Tactical and Strategic Missile Guidance

Sixth Edition

Tactical and Strategic Missile Guidance

Sixth Edition

Paul Zarchan
MIT Lincoln Laboratory
Lexington, Massachusetts

Volume 239
Progress in Astronautics and Aeronautics

Timothy C. Lieuwen, Editor-in-Chief
Georgia Institute of Technology
Atlanta, Georgia

Published by
American Institute of Aeronautics and Astronautics
12700 Sunrise Valley Drive, Suite 200
Reston, VA 20191-5807

American Institute of Aeronautics and Astronautics, Inc., Reston, Virginia

1 2 3 4 5

Cover art courtesy of Air University Press, Air Force Research Institute. From an illustration by Daniel Armstrong for Colonel Mike Corbett, USAF, Retired, and Paul Zarchan, "The Role of Airpower in Active Missile Defense," *Air and Space Power Journal* 24, no. 2 (Summer 2010): 57–71.

ISBN 978-1-60086-894-8

To
Maxine, Adina, Ari, and Ronit

TABLE OF CONTENTS

Software download information can be found at the end of the book on the Supporting Materials page.

ACKNOWLEDGMENTS

Although many of the devices discussed in *Tactical and Strategic Missile Guidance* work in a vacuum, I do not. In this section I would like to take the opportunity to acknowledge those individuals who contributed in some way to the content and publication of this text.

First, I would like to thank Dr. Donald C. Fraser, former Executive Vice President of The Charles Stark Draper Laboratory, Inc. (CSDL) for recommending me as an author for AIAA's Progress in Astronautics and Aeronautics book series. He knew of my desire to write a book on missile guidance and through his efforts the dream is now a reality. During the last 16 years, Mr. John M. Elwell, former Guidance, Navigation & Control Director at CSDL, provided me with numerous technical opportunities that expanded my missile guidance background. The technical thread running throughout the text is based on those technical opportunities. The highly competent staff and superb technical environment of CSDL provided the atmosphere that not only made this project possible but also fun. In particular, I would like to thank CSDL technical staff members Dr. Owen Deutsch, Mr. Steven L. Nelson, and Dr. Richard E. Phillips for our many valuable technical discussions. Much of the material in the strategic interceptor chapters was a consequence of our valuable technical interchanges. Some of the material in several of the chapters pertaining to tactical missile guidance was based on knowledge gained during my employment at Raytheon. Here, I not only learned tactical interceptor guidance and control fundamentals from the people who first made radar homing missiles a reality, but also learned of and developed an appreciation for the hard work involved in converting theory to something that works.

Next, I would like to acknowledge those associated with AIAA responsible for the inception and completion of this project. I would like to thank Dr. Allen E. Fuhs, former AIAA Progress in Astronautics and Aeronautics Editorial Board Member, for inviting me to participate in this series and then helping me formulate an outline from which a manuscript could be written. His kind and constructive review of the initial draft led to numerous changes, all of which improved the overall quality of the text. Mr. John Newbauer, former AIAA Administrator of Scientific Publications, provided me with excellent examples of the standards expected of a Progress series volume and, in addition, helped outline the preface. Mrs. Jeanne Godette, former AIAA Director of Book Publications, and her very competent staff not only taught me a great deal about grammar but, in addition, moved this project forward very rapidly.

Finally, I wish to express my love and gratitude to my wife Maxine for having patience and understanding, while our home fell into a state of disrepair, for not complaining while I put in the many hours required in writing this text.

PREFACE TO THE SIXTH EDITION

The sixth edition contains six new chapters on important topics related to improving missile guidance system performance and understanding key design concepts and tradeoffs. In addition, at the request of many readers, all of the FORTRAN source code that appeared in the first five editions of *Tactical and Strategic Missile Guidance* has been converted to MATLAB. Interested readers can obtain all MATLAB source code—and the equivalent FORTRAN versions—in electronic form on the AIAA Web site as mentioned at the back of this book on the Supporting Materials page.

The first new chapter, Chapter 29, presents two new applications of the method of adjoints for mixed continuous-discrete systems. The first application involves multiple samplers with each of the samplers operating at a different sampling rate. The second application involves taking the adjoint of a three-state discrete Kalman filter in the homing loop.

The second new chapter, Chapter 30, introduces a new guidance law that can be used to shape the interceptor trajectory against a stationary target. The unique advantage of this new guidance approach over existing guidance laws is that time-to-go information is not required. The second part of this chapter considers both the problem of finding the minimum achievable miss distance of a radar homing missile and exploring practical techniques that can be used to achieve the minimum possible miss.

The third new chapter, Chapter 31, introduces the differential game guidance law with bounded controls and demonstrates its performance improvements over conventional guidance laws under challenging conditions in which the missile-to-target acceleration advantage is very low.

The fourth new chapter, Chapter 32, introduces techniques for graphically presenting strategic information on successful intercepts of an impulsive ballistic target being pursued by an impulsive interceptor. The three graphical techniques discussed are the operational area method, the launch area denied method, and the defended area method. Examples are presented illustrating the utility of each method.

The fifth new chapter, Chapter 33, examines two filtering options for the boost phase intercept of a strategic target. The two filters compared are a template-based Kalman filter that has perfect *a priori* information and a linear three-state polynomial Kalman filter that does not require such information. The filters are compared under both ideal and realistic conditions. The results of this chapter may surprise some readers.

Finally, the sixth new chapter, Chapter 34, addresses some of the guidance and control issues involved in enabling an air-launched interceptor carrying a highly maneuverable kinetic kill vehicle to perform an exoatmospheric intercept of a boosting threat target that is capable of traveling many thousands of kilometers. This chapter takes the reader through the first iteration of the multi-iteration design process in order to show how much divert and acceleration may be required by the kinetic kill vehicle so that it can hit the target. Simplified examples are presented to indicate how conventional guidance and filtering techniques can be used as a starting point in the iterative design process for this important problem in missile defense.

These six new chapters are based on material from the AIAA continuing education short course "Fundamentals of Tactical and Strategic Missile Guidance" that I have been teaching since 1990. The new topics are treated from both an analytical and a simulation point of view so that readers with different backgrounds and learning styles can benefit from the new material.

Readers of previous editions of *Tactical and Strategic Missile Guidance* will notice that even though the sixth edition has six new chapters, the book is approximately the same size as the fifth edition. This was achieved by deleting fifth edition Chapters 10 and 27; half of Chapters 16, and 29, and Appendix A; and all of Appendices B – E. In addition, Chapter 20 has been rewritten in response to questions by readers. Although the material has been deleted from the sixth edition, interested readers can find the fifth edition chapters and appendices in PDF format on AIAA's Web site.

On a personal note, it continues to be very gratifying for me to learn that many people working with or needing to learn about missile guidance have found *Tactical and Strategic Missile Guidance* useful. Over the years, many readers have contacted me and have asked questions when the book's material was not clear to them. Material in the existing chapters has been clarified so that all will benefit from my interaction with the readers. It is still my hope that this sixth edition will be of value not only to new readers, but will also be worthwhile to those who have read previous editions.

Paul Zarchan
MIT Lincoln Laboratory
February 2012

INTRODUCTION

The requirements for tactical missile guidance systems were born at the end of World War II as a result of the highly effective kamikaze attacks on U.S. vessels. After the war it was clear that naval guns using unguided shells were not adequate for shooting down hostile aircraft making suicidal attacks against U.S. ships. To counter kamikaze-like threats, the U.S. Navy initiated the development of the Lark guided missile in 1944. Approximately six years of hard work was required before the first successful intercept, made by the Lark against an unmanned aircraft, occurred on Dec. 2, 1950. A very readable history of this significant event, told by one of the engineering participants, can be found in a work by M. W. Fossier ("The Development of Radar Homing Missiles," *Journal of Guidance, Control, and Dynamics*, Vol. 7, No. 6, 1984, pp. 641–651).

Operational endoatmospheric tactical missile systems, using the same guidance principles as those used by the Lark, have now been in existence for approximately six decades. These missiles use guidance concepts that work well not only against stationary or predictable targets but also are effective against responsive threats (such as aircraft executing evasive maneuvers) whose future position is highly uncertain. In the tactical arena, current guidance law technology appears to be adequate if the flight time is long compared to the effective time constant of the guidance system and if the missile enjoys a considerable acceleration advantage over the target. In fact, it is not uncommon for a tactical missile to have an acceleration advantage of more than five against an aircraft target. This advantage is more than adequate for a successful intercept with current guidance law technology.

Strategic ballistic missiles are different from tactical guided missiles because they travel much longer distances and are designed to intercept stationary targets whose location is known precisely. A tactical homing missile acquires the target with a seeker and then guides all the way to intercept, whereas a strategic interceptor does not require a seeker because all of its guidance is in the boost phase. Because the boost phase represents only a small fraction of the total flight of a strategic ballistic missile, the interceptor glides most of the way toward the stationary target. In this type of strategic application, precise instrumentation is necessary so that the interceptor can steer to the correct position and velocity states at the end of the boost phase. With the correct states the missile will be able to glide ballistically, without further corrective maneuvers, toward the target.

The concept of long-range strategic ballistic missiles was also born during World War II with the highly effective V-2 rocket developed by German scientists at Peenemunde. The first combat round, launched on September 6, 1944, represented the culmination of nearly 14 years of work (G. P. Kennedy, *Rockets, Missiles, and Spacecraft of the National Air and Space Museum*, Smithsonian Institution Press, Washington, DC, 1983). The range of the V-2 was only a few hundred miles, and a 2000-lb warhead compensated somewhat for its lack of accuracy. After World War II some of the German scientists from Peenemunde were brought to the United States, along with 80 missiles, in order to advance the V-2 technology. Guidance accuracy was improved by more than an order of magnitude by incorporating precise inertial sensors into the guidance system design. The short-range Redstone missile became the first U.S. ballistic missile with a highly accurate inertial guidance system. More precise and smaller inertial sensor hardware and improved guidance software were achieved with the U.S. medium-range Jupiter missile. An eyewitness account of this advance in strategic missile technology can be found in a work by W. Haeussermann ("Developments in the Field of Automatic Guidance and Control of Rockets," *Journal of Guidance, Control, and Dynamics*, Vol. 4, No. 3, 1981, pp. 225–239). Today's highly accurate strategic ballistic missiles travel at intercontinental ranges at near-orbital speeds.

In newer systems, such as those being considered by the Ballistic Missile Defense Organization (BMDO), interceptors will have to make use of and improve on yesterday's tactical and strategic missile technology. Tomorrow's interceptors will have to fly strategic distances against moving and possibly accelerating targets whose future position is highly uncertain. In these applications it is not sufficient to apply only ballistic missile technology. Some type of homing guidance system is required after the interceptor boost phase to take out the inevitable errors. However, it is not clear whether tactical guidance law technology is appropriate for the homing phase of these strategic interceptors. Unlike tactical endoatmospheric interceptors, which maneuver by generating lift, the newer exoatmospheric interceptors require fuel for the missile divert engines in order to maneuver or divert in response to guidance commands. If all of the divert fuel is consumed, the interceptor cannot maneuver. In addition, fuel is at a premium, because, according to the rocket equation, interceptor weight grows exponentially with fuel weight. Therefore, unlike tactical endoatmospheric missiles, the newer exoatmospheric interceptors require guidance laws that minimize fuel consumption. In addition, because of practical limits on achievable divert engine thrust-to-weight ratios; the newer interceptors may no longer enjoy the tactical missile's acceleration advantage over the target. In fact, strategic exoatmospheric interceptors may be working at an *acceleration disadvantage* against a boosting target!

With this perspective, this book attempts to lay the foundation for meeting today's new challenges. The principles of both tactical and strategic missile guidance are presented in a common language, notation, and perspective. The mathematics, arguments, and examples presented in the text are intended to be nonintimidating so that the designers working in the tactical world will be able to understand and appreciate the difficulty of the strategic problem and vice versa. Numerous examples are presented to illustrate all of the concepts presented in the text. In later chapters of the book, examples are presented showing elementary ways in which tactical and strategic guidance principles can be combined.

In this way the potential for a cross-fertilization of ideas, which is necessary for today's challenges in guidance technology, is increased.

The book is written with both the expert and novice in mind. Proven methods of guidance in both the tactical and strategic world are presented from several points of view. The guidance laws are usually first derived mathematically, then explained from a heuristic point of view, and finally a numerical example, along with a MATLAB listing, is included to prove that the guidance law performs as expected. The numerical example is usually chosen to highlight the strengths and weaknesses of the guidance approach. The expert or more interested reader using the MATLAB listing may not only get a deeper understanding of the principles involved, but may also explore issues beyond the scope of the text.

It might appear from glancing at the many block diagrams scattered throughout the text that the subject matter requires a control system background. *This is not true*! The block diagrams are offered as a pictorial description of differential equations. The reader can always ignore the block diagrams and just read the differential equations. The only assumption made in the text is that the reader has either an engineering or physics background. All of the numerical and mathematical techniques used in the text are explained in earlier chapters.

Chapter 1 presents the basis of most of the numerical techniques used throughout the text. The utility of Laplace transforms in solving and representing differential equations is explained. It is then shown how differential equations can be solved using the second-order numerical Runge–Kutta integration technique. MATLAB code for a representative example is presented so that the reader can better understand how numerical integration is actually applied. The code is written so that it is easy to apply to problems involving other differential equations without changing more than a few lines of code. In fact, most of this sample code is repeated throughout the text when the differential equations representing the system under consideration change.

Chapter 2 introduces the reader to the most important and widely used tactical missile guidance law: proportional navigation. A simplified engagement simulation is developed in order to show how proportional navigation operates and why it is an effective guidance law. Next, linearization techniques are used to get a deeper understanding of how proportional navigation works and to form an analytical foundation from which guidance theory can be advanced. Closed-form solutions for the required missile acceleration due to heading error and target maneuver disturbances are derived so that the reader can see how these important errors influence total system performance. It is then demonstrated that the derived closed-form solutions are not only analytically convenient but are in fact accurate indicators of the expected system performance.

Chapter 3 introduces one of the most important methods used in analyzing tactical missile guidance systems: the method of adjoints. The rules for constructing an adjoint are presented, and the necessary mathematics required to understand why adjoints are useful are included in this chapter. A numerical example involving a missile guidance system is used both as a practical application of the theory and to ensure that the reader fully understands how to apply the adjoint technique. The numerical results of the adjoint example are compared to traditional simulation results so that the reader can appreciate both the power and utility of this very elegant technique. New closed-form solutions are derived using adjoints so that the reader can also begin to appreciate the relationship between miss distance, the guidance system time constant, and the error source.

Chapter 4 reviews all of the necessary theory so that the reader can understand how measurement noise and random phenomenon influence the performance of a missile guidance system. Basic definitions are reviewed, and numerical examples are presented showing the reader how theory is used in practice. The method of adjoints, which was introduced in Chapter 3 for deterministic systems, is extended so that a missile guidance system can be analyzed if the error sources are random. A numerical example is presented showing how the method of adjoints can be used to obtain statistical performance projections in only *one computer run*. These results are compared to multiple-run Monte Carlo projections (that is, repeated simulation trials with ensemble averaging of the resultant output data).

Another popular computerized analytical technique, known as covariance analysis, is introduced in Chapter 5. It is shown that covariance analysis can also be used to yield exact statistical performance projections of a missile guidance system in the presence of random error sources in one computer run. The numerical requirements for the successful implementation of covariance analysis are discussed and compared to the adjoint method. Chapter 5 also shows how the adjoint technique can be extended to yield acceleration as well as miss distance information.

Chapter 6 presents most of the important properties of a proportional navigation guidance system. It is shown that an accurate dynamic model of the guidance system is crucial in obtaining accurate miss distance performance projections. Normalized design curves are developed for a high-order canonical guidance system so that an engineer can quickly estimate system performance given a minimum amount of information. The concept of an optimal target evasive maneuver is developed. It is shown how one can compute the maximum miss distance that a maneuvering target can induce. This type of information is important in determining the vulnerability of a guidance system. Finally, the influence of saturation and parasitic effects on system performance are considered. Design curves are presented showing how to size the missile to target acceleration advantage so that the interceptor will be effective. It is also shown in this chapter that parasitic effects such as radome place fundamental limits on the attainable speed and gain of the guidance system.

The missile seeker provides a noisy measurement of the line-of-sight angle. Chapter 7 demonstrates how simple digital fading memory filters can be used as part of a missile guidance system to provide an estimate of the line-of-sight angle and rate from the noisy measurement. Adjoint theory is again extended to handle systems with continuous and digital parts so that a mixed continuous-discrete missile guidance system can be analyzed efficiently. Some of the more important properties of fading memory filters are illustrated via numerical examples. This chapter also shows that, if range measurements are available, it is possible to estimate target acceleration with a fading memory filter.

Chapter 8 reviews proportional navigation concepts and sets up the mathematical foundation so that more advanced tactical guidance laws can be derived. Augmented proportional navigation is derived from some properties of the Schwartz inequality. It is then shown, via a numerical example, how guidance system time constants can degrade miss distance performance of both proportional and augmented proportional navigation. Recognizing the cause of the miss distance degradation, a new guidance law is derived. It is demonstrated that, as long as the missile guidance system dynamics is known, the new guidance law can effectively eliminate miss distance degradation due to a guidance system time constant.

Chapter 9 introduces a class of optimal digital noise filters known as Kalman filters. Using the theoretical Kalman filtering equations, the chapter develops, in detail, a digital Kalman filter that is very useful for missile guidance system applications. It is shown, in detail, how a Kalman filter can be used in conjunction with an optimal guidance law to improve system performance and to relax missile acceleration requirements. Various experiments are conducted in order to illustrate important filtering and guidance concepts.

Chapter 10 wraps up the discussion of tactical interceptors with consideration of the missile's operational zone. Reach considerations, based on the rocket equation and drag effects, are presented. Gravitational effects, which were previously neglected, are offered as another phenomenon that further limits the zone. Numerical examples are used to illustrate important effects considered in this chapter.

Chapter 11 introduces strategic interceptor concepts from a tactical point of view. A gravitational model, based on Newton's law of universal gravitation, is developed for strategic flight. Comparisons between a flat-Earth gravitational model and a strategic gravitational model are made. Although strategic engagement simulation models are presented in the text in a Cartesian Earth-centered coordinate system, a polar coordinate system is also introduced so that important closed-form solutions can be derived. Key formulas for velocity and flight time for an impulsive ballistic missile to travel a fixed distance, given an initial flight-path angle, are developed using the polar coordinate system. A Cartesian Earth-centered simulation is used to confirm the analytical results.

Chapter 12 shows how preliminary strategic booster sizing can be done with the rocket equation. Simplified booster sizing examples are presented in order to clarify the concepts. The rocket equation is extended so that the virtues of staging can be illustrated via a numerical example. Finally, the gravity turn maneuver is introduced as the simplest possible steering method a booster can employ in traveling from its launch point to a desired destination.

Starting from the closed-form solutions derived in Chapter 11, the concepts of Lambert steering, which is fundamental to booster and spacecraft steering, are developed in Chapter 13. A simple to understand but numerically inefficient way of solving Lambert's problem is derived. A numerical example is presented showing how the numerical solution to Lambert's problem can be implemented. A novel use of the secant method is demonstrated to speed up the solution to Lambert's problem by more than two orders of magnitude! It is then shown how the implemented solution can be modified with a simple feedback scheme to steer an interceptor, during its boost phase, to its intended target. Another subset of Lambert steering, known as general energy management (GEM) steering, is also derived and demonstrated. A numerical example highlighting the similarities and differences between Lambert and GEM steering is presented.

Chapter 14 shows elementary but fundamental methods of combining the tactical missile guidance concepts of Chapters 2–9 with the strategic notions of Chapters 11–13. Unifying numerical examples are used to illustrate the strengths and weaknesses of the combined approach. Previously derived closed-form solutions for the required missile acceleration to hit a target by a tactical interceptor are converted to strategic lateral divert formulas. Nonlinear strategic engagement simulation results are used to show that the divert formulas for prediction error, apparent target maneuver, and guidance law are not only useful because of their simplicity but are in fact accurate indicators of strategic interceptor requirements.

Chapter 15 presents some additional concepts that are very important to the strategic world. It is shown how compensating for known gravity effects in the guidance law can considerably reduce interceptor lateral divert requirements. Next, predictive guidance is introduced as the ultimate guidance law. It is shown that, if accurate *a priori* information exists concerning the target, then predictive guidance can be used to substantially alleviate interceptor lateral divert requirements. All of the homing guidance concepts introduced have assumed that strategic interceptor divert engines were effectively throttleable. A pulsed guidance law is developed assuming that guidance commands can only be issued with a few discrete burns. The performance of the pulsed guidance law is compared to that of proportional navigation.

In Chapter 16, the focus of the text switches to endoatmospheric ballistic targets. Closed-form solutions are derived and validated, based on the properties of ballistic targets, showing how the magnitude of endoatmospheric ballistic target deceleration varies with speed, altitude, and re-entry angle. It is demonstrated that unless advanced guidance techniques are used, the high deceleration levels of a ballistic target make it difficult to hit under all engagement conditions.

Advanced guidance techniques and fire control logic for endoatmospheric intercepts require knowledge of the target's ballistic coefficient. Using a simplified extended Kalman filter as an example, the challenges of estimating a target's ballistic coefficient are demonstrated in Chapter 17. Common filter design pitfalls and their engineering fixes are illustrated in easy-to-understand examples. Guidelines for making an extended Kalman filter robust to large initialization errors are presented.

Chapter 18 integrates many of the text's concepts to further explain why endoatmospheric ballistic targets are challenging. Formulas are derived showing the geometry dependence of noise-induced miss distance. The dependence of the minimum achievable guidance system time constant on radome slope, geometry, and missile aerodynamic properties is shown. Numerical examples are presented in order to highlight missile performance differences for both aircraft and ballistic target threats.

When two aircraft are flying in close formation, they both appear within the pursuing interceptor's seeker beam. When the missile is close enough to the aircraft, one of the targets will drop out of the seeker beam, and it will appear to the missile as if the target has moved instantaneously. Large miss distances may result because of insufficient remaining homing time. Chapter 19 develops normalized design curves to both illustrate and quantify the multiple target problem. Rules of thumb are developed relating the necessary ratio of the time left for homing after resolution has taken place to the guidance system time constant.

It is well-known that the barrel roll or weave maneuver can cause large miss distances. Because tactical ballistic missiles can spiral or weave into resonance as they re-enter the atmosphere due to either mass or configurational asymmetries, the weave maneuver is of particular interest to the guidance system designer. Chapter 20 studies the influence of the target weave maneuver on a proportional navigation guidance system. This chapter demonstrates how the target weave frequency and amplitude, the missile guidance system time constant, effective navigation ratio, and acceleration capability all play an important role in determining system performance. Methods for improving system performance are also discussed.

To design a flight-control system the interceptor airframe equations must first be linearized. Chapter 21 shows how the nonlinear missile force and moment equations are related to the geometry of the missile airframe. A simple method for linearizing the force and moment equations is introduced so that transfer functions can be derived for the missile airframe. It is shown that the transfer function approximation to the airframe is an excellent match to reality for small angles of attack.

Chapter 22 illustrates how the flight-control system interacts with the guidance system. Both the open-loop and rate gyro flight-control systems are discussed in detail. It is shown that the open-loop flight-control system has the dynamics of the bare airframe and is therefore not usually acceptable in radar homing applications because of its low damping. It is demonstrated that the rate gyro flight-control system improves the system damping by using a sensor and the principles of feedback.

Chapter 23 demonstrates that by using an accelerometer and rate gyro with the flight-control system, the system damping, time constant, and open-loop crossover frequency can be controlled independently. Controlling the system damping ensures that the guidance system is not overly sensitive to radome slope effects at the high altitudes. Selecting the system time constant means that there will be adequate performance against maneuvering targets. Controlling the open-loop crossover frequency ensures that the design will be robust and not overly sensitive to unmodeled high frequency dynamics. A unifying example is presented to demonstrate how typical guidance system tradeoffs are conducted. The method of adjoints is used to show how miss distance error budgets can be generated, actuator requirements set, and missile acceleration requirements derived.

In all the guidance work done thus far, the goal has been to hit the target using the least amount of energy. In some applications, in addition to hitting the target, it may also be desirable to shape the missile trajectory near impact. For example, in antitank or antiballistic missile applications we may want to have the missile approach the target at certain strike angles to improve lethality. Chapter 24 shows how the guidance problem can be reformulated so that a new guidance law can be derived that hits the target using minimum energy and, in addition, travels on the desired trajectory.

The text has already shown that advanced guidance laws derived in Chapter 20 can be used to improve system performance against spiraling targets. Chapter 25 explores the various filtering options that can be used to estimate the weaving target states required for advanced guidance techniques. After a brief review of existing filtering techniques for the weaving target problem, it is assumed that the target weave frequency is known and an optimal linear four-state weave Kalman filter, which estimates both target acceleration and jerk, is both derived and evaluated. It is then assumed that the target weave frequency is not known but must be estimated along with target acceleration and jerk using an extended five-state Kalman filter. The chapter compares the various candidate Kalman filter options and appropriate guidance laws in terms of both performance and robustness.

So far, all of the guidance laws presented in the text were derived using an analytical technique based on the Schwartz inequality. Chapter 26 presents several alternative ways of developing guidance laws numerically. These techniques can either be used to check existing guidance laws or to derive more advanced guidance laws when the Schwartz inequality technique becomes either too cumbersome or impossible to use. An example is presented showing the advantages of a new advanced guidance law.

Previous editions of the text have shown that intentional or unintentional spiraling maneuvers on the part of a tactical ballistic missile can also make it particularly difficult for a pursuing missile to hit. The third and fourth editions have already discussed ways in which performance against weaving or spiraling targets could be improved from both a guidance and filtering point of view if the target weave frequency is known. However, if the target weave frequency is unknown it can be demonstrated that an extended Kalman filter approach only works if it is initialized close to the actual target weave frequency. Chapter 27 presents a linear Kalman-filter bank approach, originally introduced in the 1960s, for accurately estimating the target weave frequency. As with other material in the text, this chapter uses a step-by-step approach in order to explain how the filter bank approach works.

So far, all of the engagement simulations presented in this text have either been in one or two dimensions. This was done to make it easier for the reader to more easily understand all of the concepts presented in the text. Chapter 28 provides several examples on how to convert important elements of previous engagement simulation code to three dimensions in both the tactical and strategic worlds. In addition, this chapter demonstrates how additional insights for the weaving target problem can be gained in the tactical world by going to three dimensions.

Chapter 29 presents two new applications of the method of adjoints for a mixed continuous-discrete system. The first application involves multiple samplers used in a system in which measurement data is first preprocessed at one sampling rate and then used for input to a fading memory filter operating at another sampling rate. The second application involves taking the adjoint of a three-state discrete Kalman filter in the homing loop.

Chapter 30 discusses two important topics that have not yet been covered. The first topic considers the special case of shaping the interceptor trajectory against a stationary target without requiring time-to-go information. This new method is compared to optimal guidance techniques that does require time-to-go information. The next topic considers finding the minimum achievable miss distance with a radar homing missile. Practical techniques for achieving the minimum possible rms miss are discussed and examples demonstrating their utility are presented.

Chapter 31 introduces the differential game guidance law with bounded controls and demonstrates its performance improvements over conventional guidance under extreme circumstances in which the missile-to-target acceleration advantage is very low. The chapter also demonstrates how the chattering caused by the bang-bang nature of differential game guidance with bounded controls can be dramatically reduced without significant loss of performance.

Chapter 32 introduces techniques for graphically presenting strategic information on successful intercepts of an impulsive ballistic target being pursued by an impulsive interceptor. The three techniques discussed are the operational area method, the launch area denied method, and the defended area method. Examples are presented illustrating the three methods of presentation. Cases are examined showing how the size of the successful engagement area is influenced by both the interceptor velocity and the target trajectory type.

Chapter 33 examines two filtering options for a boost phase intercept problem. The two filters compared are a template-based Kalman filter that has perfect *a priori* information and a linear three-state polynomial Kalman filter that does not require such information. The filters are compared under ideal conditions and then are compared in a more realistic environment.

Finally, Chapter 34 addresses some of the guidance and control issues involved in enabling an air-launched interceptor carrying a highly maneuverable kinetic kill vehicle to perform an exoatmospheric intercept of a boosting threat missile capable of traveling many thousands of kilometers. The chapter takes the reader through part of the first iteration of the multi-iteration design process in order to show how much divert and acceleration are required by the kinetic kill vehicle to hit the target. Simplified examples are presented to indicate how conventional guidance and filtering techniques can be used as a starting point in the iterative design process for this important problem in missile defense.

This text attempts to present many of the important guidance principles involved in enabling an interceptor to hit its intended target. The utilities of these principles are explained and demonstrated with pictures, equations, and computer code. However, missing from the text is the intensity, challenge of the unknown, and plain hard work that it takes in going from a paper design to something that not only flies but also meets the system objectives. In light of the new challenges facing today's guidance engineer, it is very appropriate to quote from the work by M. W. Fossier:

> On joining Raytheon as a young engineer, I found an intensity of spirit that I had never experienced before. I felt myself carried along in what seemed almost a crusade, sharing a burning commitment to succeed against an immense challenge.
>
> As a result of this shared feeling, each triumph led to a broad-based feeling of great elation. I still vividly recall the first flight test by Lark against a low-flying drone aircraft in 1951 over the ocean at Point Mugu. The intercept was a relatively short range and was the first to be in full view of the handful of observers permitted on the beach. When the missile homed unerringly to a spectacular direct hit on the drone, the human explosion matched the one in the air. The formal celebration ended in the wee hours of the morning, but the emotional wave lasted for months.
>
> On the other hand, I can still feel the despair that resulted when a technician inadvertently connected B+ (250 V) to the filament string, blowing out every tube in an early Sparrow being prepared for flight test. At that time, the best flight test engineers we had took about a month of 12-hour days to check out a missile. The resulting delay was felt at a personal level in every corner of the organization.
>
> For years I attributed this intensity to the relative youth of most of the participants. However, I found that age has served mainly to mute their exuberance, but not their intensity, and I was forced to seek another source. My current view is that the intensity was (and is) a result of the intellectual challenge of the unknown. It is the characteristic of the engineering profession that there are always new problems waiting in the wings to replace the old ones as they are solved. The constant element is the challenge itself, which demands that we do our utmost and rewards us only when we do.

CHAPTER 1

Numerical Techniques

INTRODUCTION

The numerical techniques introduced in this chapter involve the use of Laplace transforms for manipulating and displaying differential equations and numerical integration for solving the differential equations. These techniques form the basis of all of the numerical methods used throughout the text. A numerical example will be presented that will illustrate a practical application of the use of Laplace transforms and numerical integration. Another example will be presented showing how z transforms can be used to both represent difference equations and get their solution.

LAPLACE TRANSFORMS AND DIFFERENTIAL EQUATIONS

Transform methods are often useful because certain operations in one domain are different and often simpler than operations in the other domain. For example, ordinary differential equations in the time domain become algebraic expressions in the s domain after being Laplace transformed. In control system engineering, Laplace transforms are used both as a shorthand notation and as a method for solving linear differential equations. In this text we will frequently use Laplace transform notation to represent subsystem dynamics in tactical missile guidance systems.

If we define $F(s)$ as the Laplace transform of $f(t)$, then the Laplace transform has the following definition:

$$F(s) = \int_0^\infty f(t)e^{-st}\,\mathrm{d}t$$

With this definition it is easy to show that a summation in the time domain is also a summation in the Laplace transform or frequency domain. For example, if $f_1(t)$

and $f_2(t)$ have Laplace transforms $F_1(s)$ and $F_2(s)$, respectively, then

$$\mathcal{L}[f_1(t) \pm f_2(t)] = F_1(s) \pm F_2(s)$$

Again, using the definition of the Laplace transform, it is easy to show that differentiation in the time domain is equivalent to frequency multiplication in the Laplace transform domain, or

$$\mathcal{L}\left(\frac{df(t)}{dt}\right) = sF(s) - f(0)$$

where $f(0)$ is the initial condition on $f(t)$. The Laplace transform of the nth derivative of a function is given by

$$\mathcal{L}\left(\frac{d^n f(t)}{dt^n}\right) = s^n F(s) - s^{n-1} f(0) - s^{n-2}\frac{df(0)}{dt} - \cdots$$

From the preceding equation we can see that, for zero initial conditions, the nth derivative in the time domain is equivalent to a multiplication by s^n in the Laplace transform domain.

Laplace transforms can also be used to convert the input-output relationship of a differential equation to a shorthand notation called a *transfer function representation*. For example, given the second-order equation

$$\frac{d^2 y(t)}{dt^2} + 2\frac{dy(t)}{dt} + 4y(t) = x(t)$$

with zero initial conditions, or

$$\frac{dy(0)}{dt} = 0, \qquad y(0) = 0$$

we can find the same differential equation in the Laplace transform domain to be

$$s^2 Y(s) + 2sY(s) + 4Y(s) = X(s)$$

Combining like terms in the preceding equation to get a fractional relationship between the output and input, known as a *transfer function*, yields

$$\frac{Y(s)}{X(s)} = \frac{1}{s^2 + 2s + 4}$$

Similarly, given a transfer function, we can go back to the differential equation form. Consider the second-order transfer function

$$\frac{Y(s)}{X(s)} = \frac{1 + 2s}{1 + 2s + s^2}$$

We know that, according to the chain rule, the transfer function can be expressed as

$$\frac{Y(s)}{X(s)} = \frac{E(s)}{X(s)}\frac{Y(s)}{E(s)}$$

Therefore, we can break the relationship into the following two equivalent transfer functions:

$$\frac{E(s)}{X(s)} = \frac{1}{1+2s+s^2}, \qquad \frac{Y(s)}{E(s)} = 1+2s$$

Cross multiplication results in

$$s^2E(s) + 2sE(s) + E(s) = X(s)$$

and

$$2sE(s) + E(s) = Y(s)$$

Converting the first equation to the time domain yields the second-order differential equation

$$\frac{d^2e(t)}{dt^2} + 2\frac{de(t)}{dt} + e(t) = x(t)$$

and converting the second equation yields the output relationship

$$y(t) = 2\frac{de(t)}{dt} + e(t)$$

The implication from the transfer function notation is that the initial conditions on the second-order differential equation are zero, or

$$\frac{de(0)}{dt} = 0, \qquad e(0) = 0$$

Often we will use Laplace transform notation and, for shorthand, drop the functional dependence on s in the notation [that is, F is equivalent to $F(s)$]. Similarly, when we are in the time domain, the functional dependence on t will often be dropped [that is, f is equivalent to $f(t)$]. In addition, block diagrams and program listings will frequently use the overdot notation to represent time derivatives. With this notation, each overdot represents a derivative. For example,

$$\dot{y} = \frac{dy}{dt}, \quad \ddot{y} = \frac{d^2y}{dt^2}, \quad \dddot{y} = \frac{d^3y}{dt^3}, \quad \text{etc.}$$

Therefore, converting

$$\frac{d^2e(t)}{dt^2} + 2\frac{de(t)}{dt} + e(t) = x(t)$$

TABLE 1.1 COMMON INVERSE LAPLACE TRANSFORMS

$F(s)$	$f(t)$
$\frac{K}{s}$	K
$\frac{K}{s^n}(n = 1, 2, \ldots)$	$\frac{Kt^{n-1}}{(n-1)!}$
$\frac{K}{(s-a)^n}(n = 1, 2, \ldots)$	$\frac{Kt^{n-1}e^{at}}{(n-1)!}$
$\frac{K}{s^2+a^2}$	$\frac{K\sin(at)}{a}$
$\frac{Ks}{s^2+a^2}$	$K\cos(at)$
$\frac{K}{(s-a)^2+b^2}$	$\frac{Ke^{at}\sin(bt)}{b}$
$\frac{K(s-a)}{(s-a)^2+b^2}$	$Ke^{at}\cos(bt)$

to the overdot notation yields

$$\ddot{e} + 2\dot{e} + e = x$$

Occasionally, we shall either convert time functions to Laplace transforms or vice versa, by inspection. Some common transfer functions [1], along with their time domain equivalents, appear in Table 1.1. A more extensive listing of inverse Laplace transforms can be found in [1].

NUMERICAL INTEGRATION OF DIFFERENTIAL EQUATIONS

Throughout this text we will be simulating both linear and nonlinear ordinary differential equations. Because, in general, these equations have no closed-form solutions, it will be necessary to resort to numerical integration techniques to solve or simulate these equations. Many numerical integration techniques [2] exist for solving differential equations. However, we shall use the second-order Runge–Kutta technique throughout the text because it is simple to understand, easy to program, and, most importantly, yields accurate answers for all of the examples presented in this text.

The second-order Runge–Kutta numerical integration procedure is easy to state. Given a first-order differential equation of the form

$$\dot{x} = f(x, t)$$

where t is time, we seek to find a recursive relationship for x as a function of time. With the second-order Runge–Kutta numerical technique, the value of x at the next integration interval h is given by

$$x_{K+1} = x_K + \frac{hf(x,t)}{2} + \frac{hf(x,t+h)}{2}$$

where the subscript K represents the last interval and $K + 1$ represents the new interval. From the preceding expression we can see that the new value of x is simply the old value of x plus a term proportional to the derivative evaluated at time t and another term with the derivative evaluated at time $t + h$.

The integration step size h must be small enough to yield answers of sufficient accuracy. A simple test, commonly practiced among engineers, is to find the appropriate integration step size by experiment. As a rule of thumb, the initial step size is chosen to be several times smaller than the smallest time constant in the system under consideration. The step size is then halved to see if the answers change significantly. If the new answers are approximately the same, the larger integration step size is used to avoid excessive computer running time. If the answers change substantially, then the integration interval is again halved and the process is repeated.

To see how the Runge–Kutta technique can be applied to a practical example, let us consider the problem of finding the step response of one of the second-order networks from Table 1.1. Consider the sinusoidal transfer function

$$\frac{Y}{X} = \frac{\omega}{s^2 + \omega^2}$$

where x is the input, Y the output, ω the natural frequency of the second-order network, and s the Laplace transformation notation for a derivative. Cross multiplying the numerator and denominator of the transfer function and solving for the highest derivative, as was shown in the previous section, yields the following second-order differential equation:

$$\ddot{y} = \omega x - \omega^2 y$$

where the double overdot represents two differentiations. This second-order differential equation can be represented in block diagram form as shown in Fig. 1.1. In this diagram each $1/s$ represents an integration. The outputs of each integrator are sometimes called states and are y and y dot respectively.

If x is a step input in Fig. 1.1, we can find the response y exactly using Laplace transform techniques. Recall from Table 1.1 that $1/s$ represents a step function in the Laplace transform domain. Therefore we can express the output y in the Laplace transform domain as

$$Y(s) = \frac{\omega}{s(s^2 + \omega^2)}$$

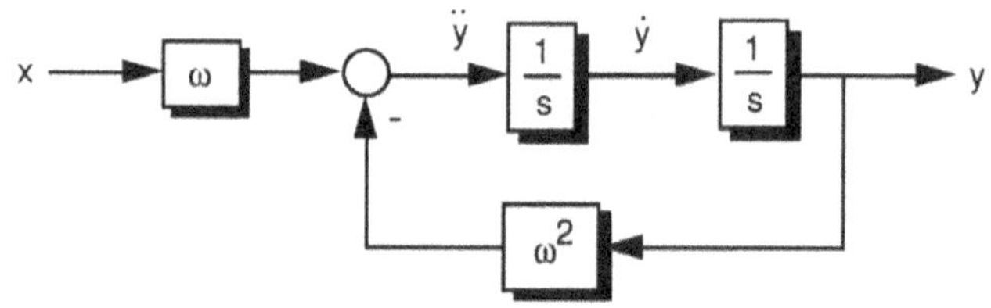

Fig. 1.1 Block diagram representation of second-order system.

Expanding the preceding expression using partial fraction expansion yields

$$Y(s) = \frac{1}{\omega}\left[\frac{1}{s} - \frac{s}{s^2 + \omega^2}\right]$$

The inverse Laplace transform of $Y(s)$ produces y in the time domain or $y(t)$. The output can be found by using Table 1.1 obtaining

$$y = \frac{1}{\omega}(1 - \cos \omega t)$$

To check the preceding theoretical closed-form solution for y, a simulation involving numerical integration was written based on the system of Fig. 1.1. A simulation of the second-order system, using the second-order Runge–Kutta integration techniques, appears in Listing 1.1. We can see from the listing that the second-order differential equation, or derivative information, appears just before the FLAG=1 statement. We come to this code twice during the integration interval: once to evaluate the derivative at time t and once to evaluate the derivative at time $t + h$. We can also see from Listing 1.1 that every 0.01 s we print out the output along with the closed-form solution. In this particular example the natural frequency ω of the second-order system is 20 rad/s.

We can see from Listing 1.1 that the integration step size h is 0.001 s. Because the simulation time is 1 s, the ratio of the simulation time to the step size is 1000. This means that 2000 passes are made to the differential equations. The resultant system transient response, due to a step input ($x = 1$), is shown in Fig. 1.2. We can see that the simulation output agrees exactly with the closed-form solution.

LISTING 1.1 SIMULATION OF SECOND-ORDER SYSTEM

```
T=0.;
S=0.;
Y=0.;
YD=0.;
X=1.;
```

```
H=.001;
S=0.;
while T <=(1.-1e-5)
  YOLD=Y;
  YDOLD=YD;
  STEP=1;
  FLAG=0;
  while STEP <=1
    if FLAG==1
      STEP=2;
      Y=Y+H*YD;
      YD=YD+H*YDD;
      T=T+H;
    end
    YDD=W*X-W*W*Y;
    FLAG=1;
  end
  FLAG=0;
  Y=.5*(YOLD+Y+H*YD);
  YD=.5*(YDOLD+YD+H*YDD);
  S=S+H;
  if S >=.000999
    S=0.;
    n=n+1;
    ArrayT(n)=T;
```

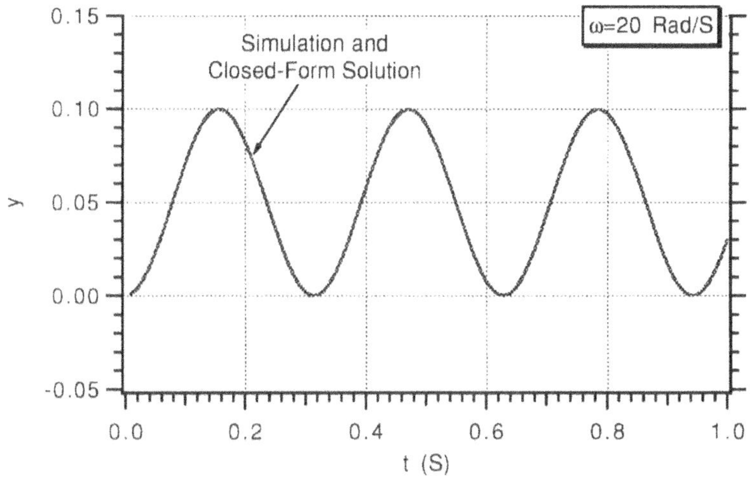

ig. 1.2 Numerically integrating differential equations yields same results as losed-form solution.

```
        ArrayY(n)=Y;
    end
end
figure
plot(ArrayT,ArrayY),grid
xlabel('Time (Sec)')
ylabel('y')
clc
output=[ArrayT',ArrayY'];
save datfil output -ascii
disp 'simulation finished'
```

Z TRANSFORMS AND DIFFERENCE EQUATIONS

We have already shown that Laplace transforms are a useful way of representing differential equations. In this text we shall also want to simulate difference equations. Z transforms can also be used as an engineering shorthand for representing the difference equations. Later in this section we will also show how Z transforms can be used to solve difference equations and sometimes check simulation results [3].

If we define $F(z)$ as the Z transform of $f(n)$, then the Z transform has the following definition:

$$F(z) = \sum_{n=0}^{\infty} f(n) z^{-n}$$

With this definition it is easy to show that a summation in the time or n domain is also a summation in the Z transform domain. For example, if $f_1(n)$ and $f_2(n)$ have Z transforms $F_1(z)$ and $F_2(z)$, respectively, then

$$Z[f_1(n) \pm f_2(n)] = F_1(z) \pm F_2(z)$$

One can show that the Z transform of a signal at time $n+1$ is a multiplication of the function in the Z transform domain by z. The Z transform of the signal at time n according to

$$Z(f_{n+1}) = zF(z) - zf(0)$$

where $f(0)$ is an initial condition. Often we will be working with systems having zero initial conditions.

A list of some common Z transforms can be found in Table 1.2. From the table we can see that there is a relationship between the sampling time T_s and time t given by

$$t = nT_s$$

TABLE 1.2 Z TRANSFORMS OF COMMON FUNCTIONS

Function	Z transform
$\delta(n)$	1
1	$z/(z-1)$
a^n	$z/(z-a)$
n	$z/(z-1)^2$
$\sin \omega n T_s$	$z \sin \omega T_s/(z^2 - 2z \cos \omega T_s + 1)$

To illustrate how Z transforms can be used to solve difference equations, let us consider a numerical example involving the fading memory filters we will be working with in Chapter 7. The simplest fading memory filter can be expressed as the difference equation

$$y_{n+1} = y_n + G(x_{n+1} - y_n)$$

where y is the filter estimate or output, x the filter input or measurement, and G the filter gain. For the first-order fading memory filter, the filter gain is a designer chosen number between zero and unity. We can find the filter response to a step input (that is, $x_{n+1} = 1$) by observing from Table 1.2 that the Z transform of a unit step function or constant is given by

$$Z(1) = z/(z-1)$$

Therefore taking the Z transform of both sides of the difference equation yields

$$zY = Y + G\left(\frac{z}{z-1} - Y\right)$$

If we bring all the terms in Y to the left-hand side of the equation, we get

$$Y(z - 1 + G) = Gz/(z-1)$$

Solving for Y produces

$$y = \frac{Gz}{(z-1)(z-a)}$$

where

$$a = 1 - G$$

Using a partial fraction expansion on the solution for Y yields

$$\frac{G}{(z-1)(z-a)} = \frac{G}{1-a}\left[\frac{1}{z-1} - \frac{1}{z-a}\right]$$

Therefore by multiplying both sides of the preceding equation by z, we obtain

$$\frac{G}{(z-1)(z-a)} = \frac{G}{1-a}\left[\frac{z}{z-1} - \frac{z}{z-a}\right]$$

Using Table 1.2 to find the inverse Z transform of the preceding expression yield

$$y_n = \frac{G}{1-a}(1-a^n)$$

Substitution of the value of a in the preceding expression yields the closed-forn solution for y as

$$y_n = 1-(1-G)^n$$

LISTING 1.2 DIFFERENCE EQUATION SIMULATION

```
G=.5;
X=1.;
TS=.1;
Y=0.;
T=0.;
N=0;
count=0;
YTHEORY=1.-(1.-G)^N;
for N=1:20
   Y=Y+G*(X-Y);
   T=N*TS;
   YTHEORY=1.-(1.-G)^N;
   count=count+1;
   ArrayT(count)=T;
   ArrayY(count)=Y;
   ArrayYTHEORY(count)=YTHEORY;
end;
figure
plot(ArrayT,ArrayY,ArrayT,ArrayYTHEORY),grid
title('Output')
xlabel('T (S)')
ylabel('Y')
clc
output=[ArrayT',ArrayY',ArrayYTHEORY'];
save datfil output -ascii
disp 'simulation finished'
```

We now have an exact expression for the filter output as a function of th number of measurements n. To test the accuracy of the preceding closed-forn

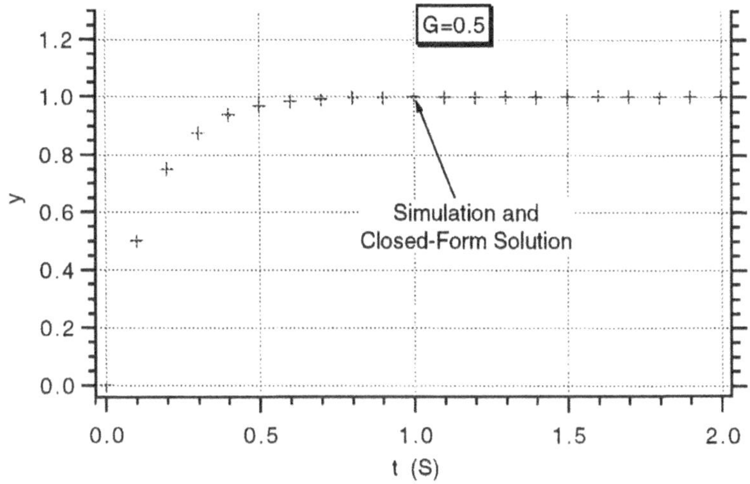

Fig. 1.3 Difference equation simulations results agree with closed-form solution.

solution for y, a simulation of the original difference equation was written and appears in Listing 1.2. We can see from the listing that unlike the previous simulation, numerical integration is not required. In this simulation we are simply solving the difference equation at each iteration of the "for loop" to get a new value for y. As we can see from the listing, the simulation solves the difference equation 20 times. The closed-form solution for y is also calculated at each iteration in order to check the validity of the simulation.

We can see from Fig. 1.3 that the filter output eventually matches the filter input. The amount of time it takes the filter output to reach 63% of its steady-state value is the filter time constant. Varying the filter gain G will change the time constant of the fading memory filter. We can also see from Fig. 1.3 that the simulation results agree with the closed-form solution.

REFERENCES

[1] Selby, S. M., *Standard Mathematical Tables—Twentieth Edition*, Chemical Rubber Co., Cleveland, OH, 1972.

[2] Press, W. H., Flannery, B. P., Teukolsky, S. A., and Vetterling, W. T., *Numerical Recipes: The Art of Scientific Computation*, Cambridge Univ. Press, London, 1986.

[3] Schwarz, R., and Friedland, B., *Linear Systems*, McGraw-Hill, New York, 1965.

CHAPTER 2

Fundamentals of Tactical Missile Guidance

INTRODUCTION

TACTICAL guided missiles apparently had their origin in Germany. For example, the Hs. 298 was one of a series of German air-to-air guided missiles developed by the Henschel Company during World War II [1]. A high-thrust first stage accelerated the missile from the carrier aircraft, whereas a low-thrust, long-burning sustainer maintained the vehicle's velocity. The Hs. 298, which was radio-controlled from the parent aircraft, was to be released either slightly above or below the target. Apparently the height differential made it easier to aim and guide the missile. This first air-to-air missile weighed 265 lb and had a range of nearly 3 miles. On December 22, 1944, three missiles were test flown from a JU 88G aircraft. All three tests resulted in failure. Although 100 of these air-to-air missiles were manufactured, none was used in combat.

The Rheintochter (R-1) was a surface-to-air missile also developed in Germany during World War II [1]. This unusual looking two-stage radio-controlled missile weighed nearly 4000 lb and had three sets of plywood fins: one for the booster and two for the sustainer. Eighty-two of these missiles flew before production was halted in December 1944. The missile was ineffective because Allied bombers, which were the R-1's intended target, flew above the range (about 20,000 ft) of this surface-to-air missile.

Although proportional navigation was apparently known by the Germans during World War II at Peenemünde, no applications on the Hs. 298 or R-1 missiles using proportional navigation were reported [2]. The Lark missile, which had its first successful test in December 1950, was the first missile to use proportional navigation. Since that time proportional navigation guidance has been used in virtually all of the world's tactical radar, infrared (IR), and television (TV) guided missiles [3]. The popularity of this interceptor guidance law is based upon its simplicity, effectiveness, and ease of implementation. Apparently, proportional navigation was first studied by C. Yuan and others at the RCA Laboratories during World War II under the auspices of the U.S. Navy [4].

The guidance law was conceived from physical reasoning and equipment available at that time. Proportional navigation was extensively studied at Hughes Aircraft Company [5] and implemented in a tactical missile using a pulsed radar system. Finally, proportional navigation was more fully developed at Raytheon and implemented in a tactical continuous wave radar homing missile [6]. After World War II, the U.S. work on proportional navigation was declassified and first appeared in the *Journal of Applied Physics* [7]. Mathematical derivations of the "optimality" of proportional navigation came more than 20 years later [8].

Keeping with the spirit of the origins of proportional navigation, we shall avoid mathematical proofs in this chapter on deriving the guidance law, but shall, instead, concentrate first on proving to the reader that the guidance technique works. Next we shall investigate some properties of the guidance law that we shall both observe and derive. Finally, we shall show how this classical guidance law provides the foundation for more advanced techniques of interceptor guidance.

WHAT IS PROPORTIONAL NAVIGATION?

Theoretically, the proportional navigation guidance law issues acceleration commands, perpendicular to the instantaneous missile-target line-of-sight, which are proportional to the line-of-sight rate and closing velocity. Mathematically, the guidance law can be stated as

$$n_c = N' V_c \dot{\lambda}$$

where n_c is the acceleration command (in $\mathrm{ft/s^2}$), N' a unitless designer-chosen gain (usually in the range of 3–5) known as the effective navigation ratio, V_c the missile-target closing velocity (in ft/s), and λ the line-of-sight angle (in rad). The overdot indicates the time derivative of the line-of-sight angle or the line-of-sight rate.

In tactical radar homing missiles using proportional navigation guidance, the seeker provides an effective measurement of the line-of-sight rate, and a Doppler radar provides closing velocity information. In tactical IR missile applications of proportional navigation guidance, the line-of-sight rate is measured, whereas the closing velocity, required by the guidance law, is "guesstimated."

In tactical endoatmospheric missiles, proportional navigation guidance commands are usually implemented by moving fins or other control surfaces to obtain the required lift. Exoatmospheric strategic interceptors use thrust vector control, lateral divert engines, or squibs to achieve the desired acceleration levels.

SIMULATION OF PROPORTIONAL NAVIGATION IN TWO DIMENSIONS

To better understand how proportional navigation works, let us consider the two-dimensional, point mass missile-target engagement geometry of Fig. 2.1. Here we

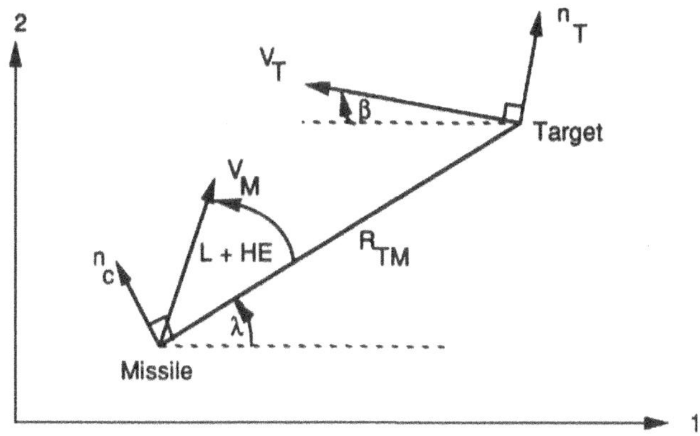

Fig. 2.1 Two-dimensional missile-target engagement geometry.

have an inertial coordinate system fixed to the surface of a flat-Earth model (that is, the 1 axis is downrange and the 2 axis can either be altitude or crossrange). Using the inertial coordinate system of Fig. 2.1 means that we can integrate components of the accelerations and velocities along the 1 and 2 directions without having to worry about additional terms due to the Coriolis effect. In this model it is assumed that both the missile and target travel at constant velocity. In addition, gravitational and drag effects have been neglected for simplicity.

We can see from the figure that the missile, with velocity magnitude V_M, is heading at an angle of $L + HE$ with respect to the line of sight. The angle L is known as the missile lead angle. The lead angle is the theoretically correct angle for the missile to be on a collision triangle with the target. In other words, if the missile is on a collision triangle, no further acceleration commands are required for the missile to hit the target. The angle HE is known as the heading error. This angle represents the initial deviation of the missile from the collision triangle.

In Fig. 2.1 the imaginary line connecting the missile and target is known as the line of sight. The line of sight makes an angle of λ with respect to the fixed reference, and the length of the line of sight (instantaneous separation between missile and target) is a range denoted R_{TM}. From a guidance point of view, we desire to make the range between missile and target at the expected intercept time as small as possible (hopefully zero). The point of closest approach of the missile and target is known as the miss distance.

The closing velocity V_c is defined as the negative rate of change of the distance from the missile to the target, or

$$V_c = -\dot{R}_{\text{TM}}$$

Therefore, at the end of the engagement, when the missile and target are in closest proximity, the sign of V_c will change. In other words, from calculus we know that the closing velocity will be zero when R_{TM} is a minimum (that is, the function is either minimum or maximum when its derivative is zero). The desired acceleration command n_c, which is derived from the proportional navigation guidance law, is perpendicular to the instantaneous line of sight.

In our engagement model of Fig. 2.1, the target can maneuver evasively with acceleration magnitude n_T. Since target acceleration n_T in the preceding model is perpendicular to the target velocity vector, the angular velocity of the target can be expressed as

$$\dot{\beta} = \frac{n_T}{V_T}$$

where V_T is the magnitude of the target velocity. The components of the target velocity vector in the Earth or inertial coordinate system can be found by integrating the differential equation given earlier for the flight-path angle of the target β and substituting in

$$V_{T1} = -V_T \cos\beta$$
$$V_{T2} = V_T \sin\beta$$

Target position components in the Earth fixed coordinate system can be found by directly integrating the target velocity components. Therefore, the differential equations for the components of the target position are given by

$$\dot{R}_{T1} = V_{T1}$$
$$\dot{R}_{T2} = V_{T2}$$

Similarly, the missile velocity and position differential equations are given by

$$\dot{V}_{M1} = a_{M1}$$
$$\dot{V}_{M2} = a_{M2}$$
$$\dot{R}_{M1} = V_{M1}$$
$$\dot{R}_{M2} = V_{M2}$$

where a_{M1} and a_{M2} are the missile acceleration components in the Earth coordinate system. To find the missile acceleration components, we must first find the components of the relative missile-target separation. This is accomplished by first defining the components of the relative missile-target separations by

$$R_{\text{TM1}} = R_{T1} - R_{M1}$$
$$R_{\text{TM2}} = R_{T2} - R_{M2}$$

We can see from Fig. 2.1 that the line-of-sight angle can be found, using trigonometry, in terms of the relative separation components as

$$\lambda = \tan^{-1} \frac{R_{\text{TM2}}}{R_{\text{TM1}}}$$

If we define the relative velocity components in Earth coordinates to be

$$V_{\text{TM1}} = V_{T1} - V_{M1}$$
$$V_{\text{TM2}} = V_{T2} - V_{M2}$$

we can calculate the line-of-sight rate by direct differentiation of the expression for line-of-sight angle. After some algebra we obtain the expression for the line-of-sight rate to be

$$\dot{\lambda} = \frac{R_{\text{TM1}} V_{\text{TM2}} - R_{\text{TM2}} V_{\text{TM1}}}{R_{\text{TM}}^2}$$

The relative separation between missile and target R_{TM} can be expressed in terms of its inertial components by application of the distance formula, as

$$R_{\text{TM}} = \left(R_{\text{TM1}}^2 + R_{\text{TM2}}^2\right)^{\frac{1}{2}}$$

Because the closing velocity is defined as the negative rate of change of the missile target separation, it can be obtained by differentiating the preceding equation, yielding

$$V_c = -\dot{R}_{\text{TM}} = \frac{-(R_{\text{TM1}} V_{\text{TM1}} + R_{\text{TM2}} V_{\text{TM2}})}{R_{\text{TM}}}$$

The magnitude of the missile guidance command n_c can then be found from the definition of proportional navigation, or

$$n_c = N' V_c \dot{\lambda}$$

Because the acceleration command is perpendicular to the instantaneous line of sight, the missile acceleration components in Earth coordinates can be found by trigonometry using the angular definitions from Fig. 2.1. The missile acceleration components are

$$a_{M1} = -n_c \sin \lambda$$
$$a_{M2} = n_c \cos \lambda$$

We have now listed all of the differential equations required to model a complete missile-target engagement in two dimensions. However, some additional equations are required for the initial conditions on the differential equations in order to complete the engagement model.

A missile employing proportional navigation guidance is not fired at the target but is fired in a direction to lead the target. The initial angle of the missile velocity vector with respect to the line of sight is known as the missile lead angle L. In essence we are firing the missile at the expected intercept point. We can see from Fig. 2.1 that, for the missile to be on a collision triangle (missile will hit the target if both continue to fly along a straight-line path at constant velocities), the theoretical missile lead angle can be found by application of the law of sines, yielding

$$L = \sin^{-1}\frac{V_T \sin(\beta + \lambda)}{V_M}$$

In practice, the missile is usually not launched exactly on a collision triangle, as the expected intercept point is not known precisely. The location of the intercept point can only be approximated because we do not know in advance what the target will do in the future. In fact, that is why a guidance system is required! Any initial angular deviation of the missile from the collision triangle is known as a heading error HE. The initial missile velocity components can therefore be expressed in terms of the theoretical lead angle L and actual heading error HE as

$$V_{M1}(0) = V_M \cos(L + HE + \lambda)$$
$$V_{M2}(0) = V_M \sin(L + HE + \lambda)$$

TWO-DIMENSIONAL ENGAGEMENT SIMULATION

To witness and understand the effectiveness of proportional navigation, it is best to simulate the guidance law and test its properties under a variety of circumstances. A two-dimensional missile-target engagement simulation was set up using the differential equations derived in the previous section. The simulation inputs are the initial location of the missile and target, speeds, flight time, and effective navigation ratio. The user can vary the level of the two error sources considered: target maneuver and heading error.

LISTING 2.1 TWO-DIMENSIONAL TACTICAL MISSILE-TARGET ENGAGEMENT SIMULATION

```
n=0;
VM = 3000.;
VT = 1000.;
XNT = 0.;
HEDEG = -20.;
XNP = 4.;
RM1 = 0.;
RM2 = 10000.;
```

```
RT1 = 40000.;
RT2 = 10000.;
BETA=0.;
VT1=-VT*cos(BETA);
VT2=VT*sin(BETA);
HE=HEDEG/57.3;
T=0.;
S=0.;
RTM1=RT1-RM1;
RTM2=RT2-RM2;
RTM=sqrt(RTM1*RTM1+RTM2*RTM2);
XLAM=atan2(RTM2,RTM1);
XLEAD=asin(VT*sin(BETA+XLAM)/VM);
THET=XLAM+XLEAD;
VM1=VM*cos(THET+HE);
VM2=VM*sin(THET+HE);
VTM1 = VT1 - VM1;
VTM2 = VT2 - VM2;
VC=-(RTM1*VTM1 + RTM2*VTM2)/RTM;
while VC >= 0
    if RTM < 1000
            H=.0002;
    else
            H=.01;
    end
    BETAOLD=BETA;
    RT1OLD=RT1;
    RT2OLD=RT2;
    RM1OLD=RM1;
    RM2OLD=RM2;
    VM1OLD=VM1;
    VM2OLD=VM2;
    STEP=1;
    FLAG=0;
    while STEP <=1
        if FLAG==1
                STEP=2;
            BETA=BETA+H*BETAD;
                RT1=RT1+H*VT1;
                RT2=RT2+H*VT2;
                RM1=RM1+H*VM1;
                RM2=RM2+H*VM2;
                VM1=VM1+H*AM1;
                VM2=VM2+H*AM2;
                T=T+H;
        end
```

```
        RTM1=RT1-RM1;
            RTM2=RT2-RM2;
            RTM=sqrt(RTM1*RTM1+RTM2*RTM2);
            VTM1=VT1-VM1;
            VTM2=VT2-VM2;
            VC=-(RTM1*VTM1+RTM2*VTM2)/RTM;
            XLAM=atan2(RTM2,RTM1);
            XLAMD=(RTM1*VTM2-RTM2*VTM1)/(RTM*RTM);
            XNC=XNP*VC*XLAMD;
            AM1=-XNC*sin(XLAM);
            AM2=XNC*cos(XLAM);
            VT1=-VT*cos(BETA);
            VT2=VT*sin(BETA);
            BETAD=XNT/VT;
        FLAG=1;
    end
    FLAG=0;
    BETA=.5*(BETAOLD+BETA+H*BETAD);
    RT1=.5*(RT1OLD+RT1+H*VT1);
    RT2=.5*(RT2OLD+RT2+H*VT2);
    RM1=.5*(RM1OLD+RM1+H*VM1);
    RM2=.5*(RM2OLD+RM2+H*VM2);
    VM1=.5*(VM1OLD+VM1+H*AM1);
    VM2=.5*(VM2OLD+VM2+H*AM2);
    S=S+H;
    if S > =.09999
            S=0.;
        n=n+1;
        ArrayT(n)=T;
        ArrayRT1(n)=RT1;
        ArrayRT2(n)=RT2;
        ArrayRM1(n)=RM1;
        ArrayRM2(n)=RM2;
        ArrayXNCG(n)=XNC/32.2;
        ArrayRTM(n)=RTM;
    end
end
RTM
figure
plot(ArrayRT1,ArrayRT2,ArrayRM1,ArrayRM2),grid
title('Two-dimensional tactical missile-target engagement simulation')
xlabel('Downrange (Ft) ')
ylabel('Altitude (Ft)')
figure
plot(ArrayT,ArrayXNCG),grid
title('Two-dimensional tactical missile-target engagement simulation')
```

```
xlabel('Time (sec)')
ylabel('Acceleration of missle (G)')
clc
output=[ArrayT',ArrayRT1',ArrayRT2',ArrayRM1',ArrayRM2',ArrayXNCG',ArrayRTM' ];
save datfil.txt output /ascii
disp '*** Simulation Complete'
```

A tactical missile-target engagement simulation appears in Listing 2.1. We can see from the listing that the missile and target differential equations are solved using the second-order Runge–Kutta numerical integration technique. As was the case in the second-order system simulation of Chapter 1, the differential equations appear before the FLAG=1 statement. The integration step size is fixed for most of the flight ($H = 0.01$ s) but is made smaller near the end of the flight ($H = 0.0002$ s when $R_{TM} < 1000$ ft) to accurately capture the magnitude of the miss distance. The program is terminated when the closing velocity changes sign, because this means that the separation between the missile and target is a minimum. At this time the missile-target separation is the miss distance: We can see from the preceding equations that the miss distance will always be positive because it is calculated from the distance formula. We can see from the listing that errors can be introduced by changing values in the data statements. Status of the missile and target location, along with acceleration and separation information, is displayed every 0.1 s. Note that the missile acceleration is written to a file datfil.txt in units of gravity.

A sample case was run in which the only disturbance was a 20-deg heading error (HEDEG = −20.). Sample trajectories for effective navigation ratios of 4 and 5 are depicted in Fig. 2.2. We can see from the figure that initially the missile is flying in the wrong direction because of the heading error. Gradually the guidance

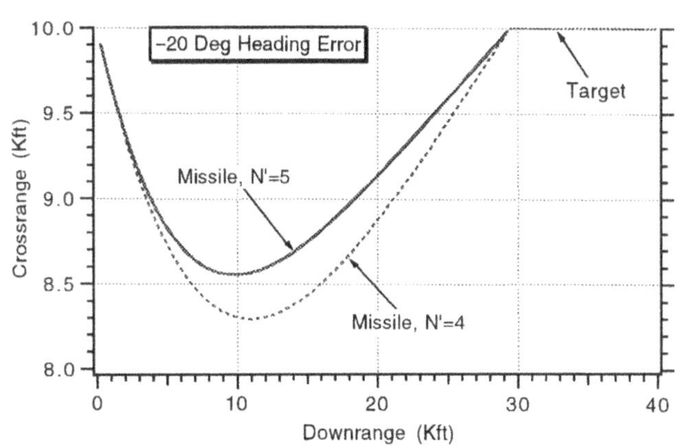

Fig. 2.2 Increasing effective navigation ratio causes heading error to be removed more rapidly.

law forces the missile to home on the target. The larger effective navigation ratio enables the missile to remove the initial heading error more rapidly, thus causing a much tighter trajectory. In both cases, proportional navigation appears to be an effective guidance law because the missile hits the target (near zero miss distance with the simulation).

The resultant missile acceleration histories, displayed in Fig. 2.3, for both cases are somewhat different. The quicker removal of heading error in the higher effective navigation ratio case ($N' = 5$) results in larger missile accelerations at the beginning of the flight and lower accelerations near the end of the flight. In both cases the acceleration profiles for the required missile acceleration to take out the heading error and to hit the target is monotonically decreasing and zero at the end of the flight. Thus, a property of a proportional navigation guidance system is to start taking out heading error as soon as possible but also gradually throughout the *entire* flight. In Chapter 15 we shall study a guidance system that tries to remove the entire heading error immediately. By increasing the effective navigation ratio, we are allowing the missile to take out heading error more rapidly.

Another sample case was run in which the only disturbance was a 3-g target maneuver (XNT = 96.6, HEDEG = 0). In this scenario the missile and target are initially on a collision triangle and flying along the downrange component of the Earth fixed coordinate system (cross-range velocity components of both interceptor and target are zero). Therefore, the target velocity vector is initially along the line of sight, and at first all 3 g of the target acceleration are perpendicular to the line of sight. As the target maneuvers, the magnitude of the target acceleration perpendicular to the line of sight diminishes due to the turning of the target. Sample missile-target trajectories for this case with effective navigation ratios

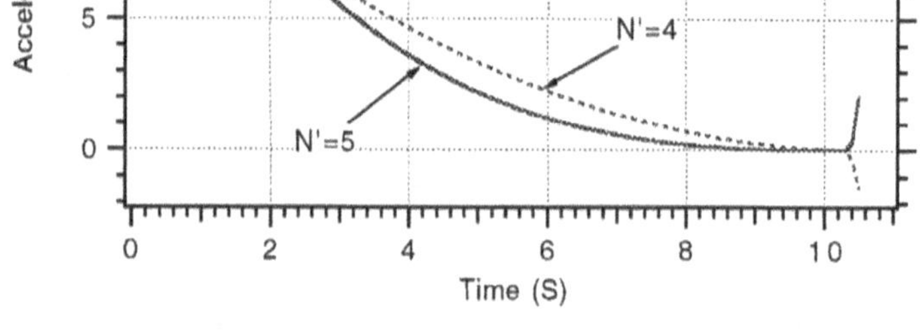

Fig. 2.3 Increasing effective navigation ratio causes more acceleration initially.

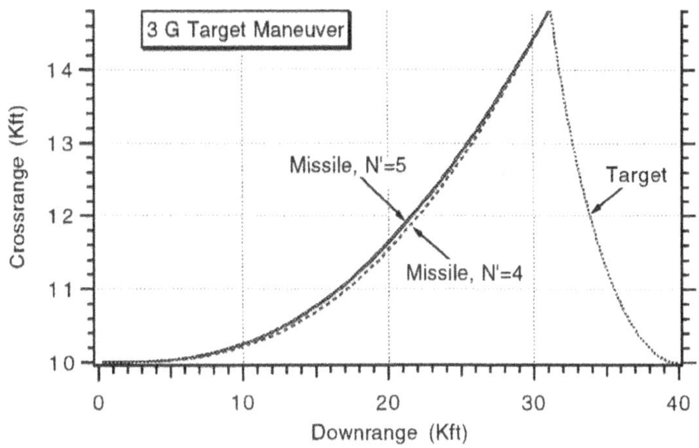

Fig. 2.4 Proportional navigation works against maneuvering target.

of 4 and 5 are depicted in Fig. 2.4. We can see that the higher effective navigation ratio causes the missile to lead the target slightly more than the lower navigation ratio case. Otherwise the trajectories are virtually identical. In both cases, the proportional navigation guidance law enabled the missile to hit the maneuvering target.

However, Fig. 2.5 shows that there are significant differences between the acceleration profiles for the maneuvering target case. Although both acceleration profiles are virtually monotonically increasing for the entire flight, the higher effective navigation ratio requires less acceleration capability of the missile.

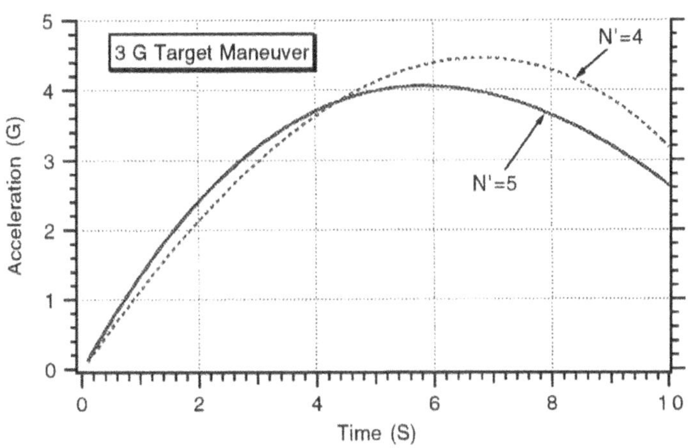

Fig. 2.5 Higher navigation ratio yields less acceleration to hit maneuvering target.

In addition, we can see that the peak acceleration required by the missile to hit the target is significantly higher than the maneuver level of the target (3 g).

In both simulation examples we have seen the effectiveness of proportional navigation guidance. First we saw that proportional navigation is able to hit a target, even if it is initially launched in the wrong direction by 20 deg. Then we observed that the guidance law was also effective in hitting a maneuvering target. In both cases certain acceleration levels were required of the missile in order for it to hit the target. The levels were dependent on the type of error source and the effective navigation ratio. If the missile does not have the acceleration required by the guidance law, a miss will result.

LINEARIZATION

Thus far our understanding of the effectiveness of proportional navigation has come from the numerical simulation results of the two-dimensional engagement simulation. It is critical for the analysis, understanding, and development of design relationships to temporarily depart from the nonlinear missile-target simulation and develop a simpler model. Therefore, we will linearize the two-dimensional engagement model in the hope of gaining more understanding. This does not mean that we will abandon the nonlinear engagement model. In fact, we will always use the nonlinear engagement model to verify the insights generated by powerful analytical techniques to be used on the linearized engagement model.

The linearization of the missile-target geometry can easily be accomplished if we define some new relative quantities as shown in Fig. 2.6. Here y is the relative separation between the missile and target perpendicular to the fixed reference.

The relative acceleration (difference between missile and target acceleration) can be written by inspection of Fig. 2.6 as

$$\ddot{y} = n_T \cos\beta - n_c \cos\lambda$$

If the flight-path angles are small (near head-on or tail chase case), the cosine terms are approximately unity, and the preceding equation becomes

$$\ddot{y} = n_T - n_c$$

Similarly, the expression for the line-of-sight angle can also be linearized using the small-angle approximation, yielding

$$\lambda = y/R_{\mathrm{TM}}$$

For a head-on case, we can approximate the closing velocity as

$$V_c = V_M + V_T$$

whereas in a tail chase case, the closing velocity can be approximated as

$$V_c = V_M - V_T$$

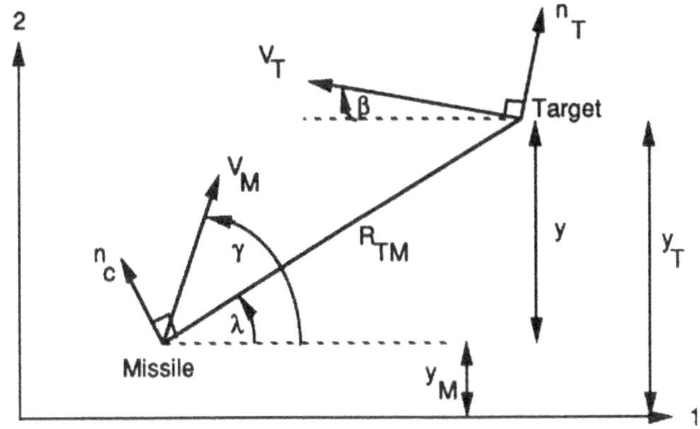

Fig. 2.6 Engagement model for linearization.

Therefore, in a linearized analysis we will treat the closing velocity as a positive constant. Because closing velocity has also been previously defined as the negative derivative of the range from the missile to target, and since the range must go to zero at the end of the flight, we can also linearize the range equation with the time-varying relationship

$$R_{\mathrm{TM}} = V_c(t_F - t)$$

where t is current time and t_F the total flight time of the engagement. Note that t_F is also now a constant. The quantity $t_F - t$ is the time to go until the end of the flight. Therefore, the range from the missile to the target is also the closing velocity multiplied by the time to go until intercept. Because range goes to zero at the end of the flight by definition, we must reexamine the definition of miss distance. The linearized miss distance is taken to be the relative separation between missile and target y at the end of the flight, or

$$\mathrm{Miss} = y(t_F)$$

Because the linearized miss is not obtained from the distance formula, it is only an approximation to the actual miss. However, we shall soon see that the miss distance approximation is very accurate.

LINEARIZED ENGAGEMENT SIMULATION

In the previous section we developed linearized equations for the missile-target engagement. In this section we will see if the resultant linearized equations give performance projections that have trends similar to those of the nonlinear engagement equations. If they do not, then there is no point in developing design relationships based on a meaningless model. If they do, then there may be a point for the interested reader to continue reading this text!

The linearized proportional navigation engagement simulation appears in Listing 2.2. In this simulation the flight time t_F is an input rather than output. We can see from the listing that the simulation only consists of two differential equations: one for relative velocity and the other for relative acceleration. These differential equations are also solved using the second-order Runge–Kutta numerical integration technique. The linearized differential equations appear in the listing before the FLAG=1 statement. Unlike the nonlinear engagement simulation, the integration step size in the linear simulation can be kept fixed for the entire flight ($H = 0.01$ s). The program is stopped when the current time equals the flight time. Nominally the program is set up without errors. Errors can be introduced by changing values in the data statements. The status of the relative position and velocity, along with missile acceleration information, is displayed every 0.1 s.

LISTING 2.2 LINEARIZED ENGAGEMENT SIMULATION

```
XNT=0.;
Y=0.;
VM=3000.;
HEDEG=-20.;
TF=10.;
XNP=4.;
YD=-VM*HEDEG/57.3;
T=0.;
H=.01;
S=0.;
n=0.;
while T<=(TF-1e-5)
    YOLD=Y;
    YDOLD=YD;
    STEP=1;
    FLAG=0;
    while STEP<=1
        if FLAG==1
           STEP=2;
           Y=Y+H*YD;
           YD=YD+H*YDD;
           T=T+H;
        end
        TGO=TF-T+.00001;
        XLAMD=(Y+YD*TGO)/(VC*TGO*TGO);
        XNC=XNP*VC*XLAMD;
        YDD=XNT-XNC;
        FLAG=1;
    end
```

```
    FLAG=0;
    Y=.5*(YOLD+Y+H*YD);
    YD=.5*(YDOLD+YD+H*YDD);
    S=S+H;
    if S>=.0999
       S=0.;
       n=n+1;
       ArrayT(n)=T;
       ArrayY(n)=Y;
       ArrayYD(n)=YD;
       ArrayXNCG(n)=XNC/32.2;
    end
end
figure
plot(ArrayT,ArrayXNCG),grid
xlabel('Time (Sec)')
ylabel('Missile Acceleration (G)')
clc
output=[ArrayT',ArrayY',ArrayYD',ArrayXNCG'];
save datfil.txt output -ascii
disp 'simulation finished'
```

To verify that the linearized engagement model is a reasonable approximation to the nonlinear engagement model, cases that were run for the nonlinear engagement model were repeated using the simulation of Listing 2.2. A sample run was made with the linearized engagement model in which the only disturbance was a −20-deg heading error (HEDEG = −20.). In this case the effective navigation ratio was 4. Acceleration profile comparisons for both the linear and nonlinear engagement models are presented in Fig. 2.7. The figure clearly shows that, even for a relatively large heading error disturbance, the resultant acceleration profiles are virtually indistinguishable. Thus, the linearized model is an excellent approximation to the nonlinear engagement model in the case of a heading error disturbance.

Another sample run was made with the linear engagement model; this time with a 3-*g* target maneuver disturbance. Figure 2.8 shows that this time the linearized model overestimates the missile acceleration requirements. The reason for the discrepancy is that the linear model assumes that the target acceleration magnitude, perpendicular to the line of sight, is always the same and equal to the magnitude of the maneuver. In reality, as the target maneuvers, the component of acceleration perpendicular to the line of sight decreases because the target is turning. Therefore, the nonlinear acceleration requirements due to a maneuvering target are somewhat less than those predicted by the linearized engagement model. However, it is important to note that the linear engagement model accurately predicts the monotonically increasing trend (for most of the flight) for the missile acceleration profile due to a target maneuver.

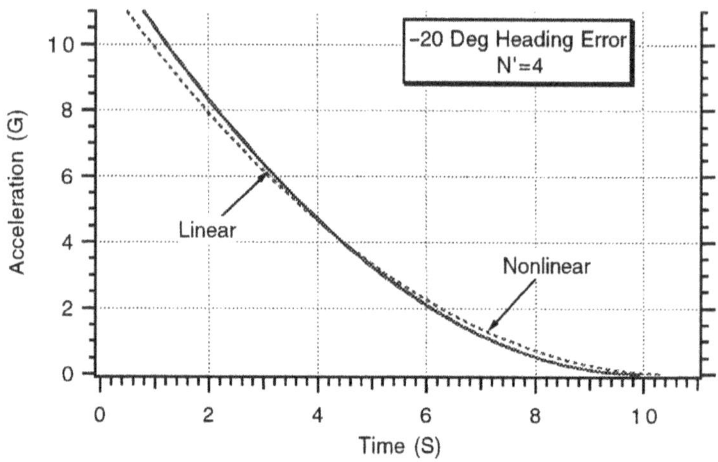

Fig. 2.7 Linearized engagement model yields accurate performance projections for heading error disturbance.

At this point we can conclude that the linearized engagement model yields performance projections of sufficient accuracy to make it worthwhile to proceed with the development of design relationships. We will test the validity of those relationships throughout the text in a variety of environments.

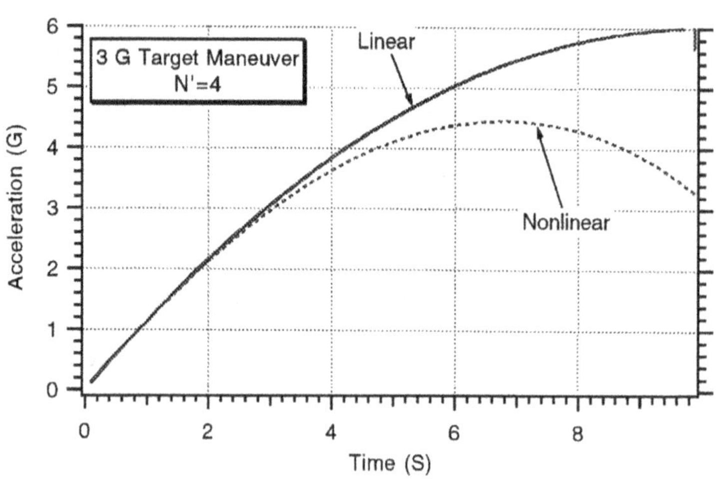

Fig. 2.8 Linear model overestimates missile acceleration due to target maneuver.

IMPORTANT CLOSED-FORM SOLUTIONS

The linearization of the engagement model is important for two reasons. First, with a linear model, powerful computerized techniques such as the method of adjoints (described in Chapters 3 and 4) can be used to analyze the missile guidance system both statistically and deterministically in one computer run. With this technique, error budgets are automatically generated so that key system drivers can be identified and a balanced guidance system design can be achieved. The linear model is also important because, under special circumstances, closed-form solutions can be obtained. These solutions can be used as system sizing aids. In addition, the form of the solutions will suggest how key parameters influence system performance.

Let us consider obtaining closed-form solutions for the two important cases we have already considered in both the linear and nonlinear engagement simulations. The first case is the missile acceleration required to remove a heading error, and the second case is the missile acceleration required to hit a maneuvering target. In the absence of target maneuver the relative acceleration (target acceleration minus missile acceleration) can be expressed as

$$\ddot{y} = -N'V_c\dot{\lambda}$$

Integrating the preceding differential equation once yields

$$\dot{y} = -N'V_c\lambda + C_1$$

where C_1 is the constant of integration. Substitution of the linear approximation to the line-of-sight angle in the preceding expression yields the following time-varying first-order differential equation:

$$\frac{\mathrm{d}y}{\mathrm{d}t} + \frac{N'y}{t_F - t} = C_1$$

As a first-order differential equation of the form

$$\frac{\mathrm{d}y}{\mathrm{d}t} + a(t)y = h(t)$$

has the solution [9–12]

$$y = \exp\left[-\int_0^t a(T)\mathrm{d}T\right]\left\{\int_0^t h(n)\exp\left[\int_0^n a(T)\mathrm{d}T\right]dn + C_2\right\}$$

we can solve the linearized trajectory differential equation exactly. Note that the first constant of integration C_1 is contained in $h(t)$ while the second constant of integration C_2 appears in the preceding equation. Both constants of integration can be found by evaluating initial conditions on y and its derivative. Let us assume that the initial condition on the first state is zero, or

$$y(0) = 0$$

and that the initial condition on the second state is related to the heading error by

$$\dot{y}(0) = -V_M HE$$

where V_M is the missile velocity and HE the heading error in radians. Under these circumstances, after much algebra, we find that the closed-form solution for the missile acceleration due to heading error is given by

$$n_c = \frac{-V_M HE\, N'}{t_F}\left(1 - \frac{t}{t_F}\right)^{N'-2}$$

where t_F is the flight time and N' the effective navigation ratio. We can see that the magnitude of the initial acceleration is proportional to the heading error and missile velocity and inversely proportional to the flight time. Doubling the velocity or heading error will double the initial missile acceleration, whereas doubling the flight time or time available for guidance will halve the initial missile acceleration. In addition, the closed-form solution for the miss distance $y(t_F)$ is zero. In other words, as long as the missile has sufficient acceleration capability, there is no miss due to heading error!

The closed-form solution for the missile acceleration response due to heading error is displayed in normalized form in Fig. 2.9. We can see that higher effective navigation ratios require more acceleration at the beginning of flight than at the end of the flight and less acceleration as the flight progresses. From a system sizing point of view, the designer usually wants to ensure that the acceleration capability of the missile is adequate at the beginning of flight so that saturation can be avoided. For a fixed missile acceleration capability, Fig. 2.9 shows how requirements are placed on minimum guidance or flight time and maximum allowable heading error and missile velocity.

Similarly, if the only disturbance is a target maneuver, the appropriate second-order differential equation becomes

$$\ddot{y} = -N' V_c \dot{\lambda} + n_T$$

with initial conditions

$$y(0) = 0$$
$$\dot{y}(0) = 0$$

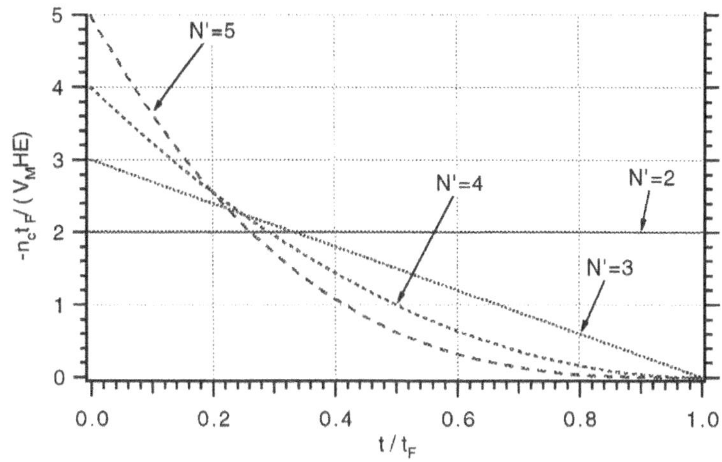

Fig. 2.9 Normalized missile acceleration due to heading error for proportional navigation guidance.

After conversion to a first-order differential equation and much algebra, the solution can be found to be

$$n_c = \frac{N'}{N'-2}\left[1 - \left(1 - \frac{t}{t_F}\right)^{N'-2}\right] n_T$$

It appears that something "magical" happens to the acceleration when the effective navigation ratio is two. Application of L'Hopital's rule eliminates the division by zero in the preceding formula and indicates that

$$\lim_{N' \to 2} n_c = -2\,\ell n\left(\frac{t_F - t}{t_F}\right)$$

This is approximately the same solution as if we simply let $N' = 2.01$ or $N' = 1.99$ in the original closed-form solution for the acceleration as a function of the effective navigation ratio. As with the heading error case, the closed-form solution indicates that the miss distance due to target maneuver is exactly zero!

Unlike the heading error case, missile acceleration due to maneuver is independent of flight time and missile velocity and only depends on the magnitude of the maneuver and the effective navigation ratio. Doubling the maneuver level of the target doubles the missile acceleration requirements.

The closed-form solution for the missile acceleration response due to target maneuver is displayed in normalized form in Fig. 2.10. We can see that higher effective navigation ratios relax the acceleration requirements at the end of the flight. Unlike the heading error response, the missile acceleration required to

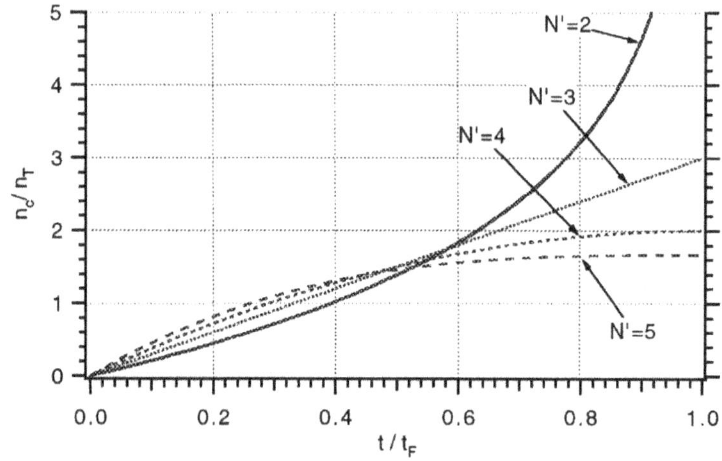

Fig. 2.10 Normalized missile acceleration due to target maneuver for proportional navigation guidance.

hit a maneuvering target increases as the flight progresses. From a system sizing point of view, the designer must ensure that the acceleration capability of the missile is adequate at the end of flight so that saturation can be avoided so that the missile can hit the target.

PROPORTIONAL NAVIGATION AND ZERO EFFORT MISS

Thus far we have seen from simulation results and closed-form solutions that proportional navigation appears to be effective, but we do not know why. Although it is possible to construct geometric arguments showing that it is very logical to issue acceleration commands proportional to the line-of-sight rate (that is, zero line-of-sight rate means we are on a collision triangle and therefore no further commands are necessary), it is not obvious what is happening. The concept of zero effort miss is not only useful in explaining proportional navigation but is also useful in deriving and understanding more advanced guidance laws.

We can define the zero effort miss to be the distance the missile would miss the target if the target continued along its present course and the missile made no further corrective maneuvers. Therefore, if the target does not maneuver, the two components, in the Earth fixed coordinate system, of the zero effort miss can be expressed in terms of the previously defined relative quantities as

$$ZEM_1 = R_{TM1} + V_{TM1}t_{go}$$
$$ZEM_2 = R_{TM2} + V_{TM2}t_{go}$$

where t_{go} is the time to go until intercept. Thus, we can see that in this case the zero effort miss is just a simple prediction (assuming constant velocities and zero acceleration) of the future relative separation between missile and target. From Fig. 2.1 we can see that the component of the zero effort miss that is perpendicular to the line of sight ZEM_{PLOS} can be found by trigonometry and is given by

$$ZEM_{PLOS} = -ZEM_1 \sin\lambda + ZEM_2 \cos\lambda$$

Expansion and simplification of the preceding equation yields

$$ZEM_{PLOS} = \frac{t_{go}(R_{TM1}V_{TM2} - R_{TM2}V_{TM1})}{R_{TM}}$$

Comparing the preceding expression to the expression for line-of-sight rate, we can see that the line-of-sight rate can be expressed in terms of the component of the zero effort miss perpendicular to the line of sight or

$$\dot{\lambda} = \frac{ZEM_{PLOS}}{R_{TM}t_{go}}$$

If we assume that the relative separation between missile and target and closing velocity are approximately related to the time to go by

$$R_{TM} = V_c t_{go}$$

then the proportional navigation guidance command can be expressed in terms of the zero effort miss perpendicular to the line sight as

$$n_c = \frac{N' ZEM_{PLOS}}{t_{go}^2}$$

Thus, we can see that the proportional navigation acceleration command that is perpendicular to the line of sight is not only proportional to the line-of-sight rate and closing velocity but is also proportional to the zero effort miss and inversely proportional to the square of time to go. We shall see later (in Chapters 8, 15, and 20) that this is a very powerful concept, as the zero effort miss can be computed by a variety of methods, including the on-line numerical integration of the assumed nonlinear differential equations of the missile and target.

SUMMARY

In this chapter we have developed and shown the results of a simple two-dimensional proportional navigation missile-target engagement simulation. Results have shown that the proportional navigation law is effective in a variety of cases. Linearization of the nonlinear missile-target geometry was shown to be an accurate approximation to the actual geometry. Closed-form solutions were derived, based on the linearized geometry, for the missile acceleration requirements due to heading error and target maneuver. From these solutions it was shown how the effective navigation ratio influences system performance. Finally, the concept of zero effort miss was introduced, and it was shown how the proportional navi-gation guidance law could be expressed in terms of this concept. Later (Chapters 8, 15, and 20), we shall develop more advanced guidance laws based upon the zero effort miss concept.

REFERENCES

[1] Kennedy, G. P., *Rockets, Missiles and Spacecraft of the National Air and Space Museum*, Smithsonian Institution Press, Washington, DC, 1983.

[2] Benecke, T., and Quick, A. W. (eds.), "History of German Guided Missile Development," *Proceedings of AGARD First Guided Missile Seminar,* 1956.

[3] Nesline, F. W., and Zarchan, P., "A New Look at Classical Versus Modern Homing Guidance," *Journal of Guidance and Control,* Vol. 4, Jan.-Feb. 1981, pp. 78-85.

[4] Yuan, C. L., "Homing and Navigation Courses of Automatic Target-Seeking Devices," RCA Labs., Rept. PTR-12C, Princeton, NJ, Dec. 1942.

[5] Bennett, R. R., and Mathews, W. E., "Analytical Determination of Miss Distance for Linear Homing Navigation Systems," Hughes Aircraft Co., TN-260, Culver City, CA, March 1952.

[6] Fossier, M. W., "The Development of Radar Homing Missiles," *Journal of Guidance, Control, and Dynamics,* Vol. 7, Nov.-Dec. 1984, pp. 641-651.

[7] Yuan, C. L., "Homing and Navigation Courses of Automatic Target-Seeking Devices," *Journal of Applied Physics,* Vol. 19, Dec. 1948, pp. 1122-1128.

[8] Bryson, A. E., and Ho, Y. C., *Applied Optimal Control,* Blaisdell, Waltham, MA, 1969.

CHAPTER 3

Method of Adjoints and the Homing Loop

INTRODUCTION

Although direct simulation is always used in evaluating missile system designs, the adjoint technique has historically been the main computerized analysis and design tool used in tactical missile guidance system design. The adjoint technique goes back at least to Vito Volterra [1], circa 1870, and was used particularly by ballisticians in connection with their theoretical studies of artillery hit dispersions [2]. The adjoint was popularized by Laning and Battin in the 1950s [3].

The adjoint technique is based on the system impulse response and can be used to analyze linear time-varying systems such as the missile homing loop. With the adjoint method, exact performance projections of any quantity at a particular time and information showing how all disturbance terms contribute to the performance are available [4, 5]. In other words, error budgets are automatically generated with the adjoint technique. Although this technique has been used mainly in missile guidance system design and analysis, its application potential is much broader.

In this chapter we shall show how to construct an adjoint model from a missile guidance system homing loop. Numerical examples will be presented that demonstrate the power and utility of the adjoint approach. Performance projection comparisons will be made from nonlinear engagement simulation results and adjoint solutions.

HOMING LOOP

It is convenient to take the linearized engagement equations of Chapter 2 and draw them in block diagram form as is shown in Fig. 3.1. This type of block diagram is known as a homing loop because it is drawn as a feedback control system. In this diagram missile acceleration is subtracted from target acceleration to form a relative acceleration. After two integrations we have relative position, which at the end of the flight is the miss distance. A division by range (or the

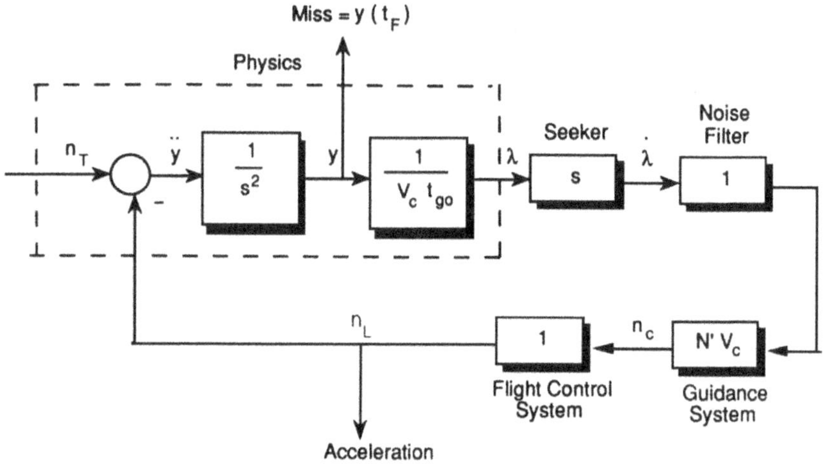

Fig. 3.1 Simplest possible proportional navigation guidance homing loop.

closing velocity multiplied by the time to go until intercept) yields the geometric line-of-sight angle where the time to go is defined as

$$t_{go} = t_F - t$$

The missile seeker, which is represented in Fig. 3.1 as a perfect differentiator, attempts to track the target. Effectively the seeker takes the derivative of the geometric line-of-sight angle, thus providing a measurement of the line-of-sight rate. The noise filter must process the noisy line-of-sight rate measurement of the seeker and provide an estimate of the line-of-sight rate. A guidance command is generated, based on the proportional navigation guidance law, from the noise filter output. In tactical aerodynamic missiles the flight-control system (which is represented by unity gain in Fig. 3.1) must, by moving control surfaces, cause the missile to maneuver in such a way that the achieved acceleration matches the desired acceleration.

Figure 3.1 presents the simplest possible proportional navigation homing loop. In this perfect homing loop, models of the seeker, noise filter, guidance, and flight-control systems have been considered to be perfect and without dynamics. Such a block diagram is known as a *zero-lag guidance system*. The miss distance will always be zero in a zero-lag proportional navigation homing loop.

Guidance system lags or subsystem dynamics will cause miss distance. As long as the lags can be represented by either linear differential or difference equations, the homing loop will still remain linear and more powerful methods of analysis, such as the method of adjoints, can be used to determine system performance and behavior.

SINGLE TIME CONSTANT GUIDANCE SYSTEM

Thus far, in our homing loop analysis, the missile has always hit the target. The strength of proportional navigation is that, in the absence of acceleration saturation effects, zero miss distance can be achieved if there are no lags within the homing loop. If the flight-control system dynamics were modeled as a single lag, or

$$\frac{n_L}{n_c} = \frac{1}{1 + sT}$$

where n_L is the achieved missile acceleration, n_c the commanded missile acceleration, and T the flight-control system time constant. Note that the relative acceleration equation in Fig. 3.1 would also have to be modified to

$$\ddot{y} = n_T - n_L$$

To determine how flight-control system time constant influences miss distance, a massive simulation experiment was conducted. Both the linear and nonlinear engagement simulations, developed in Chapter 2, were run for many different flight times. Each simulation trial had a 1-s flight-control system time constant ($T = 1$ s), an effective navigation ratio of 4, a -20-deg heading error, and a different flight time. The flight times ranged from 0.1 s to 10 s in steps of 0.1 s.

In the linearized model the miss distance $y(t_F)$ can be either positive or negative. Recall that we have already included a few lines of extra code in the nonlinear engagement simulation to also determine if the miss is positive or negative. A positive miss means that the target is above the missile, whereas a negative miss means the opposite.

The miss distance results for each run representing a different flight time, for both the linear and nonlinear engagement models, were recorded. Figure 3.2 displays miss distance as a function of flight time for both the linear and nonlinear engagement simulation results. First, the figure shows that a 1-s flight-control system time constant can have a profound influence on the miss distance. In order for the miss distance to be negligible, the flight time must be large when compared to the flight-control system time constant. In addition, we can see from Fig. 3.2 that the linearized engagement model and nonlinear engagement model results are in close proximity, which demonstrates that the linearized model accurately captures the interaction between guidance system dynamics (flight-control system time constant) and miss distance for the heading error disturbance.

Another important disturbance is *target maneuver*. We saw in Chapter 2 that the linearization of the engagement model in the case of target maneuver was not as accurate as it was for heading error. It is important to determine if the inaccuracy in linearization leads to false conclusions concerning system performance. Both the linear and nonlinear engagement simulations were run

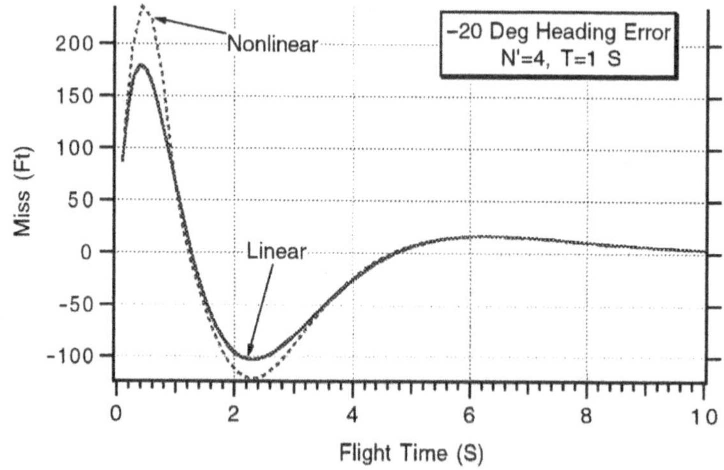

Fig. 3.2 Both models show that heading error miss approaches zero as flight time increases.

for many different flight times, each run having a 1-s flight-control system time constant, an effective navigation ratio of 4, and a 3-*g* target maneuver. Figure 3.3 displays miss distance as a function of the flight time. Again we can see that the miss due to a constant target maneuver is only negligible if the flight time is much larger than the flight-control system time constant. In addition, Fig. 3.3 shows that the linearized guidance system model accurately

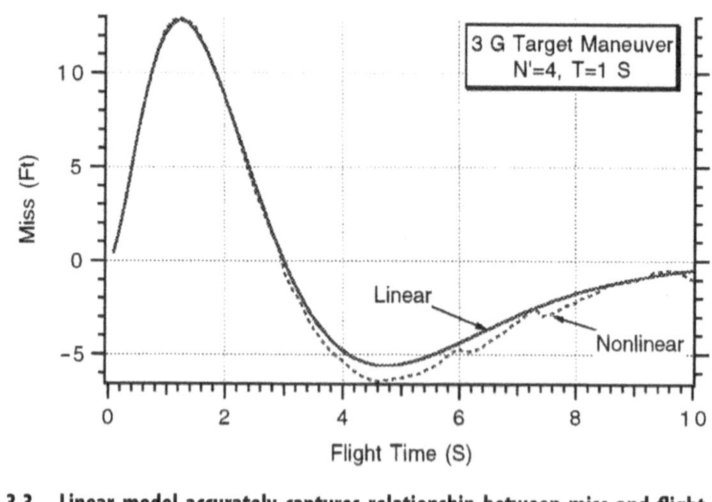

Fig. 3.3 Linear model accurately captures relationship between miss and flight time.

captures the effect of flight control-system time constant on miss distance. The jaggedness in the nonlinear results is due to the approximate way in which the miss distance is computed. At the end of the flight, the integration step size is reduced to 0.0002 s. This means that, for the case considered, where the closing velocity is 4000 ft/s, the nonlinear miss distance computation is only good to 0.8 ft (4000 * 0.0002).

In this section we have shown the very important result that, when the homing loop has guidance system dynamics, the linearized guidance system model yields very accurate miss distance performance projections for both heading error and target maneuver disturbances. Therefore, techniques that depend on the linearized engagement model for miss distance projections should also be accurate.

HOW TO CONSTRUCT AN ADJOINT

In this section we will see how the miss distance results of Figs. 3.2 and 3.3, which were generated from many simulation trials, can be obtained in one computer run using the method of adjoints. However, we must first learn how to construct an adjoint model from the original system.

For every linear deterministic system, there exists an adjoint system that can be constructed from the original system, given in block diagram form, by application of three Rules [3–6].

RULE 1 : CONVERT ALL SYSTEM INPUTS TO IMPULSES

To construct an adjoint we must have impulsive inputs in the original system. Because impulsive inputs may not exist in the original system, block diagram manipulation of the actual inputs of the original system may be necessary. For example, deterministic inputs can be converted to impulsive inputs by judicious use of integrators. Figure 3.4 shows that a step input and an integrator-driven impulse are equivalent at the integrator output. The figure also shows that an initial condition is equivalent at the integrator output to an integrator with an impulsive input.

RULE 2: REPLACE t BY $t_F - t$ IN THE ARGUMENTS OF ALL TIME-VARYING COEFFICIENTS

Often a linear system has a gain that can be expressed as a function of time, either in analytical or tabular form. Care must be taken with time-varying gains when applying the method of adjoints. Figure 3.5 shows, by example, how both a time-varying gain and a gain expressed as a tabular function of time can be converted to the adjoint domain. Notice that the adjoint of a table, which is a function of time, is the same table with the gains reversed. Otherwise, gains that are a function of time simply have t replaced by $t_F - t$ when the adjoint is taken, where t_F is the final time or time of flight.

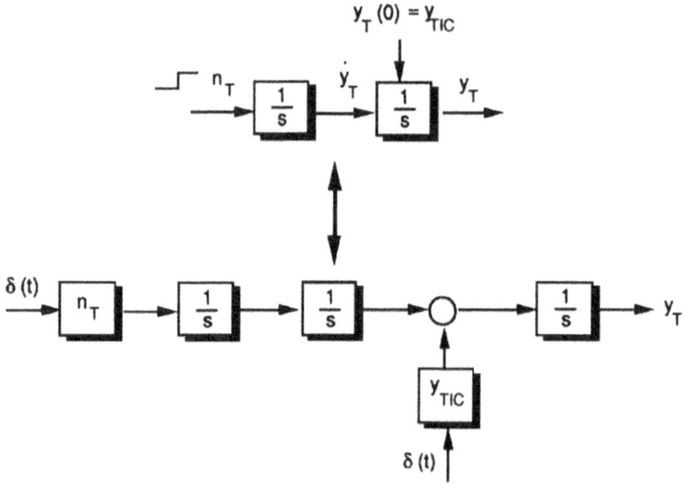

Fig. 3.4 Steps and initial conditions can be replaced by impulses.

RULE 3: REVERSE ALL SIGNAL FLOW, REDEFINING NODES AS SUMMING JUNCTIONS AND VICE VERSA

Figure 3.6 shows how summing junctions and nodes are converted in going from the original to the adjoint system. Note that all original system inputs become adjoint outputs and vice versa. This simple relationship between the two systems enables one to take an adjoint by first drawing the original block diagram and then using tracing paper to construct the adjoint model.

Function \ System	Original System	Adjoint System
Time Varying Gain	$K(t) = at + b$ $K(t) = \dfrac{1}{a(t_F - t) + b}$	$K(t_F - t) = a(t_F - t) + b$ $K(t_F - t) = \dfrac{1}{at + b}$
Table	t / K: 0 / 8; 1 / 4; 2 / 3; 3 / 9	t / K: 0 / 9; 1 / 3; 2 / 4; 3 / 8

Fig. 3.5 Taking the adjoint of a time-varying gain.

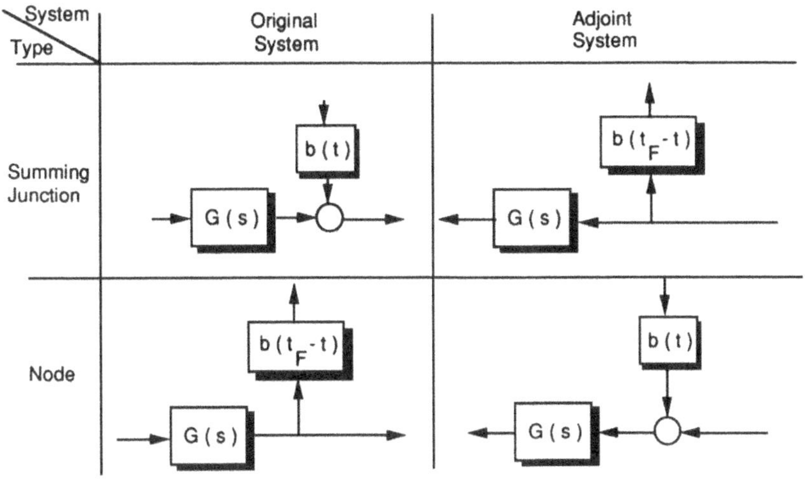

Fig. 3.6 Adjoints redefine branch points and nodes.

Figure 3.7 presents an example of a single-lag, proportional navigation homing loop in which a step target maneuver disturbance has been converted to an impulsive input by the use of an extra integrator. In addition, a heading error initial condition has also been converted to an impulsive input. The output of interest is the miss distance or $y(t_F)$. A simulation of this block diagram will yield y as a function of time $y(t)$, and the last value of y will be the miss distance $y(t_F)$. To find the miss due to a target maneuver disturbance,

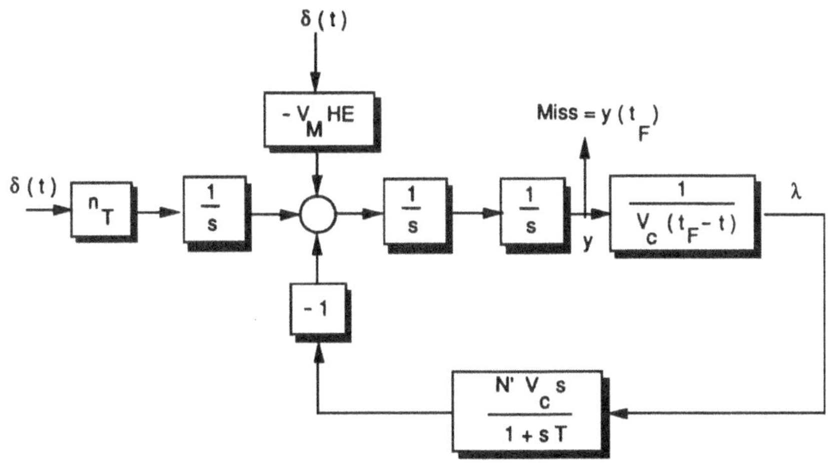

Fig. 3.7 Single-lag proportional navigation homing loop.

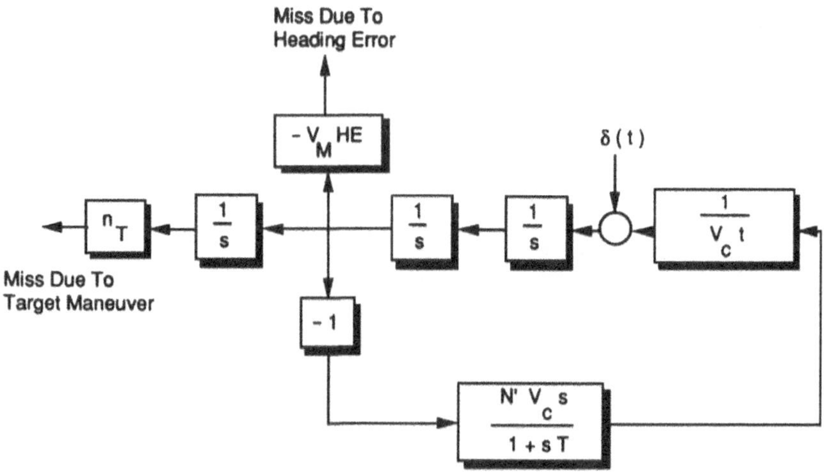

Fig. 3.8 Adjoint of homing loop.

we would have to set the heading error disturbance to zero, and to find the miss due to an initial heading error, the target maneuver disturbance would have to be set to zero.

The adjoint of this homing loop, obtained by following the rules for constructing an adjoint, is shown in Fig. 3.8. Here the original output of interest [miss distance or $y(t_F)$] becomes an impulsive input to the adjoint system, and the two original system inputs (target maneuver and heading error) become two adjoint outputs. A simulation of the adjoint block diagram will yield y as a function of flight time $y(t_F)$. This means that in an adjoint simulation we obtain the miss distances due to both a step target maneuver and initial heading error for various flight times—all obtained in *one* computer run!

ADJOINT MATHEMATICS

The impulse response of the adjoint system h^* and the impulse response of the original system h are related by

$$h^*(t_F - t_I, t_F - t_O) = h(t_O, t_I)$$

where t_I and t_O are the impulse application and observation times, respectively, of the original system. This equation means that applying an impulse at time t_I and observing the output at time t_O of the original system is equivalent to applying an impulse to the adjoint system at time $t_F - t_O$ and observing the adjoint output at time $t_F - t_I$. The importance of this fundamental relationship becomes more apparent when it is desired to observe the impulse response of the original

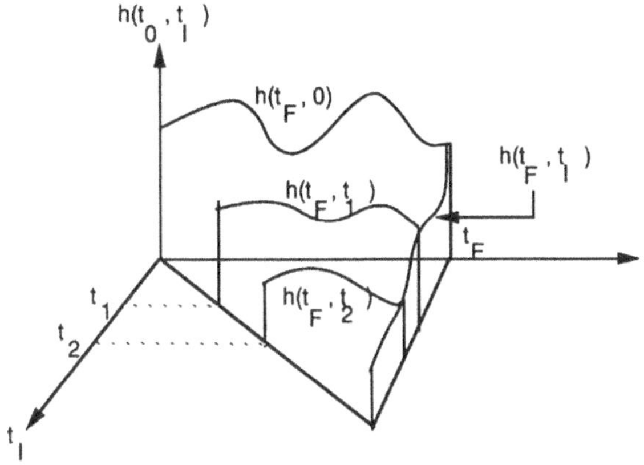

Fig. 3.9 Impulse response of original system for different application times.

system at time t_F due to various impulse application times. This means that in order to generate $h(t_F, t_I)$ it becomes necessary to simulate the system response for each impulse application time as shown in Fig. 3.9.

However, since the observation time is the final time $(t_O = t_F)$, only one adjoint response needs be generated since the fundamental adjoint relationship simplifies to

$$h^*(t_F - t_I, 0) = h(t_F, t_I)$$

Therefore, an impulse applied at any time t_I and observed only at the final time t_F in the original system is equivalent to applying an impulse at time zero in the adjoint system and monitoring the output at time $t_F - t_I$. Figure 3.10 shows that the adjoint impulse response is identical to the impulse response of the original system, except that it is generated backwards!

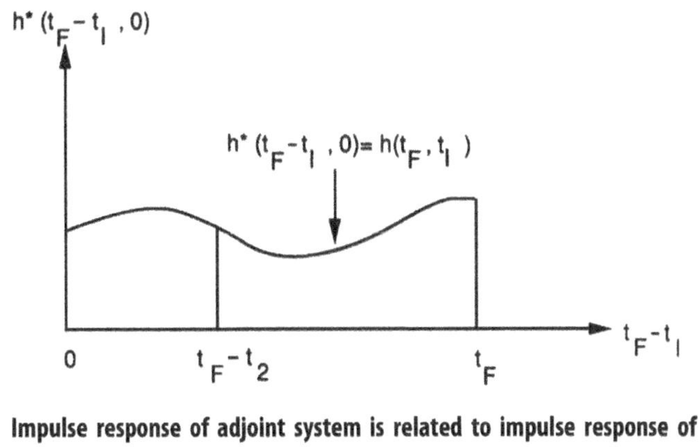

Fig. 3.10 Impulse response of adjoint system is related to impulse response of original system.

ADJOINTS FOR DETERMINISTIC SYSTEMS

To fully understand the utility of adjoints, let us consider the convolution integral, or

$$y(t) = \int_{-\infty}^{t} x(\tau)h(t, \tau)\,\mathrm{d}\tau$$

where x is the system input and h the system impulse response. For physically realizable (noncausal) systems, this integral becomes

$$y(t) = \int_{0}^{t} x(\tau)h(t, \tau)\,\mathrm{d}\tau$$

A step input of magnitude a changes the preceding equation to

$$y(t) = a\int_{0}^{t} h(t, \tau)\,\mathrm{d}\tau$$

Therefore, this equation states that the step response of a system can be found by integrating the impulse response. A closer examination reveals that this revelation is of no practical utility because the variable of integration is with respect to τ. This means that many impulse responses, each with a different application time, would have to be generated. Then the results would have to be saved and then integrated—just to get a system step response! Of course, it would be much easier to avoid the convolution integral and instead just simulate the system with the step input and then observe the output in order to get the system step response.

Let us now see if the method of adjoints can be useful in the case where the system has a step input. We can substitute the fundamental relationship between the original and adjoint system impulse responses into the convolution integral, yielding

$$y(t) = a\int_{0}^{t} h^{*}(t_F - \tau, t_F - t)\,\mathrm{d}\tau$$

Variables can be changed according to

$$x = t_F - \tau$$
$$\mathrm{d}x = -\mathrm{d}\tau$$

Hence, we obtain

$$y(t) = a\int_{t_F - t}^{t_F} h^{*}(x, t_F - t)\,\mathrm{d}x$$

If the observation time of interest is the final time ($t = t_F$), the preceding relationship simplifies to

$$y(t_F) = a \int_0^{t_F} h^*(x, 0)\, dx$$

Note that the integration in the adjoint system is with respect to the observation time rather than the impulse application time. This means that the original system step response output at time t_F can be obtained by using an impulsive input in the adjoint system at time zero and then integrating the output. The resultant one computer run adjoint solution obtains the step response value at the final time for *all* final time values! For time-invariant systems the original system step response, for all final times, could also have been obtained in one computer run. However, for time-varying systems, many original system computer runs would have been required to obtain the same information as that provided by the adjoint response.

To see further benefits from the method of adjoints, let us consider many step input disturbances to the original system as shown in Fig. 3.11. Here, not only would many computer runs be required to find the system step response for different flight times, but also information showing how each disturbance contributed to the total output would require many more computer runs. To generate this type of error budget in the original system for N disturbances, N computer runs

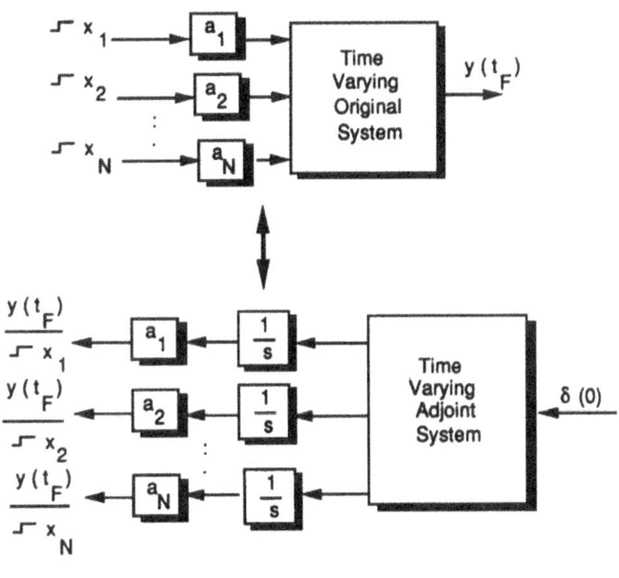

Fig. 3.11 Equivalence between adjoint and original systems for deterministic step disturbances.

would be required (each run only having that disturbance), and then superposition could be invoked (add up all responses) to get the total step response. However, original system inputs become adjoint outputs; thus, only one adjoint run would be required to get the total step response value at the final time and to also automatically generate an error budget. Only the adjoint outputs (original system inputs) have to be monitored, as shown in Fig. 3.11, and superposition allows the total output to be expressed as

$$y(t_F) = \frac{y(t_F)}{\sqrt{x_1}} + \frac{y(t_F)}{\sqrt{x_2}} + \cdots + \frac{y(t_F)}{\sqrt{x_N}}$$

Thus, when there are many deterministic disturbances to the original system, one adjoint computer run yields the system response for all final times, along with a detailed error budget showing how each disturbance influenced total system performance.

DETERMINISTIC ADJOINT EXAMPLE

To demonstrate the practical utility of adjoint theory for a system with deterministic inputs, let us reconsider the proportional navigation homing loop example of Fig. 3.7. After following the rules for constructing an adjoint, we obtain the detailed adjoint model shown in Fig. 3.12.

All of the integrator inputs and outputs are marked in Fig. 3.12 for the purpose of understanding the adjoint simulation. With the exception of the two adjoint outputs, none of the quantities shown in the adjoint model has any physical

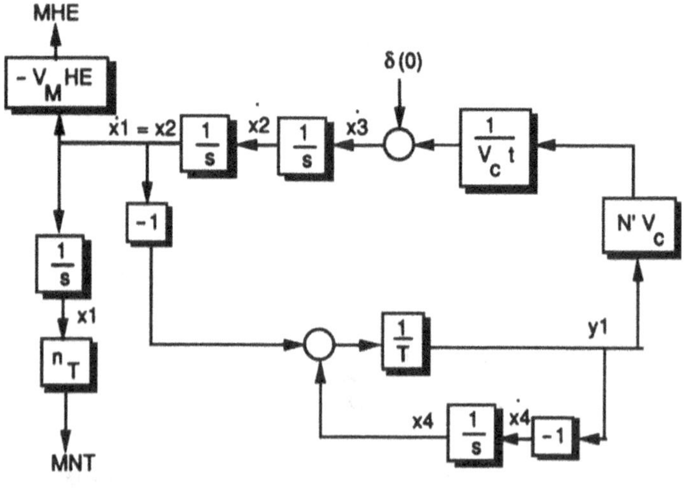

Fig. 3.12 Adjoint simulation model of single-lag guidance system.

meaning. The impulse required for adjoint initialization can also be represented as an initial condition on the $x3$ integrator. The output $x1$ is the miss distance sensitivity (multiply by n_T to get miss) due to a step target maneuver, whereas the output $x2$ is related to the miss sensitivity (multiply by $-V_M\ HE$ to get miss) due to an initial heading error.

Listing 3.1 presents the adjoint simulation of Fig. 3.12. Here the four differential equations of Fig. 3.12 (which appear before the FLAG=1 statement in the listing) are integrated using the second-order Runge–Kutta method with an integration step size of 0.01 s. The integration step size is small enough in this sample program to get fairly accurate answers. Note that in this example the effective navigation ratio is four, the guidance time constant is 1 s, there is -20 deg of heading error, and the maneuver level is 96.6 ft/s^2 or 3 *g*.

The adjoint output due to a 3-*g* step target maneuver appears in Fig. 3.13 along with results from the linearized engagement simulation (run for many different flight times). The figure shows that the adjoint simulation yields accurate miss distance projections for many different flight times in one computer run. Because the adjoint is linear, we can apply superposition to the results. Doubling the target maneuver acceleration level doubles the miss. The abscissa of Fig. 3.13 can either be interpreted as flight time or the time to go until intercept at which the target initiates its maneuver. Therefore, in this example, a 3-*g* maneuver causes nearly 12 ft of miss if the flight time is only 1 s or if the maneuver occurs at 1 s to go before intercept. The figure also shows that long flights (flight time large compared to guidance system time constant), or flights with maneuver initiation occurring at the beginning of flight (large time to go), result in virtually zero miss.

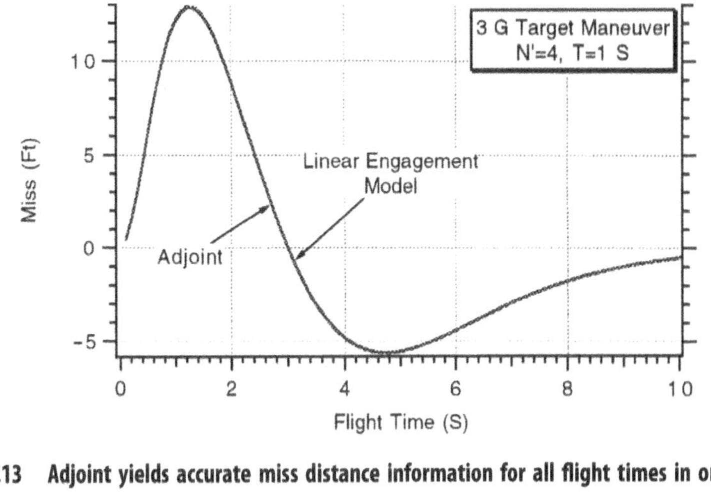

Fig. 3.13 Adjoint yields accurate miss distance information for all flight times in one run.

LISTING 3.1 SINGLE-LAG ADJOINT WITH SECOND-ORDER RUNGE-KUTTA INTEGRATION

```
n=0;
XNT=96.6;
XNP=4.;
TAU=1.;
TF=10;
VM=3000.;
HEDEG=-20.;
T=0.;
S=0.;
TP=T+.00001;
X1=0;
X2=0;
X3=1;
X4=0;
H=.01;
HE=HEDEG/57.3;
while TP <=(TF-1e-5)
    X1OLD=X1;
    X2OLD=X2;
    X3OLD=X3;
    X4OLD=X4;
    STEP=1;
    FLAG=0;
    while STEP <=1
            if FLAG==1
                STEP=2;
                X1=X1+H*X1D;
                X2=X2+H*X2D;
                X3=X3+H*X3D;
                X4=X4+H*X4D;
                TP=TP+H;
            end
            X1D=X2;
            X2D=X3;
            Y1=(X4-X2)/TAU;
            TGO=TP+.00001;
            X3D=XNP*Y1/TGO;
            X4D=-Y1;
            FLAG=1;
    end
    FLAG=0;
    X1=(X1OLD+X1)/2+.5*H*X1D;
    X2=(X2OLD+X2)/2+.5*H*X2D;
    X3=(X3OLD+X3)/2+.5*H*X3D;
```

```
    X4=(X4OLD+X4)/2+.5*H*X4D;
    S=S+H;
    if S>=.0999
            S=0.;
            n=n+1;
            ArrayTP(n)=TP;
            ArrayXMNT(n)=XNT*X1;
            ArrayXMHE(n)=-VM*HE*X2;
    end
end
figure
plot(ArrayTP,ArrayXMNT),grid
xlabel('Flight Time (Sec)')
ylabel('Target Maneuver Miss (Ft)')
figure
plot(ArrayTP,ArrayXMHE),grid
xlabel('Flight Time (Sec)')
ylabel('Heading Error Miss (Ft)')
clc
output=[ArrayTP',ArrayXMNT',ArrayXMHE'];
save datfil.txt output -ascii
disp 'simulation finished'
```

Another adjoint output, from the same simulation trial, also yields miss distance information for a heading error disturbance. These results, along with linearized engagement simulation results (run for many different flight times) appear in Fig. 3.14. We can see that the adjoint also yields accurate performance projections in the case of the heading error disturbance.

Therefore, we can see that a great deal of information is available from one adjoint run representing a system with many deterministic disturbances. Miss distance sensitivity data for a variety of disturbances and for all flight times can be obtained in a single adjoint run. We can also tell from the adjoint output when the guidance system is most sensitive to an error source. In addition, the shape of the adjoint output also provides information concerning system behavior. For example, if the adjoint output for a deterministic input does not approach zero as the flight time increases, then we know that the missile cannot guide effectively.

ADJOINT CLOSED-FORM SOLUTIONS [7, 8]

If there are no dynamics in a proportional navigation homing loop, then the resultant miss distance should always be zero. Guidance system dynamics cause miss distance. Under special circumstances it is possible to obtain closed-form solutions for the miss distance when there are dynamics in the homing loop.

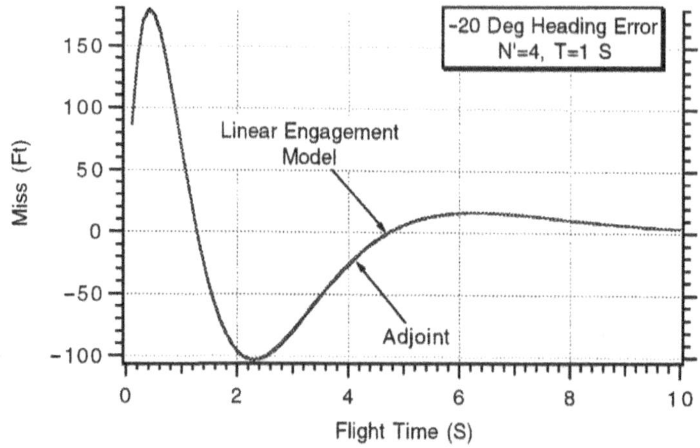

Fig. 3.14 Adjoint also yields accurate heading-error-induced miss information for all flight times in one run.

These closed-form solutions can be used to gain insight into the effectiveness of homing and also be used to check the accuracy of computerized adjoints. In addition, we shall see in Chapter 6 that the normalization factors developed from simpler systems are also valid for more complex systems.

Consider the single time constant proportional navigation guidance system shown in Fig. 3.15a. In this guidance system there are two disturbances (the target maneuver and the heading error) that are represented as impulsive inputs so that we can easily take the adjoint later. For convenience, we can use block diagram manipulation to remove the two integrations from the main loop and place them before the two inputs. After some work we obtain Fig. 3.15b.

For conceptual purposes we may be interested in guidance system dynamics other than first order. Therefore we will use shorthand notation and express the guidance system dynamics as W. For a single-lag guidance system, W can be expressed as

$$W_{\substack{\text{Single}\\ \text{Lag}}} = \frac{N'}{s(1+sT)}$$

and in a perfect or zero-lag guidance system W is given by

$$W_{\substack{\text{Zero}\\ \text{Lag}}} = \frac{N'}{s}$$

In general W can represent guidance system dynamics of any order.

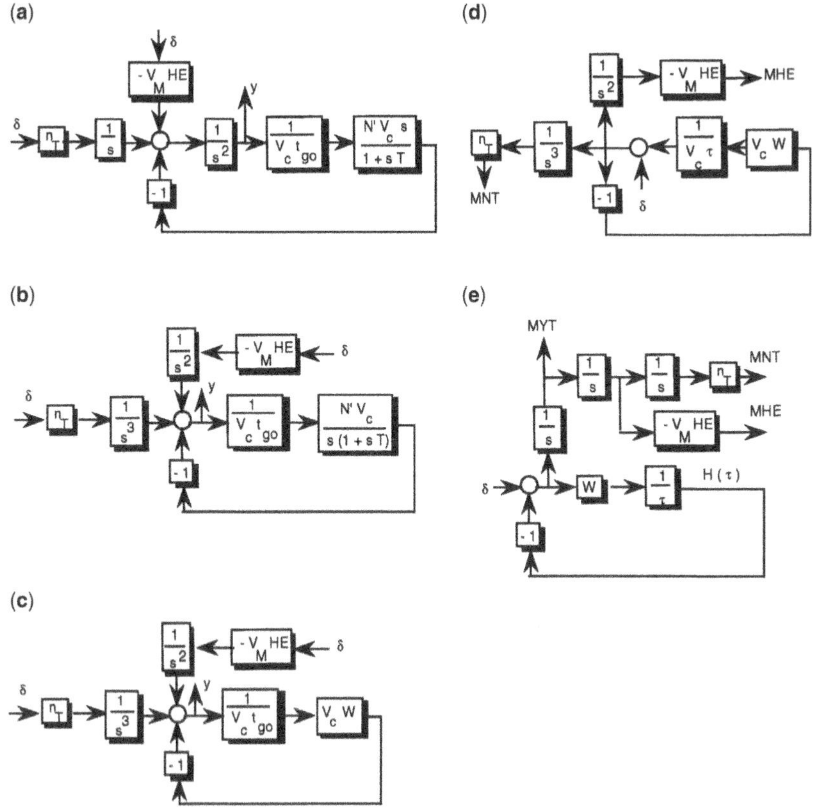

Fig. 3.15 Methodology in getting generalized homing loop adjoint: a) single time constant proportional navigation guidance system; b) block diagram manipulation; c) replace guidance transfer function with *W*; d) take adjoint for miss distance; e) reorient block diagram.

Using W to represent the guidance system dynamics, a more generalized proportional navigation guidance system is displayed in Fig. 3.15c. The miss distance adjoint of the generalized guidance system can be obtained by inspection (that is, by following the rules of adjoints outlined in this chapter) and applying an impulse at time zero where y was in the original system. The resultant generalized adjoint block diagram appears in Fig. 3.15d. Using block diagram manipulation and reorienting the figure yields the final generalized adjoint shown in Fig. 3.15e. We can see that the miss due to a step target maneuver MNT is after three integrations while the miss due to heading error MHE (that is, step in target velocity) is after two integrations. Therefore MYT, which is after one integration, must represent a step in target displacement. This important error source will be investigated in detail in Chapter 19.

For convenience, $H(\tau)$ is also indicated as an adjoint signal of interest. From the convolution integral we can relate the adjoint output to the input by

$$H(\tau) = \frac{1}{\tau}\int W(x)[\delta(\tau - x) - H(\tau - x)]\,\mathrm{d}x$$

Converting from the time to the frequency domain (Laplace transform notation), we can express the preceding relationship as

$$\frac{-\mathrm{d}H(s)}{\mathrm{d}s} = W(s)[1 - H(s)]$$

because of a Laplace transform identity. Recall also that

$$\frac{\mathrm{d}}{\mathrm{d}s}[1 - H(s)] = \frac{-\mathrm{d}H(s)}{\mathrm{d}s}$$

For convenience let us allow $H(s)$ to be replaced by H and $W(s)$ to be replaced by W. Therefore, substitution yields

$$\int \frac{\mathrm{d}(1 - H)}{1 - H} = \int W\,\mathrm{d}s$$

Finally, integrating both sides of the preceding equations yields the important conclusion

$$1 - H = c\exp\left(\int W\,\mathrm{d}s\right)$$

where c is a constant of integration. We can evaluate c by first recognizing from Fig. 3.15e that a miss due to a unit step in target displacement MYT can be expressed in Laplace transform notation as

$$MYT(s) = \frac{1 - H(s)}{s}$$

We know that in the time domain the miss due to a unit step target displacement at flight time zero is unity. Therefore, using the initial value theorem, which relates the time domain with the Laplace transform domain, we can say that

$$MYT(0) = 1 = \lim_{s\to\infty} s\left(\frac{1 - H}{s}\right)$$

Therefore c is chosen to make

$$\lim_{s\to\infty}\left[c\exp\left(\int W\,ds\right)\right] = 1$$

Because the simplest possible guidance system has at least one integration, the constant of integration becomes unity ($c = 1$).

To demonstrate that we have enough analytical tools to find closed-form solutions under special circumstances, let us find the miss due to a step target maneuver for a single-lag guidance system. As mentioned before, the guidance system transfer function in this case is

$$W(s) = \frac{N'}{s(1+sT)}$$

where N' is the effective navigation ratio and T the guidance system time constant. Because

$$1 - H = \exp\left(\int W\,ds\right)$$

we get after integration

$$1 - H(s) = \left[s\Big/\left(s+\frac{1}{T}\right)\right]^{N'}$$

Therefore, the miss due to a step maneuver of magnitude n_T is given by

$$\frac{MNT}{n_T}(s) = \frac{1-H(s)}{s^3} = \frac{1}{s^3}\left[s\Big/\left(s+\frac{1}{T}\right)\right]^{N'}$$

For an effective navigation ratio of 4, the miss, in Laplace transform notation, is given by

$$\left.\frac{MNT}{n_T}\right|_{N'=4}(s) = \frac{s}{(s+1/T)^4}$$

Taking the inverse Laplace transform of the preceding expression yields the miss due to a step target maneuver in the adjoint time domain as

$$\left.\frac{MNT}{n_T}\right|_{N'=4}(\tau) = \tau^2 e^{-\tau/T}\left(0.5 - \frac{\tau}{6T}\right)$$

where τ is adjoint time and can be interpreted as either time of flight (t_F) or time to go at which the maneuver occurs.

The single-lag adjoint simulation of Listing 3.1 was run for a case in which the target maneuver level was 3 g, the guidance system time constant was 1 s, and the effective navigation ratio was 4. Figure 3.16 displays the adjoint simulation results

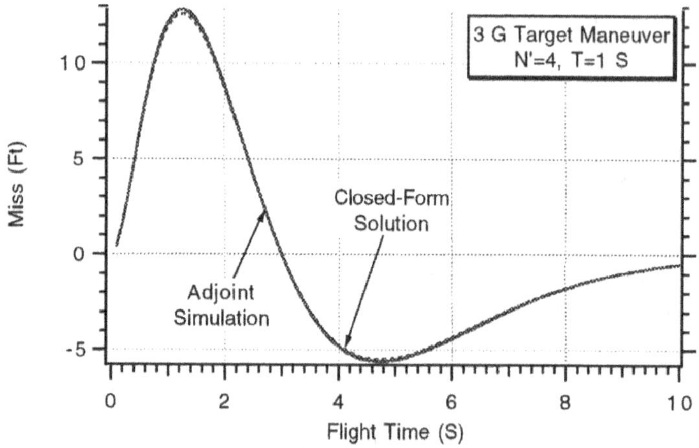

Fig. 3.16 Closed-form solution agrees with computerized adjoint for step target maneuver.

along with the closed-form solution. We can see from the figure that both the adjoint results and closed-form solution results are virtually identical, which proves that it is possible for theory and simulation to agree.

The nonlinear engagement simulation of Chapter 2 was rerun for the same case (3-g target maneuver, $N' = 4$, and $T = 1$ s) for many different flight times. The nonlinear results and the closed-form solution results are compared in Fig. 3.17. We can see that the miss distance projections from the closed-form solution are in excellent agreement with the nonlinear results.

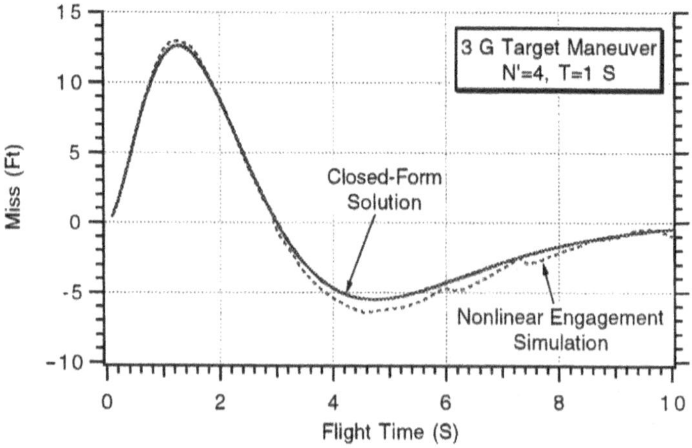

Fig. 3.17 Miss distance formula and nonlinear simulation results are in close agreement.

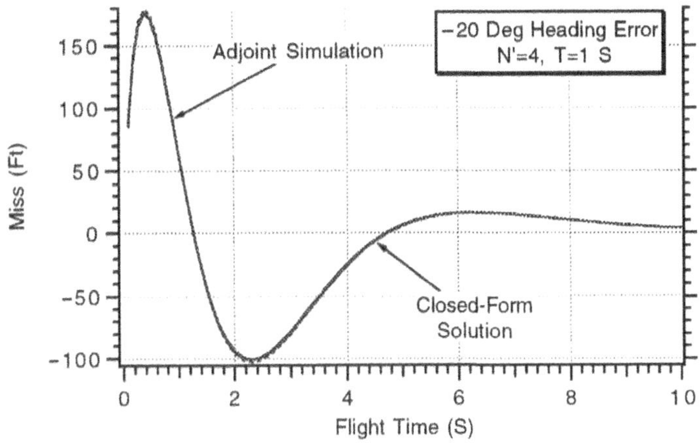

Fig. 3.18 Closed-form solution agrees with computerized adjoint for heading error.

Similarly, a closed-form expression can be found for the miss due to an initial heading error. The generalized formula for the heading error miss is given by

$$\frac{MHE}{HE}(s) = \frac{-V_M[1 - H(s)]}{s^2}$$

Following a similar procedure to that of the target maneuver case and assuming an effective navigation ratio of 4 for a single-lag guidance system, we obtain a closed-form solution for the heading error miss:

$$\left.\frac{MHE}{HE}\right|_{N'=4}(\tau) = -V_M \tau e^{-\tau/T}\left(1 - \frac{\tau}{T} + \frac{\tau^2}{6T^2}\right)$$

The adjoint simulation results for a -20-deg heading error and the closed-form solution results are displayed in Fig. 3.18. Again we can see that the adjoint simulation results are in complete agreement with the closed-form solution.

NORMALIZATION

In the previous section we showed how closed-form solutions for the heading error and target maneuver could be derived for a single time constant guidance system. Specific solutions were derived for the case in which the effective navigation ratio was 4. Following the same procedure outlined in the previous section, closed-form solutions were derived for the miss due to a target maneuver when the effective navigation ratio varied between 3 and 5. The solutions are

$$\left.\frac{\text{Miss}}{n_T}\right|_{N'=3} = 0.5t_F^2 e^{-t_F/T}$$

$$\left.\frac{\text{Miss}}{n_T}\right|_{N'=4} = t_F^2 e^{-t_F/T}\left(0.5 - \frac{t_F}{6T}\right)$$

$$\left.\frac{\text{Miss}}{n_T}\right|_{N'=5} = t_F^2 e^{-t_F/T}\left(0.5 - \frac{t_F}{3T} + \frac{t_F^2}{24T^2}\right)$$

where n_T is the maneuver level of the target (in ft/s²), t_F is the flight time (in s), T is the guidance system time constant (in s), and Miss is the miss distance (in ft).

In a similar way, miss distance formulas can be derived for heading error in a single time constant guidance system. The formulas are

$$\left.\frac{\text{Miss}}{-V_M HE}\right|_{N'=3} = t_F e^{-t_F/T}\left(1 - \frac{t_F}{2T}\right)$$

$$\left.\frac{\text{Miss}}{-V_M HE}\right|_{N'=4} = t_F e^{-t_F/T}\left(1 - \frac{t_F}{T} + \frac{t_F^2}{6T^2}\right)$$

$$\left.\frac{\text{Miss}}{-V_M HE}\right|_{N'=5} = t_F e^{-t_F/T}\left(1 - 1.5\frac{t_F}{T} + \frac{t_F^2}{2T^2} + \frac{t_F^3}{24T^3}\right)$$

where V_M is the missile velocity (in ft/s) and HE the heading error (in rad).

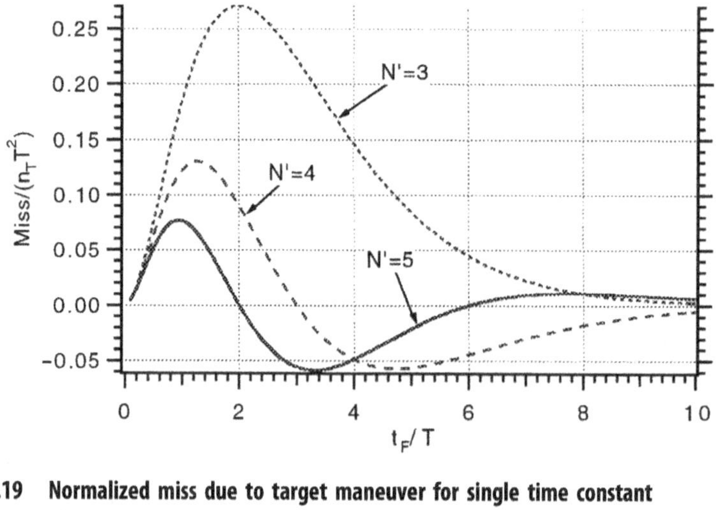

Fig. 3.19 Normalized miss due to target maneuver for single time constant guidance system.

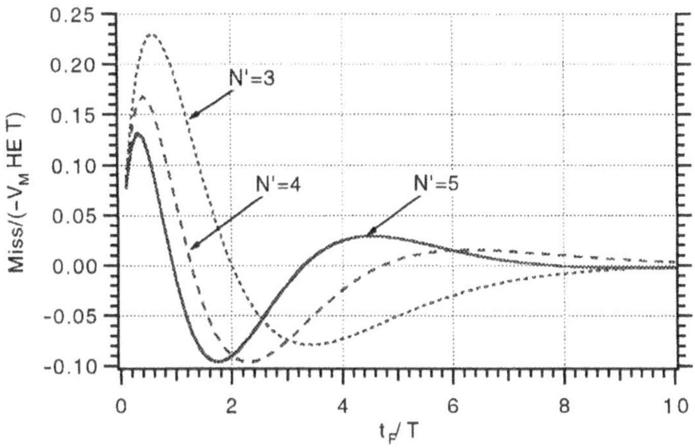

Fig. 3.20 Normalized miss due to heading error for single time constant guidance system.

The closed-form solutions for both the target maneuver and heading error miss can be normalized for conciseness. For example, Fig. 3.19 displays the target maneuver miss sensitivity, in normalized form, for various effective navigation ratios. From the normalization factor it becomes obvious that, as the ratio of the flight time to the guidance system time constant (t_F/T) becomes large, the miss eventually goes to zero. From the normalization on the ordinate it becomes obvious that, for a given ratio of flight time to guidance time constant, doubling the guidance system time constant quadruples the miss!

Figure 3.20 displays the heading error miss sensitivity, in normalized form, for various effective navigation ratios. In this case too, it is obvious from the figure that, as the ratio of the flight time to the guidance system time constant (t_F/T) becomes large, the miss eventually goes to zero. From the normalization on the ordinate, we can see that, for a given ratio of flight time to guidance time constant, doubling the guidance system time constant only doubles the heading error miss.

SUMMARY

In this chapter we have seen the power and accuracy of linearization. First we showed that the method of adjoints could be applied to the linearized homing loop. The adjoint technique permitted us to obtain miss distance performance projections as a function of flight time, in error budget form, for many inputs in a single adjoint computer run. The method was shown to be accurate when compared to linear performance projections obtained by massive simulation. It was also shown how the adjoint method could be used to derive miss distance formulas. These closed-form solutions agreed closely with results obtained by massive simulation trials of the nonlinear engagement simulation.

REFERENCES

[1] Goldsmith, J. L., "A Discussion of the Adjoint Technique of System Analysis," Raytheon, Memo SDD-74-525, Bedford, MA, Oct. 1974.

[2] Bliss, G. A., *Exterior Ballistics*, Univ. of Chicago Press, Chicago, IL, 1925.

[3] Laning, J. H., and Battin, R. H., *Random Processes in Automatic Control*, McGraw-Hill, New York, 1956.

[4] Zarchan, P., "Complete Statistical Analysis of Nonlinear Missile Guidance Systems—SLAM," *Journal of Guidance and Control*, Vol. 2, Jan.–Feb. 1979, pp. 71–78.

[5] Peterson, E. L., *Statistical Analysis and Optimization of Systems*, Wiley, New York, 1961.

[6] Zarchan, P., "Comparison of Statistical Digital Simulation Methods," *Advisory Group for Aerospace Research and Development*, AGARDograph No. 273, July 1988, pp. 2-1–2-16.

[7] Travers, P., "Interceptor Dynamics," unpublished lecture notes, Raytheon, circa 1971.

[8] Bennett, R. R., and Mathews, W. E., "Analytical Determination of Miss Distance for Linear Homing Navigation Systems," Hughes Aircraft Co., TM-260, Culver City, CA, March 1952.

CHAPTER 4

Noise Analysis

INTRODUCTION

The concept of noise is important to the guidance system engineer. For example, in a radar homing tactical missile the seeker measurement of the line-of-sight rate signal, required for the implementation of proportional navigation guidance, is not perfect but is corrupted by noise. To extract the signal from the measurement, an understanding of the concept of noise and its various properties are mandatory. In addition, in order to evaluate system performance in the presence of noise, we must first know how to simulate noise and then how to conduct and interpret experiments with repeated simulation trials. The concepts developed and illustrated in this chapter will be used throughout the text for filtering and evaluating system performance in the presence of noise or other random phenomenon.

BASIC DEFINITIONS

In this section [1] we will depart from our usual guidance discussions and start by defining some important quantities related to random variables. Random variables have unknown specific values, so they are usually quantified according to their statistical properties. One of the most important statistical properties of any random function x is its probability density function $p(x)$. This function is a measure of the likelihood of occurrence of each value of x and is defined such that

$$p(x) \geq 0$$

and

$$\int_{-\infty}^{\infty} p(x)\, \mathrm{d}x = 1$$

This means that there is a probability that x will occur, and it is certain that the value of x is somewhere between plus and minus infinity. The probability that x is between a and b can be expressed in terms of the probability density function as

$$\text{Prob}(a \le x \le b) = \int_a^b p(x)\,\mathrm{d}x$$

Another important quantity related to random variables is the distribution function. A distribution function $P(x)$ is the probability that a random variable is less than or equal to x. Therefore, if the probability density function is known, the distribution function can be found by integration as

$$P(x) = \int_{-\infty}^{x} p(u)\,\mathrm{d}u$$

The mean or expected value of x is defined by

$$m = E(x) = \int_{-\infty}^{\infty} xp(x)\,\mathrm{d}x$$

Therefore, the mean can also be thought of as the first moment of x. We can also think of the mean value of x as the sum (integral) of all values of x, each being weighted by its probability of occurrence. It can be shown that, if random variables $x_1, \ldots, x_n$ are independent, then the expectation of the sum is the sum of the expectations, or

$$E(x_1 + x_2 + \cdots + x_n) = E(x_1) + E(x_2) + \cdots E(x_n)$$

The second moment or mean squared value of x is defined as

$$E(x^2) = \int_{-\infty}^{\infty} x^2 p(x)\,\mathrm{d}x$$

Therefore, the rms of x can be obtained by taking the square root of the preceding equation, or

$$\text{rms} = [E(x^2)]^{1/2}$$

The variance of x, σ^2 is defined as the expected squared deviation of x from its mean value. Mathematically, the variance can be expressed as

$$\sigma^2 = E\{[x - E(x)]^2\} = E(x^2) - E^2(x)$$

We can see that the variance is the difference between the mean squared value of x and the square of the mean of x. If we have independent random variables $x_1, \ldots,$ x_n then the variance of the sum can be shown to be the sum of the variances, or

$$\sigma^2 = \sigma_1^2 + \sigma_2^2 + \cdots + \sigma_n^2$$

The square root of the variance σ is also known as the standard deviation. In general, the rms value and standard deviation are not the same unless the random process under consideration has a zero mean.

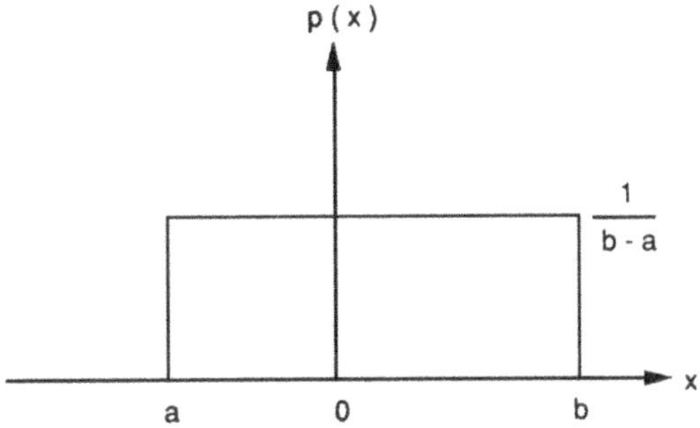

Fig. 4.1 Uniform probability distribution.

An example of a probability density function is the *uniform distribution*, which is depicted in Fig. 4.1. With this probability density function all values of x between a and b are equally likely to occur. An important practical example of the uniform distribution, which should be familiar to any engineer who has programmed on a personal computer, is the BASIC language random number generator (RND). The BASIC RND [or the MATLAB rand(1)] statement supplies a uniformly distributed random number, on each call, between 0 and 1. Soon we will see how random numbers with different probability density functions can be constructed from random numbers following the uniform distribution. From our previous definitions we can see that the mean value of a uniform distribution is

$$m = \int_{-\infty}^{\infty} xp(x)\,dx = \frac{1}{b-a}\int_a^b x\,dx = \frac{b+a}{2}$$

This makes sense, as the expected or mean value is halfway between a and b. The variance of a uniform distribution can also be found from our previous definitions and can be shown to be

$$\sigma^2 = E(x^2) - m^2 = \frac{b^3-a^3}{3(b-a)} - \left(\frac{b+a}{2}\right)^2 = \frac{(b-a)^2}{12}$$

This means that, if the random numbers from a uniform distribution vary from 0 to 1, the mean of the resultant set of numbers should be $1/2$ and the variance should be $1/12$. We will use this property of a uniform distribution later in this chapter for constructing random numbers with different probability density functions.

Another important probability density function is the *Gaussian or normal distribution.* In the missile homing loop we shall often treat the sensor noise disturbances as having a Gaussian distribution. The probability density function for this distribution is shown in Fig. 4.2 and is given by the formula

$$p(x) = \exp\left[-\frac{(x-m)^2}{2\sigma^2}\right] \Big/ [\sigma(2\pi)^{0.5}]$$

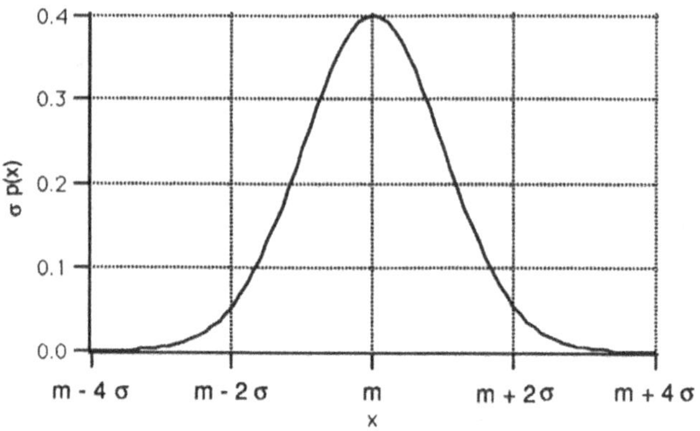

Fig. 4.2 Gaussian or normal probability density function.

where m and σ are parameters. By using our basic definitions it is easy to show that the expected or mean value of a Gaussian distribution is given by

$$E(x) = m$$

and its variance is

$$E(x^2) - m^2 = \sigma^2$$

Therefore, m and σ in the expression for the Gaussian probability density function correspond to the mean and standard deviation, respectively.

We can see from Fig. 4.2 that this bell-shaped distribution is virtually zero after three standard deviations ($\pm 3\sigma$). Integration of the probability density function, to find the distribution function, shows that there is a 68% probability that the Gaussian random variable is within one standard deviation ($\pm \sigma$) of the mean, 95% probability it is within two standard deviations of the mean, and 99% probability that it is within three standard deviations of the mean.

It can be shown that the resultant probability density function of a sum of Gaussian distributed random variables is also Gaussian. In addition, under certain circumstances, it can also be shown that the sum of independent random variables, regardless of individual density function, tends toward Gaussian as the number of random variables gets larger (an illustration of this phenomenon will be illustrated in the next section). That is in fact why so many random variables are Gaussian distributed.

GAUSSIAN NOISE EXAMPLE

To simulate noise or random events we have to know how to generate, via the computer, pseudorandom numbers with the appropriate probability density function. The FORTRAN language, for example, does not come with a random number generator. However, many microcomputer implementations of FORTRAN provide extensions from which noise, with the desired probability density function, can be constructed. For example, in Macintosh Absoft FORTRAN, the irand(0) statement produces a uniformly distributed integer between 0 and 2^{31}. It can be shown from the central limit theorem that the addition of many uniformly distributed variables produces a Gaussian distributed variable.

The first step in constructing random numbers with the desired probability density function is to normalize the uniform noise generator so that random numbers between 0 and 1 are produced. The second step is to add 12 uniformly distributed random variables and subtract 6 in order to obtain a zero-mean Gaussian variable with unity standard deviation (as the variance of one uniformly distributed random variable is $1/12$, the variance of 12 must be 1). The MATLAB listing for the generation of 100 Gaussian-distributed random numbers with zero mean and unity variance is shown in Listing 4.1. It is important to note that we are not making use of the MATLAB statement randn, which automatically generates a Gaussian random number.

LISTING 4.1 MATLAB GAUSSIAN RANDOM NUMBER GENERATOR (THE HARD WAY)

```
count=0;
N=100;
for I=1:N
        SUM=0;
        for J=1:12
            RAND=rand(1);
            SUM=SUM+RAND;
        end;
        X=SUM-6;
        count=count+1;
        ArrayI(count)=I;
        ArrayX(count)=X;
end;
figure
plot(ArrayI,ArrayX),grid
title('One hundred random numbers with Gaussian distribution')
xlabel('Number')
ylabel('Value')
clc
output=[ArrayI',ArrayX'];
save datfil output -ascii
disp 'simulation finished'
```

Figure 4.3 displays the values of each of the 100 random numbers, generated via the program of Listing 4.1, in graphic form. A quick glance at the plot indicates that the data appear to have approximately zero mean. The standard deviation of the data can be "eyeballed" by looking at the maximum excursions (99% chance

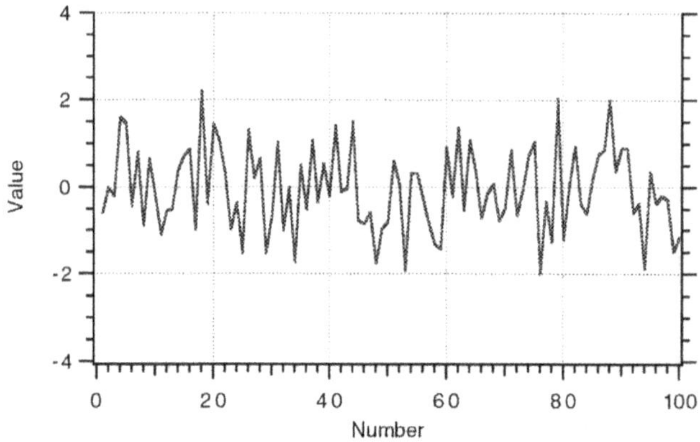

Fig. 4.3 One hundred random numbers with Gaussian distribution.

that data is within the 3σ values) and using the simplified relationship

$$\sigma_{\text{approx}} = (\text{peak to peak})/6 = 4/6 = 0.67$$

Thus, the eyeballed value of σ does not quite meet the theoretical expectations of unity standard deviation.

To get an idea of the resultant probability density function of the computer-generated 100 random numbers, another MATLAB program was written and appears in Listing 4.2. Essentially each random number is placed in a bin in order to calculate the frequency of occurrence and hence the probability density function [2]. Also included in the listing, for comparative purposes, is the theoretical formula for the probability density function of a zero-mean, unity variance, Gaussian distribution.

LISTING 4.2 MATLAB PROGRAM USED TO GENERATE PROBABILITY DENSITY FUNCTION

```
% Preallocation
H=zeros(1,10000);
X=zeros(1,10000);
count=0;
XMAX=6;
XMIN=-6;
RANGE=XMAX-XMIN;
TMP=1./sqrt(6.28);
N=100;
BIN=50;
for I=1:N
```

```
        SUM=0;
        for J=1:12
           RAND=rand(1);
           SUM=SUM+RAND;
        end;
        X(I)=SUM-6;
end;
for I=1:BIN
        H(I)=0;
end;
% FIX Round towards zero.
%       FIX(X) rounds the elements of X to the nearest integers
%       towards zero.
for I=1:N
       K=fix(((X(I)-XMIN)/RANGE)*BIN)+.99;
       if K < 1, K=1; end;
       if K > BIN, K=BIN; end;
%CORRECTION HERE >>>
        K=round(K);
       H(K)=H(K)+1;
end;
for K=1:BIN
        PDF=(H(K)/N)*BIN/RANGE;
        AB=XMIN+K*RANGE/BIN;
        TH=TMP*exp(-AB*AB/2.);
        count=count+1;
        ArrayAB(count)=AB;
        ArrayPDF(count)=PDF;
        ArrayTH(count)=TH;
end;
figure
plot(ArrayAB,ArrayPDF,ArrayAB,ArrayTH),grid
title('Sample Gaussian distribution')
xlabel('X')
ylabel('Probability Density Function')
clc
output=[ArrayAB',ArrayPDF',ArrayTH'];
save datfil output -ascii
disp 'simulation finished'
```

Figure 4.4 presents the calculated probability density function in graphic form. Superimposed on the figure is a plot of the theoretical Gaussian distribution. The figure indicates that, with a sample size of only 100 random numbers, it is not immediately obvious that the computer-generated probability density function follows a Gaussian distribution.

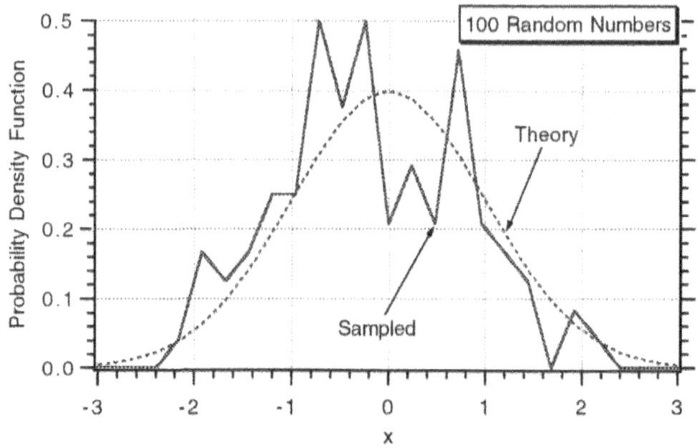

Fig. 4.4 Sampled Gaussian distribution does not closely follow theory for 100 random numbers.

Increasing the sample size from 100 random numbers to 1000 random numbers will clarify the "goodness" of the computer-generated random numbers. Figure 4.5 displays each of the 1000 random numbers generated. The figure demonstrates that, although the mean of the random numbers is still about zero, we now have larger excursions (numbers vary between $\pm 3\sigma$) and the approximate value of the standard deviation is

$$\sigma_{\text{approx}} = (\text{peak to peak})/6 = 6/6 = 1$$

which is the theoretically correct value.

Finally, Fig. 4.6 shows that when the sample size is increased to 1000 numbers, the resultant probability density function closely follows the theoretical

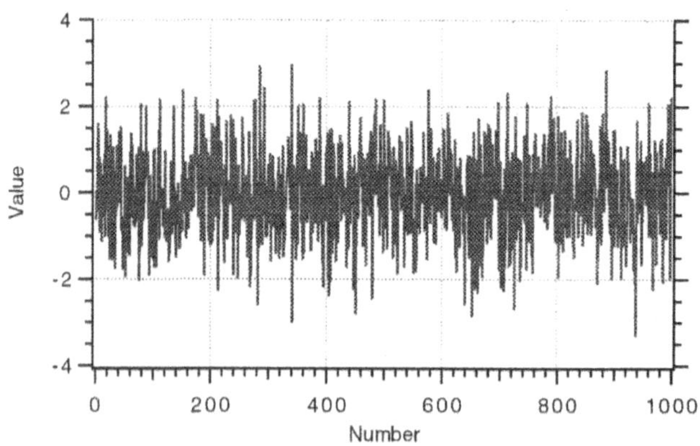

Fig. 4.5 One thousand random numbers with Gaussian distribution.

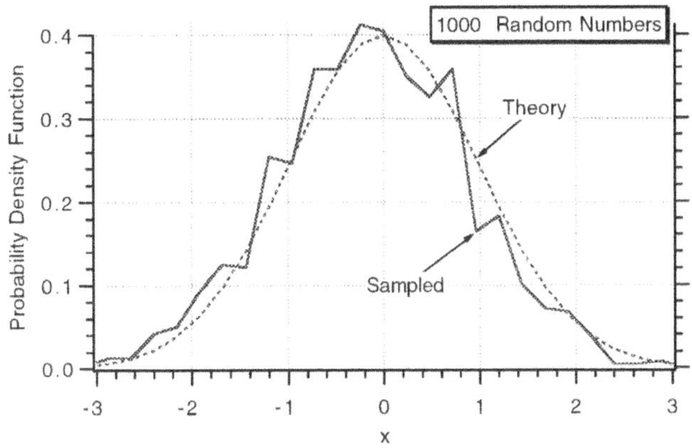

Fig. 4.6 Sampled Gaussian distribution more closely follows theory for 1000 random numbers.

bell-shaped curve. Thus, we have seen how a Gaussian distribution can be constructed from the summation of 12 uniformly distributed random variables. This is a practical application of the *central limit theorem*. In practice, to save computer running time, we can add fewer than 12 uniformly distributed numbers to get a Gaussian-distributed random number. Throughout the rest of this book we will add only six uniformly distributed numbers to get the desired Gaussian distribution.

COMPUTATIONAL ISSUES

Often, from simulation outputs, we wish to compute some of the basic random variable properties (such as mean, variance, and so on). Stated more mathematically, we wish to compute these basic random variable properties from a finite set of data x_i when only n samples are available. The discrete equivalent of the previously presented formulas for basic random variable properties are presented in the following equations:

$$\text{mean} = \sum_{i=1}^{n} x_i/n$$

$$\text{mean square} = \sum_{i=1}^{n} x_i^2/(n-1)$$

$$\text{standard deviation} = \left\{ \left[\sum_{i=1}^{n} (x_i - \text{mean})^2 \right] \Big/ (n-1) \right\}^{1/2}$$

We can see from these equations that integrals from the theoretical or continuous formulas have been replaced with summations in their discrete equivalents. In order for the theoretical and calculated random variable properties to be equal, the number of samples in the discrete computations must be infinite. Because the sample size is finite, the discrete or calculated formulas are approximations. In fact, the answers generated from these formulas have statistics of their own.

Recognizing that simulation outputs based upon random inputs can vary, the Monte Carlo approach [3] will often be used in this text to obtain system performance. The *Monte Carlo method* is approximate and is simply repeated simulation trials plus postprocessing of the resultant data in order to do ensemble averaging (using the preceding formulas) to get the mean and standard deviation. Usually a large number of simulation trials are required in order to provide confidence in the accuracy of the results. Because of its simplicity and generality, however, the Monte Carlo approach is probably the most popular computerized method of statistical analysis.

To demonstrate that our computed statistics are not precise and in fact are random variables with statistics, a MATLAB simulation of the Gaussian noise was generated, and Listing 4.3 shows the computation of the sampled standard deviation. The number of i samples used in the program computation was made a parameter in the study and varied from 1 to 100.

LISTING 4.3 MATLAB PROGRAM FOR COMPUTING SAMPLED STANDARD DEVIATION (THE HARD WAY)

```
% Preallocation
Z=zeros(1,100);
count=0;
Z1=0;
for I=1:100
    SUM=0;
    for J=1:12
        RAND=rand(1);
        SUM=SUM+RAND;
    end;
    X=SUM-6;
    Z(I)=X;
    Z1=Z(I)+Z1;
    XMEAN=Z1/I;
end;
SIGMA=0;
Z1=0;
for I=1:100
    Z1=(Z(I) -XMEAN) ^2+Z1;
```

```
    if I == 1
        SIGMA=0;
    else
        SIGMA=sqrt(Z1/(I-1));
    end;
    count=count+1;
    ArrayI(count)=I;
    ArraySIGMA(count)=SIGMA;
end;
figure
plot(ArrayI,ArraySIGMA),grid
title('Sampled Standard Deviation')
xlabel('Number of Samples')
ylabel('Calculated Standard Deviation')
clc
disp '*** Simulation Complete'
output=[ArrayI',ArraySIGMA'];
save datfil.txt output -ascii
disp 'simulation finished'
```

Figure 4.7 shows that the computed standard deviation (actual standard deviation is unity) obtained from the MATLAB program is a function of the sample size used. Large errors in the standard deviation estimate occur when there are less than 20 samples. The accuracy of the computation improves significantly when many samples are used in computing the standard deviation. In this example, we need more than 100 samples for the computed standard deviation to be within 5% of the theoretical value of unity. When we begin to evaluate system performance, when the inputs are random, we will take this information into account

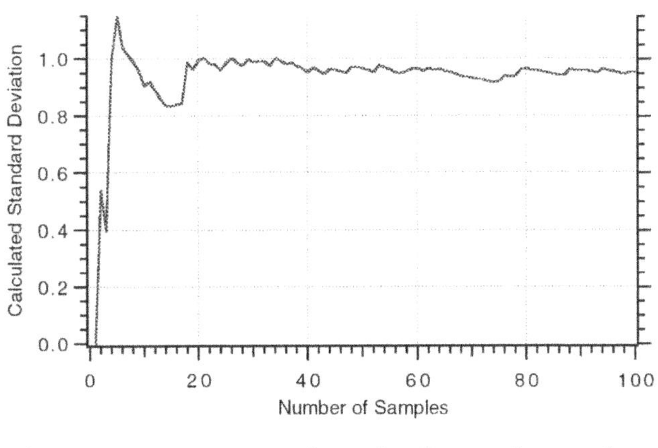

Fig. 4.7 Large errors occur when only a few samples are taken.

in determining how many simulation trials (Monte Carlo runs) will be required to get reasonably accurate results. Usually we will consider 50 runs to be sufficient in the tradeoff between computer running time and numerical accuracy.

MORE BASIC DEFINITIONS

A few more definitions [1] are required before we can build up the tools required for the analysis of noise-driven systems. Thus far we have discussed the second-order statistics of random processes. However, in practice, we are limited to even less information than that given by the probability density function. Often, only the first moment of these random processes is measured. One such moment is the *autocorrelation function*, which is defined by

$$\phi_{xx}(t_1, t_2) = E[x(t_1)x(t_2)]$$

The Fourier transform of the autocorrelation function is called the *power spectral density* and is defined as

$$\Phi_{xx} = \int_{-\infty}^{\infty} \phi_{xx}(\tau)e^{-j\omega\tau}\,d\tau$$

where the power spectral density, using these definitions, has dimensions of unit squared per Hertz. In all of the statistical work presented throughout this text, the power spectral density will have those units.

One simple and useful form for the power spectral density is that of white noise, in which the power spectral density is constant, or

$$\Phi_{xx} = \Phi_0 \qquad \text{(white noise)}$$

The autocorrelation function for white noise is a delta function given by

$$\phi_{xx} = \Phi_0\delta(\tau) \qquad \text{(white noise)}$$

Although white noise is not physically realizable, it can be used to serve as an invaluable approximation to situations in which a disturbing noise is wideband compared to the system bandwidth. In addition, white noise is useful for analytical operations because of the impulsive nature of the autocorrelation function (it makes integrals disappear).

RESPONSE OF LINEAR SYSTEM TO WHITE NOISE

Often we are interested in finding the response of a linear system to noise. If the system is linear, with impulse response $h(t, \tau)$, the output $y(t)$, can be expressed in terms of the input $x(t)$ via the convolution integral, or

$$y(t) = \int_{-\infty}^{t} x(\tau)h(t, \tau)\,d\tau$$

Squaring both sides of the preceding equation yields

$$y^2(t) = \int_{-\infty}^{t} x(\tau_1)h(t,\tau_1)\,\mathrm{d}\tau_1 \int_{-\infty}^{t} x(\tau_2)h(t,\tau_2)\,\mathrm{d}\tau_2$$

If $x(t)$ is random, we can take expectations of both sides, or

$$E[y^2(t)] = \int_{-\infty}^{t}\int_{-\infty}^{t} h(t,\tau_1)h(t,\tau_2)E[x(\tau_1)x(\tau_2)]\,\mathrm{d}\tau_1\,\mathrm{d}\tau_2$$

In addition, if the input $x(t)$ is white noise with power spectral density Φ, the double integral of the preceding equation can be simplified because of the impulsive nature of the autocorrelation function, or

$$E[x(\tau_1)x(\tau_2)] = \Phi\delta(\tau_1 - \tau_2)$$

Substitution yields

$$E[y^2(t)] = \Phi\int_{-\infty}^{t} h^2(t,\tau)\,\mathrm{d}\tau$$

Therefore, the mean square response of a linear system driven by white noise with power spectral density Φ (where Φ has the dimensions of unit^2/Hz) is proportional to the integral of the square of the impulse response. The preceding equation is a general relationship and is valid for both time-varying and time-invariant linear systems driven by white noise.

LOW-PASS-FILTER EXAMPLE

To illustrate the utility of the mean square response equation, let us find the response of a low-pass filter to white noise as shown in Fig. 4.8. Here the input x is white noise with power spectral density Φ. As this system is time-invariant and physically realizable (noncausal), the fundamental noise relationship simplifies to

$$E[y^2(t)] = \Phi\int_{0}^{t} h^2(\tau)\,\mathrm{d}\tau$$

We can find the system impulse response in the preceding integral by first recognizing that the transfer function of the low-pass filter, shown in Fig. 4.8, is given by

$$H(s) = \frac{1}{1+sT}$$

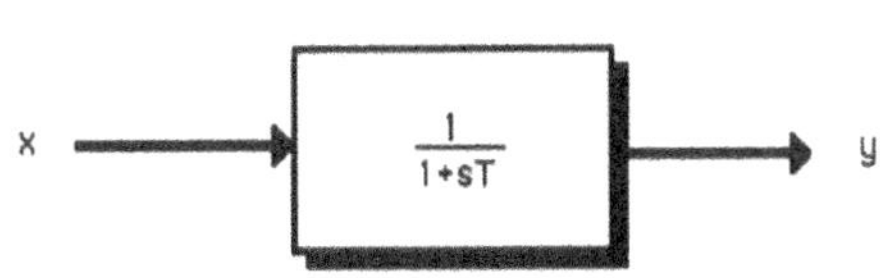

Fig. 4.8 Low-pass filter with white noise input.

Therefore, its inverse Laplace transform is

$$h(t) = \mathcal{L}^{-1}[H(s)] = \frac{e^{-t/T}}{T}$$

Substitution yields

$$E[y^2(t)] = \frac{\Phi}{T^2}\int_0^t e^{-2\tau/T}\,\mathrm{d}\tau$$

Evaluation of the upper and lower limits results in the final answer:

$$E[y^2(t)] = \frac{\Phi(1 - e^{-2t/T})}{2T}$$

In the steady state, the exponential term drops out, yielding

$$E[y^2(\infty)] = \frac{\Phi}{2T}$$

We can write a MATLAB program to simulate the problem and to see how the theoretical results and simulation results agree. To do this we must be able to simulate Gaussian white noise. In MATLAB Gaussian random numbers can also be generated by using `randn`. Because the MATLAB Gaussian distributed random numbers are independent, the resultant Gaussian random numbers will look white to the low-pass filter if its bandwidth is much greater than the filter bandwidth. In a simulation of the continuous system of Fig. 4.8, the MATLAB Gaussian noise generator is called every integration interval h. Because integration interval is always chosen to be at least several times smaller than the smallest system time constant $T/h \ll T$ in order to get correct answers with numerical integration techniques), the noise will look white to the system.

The standard deviation of the pseudowhite noise (actual white noise has infinite standard deviation) is related to the desired white noise spectral density Φ and integration interval h according to [3]

$$\sigma = \sqrt{\frac{\Phi}{h}}$$

where Φ has dimensions of units squared per Hertz. The MATLAB simulation listing of this white-noise-driven low-pass filter is shown in Listing 4.4. We can see from the listing that the Gaussian noise with unity standard deviation is modified to get the desired pseudowhite noise spectral density ($\Phi = 1$). The approximate white noise enters the system at every integration interval. A sample output for a correlation time of 0.2 s is shown in Fig. 4.9. Also shown in the listing and figure are the theoretical results obtained from the previously derived formula for the output standard deviation of a white-noise-driven

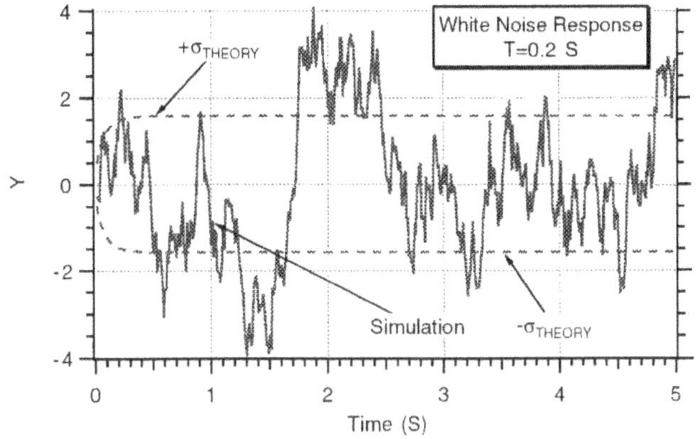

Fig. 4.9 Low-pass filter output agrees with theory.

low-pass filter, which is

$$\sigma_{\text{theory}} = \pm\sqrt{\frac{\Phi[1 - e^{-2t/T}]}{2T}}$$

We can see from this figure that the simulation results, based upon the MATLAB listing, agree with theory in the sense that the simulation results lay within the $\pm\sigma$ bounds approximately 68% of the time. Therefore, we can say that the experimental and theoretical results are in agreement.

Increasing the correlation time constant increases the smoothing action of the low-pass filter. Figure 4.10 shows the filter output when the correlation time constant is increased from 0.2 s to 1 s. Here we can see that the larger filter time

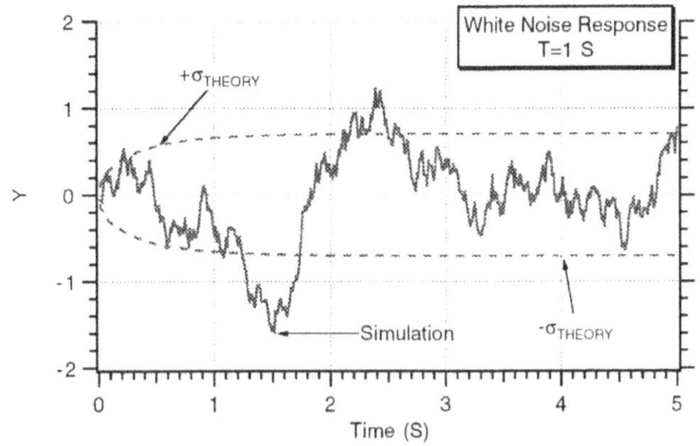

Fig. 4.10 Increasing filter time constant reduces the noise transmission.

constant not only provides more filtering action but also tends to correlate the noise. In other words, the randomness of the noise is starting to disappear as the filter time constant increases. Again, this figure shows that the simulated results appear to be within the $\pm\sigma$ bounds about 68% of the time.

It is important to note that both the simulated time domain results and the theoretical second-order statistical results provide invaluable visual information. The use of both theory and simulation can be used to not only verify results but also to provide a deeper understanding of the processes involved.

LISTING 4.4 MATLAB SIMULATION OF LOW-PASS FILTER DRIVEN BY WHITE NOISE

```
TAU=0.2;
PHI=1.0;
count=0;
T=0;
H=.01;
SIG=sqrt(PHI/H);
Y=0;
while T <= 5.0
        X=SIG*randn;
        YOLD=Y;
        STEP=1;
        FLAG=0;
        while STEP <=1
          if FLAG ==1
                  STEP=2;
                  Y=Y+H*YD;
          T=T+H;
          end
          YD=(X-Y)/TAU;
          FLAG=1;
        end
        FLAG=0;
        Y=(YOLD+Y)/2.+.5*H*YD;
        SIGPLUS=sqrt(PHI*(1.-exp(-2.*T/TAU))/(2.*TAU));
        SIGMINUS=-SIGPLUS;
        count=count+1;
        ArrayT(count)=T;
        ArrayY(count)=Y;
        ArraySIGPLUS(count)=SIGPLUS;
        ArraySIGMINUS(count)=SIGMINUS;
end
figure
plot(ArrayT,ArrayY,ArrayT,ArraySIGPLUS,ArrayT,ArraySIGMINUS),grid
title('Simulation of low-pass filter driven by white noise')
```

```
xlabel('Time (S)')
ylabel('Y')
clc
output=[ArrayT',ArrayY',ArraySIGPLUS',ArraySIGMINUS'];
save datfil.txt output -ascii
disp 'simulation finished'
```

ADJOINTS FOR NOISE-DRIVEN SYSTEMS

In the previous chapter we saw that the method of adjoints could be very useful in analyzing linear time-varying deterministic systems. We shall now demonstrate that adjoints can also be of great utility in analyzing linear time-varying systems driven by white noise [4, 5]. It was shown earlier that the mean square response of a linear time-varying system driven by white noise is given by

$$E[y^2(t)] = \Phi \int_0^t h^2(t, \tau)\, d\tau$$

where Φ is the white noise power spectral density and $h(t, \tau)$ the impulse response of the linear time-varying system. In this case τ is the impulse application time and t the observation time. The previous section presented a simple example demonstrating the practical utility of this equation. For time-varying systems, however, this equation is not as useful because the integration is with respect to the impulse application time τ. As with the deterministic case, this means that many computer runs would have to be made, each having a different impulse application time, in order to evaluate the preceding equation.

If we go back to the fundamental relationship between the original and adjoint systems, we can say that

$$E[y^2(t)] = \Phi \int_0^t h^*[(t_F - \tau, t_F - t)]^2\, d\tau$$

After making the substitution

$$x = t_F - \tau$$
$$dx = -d\tau$$

we obtain

$$E[y^2(t)] = \Phi \int_{t_F - t}^{t_F} [h^*(x, t_F - t)]^2\, dx$$

If the final time is of interest ($t = t_F$), the preceding equation can be rewritten as

$$E[y^2(t_F)] = \Phi \int_0^{t_F} [h^*(x, 0)]^2\, dx$$

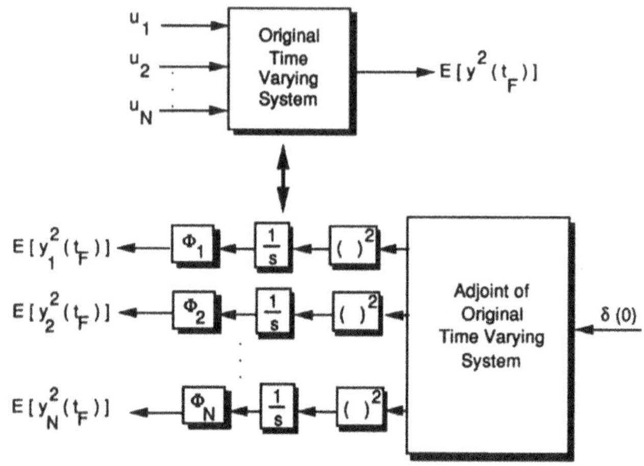

Fig. 4.11 Equivalence between adjoint and original systems for stochastic inputs.

Since the integration is now with respect to the observation time, this new equation is quite useful. Therefore, we can find the mean square response of a linear, time-varying system driven by white noise by squaring and integrating the output of the impulsively driven adjoint system *in one computer run!*

The benefits of the adjoint approach become even more dramatic when we consider many white noise inputs to the original system as shown in Fig. 4.11.

As with the deterministic inputs, white noise inputs to the original system become outputs in the adjoint system. Therefore, by superposition, one adjoint run yields an exact statistical analysis of the noise-driven system plus a statistical error budget showing how each white noise error source contributed to the total performance projection. In Fig. 4.11 the total mean square response is computed from

$$E[y^2(t)] = E[y_1^2(t)] + E[y_2^2(t)] + \cdots + E[y_N^2(t)]$$

SHAPING FILTERS AND RANDOM PROCESSES

Thus far we have seen the importance of target maneuver on system performance. In this section we will show how shaping filters can be used to accurately represent aircraft evasive maneuvers [6, 7]. The purpose of the shaping filter approach is to allow us to use efficient and effective means of performance analysis such as the method of adjoints.

The concept of *shaping filter* has been used for many years in the analysis of physical systems because it allows a system with a random input to be replaced by an augmented system (the original system plus the shaping filter) excited only by white noise. An example of this is the replacement of the random telegraph signal by white noise through a simple lag network. This approach is generally applied to problems where mean square values of outputs are of prime importance. In such cases only second-order statistics are important, and rather complex input processes can sometimes be represented by very simple shaping filters. This is due to the fact that random processes that have the same mean and autocorrelation are mathematically equivalent. This is true even though the associated probability density functions of the random processes may be quite different. In other words, a random phenomenon and its shaping filter equivalent are indistinguishable as far as their second-order statistics are concerned. The concept of shaping filter can also be applied to the statistical representation of signals with known form but random starting time. Consider a signal of known form $h(t)$ with random starting time so that the resultant signal $x(t)$ is given by

$$x(t) = h(t - T)u(t - T)$$

where the probability density function of T is given by $p_T(t)$, and $u(t)$ is the unit step function. Note that, although $h(t)$ is deterministic, $x(t)$ is random because of the random starting time. It can be shown that the white-noise-driven shaping network of Fig. 4.12 has the same mean and autocorrelation functions as those of the preceding equation. Here we can see that the white noise has a power spectral density equal to the probability density function of the random starting time and that the inverse Laplace transform of the shaping filter is equal to the deterministic signal.

The output of the shaping network and the actual random process are equivalent in terms of second-order statistics. If either process is passed through a linear physical system, the outputs would be indistinguishable if second-order statistics are being observed (that is, mean square values).

Consider a step target maneuver that has a starting time that is uniformly distributed over the flight time. Mathematically speaking, the maneuver can be modeled as a constant signal of magnitude n_T, which starts at time T, or

$$x(t) = n_T u(t - T)$$

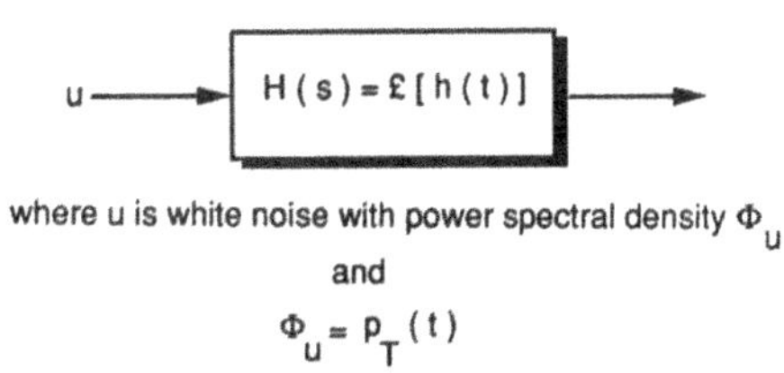

Fig. 4.12 Shaping network representation of deterministic signal with random starting time.

where $u(t - T)$ is a unit step function defined by

$$u(t-T) = 0 \quad \text{for } t < T$$
$$= 1 \quad \text{otherwise}$$

Let us assume that the initiation of the maneuver is equally likely to occur anywhere during the flight. More precisely, we can say that the starting time T is uniformly distributed over the flight time t_F, with probability density function

$$p_T(t) = 1/t_F \quad \text{for } 0 \leq t \leq t_F$$
$$= 0 \qquad \text{otherwise}$$

Therefore, the autocorrelation function of this signal with random starting time is given by

$$\phi_{xx}(t_1, t_2) = \int_{-\infty}^{\infty} x(t_1)x(t_2)p_T(T)\,\mathrm{d}T$$

or

$$\phi_{xx}(t_1, t_2) = \int_0^{t_F} n_T u(t_1 - T) n_T u(t_2 - T) \frac{dT}{t_F}$$

Assuming that

$$0 < t_1 < t_2 < t_F$$

the autocorrelation function simplifies to

$$\phi_{xx}(t_1, t_2) = \frac{n_T^2}{t_F} \int_0^{t_1} \mathrm{d}T$$

It is important to note that the output autocorrelation function of a linear time-invariant system with impulse response $h(t)$ driven by white noise can be expressed as

$$\phi_{yy}(t_1, t_2) = \int_{-\infty}^{t_1} h(t_1 - \tau_1) \int_{-\infty}^{t_2} h(t_2 - \tau_2)\phi_{uu}(\tau_1, \tau_2)\,\mathrm{d}\tau_1\,\mathrm{d}\tau_2$$

The autocorrelation function of the white noise input is

$$\phi_{uu}(\tau_1, \tau_2) = \Phi_u(\tau_1)\delta(\tau_1 - \tau_2)$$

where the spectral density of the white noise Φ_u is a function of time. Substitution of the white noise autocorrelation function into the preceding integral equation eliminates one of the integrals. After some manipulations and assuming $t_1 < t_2$, we obtain

$$\phi_{yy}(t_1, t_2) = \int_{-\infty}^{t_1} \Phi_u(\tau_1)h(t_1 - \tau_1)h(t_2 - \tau_1)\,\mathrm{d}\tau_1$$

If the spectral density takes on values of

$$\phi_u(t) = \phi_u \quad 0 \le t \le t_F$$
$$= 0 \quad \text{otherwise}$$

and we assume that

$$0 < t_1 < t_2 < t_F$$

then we can say that

$$\phi_{yy}(t_1, t_2) = \phi_u \int_0^{t_1} h(t_1 - \tau_1) h(t_2 - \tau_2)\, \mathrm{d}\tau_1$$

Therefore, we can say that the two different expressions for the autocorrelation function are equivalent if

$$\Phi_u = n_T^2 / t_F$$

and

$$h(t) = 1$$

In summary, a step maneuver of amplitude n_T, where starting time is uniformly distributed over the flight time t_F has the same autocorrelation function as a linear network with transfer function

$$H(s) = \frac{1}{s}$$

driven by white noise with power spectral density

$$\Phi_u = n_T^2 / t_F \quad 0 < t < t_F$$
$$= 0 \quad \text{otherwise}$$

The uniformly distributed step maneuver is shown in Fig. 4.13 and its shaping filter equivalent is shown in Fig. 4.14.

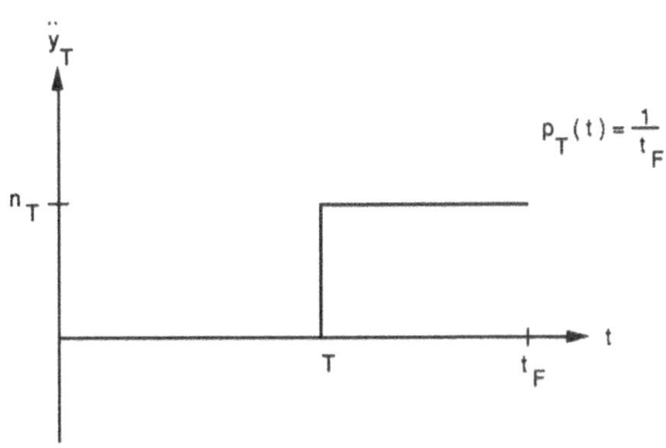

Fig. 4.13 Step maneuver with uniformly distributed starting time.

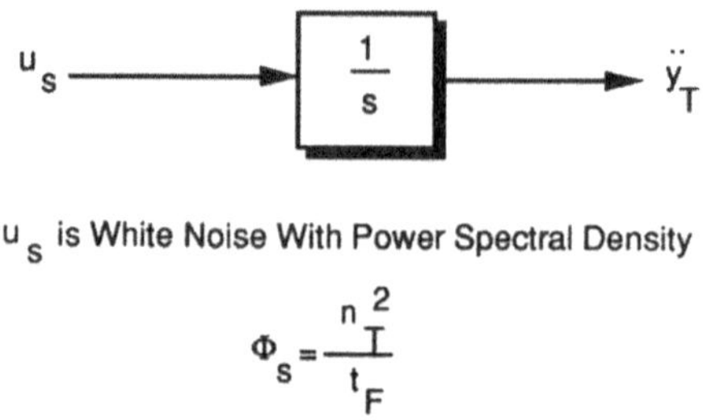

Fig. 4.14 Shaping filter equivalent of random starting time step maneuver.

EXAMPLE OF A STOCHASTIC ADJOINT

To show how the adjoint can also be used to analyze linear systems driven by stochastic or random disturbances, let us revisit the single-lag proportional navigation homing loop. However, this time we will consider a target maneuver with a random starting time (starting time that is uniformly distributed over the flight time) as the error source. A single-lag proportional navigation homing loop with the stochastic input is shown in Fig. 4.15.

In Fig. 4.15 the target maneuver is a constant from flight to flight (either plus or minus n_T). However, on a given flight its initiation time is equally likely to occur anywhere during the flight (uniformly distributed over the flight time). A Monte Carlo simulation of Fig. 4.15 with the random target maneuver appears in Listing 4.5. We can see from the listing that there are two main loops. The outer loop varies the flight time from 1 to 10 s in increments of 1 s. The inner loop performs 50 sets of runs on a particular case. In each of these cases the starting time of the maneuver is chosen from a uniform distribution. After each Monte Carlo set, the standard deviation and mean of the 50 miss distances are computed according to the formulas developed in this chapter.

A case was run for the single-time-constant guidance system in which the time constant was set to 1 s and the effective navigation ratio was set to 3. Fifty-run Monte Carlo sets for 10 different flight times were run for this single case, which actually encompassed a total of 500 runs (50 × 10). The standard deviation of the miss for each flight time was calculated and is displayed as a function of flight time in Fig. 4.16. We can see that the miss distance is small for both small and large flight times. The miss is small for short flight times because the miss distance does not have enough time to develop. At the larger flight times

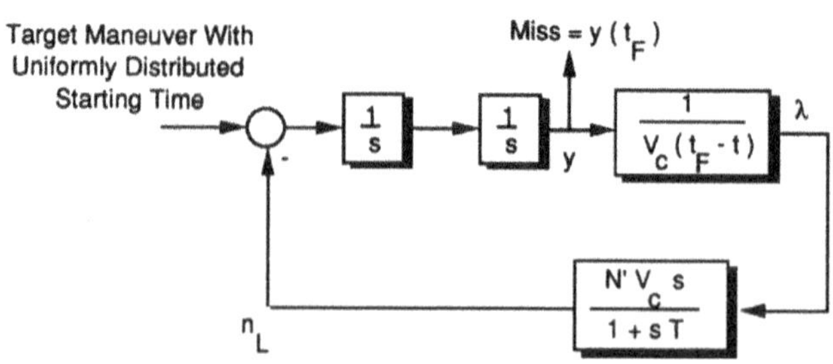

Fig. 4.15 Single-lag homing loop with stochastic inputs.

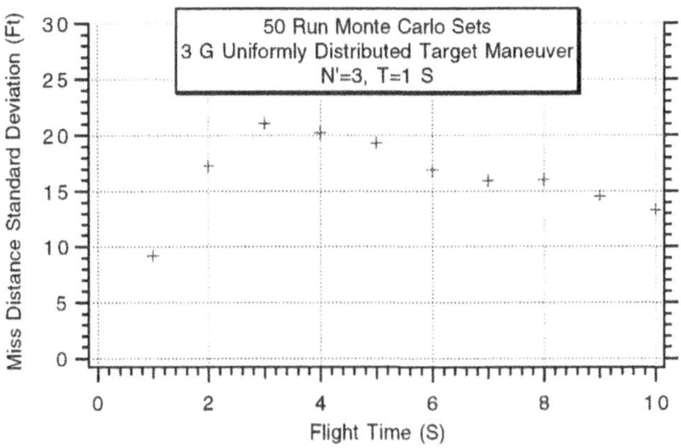

Fig. 4.16 Monte Carlo results for uniformly distributed target maneuver.

there is a good chance that the target maneuver initiation time will be at a long time to go, relative to the guidance system time constant, and will therefore induce a smaller miss distance.

We have seen in the previous section that the shaping filter equivalent for a uniformly distributed target maneuver is white noise, with spectral density Φ_s, through an integrator. The spectral density of the white noise is related to the maneuver level and the flight time according to

$$\Phi_s = \frac{n_T^2}{t_F}$$

LISTING 4.5 MONTE CARLO SIMULATION OF HOMING LOOP WITH RANDOM TARGET MANEUVER

```
% Preallocation
Z=zeros(1,1000);
TF=zeros(1,10);
I=zeros(1,100);
ArrayTF=zeros(1,10);
ArraySIGMA=zeros(1,10);
ArrayXMEAN=zeros(1,10);
count=0;
VC=4000;
XNT=96.6;
VM=3000;
XNP=3;
TAU=1;
```

```
RUN=50
for TF=1:10
          Z1=0;
          for I=1:RUN
             SUM=uniform;
             TSTART=TF*SUM;
             PZ=uniform;
             PZ=PZ-.5;
             if PZ > 0
                         COEF=1;
             else
                         COEF=-1;
             end;
         Y=0;
         YD=0;
         T=0;
         H=.01;
         S=0;
         XNC=0;
         XNL=0;
         while T < =(TF - 1e-5)
          if T < TSTART
             XNT=0;
          else
             XNT=COEF*96.6;
          end;
          YOLD=Y;
          YDOLD=YD;
          XNLOLD=XNL;
          STEP=1;
          FLAG=0;
          while STEP < =1
          if FLAG==1
             Y=Y+H*YD;
             YD=YD+H*YDD;
             XNL=XNL+H*XNLD;
             T=T+H;
             STEP=2;
         end
         TGO=TF-T+.00001;
         RTM=VC*TGO;
         XLAMD=(RTM*YD+Y*VC)/(RTM^2);
         XNC=XNP*VC*XLAMD;
         XNLD=(XNC-XNL)/TAU;
         YDD=XNT-XNL;
         FLAG=1;
```

```
        end;
        FLAG=0;
        Y=.5*(YOLD+Y+H*YD);
        YD=.5*(YDOLD+YD+H*YDD);
        XNL=.5*(XNLOLD+XNL+H*XNLD);
        S=S+H;
    end;
    Z(I)=Y;
    Z1=Z(I)+Z1;
    XMEAN=Z1/I;
            end;
            SIGMA=0;
            Z1=0;
            for I=1:RUN
            Z1=(Z(I)-XMEAN)^2+Z1;
            if I==1
            SIGMA=0;
            else
            SIGMA=sqrt(Z1/(I-1));
            end
            end;
            count=count+1;
            ArrayTF(count)=TF;
            ArraySIGMA(count)=SIGMA;
            ArrayXMEAN(count)=XMEAN;
end;
figure
plot(ArrayTF,ArraySIGMA,'c+')
title('Shaping filter Monte Carlo results')
xlabel('Time')
ylabel('Standard Deviation / Mean')
axis([00,10,00,30])
clc
output=[ArrayTF',ArraySIGMA',ArrayXMEAN'];
save datfil.txt output  -ascii
disp 'simulation finished'
```

An adjoint model can be constructed from the original system by following the rules for constructing stochastic adjoints [4, 5]. The only additional rule for stochastic systems is that all stochastic inputs to the original system must be modeled as white noise inputs, which then become outputs in the adjoint system. Since the input to the original system can be modeled as white noise through an integrator, the adjoint model will reverse the signal flow and square and integrate the output. The resultant adjoint model is shown in Fig. 4.17. The impulsive input can be replaced by an initial condition of unity on integrator $x3$.

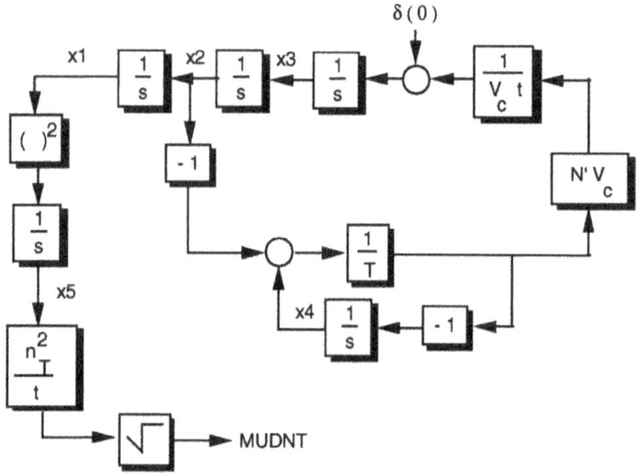

Fig. 4.17 Adjoint model for stochastic example.

An adjoint simulation of the model in Fig. 4.17 appears in Listing 4.6. The adjoint program only has to run once to find the standard deviation of the miss as a function of the flight time.

The adjoint simulation was run using the input parameters shown in Listing 4.6. The adjoint results for this example are shown in Fig. 4.18. Superimposed on the plot are the Monte Carlo results previously generated. We can see that the adjoint and Monte Carlo results are in close proximity, thus experimentally confirming the shaping filter approach and demonstrating the utility

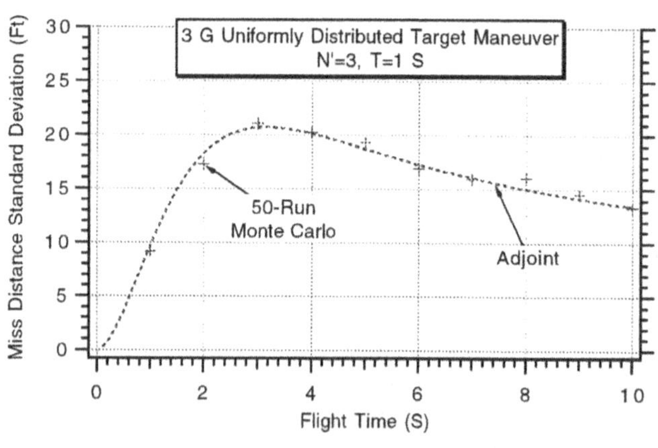

Fig. 4.18 Shaping filter adjoint and Monte Carlo results are in close agreement.

of stochastic adjoints. One single adjoint run gave results that were the equivalent of ten 50-run Monte Carlo sets.

LISTING 4.6 ADJOINT MODEL USING SHAPING FILTER APPROACH

```
count=0;
XNT = 96.6;
XNP = 3;
TAU = 1;
TF = 10;
T=0;
S=0;
TP=T+.00001;
X1=0;
X2=0;
X3=1;
X4=0;
X5=0.;
H=.01;
while TP < = (TF - 1e-5)
      STEP=1;
      FLAG=0;
      S=S+H;
      X1OLD=X1;
      X2OLD=X2;
      X3OLD=X3;
      X4OLD=X4;
      X5OLD=X5;
      while STEP < =1
        if FLAG==1
                STEP=2;
                X1=X1+H*X1D;
                X2=X2+H*X2D;
                X3=X3+H*X3D;
                X4=X4+H*X4D;
                X5=X5+H*X5D;
                TP=TP+H;
        end;
        X1D=X2;
        X2D=X3;
        Y1=(X4-X2)/TAU;
        TGO=TP+.00001;
        X3D=XNP*Y1/TGO;
        X4D=-Y1;
        X5D=X1*X1;
```

```
            FLAG=1;
        end;
        FLAG=0;
        X1=(X1OLD+X1)/2+.5*H*X1D;
        X2=(X2OLD+X2)/2+.5*H*X2D;
        X3=(X3OLD+X3)/2+.5*H*X3D;
        X4=(X4OLD+X4)/2+.5*H*X4D;
        X5=(X5OLD+X5)/2+.5*H*X5D;
        S=S+H;
        if S > =.000999
            S=0.;
            XMUDNT=XNT*sqrt(X5/TGO);
            count=count+1;
            ArrayTP(count)=TP;
            ArrayXMUDNT(count)=XMUDNT;
        end;
end
figure
plot(ArrayTP, ArrayXMUDNT),grid
title('Adjoint model using shaping filter approach')
xlabel('Flight Time (S)')
ylabel('Miss Dist Standard Deviation (Ft)')
axis([00,10,00,30])
clc
output=[ArrayTP',ArrayXMUDNT'];
save datfil.txt output /ascii
disp('Simulation Complete')
```

CLOSED-FORM SOLUTION FOR RANDOM TARGET MANEUVER

If we closely investigate Fig. 4.17 we can see that the miss due to a target maneuver with uniformly distributed starting time can be found by squaring the sensitivity due to a step target maneuver and then integrating and taking the square root of the resultant adjoint signal. If we define the miss due to a step target maneuver as *MNT*, the miss due to a uniformly distributed target maneuver can be expressed mathematically as

$$MUDNT = n_T\sqrt{\frac{1}{t_F}\int_0^{t_F} MNT^2\,\mathrm{d}\tau}$$

where n_T is the target maneuver level and t_F the flight time. For a single-lag guidance system with an effective navigation ratio of 3, we have already shown in Chapter 3 that the miss due to a step target maneuver is given by

$$MNT|_{N'=3} = 0.5t_F^2 e^{-t_F/T}$$

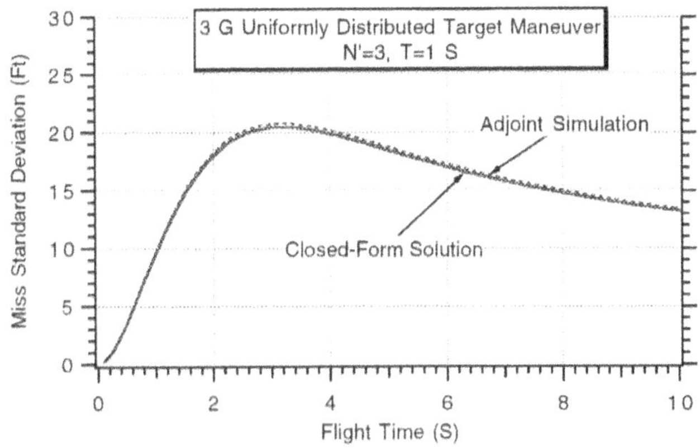

Fig. 4.19 Closed-form solution and adjoint simulation results agree.

where T is the guidance system time constant. Substitution of the step target maneuver solution into the expression for the uniformly distributed target maneuver yields

$$MUDNT|_{N'=3} = n_T \sqrt{\frac{1}{t_F} \int_0^{t_F} 0.25\tau^4 e^{-2\tau/T}\, d\tau}$$

After integration by parts and much algebra we obtain the closed-form expression, valid for an effective navigation ratio of 3, for the uniformly distributed target maneuver:

$$MUDNT|_{N'=3} = \frac{n_T}{4} \sqrt{\frac{T^5}{t_F}[3 - e^{-2x}(2x^4 + 4x^3 + 6x^2 + 6x + 3)]}$$

where x is defined as normalized time, or

$$x = \frac{t_F}{T}$$

Figure 4.19 displays the closed-form solution for the case in which the guidance system time constant is 1 s. Superimposed on the plot are the previously presented adjoint results for a uniformly distributed target maneuver. We can see from the close proximity of the two curves that both solutions are in close agreement.

SUMMARY

Starting from basic definitions of random variables, we have shown how to simulate random phenomena and properly interpret the simulation results.

Throughout this chapter we have shown two ways of doing problems: the theoretical way, which only works under certain circumstances, and simulation, which is always valid. Numerical examples have been presented that not only demonstrate that theory and simulation agree but also show how each method offers new insights. Finally, it was shown how the method of adjoints can be extended to evaluate system performance in the presence of random disturbances.

REFERENCES

[1] Gelb, A., *Applied Optimal Estimation*, MIT Press, Cambridge, MA, 1974.

[2] Zarchan, P., "Engineering Tips For Plotting," *MacTutor*, Vol. 4, Feb. 1988, pp. 78–82.

[3] Zarchan, P., "Comparison of Statistical Digital Simulation Methods," *Advisory Group For Aerospace Research and Development*, AGARDograph No. 273, July 1988, pp. 2-1–2-16.

[4] Laning, J. H., and Battin, R. H., *Random Processes in Automatic Control*, McGraw-Hill, New York, 1956.

[5] Zarchan, P., "Complete Statistical Analysis of Nonlinear Missile Guidance Systems—SLAM," *Journal of Guidance and Control*, Vol. 2, Jan.–Feb. 1979, pp. 71–78.

[6] Zarchan, P., "Representation of Realistic Evasive Maneuvers by the Use of Shaping Filters," *Journal of Guidance and Control*, Vol. 2, July–Aug. 1979, pp. 290–295.

[7] Fitzgerald, R. J., "Shaping Filters for Disturbances with Random Starting Times," *Journal of Guidance and Control*, Vol. 2, March–April 1979, pp. 152–154.

CHAPTER 5

Covariance Analysis and the Homing Loop

BACKGROUND

Covariance analysis is another useful computerized tool that can be used to analyze time-varying linear systems driven by random inputs. *Covariance analysis*, like the adjoint technique, is an exact method of analysis that is restricted to linear systems. With this method, the covariance matrix of the system state vector is propagated as a function of time by the direct integration of a nonlinear matrix differential equation. Exact statistical performance projections of any state or combination of states as a function of time can be obtained with this technique. Covariance analysis is quite popular in problems associated with inertial navigation and optimal estimation. We shall show that the covariance analysis technique can also be used to get exact statistical performance projections in a missile guidance system.

THEORY

So far we are accustomed to writing computer programs directly from inspection of the system block diagram. To apply covariance analysis, we must first change our method of operation and convert the system block diagram to state space notation or an equivalent set of first-order differential equations expressed in matrix form.

The dynamics of any linear system driven by white noise inputs can be converted to the following first-order vector differential equation:

$$\dot{\boldsymbol{x}}(t) = \boldsymbol{F}(t)\boldsymbol{x}(t) + \boldsymbol{u}(t)$$

where $\boldsymbol{x}(t)$ is the system state vector, $\boldsymbol{F}(t)$ is the system dynamics matrix, and $\boldsymbol{u}(t)$ is a white noise vector with spectral density matrix $\boldsymbol{Q}(t)$, or

$$\boldsymbol{Q}(t) = E[\boldsymbol{u}(t)\boldsymbol{u}^T(t)]$$

The matrix differential equation for the propagation of the covariance of this general system is [1, 2]

$$\dot{\boldsymbol{X}}(t) = \boldsymbol{F}(t)\boldsymbol{X}(t) + [\boldsymbol{F}(t)\boldsymbol{X}(t)]^T + \boldsymbol{Q}(t)$$

where the covariance matrix $\boldsymbol{X}(t)$ is related to the state $\boldsymbol{x}(t)$ according to

$$\boldsymbol{X}(t) = \boldsymbol{E}[\boldsymbol{x}(t)\boldsymbol{x}^T(t)]$$

The diagonal elements of the covariance matrix represent the variances of the state variables if the disturbance processes are zero mean. The off-diagonal elements of the covariance matrix represent the degree of correlation between the various state variables.

LOW-PASS FILTER EXAMPLE

To demonstrate the application of covariance analysis, let us revisit the example of Chapter 4 in which a low-pass filter with a white noise input has been redrawn in block diagram form as shown in Fig. 5.1. In this example the input u_s is white noise with power spectral density Φ_S, T is the time constant of the low-pass filter, and x is the filter output. We want to find the variance of x as a function of time.

By inspection of Fig. 5.1, we can write the first-order differential equation of the low-pass filter in state space form as

$$\dot{x} = -\frac{x}{T} + \frac{u_S}{T}$$

Therefore, for this example, the system dynamic matrix and spectral density matrix are time-invariant scalars and can be written by inspection of the preceding differential equation as

$$\boldsymbol{F} = \frac{-1}{T}$$

$$\boldsymbol{Q} = \frac{\Phi_S}{T^2}$$

The differential equation for the propagation of the covariance simplifies to the linear equation

$$\dot{X} = \frac{-2}{T}X + \frac{\Phi_S}{T^2}$$

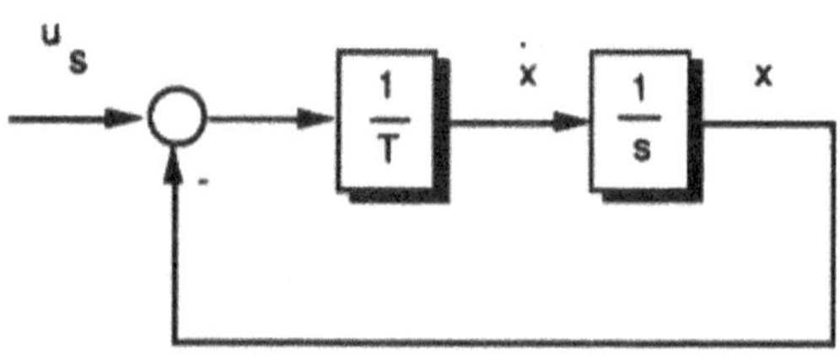

Fig. 5.1 Low-pass filter with white noise input.

The solution to the preceding linear covariance analysis differential equation can be found, using standard differential equation solution techniques, to be

$$X = \frac{\Phi_S}{2T}[1 - e^{-2t/T}]$$

Because X represents the mean square value of x, we recognize that the preceding answer is identical to the answer obtained in Chapter 4 using the impulse response technique.

NUMERICAL CONSIDERATIONS

In all of the systems simulated in the text, the second-order Runge–Kutta numerical integration method is used to solve the necessary differential equations. Although more accurate numerical integration techniques exist, the second-order Runge–Kutta technique is adequate for getting the correct answers. When the equations associated with covariance analysis are solved numerically, higher order integration methods are required to get the desired accuracy.

Let us again consider the second-order network simulation of Chapter 1 (that is, Fig. 1.1 and Listing 1.1) in which the second-order Runge–Kutta numerical integration technique was used. In that simulation the integration step size was made very small (h = 0.001 s). If we arbitrarily increase the integration step size, we can see from Fig. 5.2 that the accuracy of the answers begin to degrade.

It is apparent from Fig. 5.2 that h = 0.01 s is about the largest the integration step size can be made without degrading accuracy. This is not surprising because the natural frequency of the second-order system in this example is 20 rad/s. This means that the approximate time constant of the system under consideration is 0.05 s (1/20 = 0.05). With second-order Runge–Kutta numerical integration

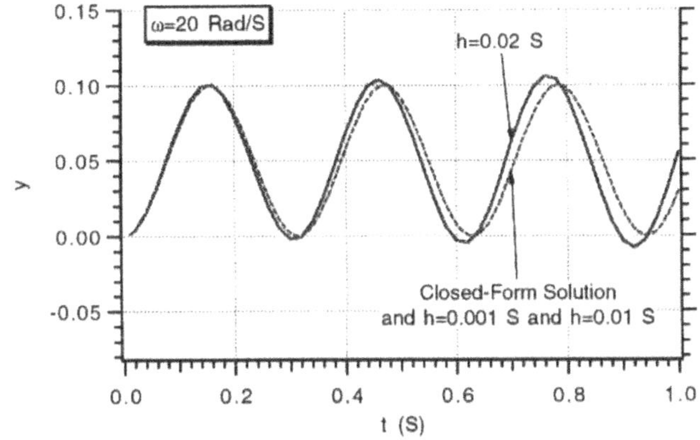

Fig. 5.2 Increasing integration step size eventually degrades accuracy.

we must make the integration interval at least five times smaller than the natural frequency of the network in order to get accurate answers. Making the integration step size equal to or larger than 0.02 s means that we will be missing the effect of the system's high bandwidth.

Better accuracy can be achieved with the fourth-order Runge–Kutta numerical integration technique [3]. Given a first-order differential equation of the form

$$\dot{x} = f(x, t)$$

where t is time, we want to find a numerical integration recursive relationship for x as a function of time. With the fourth-order Runge–Kutta numerical integration technique the value of x at the next integration interval h is given by

$$x_{k+1} = x_k + \frac{h}{6}[K_0 + 2K_1 + 2K_2 + K_3]$$

where

$$\begin{aligned} K_0 &= f(x_k, t_k) \\ K_1 &= f(x_k + 0.5K_0, t + 0.5h) \\ K_2 &= f(x_k + 0.5K_1, t + 0.5h) \\ K_3 &= f(x_k + K_2, t + h) \end{aligned}$$

From the preceding expressions we can see that the new value of x is simply the old value of x plus terms proportional to the derivative evaluated at various times between t and $t + h$. Using the relationships for the fourth-order Runge–Kutta integration technique, we can write a program to simulate the second-order

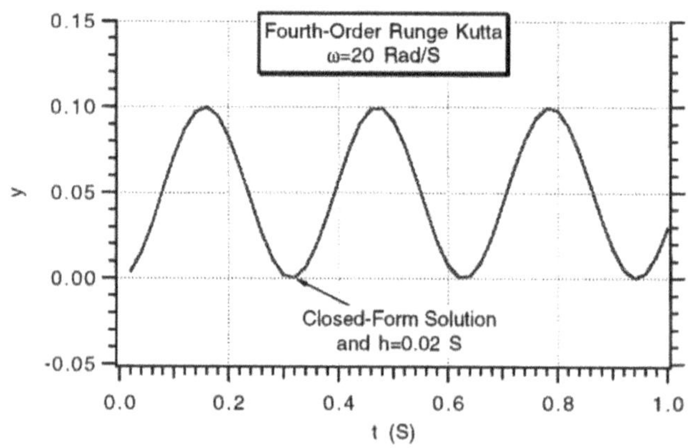

Fig. 5.3 Fourth-order Runge–Kutta integration yields adequate accuracy with larger integration step sizes.

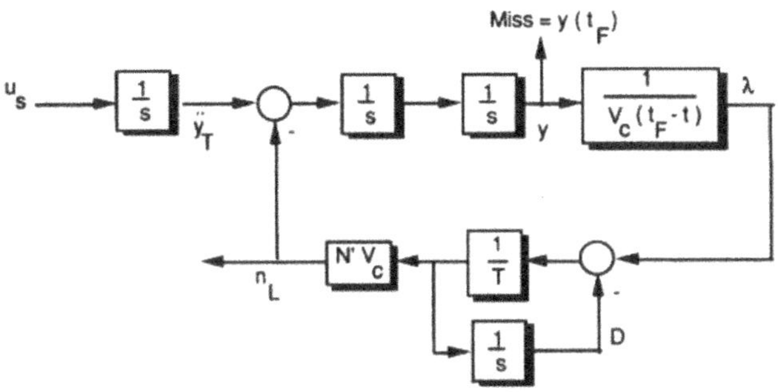

Fig. 5.4 Single-lag homing loop with random target maneuver.

network as shown in Listing 5.1. We can see that the structure of this program is identical to the one of Listing 1.1 (that is, differential equations appear just before the FLAG = 1 statement) except extra steps have been added to the integration procedure.

Figure 5.3 shows that when we simulate the second-order network with Listing 5.1 using fourth-order rather than second-order Runge–Kutta numerical integration we can use a larger integration step size to get the same accuracy. In this example, a step size of 0.02 s was adequate for getting the correct solution.

HOMING LOOP EXAMPLE

To demonstrate the utility of covariance analysis for a more relevant example, let us revisit the single-lag homing loop example of Chapter 4 in which the random error source is a uniformly distributed target maneuver. The homing loop model of Fig. 4.15 is redrawn in Fig. 5.4 for convenience. In Fig. 5.4 the uniformly distributed target maneuver has been replaced by its shaping filter equivalent, which is white noise through an integrator. The spectral density of the white noise input u_s is given by Φ_S, which was shown in Chapter 4 to be

$$\Phi_S = \frac{n_T^2}{t_F}$$

LISTING 5.1 SIMULATION OF SECOND-ORDER SYSTEM WITH FOURTH-ORDER RUNGE–KUTTA INTEGRATION TECHNIQUE

```
% Preallocation
clear
K0=zeros([1,6]);
K1=zeros([1,6]);
K2=zeros([1,6]);
```

```
K3=zeros([1,6]);
count=0;
XIN=1.;
W=20.;
Y=0.;
YD=0.;
T=0.;
H=.01;
S=0.;
while ~(T >= 1.)
	S=S+H;
	YOLD=Y;
	YDOLD=YD;
	STEP=1;
	while ( (STEP == 1) | (STEP == 2) | (STEP == 3))
	  YDD=W*XIN-W*W*Y;
	  if (STEP == 1)
			STEP=2;
			K0(1,1)=YD;
			K0(1,2)=YDD;
			TNEW=T+.5*H;
			Y=YOLD+.5.*H.*K0(1,1);
			YD=YDOLD+.5.*H.*K0(1,2);
	  elseif (STEP ==2)
			STEP=3;
			K1(1,1)=YD;
			K1(1,2)=YDD;
			TNEW=T+.5*H;
			Y=YOLD+.5.*H.*K1(1,1);
			YD=YDOLD+.5.*H.*K1(1,2);
	  else
			STEP=4;
			K2(1,1)=YD;
			K2(1,2)=YDD;
			TNEW=T+H;
			Y=YOLD+H.*K2(1,1);
			YD=YDOLD+H.*K2(1,2);
	  end
	end
	YDD=W*XIN-W*W*Y;
	K3(1,1)=YD;
	K3(1,2)=YDD;
	T=TNEW;
	Y=YOLD+H.*(K0(1,1)+2.*(K1(1,1)+K2(1,1))+K3(1,1))./6;
	YD=YDOLD+H.*(K0(1,2)+2.*(K1(1,2)+K2(1,2))+K3(1,2))./6;
	count=count+1;
```

```
        ArrayT(count)=T;
        ArrayY(count)=Y;
end
figure
plot(ArrayT,ArrayY)
xlabel('Time (s) ')
ylabel('Y')
title('Forth-order Runge-Kutta: Second Order Network')
output=[ArrayT', ArrayY'];
save datfil.txt output /ascii
disp 'simulation finished'
```

where n_T is the magnitude of the target maneuver and t_F is the flight time over which the maneuver is equally likely to occur.

To apply covariance analysis to the homing loop of Fig. 5.4, we must convert this block diagram to state space form. To perform the conversion, the homing loop equations must first be expressed as a set of first-order linear differential equations or

$$\ddot{y} = \ddot{y}_T - N'V_c\dot{D} = \ddot{y}_T - \frac{N'V_c}{T}\left[\frac{y}{V_c(t_F - t)} - D\right]$$

$$\dddot{y}_T = u_s$$

$$\dot{D} = \frac{1}{T}\left[\frac{y}{V_c(t_F - t)} - D\right]$$

Since the preceding set of first-order differential equations are functions of the states, they can be written in state space form by inspection as

$$\begin{bmatrix} \dot{y} \\ \ddot{y} \\ \dddot{y}_T \\ \dot{D} \end{bmatrix} = \begin{bmatrix} 0 & 1 & 0 & 0 \\ \dfrac{-N'}{T(t_F - t)} & 0 & 1 & \dfrac{N'V_c}{T} \\ 0 & 0 & 0 & 0 \\ \dfrac{1}{TV_c(t_F - t)} & 0 & 0 & \dfrac{-1}{T} \end{bmatrix} \begin{bmatrix} y \\ \dot{y} \\ \ddot{y}_T \\ D \end{bmatrix} + \begin{bmatrix} 0 \\ 0 \\ u_S \\ 0 \end{bmatrix}$$

By comparing the preceding matrix differential equation with the generalized state space equation we can see that the state vector for this example is

$$\boldsymbol{x} = \begin{bmatrix} y \\ \dot{y} \\ \ddot{y}_T \\ D \end{bmatrix}$$

and the system dynamic matrix is given by

$$F = \begin{bmatrix} 0 & 1 & 0 & 0 \\ \dfrac{-N'}{T(t_F - t)} & 0 & 1 & \dfrac{N'V_c}{T} \\ 0 & 0 & 0 & 0 \\ \dfrac{1}{TV_c(t_F - t)} & 0 & 0 & \dfrac{-1}{T} \end{bmatrix}$$

From the homing loop state space equation we can also see that $\boldsymbol{u}(t)$ is

$$\boldsymbol{u} = \begin{bmatrix} 0 \\ 0 \\ u_S \\ 0 \end{bmatrix}$$

and therefore the spectral density matrix $\boldsymbol{Q}(t)$ becomes

$$\boldsymbol{Q} = \begin{bmatrix} 0 & 0 & 0 & 0 \\ 0 & 0 & 0 & 0 \\ 0 & 0 & \Phi_S & 0 \\ 0 & 0 & 0 & 0 \end{bmatrix}$$

where Φ_S has been previously defined.

Integration of the covariance analysis nonlinear matrix differential equation yields statistical information for all of the states. For this homing loop example, the standard deviation of the relative trajectory y can be found by taking the square root of the first diagonal element of the covariance matrix X or

$$\sigma_y(t) = \sqrt{X(1,1)}$$

The source code listing of the homing loop covariance analysis program appears in Listing 5.2. We can see from the listing that the fourth-order Runge–Kutta integration is used to get the necessary accuracy. From Listing 5.2 we can see that the only error source in the guidance system is a 3-g uniformly distributed target maneuver.

LISTING 5.2 HOMING LOOP COVARIANCE ANALYSIS PROGRAM

```
clear
F=zeros([4,4]);
X=zeros([4,4]);
XOLD=zeros([4,4]);
Q=zeros([4,4]);
K0=zeros([4,4]);
K1=zeros([4,4]);
```

```
K2=zeros([4,4]);
K3=zeros([4,4]);
XD=zeros([4,4]);
A=zeros([1,4]);
AXAT=zeros([1,1]);
count=0;
T=0.;
TNEW=T;
S=0.;
H=.01;
XNP=3.;
TAU=1.;
XNT=96.6;
VC=4000.;
TF=10.;
TGO=TF-T+.00001;
PHIS=XNT*XNT/TF;
F(1,2)=1.;
F(2,1)=-XNP/(TAU*TGO);
F(2,3)=1.;
F(2,4)=XNP*VC/TAU;
F(4,1)=1./(TAU*VC*TGO);
F(4,4)=-1./TAU;
Q(3,3)=PHIS;
while ~(T >= (TF-.0001))
        S=S+H;
        XOLD=X;
        STEP=1;
        while ( (STEP == 1) | (STEP == 2) | (STEP == 3))
           TGO=TF-TNEW+.00001;
           F(2,1)=-XNP/(TAU*TGO);
           F(4,1)=1./(TAU*VC*TGO);
           XD=(F*X)+(F*X)'+Q;
           if (STEP == 1)
                    STEP=2;
                    K0=XD;
                    TNEW=T+.5*H;
                    X=XOLD+.5.*H.*K0;
           elseif (STEP ==2)
                    STEP=3;
                    K1=XD;
                    TNEW=T+.5*H;
                    X=XOLD+.5.*H.*K1;
           else
                    STEP=4;
                    K2=XD;
```

```
                TNEW=T+H;
                X=XOLD+H.*K2;
        end
    end
    TGO=TF-TNEW+.00001;
    F(2,1)=-XNP/(TAU*TGO);
    F(4,1)=1./(TAU*VC*TGO);
    XD=(F*X)+(F*X)'+Q;
    K3=XD;
    T=TNEW;
    X=XOLD+H.*(K0+2.*(K1+K2)+K3)./6;
    if S > =.09999
        S=0.;
        A(1,1)=XNP/(TAU*TGO);
        A(1,2)=0.;
        A(1,3)=0.;
        A(1,4)=-XNP*VC/TAU;
        AXAT=A*X*A';
        SIGY=sqrt(X(1,1));
        SIGNL=sqrt(AXAT(1,1));
        count=count+1;
        ArrayT(count)=T;
        ArraySIGY(count)=SIGY;
        ArraySIGNLG(count)=SIGNL/32.2;
     end
end
figure
plot(ArrayT,ArraySIGY)
xlabel('Time (s) ')
ylabel('Standard Deviation of Relative Position (Ft)')
figure
plot(ArrayT,ArraySIGNLG)
xlabel('Time (s) ')
ylabel('Standard Deviation of Acceleration (G)')
axis([0 10 0 20])
clc
output=[ArrayT', ArraySIGY',ArraySIGNLG'];
save datfil.txt output /ascii
disp 'simulation finished'
SIGY
```

The homing loop covariance analysis program of Listing 5.2 was run, and Fig. 5.5 presents the resultant standard deviation of the relative separation between the missile and target [that is, square root of first diagonal element of covariance matrix represents $\sigma_y(t)$] for the entire 10-s flight. At the end of the flight, the relative separation between the missile and target is the miss distance

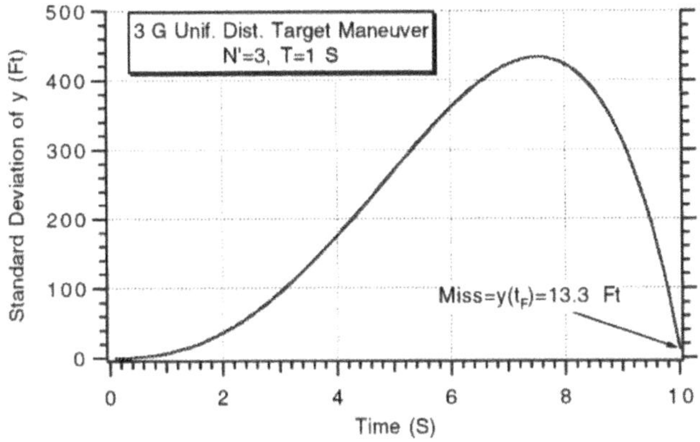

Fig. 5.5 Covariance analysis miss distance results agree with adjoint.

[that is, $\sigma_{\text{Miss}} = \sigma_y(t_F)$]. In this example the covariance analysis program indicates that the standard deviation of the miss distance is 13.3 ft, which is identical to the adjoint results of Chapter 4 (see Fig. 4.18). Unlike the adjoint technique, covariance analysis does not provide miss distance error budget information for all different flight times in a single computer run. However, as can be seen from Fig. 5.5, covariance analysis does provide relative trajectory information at all times for a given flight time. If many random error sources are present, one covariance analysis computer run yields a total statistical performance projection. If learning how each error source contributed to the total performance is desired, additional computer runs must be made—each one run with one error source at a time!

Covariance analysis also has the capability of providing information concerning other quantities in the same computer run. For example, covariance analysis could also show us how the standard deviation of the missile acceleration varies with time in the same computer run. However, we must first express the missile acceleration as a function of the states. From Fig. 5.4 we can see that the missile acceleration is related to the states according to

$$n_L = \frac{N'V_c}{T}\left[\frac{y}{V_c(t_F - t)} - D\right]$$

or more concisely, in matrix form we can say that

$$n_L = \boldsymbol{A}\boldsymbol{x}$$

where $\boldsymbol{x}$ is the system state vector, and for this example $\boldsymbol{A}$ is given by

$$\boldsymbol{A} = \left[\frac{N'}{T(t_F - t)} \quad 0 \quad 0 \quad \frac{-N'V_c}{T}\right]$$

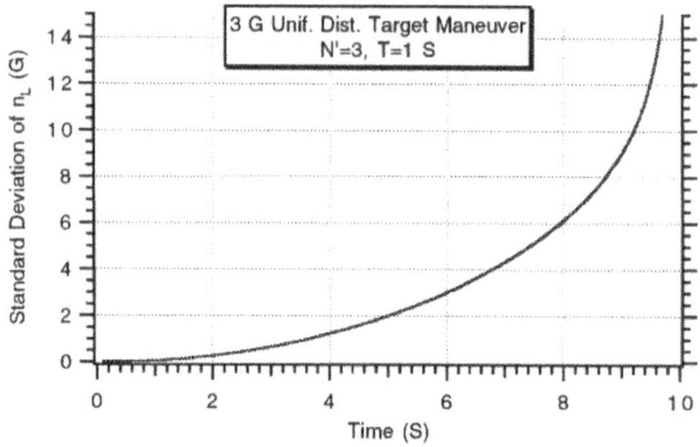

Fig. 5.6 Covariance analysis also provides acceleration profile information.

Therefore the variance of the missile acceleration is given by

$$E[n_L n_L^T] = \sigma_{n_L}^2 = \boldsymbol{AXA}^T$$

where $\boldsymbol{X}$ is the covariance matrix and the standard deviation of the acceleration is simply the square root of the preceding expression. Figure 5.6 displays the resultant missile acceleration profile, using the preceding expression in the covariance analysis program, for the entire flight. We can see that the standard deviation of the missile acceleration is monotonically increasing for the 10-s flight. If we make the Gaussian assumption, we can infer that 68% of the time 15 g (that is, 483 ft/s^2) or 5 times the acceleration of the target is required to avoid acceleration saturation. In more pessimistic terms we can also say that if the missile only has a 15-g capability there is a 32% probability that the missile will acceleration saturate for this example. This example demonstrates that although covariance analysis does not provide all of the information of the adjoint, it does provide extra useful information that can be used to access system performance. In addition, the covariance analysis technique can be used to provide an independent check of the accuracy of an adjoint simulation.

ACCELERATION ADJOINT

We state in Chapter 3 that the impulse response of the original system and adjoint system are related according to

$$h^*(t_F - t_I, t_F - t_o) = h(t_o, t_I)$$

where h denotes the impulse responses of the original system and h^* is the impulse response of the adjoint system. This important relationship means that putting an impulse into the original system at time t_I and observing the output at time t_o is identical to putting an impulse into the adjoint system at time $t_F - t_o$ and observing the output at time $t_F - t_I$, where t_F is the final time or flight time. In all of the adjoint applications discussed so far, the observation time was always the final time t_F since we were only interested in the miss distance. If all disturbances occur at time zero in the original system but the observation time is not the final time, the fundamental adjoint relationship simplifies to

$$h^*(t_F, t_F - t_o) = h(t_o, 0)$$

The preceding relationship means that applying an impulse to the original system at time zero and observing the output at time t_0 is equivalent to initiating the impulse at time $t_F - t_o$ in the adjoint system and observing the output at time t_F. *In other words, if we would like to develop other types of adjoints it is only necessary to change the impulse application time and the location of the impulse application. The adjoint block diagram remains unchanged!*

Figure 5.7 is the adjoint block diagram of the single-lag homing loop of Fig. 5.4. This adjoint diagram is identical to Fig. 4.17 (adjoint model in Chapter 4) except that it is noted that certain initial conditions are used if a miss distance adjoint is desired and other initial conditions are used for an acceleration adjoint. If a miss distance adjoint is being run, an initial condition of unity is

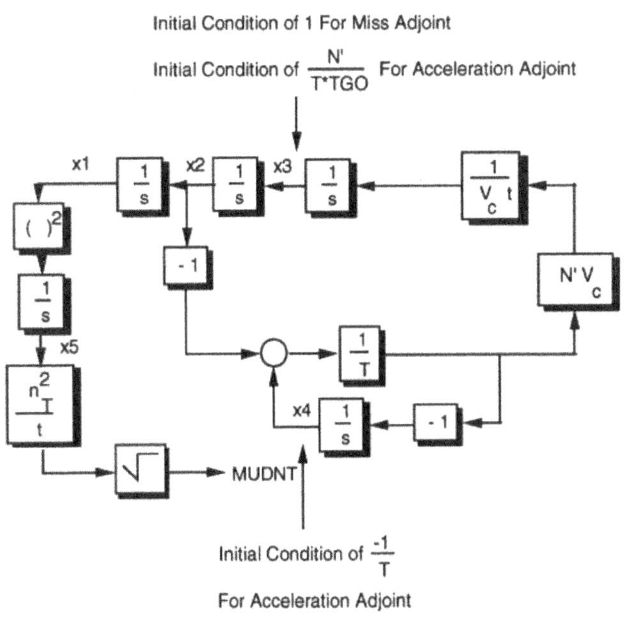

Fig. 5.7 Adjoint model for miss distance and acceleration.

applied at time zero on the $x3$ integrator. If an acceleration adjoint is required, initial conditions are applied at time $t_F - t_o$ to the $x3$ and $x4$ integrators. In this case time $t_F - t_o$ corresponds to the time to go until intercept in which we desire to observe the acceleration in the original system. In other words, if we desire to observe acceleration in the original system at time 8 s for a 10-s flight, that is the same as observing the acceleration at 2 s to go (that is, $10 - 8 = 2$). Therefore the impulse (or initial conditions on appropriate integrators) is applied at time 2 s in the adjoint system and the output is observed at 10 s in the adjoint system. Changing the observation time in the adjoint system corresponds to observing acceleration at 2 s to go for different flight times in the original system.

The source code listing for the adjoint program, which can be used for both miss distance and acceleration computation, appears in Listing 5.3. In this program TINT represents the time to go in the original system in which we wish to observe the quantity of interest. If we want to compute miss distance, MISS should be set to 1 and TINT set to zero. For an acceleration adjoint, MISS should be set to 0 and TINT set to a number representing the time to go at which we want to observe the acceleration. The adjoint program of Listing 5.3 is set to run as an acceleration adjoint in which acceleration levels correspond to 0.5 s to go in the original system (TINT = 0.5).

The preceding acceleration adjoint program was run for values of observation time in the original system corresponding to 0.5, 1, and 2 s to go (TINT = 0.5, 1, 2), and the results for the three adjoint runs are displayed in Fig. 5.8. We can interpret the abscissa of the plot as either adjoint time or flight time. The curve representing acceleration at 0.5 s to go, TINT = 0.5 (labeled $t_{go} = 0.5$), indicates that the standard deviation of the missile acceleration is 12 g at 0.5 s to go for a 10-s flight, approximately 12 g at 0.5 s to go for a 6-s flight, and approximately 4 g for a 2-s flight. The missile acceleration values for a 10-s flight at observation times to go of 0.5, 1, and

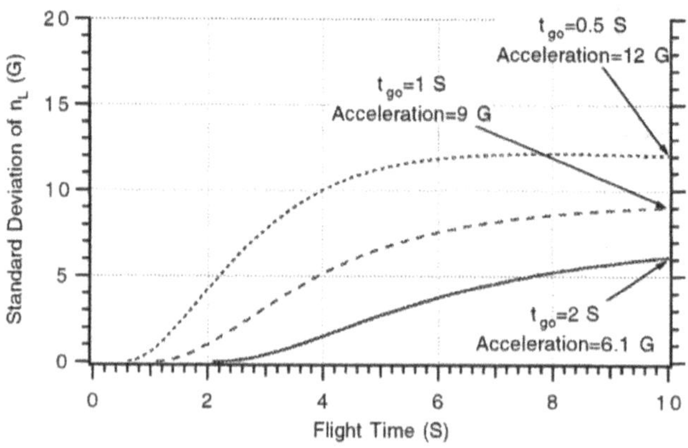

Fig. 5.8 Adjoint and covariance analysis acceleration results agree.

2 s (12 *g*, 9 *g*, and 6.1 *g*) respectively agree exactly with the single run covariance analysis results of Fig. 5.6.

LISTING 5.3 ACCELERATION AND MISS DISTANCE ADJOINT PROGRAM

```
count=0;
XNT=96.6;
XNP=3.;
TAU=1.;
TF=10.;
T=0.;
S=0.;
TINT=.5;
MISS=0;
TP=T+.00001+TINT;
X1=0;
X2=0;
X5=0.;
if MISS==1
        X3=1.;
        X4=0.;
else
        X3=XNP/(TAU*TINT);
        X4=-1./TAU;
end
H=.01;
while TP<=(TF - 1e-5)
        X1OLD=X1;
        X2OLD=X2;
        X3OLD=X3;
        X4OLD=X4;
        X5OLD=X5;
        STEP=1;
        FLAG=0;
        while STEP<=1
           if FLAG==1
                   X1=X1+H*X1D;
                   X2=X2+H*X2D;
                   X3=X3+H*X3D;
                   X4=X4+H*X4D;
                   X5=X5+H*X5D;
                   TP=TP+H;
                   STEP=2;
           end
           X1D=X2;
```

```
            X2D=X3;
            Y1=(X4-X2)/TAU;
            TGO=TP+.00001;
            X3D=XNP*Y1/TGO;
            X4D=-Y1;
            X5D=X1*X1;
            FLAG=1;
        end;
        FLAG=0;
        X1=(X1OLD+X1)/2+.5*H*X1D;
        X2=(X2OLD+X2)/2+.5*H*X2D;
        X3=(X3OLD+X3)/2+.5*H*X3D;
        X4=(X4OLD+X4)/2+.5*H*X4D;
        X5=(X5OLD+X5)/2+.5*H*X5D;
        S=S+H;
        if S > =.099999
          S=0.;
          XMUDNT=XNT*sqrt(X5/TGO);
          if MISS==0
                XMUDNT=XMUDNT/32.2;
          end
          count=count+1;
          ArrayTP(count)=TP;
          ArrayXMUDNT(count)=XMUDNT;
        end
end
%figure
plot(ArrayTP,ArrayXMUDNT),grid
xlabel('Flight Time (Sec)')
ylabel('Acceleration (G) ')
clc
output=[ArrayTP',ArrayXMUDNT'];
save datfil.txt output /ascii
disp 'simulation finished'
```

SUMMARY

In this chapter we have shown how the covariance analysis technique can be applied to a missile guidance system. The fourth-order Runge–Kutta numerical integration technique was required in order to obtain performance projections of the desired accuracy. Although covariance analysis techniques do not yield error budget information as does the adjoint technique, exact performance projects can be obtained for all quantities of interest in a single run. It was also shown how the adjoint technique could be extended to yield acceleration as well as miss distance information.

REFERENCES

[1] Gelb, A., *Applied Optimal Estimation*, MIT Press, Cambridge, MA, 1974.

[2] Bryson, A. E., and Ho, Y. C., *Applied Optimal Control*, Blaisdell, Waltham, MA, 1969.

[3] Press, W. H., Flannery, B. P., Teukolsky, S. A., and Vetterling, W. T., *Numerical Recipes: The Art of Scientific Computation*, Cambridge, University Press, London, UK 1986.

CHAPTER 6

Proportional Navigation and Miss Distance

INTRODUCTION

The relationship between proportional navigation and miss distance will be investigated more extensively in this chapter. First we will demonstrate, via numerical examples, that it is important to have an accurate guidance system model in order to get performance projections that are meaningful and not overly optimistic. Normalized design curves will be presented that allow an analyst to rapidly predict system behavior given a minimum of information. Curves of this type are invaluable in preliminary system design. The influence of optimal target maneuvers on system performance will be evaluated to highlight potential guidance system weaknesses. Finally, the influence of saturation and parasitic effects will be demonstrated to help the designer place realistic bounds on achievable system performance.

SYSTEM ORDER [1]

Thus far, the work presented has concerned itself with either a zero- or single-lag guidance system. We have seen that, if the flight time is *not* significantly larger than the guidance system time constant, then the difference between the performance of a zero- and single-lag guidance system can be significant. Both the single- and zero-lag guidance systems are convenient analytical models but do not quite match reality. It is important to determine if a higher-order guidance system representation would influence system performance. To separate time constant and system order effects, it is convenient to use a binomial representation of the guidance system:

$$\frac{n_L}{\dot{\lambda}} = (N'V_c)\Big/\left[\left(\frac{1+sT}{n}\right)^n\right]$$

In the preceding representation, T is the effective guidance system time constant and n the system order. If $n = 1$, then the binomial expansion reduces to the single-lag guidance system, which we have already previously studied. This particular form of the binomial representation is useful, although not especially realistic, because an expansion of the guidance system denominator always yields

$$\left(\frac{1+sT}{n}\right)^n = 1 + sT + \cdots$$

which means that T is always the approximate time constant of the guidance system, regardless of system order.

Figure 6.1 shows how the miss distance due to a 3-g target maneuver varies with flight time and system order for a binomial guidance system in which the effective navigation ratio is 4 and the effective guidance system time constant is 1 s. We can see that the performance projections resulting from a single-lag guidance system model are a serious underestimate of the influence of target maneuver on miss when the flight time is not an order of magnitude greater than the guidance system time constant. The importance of system order and its influence on system performance becomes less important as system order increases. The experiment conveys the importance of accurately modeling the guidance system (which is generally not a binomial) under consideration if accurate performance projections are required.

An experiment was also conducted to determine if, in the presence of guidance system dynamics, the linearized model of the homing loop still gives accurate performance projections. Fifth-order binomial guidance system models were included in both the linearized and nonlinear engagement simulations. Cases were run for both simulations in the case of a 3-g target maneuver disturbance for various flight times, and the resultant miss distances were monitored.

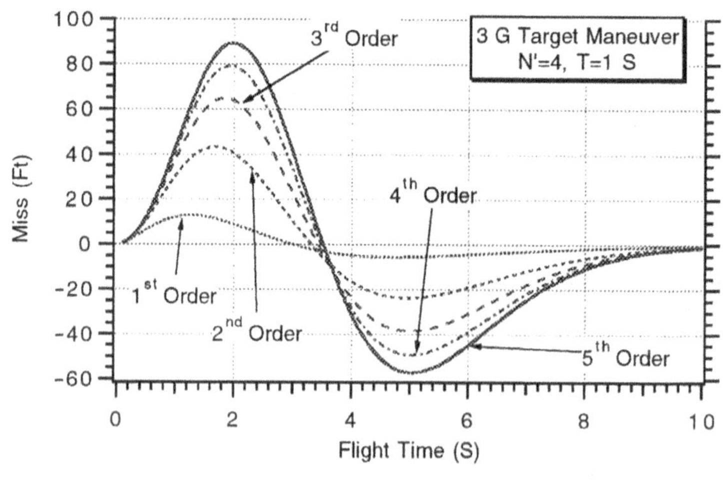

Fig. 6.1 System order has a profound influence on system performance.

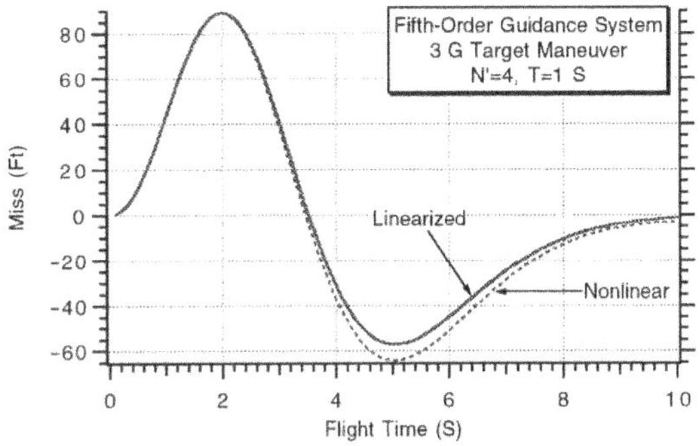

Fig. 6.2 Linearized guidance system model gives very accurate performance projections.

Figure 6.2 shows that the linearized model of the homing loop gives very accurate performance projections. Thus, we can feel confident in using our linearized guidance system model for studies involving a binomial representation of the guidance system.

Normally a missile guidance system is represented by n different time constants for an n-state system. If the time constants are widely separated, then the slowest time constant will usually dictate system performance. If the time constants are closely spaced, one must evaluate the guidance system to get accurate performance projections.

DESIGN RELATIONSHIPS

We have just seen that target maneuver can play a major role in determining missile system performance. Target maneuver can induce very large miss distances if the effective guidance time constant is too large or if the flight time is very short. In addition, we have seen in Chapter 2 that target maneuver induces large missile acceleration levels. This may lead to acceleration saturation, which will significantly further increase the induced miss distance. The purpose of this section is to quantify the influence of target maneuver on system performance in a form that will be of value to an analyst in preliminary system design.

In the previous section we established that the linearized model of the guidance system gave accurate performance projections in terms of miss distance induced by target maneuver. Performance

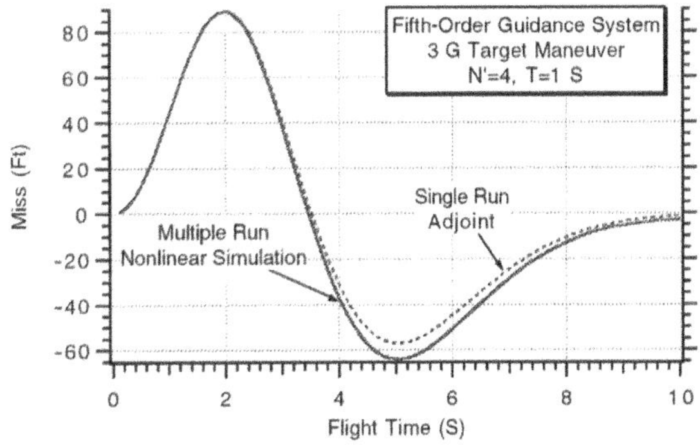

Fig. 6.3 One adjoint run gives the same result as many runs with nonlinear engagement simulation.

projections were obtained by running both the linear and nonlinear engagement simulation for many times of flight, and the resultant data were plotted. The same results could have been obtained by making one adjoint run as shown in Fig. 6.3.

In Chapter 3 a closed-form solution was developed for the miss distance induced by target maneuver in a single time constant representation of the guidance system. Although the miss distance formula will change for varying effective navigation ratios and canonic system form and order, the normalization for miss due to target maneuver will be the same. In this section we will use the method of adjoints to develop design curves that may be of use in preliminary system sizing. We will choose a guidance system form that has only one parameter: the guidance system time constant.

The model to be used for the development of normalized design curves is the fifth-order binomial proportional navigation system. This guidance system, which is depicted in Fig. 6.4, has guidance system transfer function

$$\frac{n_L}{\lambda} = (N'V_c s)\bigg/\left[\left(\frac{1+sT}{5}\right)^5\right]$$

where T is the guidance system time constant. In this canonic model, one time constant represents the seeker, another represents the noise filter, and the three time constants represents the flight-control system. Hopefully, the simplicity of this model will shed some light on fundamental issues and be of value for other guidance system forms.

Figure 6.5 presents the adjoint model of the fifth-order binomial guidance system. The adjoint model consists of three outputs that are related to three input disturbances in the original system. The miss due to a step target maneuver is represented by MNT, the miss due to a ramp target maneuver is represented by MNTD, and the miss due to a parabolic maneuver is denoted MNTDD. In the figure each integrator output is denoted by variables $x1$ to $x10$. The impulse

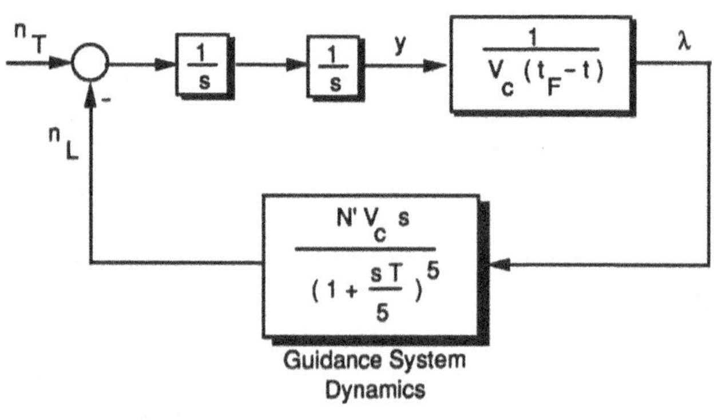

Fig. 6.4 Fifth-order binomial guidance system.

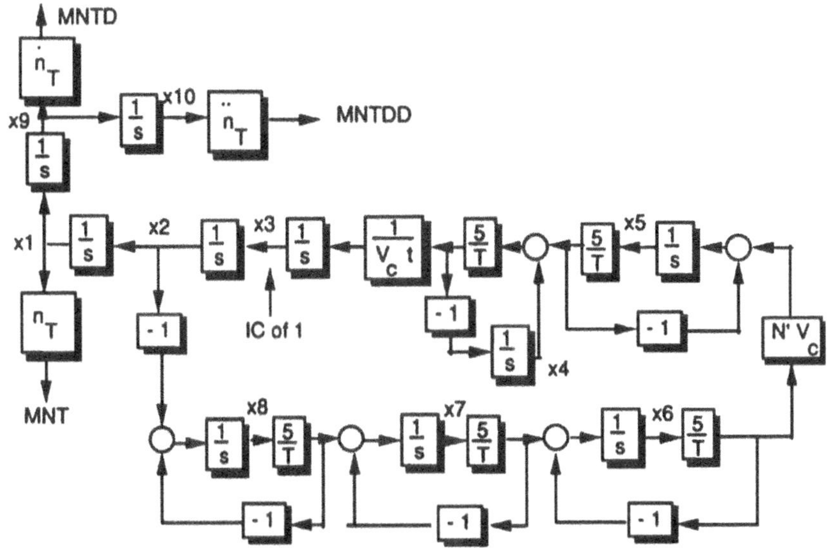

Fig. 6.5 Adjoint of fifth-order binomial guidance system.

needed to start a miss distance adjoint is represented by a unity initial condition on integrator $x3$.

An adjoint simulation can be derived from the model of Fig. 6.5. Listing 6.1 presents a MATLAB adjoint program of this fifth-order binomial guidance system. We can see from the listing that the nominal value of the target maneuver is $1g$, the value of target jerk XNTD is $1\ g/s$, and the value of target yank XNTDD is $1g/s^2$. As in our other simulations, the differential equations describing the adjoint system can be found before the FLAG=1 statement. All integrator initial conditions are zero, except for integrator $x3$. We can see from the listing that this integrator has a unity initial condition in order to make a miss distance adjoint. We can also see from the listing that a small number is added to adjoint time so that we can avoid a division by zero. This is a practical way of applying L'Hopital's rule.

The adjoint program is set up to generate normalized results by choosing the guidance system time constant TAU to be unity and the step target maneuver disturbance XNT to be $1\ g$ or $32.2\ ft/s^2$. The value of closing velocity is not important, as there is a cancellation of this term in the guidance loop. Normalized adjoint results can be generated by running the program once for a value of unity guidance system time constant. The normalization factors, derived in Chapter 3 for a single-lag guidance system, are also valid for the fifth-order binomial guidance system. Therefore, the adjoint program only has to be rerun for each effective navigation ratio XNP. For example, Fig. 6.6 presents the normalized system response to a step in target acceleration. The abscissa can be interpreted as

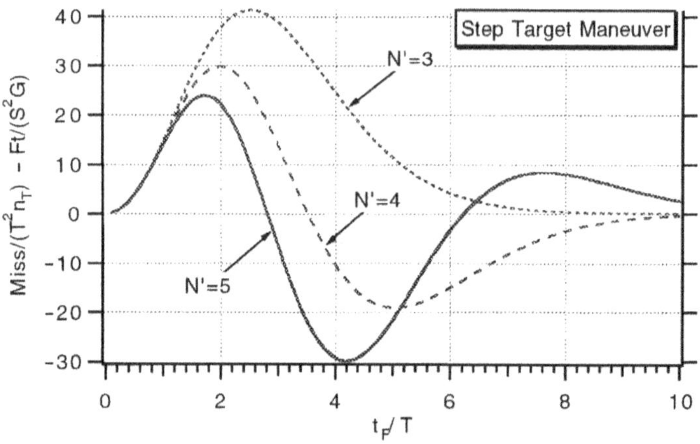

Fig. 6.6 Normalized miss due to step target maneuver.

either normalized time of flight for a step maneuver occurring at the beginning of flight or the normalized time to go at which the disturbance occurs. We can see from Fig. 6.6 that for long normalized flight times the miss approaches zero and for small normalized flight times the miss can be quite large. Increasing the effective navigation ratio tends to reduce the miss for small normalized flight times and increases the miss at the larger normalized flight times.

LISTING 6.1 ADJOINT OF FIFTH-ORDER BINOMIAL GUIDANCE SYSTEM

```
XNT=32.2;
XNP=3.;
TAU=1.;
TF=10.;
VC=4000.;
XNTD=32.2;
XNTDD=32.2;
T=0.;
S=0.;
TP=T+.00001;
X1=0;
X2=0;
X3=1;
X4=0;
X5=0.;
X6=0.;
X7=0.;
X8=0.;
X9=0.;
X10=0.;
```

```
H=.01;
n=0.;
while TP<=(TF-1e-5)
   X1OLD=X1;
   X2OLD=X2;
   X3OLD=X3;
   X4OLD=X4;
   X5OLD=X5;
   X6OLD=X6;
   X7OLD=X7;
   X8OLD=X8;
   X9OLD=X9;
   X10OLD=X10;
   STEP=1;
   FLAG=0;
   while STEP<=1
      if FLAG==1
         STEP=2;
         X1=X1+H*X1D;
         X2=X2+H*X2D;
         X3=X3+H*X3D;
         X4=X4+H*X4D;
         X5=X5+H*X5D;
         X6=X6+H*X6D;
         X7=X7+H*X7D;
         X8=X8+H*X8D;
         X9=X9+H*X9D;
         X10=X10+H*X10D;
         TP=TP+H;
      end
      X1D=X2;
      X2D=X3;
      Y1=5.*(5.*X5/TAU+X4)/TAU;
      TGO=TP+.00001;
      X3D=Y1/(VC*TGO);
      X4D=-Y1;
      X5D=-5.*X5/TAU+5.*X6*XNP*VC/TAU;
      X6D=-5.*X6/TAU+5.*X7/TAU;
      X7D=-5.*X7/TAU+5.*X8/TAU;
      X8D=-5.*X8/TAU-X2;
      X9D=X1;
      X10D=X9;
      FLAG=1;
   end
   FLAG=0;
   X1=.5*(X1OLD+X1+H*X1D);
   X2=.5*(X2OLD+X2+H*X2D);
```

```
    X3=.5*(X3OLD+X3+H*X3D);
    X4=.5*(X4OLD+X4+H*X4D);
    X5=.5*(X5OLD+X5+H*X5D);
    X6=.5*(X6OLD+X6+H*X6D);
    X7=.5*(X7OLD+X7+H*X7D);
    X8=.5*(X8OLD+X8+H*X8D);
    X9=.5*(X9OLD+X9+H*X9D);
    X10=.5*(X10OLD+X10+H*X10D);
    S=S+H;
    if S>=.0999
        S=0.;
        n=n+1;
        ArrayTP(n)=TP;
        ArrayXMNT(n)=XNT*X1;
        ArrayXMNTD(n)=XNTD*X9;
        ArrayXMNTDD(n)=XNTDD*X10;
    end
end
figure
plot(ArrayTP,ArrayXMNT),grid
xlabel('Normalized Flight Time (Sec)')
ylabel('Missile Miss Due To Step Maneuver(Ft/G-Sec^2)')
figure
plot(ArrayTP,ArrayXMNTD),grid
xlabel('Normalized Flight Time (Sec)')
ylabel('Missile Miss Due To Ramp Maneuver(Ft/G-Sec^3/S)')
figure
plot(ArrayTP,ArrayXMNTDD),grid
xlabel('Normalized Flight Time (Sec)')
ylabel('Missile Miss Due To Parabolic Maneuver(Ft/G-Sec^4/S^2)')
clc
output=[ArrayTP',ArrayXMNT',ArrayXMNTD',ArrayXMNTD'];
save datfil.txt output -ascii
disp 'simulation finished'
```

To illustrate the use of the normalized miss distance results of Fig. 6.6, let us consider a numerical example. If the guidance time constant is 0.5 s, and the time of flight is 2.5 s, then the normalized flight time is 5, or

$$t_F/T = 2.5/0.5 = 5$$

For an effective navigation ratio of 3, the normalized miss can be read from Fig. 6.6 as 12, or

$$\frac{\text{Miss}}{T^2 n_T} = 12$$

To compute the actual miss distance in this example, we must assume a target maneuver level n_T. With a 4-g maneuver level the actual miss distance becomes

$$\text{Miss} = 12T^2 n_T = 12 * 0.5^2 * 4 = 12\,\text{ft}$$

Increasing the guidance system time constant can substantially influence the miss distance. For example, if we increase the guidance system time constant from 0.5 s to 1 s, the normalized flight time becomes

$$t_F/T = 2.5/1 = 2.5$$

Keeping the effective navigation ratio to 3 yields a new normalized miss of approximately 42, or

$$\frac{\text{Miss}}{T^2 n_T} = 42$$

which means that for a 4-g maneuver the actual miss is

$$\text{Miss} = 42T^2 n_T = 42 * 1^2 * 4 = 168\,\text{ft}$$

In other words, for this case doubling the guidance system time constant increased the miss distance by more than an order of magnitude!

By integrating the miss due to a step target maneuver in the adjoint program, we can also find the normalized miss due to a ramp maneuver. Figure 6.7 presents normalized miss distance results for a *ramp maneuver disturbance*. Again, the abscissa has the same interpretation as before. Here we can see that the effective navigation ratio must be greater than 3 for the miss to approach zero for long flight times. This means that, if the actual maneuver is a ramp, we need an effective navigation ratio of more than 3 to hit the target. In addition, we can see from

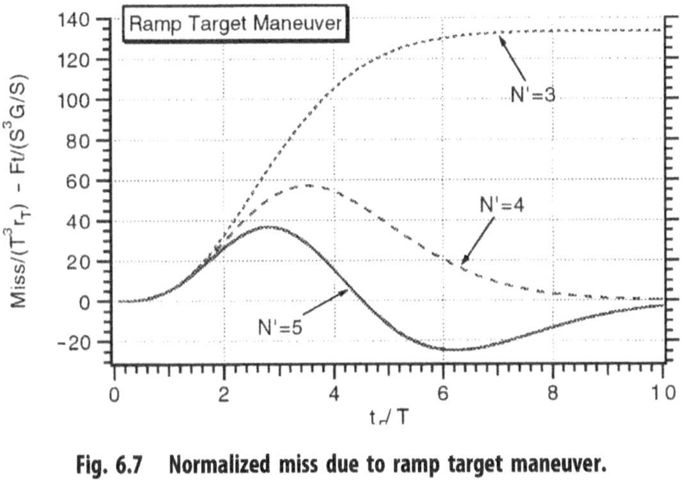

Fig. 6.7 Normalized miss due to ramp target maneuver.

Fig. 6.7 that the normalization on the ordinate is different from that in the case of a step target maneuver.

If we had a ramp maneuver that reached the 4-g level in 2.5 s, then its acceleration rate would be

$$r_T = n_T/t_F = 4/2.5 = 1.6\,g/s$$

With the same inputs as before (that is, $T = 0.5$ s), we can read the normalized miss from Fig. 6.7 as

$$\frac{\text{Miss}}{T^3 r_T} = 120$$

Therefore, the actual miss distance is

$$\text{Miss} = 120T^3 r_T = 120 * 0.5^3 * 1.6 = 24\,\text{ft}$$

We can see that, although the ramp maneuver only reaches the 4-g level by the end of the flight, its influence on miss distance, for this example, is much greater than that of the step maneuver. Increasing the effective navigation ratio to 4 reduces the normalized miss to

$$\frac{\text{Miss}}{T^3 r_T} = 40$$

which reduces the actual miss to

$$\text{Miss} = 40 * 0.5^3 * 1.6 = 8\,\text{ft}$$

This numerical example illustrates the need for larger effective navigation ratios in a proportional navigation guidance system for nonconstant target maneuvers.

Integrating the adjoint ramp maneuver output yields the miss due to a parabolic maneuver. Figure 6.8 presents the normalized miss distance induced by a parabolic target maneuver. Here we can see that an effective navigation ratio of 5 is required for the miss to go to zero for long flight times.

For consistency, let us consider a case in which all numerical values are related to the previous cases considered. If the parabolic maneuver reaches the 4-g level in 2.5 s, the acceleration jerk will be

$$p_T = n_T/t^2 = 4/2.5^2 = 0.64\,g/s^2$$

For the same inputs as before, the miss for a navigation ratio of 3 becomes

$$\text{Miss}|_{N''=3} = 300T^4 p_T = 300 * 0.5^4 * 0.64 = 12\,\text{ft}$$

Increasing the navigation ratio reduces the miss, as we can see from

$$\text{Miss}|_{N'=4} = 150 * 0.5^4 * 0.64 = 6\,\text{ft}$$
$$\text{Miss}|_{N'=5} = 75 * 0.5^4 * 0.64 = 3\,\text{ft}$$

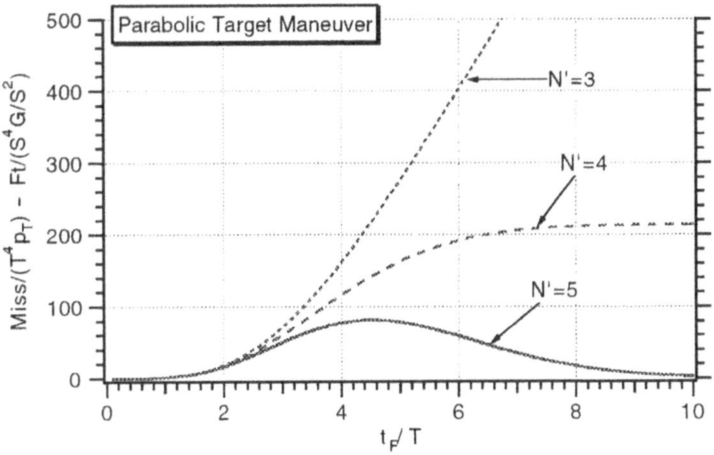

Fig. 6.8 Normalized miss due to parabolic target maneuver.

In summary, we can say that increasing the effective navigation ratio and decreasing the guidance system time constant work in the direction of reducing the miss due to target maneuver. We shall see in Chapter 18 that increasing the navigation ratio also increases the miss due to noise and parasitic effects. Variable maneuver levels such as ramps and parabolas may cause more miss than constant maneuvers because the navigation ratio may not be set at a high enough level.

OPTIMAL TARGET EVASIVE MANEUVERS

We have seen that the effective navigation ratio has a strong influence on missile guidance system performance against all types of target maneuvers. Let us consider the influence of step target maneuver on a fifth-order binomial guidance system in more detail [2–6]. From a target's point of view an optimal maneuver is one that induces the most miss distance. Figure 6.9 shows that, when the effective navigation ratio is 3, the normalized miss distance curve has a maxima at a normalized flight time of 2.5. Since flight time and time to go are interchangeable for this system, we can interpret the abscissa of Fig. 6.9 as normalized time to go. Therefore, as shown in Fig. 6.9, the target can induce the most miss distance by first executing $+n_T g$ (normalized time to go is large) and then rolling 180 deg at a normalized time to go of 2.5 so that the target will be executing $-n_T g$. As far as the missile is concerned, the target appears to be executing a maneuver of magnitude $2n_T$! The optimality of this maneuver is proven mathematically in [2] using a combination of optimal control theory and adjoint theory.

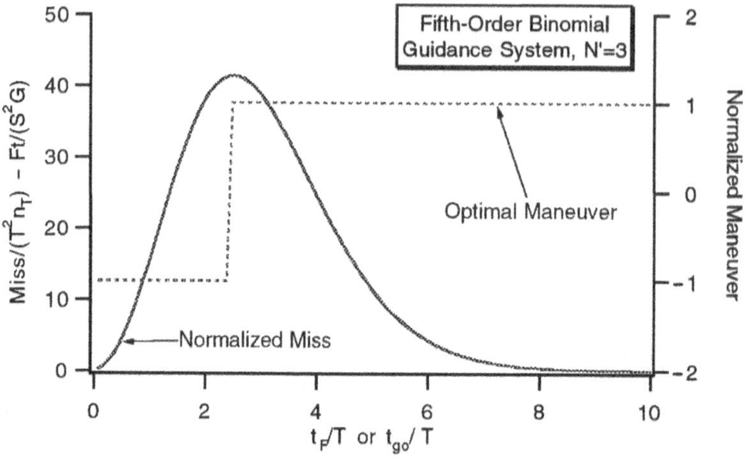

Fig. 6.9 Optimal maneuver policy for effective navigation ratio of 3.

We can see from Fig. 6.9 that the induced miss distance caused by this optimal maneuver will be

$$\text{Miss}|_{\text{opt},N'=3} = 41.4 * (2n_T)T^2 = 82.8T^2 n_T$$

For a maneuver level of 4 g and a guidance system time constant of 0.5 s, the largest miss distance the target maneuver can induce for an effective navigation ratio of 3 is

$$\text{Miss}|_{\text{opt},N'=3} = 82.8 * 0.5^2 * 4 = 82.8\,\text{ft}$$

This miss distance is considerably larger than 12 ft, which was previously obtained with a step target maneuver occurring at a normalized time to go of 5 s.

If the effective navigation ratio is 4, the miss response has maxima indicated in Fig. 6.10. Therefore, in this case, the optimal maneuver policy is for the target to begin with a maneuver level of $-n_T\,g$ until $t_{go}/T = 5$, then rolling 180 deg in order to execute $+n_T\,g$ and then finally at $t_{go}/T = 2$, rolling another 180 deg in order to execute $-n_T\,g$. We can see from Fig. 6.10 that the induced miss in this case will be

$$\text{Miss}|_{\text{opt},N'=4} = (19.1 + 29.8) * (2n_T)T^2 = 97.8T^2 n_T$$

If the navigation ratio is 5, the miss response has three maxima, as shown in Fig. 6.11. The optimal maneuver strategy is superimposed on this figure, and the optimal induced miss turns out to be

$$\text{Miss}|_{\text{opt},N'=5} = (8.4 + 29.9 + 24) * (2n_T)T^2 = 124.6T^2 n_T$$

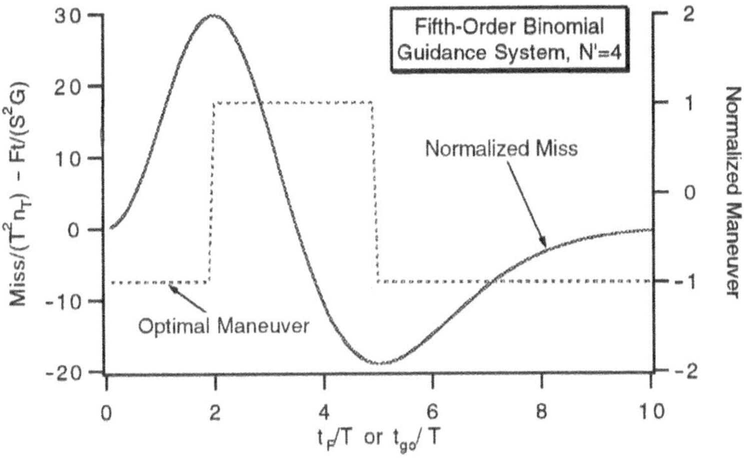

Fig. 6.10 Optimal maneuver policy for effective navigation ratio of 4.

It is interesting to note that, as we increase the effective navigation ratio, the optimal miss due to a step target maneuver also increases because of the increased number of maxima in the miss distance sensitivity curve.

The concept of an optimal maneuver is useful in that it identifies the largest possible miss distance that the target can induce and possibly aid in the selection of the missile guidance system time constant. Of course, this optimal maneuver assumes unrealistically that the target has precise knowledge of the time to go

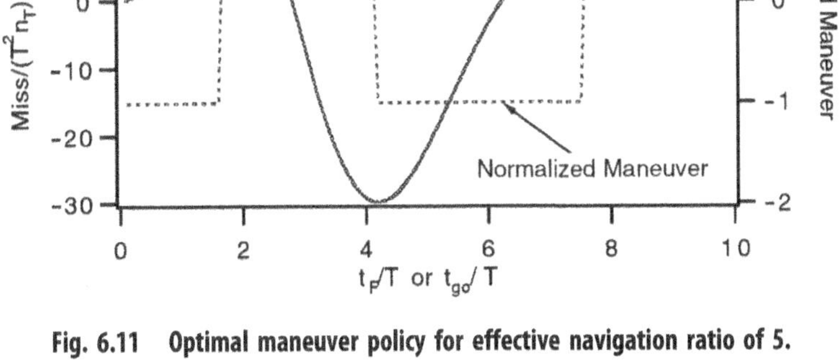

Fig. 6.11 Optimal maneuver policy for effective navigation ratio of 5.

until intercept and of the missile guidance system dynamics. It is readily apparent from the preceding miss distance curves that the missile guidance time constant must be minimized if the miss distance due to target maneuver is to be kept small. However, we shall see in Chapter 18 that noise and parasitic effect place a practical lower limit on the minimum achievable guidance system time constant.

PRACTICAL EVASIVE MANEUVERS

In the previous section we showed that optimal target evasive maneuvers could induce very large miss distances if *a priori* information concerning the missile guidance system was available. In this section we shall demonstrate that when *a priori* information is lacking, practical periodic target evasive maneuvers can also generate very large miss distances. Two such practical evasive periodic maneuver policies are the *barrel roll* and the *Vertical-S*.

The barrel roll can be described in one dimension as a sinusoid with radian frequency ω or period T and amplitude n_T as given by

$$\ddot{y}_T = n_T \sin \omega t = n_T \sin \frac{2\pi}{T} t$$

From the preceding relationship we can see that the barrel roll only yields maximum acceleration levels some of the time. With the Vertical-S maneuver, however, the aircraft is always at maximum acceleration and the sign of the acceleration is periodically reversed by rolling the aircraft through 180 deg. With a theoretically infinite roll rate, this maneuver policy can be approximated by a periodic square wave in one dimension. The barrel roll and Vertical-S maneuver policies do not require information about the missile guidance system. The amplitudes of both target maneuvers are chosen to reflect the maximum acceleration capability of the aircraft, whereas the frequencies of both maneuvers are chosen to be physiologically possible for a human pilot and robust enough to cause any missile guidance system problems.

Both the barrel roll and Vertical-S maneuver policies are illustrated in Fig. 6.12 where, for illustrative purposes, it is assumed that the maneuver amplitude n_T is 4 g and the maneuver frequency ω is 1 rad/s. Figure 6.12 confirms that the effective maneuver period T is 6.28 s.

For comparative purposes, the preceding maneuver policies were evaluated on the fifth-order proportional navigation binomial guidance system. The guidance system time constant was 0.5 s, the target maneuver amplitude was 4 g, and the effective navigation ratio considered was three so that comparisons could be made with the optimal evasive maneuver induced miss distances. With these numbers it was shown in the previous section that the optimal miss distance was 82.8 ft. Figure 6.13 shows how the miss distances vary with flight time for both the barrel roll and Vertical-S maneuver policies. It is apparent that the Vertical-S maneuver generates the largest miss distances because the target is

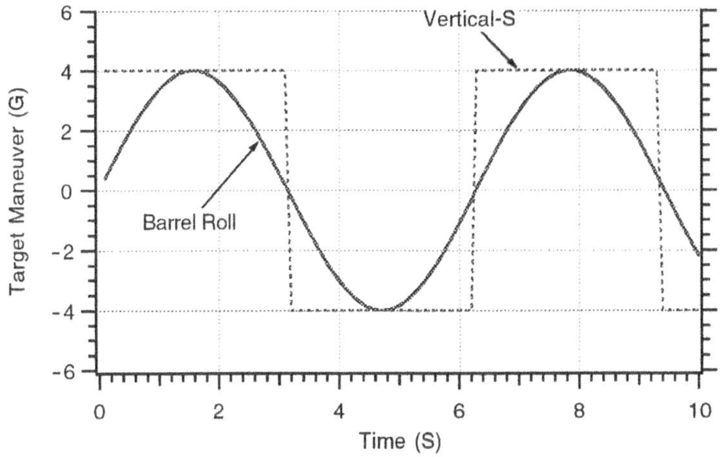

Fig. 6.12 More realistic evasive maneuver policies.

always at maximum acceleration. We can also see from Fig. 6.13 that, on the average, the miss distances for both maneuver policies are quite high. For this example the Vertical-S maneuver yields miss distances which approach that of the optimal maneuver when the flight times are 0.5, 1.8, 4.4, or 7.5 s. It appears that if the pilot is not lucky and the flight time is 1.0, 3.4, 6.5, or 9.5 s, the miss distance will be quite small. The appendix shows that when we move to three dimensions, the peak miss also represents the average miss.

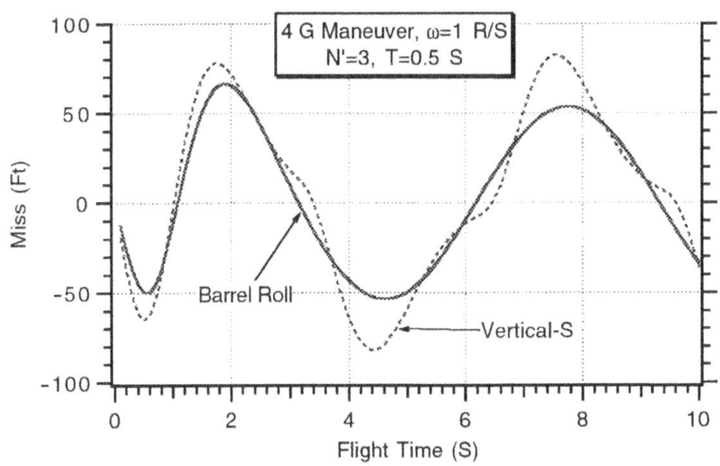

Fig. 6.13 Realistic maneuvers can induce very large miss distances.

More details on the influence of practical evasive maneuver strategies on miss distance can be found in [5]. In addition, [5] derives the shaping filter equivalent for many practical maneuvers, so that the method of adjoints can be used to assess system performance in a single computer run.

SATURATION [2]

Thus far we have seen normalized miss distance curves for a fifth-order proportional navigation binomial guidance system. The results presented have implicitly assumed that the missile had adequate acceleration capability in order to guide and hit the target. If adequate acceleration capability is not available, the missile acceleration *saturates*, which results in additional miss distance. In endoatmospheric interceptors, angle-of-attack constraints limit maximum achievable accelerations at high altitudes, whereas the missile structure limits achievable acceleration levels at the lower altitudes. The lateral engine thrust-to-weight ratio limits the acceleration level in exoatmospheric interceptors.

The basic homing loop can be modified and made nonlinear to account for acceleration saturation as shown in Fig. 6.14. In this figure the guidance system is also represented as a fifth-order binomial guidance system with guidance time constant T. Two of the time constants are devoted to the seeker and noise filter, and the other three time constants are devoted to the flight-control system. The acceleration limit is on the acceleration command n_c, and the resultant acceleration command is denoted n_{CLIM}.

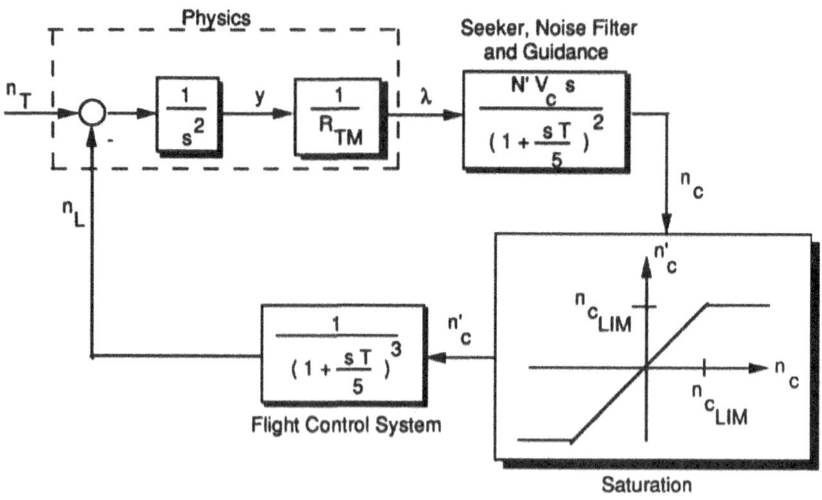

Fig. 6.14 Homing loop with acceleration saturation.

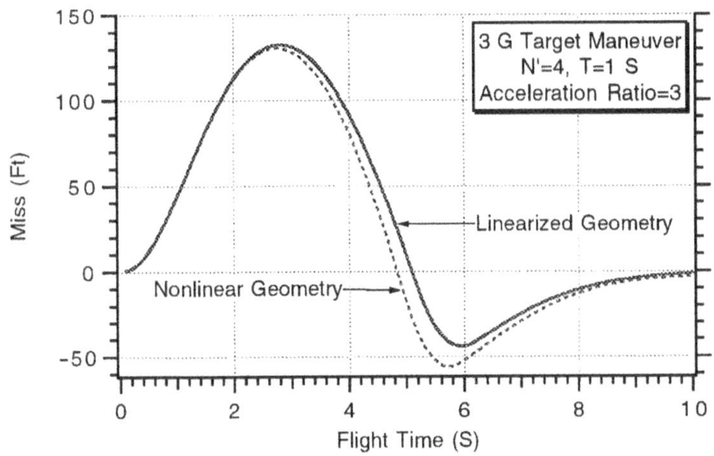

Fig. 6.15 Linearized geometry model is adequate for investigating saturation effects.

The first question that must be answered again is whether or not linearizing the geometry in the presence of the nonlinear acceleration saturation model is adequate for capturing important miss distance effects. For convenience, let us define the acceleration ratio as the ratio between the missile acceleration limit to the maneuver level of the target. Figure 6.14 will only be linear if the acceleration ratio is infinity.

Figure 6.15 represents the results of running both the nonlinear engagement simulation, with saturation effects modeled, and an engagement simulation of the model shown in Fig. 6.14. In the case considered, the guidance time constant was 1 s, the effective navigation ratio was 4, and the level of the target maneuver was 3 *g*. The acceleration ratio considered was 3. This means that since the target maneuver level is 3 *g*, the effective acceleration limit of the missile is 9 *g*. We can see from Fig. 6.15 that the miss distance results for both the linearized geometry model and nonlinear geometry models are virtually identical. Therefore, we can conclude that the linearized geometry model is adequate for investigating saturation effects. By comparing Fig. 6.15 with the nonsaturation case of Fig. 6.6, we can also conclude that even with a missile-to-target acceleration advantage of 3, considerable miss distance is contributed by saturation, especially for the shorter flight times.

Using the missile to target acceleration capability (n_{CLIM}/n_T) and the normalization factors for miss due to a constant target maneuver, we can derive normalized miss distance curves by the method of brute force (running engagement model with nonlinear saturation effect for many different flight times and noting the miss distance). In other words, we can generate normalized design curves by simulating all of the possibilities. We can then infer performance by

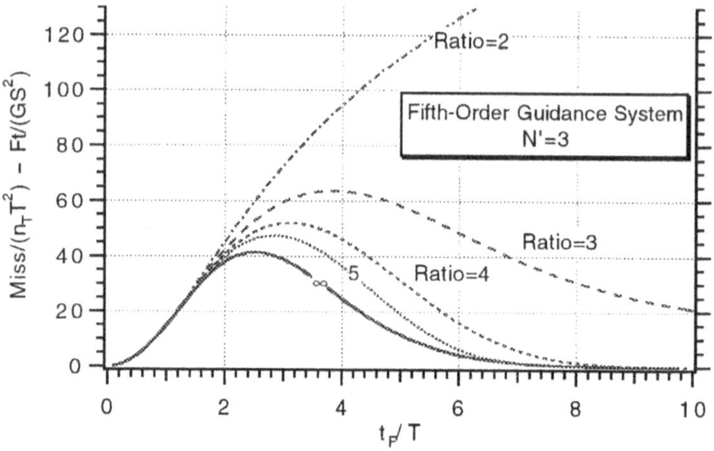

Fig. 6.16 Normalized miss due to target maneuver with saturation effects ($N' = 3$).

making extrapolations from the normalized design curves. Figure 6.16 presents the normalized miss distance due to a step target maneuver when the effective navigation ratio is 3. This figure shows that miss distance always increases with increasing flight time if the acceleration ratio is only 2. For acceleration ratios of 4 or more, the miss is virtually 0 for flight times approximately 10 times greater than the guidance time constant. An acceleration ratio of about 5 closely follows the infinite ratio or linear case.

Let us do a numerical example in order to clarify the use of these curves. If the guidance time constant is 0.5 s and the flight time is 2.5 s, we get a normalized flight time of

$$t_F/T = 2.5/0.5 = 5$$

Assuming a target maneuver level of 4 *g*, we can then calculate the miss distances for various acceleration limits as

$$\text{Miss}|_{\infty g} = 12.0T^2 n_T = 12.0 * 0.5^2 * 4 = 12.0\,\text{ft}$$

$$\text{Miss}|_{20g} = 18.9T^2 n_T = 18.9 * 0.5^2 * 4 = 18.9\,\text{ft}$$

$$\text{Miss}|_{16g} = 31.1T^2 n_T = 31.1 * 0.5^2 * 4 = 31.1\,\text{ft}$$

$$\text{Miss}|_{12g} = 58.1T^2 n_T = 58.1 * 0.5^2 * 4 = 58.1\,\text{ft}$$

$$\text{Miss}|_{8g} = 112T^2 n_T = 112 * 0.5^2 * 4 = 112\,\text{ft}$$

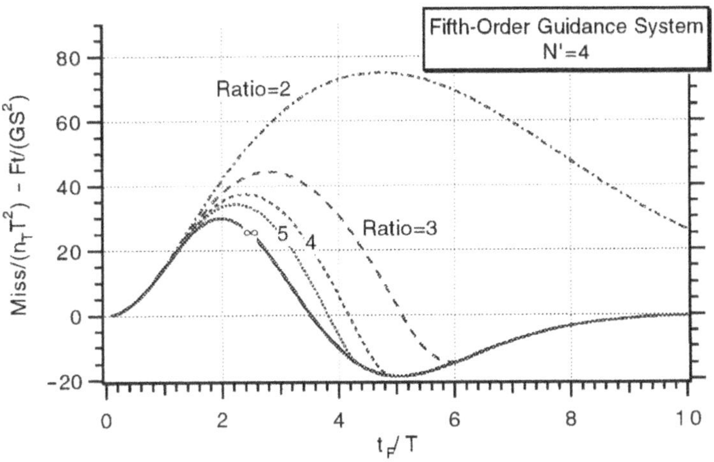

Fig. 6.17 Normalized miss due to target maneuver with saturation effects ($N' = 4$).

Therefore, we can see that the miss goes up by nearly a factor of 5 from the linear case if the missile-to-target acceleration advantage is only 3 and by a factor of 10 if the acceleration advantage is only 2.

Increasing the effective navigation ratio tends to reduce the acceleration requirements as shown in the normalized curves of Figs. 6.17 and 6.18.

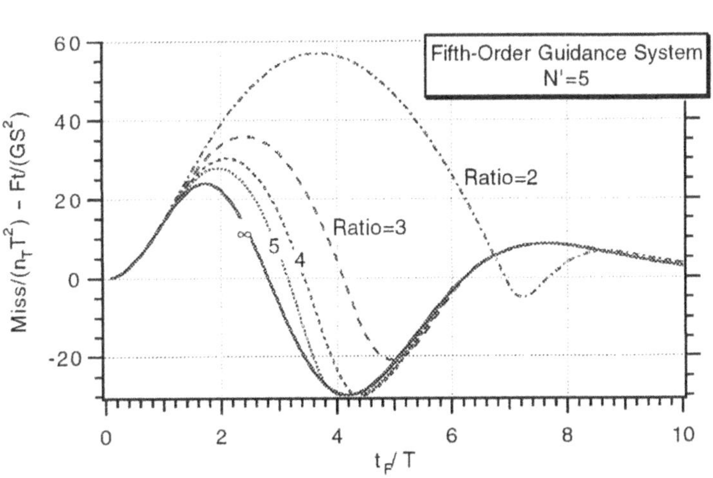

Fig. 6.18 Normalized miss due to target maneuver with saturation effects ($N' = 5$).

PARASITIC EFFECTS [7–10]

Thus far, from all of the results presented it would appear that the guidance system designer has an easy job, since all the graphs indicate that smaller time constants and larger effective navigation ratios appear to improve system performance. Actually, parasitic or unwanted feedback paths within the homing loop will work in the direction of larger time constants and smaller effective navigation ratios to get acceptable performance. One of the most serious unwanted feedback paths is created in tactical radar homing missile applications by the missile radome. The radome causes a refraction or bending of the incoming radar wave, which in turn gives a false indication of the target location. A parameter associated with missile maneuverability, which has a significant interaction with radome effects, is the turning rate time constant T_α. If we consider the basic geometry of Fig. 6.19, the turning rate time constant can be defined as the amount of time it takes to turn the missile flight-path angle γ through an equivalent angle of attack α, or

$$T_\alpha = \frac{\alpha}{\dot{\gamma}}$$

where the angle of attack and the flight-path angle are defined in Fig. 6.19. Generally the turning rate time constant increases with increasing missile altitude and decreasing missile velocity.

To see how the turning rate time constant enters into the homing loop, we must see how it is related to other important quantities. From Fig. 6.19 we can see that the missile pitch angle θ can be expressed as

$$\theta = \gamma + \alpha$$

Taking derivatives of both sides of the equation yields

$$\dot{\theta} = \dot{\gamma} + \dot{\alpha} = \dot{\gamma} + \frac{s\alpha\dot{\gamma}}{\dot{\gamma}}$$

Since the missile acceleration is perpendicular to the missile velocity, we can say that

$$n_L = V_M\dot{\gamma}$$

Therefore, we can express the missile pitch rate in terms of the missile acceleration as

$$\dot{\theta} = \frac{n_L}{V_M} + sT_\alpha\frac{n_L}{V_M}$$

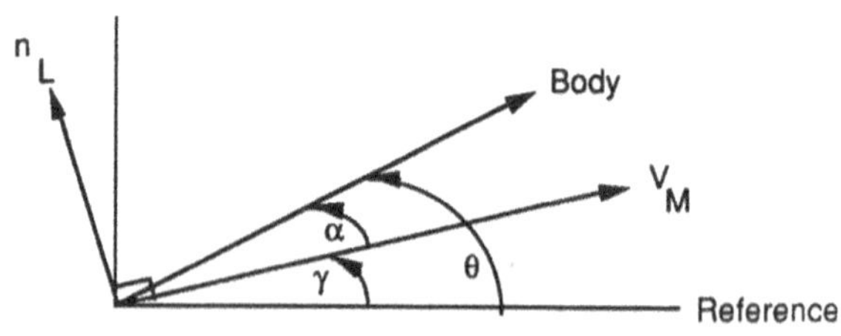

Fig. 6.19 Geometry for turning rate time constant.

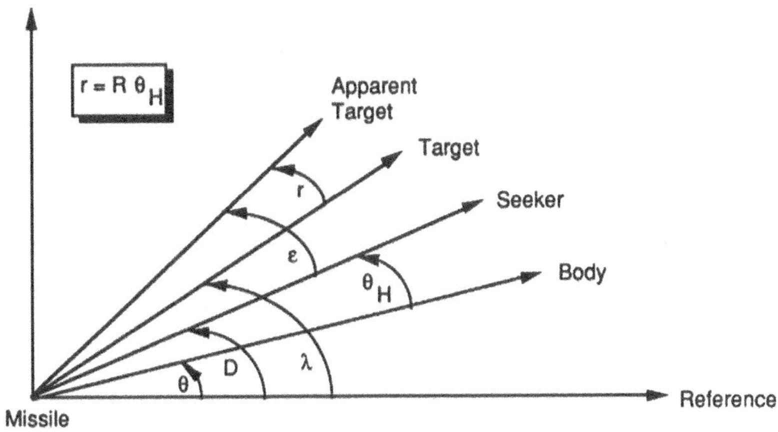

Fig. 6.20 Basic geometry for radome analysis.

Dividing both sides by the missile acceleration yields the *missile pitch rate transfer function*,

$$\frac{\dot{\theta}}{n_L} = \frac{1}{V_M}(1 + T_\alpha S)$$

This aerodynamic transfer function shows that there is a missile body rate whenever the missile is accelerating.

Now we need to see how the missile aerodynamic transfer function interacts with the radome slope. Consider the basic geometry of Fig. 6.20 in which the seeker is not pointed at the actual target because of seeker dynamics and radome effects.

The radome refraction angle r varies with the missile gimbal angle θ_H. For preliminary analysis it is usually assumed that the refraction angle is linearly proportional to the gimbal angle, or

$$r = R\theta_H$$

where R is constant known as the *radome slope*. The radome slope is a function of the radome material, radome diameter, and fineness ratio, and the wavelength of the incoming signal. From Fig. 6.20 we can see that it is possible to express the missile boresight error ϵ as

$$\epsilon = \lambda - \theta - \theta_H + r = \lambda - \theta - \theta_H + R\theta_H$$

A block diagram of the homing loop with the radome unwanted feedback path is indicated in Fig. 6.21. We can see that without radome effects ($R = 0$) we would have a fifth-order binomial guidance system transfer function. The missile aerodynamic transfer function [11] provides the unwanted feedback path in the guidance system.

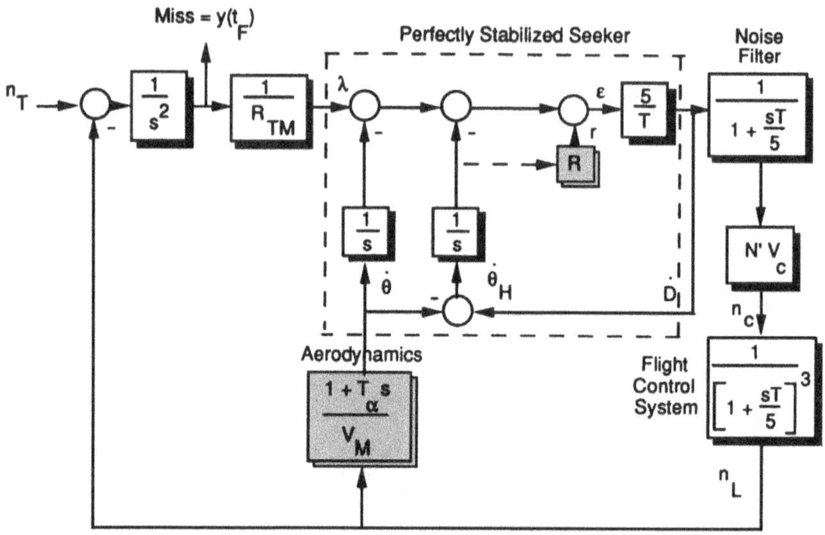

Fig. 6.21 Fifth-order binomial model of guidance system with radome effects.

Listing 6.2 presents an engagement simulation with the fifth-order binomial model, including radome effects, of Fig. 6.21. The simulation is set to run multiple

LISTING 6.2 ENGAGEMENT SIMULATION WITH RADOME EFFECTS

```
VC=4000.;
XNT=32.2;
YIC=0.;
VM=3000.;
HEDEG=0.;
TAU=.5;
XNP=3.;
TA=0.;
R=-.01;
n=0.;
for TF=.1:.1:10
        Y=YIC;
        YD=-VM*HEDEG/57.3;
        YDIC=YD;
        XNL=0.;
        ELAMDH=0.;
        X4=0.;
        X5=0.;
        TH=0.;
```

```
THH=0.;
T=0.;
H=.01;
while T<=(TF-1e-5)
   YOLD=Y;
   YDOLD=YD;
   XNLOLD=XNL;
   ELAMDHOLD=ELAMDH;
   X4OLD=X4;
   X5OLD=X5;
   THOLD=TH;
   THHOLD=THH;
   STEP=1;
   FLAG=0;
   while STEP<=1
   if FLAG==1
                        STEP=2;
                        Y=Y+H*YD;
                        YD=YD+H*YDD;
                        XNL=XNL+H*XNLD;
                        ELAMDH=ELAMDH+H*ELAMDHD;
                        X4=X4+H*X4D;
                        X5=X5+H*X5D;
                        TH=TH+H*THD;
                        THH=THH+H*THHD;
                        T=T+H;
   end
   TGO=TF-T+.00001;
   XLAM=Y/(VC*TGO);
   EPS=XLAM-TH-THH+R*THH;
   DD=5.*EPS/TAU;
   ELAMDHD=5.*(DD-ELAMDH)/TAU;
   XNC=XNP*VC*ELAMDH;
   X4D=5.*(XNC-X4)/TAU;
   X5D=5.*(X4-X5)/TAU;
   XNLD=5.*(X5-XNL)/TAU;
   THD=XNL/VM+TA*XNLD/VM;
   THHD=DD-THD;
   YDD=XNT-XNL;
   FLAG=1;
   end
   FLAG=0;
   Y=.5*(YOLD+Y+H*YD);
   YD=.5*(YDOLD+YD+H*YDD);
   XNL=.5*(XNLOLD+XNL+H*XNLD);
   ELAMDH=.5*(ELAMDHOLD+ELAMDH+H*ELAMDHD);
```

```
            X4=.5*(X4OLD+X4+H*X4D);
            X5=.5*(X5OLD+X5+H*X5D);
            TH=.5*(THOLD+TH+H*THD);
            THH=.5*(THHOLD+THH+H*THHD);
        end
        n=n+1;
        ArrayTF(n)=TF;
        ArrayY(n)=Y;
end
figure
plot(ArrayTF,ArrayY),grid
xlabel('Flight Time (Sec)')
ylabel('Miss (Ft)')
clc
output=[ArrayTF',ArrayY'];
save datfil.txt output -ascii
disp 'simulation finished'
```

cases with the flight time as a parameter so that miss distance sensitivity curves can be generated by brute force. Again, the differential equations representing the guidance system of Fig. 6.21 appear before the FLAG=1 statement.

To see how the turning rate time constant influences system performance, a case was run for a 1-*g* target maneuver disturbance in which the guidance system time constant was 0.5 s, the radome slope was -0.01, and the effective navigation ratio was 3. The turning rate time constant was varied from 0 to 10 s. Figure 6.22 shows that when the turning rate time constant is zero the

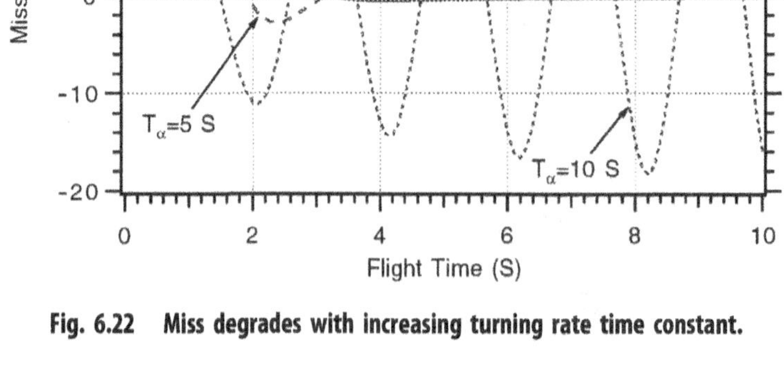

Fig. 6.22 Miss degrades with increasing turning rate time constant.

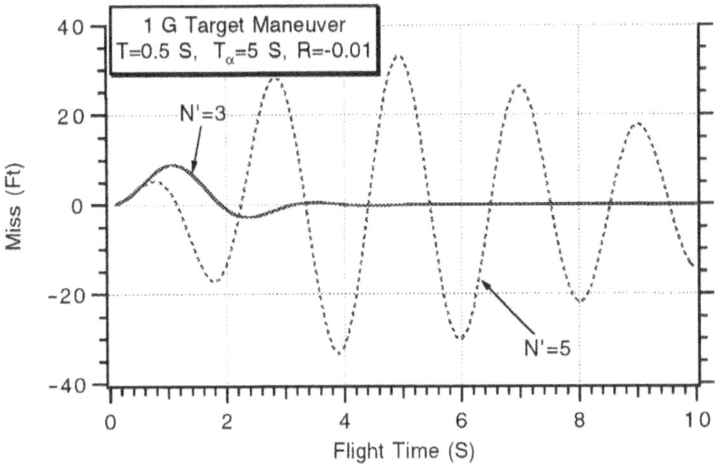

Fig. 6.23 Increasing effective navigation ratio has destabilizing effect in presence of negative radome slope.

miss distance response is virtually identical to the case in which there are no parasitic paths in the homing loop (compare with Fig. 6.6, for example). When the turning rate time constant is increased to 5 s, the miss distance response begins to become more oscillatory, but the miss distances are still small and tend to zero as the flight time increases. Finally, when the turning rate time constant is increased to 10 s, the miss distance response becomes unstable. Thus, we can see that we have to be concerned about radome effects from both a miss distance and stability point of view.

As was mentioned previously, the magnitude of the effective radome slope is determined by the physical characteristics of the radome and the wavelength of the incoming signal. For a given radome, the guidance designer has only two parameters (that is, guidance system time constant and effective navigation ratio) under control to get acceptable performance and meet stability requirements. Figure 6.23 shows how miss distance due to a 1-g target maneuver varies with flight time for two different values of effective navigation ratio in the presence of a negative radome slope ($R = -0.01$). We can see that the higher effective navigation ratio has a destabilizing effect. This is not unreasonable because we are essentially increasing the guidance system gain. Thus, the guidance system designer desires to keep the effective navigation ratio as small as possible to meet the stability requirements and yet large enough so that homing will be effective.

Figure 6.24 shows that, in the presence of a large effective navigation ratio and negative radome slope, increasing the guidance system time constant from 0.5 s to 0.75 s has a stabilizing effect.

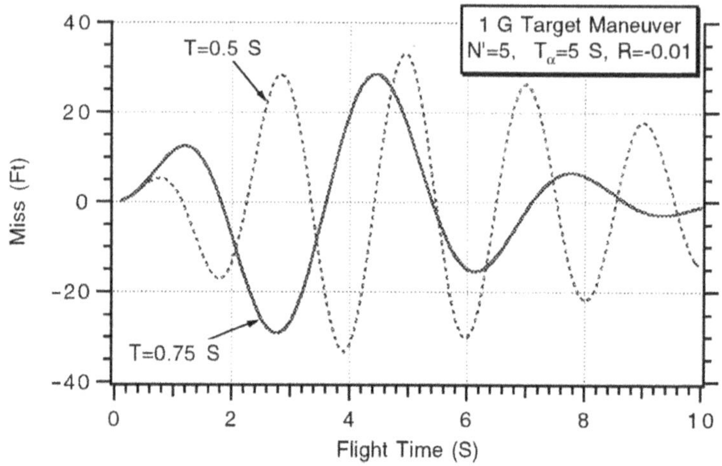

Fig. 6.24 Increasing guidance system time constant has stabilizing effect in presence of negative radome slope.

In tactical missile design the guidance system time constant is generally made larger at the higher altitudes because the turning rate time constant is largest at the higher altitudes. Of course, the penalty for such a decision is that miss distances tend to increase with increasing guidance system time constant. Therefore, the guidance system designer attempts to make the guidance system time constant as small as possible subject to meeting guidance system stability requirements.

THRUST VECTOR CONTROL

We saw in the previous section that, if the turning rate time constant of a tactical aerodynamic missile was large, radome effects caused stability problems and miss distance deterioration. This problem is not confined to only tactical aerodynamic missiles. Consider a missile that operates outside the atmosphere and uses thrust vector control to maneuver. Figure 6.25 presents a diagram of a thrust vector controlled missile with all important angles indicated. The missile acceleration n_L needed to maneuver in accordance with guidance commands is obtained from the component of the thrust T perpendicular to the missile body.

For simplicity we are neglecting the fact that, if the thrust does not go through the center of gravity, the missile will tumble. The rate of change of the missile flight-path angle γ is related to the missile acceleration and velocity according to

$$\dot{\gamma} = \frac{n_L}{V_M}$$

where n_L is the missile acceleration and V_M the missile velocity. From Fig. 6.25 we can see that the flight-path rate can also be expressed as

$$\dot{\gamma} = \frac{n_L}{V_M} = \frac{Tg\sin\alpha}{WV_M}$$

where T is thrust (in lb), g is the gravitational acceleration (in ft/s^2), α is the angle of attack, and W is the missile weight. For small angles of attack we can approximate the flight-path rate to be

$$\dot{\gamma} = \frac{Tg\alpha}{WV_M}$$

Recalling that the turning rate time constant is the ratio of the angle of attack to the flight-path rate, we obtain

$$T_\alpha = \frac{\alpha}{\dot{\gamma}} = \frac{WV_M}{Tg}$$

This means that the effective turning rate time constant for a thrust vector controlled missile is proportional to the missile weight and velocity and inversely proportional to the thrust.

To illustrate the importance of turning rate time constant to a thrust vector controlled missile, let us work a numerical example. Consider a missile traveling at 20,000 ft/s and requiring a 5-deg angle of attack in order to generate 5 g of acceleration. To generate 5 g of acceleration at 5-deg angle of attack, the missile must have a thrust-to-weight ratio given by

$$\frac{T}{W} = \frac{n_L/g}{\alpha} = \frac{5}{5/57.3} = 57.3$$

This means that the effective turning rate time constant is

$$T_\alpha = \frac{V_M}{(T/W)g} = \frac{20,000}{57.3 * 32.2} = 10.8\,\text{s}$$

The turning rate time constant in this example is quite large compared to values indicated in the previous section. However, because the thrust

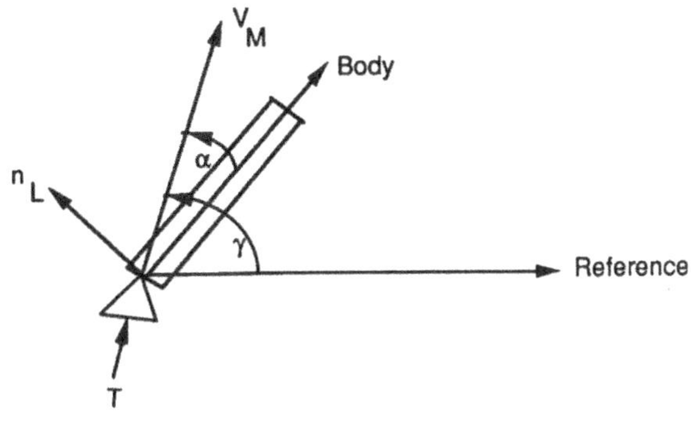

Fig. 6.25 Important angles in thrust vector control.

vector controlled missile operates outside the atmosphere, the shape of the missile nose can be made near-hemispherical. This means that the effective radome slope will be close to zero. For a thrust vector controlled missile the guidance system designer must pay close attention to the product of the radome slope and turning rate time constant to ensure adequate stability margins in the resultant design. If the design yields unacceptable stability margins, the guidance system time constant must be increased to yield a workable design.

SUMMARY

In this chapter we have shown how system order, optimal target maneuvers, saturation, and parasitic effects all influence miss distance. Miss distance design curves were presented to aid the guidance system designer in predicting preliminary system performance. These curves could also be used to ensure that the interceptor had adequate acceleration capability. Examples were presented showing the conflicting tradeoffs the guidance system designer must confront in choosing acceptable guidance system parameters.

REFERENCES

[1] Nesline, F. W., and Zarchan, P., "Miss Distance Dynamics in Homing Missiles," *Proceedings of AIAA Guidance and Control Conference*, AIAA, New York, Aug. 1984.

[2] Shinar, J., and Steinberg, D., "Analysis of Optimal Evasive Maneuvers Based On a Linearized Two-Dimensional Model," *Journal of Aircraft*, Vol. 14, Aug. 1977, pp. 795–802.

[3] Bennett, R. R., and Mathews, W. E., "Analytical Determination of Miss Distance for Linear Homing Navigation Systems," Hughes Aircraft Co., TM-260, Culver City, CA, March 1952.

[4] Travers, P., *Interceptor Dynamics*, unpublished lecture notes, Raytheon Co., circa 1971.

[5] Zarchan, P., "Representation of Realistic Evasive Maneuvers by the Use of Shaping Filters," *Journal of Guidance and Control*, Vol. 2, July–Aug. 1979, pp. 290–295.

[6] Howe, R. M., "Guidance," *System Engineering Handbook*, edited by R. E. Machol, W. P. Tanner Jr., and S. N. Alexander, McGraw-Hill, New York, 1965, Chap. 19.

[7] Nesline, F. W., and Zarchan, P., "Radome Induced Miss Distance in Aerodynamically Controlled Homing Missiles," *Proceedings of AIAA Guidance and Control Conference*, AIAA, New York, Aug. 1984.

[8] Youngren, F. R., "Minimizing Boresight Errors in Aerodynamic Radomes," *Electronic Design*, Dec. 20, 1961, pp. 152–157.

[9] Peterson, E. L., *Statistical Analysis and Optimization of Systems*, Wiley, New York, 1961.

[10] Eichblatt, E. (ed.), *Test and Evaluation of the Tactical Missile*, Vol. 119, Progress in Astronautics and Aeronautics, AIAA, Washington, DC, 1989, p. 415.

[11] Hemsch, M. J., and Nielsen, J. N. (eds.), *Tactical Missile Aerodynamics*, Vol. 104, Progress in Astronautics and Aeronautics, AIAA, New York, 1986, p. 858.

CHAPTER 7

Digital Fading Memory Noise Filters in the Homing Loop

INTRODUCTION

Thus far, we have assumed in our analysis that the geometric line-of-sight rate was available for guidance purposes. Actually, the seeker measurement of the line-of-sight angle is corrupted by noise. Therefore, in order to derive the line-of-sight rate estimate required by proportional navigation guidance, it is necessary to use a digital noise filter in an onboard guidance system. Although we shall study optimal digital noise filters in Chapter 9, we shall first consider simple constant gain filters, known as *fading memory filters*, to derive the line-of-sight rate estimate. We will investigate, by example, some of the properties of digital fading memory filters and their influence on system performance. Fading memory filters will serve as the foundation for more advanced digital filters, known as *Kalman filters*.

FADING MEMORY FILTERS [1]

A simple digital noise filter is known as a fading memory filter. This filter is recursive and weights new measurements more heavily than older measurements. First-, second-, and third-order fading memory filters and their gains are tabulated in recursive form in Table 7.1. We can see from the table that the filter estimate is essentially the old estimate plus a gain times a residual (difference between current measurement and previous estimate). Table 7.1 also shows that the fading memory filter gains are constant and are a function of only one parameter β. This parameter is associated with the memory length of the filter and is a constant between zero and unity. Increasing β tends to decrease the bandwidth of the filter and enables the filter to "remember" more about previous measurements.

We can see from Table 7.1 that the fading memory filter assumes a polynomial model for the actual process. If the polynomial process of the filter is an underestimate of the polynomial degree of the actual process, then there will be a filter truncation error. The lowest order filter does the best job of removing the

TABLE 7.1 DIFFERENT ORDER DIGITAL FADING MEMORY FILTERS

Filter	Gaints
$\hat{x}_n = \hat{x}_{n-1} + G[x_n^* - \hat{x}_{n-1}]$	$G = 1 - \beta$
$\hat{x}_n = \hat{x}_{n-1} + \hat{\dot{x}}_{n-1}T_s + G[x_n^* - (\hat{x}_{n-1} + \hat{\dot{x}}_{n-1}T_s)]$	$G = 1 - \beta^2$
$\hat{\dot{x}}_n = \hat{\dot{x}}_{n-1} + \frac{H}{T_s}[x_n^* - (\hat{x}_{n-1} + \hat{\dot{x}}_{n-1}T_s)]$	$H = (1 - \beta)^2$
$\hat{x}_n = \hat{x}_{n-1} + \hat{\dot{x}}_{n-1}T_s + 0.5\hat{\ddot{x}}_{n-1}T_s^2 + G[x_n^* - (\hat{x}_{n-1} + \hat{\dot{x}}_{n-1}T_s + 0.5\hat{\ddot{x}}_{n-1}T_s^2)]$	$G = 1 - \beta^3$
$\hat{\dot{x}}_n = \hat{\dot{x}}_{n-1} + \hat{\ddot{x}}_{n-1}T_s + \frac{H}{T_s}[x_n^* - (\hat{x}_{n-1} + \hat{\dot{x}}_{n-1}T_s + 0.5\hat{\ddot{x}}_{n-1}T_s^2)]$	$H = 1.5(1 - \beta)^2(1 + \beta)$
$\hat{\ddot{x}}_n = \hat{\ddot{x}}_{n-1} + \frac{2K}{T_s^2}[x_n^* - (\hat{x}_{n-1} + \hat{\dot{x}}_{n-1}T_s + 0.5\hat{\ddot{x}}_{n-1}T_s^2)]$	$K = 0.5(1 - \beta)^3$

noise from the signal. However, it also has the potential for having the most truncation error. The filter designer must select the appropriate filter order to trade off filter variance reduction vs truncation error buildup. Fading memory filters are quite popular in radar tracking applications [2, 3] but, as we shall see, can be made to work in tactical missile homing applications as well.

FADING MEMORY FILTER IN HOMING LOOP

Figure 7.1 shows an example of how a second-order fading memory filter can be included in the homing loop. In this loop the actual line-of-sight angle λ is sampled with noise added every T_s seconds, thus providing an idealized seeker model. Estimates of the line-of-sight angle and rate are made with a digital two-state fading memory filter's measurement of the noisy line-of-sight angle λ_k^*. As was mentioned in Chapter 1, the notation z^{-1} is Z transform notation for a pure delay of T_s seconds. A guidance command is generated, using the proportional navigation guidance law from the estimated line-of-sight rate. The resultant command is passed through a "hold" network that converts the digital signal to a continuous signal for the flight-control system. The diagram shows a unity gain for an idealized representation of the flight-control system.

Listing 7.1 is a MATLAB engagement simulation of the homing loop shown in Fig. 7.1. Zero-mean Gaussian noise, independent from sample to sample, with standard deviation, SIGNOISE, is added to the measured line-of-sight angle every T_s seconds. We can see from the listing that the program consists of two separate parts. The first part, which represents the real world, consists of differential equations and the second-order Runge–Kutta numerical integration technique,

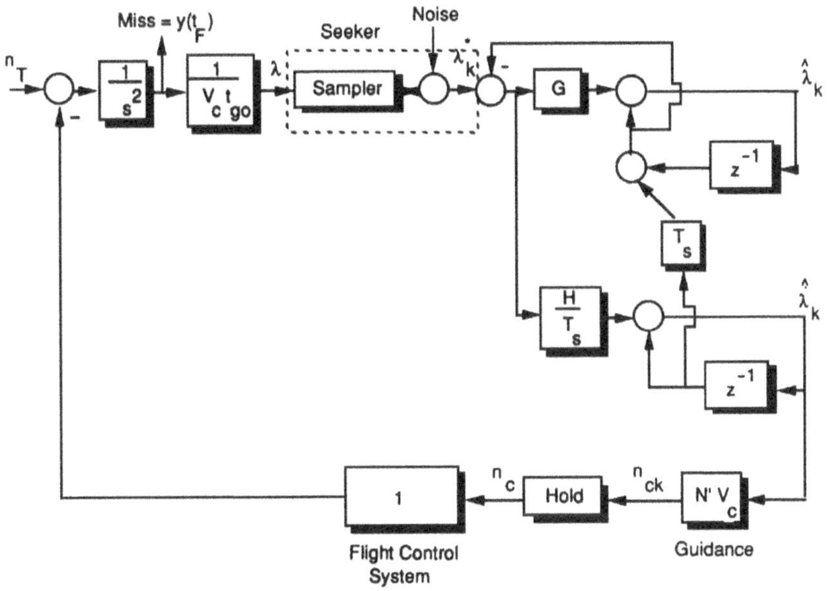

Fig. 7.1 Second-order fading memory filter in homing loop.

and the second part, which represents an onboard guidance system, has the difference equations for the second-order digital fading memory filter. We solve the differential equations every H seconds, and the difference equations are solved every T_s seconds. It is important to note that the ratio T_s/H must be a large integer so that effects in between sampling instants are treated properly and accurately.

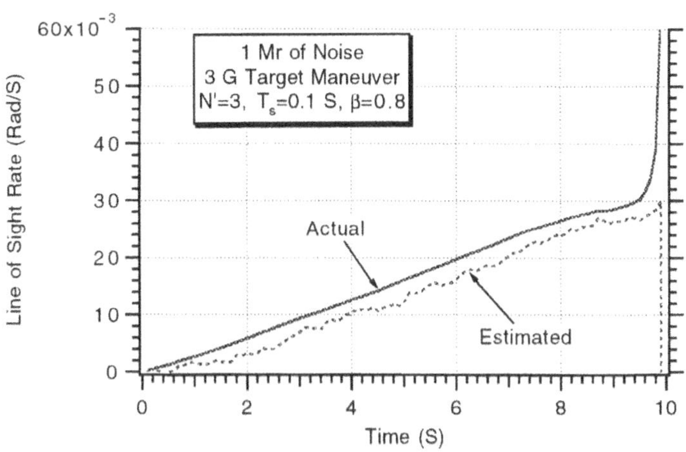

Fig. 7.2 Filter is sluggish and lags signal when $\beta = 0.8$.

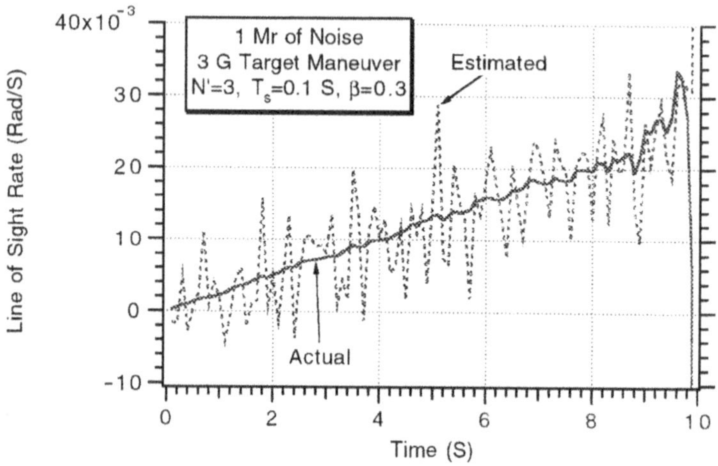

Fig. 7.3 Decreasing β increases noise transmission of fading memory filter.

The engagement simulation was exercised and a nominal case was run in which β of the fading memory filter was set to 0.8. Figure 7.2 compares the actual line-of-sight rate to the filter estimate of the derivative of the measurement. We can see that the filter estimate of the line-of-sight rate is smooth but lags the actual line-of-sight rate, indicating that the filter is sluggish.

Figure 7.3 indicates that we can effectively increase the bandwidth of the fading memory filter by decreasing β. Here we can see that the line-of-sight rate estimate no longer lags the actual signal when β is reduced from 0.8 to 0.3. However, we can see from the figure that the noisiness of the line-of-sight rate estimate is the price paid for reducing β. In other words, decreasing β increases the fading memory filter's noise transmission.

The results presented thus far are for a single flight with a particular noise stream. Answers will change for another flight with a different noise stream. To get accurate performance projection in terms of miss distance, we must run the program in the Monte Carlo mode. That is, repeated simulation trials must be conducted for each flight time of interest. The resultant miss distance data must be postprocessed, as was done in Chapter 4 when dealing with the random target maneuver, to calculate the mean and standard deviation of the resultant miss distances. Listing 7.2 presents a modification to the engagement simulation of Listing 7.1. Here two loops are added to the program. One loop executes 50 simulation trials (RUN = 50) for each flight time of interest, and the other loop selects different flight times (TF ranges from 0.5 to 10 s in increments of 0.5 s). In other words, the simulation of Listing 7.2 runs Monte Carlo sets for engagements in which the flight time is a parameter. Postprocessing of the resultant data is conducted at the end of the first "for loop" in accordance with the formulas and routines developed in Chapter 4.

LISTING 7.1 ENGAGEMENT SIMULATION WITH SECOND-ORDER FADING MEMORY FILTER

```
count=0;
VC=4000;
XNT=96.6;
YIC=0;
VM=3000;
HEDEG=0;
BETA=0.3;
XNP=3;
SIGNOISE=.001;
TF=10;
TS=.1;
NOISE=1;
Y=YIC;
YD=-VM*HEDEG/57.3;
YDIC=YD;
T=0;
H=.01;
S=0;
GFILTER=1.-BETA^2;
HFILTER=(1.-BETA)^2;
XLAMH=0;
XLAMDH=0;
XNC=0;
while T <= (TF - 1e-5)
        YOLD=Y;
        YDOLD=YD;
        STEP=1;
        FLAG=0;
        while STEP <=1
          if FLAG==1
        Y=Y+H*YD;
                    YD=YD+H*YDD;
                    T=T+H;
                    STEP=2;
          end;
          TGO=TF-T+1e-5;
          RTM=VC*TGO;
          XLAM=Y/(VC*TGO);
          XLAMD=(RTM*YD+Y*VC)/(RTM^2);
          YDD=XNT-XNC;
          FLAG=1;
        end;
        FLAG=0;
        Y=.5*(YOLD+Y+H*YD);
        YD=.5*(YDOLD+YD+H*YDD);
```

```
        S=S+H;
        if S > =(TS - 1e-5)
            S=0.;
            if NOISE==1
                    XLAMNOISE=SIGNOISE*randn;
            else
                    XLAMNOISE=0.;
            end;
            RES=XLAM-(XLAMH+TS*XLAMDH)+XLAMNOISE;
            XLAMH=GFILTER*RES+XLAMH+TS*XLAMDH;
            XLAMDH=HFILTER*RES/TS+XLAMDH;
            XNC=XNP*VC*XLAMDH;
            count=count+1;
            ArrayT(count)=T;
            ArrayY(count)=Y;
            ArrayXNC(count)=XNC;
            ArrayXLAMD(count)=XLAMD;
            ArrayXLAMDH(count)=XLAMDH;
        end;
end;
figure
plot(ArrayT,ArrayXLAMD,ArrayT,ArrayXLAMDH),grid
title('Decreasing beta increase noise transmission of fading memory filter')
xlabel('Time (S)')
ylabel('Line of Sight Rate (Rad/S) ')
axis([0 10 -.01 .06])
output=[ArrayT',ArrayY',ArrayXNC',ArrayXLAMD',ArrayXLAMDH'];
save datfil.txt output /ascii
disp('Simulation Complete')
```

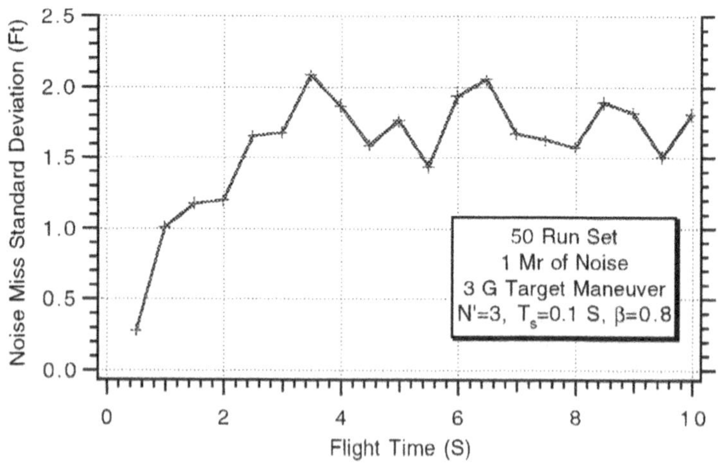

Fig. 7.4 Standard deviation of miss for various flight times.

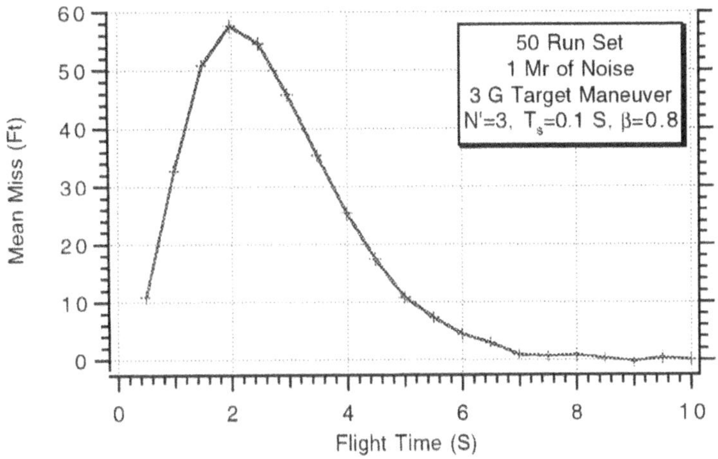

Fig. 7.5 Mean of miss for various flight times.

A nominal case was considered in which there was a constant 3-g target maneuver and 1 milliradian (mr) of measurement noise. A filter fading memory factor of 0.8 and a sampling time of 0.1 s were selected for the nominal case. A 50-run Monte Carlo set was run for 20 different values of flight time with the program of Listing 7.2 for a total of 1000 runs! The standard deviation and mean miss were computed for each of the 50-run Monte Carlo sets, and the results are displayed in Figs. 7.4 and 7.5. In this experiment there are only two disturbances. The target maneuver is deterministic (always 3 g), and the noise is a zero-mean random process. Therefore, we can assume that the standard deviation of the miss must be due to the noise, and the mean of the miss must be due to the target maneuver. Figures 7.4 and 7.5 show that for the value of β selected the noise-induced miss is small compared to the target-maneuver-induced miss for most flight times. This is not surprising because we know that the fading memory filter with $\beta = 0.8$ is sluggish.

LISTING 7.2 MONTE CARLO VERSION OF FADING MEMORY FILTER IN HOMING LOOP

```
%Preallocation
Z=zeros(size(1:1000));
I=zeros(size(1:50));
TF=zeros(size(1:50));
count=0;
VC=4000;
XNT=96.6;
YIC=0;
VM=3000;
```

```
HEDEG=0;
BETA=.8;
XNP=3;
SIGNOISE=.001;
TS=.1;
RUN=50;
NOISE=1;
for TF=.5:.5:10.0,
      Z1=0;
      for I=1:RUN
        Y=YIC;
        YD=-VM*HEDEG/57.3;
        YDIC=YD;
        T=0.;
        H=.01;
        S=0.;
        GFILTER=1.-BETA^2;
        HFILTER=(1.-BETA)^2;
        XLAMH=0.;
        XLAMDH=0.;
        XNC=0.;
        while T <= (TF - 1e-5)
                YOLD=Y;
                YDOLD=YD;
                STEP=1;
                FLAG=0;
                while STEP <=1
                        if FLAG==1
                        Y=Y+H*YD;
                                YD=YD+H*YDD;
                                T=T+H;
                                STEP=2;
                        end;
                        TGO=TF-T+.00001;
                        RTM=VC*TGO;
                        XLAM=Y/(VC*TGO);
                        XLAMD=(RTM*YD+Y*VC)/(RTM^2);
                        YDD=XNT-XNC;
                        FLAG=1;
                end;
                FLAG=0;
                Y=.5*(YOLD+Y+H*YD);
                YD=.5*(YDOLD+YD+H*YDD);
                S=S+H;
                if S > =(TS - 1e-5)
                S=0.;
```

```
                    if NOISE==1,
                            XLAMNOISE=gaussc7(SIGNOISE);
                    else
                            XLAMNOISE=0;
                    end;
                    RES=XLAM-(XLAMH+TS*XLAMDH)+XLAMNOISE;
                    XLAMH=GFILTER*RES+XLAMH+TS*XLAMDH;
                    XLAMDH=HFILTER*RES/TS+XLAMDH;
                    XNC=XNP*VC*XLAMDH;
                    end;
            end;
            Z(I)=Y;
            Z1=Z(I)+Z1;
            XMEAN=Z1/I;
        end;
        SIGMA=0;
        Z1=0;
        for I=1:RUN,
        Z1=(Z(I)-XMEAN)^2+Z1;
        if I==1,
            SIGMA=0;
        else
            SIGMA=sqrt(Z1/(I-1));
        end;
        end;
        count=count+1;
        ArrayTF(count)=TF;
        ArraySIGMA(count)=SIGMA;
        ArrayXMEAN(count)=XMEAN;
end;
figure
plot(ArrayTF',ArraySIGMA'),grid
title('Standard deviation of miss for various flight times')
xlabel('Flight Time (S)')
ylabel('Noise Miss Standard Deviation (Ft) ')
axis([00,10,00,4])
figure
plot(ArrayTF',ArrayXMEAN'),grid
title('Mean of miss for various flight times')
xlabel('Flight Time (S)')
ylabel('Noise Miss Standard Deviation (Ft) ')
axis([00,10,00,60])
clc
output=[ArrayTF',ArraySIGMA',ArrayXMEAN'];
save datfil.txt output /ascii
disp('Simulation Complete')
```

To generate the data of Figs. 7.4 and 7.5, 1000 runs had to be made! One thousand run sets will have to be made each time a parameter of interest is changed. In addition, we were able to separate the contributions to the miss from the measurement noise and the target maneuver in the 1000 run set *only because* the noise was random and the target maneuver was deterministic. In a system with many deterministic and stochastic inputs, one would have to run 1000-run sets for one disturbance at a time in order to generate a miss distance error budget. Fortunately the adjoint technique allows us to get error budget information of this type in only *one* run! Because the system under consideration is a mixed continuous discrete system, we have to extend the rules for adjoints covered in Chapters 3 and 4.

MIXED CONTINUOUS DISCRETE ADJOINT THEORY [4, 5]

The rules for constructing an adjoint of a mixed continuous discrete system are simple and are similar to the adjoint rules for continuous systems. Given a linear time-varying discrete system with impulse response H_D in which the ratio of the time of flight to the sampling time is an integer given by

$$N = t_F/T_s$$

there exists an adjoint system with impulse response H_D^*. One can construct a mixed continuous discrete adjoint from the original system using the rules of Chapters 3 and 4 and the additional rule given in the next subsection.

REPLACE *n* BY *N* − *n* IN THE ARGUMENTS OF ALL VARIABLE COEFFICIENTS

Therefore, the rules for continuous and mixed continuous discrete adjoints are identical in that the signal flow of the original system is reversed and the time-varying gains in the original system are generated backwards for the adjoint system. In mixed continuous discrete systems the adjoint rules get slightly more complicated because additional elements are required for interfacing the continuous and discrete parts of the system. For example, a sampler or analog-to-digital converter is required as shown in Fig. 7.6, to make the connection from a continuous system to a discrete system.

The input/output characteristics of a sampler can easily be illustrated. For example, Fig. 7.7 shows that, if the input signal to the sampler is continuous,

Fig. 7.6 Connecting continuous and discrete systems.

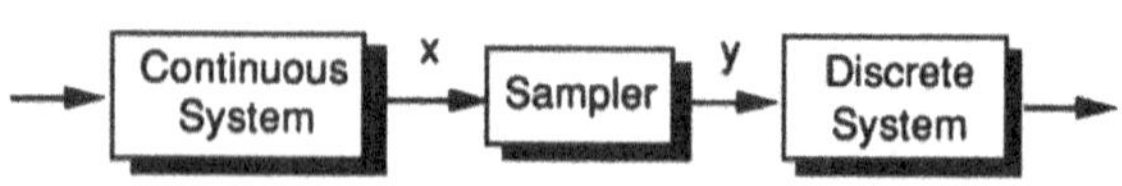

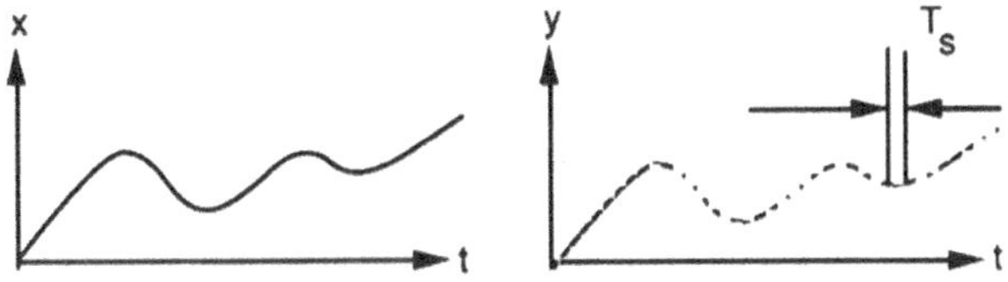

Fig. 7.7 Effect of sampling on continuous signal.

the output signal has the same shape but is defined only at each sampling instant by a number. These numbers, or sample points, are spaced T_s seconds apart.

Applying adjoint theory to mixed continuous discrete systems requires taking the adjoint of a sampler. The adjoint of a sampler is depicted in Fig. 7.8. Here the s block represents a pure derivative and the "hold" block will soon be defined. The z^{-1} block is Z transform notation and represents a pure delay of T_s seconds.

A hold network or digital-to-analog converter is required to connect signal flow from a discrete network to a continuous network as shown in Fig. 7.9.

If the input to the hold is a set of numbers, Fig. 7.10 illustrates the proper input/output characteristics of the hold network. Here we can see that, after a discrete signal has been "held," it becomes continuous.

The adjoint of a hold is shown in Fig. 7.11. Here the $1/s$ term is the Laplace transform representation of an integrator. Again, the z^{-1} term represents a pure delay of T_s seconds.

We now have enough rules to enable us to take the adjoint of a mixed continuous discrete system. Consider the model of Fig. 7.12. In this model there are three continuous linear time-varying networks with impulse responses, H_{C1}, H_{C2}, and H_{C3}, respectively. White noise u_C with spectral density Φ_C enters the continuous portion of the system through the shaping network H_{C3}. In addition, a step disturbance of magnitude a also enters the system through the shaping network H_{C1}. The step input has been represented as an impulse through an integrator so that adjoint theory can be applied to this error source.

In this example we are interested in observing the continuous quantity y at the final time t_F. The output of the network H_{C2} is sampled and sent through a discrete network with impulse response H_{D1}. Zero-mean Gaussian noise with variance σ_D^2 enters the discrete portion of the system through shaping network H_{D2}.

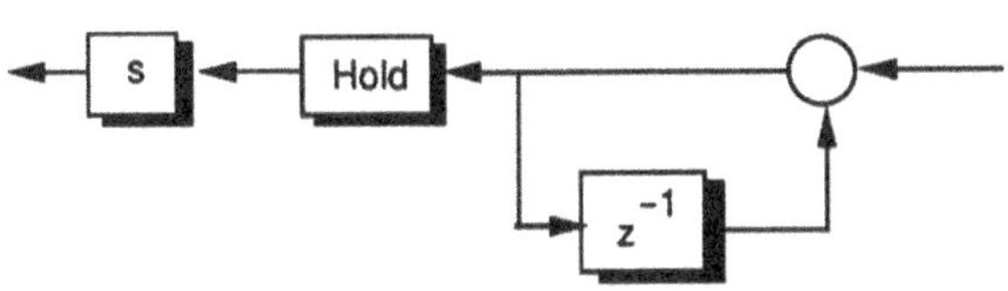

Fig. 7.8 Adjoint of a sampler.

Fig. 7.9 Connecting discrete and continuous systems.

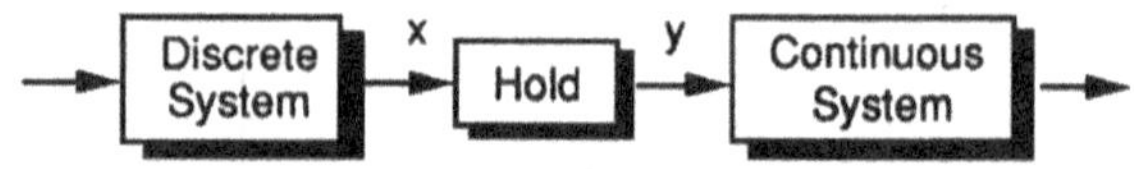

After the resultant signal goes through the discrete network H_{D1}, the output is held and fed back to the continuous network H_{C1}, thus completing the loop. In this example we seek to find $y(t_F)$ due to each of the disturbances. Adjoint theory can readily be applied to this example.

Following the rules of adjoints we can obtain the adjoint system of Fig. 7.13. Although this adjoint model is driven by an impulse, we have seen in Chapters 3–5 that it is not necessary to simulate an impulse. The impulse becomes initial conditions on integrators in its forward path. The impulse is applied at time zero in the adjoint system because the output of interest in the original system is at the final time. The adjoint model shows a differentiator appearing before H^*_{C2}. Again, one need not simulate the differentiator but just use block diagram manipulation to eliminate it (that is, feed it through H^*_{C2}).

The outputs of the adjoint model represent output sensitivities of the system. They are referred to as sensitivities because a change in their levels does not necessitate a rerunning of the adjoint simulation. As can be seen from Fig. 7.13, the new outputs can be calculated by inspection. Note that, in order to find $y(t_F)$ due to a continuous random disturbance, we square and integrate a continuous signal (that is, output of H^*_{C3}). To find $y(t_F)$ due to a discrete random disturbance, we square and sum a discrete signal (that is, output of H^*_{D2}).

USING ADJOINTS TO EVALUATE FILTER PERFORMANCE

Mixed continuous discrete adjoint theory can be applied to the engagement model of Fig. 7.1, a sample data homing loop containing a two-state fading memory filter.

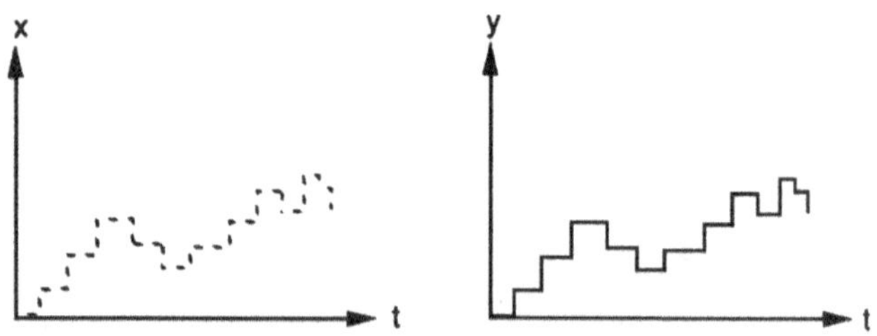

Fig. 7.10 Effects of holding a discrete signal.

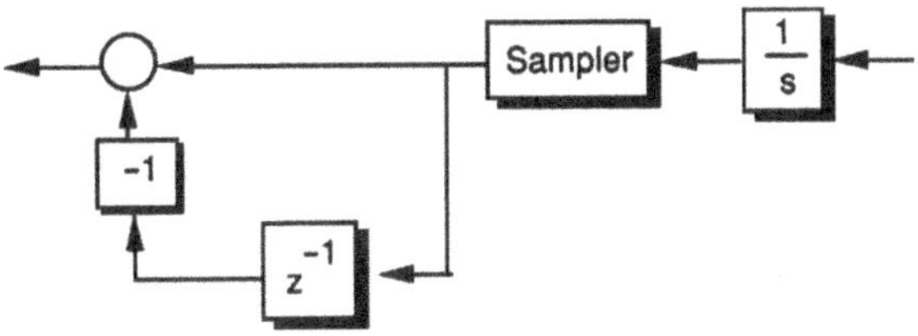

Fig. 7.11 Adjoint of hold.

Recall that in this example there are two disturbances: a deterministic target maneuver and measurement noise on the line-of-sight angle. However, before we take the complete adjoint, let us realize that when the adjoint of a "sampler" is taken we will have a pure differentiator in the homing loop. It is desirable, for simulation reasons, to eliminate the differentiator by block diagram manipulation. This can easily be done by modifying the original system to have an extra integrator before the sampler. This can be accomplished by first generating the line-of-sight rate and then integrating it to get line-of-sight angle. First we must realize that the line-of-sight angle can be expressed as

$$\lambda = \frac{y}{R_{TM}} = \frac{y}{V_c t_{go}}$$

Taking the derivative of the preceding expression, using the quotient rule, and expressing the result in block diagram form we obtain Fig. 7.14.

The resultant adjoint block diagram for the entire homing loop, following the mixed continuous discrete adjoint rules discussed in the previous section, appears

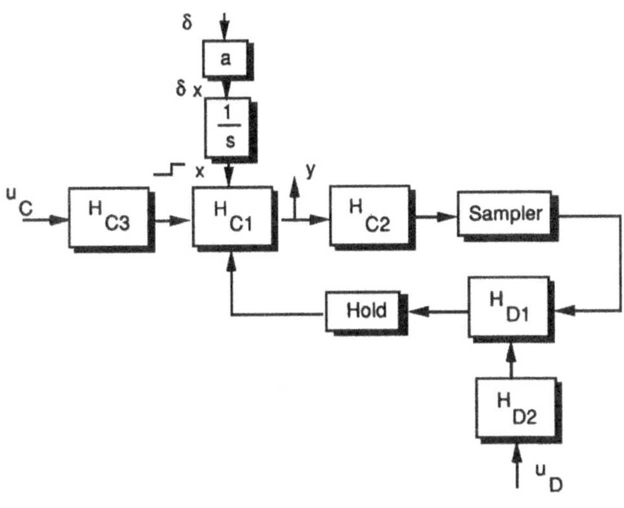

Fig. 7.12 Model of mixed continuous discrete system.

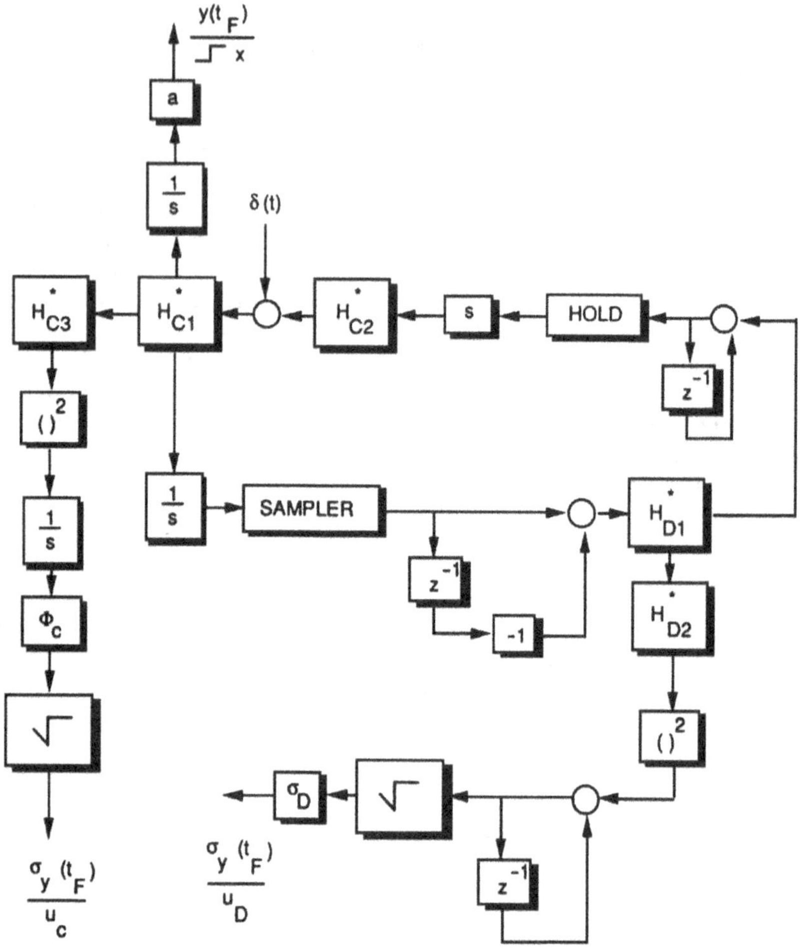

Fig. 7.13 Adjoint of mixed continuous discrete system.

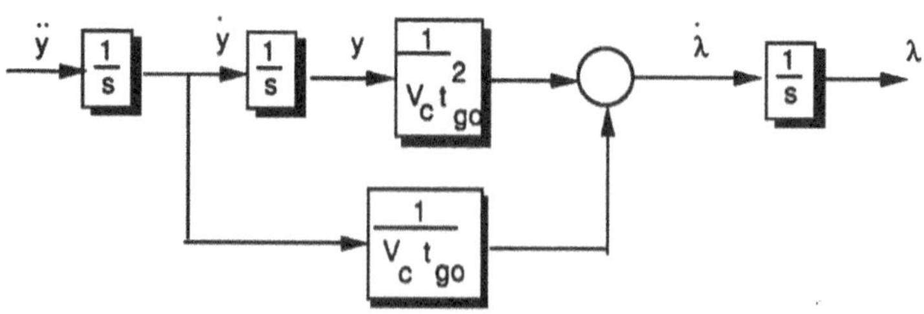

Fig. 7.14 Block diagram for line-of-sight rate.

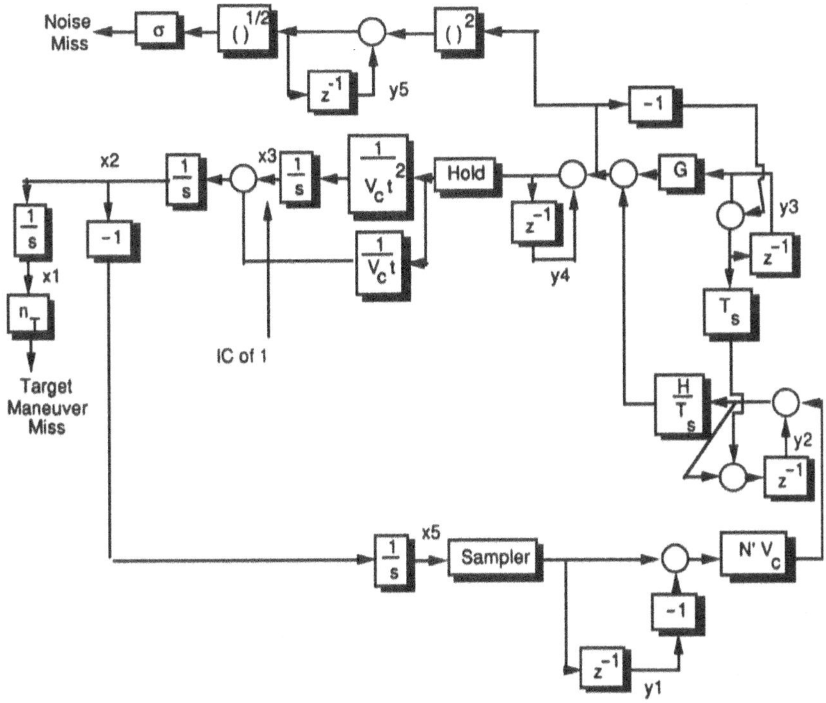

Fig. 7.15 Adjoint of second-order fading memory filter in homing loop.

in Fig. 7.15. The two disturbances of the original system become adjoint outputs, whereas the miss distance output of the original system becomes an impulsive input (or initial condition on integrator $x3$) in the adjoint system. Note that, because the noise is digital, the adjoint noise miss distance sensitivity is obtained by squaring and summing the appropriate signal.

Listing 7.3 presents the MATLAB adjoint program for the engagement model of Fig. 7.15 in which the homing loop contains a second-order fading memory filter. As with the original engagement simulation presented in this chapter, the adjoint program also consists of two sections: one for the differential equations and the other for the difference equations. Care must also be taken in the adjoint program to ensure that the ratio of the sampling interval to the integration interval be a large integer.

A single adjoint run was made for the nominal case considered at the beginning of the chapter ($\beta = 0.8$). The target maneuver miss and noise miss outputs are plotted separately vs adjoint or flight time in Figs. 7.16 and 7.17. Superimposed on these *single run* adjoint results are the standard deviation and mean of the Monte Carlo miss distance results, obtained with *1000 runs* (50 run sets for 20

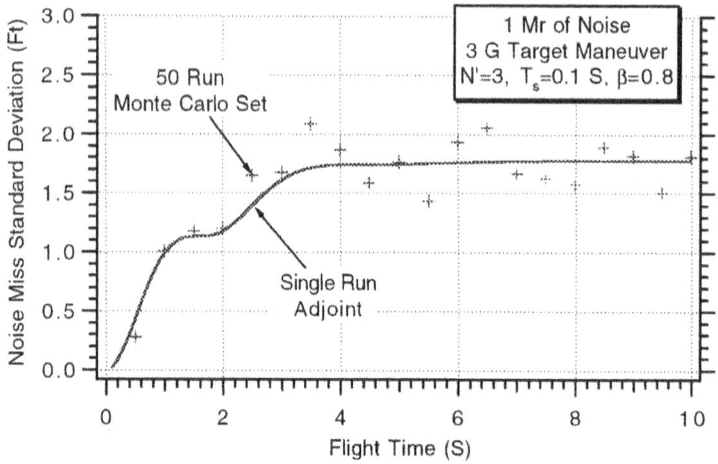

Fig. 7.16 Adjoint noise miss projections are in agreement with Monte Carlo results.

flight times). We can see that both methods yield approximately the same answers. If there were more error sources, the miss distance performance projections could still have been obtained from the same adjoint run by monitoring additional outputs. Thus, we can see that the adjoint technique is a very powerful method for efficiently generating miss distance error budgets.

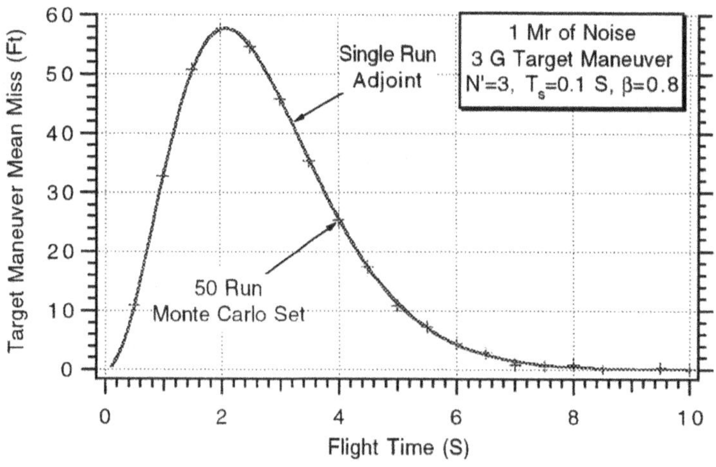

Fig. 7.17 Adjoint target maneuver miss projections are in agreement with Monte Carlo results.

LISTING 7.3 ADJOINT ENGAGEMENT SIMULATION WITH TWO-STATE FADING MEMORY FILTER

```
count=1;
XNT=96.6;
XNP=3.;
TF=10.;
TS=.1;
BETA=.8;
SIGNOISE=.001;
VC=4000.;
T=0.;
S=0.;
TP=T+.00001;
X1=0;
X2=0;
X3=1;
X5=0.;
Y1OLD=0.;
Y2OLD=0.;
Y3OLD=0.;
Y4OLD=0.;
Y5OLD=0.;
H=.01;
GFILTER=1.-BETA^2;
HFILTER=(1.-BETA)^2;
while TP<=(TF-1e-5)
      X1OLD=X1;
      X2OLD=X2;
      X3OLD=X3;
      X5OLD=X5;
      STEP=1;
      FLAG=0;
      while STEP<=1
        if FLAG==1
        STEP=2;
        X1=X1+H*X1D;
        X2=X2+H*X2D;
        X3=X3+H*X3D;
        X5=X5+H*X5D;
        TP=TP+H;
        end
        TGO=TP+.00001;
        X1D=X2;
        X2D=X3+Y4OLD/(VC*TGO);
```

```
            X3D=(Y4OLD)/(VC*TGO*TGO);
            X5D=-X2;
            FLAG=1;
        end
        FLAG=0;
        X1=(X1OLD+X1)/2+.5*H*X1D;
        X2=(X2OLD+X2)/2+.5*H*X2D;
        X3=(X3OLD+X3)/2+.5*H*X3D;
        X5=(X5OLD+X5)/2+.5*H*X5D;
        S=S+H;
        if S>=(TS-.0001)
            S=0.;
            TEMP1=(X5-Y1OLD)*XNP*VC;
            TEMP2=HFILTER*(Y2OLD+TEMP1)/TS+GFILTER*Y3OLD;
            Y1NEW=X5;
            Y2NEW=TEMP1+Y2OLD+TS*(Y3OLD-TEMP2);
            Y3NEW=Y3OLD-TEMP2;
            Y4NEW=Y4OLD+TEMP2;
            Y5NEW=Y5OLD+TEMP2*TEMP2;
            Y1OLD=Y1NEW;
            Y2OLD=Y2NEW;
            Y3OLD=Y3NEW;
            Y4OLD=Y4NEW;
            Y5OLD=Y5NEW;
            XMNOISE=SIGNOISE*sqrt(Y5NEW);
            XMNT=XNT*X1;
            count=count+1;
            ArrayTP(count)=TP;
            ArrayXMNT(count)=XMNT;
            ArrayXMNOISE(count)=XMNOISE;
        end
end
figure
plot(ArrayTP,ArrayXMNT),grid
xlabel('Flight Time (Sec)')
ylabel('Target Maneuver Miss (Ft)')
figure
plot(ArrayTP,ArrayXMNOISE),grid
xlabel('Flight Time (Sec)')
ylabel('Noise Miss (Ft)')
clc
output=[ArrayTP',ArrayXMNT',ArrayXMNOISE'];
save datfil.txt output  -ascii
disp 'simulation finished'
```

SOME PROPERTIES OF FADING MEMORY FILTERS

The filter parameter β determines how much the filter will remember about past measurements, which in turn will determine the filter bandwidth or speed of response. Higher values of β yield a filter that remembers a great deal about the past. This type of filter will have low bandwidth and slow speed of response. Low values of β result in a high bandwidth fast filter. Figures 7.18 and 7.19 show, based on adjoint simulation results, how miss distance varies with the fading memory filter parameter β. It is not surprising that Fig. 7.18 shows dramatically improved miss distance results for the wider bandwidth filter (faster guidance systems yield smaller miss distances due to target maneuver). However, it is surprising that Fig. 7.19 shows that there is slightly less miss distance due to line-of-sight angle noise for the faster filter, even though we know that the filter has more noise transmission. In general, reducing the guidance system time constant will reduce the miss for most disturbances in a proportional navigation guidance system in the absence of parasitic and saturation effects. Ultimately excessive noise transmission will lead to saturation and increased miss distance.

Changing the sampling time also effects filter and system performance. Increasing the sampling rate (lower values for T_s) means that the filter has more information per unit time. Therefore, increasing the sampling rate should be beneficial. Figure 7.20 shows, from single flight results, that increasing the sampling rate (reduce T_s from 0.1 s to 0.05 s) removes the previously noted lag in the line-of-sight rate estimate when β is 0.8 (see Fig. 7.2 for comparison). The noise transmission appears to be about the same, or slightly reduced, from the case when the sampling time was 0.1 s.

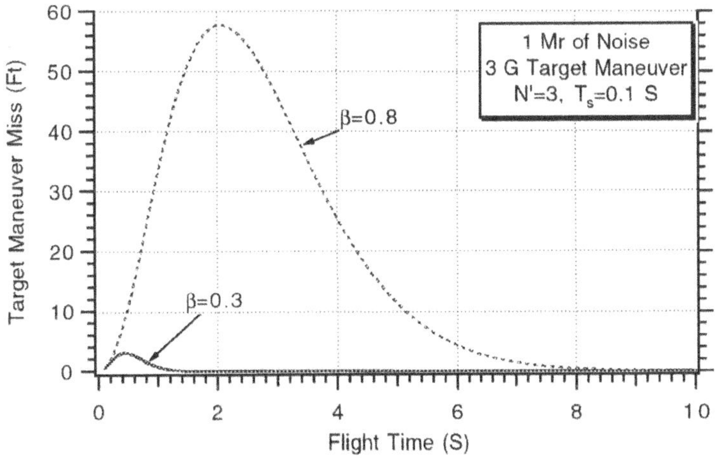

Fig. 7.18 Faster fading memory filter yields less miss due to target maneuver.

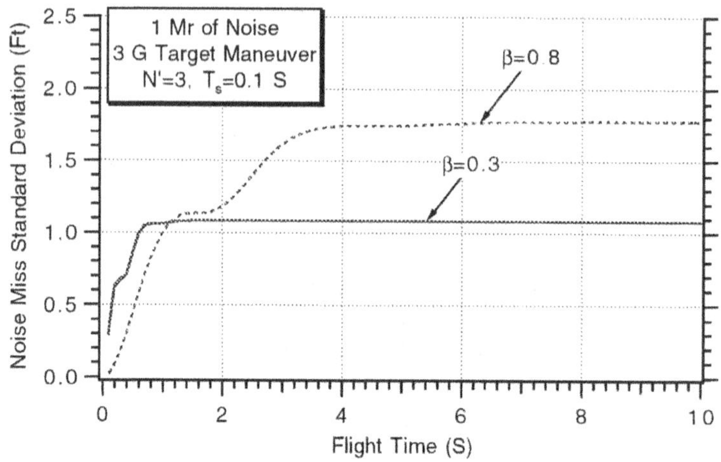

Fig. 7.19 Faster noise filter yields less miss due to line-of-sight angle noise.

From a system performance point of view, increasing the sampling rate should also be beneficial. In essence, we are speeding up the guidance system, which means for the inputs previously considered, miss should decrease. Adjoint results, which are displayed in Fig. 7.21, confirm that doubling the sampling rate (reducing T_s from 0.1 s to 0.05 s) dramatically reduces the miss sensitivity due to target maneuver.

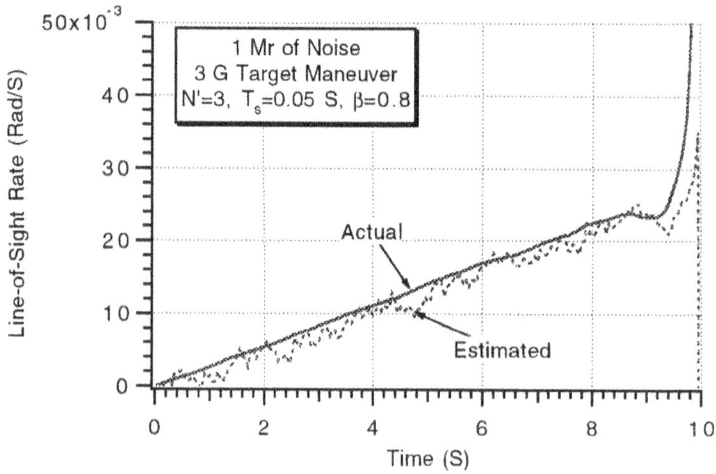

Fig. 7.20 Increasing sampling rate makes fading memory filter faster.

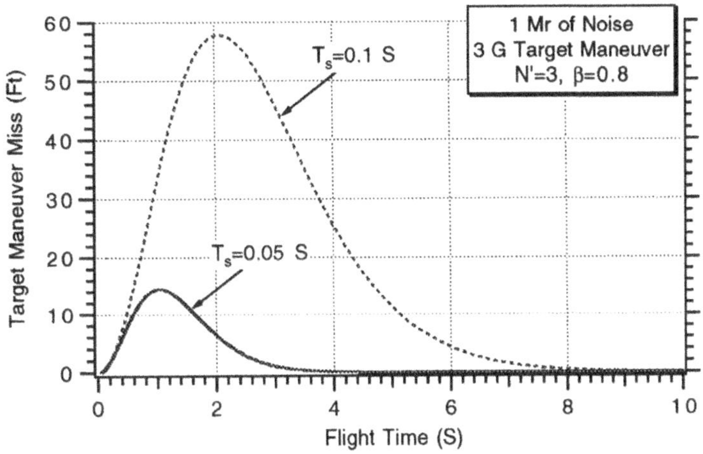

Fig. 7.21 Increasing sampling rate reduces miss due to target maneuver.

Figure 7.22 also confirms that the miss due to noise decreases with increasing sampling rate. Usually, increased system costs are associated with higher sampling rates. Therefore, financial considerations usually place an upper limit on practical achievable sampling rates.

It is important to note that in the preceding experiment the noise standard deviation remained constant as the data rate changed. In many systems the noise spectral density remains constant and so the noise standard deviation

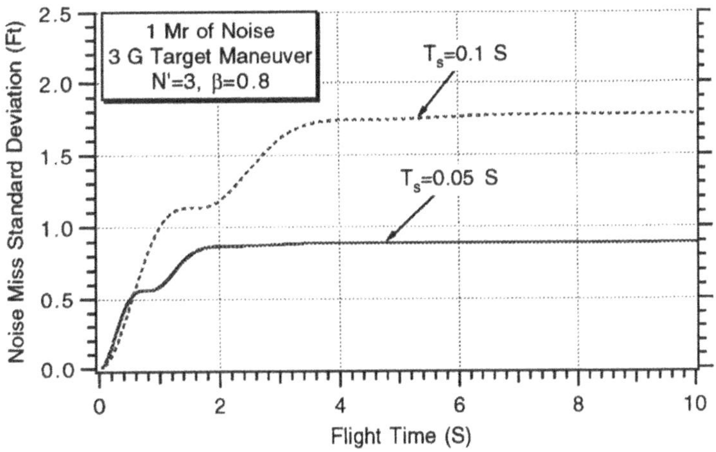

Fig. 7.22 Increasing sampling rate reduces miss due to noise.

changes with changing data rate. The interested reader is referred to the appendix for a more complete discussion of this topic.

ESTIMATING TARGET MANEUVER

In Chapter 8 we will investigate more advanced guidance laws. To implement more advanced guidance laws, we must have knowledge of all of the target states. In other words, we must know what the target is doing. Mathematically stated, we would like to be able to estimate the current maneuver level of the target based on a noisy measurement of the line-of-sight angle. Theoretically it is impossible, without additional measured or *a priori* information, to estimate the maneuver level of the target based on angle-only measurements from a single sensor. However, many tactical radar homing missiles also measure range and range rate in addition to the line-of-sight angle, which makes target acceleration estimation possible.

Figure 7.23 presents a guidance system that uses a three-state fading memory filter to estimate target acceleration from a measurement of the line-of-sight angle,

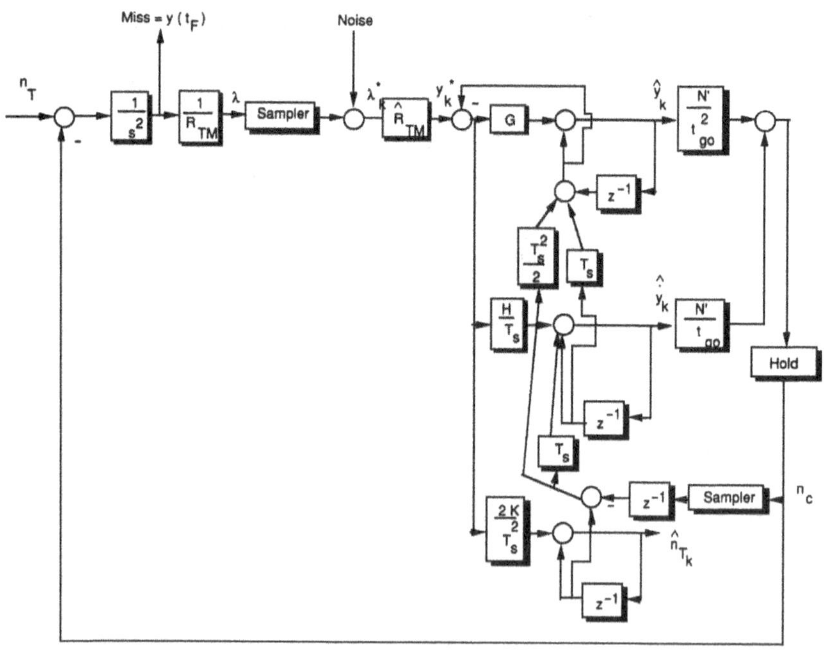

Fig. 7.23 Estimating target maneuver with three-state fading memory filter.

range, and closing velocity. The noisy measurement of the line-of-sight angle is multiplied by a range measurement to get a pseudomeasurement of relative position y_k^*. The filter then estimates the derivatives of the measurement. Using knowledge of the missile acceleration, which is assumed to be known precisely, it is then possible to estimate target acceleration from a relative acceleration as shown in Fig. 7.23. With this type of guidance system we also need time-to-go information, which can be obtained from the range and range rate measurements, to implement either the proportional or augmented proportional navigation guidance law.

Listing 7.4 presents a MATLAB engagement simulation with the three-state fading memory filter as shown in Fig. 7.23. Note that the three-state filter gains are different from the two-state filter gains.

LISTING 7.4 ENGAGEMENT SIMULATION WITH THREE-STATE FADING MEMORY FILTER

```
count=0;
VC=4000.;
XNT=96.6;
YIC=0.;
VM=3000.;
HEDEG=0.;
BETA=.8;
XNP=3.;
SIGNOISE=.001;
TF=10.;
TS=.1;
NOISE=1;
Y=YIC;
YD=-VM*HEDEG/57.3;
YDIC=YD;
T=0.;
H=.01;
S=0.;
GFILTER=1.-BETA^3;
HFILTER=1.5*((1.-BETA)^2)*(1.+BETA);
KFILTER=.5*((1.-BETA)^3);
YH=0.;
YDH=0.;
XNTH=0.;
XNC=0.;
while T <= (TF - 1e-5)
        YOLD=Y;
        YDOLD=YD;
        STEP=1;
```

```
        FLAG=0;
        while STEP <=1
            if FLAG==1
        Y=Y+H*YD;
        YD=YD+H*YDD;
        T=T+H;
        STEP=2;
        end
        TGO=TF-T+.00001;
        RTM=VC*TGO;
        XLAM=Y/(VC*TGO);
        XLAMD=(RTM*YD+Y*VC)/(RTM^2);
        YDD=XNT-XNC;
        FLAG=1;
        end
        FLAG=0;
        Y=.5*(YOLD+Y+H*YD);
        YD=.5*(YDOLD+YD+H*YDD);
        S=S+H;
        if S>=(TS - 1e-5)
            S=0.;
            if NOISE==1,
                        XLAMNOISE=SIGNOISE*randn;
            else
                        XLAMNOISE=0.;
            end;
            YSTAR=RTM*(XLAM+XLAMNOISE);
            RES=YSTAR-YH-TS*YDH-.5*TS*TS*(XNTH-XNC);
            YH=GFILTER*RES+YH+TS*YDH+.5*TS*TS*(XNTH-XNC);
            YDH=HFILTER*RES/TS+YDH+TS*(XNTH-XNC);
            XNTH=2.*KFILTER*RES/(TS*TS)+XNTH;
            XLAMDH=(YH+YDH*TGO)/(VC*TGO*TGO);
            XNC=XNP*VC*XLAMDH;
            count=count+1;
            ArrayT(count)=T;
            ArrayY(count)=Y;
            ArrayXNCG(count)=XNC/32.2;
            ArrayXLAMD(count)=XLAMD;
            ArrayXLAMDH(count)=XLAMDH;
            ArrayXNTG(count)=XNT/32.2;
            ArrayXNTHG(count)=XNTH/32.2;
        end
end
figure
plot(ArrayT,ArrayXLAMD,ArrayT,ArrayXLAMDH),grid
```

```
xlabel('Time (S)')
ylabel('Line of Sight Rate (Rad/S) ')
axis([0 10 0 .05])
figure
plot(ArrayT,ArrayXNTG,ArrayT,ArrayXNTHG),grid
xlabel('Time (S)')
ylabel('Acceleration (G) ')
clc
output=[ArrayT',ArrayY',ArrayXNCG',ArrayXLAMD',ArrayXLAMDH',ArrayXNTG',
ArrayXNTHG'];
save datfil.txt output /ascii
disp 'simulation finished'
```

A nominal case was run with the simulation of Listing 7.4 in which the fading memory factor of the filter was 0.8 and the sampling time was 0.1 s. Figure 7.24 compares the line-of-sight rate estimate of the filter with the actual line-of-sight rate for the nominal case. We can see that the filter estimate follows the geometric line-of-sight rate without excessive noise transmission.

Figure 7.25 shows, for the same case, the filter estimate of the target maneuver. Superimposed on the figure is the actual maneuver. We can see that for this case it takes the filter about 5 s to get a reasonable estimate of the maneuver level. A faster filter would have a smaller transient period but much more noise transmission. Estimates of the quality shown in Fig. 7.25 are sufficient for improving guidance system performance.

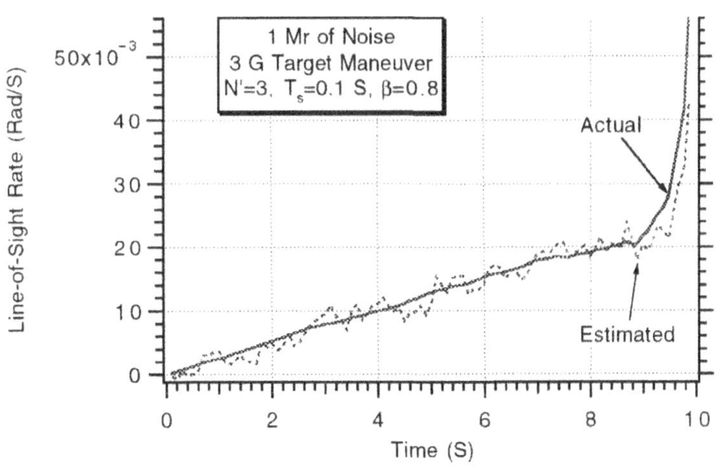

Fig. 7.24 Three-state filter yields excellent estimate of line-of-sight rate.

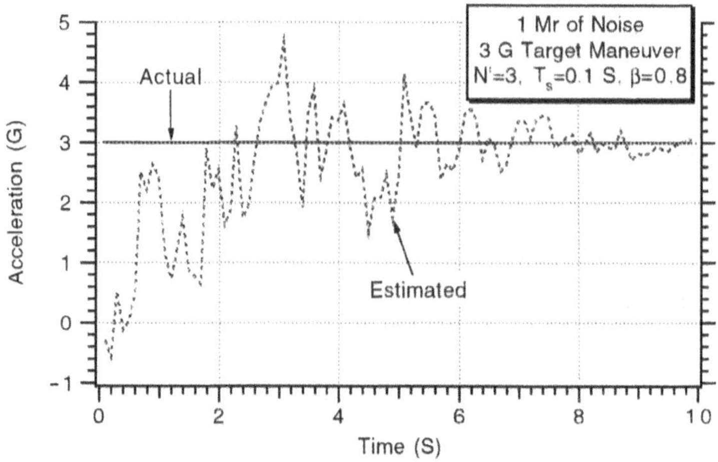

Fig. 7.25 Three-state fading memory filter is able to estimate target maneuver.

SUMMARY

In this chapter it was shown how a simple constant-gain, digital noise filter, known as a fading memory filter, could be implemented in a missile guidance system. It was shown that both filter bandwidth and sampling rate are important parameters in determining overall system performance. The method of adjoints was extended so that it could be used to yield performance projections of a missile guidance system with a digital noise filter. Experiments confirmed that Monte Carlo simulation results were in complete agreement with single-run adjoint performance projections. Finally, it was shown how a fading memory filter could be utilized to provide target acceleration estimates.

REFERENCES

[1] Morrison, N., *Introduction to Sequential Smoothing and Prediction*, McGraw-Hill, New York, 1969.

[2] Bar-Shalom, Y., and Li, X. R., *Estimation and Tracking Principles, Techniques, and Software*, Artech House, Boston, MA, 1993.

[3] Kalata, P. R., "The Tracking Index: A Generalized Parameter for Alpha-Beta-Gamma Target Trackers," *IEEE Transactions Aerospace and Electronic Systems*, AES-20, March 1994, pp. 174–182.

[4] Moroney, R., "The Adjoint of Mixed Continuous/Discrete Systems," Raytheon Memo RM-69-1, Jan. 1969.

[5] Zarchan, P., and Warren, R. S., "Discrete Adjoint Simulation," Raytheon Rept, BR-5440-1, Oct. 1969.

CHAPTER 8

Advanced Guidance Laws

INTRODUCTION

Thus far we have used proportional navigation as an interceptor guidance law because it is easy to implement and is very effective. In fact, proportional navigation is used extensively in the tactical missile world. However, there are other more advanced guidance laws. These advanced guidance laws relax the interceptor acceleration requirements and also yield smaller miss distances. The price paid for these more advanced guidance laws is that more information, such as time to go and missile-target range, is required for their successful implementation. The concept of *zero effort miss*, originally introduced in Chapter 2, will be used to develop and understand new guidance laws. The zero effort miss concept will also be important when we move to the strategic world and encounter predictive guidance. The Schwartz inequality will be used to derive optimal guidance laws analytically.

REVIEW OF PROPORTIONAL NAVIGATION

The basic homing loop for a zero-lag proportional navigation guidance system, which first appeared in Chapter 2, is repeated for convenience in Fig. 8.1. In this zero-lag loop, the seeker, noise filter, and flight-control system dynamics have been neglected. As can be seen from the figure, the proportional navigation guidance law can be expressed as

$$n_c = N'V_c\dot{\lambda}$$

where N' is a gain known as the effective navigation ratio, V_c the closing velocity, and the λ line-of-sight angle.

We have already shown in Chapter 2 that closed-form solutions for the required missile acceleration exist for this zero-lag guidance system. The resultant formula for the missile acceleration n_c due to a step target maneuver was derived

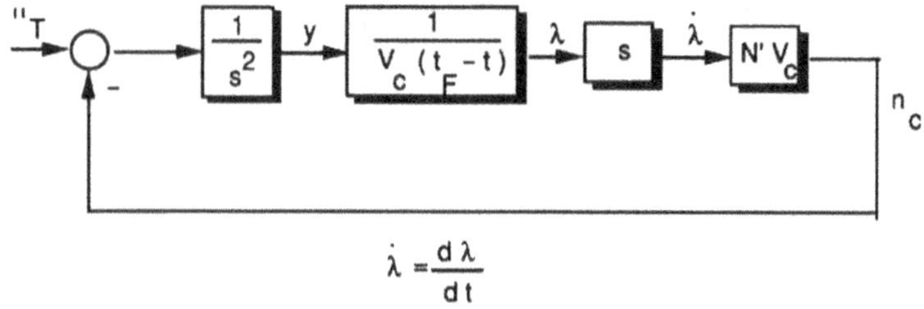

Fig. 8.1 Zero-lag proportional navigation homing loop.

from the first-order time-varying proportional navigation homing loop differential equation originally presented in Chapter 2. The solution, which is repeated here for convenience, is given by

$$n_c = \frac{N'}{N'-2}\left[1-\left(1-\frac{t}{t_F}\right)^{N'-2}\right]n_T$$

where t_F is the time of flight, t the time, and n_T the target maneuver level. We can see from the closed-form solution that the required missile acceleration is directly proportional to the target maneuver acceleration level. Doubling the target acceleration level also doubles the missile acceleration requirements.

To convey the maximum amount of information concisely, the closed-form solution for the missile acceleration induced by target maneuver is normalized and displayed in Fig. 8.2 for different values of the effective navigation ratio. We can see that, regardless of the effective navigation ratio, the required missile acceleration induced by a target maneuver is largest at the end of the flight. Increasing the effective navigation ratio tends to reduce the maximum missile acceleration requirement. Of course we have already seen that, due to parasitic effects and possibly noise considerations, there is a practical upper limit on maximum allowable values for the effective navigation ratio.

The missile's maximum required acceleration, which occurs at the end of the flight $(t = t_F)$, can be obtained from the closed-form solution as

$$n_{c\,\max}\big|_{\mathrm{PN}} = \frac{N' n_T}{N'-2}$$

Therefore, for an effective navigation ratio of 3, the missile requires three times the acceleration capability of the target for a successful intercept. However, increasing the effective navigation ratio from 3 to 5 reduces the required missile acceleration advantage from 3 to 1.67. Of course, other disturbances plus system dynamics will work in the direction of increasing the required missile acceleration advantage.

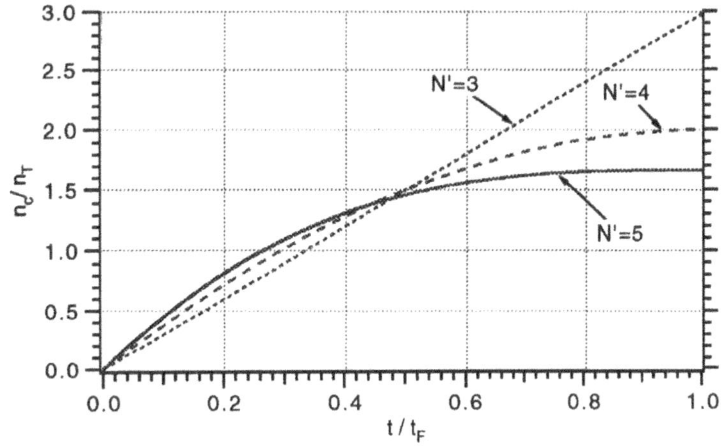

Fig. 8.2 Normalized missile acceleration due to target maneuver for proportional navigation guidance.

AUGMENTED PROPORTIONAL NAVIGATION

More advanced guidance laws can be developed from the zero-lag homing loop model of Fig. 8.1. First we note that the line-of-sight angle can also be expressed as

$$\lambda = \frac{y}{R_{\rm TM}} = \frac{y}{V_c(t_F - t)}$$

where y is the relative missile-target separation and $R_{\rm TM}$ the range from the missile to the target. We can find the line-of-sight rate by taking the derivative of the preceding expression, using the quotient rule, obtaining

$$\dot{\lambda} = \frac{y + \dot{y}t_{\rm go}}{V_c t_{\rm go}^2}$$

where $t_{\rm go}$ is the time to go until intercept and can be defined as

$$t_{\rm go} = t_F - t$$

Thus, we can also express the proportional navigation guidance law as the mathematically equivalent expression

$$n_c = N' V_c \dot{\lambda} = \frac{N'(y + \dot{y}t_{\rm go})}{t_{\rm go}^2}$$

The expression in the parentheses of the preceding equation represents the future separation between missile and target. More simply, the expression in parentheses is the miss distance that would result if the missile made no further corrective acceleration and the target did not maneuver. This expression is referred to as the zero effort miss, *ZEM*. Therefore, we can also think of proportional navigation as a guidance law in which acceleration commands are issued inversely proportional to the square of time to go and directly proportional to the zero effort miss.

If the target maneuvers, the zero effort miss must be augmented by an additional term. The new equation for the zero effort miss, in the presence of a constant target maneuver, is simply

$$ZEM_{\text{TGT MVR}} = y + \dot{y}t_{go} + 0.5n_T t_{go}^2$$

where n_T is the target maneuver acceleration level. Therefore, a perfectly plausible guidance law, in the presence of target maneuver, would be

$$n_c|_{\text{APN}} = \frac{N' ZEM_{\text{TGT MVR}}}{t_{go}^2} = N' V_c \dot{\lambda} + \frac{N' n_T}{2}$$

This new guidance law, known as augmented proportional navigation, is proportional navigation with an extra term to account for the maneuvering target [1].

A zero-lag augmented proportional navigation homing loop is shown in block diagram form in Fig. 8.3. The additional target maneuver term, required by the guidance law, appears as a feedforward term in the homing loop block diagram. As with the proportional navigation guidance law, we can also obtain closed-form solutions for the required missile acceleration due to a constant target maneuver for the zero-lag homing loop depicted in Fig. 8.3 The resultant solution for the required missile acceleration is

$$n_c|_{\text{APN}} = 0.5 n_T N' \left(1 - \frac{t}{t_F}\right)^{N'-2}$$

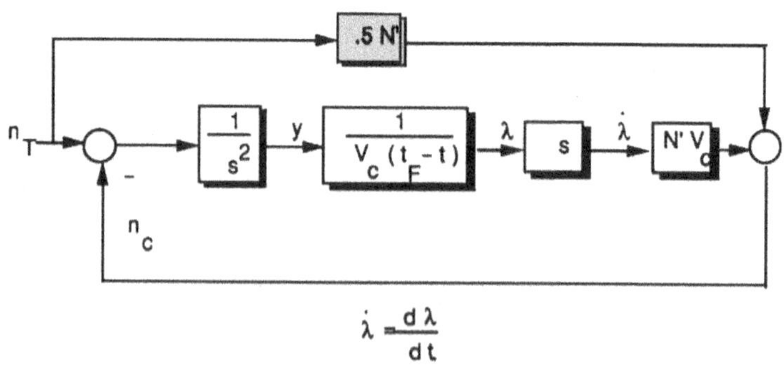

Fig. 8.3 Augmented proportional navigation homing loop.

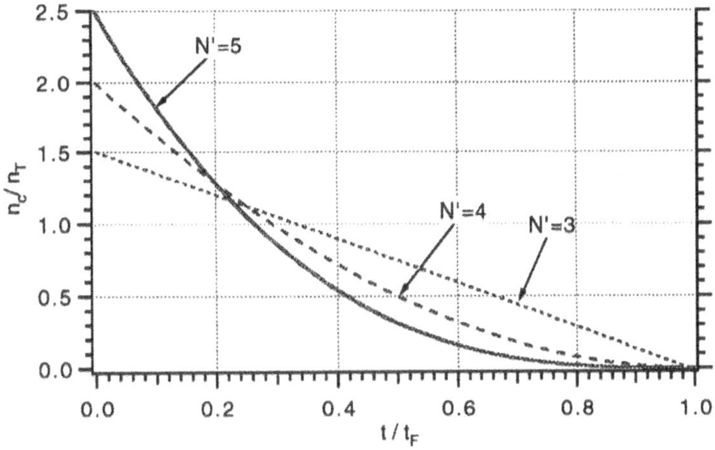

Fig. 8.4 Normalized acceleration for augmented proportional navigation to hit a maneuvering target.

The closed-form solution for the missile acceleration required to hit a maneuvering target with augmented proportional navigation is displayed in normalized form in Fig. 8.4. Here we can see that the required missile acceleration decreases monotonically with time, regardless of the effective navigation ratio, rather than increasing monotonically with time as was the case with proportional navigation. Increasing the effective navigation ratio increases the maximum acceleration at the beginning of the flight but also reduces the time at which the acceleration decays to negligible levels.

The maximum acceleration required by augmented proportional navigation to hit a maneuvering target is

$$n_{c\max}\big|_{\text{APN}} = 0.5N'n_T$$

This means that, for a navigation ratio of 3, augmented proportional navigation requires half the acceleration of the missile than with proportional navigation guidance. However, for an effective navigation ratio of 5, augmented proportional navigation requires a larger maximum acceleration compared with proportional navigation guidance.

Comparative plots of proportional and augmented proportional navigation missile acceleration profiles for different values of effective navigation ratio due to a target maneuver appear in Figs. 8.5–8.7. Figure 8.5 shows that, with an effective navigation ratio of 3, augmented proportional navigation requires less acceleration capability of the missile than proportional navigation. This figure also indicates that augmented proportional navigation requires much less total acceleration than proportional navigation. This is not surprising because augmented

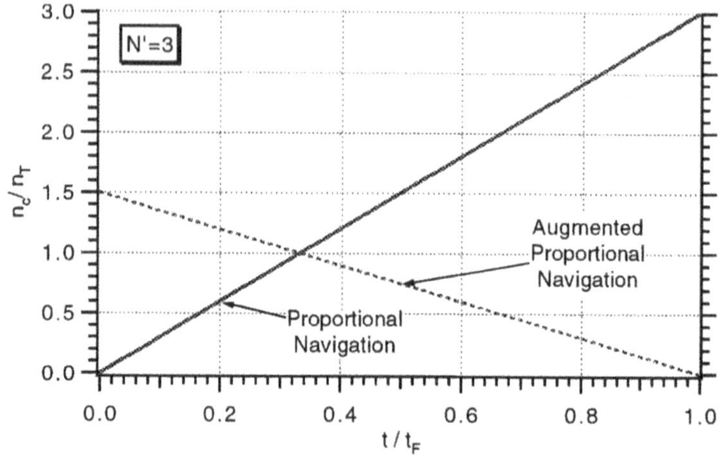

Fig. 8.5 Augmented proportional navigation requires less acceleration capability of missile for {$N' = 3$}.

proportional navigation is making use of extra information, namely, knowledge of the target maneuver. It is reasonable that this knowledge should enable the missile to maneuver in a more efficient manner.

Figure 8.6 shows that for an effective navigation ratio of 4 the maximum acceleration required by both guidance laws is the same. The total acceleration begins

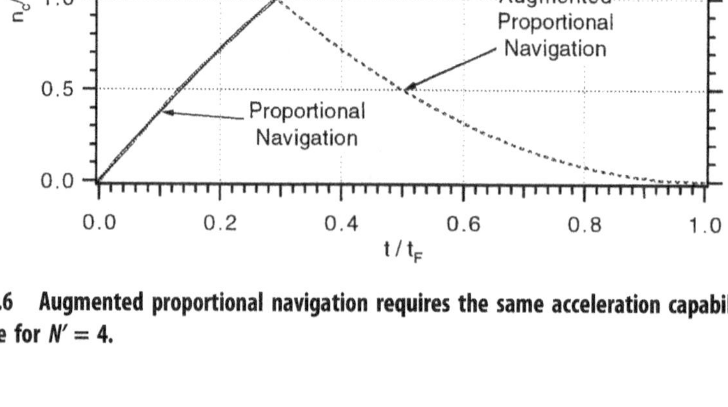

Fig. 8.6 Augmented proportional navigation requires the same acceleration capability of missile for $N' = 4$.

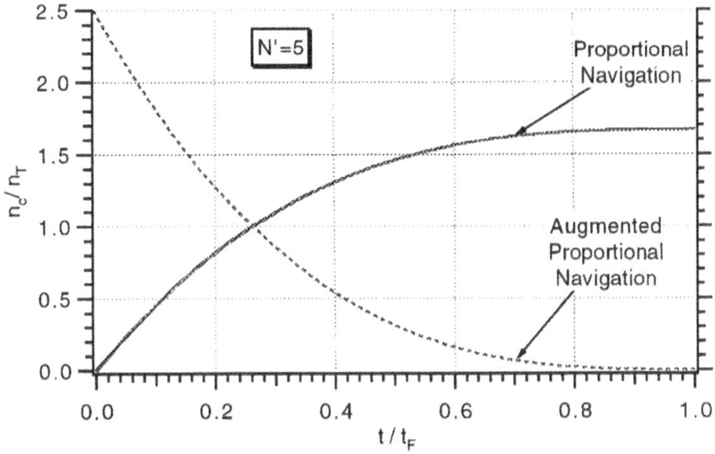

Fig. 8.7 Augmented proportional navigation requires more acceleration capability of missile for $N' = 5$.

to be less with augmented proportional navigation at a normalized time of 0.3. This means that 70% of the time-augmented proportional navigation requires less acceleration than proportional navigation to hit a maneuvering target.

Figure 8.7 shows that, when the effective navigation ratio is 5, augmented proportional navigation requires a larger maximum acceleration capability of the missile than does proportional navigation. However, about 75% of the time-augmented proportional navigation requires less missile acceleration than proportional navigation.

It appears from Figs. 8.6 and 8.7 that augmented proportional navigation does not relax the maximum missile acceleration requirements imposed by proportional navigation guidance when the effective navigation ratio is greater than or equal to 4. However, in these cases, augmented proportional navigation appears to require less total acceleration than proportional navigation for most of the flight.

To quantify this observation more precisely, we need a performance index other than maximum acceleration. One possibility is to consider the total acceleration required or to find the area under the acceleration curve. We shall see in Chapter 14 that strategic missiles require fuel to maneuver since they operate outside the atmosphere (that is, they cannot generate lift by moving control surfaces). In these cases the missile maneuverability is referred to as a lateral divert capability. Lateral divert is directly related to the amount of fuel required by the interceptor to implement the guidance law and effect an intercept outside the atmosphere. The lateral divert is in fact the total area under the absolute value of the acceleration curve. Because missile acceleration is always positive, we can find the lateral divert requirements for proportional navigation by integrating the closed-form solution for the required missile acceleration, or

$$\Delta V_{\mathrm{PN}} = \int_0^{t_F} n_c|_{\mathrm{PN}}\,\mathrm{d}t = \int_0^{t_F} \frac{N'}{N'-2}\left[1-\left(1-\frac{t}{t_F}\right)^{N'-2}\right] n_T\,\mathrm{d}t$$

After some algebra we obtain

$$\Delta V_{\mathrm{PN}} = \frac{N' n_T t_F}{N'-1}$$

Thus, we can see that increasing the effective navigation ratio makes the lateral divert requirements smaller. Following the same procedure we can express the lateral divert required for augmented proportional navigation. First we must set up the integral as

$$\Delta V_{\mathrm{APN}} = \int_0^{t_F} n_c|_{\mathrm{APN}}\,\mathrm{d}t = \int_0^{t_F} 0.5 n_T N'\left(1-\frac{t}{t_F}\right)^{N'-2}\mathrm{d}t$$

Integration and simplification yields

$$\Delta V_{\mathrm{APN}} = \frac{0.5 N' n_T t_F}{N'-1}$$

Figure 8.8 presents a comparative plot of the total energy or lateral divert required by the interceptor as a function of the effective navigation ratio for

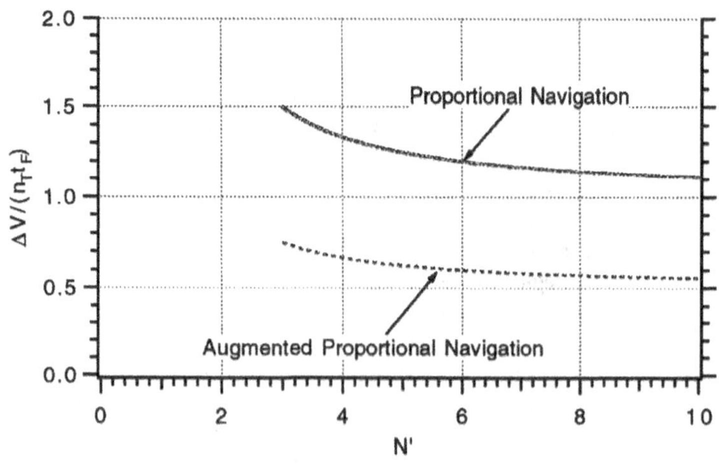

Fig. 8.8 Augmented proportional navigation has reduced divert requirement.

both guidance laws. The figure shows that the lateral divert requirements decrease with increasing effective navigation ratio for both guidance laws. We can also see from the formulas and figure that augmented proportional navigation always has one-half the lateral divert requirements of proportional navigation, regardless of the effective navigation ratio. Therefore, for strategic applications, augmented proportional navigation is a more fuel-efficient guidance law than proportional navigation.

DERIVATION OF AUGMENTED PROPORTIONAL NAVIGATION [4]

Thus far we have given a heuristic argument for the augmented proportional navigation guidance law. This is a good approach if the desire is to understand a guidance law concept, but it is not quite adequate for developing more advanced and complex laws.

Our model of the guidance system, for guidance law development, is shown in Fig. 8.9. In this zero-lag model we are saying that relative acceleration is the difference between target acceleration n_T and missile acceleration n_c.

We seek to find a guidance law that is a function of the system states. There are an infinite number of possible guidance laws; thus, it is necessary to state in mathematical terms what the guidance law should do. Certainly we would like to hit the target! Therefore, one feature of the guidance law should be a zero miss distance requirement. In addition, we would like to hit the target in an efficient manner. In other words, we desire to use minimal total acceleration. A popular and mathematically convenient way of stating the guidance problem to be solved is that we desire to achieve zero miss subject to minimizing the integral of the square of the acceleration command, or

$$y(t_F) = 0 \qquad \text{subject to minimizing} \int_0^{t_F} n_c^2(t)\, \mathrm{d}t$$

Unfortunately, if we minimized a more meaningful performance index such as the integral of the absolute value of n_c, the solution would be mathematically intractable. Typically this type of problem with a quadratic performance index is solved using techniques from optimal control theory [2, 3]. However, this class of problem can be solved more easily using the Schwartz inequality [4]. Before we begin, let us review a few fundamentals. A system of linear differential equations can always be represented in the following state space form:

$$\dot{x} = Fx + Gu$$

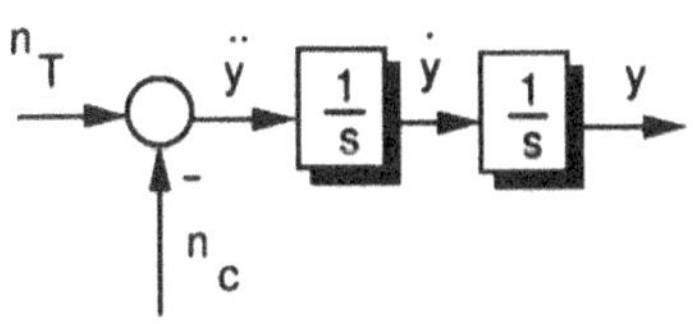

Fig. 8.9 Model for guidance law derivation.

The system of Fig. 8.9 can be expressed in state space form as

$$\begin{bmatrix} \dot{y} \\ \ddot{y} \\ \dot{n}_T \end{bmatrix} = \underbrace{\begin{bmatrix} 0 & 1 & 0 \\ 0 & 0 & 1 \\ 0 & 0 & 0 \end{bmatrix}}_{F} \begin{bmatrix} y \\ \dot{y} \\ n_T \end{bmatrix} + \underbrace{\begin{bmatrix} 0 \\ -1 \\ 0 \end{bmatrix}}_{G} n_c$$

where F is the 3×3 system dynamics matrix and $\boldsymbol{G}$ is the 3×1 vector. The solution to the state space vector differential equation is given at the final time t_F by the vector relationship [5].

$$\boldsymbol{x}(t_F) = \Phi(t_F - t)\boldsymbol{x}(t) + \int_t^{t_F} \Phi(t_F - \lambda)\boldsymbol{G}(\lambda)\boldsymbol{u}(\lambda)\,\mathrm{d}\lambda$$

where Φ is the fundamental matrix and is related to F according to

$$\Phi(t) = \mathcal{L}^{-1}[(sI - F)^{-1}]$$

where $\mathcal{L}^{-1}$ is the inverse Laplace transform. This means that in order to find the fundamental matrix we must first invert the matrix $sI - F$ and then find the inverse Laplace transform of the resultant matrix expression.

For the model of Fig. 8.9 the fundamental matrix is found to be

$$\Phi(t) = \begin{bmatrix} 1 & t & 0.5t^2 \\ 0 & 1 & t \\ 0 & 0 & 1 \end{bmatrix}$$

Using the preceding fundamental matrix in the solution for the state space vector differential equation and only looking at the first state, we get

$$y(t_F) = y(t) + \dot{y}(t)(t_F - t) + 0.5n_T(t_F - t)^2 - \int_t^{t_F} (t_F - \lambda)n_c(\lambda)\,\mathrm{d}\lambda$$

For convenience let us define the terms

$$f_1(t_F - t) = y(t) + \dot{y}(t)(t_F - t) + 0.5n_T(t_F - t)^2$$

and

$$h_1(t_F - \lambda) = t_F - \lambda$$

Then we can say that

$$y(t_F) = f_1(t_F - t) - \int_t^{t_F} h_1(t_F - \lambda)n_c(\lambda)\,\mathrm{d}\lambda$$

For the condition in which we have zero miss distance $[y(t_F) = 0]$, we can rewrite the preceding equation as

$$f_1(t_F - t) = \int_t^{t_F} h_1(t_F - \lambda) n_c(\lambda)\, \mathrm{d}\lambda$$

If we apply the Schwartz inequality to the preceding expression, we get the relationship

$$f_1^2(t_F - 1) \le \int_t^{t_F} h_1^2(t_F - \lambda)\, \mathrm{d}\lambda \int_t^{t_F} n_c^2(\lambda)\, \mathrm{d}\lambda$$

Expressing the preceding inequality in terms of the desired acceleration command, we get

$$\int_t^{t_F} n_c^2(\lambda)\, \mathrm{d}\lambda \ge \frac{f_1^2(t_F - t)}{\int_t^{t_F} h_1^2(t_F - \lambda)\, \mathrm{d}\lambda}$$

The integral of the square of the commanded acceleration will be minimized when the equality sign holds in the preceding inequality. According to the Schwartz inequality, the equality sign holds when

$$n_c(\lambda) = k h_1(t_F - \lambda)$$

This means that the integral of the squared acceleration is minimized when

$$k = \frac{f_1(t_F - t)}{\int_t^{t_F} h_1^2(t_F - \lambda)\, \mathrm{d}\lambda}$$

Therefore, the commanded acceleration is given by

$$n_c = \left[\frac{f_1(t_F - t)}{\int_t^{t_F} h_1^2(t_F - \lambda)\, \mathrm{d}\lambda}\right] h_1(t_F - t)$$

Substitution yields the feedback control guidance law

$$\frac{n_c = 3(y + \dot{y} t_{go} + 0.5 n_T t_{go}^2)}{t_{go}^2}$$

where

$$t_{go} = t_F - t$$

We can see that the "optimal" guidance law is simply augmented proportional navigation with an effective navigation ratio of 3. The effective navigation ratio turns out to be 3 because we are minimizing the integral of the square of the acceleration. If we were to minimize another function of acceleration, we would get a different answer for the optimal effective navigation ratio. It is still important to note that the optimal guidance law is proportional to the zero effort miss and inversely proportional to the square of time to go.

INFLUENCE OF TIME CONSTANTS

Thus far we have seen that augmented proportional navigation may offer considerable advantages, in terms of required missile acceleration needed to affect an intercept, over the proportional navigation guidance law. It has been demonstrated that, under certain circumstances, augmented proportional navigation may be considered to be an optimal guidance law for a zero-lag guidance system. Let us see how augmented proportional navigation performs when there is a guidance system lag.

Consider a case where the flight time is 10 s, the missile has an effective navigation ratio of 4, and there is a 3-*g* target maneuver. Figure 8.10 displays the resultant commanded acceleration profile for both proportional and augmented proportional navigation. Because the effective navigation ratio is 4, we can compare these results directly with the normalized zero-lag guidance system results of Fig. 8.6. Figure 8.10 indicates that the lag does not change the value of the maximum value of acceleration for both guidance laws. In addition, the lag does not change the fact that augmented proportional navigation requires less acceleration than proportional navigation about 70% of the time. The lag does slightly alter the shape of both acceleration profiles in the sense that the curves are not completely monotonically decreasing (APN) or monotonically increasing (PN).

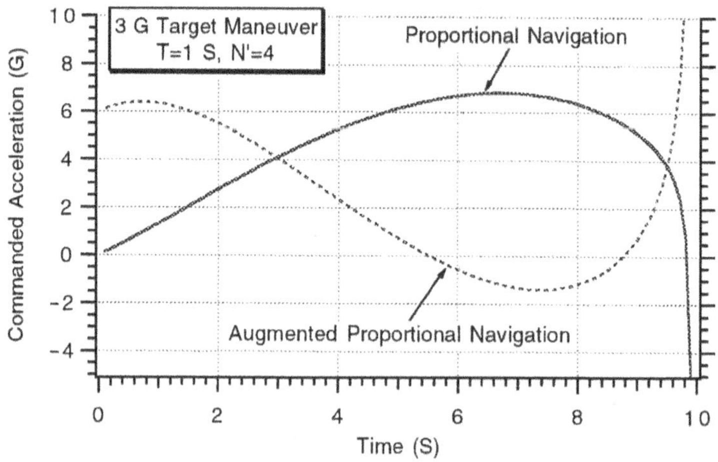

Fig. 8.10 Guidance law acceleration requirements in presence of single-lag guidance system.

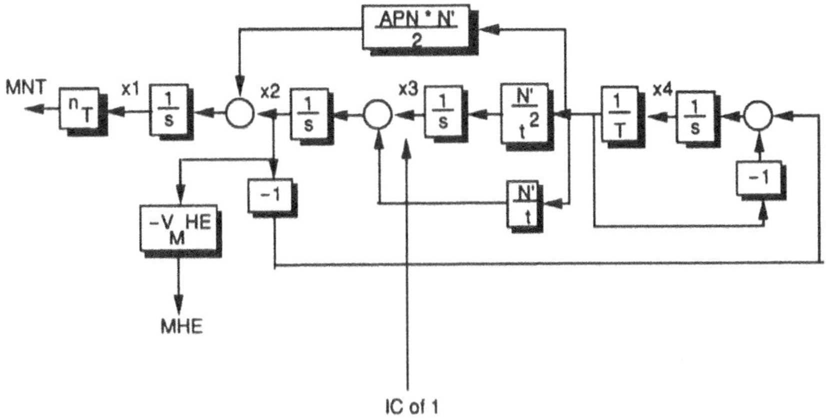

Fig. 8.11 Adjoint for investigating guidance laws in single-lag guidance system.

Having seen that the lag does not change trends in acceleration, let us use the method of adjoints to perform a miss distance sensitivity analysis for both guidance laws in the presence of the lag. The adjoint block diagram of a single time constant system appears in Fig. 8.11. In this diagram we have proportional navigation if APN = 0 and augmented proportional navigation if APN = 1.

The MATLAB listing for the adjoint simulation appears in Listing 8.1. The guidance system time constant is represented in MATLAB by TAU. The listing shows that an initial condition of unity is applied to the $x3$ integrator to make

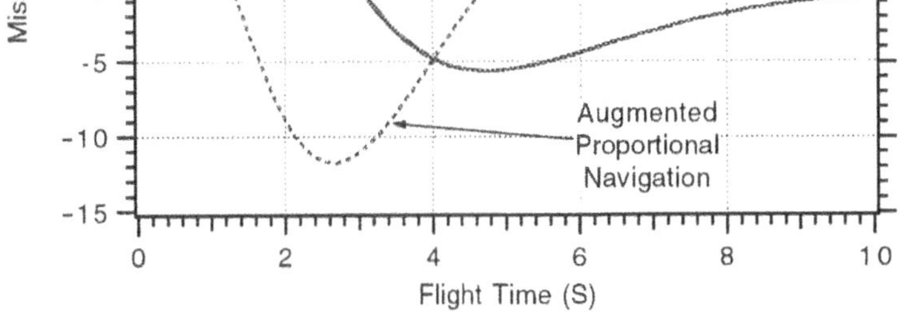

Fig. 8.12 Both guidance laws are comparable in terms of miss distance.

a miss distance adjoint. The four adjoint differential equations can be found before the FLAG=1 statement.

Two adjoint runs were made in which proportional navigation (APN = 0) and augmented proportional navigation (APN = 1) were used. The error source was a 3-*g* target maneuver in the presence of a 1-s flight-control system time constant. The value of the effective navigation ratio was 4. We can see from Fig. 8.12 that neither guidance law is superior from a miss distance point of view (assuming the system is linear and we do not have acceleration saturation). Augmented proportional navigation yields smaller miss distances for shorter flight times, whereas proportional navigation yields smaller miss distances for longer flight times.

LISTING 8.1 ADJOINT SIMULATION OF SINGLE-LAG GUIDANCE SYSTEM

```
XNT=96.6;
XNP=4.;
TAU=1.;
TF=10.;
VM=3000.;
HEDEG=-20.;
APN=1;
T=0.;
S=0.;
TP=T+.00001;
X1=0;
X2=0;
X3=1;
X4=0;
H=.01;
HE=HEDEG/57.3;
n=0.;
while TP<=(TF-1e-5)
        X1OLD=X1;
        X2OLD=X2;
        X3OLD=X3;
        X4OLD=X4;
        STEP=1;
        FLAG=0;
        while STEP<=1
           if FLAG==1
           STEP=2;
           X1=X1+H*X1D;
           X2=X2+H*X2D;
           X3=X3+H*X3D;
           X4=X4+H*X4D;
```

```
            TP=TP+H;
            end
            X1D=X2+X4*XNP*APN/(2.*TAU);
            X2D=X3+XNP*X4/(TAU*TP);
            X3D=XNP*X4/(TAU*TP*TP);
            X4D=-X4/TAU-X2;
            FLAG=1;
        end
        FLAG=0;
        X1=(X1OLD+X1)/2+.5*H*X1D;
        X2=(X2OLD+X2)/2+.5*H*X2D;
        X3=(X3OLD+X3)/2+.5*H*X3D;
        X4=(X4OLD+X4)/2+.5*H*X4D;
        S=S+H;
        if S>=.0999
            S=0.;
            n=n+1;
            ArrayTP(n)=TP;
            ArrayXMNT(n)=XNT*X1;
            ArrayXMHE(n)=-VM*HE*X2;
        end
end
figure
plot(ArrayTP,ArrayXMNT),grid
xlabel('Flight Time (Sec)')
ylabel('Target Maneuver Miss (Ft)')
clc
output=[ArrayTP',ArrayXMNT',ArrayXMHE'];
save datfil.txt output  -ascii
disp 'simulation finished'
```

Of course in a zero-lag system both guidance laws would always yield zero miss distance. So we can see that, although the lag does not significantly alter the acceleration signature, it does cause miss distance!

OPTIMAL GUIDANCE [1, 3, 4]

We have observed that by making use of target acceleration information we could derive a guidance law to reduce the missile acceleration requirements. It has been demonstrated in the previous example that guidance system lags cause miss distance. Generally, larger guidance system time constants yield larger miss distances (except for parasitic effects and some types of noise disturbances). If we had knowledge of the guidance system dynamics, could we derive a guidance law to eliminate miss distance? Mathematically speaking, the answer is yes!

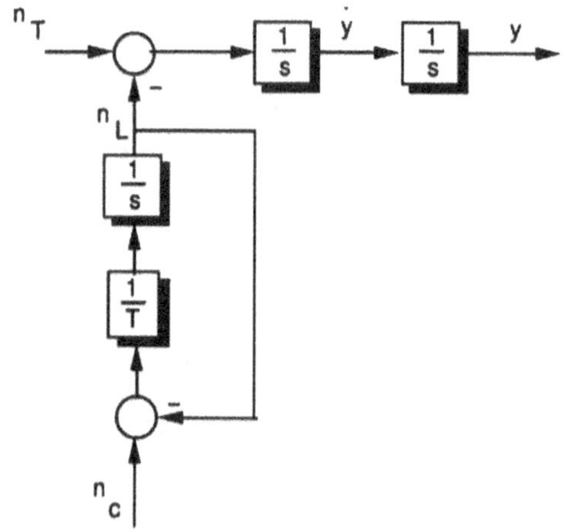

Fig. 8.13 Single-lag model for guidance law development.

A single-lag guidance system model for guidance law development is presented in Fig. 8.13. This model is identical to the one of Fig. 8.9, except that the guidance system dynamics has been represented by a single lag, or

$$\frac{n_L}{n_c} = \frac{1}{1+sT}$$

where n_c is the commanded acceleration, n_L the achieved acceleration, and T the guidance system or flight-control system time constant.

Figure 8.13 can be expressed in state space form as

$$\begin{bmatrix} \dot{y} \\ \ddot{y} \\ \dot{n}_T \\ \dot{n}_L \end{bmatrix} = \begin{bmatrix} 0 & 1 & 0 & 0 \\ 0 & 0 & 1 & -1 \\ 0 & 0 & 0 & 0 \\ 0 & 0 & 0 & \frac{-1}{T} \end{bmatrix} \begin{bmatrix} y \\ \dot{y} \\ n_T \\ n_L \end{bmatrix} + \begin{bmatrix} 0 \\ 0 \\ 0 \\ \frac{1}{T} \end{bmatrix} n_c$$

Therefore, the fundamental matrix can be found to be

$$\Phi(t) = \begin{bmatrix} 1 & t & 0.5t^2 & -tT + T^2(1 - e^{-t/T}) \\ 0 & 1 & t & -T(1 - e^{-t/T}) \\ 0 & 0 & 1 & 0 \\ 0 & 0 & 0 & e^{-t/T} \end{bmatrix}$$

Recall that we still seek a guidance law that yields zero miss subject to minimizing the integral of the square of the commanded acceleration, or

$$y(t_F) = 0 \qquad \text{subject to minimizing} \int_0^{t_F} n_c^2(t)\,\mathrm{d}t$$

Using a procedure similar to that used in the previous section, we can derive the important quantities

$$f_1(t_F - t) = y + \dot{y}(t_F - t) + 0.5 n_T (t_F - t)^2 - T^2 n_L \left[e^{-(t_F - t)/T} + \frac{(t_F - t)}{T} - 1 \right]$$

$$h_1(t_F - \lambda) = t_F - \lambda - T[1 - e^{-(t_F - \lambda)/T}]$$

Calculating

$$\int_t^{t_F} h_1^2(t_F - \lambda)\, d\lambda = T^3\left(0.5 - 0.5e^{-2t_{go}/T} - \frac{2t_{go}e^{-t_{go}/T}}{T} - \frac{t_{go}^2}{T^2} + \frac{t_{go}}{T} + \frac{t_{go}^3}{3T^3}\right)$$

and defining

$$x = \frac{t_{go}}{T}$$

we obtain the optimal guidance law

$$n_c = \frac{N'}{t_{go}^2}[y + \dot{y}t_{go} + 0.5n_T t_{go}^2 - n_L T^2(e^{-x} + x - 1)]$$

where the bracketed quantity is the zero effort miss, and the effective navigation ratio is no longer a constant but is related to the guidance system time constant and time to go by the relationship

$$N' = \frac{6x^2(e^{-x} - 1 + x)}{2x^3 + 3 + 6x - 6x^2 - 12xe^{-x} - 3e^{-2x}}$$

The effective navigation ratio for the optimal guidance law is displayed in normalized form in Fig. 8.14. We can see that at the beginning of the flight (long time to go) the effective navigation ratio is approximately constant and is approaching 3. As we get closer to intercept (small time to go), the effective navigation ratio grows considerably.

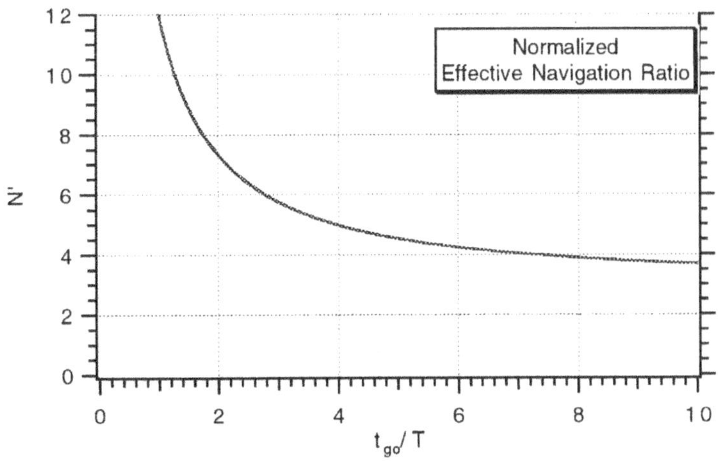

Fig. 8.14 Normalized effective navigation ratio for optimal guidance law.

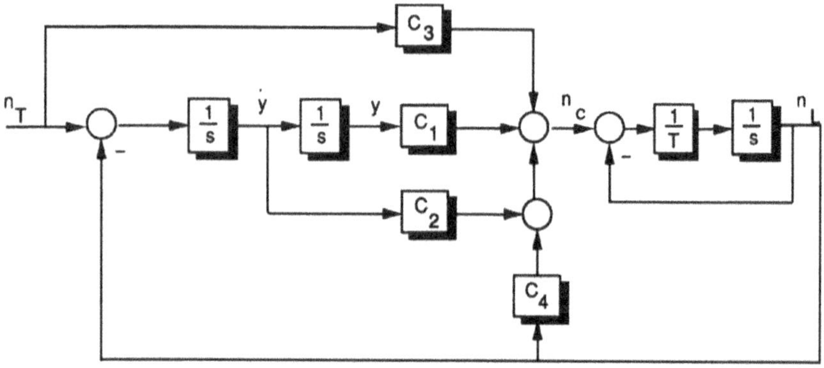

Fig. 8.15 Theoretical optimal single time constant guidance system.

In Fig. 8.15 a theoretically optimal guidance law is implemented in a single-lag guidance system. It is assumed that precise knowledge of the target and missile acceleration is available. The only error disturbance shown in this guidance system is target maneuver n_T.

The guidance law has been represented with control gains $C_1 - C_4$. These gains are functions of the time to go to intercept and the guidance system time constant. They are defined as

$$C_1 = \frac{N'}{t_{\text{go}}^2}$$

$$C_2 = \frac{N'}{t_{\text{go}}}$$

$$C_3 = 0.5N'$$

$$C_4 = \frac{-N'(e^{-x} + x - 1)}{x^2}$$

where N' is the optimal effective navigation ratio, which has been defined previously.

A case was run for a 1-s guidance system time constant in which the flight time was 10 s and the error disturbance was a 3-*g* target maneuver. Figure 8.16 shows the acceleration profile for proportional navigation ($N' = 4$), augmented proportional navigation ($N' = 4$), and optimal guidance. If we compare this figure with Fig. 8.5 we can see that the optimal guidance acceleration profile appears to be identical to the augmented proportional navigation acceleration profile *for a zero-lag guidance system and an effective navigation ratio of 3!* This means that the guidance law is dynamically canceling out the guidance system dynamics.

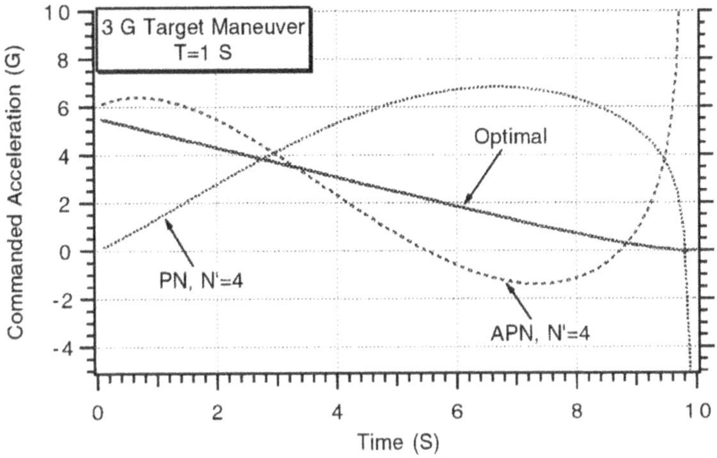

Fig. 8.16 Acceleration comparison for various guidance laws.

If the optimal guidance law were attempting to make the single-lag guidance system appear to be a zero-lag augmented proportional navigation guidance system, then the miss distance should be zero—just as it is in a zero-lag system. To test this theory, an adjoint block diagram of a single-lag optimal guidance system was constructed from Fig. 8.15 and appears in Fig. 8.17. The control

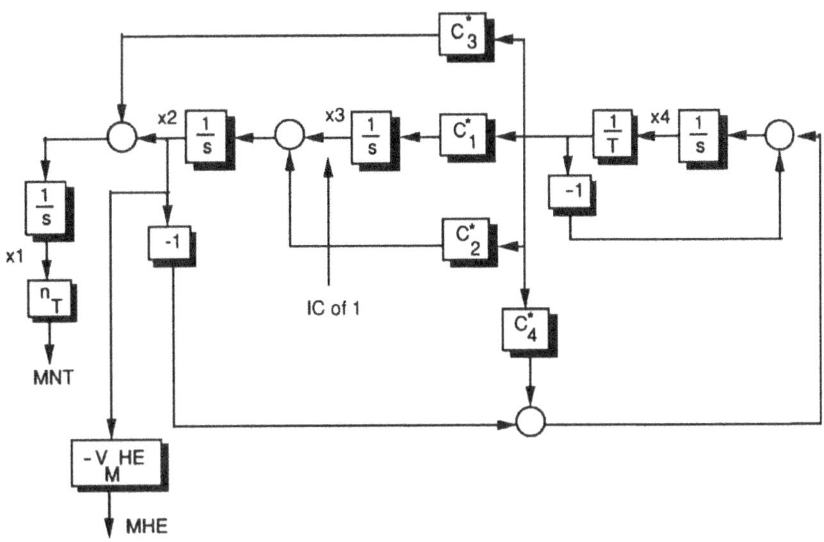

Fig. 8.17 Adjoint of theoretical optimal single time constant guidance system.

gains become C^* because they must be reversed in time according to adjoint theory. The miss sensitivities due to target maneuver MNT and heading error MHE are indicated in the figure.

The MATLAB listing of the adjoint simulation of the optimal single time constant guidance system with various guidance law options appears in Listing 8.2. We can see from the listing that the parameter APN determines the guidance law used. APN = 0 denotes proportional navigation, APN = 1 represents augmented proportional navigation, and APN = 2 defines optimal guidance.

LISTING 8.2 ADJOINT SIMULATION OF OPTIMAL GUIDANCE SYSTEM

```
XNT=96.6;
XNP=4.;
TAU=1.;
TF=10.;
VM=3000.;
HEDEG=-20.;
APN=0;
T=0.;
S=0.;
TP=T+.00001;
X1=0.;
X2=0.;
X3=1.;
X4=0.;
XNPP=0.;
H=.01;
HE=HEDEG/57.3;
n=0.;
while TP<=(TF-1e-5)
        X1OLD=X1;
        X2OLD=X2;
        X3OLD=X3;
        X4OLD=X4;
        STEP=1;
        FLAG=0;
        while STEP<=1
          if FLAG==1
          STEP=2;
          X1=X1+H*X1D;
          X2=X2+H*X2D;
          X3=X3+H*X3D;
          X4=X4+H*X4D;
          TP=TP+H;
          end
```

```
            TGO=TP+.00001;
            if APN==0
            C1=XNP/(TGO*TGO);
            C2=XNP/TGO;
            C3=0.;
            C4=0.;
            elseif APN==1
            C1=XNP/(TGO*TGO);
            C2=XNP/TGO;
            C3=.5*XNP;
            C4=0.;
            else
            X=TGO/TAU;
            TOP=6.*X*X*(exp(-X)-1.+X);
            BOT1=2*X*X*X+3.+6.*X-6.*X*X;
            BOT2=-12.*X*exp(-X)-3.*exp(-2.*X);
            XNPP=TOP/(.0001+BOT1+BOT2);
            C1=XNPP/(TGO*TGO);
            C2=XNPP/TGO;
            C3=.5*XNPP;
            C4=-XNPP*(exp(-X)+X-1.)/(X*X);
            end
            X1D=X2+C3*X4/TAU;
            X2D=X3+C2*X4/TAU;
            X3D=C1*X4/TAU;
            X4D=-X4/TAU-X2+C4*X4/TAU;
            FLAG=1;
        end
        FLAG=0;
        X1=(X1OLD+X1)/2+.5*H*X1D;
        X2=(X2OLD+X2)/2+.5*H*X2D;
        X3=(X3OLD+X3)/2+.5*H*X3D;
        X4=(X4OLD+X4)/2+.5*H*X4D;
        S=S+H;
        if S>=.0999
            S=0.;
            n=n+1;
            ArrayTP(n)=TP;
            ArrayXMNT(n)=XNT*X1;
            ArrayXMHE(n)=-VM*HE*X2;
        end
end
figure
plot(ArrayTP,ArrayXMNT),grid
xlabel('Flight Time (Sec)')
ylabel('Target Maneuver Miss (Ft)')
```

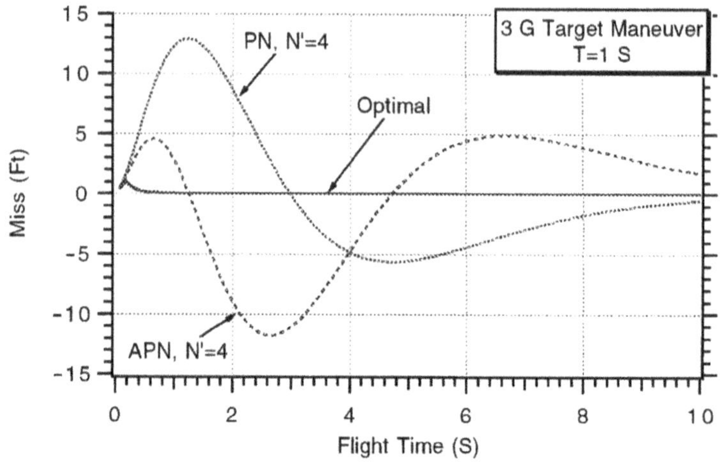

Fig. 8.18 Optimal guidance system does not have miss distance.

```
clc
output=[ArrayTP',ArrayXMNT',ArrayXMHE'];
save datfil.txt output -ascii
disp 'simulation finished'
```

Figure 8.18 shows the miss sensitivity of all the guidance laws to a 3-g step target maneuver in the presence of a single-lag guidance system with a time constant of 1 s. We can see that the optimal guidance law always yields zero miss distance. Therefore, as predicted, the optimal guidance system is attempting to make the single time constant guidance system appear to be a zero-lag augmented proportional navigation guidance system with an effective navigation ratio of 3. The interested reader is referred to the appendix to see how the optimal guidance system performs when time to go information is degraded.

SUMMARY

In this chapter we have shown how some advanced guidance laws can be derived both heuristically and mathematically. The method of adjoints was used to show the performance advantages of the more advanced guidance laws. In practice, one must also test the advanced guidance concepts in the presence of parasitic effects to ensure that performance is still better than proportional navigation [1].

REFERENCES

[1] Nesline, F. W., and Zarchan, P., "A New Look at Classical Versus Modern Homing Guidance," *Journal of Guidance and Control*, Vol. 4, No. 1, Jan.–Feb. 1981, pp. 78–85.

[2] Bryson, A. E., and Ho, Y. C., *Applied Optimal Control*, Blaisdell, Waltham, MA, 1969.

[3] Cottrell, R. G., "Optimal Intercept Guidance for Short-Range Tactical Missiles," *AIAA Journal*, Vol. 9, July 1971, pp. 1414–1415.

[4] Kliger I., "A Simple Derivation of Certain Optimal Control Laws," Raytheon, Memo SAD-1230, Bedford, MA, Nov. 1970.

[5] Gelb, A., *Applied Optimal Estimation*, MIT Press, Cambridge, MA, 1974.

CHAPTER 9

Kalman Filters and the Homing Loop

INTRODUCTION

Kalman filtering concepts can be used in the homing loop to produce an optimal digital noise filter. The filter is considered optimal because the errors in the estimates of the system states are minimized in the least-squares sense. We shall see that, for the filter to be truly optimal, the statistics of the measurement and process noise must be known. If this information is lacking or inaccurate, the filter performance will degrade. However, we shall also see that in homing loop applications the Kalman filter cannot only perform very well with significant errors in the knowledge of the statistics, but it may even be desirable to lie to the filter to achieve a desired filter bandwidth. Finally, we shall demonstrate that when Kalman filtering concepts are used it is possible to apply advanced guidance techniques and substantially improve system performance.

THEORETICAL EQUATIONS [1]

For linear systems Kalman filters can provide optimal estimators in the least-squares sense. To apply Kalman filtering theory, our model of the real world must be described by a matrix differential equation of the form

$$\dot{\boldsymbol{x}} = F\boldsymbol{x} + G\boldsymbol{u} + \omega$$

where $\boldsymbol{x}$ is a column vector describing the states of the system, F the system dynamics matrix, $\boldsymbol{u}$ a known control vector, and ω a white noise process. There is a process noise matrix Q that is related to the process noise vector according to

$$Q = E[\boldsymbol{w}\,\boldsymbol{w}^T]$$

In other words, Q is the expectation of the white process noise times its transpose. The filter will be optimal if the measurements available are linearly related to the states according to

$$z = Hx + v$$

where z is the measurement vector, H the measurement matrix, and v the white noise measurement. The measurement noise matrix R is related to the measurement noise vector v according to

$$R = E[v\,v^T]$$

The preceding relationships are valid for continuous systems. Since we are not taking measurements continuously but plan to receive information every T_s seconds, we need to discretize our system model. The fundamental matrix Φ is related to the system dynamics matrix according to

$$\Phi(t) = \mathcal{L}^{-1}\{[sI - F]^{-1}\}$$

where I is the identity matrix and $\mathcal{L}^{-1}$ the inverse Laplace transform. For discrete systems we can say that the discrete transition matrix is given by

$$\Phi_K = \Phi(T_s)$$

where T_s is the sampling time. In other words, the discrete fundamental matrix is simply the continuous fundamental matrix evaluated at the sampling time. The discrete form of the measurement equation is now

$$z_k = H\boldsymbol{x}_k + v_k$$

and

$$R_k = \sigma_n^2$$

where σ_n^2 is the variance of the measurement noise. The resultant form of the discrete Kalman filter is given by the recursive relationship in matrix form

$$\hat{\boldsymbol{x}}_k = \Phi_k\hat{\boldsymbol{x}}_{k-1} + G_k\boldsymbol{u}_{k-1} + K_k(z_k - \boldsymbol{H}\Phi_k\hat{\boldsymbol{x}}_{k-1} - \boldsymbol{H}G_k\boldsymbol{u}_{k-1})$$

where $\boldsymbol{G}_k$ is obtained from

$$\boldsymbol{G}_k = \int_0^{T_s} \Phi(\tau)\boldsymbol{G}\,\mathrm{d}\tau$$

and K_k represents the Kalman gain matrix. The Kalman gains are computed, while the filter is operating, from the matrix Ricatti equations. The Ricatti equations are a set of recursive matrix equations given by

$$M_k = \Phi_k P_{k-1}\Phi_k^T + Q_k$$

$$K_k = M_k H^T[HM_kH^T + R_k]^{-1}$$

$$P_k = (I - K_kH)M_k$$

where P_k is a covariance matrix representing errors in the state estimates before an update, and M_k is the covariance matrix representing errors in the state estimates after an update. The discrete process noise matrix Q_k can be found from the continuous process noise matrix Q and the fundamental matrix according to

$$Q_k = \int_0^{T_s} \Phi(\tau) Q \Phi^T(\tau)\, d\tau$$

To start the Ricatti equations, we need an initial covariance matrix P_0. This matrix represents the initial uncertainty in the error in the estimate. Choosing appropriate values for this initial matrix is in itself an art [1].

APPLICATION TO HOMING LOOP

To demonstrate the utility of Kalman filtering, let us consider the zero-lag homing loop model of Fig. 9.1. In this guidance system we measure noisy relative position y^* and are attempting to estimate relative position, relative velocity, and target acceleration. In our model the missile acceleration n_c is assumed to be known, and the target acceleration is considered to be modeled as a white noise through an integrator. We have shown in Chapter 4 mathematically that the shaping filter equivalent of a target maneuver with constant amplitude but random starting time (where the starting time is uniformly distributed over the flight time) is white noise through an integrator. According to the results of Chapter 4, the spectral density of this white noise process is given by

$$\Phi_s = n_T^2 / t_F$$

where n_T is the maneuver level and t_F the flight time. In Chapter 4 we also showed via a numerical experiment that this model is statistically equivalent to a maneuver of constant amplitude whose starting time is equally likely to occur anywhere during the flight.

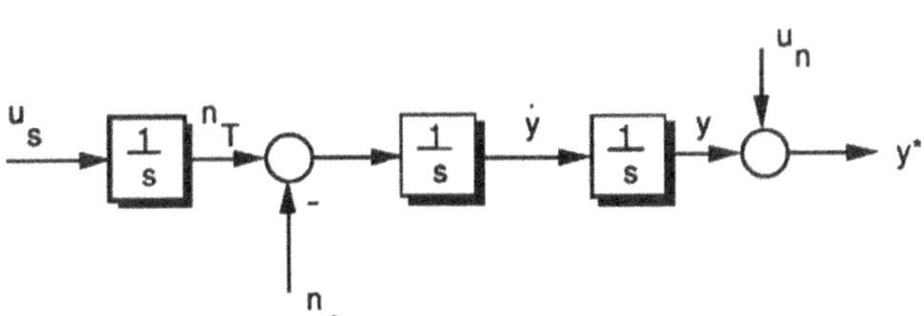

Fig. 9.1 Homing loop model for Kalman filter development.

We can express the model of Fig. 9.1 in state space form as

$$\begin{bmatrix} \dot{y} \\ \ddot{y} \\ \dot{n}_T \end{bmatrix} = \underbrace{\begin{bmatrix} 0 & 1 & 0 \\ 0 & 0 & 1 \\ 0 & 0 & 0 \end{bmatrix}}_{F} \begin{bmatrix} y \\ \dot{y} \\ n_T \end{bmatrix} + \underbrace{\begin{bmatrix} 0 \\ -1 \\ 0 \end{bmatrix}}_{G} n_c + \underbrace{\begin{bmatrix} 0 \\ 0 \\ u_s \end{bmatrix}}_{w}$$

In the previous section we showed that the fundamental matrix could be found from the system dynamics matrix. After some computation the fundamental matrix for the model of Fig. 9.1 turns out to be

$$\Phi(t) = \begin{bmatrix} 1 & t & 0.5t^2 \\ 0 & 1 & t \\ 0 & 0 & 1 \end{bmatrix}$$

or in discrete form

$$\Phi_k = \begin{bmatrix} 1 & T_s & 0.5T_s^2 \\ 0 & 1 & T_s \\ 0 & 0 & 1 \end{bmatrix}$$

The measurement equation can also be expressed in discrete form as

$$y_k^* = \underbrace{[100]}_{H} \begin{bmatrix} y_k \\ \dot{y}_k \\ n_{T_k} \end{bmatrix} + \boldsymbol{u}_k$$

where the variance of $\boldsymbol{u}_k$, known as R_k, is given by σ_n^2. The discrete form of $\boldsymbol{G}$ can be found from

$$\boldsymbol{G}_k = \int_0^{T_s} \Phi(\tau)\boldsymbol{G}(\tau)\,\mathrm{d}\tau = \begin{bmatrix} -0.5T_s^2 \\ -T_s \\ 0 \end{bmatrix}$$

The Kalman filter for the model of Fig. 9.1 can now be expressed in matrix form as

$$\begin{bmatrix} \hat{y}_k \\ \hat{\dot{y}}_k \\ \hat{n}_{Tk} \end{bmatrix} = \begin{bmatrix} 1 & T_s & 0.5T_s^2 \\ 0 & 1 & T_s \\ 0 & 0 & 1 \end{bmatrix} \begin{bmatrix} \hat{y}_{k-1} \\ \hat{\dot{y}}_{k-1} \\ \hat{n}_{T_{k-1}} \end{bmatrix} + \begin{bmatrix} -0.5T_s^2 \\ -T_s \\ 0 \end{bmatrix} n_{c_{k-1}}$$

$$+ \begin{bmatrix} K_1 \\ K_2 \\ K_3 \end{bmatrix} \left[y_k^* - [100] \begin{bmatrix} 1 & T_s & 0.5T_s^2 \\ 0 & 1 & T_s \\ 0 & 0 & 1 \end{bmatrix} \begin{bmatrix} \hat{y}_{k-1} \\ \hat{\dot{y}}_{k-1} \\ \hat{n}_{T_{k-1}} \end{bmatrix} - [100] \begin{bmatrix} -0.5T_s^2 \\ -T_s \\ 0 \end{bmatrix} n_{c_{k-1}} \right]$$

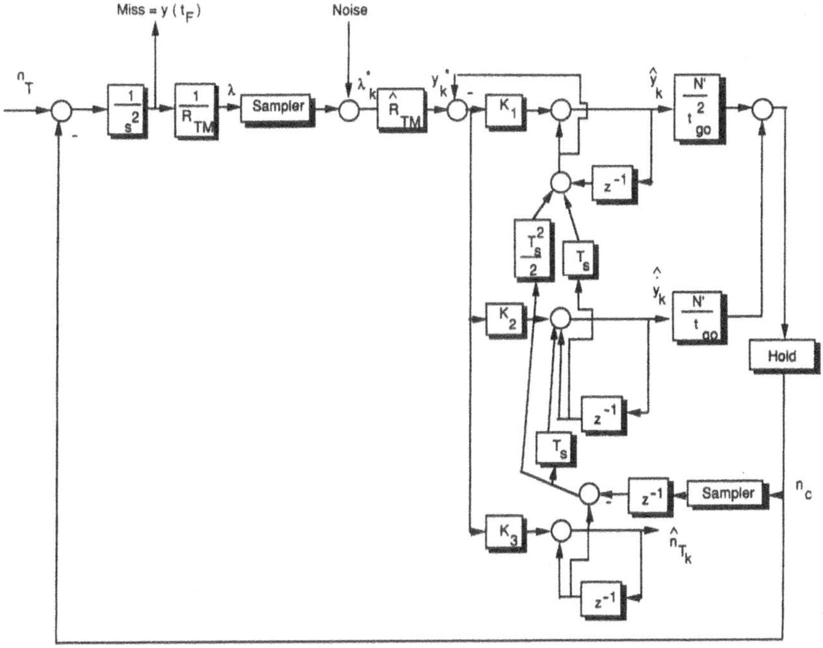

Fig. 9.2 Kalman filter as part of homing loop.

This Kalman filter is shown in block diagram form as part of the homing loop in Fig. 9.2. In this diagram z^{-1} represents a pure delay so that $z^{-1}y_k$ means y_{k-1}. In our model the measurement of the line-of-sight angle λ_k^* is corrupted by noise. We create a pseudomeasurement of relative position y_k^* by a multiplication of the line-of-sight angle measurement by our estimate or measurement of the range from missile to target. The Kalman filter then provides optimal estimates of relative position, relative velocity, and target acceleration. In this model we are using proportional navigation guidance where the guidance command is related to the state estimates according to

$$n_{ck}|_{\text{PN}} = \frac{N'}{t_{\text{go}}^2}\hat{y}_k + \frac{N'}{t_{\text{go}}}\hat{\dot{y}}_k$$

It is easy to show that this command is mathematically equivalent to the more recognizable form of proportional navigation, or

$$n_{ck}|_{\text{PN}} = N' V_c \hat{\dot{\lambda}}$$

KALMAN GAINS

In order for the Kalman filter to operate, we need to first compute the filter gains K_k. These gains are obtained from a set of recursive equations known as the matrix Ricatti equations, which were stated in the first section. The first of the Ricatti equations is

$$M_k = \Phi_k P_{k-1} \Phi_k^T + Q_k$$

where M_k represents the covariance matrix of errors in the estimates after up-dates. For the three-state system of Fig. 9.2, this matrix can be expanded in scalar form by multiplying out the matrices and by recognizing that M_k is symmetric. Substitution of the necessary matrices yields

$$M = \begin{bmatrix} 1 & T_s & 0.5T_s^2 \\ 0 & 1 & T_s \\ 0 & 0 & 1 \end{bmatrix} \begin{bmatrix} P_{11} & P_{12} & P_{13} \\ P_{12} & P_{22} & P_{23} \\ P_{13} & P_{23} & P_{33} \end{bmatrix} \begin{bmatrix} 1 & 0 & 0 \\ T_s & 1 & 0 \\ 0.5T_s^2 & T_s & 1 \end{bmatrix} + \Phi_s \begin{bmatrix} \frac{T_s^5}{20} & \frac{T_s^4}{8} & \frac{T_s^3}{6} \\ \frac{T_s^4}{8} & \frac{T_s^3}{3} & \frac{T_s^2}{2} \\ \frac{T_s^3}{6} & \frac{T_s^2}{2} & T_s \end{bmatrix}$$

The second Ricatti equation is used to obtain the Kalman gains. It appears from the second Ricatti equation,

$$K_k = M_k H^T [H M_k H^T + R_k]^{-1}$$

that it is necessary to take a matrix inverse. However, for the model of Fig. 9.1, the R_k matrix is 1×1; hence, we can take the scalar inverse by inspection and obtain

$$K_1 = \frac{M_{11}}{M_{11} + \sigma_N^2}$$

$$K_2 = \frac{M_{12}}{M_{11} + \sigma_N^2}$$

$$K_3 = \frac{M_{13}}{M_{11} + \sigma_N^2}$$

Finally, the third Ricatti equation is used to obtain the covariance matrix of the errors in the estimates before an update. The third equation,

$$P_k = (I - K_k H) M_k$$

can easily be expanded to

$$P = \begin{bmatrix} (1-K_1)M_{11} & (1-K_1)M_{12} & (1-K_1)M_{13} \\ -K_2M_{11}+M_{12} & -K_2M_{12}+M_{22} & -K_2M_{13}+M_{23} \\ -K_3M_{11}+M_{13} & -K_3M_{12}+M_{23} & -K_3M_{13}+M_{33} \end{bmatrix}$$

NUMERICAL EXAMPLES

To start the Ricatti equations, we need an initial covariance matrix P_0. A particularly useful form for the homing loop model considered is

$$P_0 = \begin{bmatrix} \sigma^2_{\text{noise}} & 0 & 0 \\ 0 & \left[\dfrac{V_M HE}{57.3}\right]^2 & 0 \\ 0 & 0 & n_T^2 \end{bmatrix}$$

where only diagonal elements are used. The initial value of the error in the estimate of position is simply the variance of the measurement noise. The initial guess in the velocity error estimate is related to missile velocity and the expected heading error. Finally, our initial value in the uncertainty in target acceleration is represented by the magnitude of the maximum possible acceleration. This is by no means the only way to initialize the covariance matrix, but it is not bad.

Listing 9.1 presents a MATLAB listing of a program used to solve the Ricatti equations recursively for the Kalman gains. In this program it is assumed that the angular measurement noise is 1 milliradian (mr). This noise must be converted to a positional noise by the multiplication of range. The process noise model is considered to be a target maneuver of amplitude 3 g, with starting time that is uniformly distributed over the 10-s flight time. We can see from the listing that the Ricatti equations have been expanded to scalar form and that the symmetry property of the Ricatti equations has been exploited.

Figure 9.3 displays the three Kalman gain profiles resulting from solving the Ricatti equations with initial conditions as shown in Listing 9.1. We can see that, unlike the constant-gain digital fading memory filter, the Kalman filter has time-varying gains. After an initial transient period, the gains appear to be monotonically increasing. This means that, after awhile, the filter bandwidth is continually increasing.

To see how the filter is performing, we must not only look at the filter gains but must also investigate the accuracy of the various state estimates. The covariance matrix has information on the accuracy of the state estimates if the filter's model of the real world is accurate. If the filter model is not matched to the real world, then the performance projections offered by the covariance matrix are not particularly useful. One way of getting more meaningful performance projections is by placing the three-state Kalman filter in the homing loop.

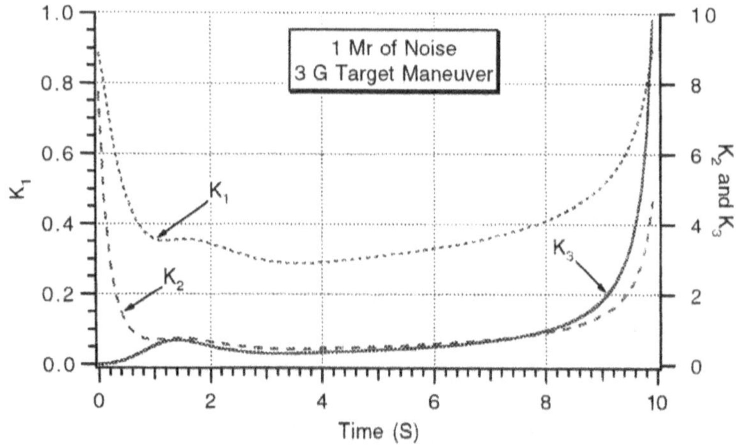

Fig. 9.3 Kalman gain profiles for nominal case.

Listing 9.2 presents an engagement simulation, based upon the model of Fig. 9.2, with the three-state Kalman filter included. In the nominal case we are not using the estimate of the target acceleration for guidance purposes. A careful examination of the listing shows that the simulation is divided into continuous and discrete parts. In the continuous section we are integrating the differential equations for the relative velocity and acceleration using the second-order Runge–Kutta numerical integration technique. In the discrete section we are solving the Ricatti equations for the Kalman gains and using the recursive Kalman filter to generate state estimates. We go to the continuous section every integration interval H, and we go to the discrete section every sampling interval T_s. For the simulation to work properly, T_s/H must be an integer.

LISTING 9.1 LISTING OF MATLAB PROGRAM TO SOLVE RICATTI EQUATIONS

```
VC=4000.;
XNT=96.6;
VM=3000.;
HEDEG=20.;
SIGRIN=.001;
TS=.1;
TF=10.;
TS2=TS*TS;
TS3=TS2*TS;
TS4=TS3*TS;
TS5=TS4*TS;
PHIN=XNT*XNT/TF;
```

```
RTM=VC*TF;
SIGNOISE=SIGRIN;
SIGPOS=RTM*SIGNOISE;
SIGN2=SIGPOS^2;
P11=SIGN2;
P12=0.;
P13=0.;
P22=(VM*HEDEG/57.3)^2;
P23=0.;
P33=XNT*XNT;
T=0.;
H=.01;
S=0.;
n=0;
while T < =(TF-1e-5)
        TGO=TF-T+.000001;
        RTM=VC*TGO;
        SIGNOISE=SIGRIN;
        SIGPOS=RTM*SIGNOISE;
        SIGN2=SIGPOS^2;
        M11=P11+TS*P12+.5*TS2*P13+TS*(P12+TS*P22+.5*TS2*P23);
        M11=M11+.5*TS2*(P13+TS*P23+.5*TS2*P33)+TS5*PHIN/20.;
        M12=P12+TS*P22+.5*TS2*P23+TS*(P13+TS*P23+.5*TS2*P33)+TS4*PHIN/8.;
        M13=P13+TS*P23+.5*TS2*P33+PHIN*TS3/6.;
        M22=P22+TS*P23+TS*(P23+TS*P33)+PHIN*TS3/3.;
        M23=P23+TS*P33+.5*TS2*PHIN;
        M33=P33+PHIN*TS;
        K1=M11/(M11+SIGN2);
        K2=M12/(M11+SIGN2);
        K3=M13/(M11+SIGN2);
        P11=(1.-K1)*M11;
        P12=(1.-K1)*M12;
        P13=(1.-K1)*M13;
        P22=-K2*M12+M22;
        P23=-K2*M13+M23;
        P33=-K3*M13+M33;
        n=n+1;
        ArrayT(n)=T;
        ArrayK1(n)=K1;
        ArrayK2(n)=K2;
        ArrayK3(n)=K3;
        T=T+TS;
end

figure
plot(ArrayT,ArrayK1),grid
```

```
xlabel('Flight Time (Sec)')
ylabel('K1')
clc
output=[ArrayT',ArrayK1',ArrayK2',ArrayK3'];
save datfil.txt output -ascii
disp 'simulation finished'
```

Figure 9.4 shows that in the nominal case the Kalman filter accurately estimates the 3-g target maneuver after about 3 s. This is consistent with Fig. 9.3, which also shows that it takes about 3 s for the Kalman gains to go through their initial transient period. Note that, after about 5 s, the error in the estimate of target acceleration has been stabilized and is quite small, as shown in Fig. 9.4. The filter's internal prediction of how well it is estimating target acceleration can be found by taking the square root of the third diagonal element in the covariance matrix. Figure 9.5 shows that the single flight error in the estimate of target acceleration agrees with the covariance matrix predictions in the sense that it is within the theoretical bounds approximately 68% of the time.

Thus far the filter knows the truth about the real world in the sense that it knows the measurement and process noise statistics exactly. In practice, because these statistics are never known *a priori*, one adjusts the bandwidth of the filter to a desirable level based on other considerations. For example, if we tell the filter that there is 10 mr of angle noise rather than 1 mr, the first Kalman gain value is approximately halved, as shown in Fig. 9.6. As the filter gain is decreasing, thus the filter bandwidth must also be decreasing. This means that, when the filter thinks there is more measurement noise, it does more filtering or slows down (lower bandwidth).

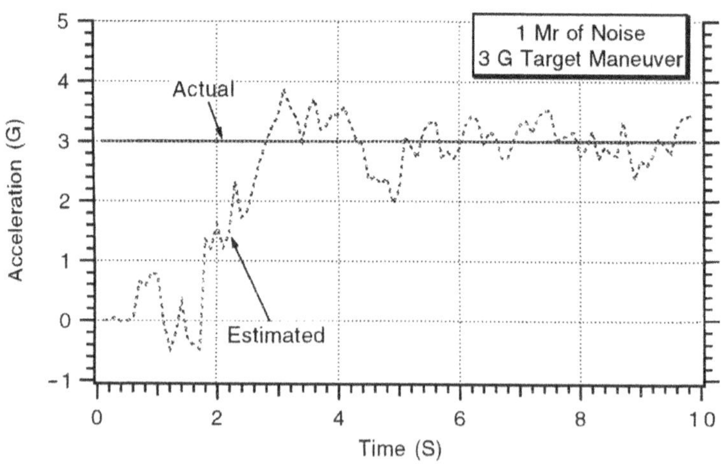

Fig. 9.4 Kalman filter estimate of target maneuver for nominal case.

LISTING 9.2 MATLAB LISTING OF KALMAN FILTER IN HOMING LOOP

```
count=0;
VC=4000.;
XNT=96.6;
YIC=0.;
VM=3000.;
HEDEG=0.;
HEDEGFIL=20.;
XNP=3.;
SIGRIN=.001;
TS=.1;
APN=0.;
TF=10.;
Y=YIC;
YD=-VM*HEDEG/57.3;
YDIC=YD;
TS2=TS*TS;
TS3=TS2*TS;
TS4=TS3*TS;
TS5=TS4*TS;
PHIN=XNT*XNT/TF;
RTM=VC*TF;
SIGNOISE=SIGRIN;
SIGPOS=RTM*SIGNOISE;
SIGN2=SIGPOS^2;
P11=SIGN2;
P12=0.;
P13=0.;
P22=(VM*HEDEGFIL/57.3)^2;
P23=0.;
P33=XNT*XNT;
T=0.;
H=.01;
S=0.;
YH=0.;
YDH=0.;
XNTH=0.;
XNC=0.;
while T <= (TF - 1e-5)
      YOLD=Y;
      YDOLD=YD;
      STEP=1;
      FLAG=0;
      while STEP <=1
        if FLAG==1
```

```
        Y=Y+H*YD;
        YD=YD+H*YDD;
        T=T+H;
        STEP=2;
           end
           TGO=TF-T+.00001;
           RTM=VC*TGO;
           XLAM=Y/(VC*TGO);
           XLAMD=(RTM*YD+Y*VC)/(RTM^2);
           YDD=XNT-XNC;
           FLAG=1;
end
FLAG=0;
           Y=.5*(YOLD+Y+H*YD);
           YD=.5*(YDOLD+YD+H*YDD);
           S=S+H;
           if S>=(TS - 1e-5)
           S=0.;
           TGO=TF-T+.000001;
           RTM=VC*TGO;
           SIGNOISE=SIGRIN;
           SIGPOS=RTM*SIGNOISE;
           SIGN2=SIGPOS^2;
           M11=P11+TS*P12+.5*TS2*P13+TS*(P12+TS*P22+.5*TS2*P23);
           M11=M11+.5*TS2*(P13+TS*P23+.5*TS2*P33)+TS5*PHIN/20.;
M12=P12+TS*P22+.5*TS2*P23+TS*(P13+TS*P23+.5*TS2*P33)+TS4*PHIN/8.;
           M13=P13+TS*P23+.5*TS2*P33+PHIN*TS3/6.;
           M22=P22+TS*P23+TS*(P23+TS*P33)+PHIN*TS3/3.;
           M23=P23+TS*P33+.5*TS2*PHIN;
           M33=P33+PHIN*TS;
           K1=M11/(M11+SIGN2);
           K2=M12/(M11+SIGN2);
           K3=M13/(M11+SIGN2);
           P11=(1.-K1)*M11;
           P12=(1.-K1)*M12;
           P13=(1.-K1)*M13;
           P22=-K2*M12+M22;
           P23=-K2*M13+M23;
           P33=-K3*M13+M33;
              XLAMNOISE=SIGNOISE*randn;
           YSTAR=RTM*(XLAM+XLAMNOISE);
           RES=YSTAR-YH-TS*YDH-.5*TS*TS*(XNTH-XNC);
           YH=K1*RES+YH+TS*YDH+.5*TS*TS*(XNTH-XNC);
           YDH=K2*RES+YDH+TS*(XNTH-XNC);
           XNTH=K3*RES+XNTH;
           XLAMDH=(YH+YDH*TGO)/(VC*TGO*TGO);
           XNC=XNP*VC*XLAMDH+APN*.5*XNP*XNTH;
```

```
            ERRNT=XNT-XNTH;
            SP33=sqrt(P33);
            SP33P=-SP33;
            count=count+1;
               ArrayT(count)=T;
               ArrayXNTG(count)=XNT/32.2;
            ArrayXNTHG(count)=XNTH/32.2;
            ArrayERRNTG(count)=ERRNT/32.2;
            ArraySP33G(count)=SP33/32.2;
            ArraySP33PG(count)=SP33P/32.2;
            ArrayY(count)=Y;
            ArrayXNCG(count)=XNC/32.2;
            end
end
figure
plot(ArrayT,ArrayXNTG,ArrayT,ArrayXNTHG),grid
xlabel('Time (S)')
ylabel('Acceleration (G) ')
figure
plot(ArrayT,ArrayERRNTG,ArrayT,ArraySP33G,ArrayT,ArraySP33PG),grid
xlabel('Time (S)')
ylabel('Error in Acceleration (G) ')
clc
output=[ArrayT',ArrayY',ArrayXNCG',ArrayXNTG',ArrayXNTHG',ArrayERRNTG',
ArraySP33G',ArraySP33PG'];
save datfil.txt output /ascii
disp 'simulation finished'
```

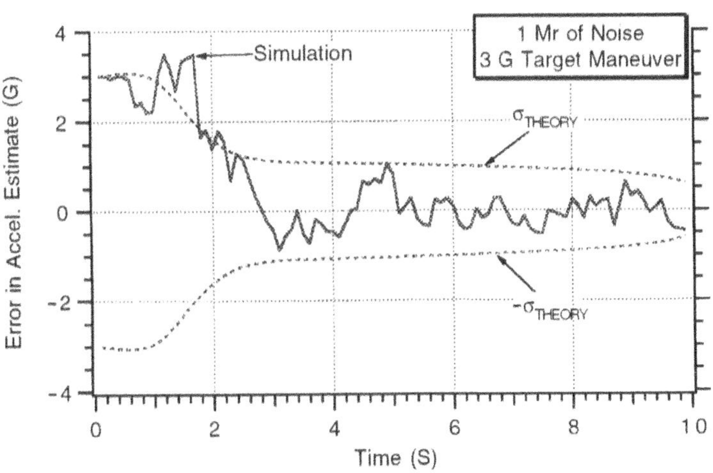

Fig. 9.5 Kalman filter prediction of performance is excellent.

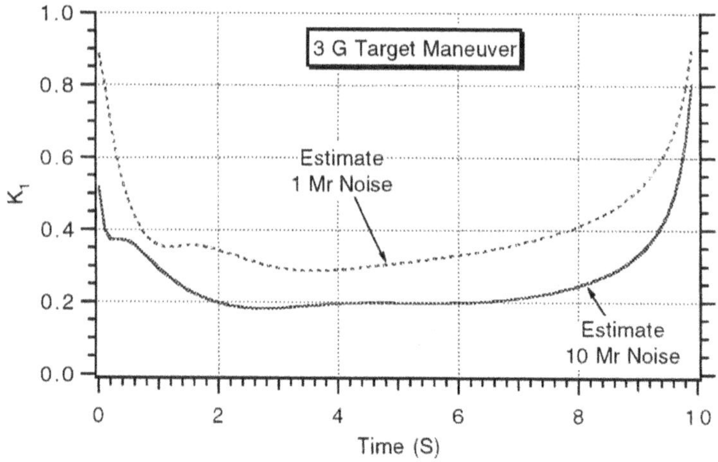

Fig. 9.6 Increasing measurement noise estimate decreases Kalman gain.

An experiment was conducted to illustrate the impact that filter bandwidth has on the resultant estimate. In this experiment the actual noise level was kept at 1 mr, whereas the filter estimate of the measurement noise statistics was changed from 1 mr (matched case, filter assumption correct) to 10 mr (mismatched case, filter assumption wrong). Figure 9.7 shows that when the filter is matched its estimate of target maneuver becomes very good at about 3 s. If the filter thinks there is 10 mr of measurement noise, it takes about 6 s for the filter

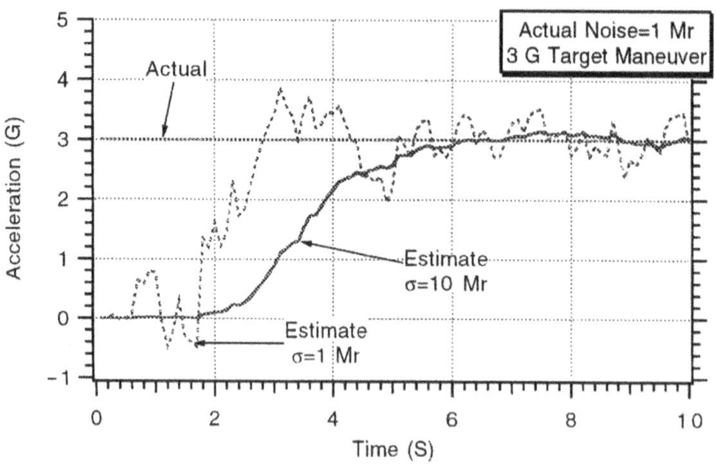

Fig. 9.7 Filter becomes sluggish when it thinks there is more noise.

estimates to become very good. Thus, telling the filter that there is more measurement noise (even if there is not) is a practical way of making the filter more sluggish or decreasing its bandwidth.

If we tell the filter that there is either 1 mr or 10 mr of measurement noise but actually turn the real measurement noise off, we have a deterministic case. We can then make flight time a parameter for the case in which there is a 3-*g* target maneuver and evaluate system miss distance caused by the target maneuver. Figure 9.8 shows that the faster system ($\sigma = 1$ mr) has less miss distance induced by target maneuver than the slower system ($\sigma = 10$ mr). However, in both cases we can see that the guidance system time constant must be very small since the miss distance sensitivity to target maneuver is small and rapidly approaches zero.

If we run our simulation with measurement noise only ($\sigma = 1$ mr and no target maneuver), we must operate in the Monte Carlo mode. Fifty-run Monte Carlo sets were made for 20 different values of flight time for a total of 1000 runs. Figure 9.9 shows how the standard deviation of the noise-induced miss varies with flight time for a case in which the filter is optimal ($\sigma = 1$ mr) and one in which the filter bandwidth has been intentionally decreased ($\sigma = 10$ mr). We can see from Fig. 9.9 that decreasing the filter bandwidth (telling the filter that there is more measurement noise) decreases the system miss distance due to the actual measurement noise (1 mr). This behavior is opposite to that of the two-state digital fading memory filter (see Fig 7.19) in which decreasing the filter bandwidth always appeared to increase miss distance! Of course, the constant gain digital fading memory filter bandwidth was fixed, whereas the Kalman filter bandwidth is time-varying. By comparing Figs. 9.8 and 9.9 we can see that the guidance system designer has a juggling act. Increasing the filter bandwidth reduces the miss due to target maneuver

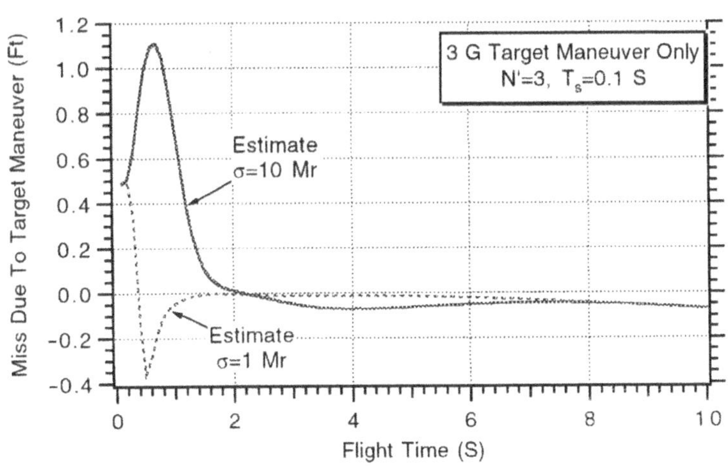

Fig. 9.8 Kalman filter guidance system has small sensitivity to target maneuver.

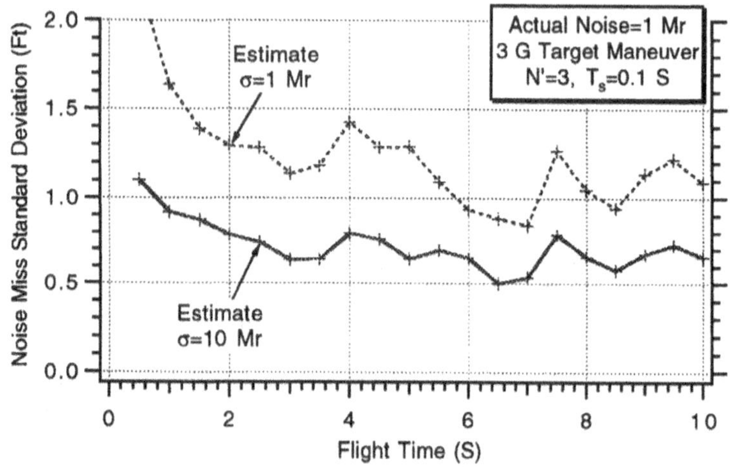

Fig. 9.9 Decreasing filter bandwidth decreases noise-induced miss.

while increasing the miss due to noise. The optimal practical filter band width is dependent on the levels of the input disturbances.

The sampling time can also have a profound effect on filtering properties and system performance. Figure 9.10 shows that increasing the sampling time T_s from 0.1 to 0.5 s (or decreasing sampling rate from 10 to 2 Hz) increases the Kalman gain. We saw from Chapter 7 on digital fading memory filters that decreasing

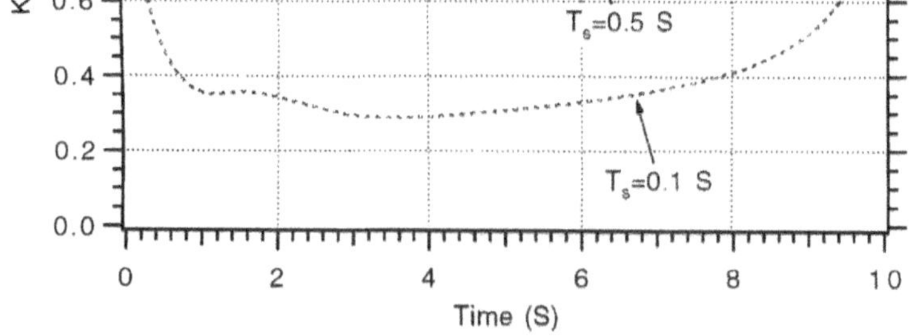

Fig. 9.10 Kalman gain increases with decreasing sampling rate.

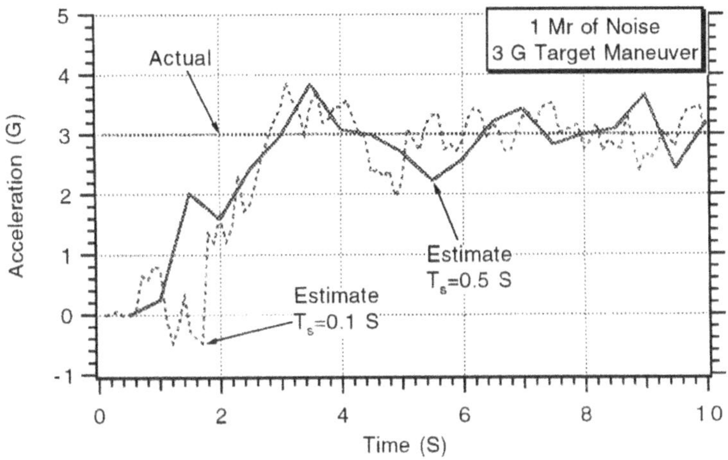

Fig. 9.11 Kalman filter bandwidth appears to be independent of sampling rate.

the sampling rate tends to decrease the total system bandwidth. Thus, the Kalman filter is attempting to increase its bandwidth to compensate for the decrease in system bandwidth due to sampling at a lower rate.

Figure 9.11 shows that the filter estimate of target acceleration for both sampling times is about the same. This means that the filter has successfully compensated for the effective decrease in system bandwidth due to a decrease in the sampling rate.

Although Kalman filter performance appears to be approximately independent of sampling rate, system performance is not! If we remove the actual measurement noise from the simulation and run with target maneuver only for different flight times, we can generate miss distance curves. Figure 9.12 shows how the target-maneuver-induced miss varies with the sampling rate. We can see that the miss for $T_s = 0.5$ s is much greater than the miss for $T_s = 0.1$ s for flight times less than 2 s.

The simulation was also run with measurement noise only in the Monte Carlo mode. Figure 9.13 shows that decreasing the sampling rate also increases the measurement-noise-induced miss. Generally, hardware costs increase with higher sampling rates. Therefore, an important job of the guidance system designer is to set a limit on the sampling rate to get both acceptable cost and adequate performance.

It is important to note that in the preceding experiment the noise standard deviation remained constant when the data rate changed. In many systems the noise spectral density remains constant and so the noise standard deviation must change as the data rate changes. The interested reader is referred to the appendix for a more complete discussion of this topic.

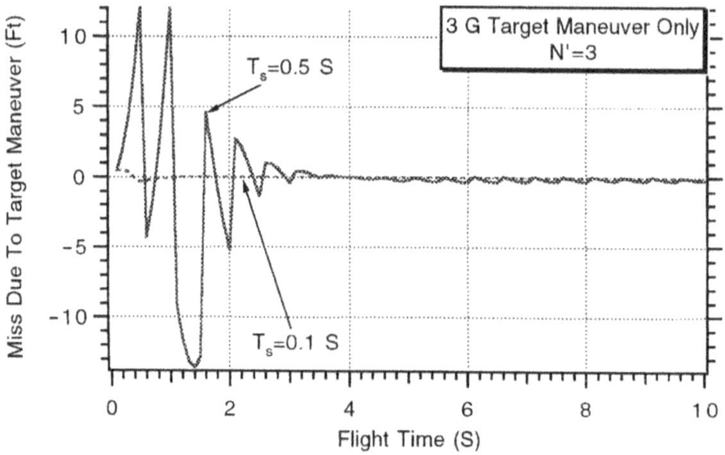

Fig. 9.12 Target maneuver miss increases with decreasing sampling rate.

EXPERIMENTS WITH OPTIMAL GUIDANCE [2]

In Chapter 8 we derived an optimal guidance law that attempted to cancel out the guidance system dynamics and, in addition, we relaxed the missile acceleration requirements. In this section we will show how an optimal guidance system might be implemented and provide a numerical example to illustrate how such a system might perform in the presence of measurement noise.

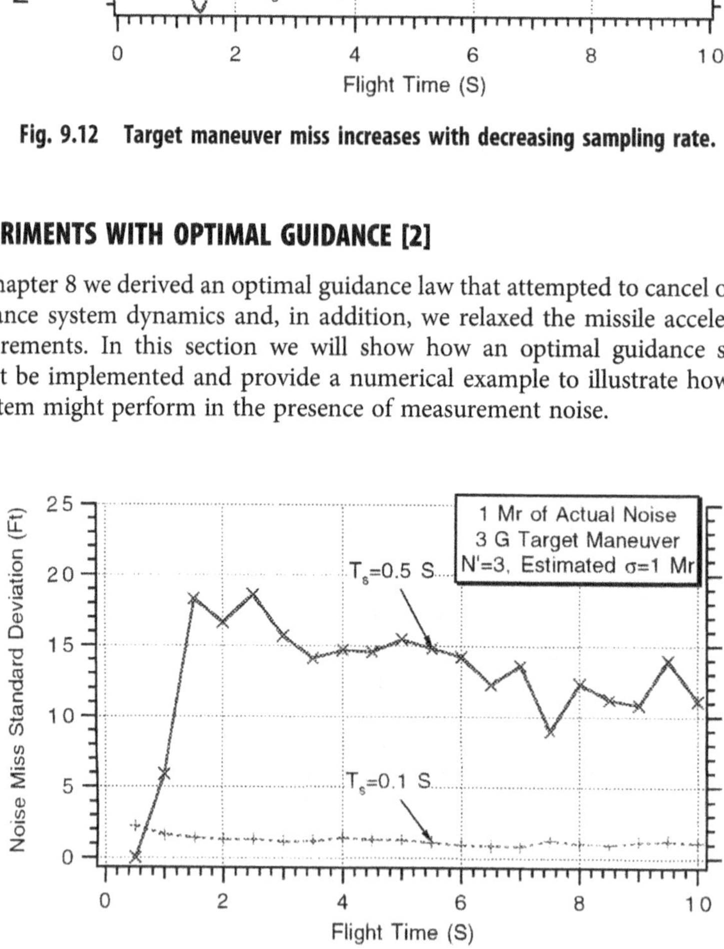

Fig. 9.13 Measurement noise miss increases with decreasing sampling rate.

Listing 9.3 presents a MATLAB Monte Carlo simulation of an optimal guidance system with a three-state digital Kalman filter and a single-lag representation of the flight-control system. The filter structure is identical to the one shown if Fig. 9.2, except that the achieved missile acceleration n_L rather than the commanded acceleration n_c is fed back into the filter. The filter estimates relative position and velocity, which can be converted into a line-of-sight rate estimate as shown in the listing. In addition, the filter estimates the target maneuver level. The achieved missile acceleration is assumed to be known perfectly. This quantity is fed into the filter and, in addition, is used as part of an optimal guidance law as was discussed in Chapter 8 and can be seen in Listing 9.3.

A 50-run Monte Carlo set was made with the engagement model of Listing 9.3. In the nominal case the flight-control system time constant was set to 0.5 s, the effective navigation ratio was 3, and the sampling time was 0.1 s. The nominal error disturbances, as can be seen from Listing 9.3, consist of 1 mr of measurement noise and a constant 3-*g* target maneuver occurring at the beginning of flight. Figures 9.14 and 9.15 show 50-run Monte Carlo results for the standard deviation and mean miss distances for this case as a function of the flight time. Both figures show results for proportional navigation (APN = 0) and an optimal guidance law (APN = 2). Because there is one random disturbance and one deterministic disturbance, we can interpret the standard deviation of the miss to be the noise-induced miss and the mean of the miss to be the target-maneuver-induced miss. Both figures clearly show that, for the case in which the guidance time constant is 0.5 s, optimal guidance yields smaller miss distances, even in the presence of measurement noise errors. The differences between the guidance laws is greatest for the smaller flight times. If the ratio of the flight time to the guidance system time constant is large, proportional navigation is known to be an effective guidance law. Thus, the optimal guidance law, discussed in a deterministic setting in Chapter 8, can be implemented and made to work successfully in a more realistic setting. Optimal guidance is yielding superior performance to proportional navigation because it is attempting to cancel out dynamically the flight-control system time constant. We can see from both figures that optimal guidance performance, unlike that of proportional navigation, is approximately independent of flight time.

LISTING 9.3 MONTE CARLO ENGAGEMENT SIMULATION TO TEST OPTIMAL GUIDANCE

```
%Preallocation
Z=zeros(size(1:1000));
I=zeros(size(1:50));
TF=zeros(size(1:50));
count=0;
VC=4000.;
XNT=96.6;
YIC=0.;
```

```
VM=3000.;
HEDEG=20.;
XNP=3.;
SIGNOISE=.001;
TS=.1;
TAU=.5;
NOISE=1;
RUN=50;
APN=0;
XLIM=999999;
for TF=.5:.5:10.0,
        Z1=0;
        for I=1:RUN
           Y=YIC;
           YD=0;
           TS2=TS*TS;
           TS3=TS2*TS;
           TS4=TS3*TS;
           TS5=TS4*TS;
           PHIN=XNT*XNT/TF;
           RTM=VC*TF;
           SIGPOS=RTM*SIGNOISE;
           SIGN2=SIGPOS^2;
           P11=SIGN2;
           P12=0.;
           P13=0.;
           P22=(VM*HEDEG/57.3)^2;
           P23=0.;
           P33=XNT*XNT;
           T=0.;
           H=.01;
           S=0.;
           YH=0.;
           YDH=0.;
           XNTH=0.;
           XNC=0.;
           XNL=0.;
           while T<= (TF - 1e-5)
                  YOLD=Y;
                  YDOLD=YD;
                  XNLOLD=XNL;
                  STEP=1;
                  FLAG=0;
                  while STEP <=1
                         if FLAG==1
                  Y=Y+H*YD;
```

```
YD=YD+H*YDD;
XNL=XNL+H*XNLD;
T=T+H;
STEP=2;
end;
TGO=TF-T+.00001;
RTM=VC*TGO;
XLAM=Y/(VC*TGO);
XLAMD=(RTM*YD+Y*VC)/(RTM^2);
XNLD=(XNC-XNL)/TAU;
YDD=XNT-XNL;
FLAG=1;
end;
FLAG=0;
Y=.5*(YOLD+Y+H*YD);
YD=.5*(YDOLD+YD+H*YDD);
XNL=.5*(XNLOLD+XNL+H*XNLD);
S=S+H;
if S>=(TS - 1e-5)
S=0.;
TGO=TF-T+.000001;
RTM=VC*TGO;
SIGPOS=RTM*SIGNOISE;
SIGN2=SIGPOS^2;
M11=P11+TS*P12+.5*TS2*P13+TS*(P12+TS*P22+...
                                          .5*TS2*P23);
M11=M11+.5*TS2*(P13+TS*P23+.5*TS2*P33)...
                              +TS5*PHIN/20.;
M12=P12+TS*P22+.5*TS2*P23+TS*(P13+TS*P23+...
                              .5*TS2*P33)+TS4*PHIN/8.;
M13=P13+TS*P23+.5*TS2*P33+PHIN*TS3/6.;
M22=P22+TS*P23+TS*(P23+TS*P33)+PHIN*TS3/3.;
M23=P23+TS*P33+.5*TS2*PHIN;
M33=P33+PHIN*TS;
K1=M11/(M11+SIGN2);
K2=M12/(M11+SIGN2);
K3=M13/(M11+SIGN2);
P11=(1.-K1)*M11;
P12=(1.-K1)*M12;
P13=(1.-K1)*M13;
P22=-K2*M12+M22;
P23=-K2*M13+M23;
P33=-K3*M13+M33;
          if NOISE==1,
                    XLAMNOISE=SIGNOISE*randn;
          else
```

```
                        XLAMNOISE=0;
                        end;
                YSTAR=RTM*(XLAM+XLAMNOISE);
                RES=YSTAR-YH-TS*YDH-.5*TS*TS*(XNTH-XNC);
                YH=K1*RES+YH+TS*YDH+.5*TS*TS*(XNTH-XNC);
                YDH=K2*RES+YDH+TS*(XNTH-XNC);
                XNTH=K3*RES+XNTH;
                XLAMDH=(YH+YDH*TGO)/(VC*TGO*TGO);
                if APN==0
                XNC=XNP*VC*XLAMDH;
                elseif APN==1
                XNC=XNP*VC*XLAMDH+APN*.5*XNP*XNTH;
                else
                X=TGO/TAU;
                TOP=6.*X*X*(exp(-X)-1.+X);
                BOT1=2*X*X*X+3.+6.*X-6.*X*X;
                BOT2=-12.*X*exp(-X)-3.*exp(-2.*X);
                XNPP=TOP/(.0001+BOT1+BOT2);
                XNEW=XNPP*XNL*(EXP(-X)+X-1.)/(X*X);
                        XNC=XNPP*VC*XLAMDH+.5*XNPP*XNTH-XNEW;
                end;
                if XNC>XLIM
                        XNC=XLIM;
                elseif XNC<-XLIM
                        XNC=-XLIM;
                end;
        end;
      end;
      Z(I)=Y;
        Z1=Z(I)+Z1;
        XMEAN=Z1/I;
      end;
      SIGMA=0;
      Z1=0;
      for I=1:RUN,
   Z1=(Z(I)-XMEAN)^2+Z1;
   if I==1,
     SIGMA=0;
   else
     SIGMA=sqrt(Z1/(I-1));
   end;
 end;
count=count+1;
ArrayTF(count)=TF;
ArraySIGMA(count)=SIGMA;
ArrayXMEAN(count)=XMEAN;
```

```
end;
figure
plot(ArrayTF',ArraySIGMA'),grid
title('Standard deviation of miss for various flight times')
xlabel('Flight Time (S)')
ylabel('Noise Miss Standard Deviation (Ft) ')
axis([00,10,00,4])
figure
plot(ArrayTF',ArrayXMEAN'),grid
title('Mean of miss for various flight times')
xlabel('Flight Time (S)')
ylabel('Mean Miss (Ft) ')
axis([00,10,00,60])
clc
output=[ArrayTF',ArraySIGMA',ArrayXMEAN'];
save datfil.txt output /ascii
disp('Simulation Complete')
```

Another performance advantage of optimal guidance is that it is supposed to relax the missile acceleration requirements. The previous case was rerun for a 10-s flight. This flight time was chosen because the performance of both proportional navigation and optimal guidance is about the same, from a miss distance point of view, as can be seen from Figs. 9.14 and 9.15. The reason for this is that the 10-s flight time is large compared to the 0.5-s flight-control system time constant. However, in the new case to be run the engagement simulation was made

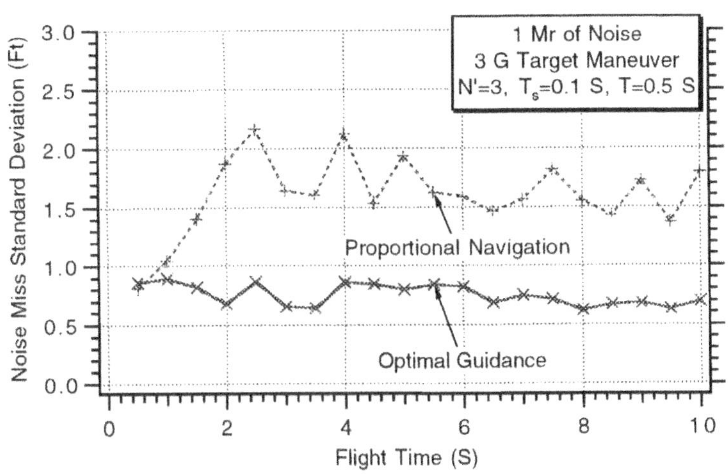

Fig. 9.14 Optimal guidance yields smaller noise-induced miss in presence of large guidance system time constant.

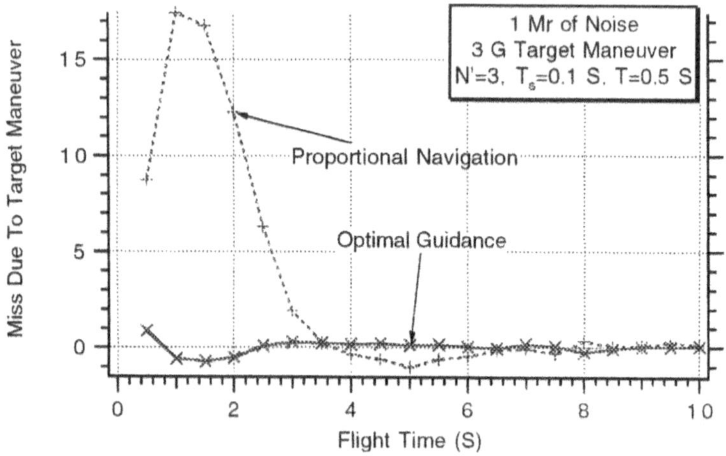

Fig. 9.15 Optimal guidance yields smaller target-maneuver-induced miss in presence of large guidance system time constant.

nonlinear in the sense that missile acceleration saturation effects were included. The missile commanded acceleration limit was made a parameter in the study. Figure 9.16 displays the mean miss distance vs the acceleration limit for a case in which there was a 3-*g* target maneuver and 1 mr of measurement noise. We can see from the figure that the acceleration requirements for optimal guidance are clearly relaxed.

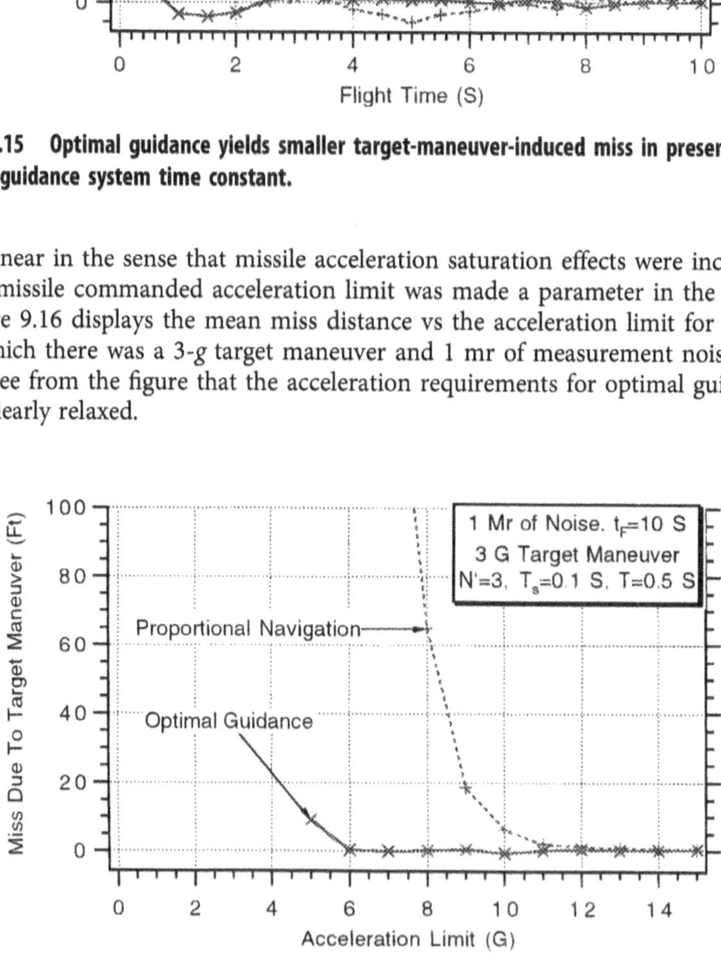

Fig. 9.16 Optimal guidance reduces missile acceleration requirements.

SUMMARY

In this chapter we have shown how both Kalman filtering and optimal guidance concepts could be applied to a missile guidance system. It was shown, via a numerical example, that when these concepts were applied there were substantial performance benefits and a relaxing of missile acceleration requirements. However, range and time-to-go information must be available for Kalman filtering and optimal guidance to work. If the required information is lacking or inaccurate, the performance of this type of guidance system may degrade to the point where its performance is worse than that of a conventional proportional navigation guidance system [2].

REFERENCES

[1] Gelb, A., *Applied Optimal Estimation*, MIT Press, Cambridge, MA, 1974.

[2] Nesline, F. W., and Zarchan, P., "A New Look at Classical Versus Modern Homing Guidance," *Journal of Guidance and Control*, Vol. 4, Jan.–Feb. 1981, pp. 78–85.

CHAPTER 10

Tactical Zones

INTRODUCTION

In the material presented so far it has been assumed that the missile can reach the target and generate sufficient lift to maneuver. For purposes of understanding fundamental guidance issues, we have neglected the fact that the ability of a tactical aerodynamic missile to maneuver is dependent upon its speed, physical characteristics (such as, wing size), and altitude. In addition, we have also assumed impulsive constant velocity missiles and have not taken into consideration that a missile must burn propellant in order to get up to speed. The resultant missile total weight is directly related to its payload weight and design speed (such as, propellant weight). Actually, it will soon become obvious that in some tactical missiles most of the weight is the propellant weight. Finally, we have also neglected the fact that while an aerodynamic missile is coasting the missile speed diminishes due to atmospheric drag. In this chapter we shall briefly address previously neglected issues and show how to modify previous computations to account for these important effects.

VELOCITY COMPUTATION

A tactical missile gets up to speed by burning propellant. If the missile is launched from the air, it already has a large initial speed. However, if the missile is launched from the ground, it needs more propellant to reach the same speed because it is starting from rest. In addition, a ground-based missile needs additional propellant because it must travel through more of the denser atmosphere. Figure 10.1 shows a typical weight and thrust profile for a boost-coast missile. The initial or total weight of the missile is denoted W_T, and its final weight, after the propellant is expended, is the glide weight W_G. The glide weight consists of the missile structure, electronics, and warhead. While propellant is being consumed, the thrust is assumed to be constant, with magnitude T.

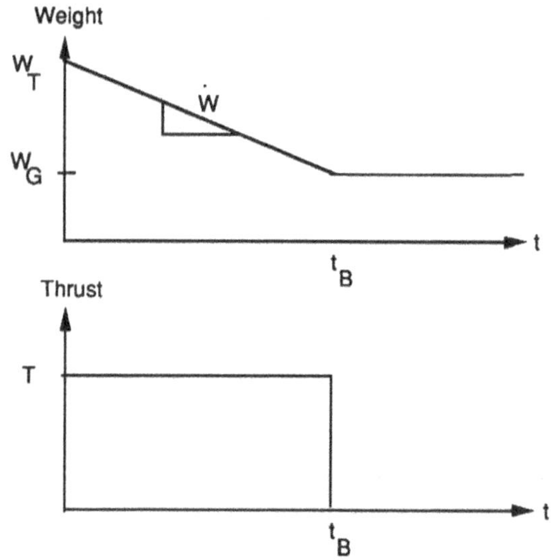

Fig. 10.1 Boost-coast thrust-weight profile.

We can find the magnitude of the missile velocity after all of the propellant is consumed from basic physics. Applying Newton's second law yields [1, 2]

$$F = ma = m\frac{\mathrm{d}V}{\mathrm{d}t} = T$$

The change in velocity with respect to time can be expressed in terms of thrust and weight as

$$\frac{\mathrm{d}V}{\mathrm{d}t} = \frac{T}{m} = \frac{Tg}{W}$$

Recognizing that as long as the missile is burning propellant ($0 < t < t_B$), we can express the instantaneous missile weight as

$$W = \dot{W}t + W_T$$

where the derivative of the weight is negative (weight is decreasing). Thus, we can find an expression for the change in velocity due to the burning of propellant by direct integration, or

$$\int_{V_0}^{V_1} \mathrm{d}V = Tg\int_0^{t_B} \frac{\mathrm{d}t}{W}$$

Substitution of the expression for the missile weight into the integral yields

$$V_1 - V_0 = \Delta V = Tg\int_0^{t_B} \frac{\mathrm{d}t}{\dot{W}t + W_T}$$

Evaluation of the integral yields

$$\Delta V = \frac{-Tg}{\dot{W}}\ell n\frac{W_T}{\dot{W}t_B + W_T}$$

Thus, the change in velocity depends only on the missile total weight, glide weight, thrust magnitude, and rate at which the propellant is burning. The preceding velocity formula, also known as the rocket equation, represents the maximum change in velocity we can impart. Practical effects such as gravity and atmospheric drag will usually work in the direction of decreasing ΔV. The preceding expression can be made more concise and useful by specifying fuel effectiveness in terms of a parameter known as the specific impulse, I_{sp}. It is a positive number in units of seconds and is related to the thrust and change in missile weight according to

$$I_{sp} = \frac{-T}{\dot{W}}$$

More fuel-efficient missiles have higher values of specific impulse. Typically, for tactical missiles, the specific impulse has values ranging from 200 s to 300 s. By substituting the specific impulse definition into the velocity change formula, we get

$$\Delta V = I_{sp} g \, \ell n \frac{W_T}{W_G}$$

Now the change in missile velocity during a burn depends only on the total weight, glide weight, and specific impulse. However, the total missile weight is the sum of the glide weight and the propellant weight, or

$$W_T = W_P + W_G$$

The fuel mass fraction mf is defined as the ratio of the missile propellant weight to the total missile weight, or

$$mf = \frac{W_P}{W_T}$$

Because a ground-to-air missile requires more fuel than an air-to-air missile to reach the same speed, it would have a larger fuel mass fraction value. We can now express the change in missile velocity in terms of the specific impulse and fuel mass fraction, or

$$\Delta V = I_{sp} g \, \ell n \frac{1}{1 - mf}$$

Using the preceding equation, Fig. 10.2 shows how the change in missile velocity varies with fuel mass fraction and specific impulse. We can see from the figure that, if the fuel mass fraction is 0.3 and the specific impulse is 200 s, the change in velocity is about 2300 ft/s. This means that, if the missile were launched from the ground, its final speed, in the absence of drag and gravitational effects, would be 2300 ft/s. If the missile with the same fuel mass fraction were launched from an aircraft traveling at 1000 ft/s, its final speed would be 3300 ft/s.

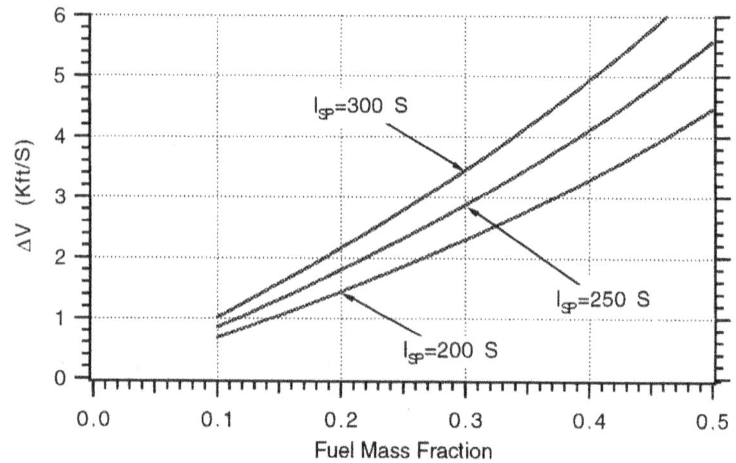

Fig. 10.2 Increasing fuel mass fraction or specific impulse increases missile speed.

DRAG [3]

Tactical missiles work within the atmosphere. Aerodynamic drag causes the missile to slow down and have less maneuver capability. The drag F_{drag} can be expressed as

$$F_{\text{drag}} = QS_{\text{ref}}C_{D0}$$

where Q is the dynamic pressure, S_{ref} a reference area, and C_{D0} the zero-lift drag. The dynamic pressure is a function of the air density ρ and velocity V and is given by

$$Q = \frac{\rho V^2}{2}$$

In the English system of units used throughout the text, air density is measured in slug per cubic foot (slug/ft^3). The reference area is the cross-sectional area of the missile body and is therefore related to the physical characteristics of the missile. The zero-lift drag is a function of the missile speed and aerodynamic shape. Since the air density decreases with altitude, the influence of drag is greatest at the lower altitudes. For analytical reasons it is convenient to use an exponential approximation to the atmosphere. One such approximation below 30,000 ft altitude is given by

$$\rho = 0.002378\, e^{-h/30{,}000} \qquad (h < 30{,}000 \text{ ft})$$

whereas above 30,000 ft the exponential approximation becomes

$$\rho = 0.0034\, e^{-h/22{,}000} \qquad (h \geq 30{,}000 \text{ ft})$$

where h is measured in feet.

To check the validity of the exponential approximations, the 1962 U.S. standard atmosphere is displayed as a function of altitude in Fig. 10.3. Superimposed on the figure are the exponential approximations. The solid curve represents the actual data points for the standard U.S. atmosphere, whereas the dashed curve represents the preceding exponential approximation. We can see that the exponential approximation is quite accurate.

In the absence of induced drag effects, the drag F_{drag} can be expressed in terms of Newton's second law as

$$F = ma = -F_{\text{drag}} = m\frac{\mathrm{d}V}{\mathrm{d}t}$$

Therefore, the rate of change of velocity can be found from

$$\frac{\mathrm{d}V}{\mathrm{d}t} = \frac{-F_{\text{drag}}}{m} = \frac{-QS_{\text{ref}}C_{D0}}{m} = \frac{-0.5\rho V^2 S_{\text{ref}} C_{D0}}{m}$$

Setting up the integral in a useful form for integration, we get

$$\int_{V_0}^{V_1} \frac{\mathrm{d}V}{V^2} = \int_{t_0}^{t_1} \frac{-0.5\rho S_{\text{ref}} C_{D0}}{m}\, \mathrm{d}t$$

If we assume that the air density does not change (constant altitude) and that the zero-lift drag is constant, we can get a closed-form expression for the new velocity

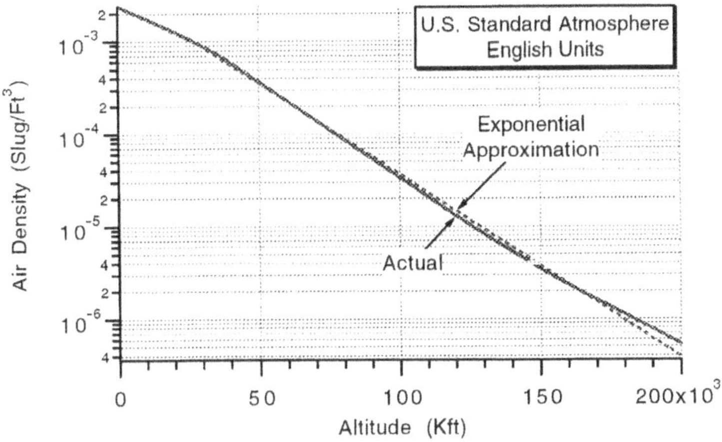

Fig. 10.3 Exponential approximation for air density is very accurate.

due to drag, or

$$\frac{1}{V_1} = \frac{\rho S_{\text{ref}} C_{D0} \Delta t}{2m} + \frac{1}{V_0}$$

where V_0 is the initial velocity, V_1 is the velocity Δt seconds later, and Δt is defined as

$$\Delta t = t_1 - t_0$$

If we define the initial drag deceleration D_0 to be

$$D_0 = \frac{\rho S_{\text{ref}} C_{D0} V_0^2}{2m}$$

and a characteristic time T_0 as

$$T_0 = V_0 / D_0$$

then we can express the velocity Δt seconds later in the simpler form

$$V_1 = \frac{V_0 T_0}{T_0 + \Delta t}$$

Integrating again yields the total distance R covered in Δt seconds as

$$R = V_0 T_0 \ell n \left(1 + \frac{\Delta t}{T_0}\right)$$

We now have sufficient information to perform some preliminary calculation concerning the effects of drag on velocity loss and range covered. However, it is first important to see how the magnitude of the drag deceleration is influenced by the zero-lift drag and altitude. Often it is convenient to combine the weight, reference area, and zero-lift drag into an expression known as the ballistic coefficient β, which is defined as

$$\beta = \frac{W}{C_{D0} S_{\text{ref}}}$$

where β is in units of pounds per square feet. Figure 10.4 shows how the drag deceleration varies with altitude for different values of ballistic coefficient. We can see that increasing the ballistic coefficient (reducing zero-lift drag) or increasing the altitude reduces the drag deceleration. It is also apparent that, for a velocity of 3000 ft/s, a ballistic coefficient of 500 lb/ft^2 yields a drag deceleration of about 22 g at sea level, 10 g at 25 kft, and 3.5 g at 50 kft. In addition, Fig. 10.4 shows that, as altitude increases, the drag deceleration becomes less dependent on the ballistic coefficient. Eventually, at the higher altitudes, there is no deceleration due to drag.

If we still assume a missile at sea level with an initial velocity of 3000 ft/s, missile speed decreases due to drag as a function of time for different ballistic

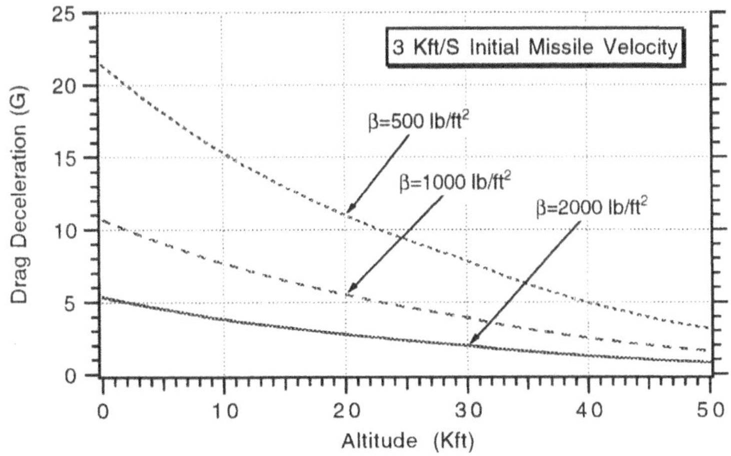

Fig. 10.4 Drag deceleration decreases with increasing altitude and increasing ballistic coefficient.

coefficients, as shown in Fig. 10.5. The velocity drops to about half of its original value in only 4.5 s for $\beta = 500\ \mathrm{lb/ft^2}$, in about 9 s for $\beta = 1000\ \mathrm{lb/ft^2}$, and in about 17.5 s for $\beta = 2000\ \mathrm{lb/ft^2}$.

Figure 10.6 shows the ranges covered for the cases given earlier. At sea level the range covered until the missile velocity drops to half of its value is about 9 kft for $\beta = 500\ \mathrm{lb/ft^2}$, about 20 kft for $\beta = 1000\ \mathrm{lb/ft^2}$, and about 35 kft for

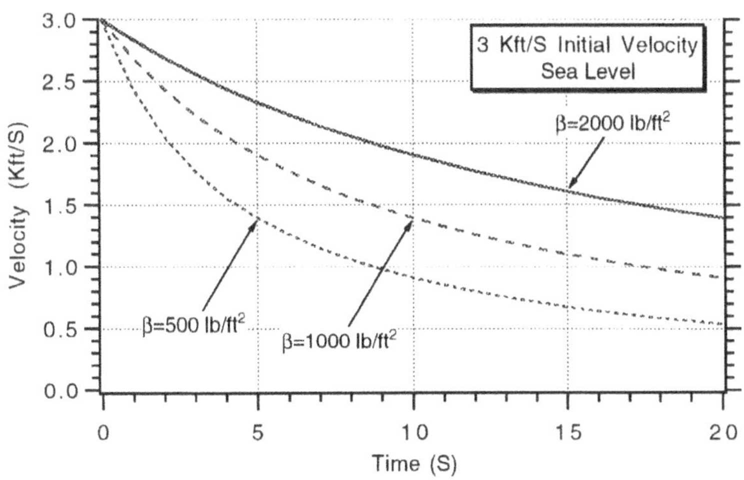

Fig. 10.5 Velocity drops faster with smaller ballistic coefficient.

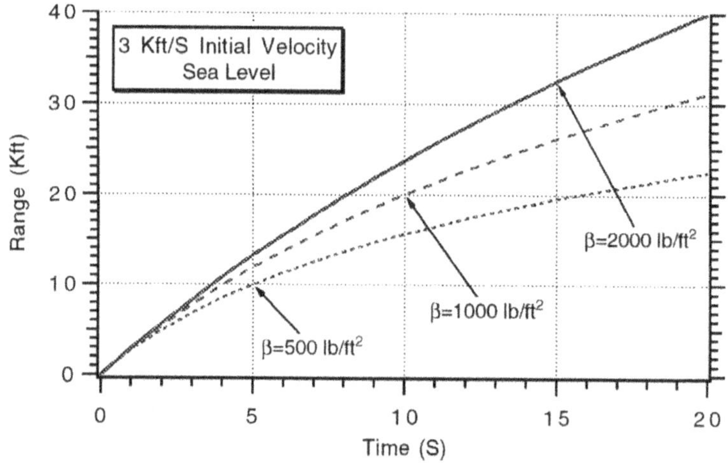

Fig. 10.6 Zone of effectiveness at sea level is not large.

$\beta = 2000 \text{ lb/ft}^2$. If the missile is not considered to be effective after it has dropped more than half of its velocity, then we can consider these values to be kinematic zones of effectiveness at sea level. The missile will have longer kinematic reach if its ballistic coefficient is higher.

Figure 10.7 shows the velocity loss at 50 kft altitude. We can see that increasing the altitude reduces the missile's velocity loss due to drag and increases the effective range of the missile. We can see from Fig. 10.7 that at 50 kft it takes

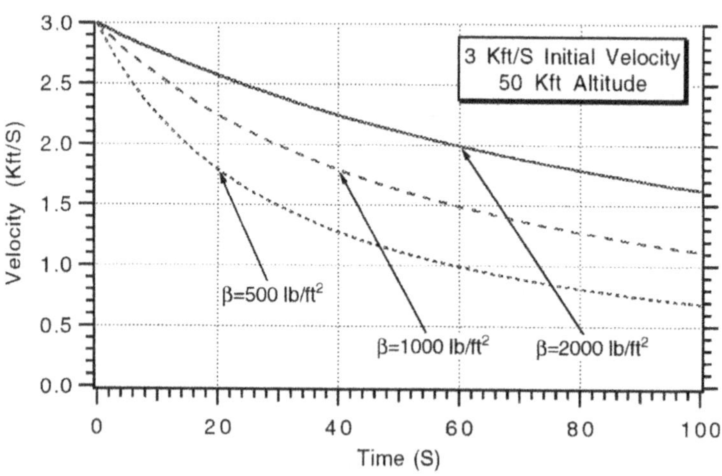

Fig. 10.7 Velocity drops much slower at 50 kft altitude.

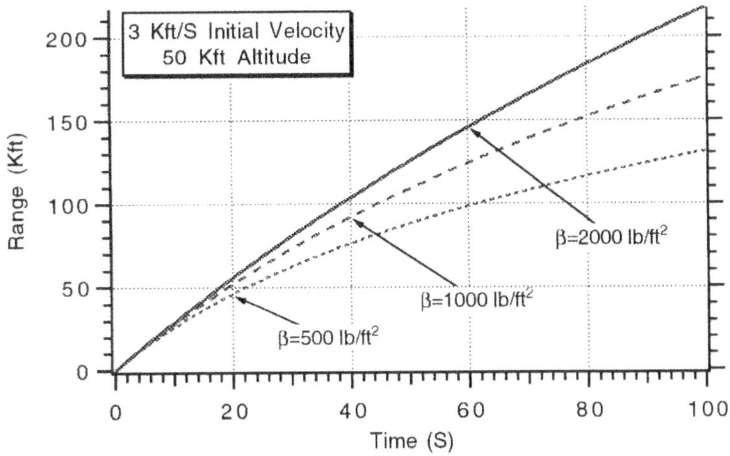

Fig. 10.8 Zone of effectiveness is much greater at 50 kft altitude.

much longer for the velocity to drop to half of its original value. At 50 kft the velocity drops to about half of its original value in about 30 s (vs 4.5 s at sea level) for $\beta = 500\ \text{lb/ft}^2$, in about 60 s (vs 9 s at sea level) for $\beta = 1000\ \text{lb/ft}^2$, and in more than 100 s (vs 17.5 s at sea level) for $\beta = 2000\ \text{lb/ft}^2$.

Figure 10.8 shows the range covered as a function of time for missiles with varying ballistic coefficients at an altitude of 50 kft. We can see from the figure that the range covered until the velocity drops to half of its value is about 70 kft (vs 9 kft at sea level) for $\beta = 500\ \text{lb/ft}^2$, about 125 kft (vs 20 kft at sea level) for $\beta = 1000\ \text{lb/ft}^2$, and more than 200 kft (vs 35 kft at sea level) for $\beta = 2000\ \text{lb/ft}^2$.

In summary, increasing the ballistic coefficient (or reducing the zero-lift drag) can have a big payoff in terms of increased zone of effectiveness for aerodynamic missiles that must fly through the more dense atmosphere at low altitudes. Tactical radar homing missiles tend to have a nose with a high fineness ratio in order to make them more aerodynamically efficient. The high fineness-ratio nose also tends to exacerbate parasitic radome effects [4].

ACCELERATION [1, 2]

Just as there was a drag coefficient to determine slowdown, there is a lift coefficient C_L to determine missile maneuverability. From Newton's second law we can say that

$$F = ma = mn_L = QS_{\text{ref}}C_L$$

where n_L represents the missile's acceleration capability. Therefore, the missile acceleration capability, expressed in units of gravity, is given by

$$\frac{n_L}{g} = \frac{0.5\rho V^2 S_{\text{ref}} C_L}{W}$$

where W is missile weight in units of pounds. The lift coefficient is a function of the missile aerodynamic shape, speed, angle of attack, and wing size. Larger wings and increasing angle of attack both work in the direction of increasing the lift coefficient.

To demonstrate the sensitivity of the missile acceleration capability to the lift coefficient and altitude, it is best to consider a numerical example. Consider a missile weighing 500 lb with an 0.5-ft^2 reference area and traveling at 3000 ft/s. Figure 10.9 shows how the missile acceleration capability decreases with increasing altitude and decreasing lift coefficient. It is important to note that a missile may have an aerodynamic acceleration capability, at a given altitude, which is far in excess of its structural capability. A loading analysis is required to set practical limits on the maximum allowable commanded missile acceleration. Figure 10.9 shows an example of a missile ($C_L = 4$) which has a 40-g capability at sea level that diminishes to about a 10-g capability at 50 kft altitude. Reducing missile weight or increasing the missile reference area (but keeping weight constant) works in the direction of increasing the missile aerodynamic maneuverability.

Speed also plays an important role in determining missile aerodynamic maneuverability. Figure 10.10 shows that decreasing the missile speed significantly decreases the missile maneuverability. A missile that travels at 3000 ft/s at

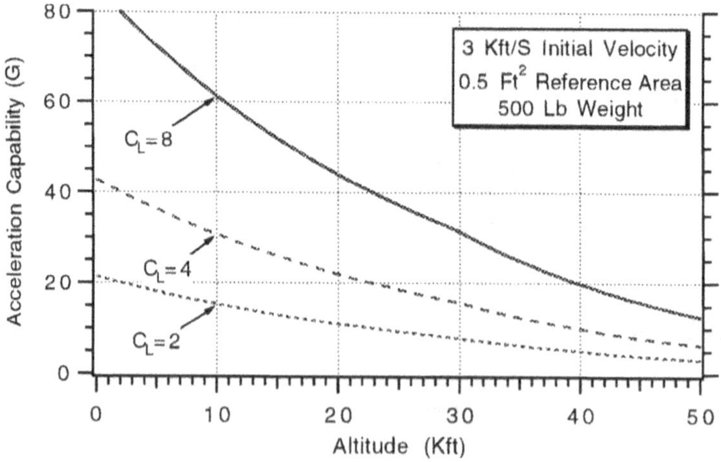

Fig. 10.9 Missile maneuverability decreases dramatically with increasing altitude.

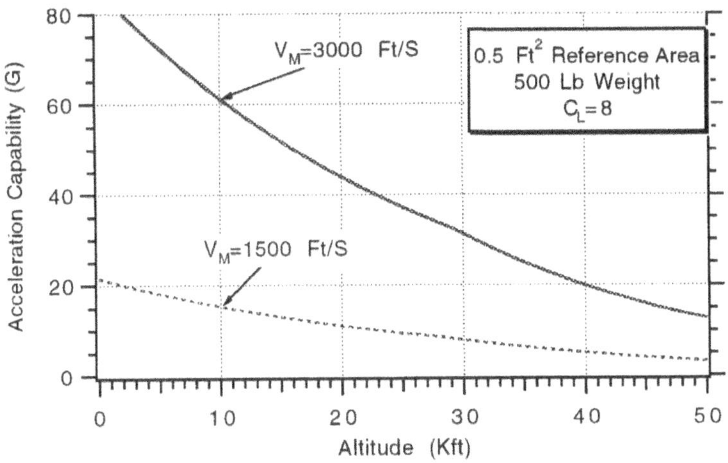

Fig. 10.10 Decreasing speed decreases maneuverability.

20 kft altitude has a maneuverability in excess of 40 *g*. Halving the missile speed more than halves its maneuverability. We have seen in previous chapters that a missile requires a certain acceleration advantage to effectively engage maneuvering targets. Therefore, for a given altitude and missile configuration, there is a minimum speed at which the missile can effectively engage a responsive threat.

GRAVITY

Thus far in our analysis we have neglected gravity and assumed a constant-altitude missile. Actually, gravity will eventually cause a coasting missile to crash to the ground. If we neglect the atmosphere and launch an impulsive 3000-ft/s missile at various flight-path angles γ, we will get different range capabilities due to gravity alone, as shown in Fig. 10.11. As expected, the 45-deg launch results in maximum range.

Atmospheric drag will of course prevent the missile from achieving the range capabilities indicated in Fig. 10.11. Listing 10.1 presents a MATLAB simulation of an impulsive missile launched at a flight-path angle GAMDEG missile in the presence of gravity and an atmosphere (zero-lift-drag). As in the previous section, the ballistic coefficient is used rather than the zero-lift drag coefficient to account for zero-lift-drag-induced slowdown effects. Lift-induced drag is neglected in this simplified analysis.

Consider a 45-deg sea-level launch of a missile that attains a velocity of 3000 ft/s instantaneously. Cases were run in which the ballistic coefficient varied from 500 lb/ft^2 to infinity (no drag). Figure 10.12 shows that drag dramatically changes the maximum range capability of the interceptor. The maximum

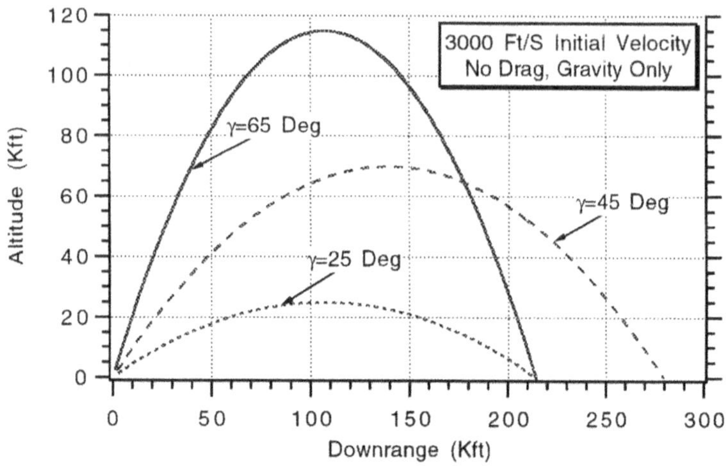

Fig. 10.11 Trajectory profiles for various launch angles.

range for a 45-deg launch angle decreases from about 300 kft (no-drag case) to about 100 kft for a ballistic coefficient of 2000 $\mathrm{lb/ft^2}$, to about 55 kft for a ballistic coefficient of 1000 $\mathrm{lb/ft^2}$, and to 30 kft for a ballistic coefficient of 500 $\mathrm{lb/ft^2}$.

Drag becomes less important at the higher altitudes. The previous case was repeated, but the initial launch altitude was increased from sea level to 50 kft. Figure 10.13 shows that the differences between the drag free trajectory and the one in which the ballistic coefficient is 2000 $\mathrm{lb/ft^2}$ is much smaller than before.

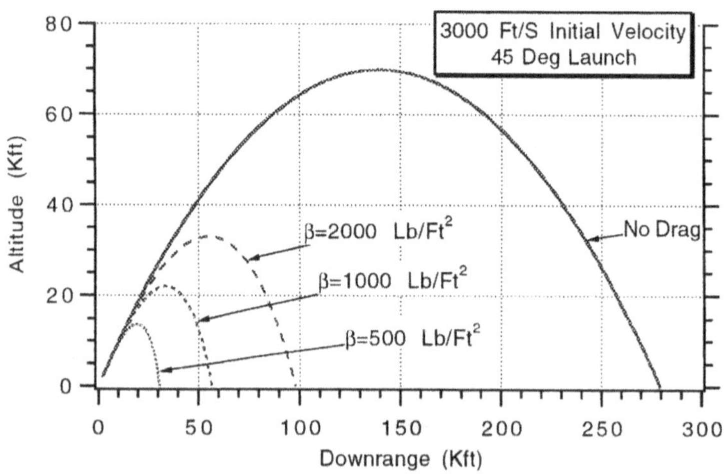

Fig. 10.12 Drag dramatically reduces range capability of missile.

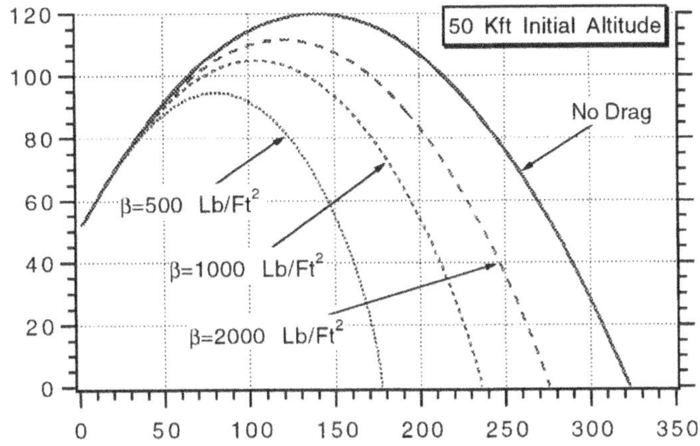

Fig. 10.13 Drag effects reduce considerably if launch altitude is high.

The maximum range without drag in this case is about 350 kft, whereas the maximum range for a ballistic coefficient of 2000 lb/ft^2 is about 275 kft. For a ballistic coefficient of 500 lb/ft^2 the maximum range reduces to 175 kft.

LISTING 10.1 TRAJECTORY SIMULATION

```
count=0;
H=.01;
VM=3000.;
BETA=1000.;
T=0.;
S=0.;
GAMDEG=45.;
VM1=VM*cos(GAMDEG/57.3);
VM2=VM*sin(GAMDEG/57.3);
RM1=0.;
RM2=0.;
while ~(T>0. & RM2<=0.)
      RM1OLD=RM1;
      RM2OLD=RM2;
      VM1OLD=VM1;
      VM2OLD=VM2;
      STEP=1;
      FLAG=0;
      while STEP <=1
        if FLAG==1
                STEP=2;
```

```
        RM1=RM1+H*VM1;
        RM2=RM2+H*VM2;
        VM1=VM1+H*AM1;
        VM2=VM2+H*AM2;
        T=T+H;
        end
        if RM2<30000.
        RHO=.002378*exp(-RM2/30000);
        else
        RHO=.0034*exp(-RM2/22000);
        end
        VM=sqrt(VM1^2+VM2^2);
        Q=.5*RHO*VM*VM;
        GAM=atan2(VM2,VM1);
        DRAG=Q*32.2/BETA;
        AM1=-DRAG*cos(GAM);
        AM2=-32.2-DRAG*sin(GAM);
        FLAG=1;
        end
        FLAG=0;
        RM1=.5*(RM1OLD+RM1+H*VM1);
        RM2=.5*(RM2OLD+RM2+H*VM2);
        VM1=.5*(VM1OLD+VM1+H*AM1);
        VM2=.5*(VM2OLD+VM2+H*AM2);
        S=S+H;
        if S>=.99999
           S=0.;
           RM1K=RM1/1000.;
           RM2K=RM2/1000.;
           count=count+1;
           ArrayT(count)=T;
           ArrayRM1K(count)=RM1K;
           ArrayRM2K(count)=RM2K;
        end
end
figure
plot(ArrayRM1K,ArrayRM2K),grid
xlabel('Downrange (Kft)')
ylabel('Altitude (Kft) ')
clc
output=[ArrayT',ArrayRM1K',ArrayRM2K'];
save datfil.txt output /ascii
disp 'simulation finished'
```

Thus, we can see that drag not only plays a role in reducing missile speed so that it has less acceleration capability but it also plays a significant role in determining the kinematic reach of the missile. For long-range ground-launched missiles, trajectory shaping is often used to get the missile to higher altitudes as quickly as possible so that range and velocity losses due to drag can be minimized.

SUMMARY

In this chapter we have considered and shown how to model previously neglected effects. A simple design formula, known as the rocket equation, was derived in order to show the influence of propellant weight on missile speed capability. In addition, we investigated how drag reduces the kinematic reach of the missile and how the atmosphere helps in providing the missile with lift to maneuver. Finally, a numerical example was presented showing how to generate flyout zones.

REFERENCES

[1] Locke, A. S., *Guidance,* Van Nostrand, Toronto, 1955.

[2] Jerger, J. J., *System Preliminary Design,* Van Nostrand, Princeton, NJ, 1960.

[3] Travers, P., "Interceptor Dynamics," Unpublished Lecture Notes, Raytheon, circa 1971.

[4] Eichblatt, E. (ed.), *Test and Evaluation of the Tactical Missile,* Vol. 119, Progress in Astronautics and Aeronautics, AIAA, Washington, DC, 1989.

CHAPTER 11

Strategic Considerations

INTRODUCTION

In all of the work presented so far we have based our models on a flat-Earth, constant-gravity model. For tactical interceptor missions, where speeds are less than 5000 ft/s, altitudes under 100 kft, and ranges covered under 100 n.mi., these assumptions are reasonable. In the strategic world where speeds are near-orbital and the distances covered are intercontinental, the flat-Earth constant-gravity assumption is not only inaccurate but can also give misleading results in terms of the size of the zone of effectiveness. However, we shall also see in subsequent chapters that the guidance lessons learned in the tactical world are still valid and give valuable insight into the requirements and effectiveness of strategic interceptors. Before proceeding with the development of models, it is worthwhile to review some of the historical background of strategic ballistic missiles [1].

BACKGROUND

Germany's V-2 was the world's first long-range ballistic missile. When a loophole in the Treaty of Versailles was found, the Wehrmacht's Ordinance Department authorized the development of this large long-range rocket and selected Artillery Captain Walter Dornberger to supervise the project. After 14 years of testing, the V-2 was ready for field use and was finally deployed in the fall of 1944. It was launched from mobile field battery positions in France and Holland. Each single-stage missile weighed nearly 30,000 lb, reached a burnout velocity of about 6000 ft/s, and had a range of approximately 230 miles. Between September 1944 and March 1945 German field units launched more than 3000 V-2 missiles. Approximately 1900 missiles were launched against Allied targets on the European continent, primarily Antwerp, Belgium. The rest fell on London and southern England.

After the war, the U.S. Army brought German V-2 engineers and enough pieces for about 80 missiles into this country. As part of Project Hermes, more than 70 V-2s were launched by the U.S. during the late 1940s and early 1950s. These rockets formed the basis for U.S. strategic ballistic missile technology and were also essential for subsequent advances in the exploration of space.

GRAVITATIONAL MODEL

In the tactical world, in the absence of thrust, drag, and lift, the flat-Earth constant-gravity assumption is easy to understand. In this mathematical model the gravitational acceleration is independent of altitude with value 32.2 ft/s^2, always in a downward direction. The tactical missile inertial coordinate system is fixed to the surface of a flat Earth and is depicted in Fig. 11.1. Here the missile has velocity V and is at a flight-path angle γ with respect to the surface of the Earth. In addition the missile is in a constant-gravity field with acceleration level g. The missile is at an initial location that is distance dr downrange from the origin of the coordinate system and at an altitude *alt* from the surface of the Earth.

The differential equations acting on the missile are

$$\dot{V}_1 = 0$$

$$\dot{V}_2 = -g$$

$$\dot{R}_1 = V_1$$

$$\dot{R}_2 = V_2$$

where V is velocity and R is range. The down-range component is denoted by 1, and the altitude component is denoted by 2. The initial conditions for velocity and position are given by

$$V_1(0) = V\cos\gamma$$
$$V_2(0) = V\sin\gamma$$
$$R_1(0) = \mathrm{d}r$$
$$R_2(0) = alt$$

Because the coordinate system is inertial, we can integrate directly in the down-range and altitude directions to get velocity from acceleration and position from velocity. In

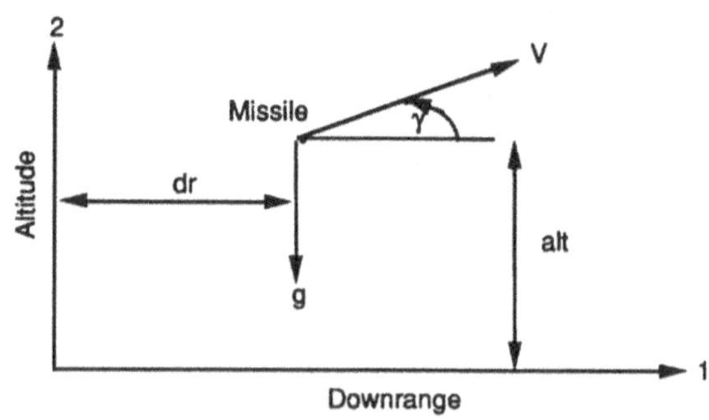

Fig. 11.1 Missile in gravity field using flat-Earth model.

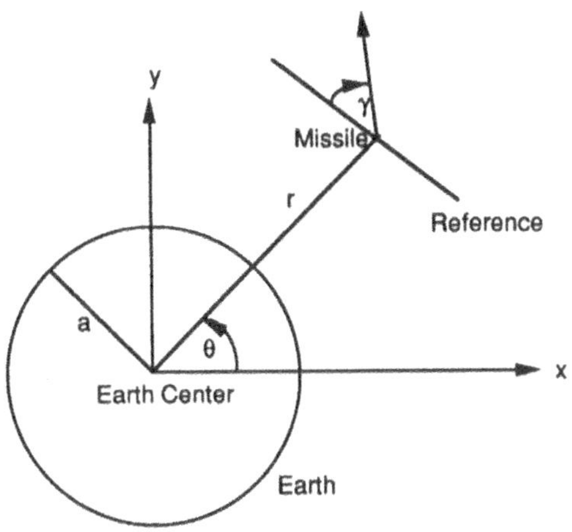

Fig. 11.2 Missile in gravity field.

this model, the gravitational acceleration is always 32.2 ft/s^2 in the negative altitude direction, regardless of altitude. Therefore, we know that this model can only be valid at the lower altitudes, since in actuality the gravitational acceleration decreases with increasing altitudes.

In general, a body in a gravitational field can be depicted in an Earth-centered coordinate system shown in Fig. 11.2. In this system the Earth is nonrotating and the gravitational acceleration acting on the missile is toward the center of the Earth. The missile has velocity V with respect to a reference that is tangent to the Earth and perpendicular to r (line from center of Earth to missile). The radius of the Earth is denoted by a in this figure.

According to Newton's law of universal gravitation, two bodies attract each other with a force that acts along a line connecting the two bodies. The force is proportional to the product of the masses of the two bodies and inversely proportional to the square of the distance between the two bodies. If one of the bodies is the Earth and the mass of the second body is negligible compared to the Earth, Newton's law of universal gravitation can be expressed in vector form as [2, 3]

$$\ddot{\boldsymbol{r}} = \frac{-gm\,\boldsymbol{r}}{r^3}$$

where $\boldsymbol{r}$ is a vector from the center of the Earth to the second body, and gm is known as the gravitational parameter with the value

$$gm = 1.4077 * 10^{16} ft^3/s^2$$

For simulation purposes and to be consistent with the work we have already done with tactical interceptors, it is natural to desire to express Newton's law in Cartesian coordinates. We shall soon see that for analytical purposes it will be more convenient to work in polar coordinates. By substituting

$$\boldsymbol{r} = x\boldsymbol{i} + y\boldsymbol{j}$$

we can express Newton's law of universal gravitation in Earth-centered inertial coordinates (x, y) as

$$\ddot{x} = \frac{-gm\,x}{(x^2 + y^2)^{1.5}}$$
$$\ddot{y} = \frac{-gm\,y}{(x^2 + y^2)^{1.5}}$$

where x and y are component distances from the center of the Earth to the body or missile. From Fig. 11.2 we can see that the initial conditions for the preceding differential equations are

$$x(0) = (a + alt_0)\cos\theta_0$$
$$y(0) = (a + alt_0)\sin\theta_0$$
$$\dot{x}(0) = V\cos\left(\frac{\pi}{2} - \gamma + \theta_0\right)$$
$$\dot{y}(0) = V\sin\left(\frac{\pi}{2} - \gamma + \theta_0\right)$$

where V is the initial missile velocity, alt_0 the initial missile altitude with respect to the surface of the Earth, γ the angle the velocity vector makes with respect to the reference, and θ_0 the initial angular location of the missile with respect to the x axis. Velocity and position components, with respect to the center of the Earth, can be found from repeated integration of the preceding differential equations. Once we have found the location of the missile with respect to the center of the Earth, it is useful to express the missile location with respect to the surface of the Earth. The instantaneous altitude of the missile can simply be found by finding the distance from the center of the Earth to the missile and then subtracting the Earth's radius, or

$$alt = (x^2 + y^2)^{0.5} - a$$

We can find the distance traveled along the surface of the Earth by referring to Fig. 11.3.

In general, the initial location of the missile can be expressed in vector notation as

$$\boldsymbol{r}_0 = x_0\boldsymbol{i} + y_0\boldsymbol{j}$$

and the future location of the missile at any arbitrary time can be expressed as

$$\boldsymbol{r} = x\boldsymbol{i} + y\boldsymbol{j}$$

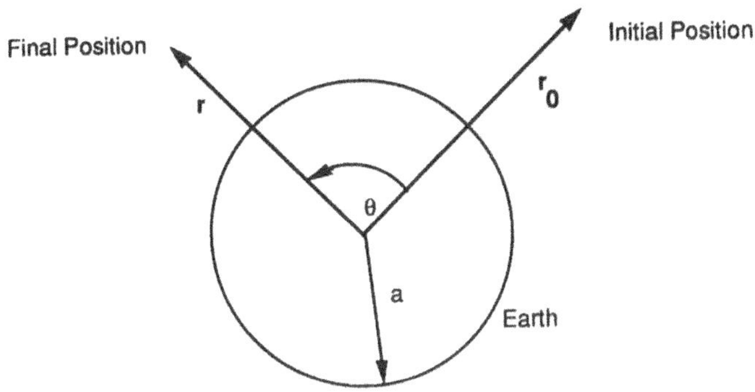

Fig. 11.3 Projecting distance missile travels on surface of Earth.

The angle between the two vectors r_0 and r can be found from the definition of the vector dot product, or

$$\theta = \cos^{-1}\frac{r_0 \cdot r}{|r_0||r|}$$

Therefore, the distance traveled, which is projected on the surface of a circular Earth, is given by

$$dist = a\theta$$

For comparative purposes, the equations of motion for a missile in a gravity field were programmed using the flat-Earth constant-gravity model and the Earth-centered coordinate system using Newton's law of universal gravitation. The MATLAB gravity field simulation appears in Listing 11.1. We can see from the listing that the position and velocity components in the flat-Earth model are denoted RT1, RT2, VT1, and VT2, respectively. In the Earth-centered system, the position and velocity components are denoted X, Y, X1, and Y1, respectively. The differential equations describing the missile in a gravity field for both coordinate systems can be found before the FLAG=1 statement. In the Earth-centered system, the missile position (x, y) is converted to a downrange and altitude so that a trajectory comparison can be made with answers obtained from the flat-Earth model.

A case was run in which an impulsive missile was launched from the surface of the Earth at a 45-deg angle. The initial missile velocity was 3000 ft/s. Figure 11.4 shows that the flat-Earth model (valid for a tactical missile) and the Earth-centered coordinate system model (valid for a strategic missile) yield the same missile trajectories. The total range traveled in both cases is about 47 n.mi., and the maximum altitude is about 12 n.mi.

Figure 11.5 shows that, when the initial speed of the impulsive missile is doubled to 6000 ft/s, we start to see some differences in the resultant missile

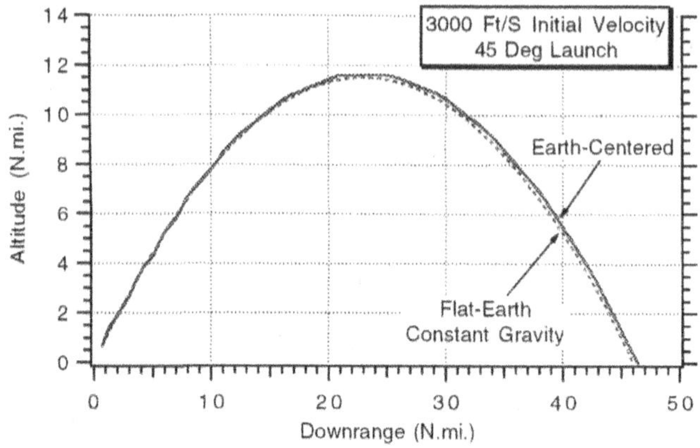

Fig. 11.4 Both models yield same answers when missile speed is small.

trajectories. In this case the missile travels about 180 n.mi. and the maximum altitude reached is about 50 n.mi. Remember that the correct answers are the ones given by the Earth-centered coordinate system differential equations. However, even in this case, the flat-Earth approximation (constant-gravity model) is fairly accurate.

Figure 11.6 shows, that when the impulsive missile speed is again doubled to 12,000 ft/s, the flat-Earth model yields large discrepancies in the resultant missile trajectory. The missile actually travels much farther than the flat-Earth model indicates, since the gravitational acceleration is reduced at the higher altitudes according to Newton's law of universal gravitation. In this case the distance traveled is more than 800 n.mi., and the peak altitude is about 220 n.mi.

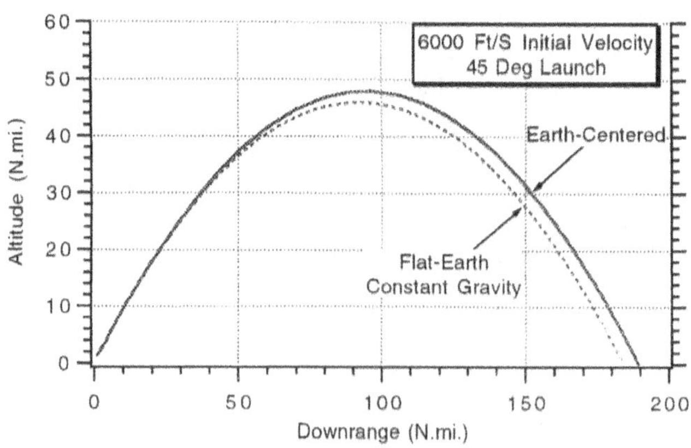

Fig. 11.5 Flat-Earth model is still fairly accurate when missile speed is doubled.

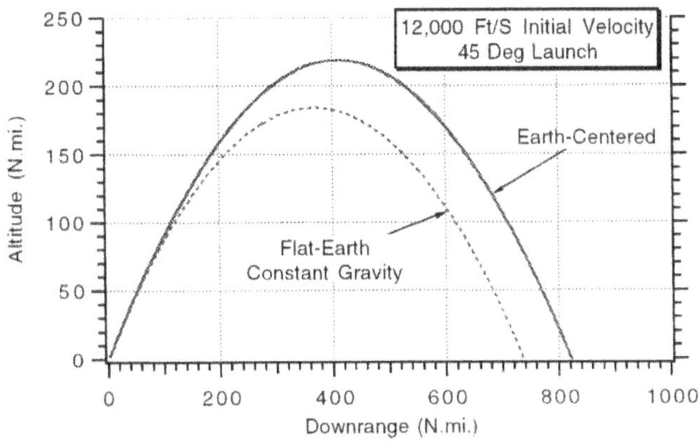

Fig. 11.6 Flat-Earth model is inaccurate when missile speed is again doubled.

LISTING 11.1 GRAVITY FIELD SIMULATION

```
count=0;
H=.01;
A=2.0926e7;
GM=1.4077e16;
GAM=45.;
ALTNM=0.;
V=3000.;
ALT=ALTNM/6076.;
ANG=0.;
VRX=V*cos(1.5708-GAM/57.3+ANG);
VRY=V*sin(1.5708-GAM/57.3+ANG);
G=32.2;
S=0.;
SCOUNT=0.;
RT1=ALT*cos(ANG);
RT2=ALT*sin(ANG);
VT1=VRX;
VT2=VRY;
X=(A+ALT)*cos(ANG);
Y=(A+ALT)*sin(ANG);
XFIRST=X;
YFIRST=Y;
X1=VRX;
Y1=VRY;
T=0.;
while ALTNM > -.0001
        RT1OLD=RT1;
```

```
RT2OLD=RT2;
VT1OLD=VT1;
VT2OLD=VT2;
XOLD=X;
YOLD=Y;
X1OLD=X1;
Y1OLD=Y1;
STEP=1;
FLAG=0;
while STEP <=1
   if FLAG==1
            STEP=2;
            RT1=RT1+H*RT1D;
            RT2=RT2+H*RT2D;
            VT1=VT1+H*VT1D;
            VT2=VT2+H*VT2D;
            X=X+H*XD;
            Y=Y+H*YD;
            X1=X1+H*X1D;
            Y1=Y1+H*Y1D;
            T=T+H;
   end
   AT1=0.;
   AT2=-G;
   RT1D=VT1;
   RT2D=VT2;
   VT1D=AT1;
   VT2D=AT2;
   TEMBOT=(X^2+Y^2)^1.5;
   X1D=-GM*X/TEMBOT;
   Y1D=-GM*Y/TEMBOT;
   XD=X1;
   YD=Y1;
   FLAG=1;
end
FLAG=0;
RT1=(RT1OLD+RT1)/2.+.5*H*RT1D;
RT2=(RT2OLD+RT2)/2.+.5*H*RT2D;
VT1=(VT1OLD+VT1)/2.+.5*H*VT1D;
VT2=(VT2OLD+VT2)/2.+.5*H*VT2D;
X=(XOLD+X)/2+.5*H*XD;
Y=(YOLD+Y)/2+.5*H*YD;
X1=(X1OLD+X1)/2+.5*H*X1D;
Y1=(Y1OLD+Y1)/2+.5*H*Y1D;
S=S+H;
if S>=1.99999
```

```
            S=0.;
            RT1NM=RT1/6076.;
            RT2NM=RT2/6076.;
            ALTNM=(sqrt(X^2+Y^2)-A)/6076.;
            R=sqrt(X^2+Y^2);
            RF=sqrt(XFIRST^2+YFIRST^2);
            CBETA=(X*XFIRST+Y*YFIRST)/(R*RF);
            BETA=acos(CBETA);
            DISTNM=A*BETA/6076.;
            count=count+1;
            ArrayT(count)=T;
            ArrayRT1NM(count)=RT1NM;
            ArrayRT2NM(count)=RT2NM;
            ArrayDISTNM(count)=DISTNM;
            ArrayALTNM(count)=ALTNM;
        end
end
figure
plot(ArrayRT1NM,ArrayRT2NM,ArrayDISTNM,ArrayALTNM),grid
xlabel('Downrange (Nmi)')
ylabel('Altitude (Nmi) ')
clc
output=[ArrayT',ArrayRT1NM',ArrayRT2NM',ArrayDISTNM',ArrayALTNM'];
save datfil.txt output /ascii
disp 'simulation finished'
```

POLAR COORDINATE SYSTEM [4, 5]

In the previous section we have shown how to accurately simulate a missile in a gravity field. The differential equations representing Newton's law of universal gravitation were first presented in vector form and then converted for simulation purposes to an Earth-centered Cartesian coordinate system. The Earth-centered coordinate system is extremely useful for simulation work because all integration can be done directly in the inertial frame. However, in order to get insight into the nature of trajectories in a gravity field and to get closed-form solutions, it is more convenient to work analytically in a polar coordinate system whose

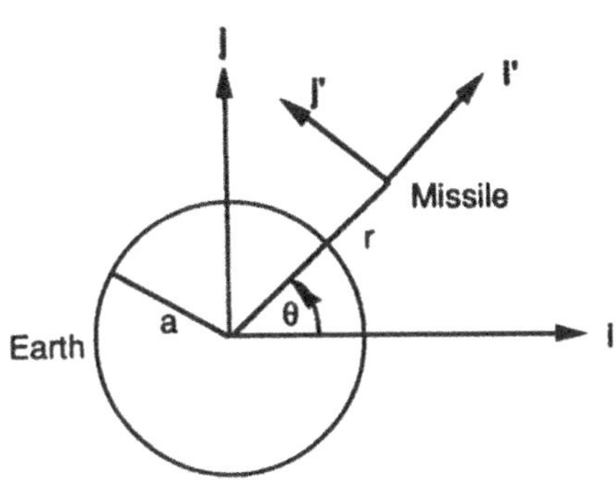

Fig. 11.7 Polar coordinate system with missile in gravity field.

origin is also at the center of the Earth. Figure 11.7 displays the polar coordinate system from which we proceed with our analysis.

In Fig. 11.7 we have defined a moving coordinate system that has the missile at the origin. The new coordinate system has an $\boldsymbol{i}'$ component along the distance vector and a $\boldsymbol{j}'$component perpendicular to $\boldsymbol{r}$. The relationship between the inertial Earth-centered coordinate system and the moving coordinate system is depicted in Fig. 11.8.

The relationship between the fixed and moving coordinate frames can be expressed mathematically as

$$\boldsymbol{i}' = \cos\theta\boldsymbol{i} + \sin\theta\boldsymbol{j}$$
$$\boldsymbol{j}' = -\sin\theta\boldsymbol{i} + \cos\theta\boldsymbol{j}$$

Since the new coordinate system is moving, we can express its rate of change with respect to the polar angle θ. Differentiating the preceding set of expressions with respect to the polar angle yields

$$\frac{\mathrm{d}\boldsymbol{i}'}{\mathrm{d}\theta} = -\sin\theta\boldsymbol{i} + \cos\theta\boldsymbol{j} = \boldsymbol{j}'$$
$$\frac{\mathrm{d}\boldsymbol{j}'}{\mathrm{d}\theta} = -\cos\theta\boldsymbol{i} - \sin\theta\boldsymbol{j} = -\boldsymbol{i}'$$

We can now find the rate of change of the new coordinate system as a function of time according to the chain rule, or

$$\frac{\mathrm{d}\boldsymbol{i}'}{\mathrm{d}t} = \frac{\mathrm{d}\theta}{\mathrm{d}t}\frac{\mathrm{d}\boldsymbol{i}'}{\mathrm{d}\theta} = \dot{\theta}\boldsymbol{j}'$$
$$\frac{\mathrm{d}\boldsymbol{j}'}{\mathrm{d}t} = \frac{\mathrm{d}\theta}{\mathrm{d}t}\frac{\mathrm{d}\boldsymbol{j}'}{\mathrm{d}\theta} = -\dot{\theta}\boldsymbol{i}'$$

We now have sufficient information so that we can take derivatives of vectors. The distance vector $\boldsymbol{r}$ can be expressed in the moving coordinate system as

$$\boldsymbol{r} = r\boldsymbol{i}'$$

Taking the derivative of the preceding expression yields

$$\dot{\boldsymbol{r}} = r\dot{\boldsymbol{i}}' + \dot{r}\boldsymbol{i}'$$

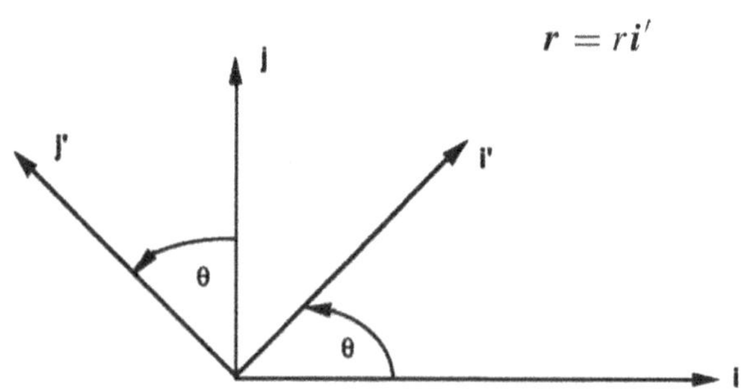

Fig. 11.8 Relationship between fixed and moving coordinate frames.

However, we have just shown that

$$\dot{i}' = -\dot{\theta}j'$$
$$\dot{j}' = \dot{\theta}i'$$

Therefore, substitution yields the radial velocity expression

$$\dot{r} = r\dot{\theta}j' + \dot{r}i'$$

Taking the derivative once more yields

$$\ddot{r} = (\ddot{r} - r\dot{\theta}^2)i' + (r\ddot{\theta} + 2\dot{r}\dot{\theta})j'$$

We know that gravitational acceleration is along i' and there is no acceleration along j'. Therefore, the preceding vector differential equation can be expressed as the following two scalar differential equations:

$$\frac{-gm}{r^2} = \ddot{r} - r\dot{\theta}^2$$
$$0 = r\ddot{\theta} + 2\dot{r}\dot{\theta}$$

Since

$$\frac{d}{dt}(r^2\dot{\theta}) = 2r\dot{r}\dot{\theta} + r^2\ddot{\theta}$$

we can say that

$$\frac{d}{d}(r^2\dot{\theta}) = 0$$

Integration yields a constant of integration that must be a moment arm times a tangential velocity, or

$$r^2\dot{\theta} = (a + alt)V\cos\gamma$$

In summary, the differential equations describing a missile in a gravity field can also be expressed in polar coordinates as

$$\ddot{r} - r\dot{\theta}^2 + \frac{gm}{r^2} = 0$$
$$r^2\dot{\theta} = (a + alt)V\cos\gamma$$

where the initial conditions are

$$r(0) = a + alt$$
$$\theta(0) = 0$$
$$\dot{r}(0) = V\sin\gamma$$

A MATLAB simulation was set up to demonstrate that the polar and Cartesian Earth-centered differential equations are equivalent. Listing 11.2 presents the gravity field simulation for both coordinate systems. The position and velocity components in the Cartesian system appear in the listing as X, Y, X1, and Y1, respectively. The range, its derivative, and the polar angle in the polar coordinate system are denoted by R0, R1, and PSI in the listing. The differential equations for both the Cartesian and polar coordinate systems appear before the FLAG=1 statement in the listing.

LISTING 11.2 GRAVITY FIELD SIMULATION WITH DIFFERENT COORDINATE SYSTEMS

```
count=0;
H=.01
A=2.0926e7;
GM=1.4077e16;
GAM=45.;
ALTNM=0.;
V=24000.;
ANGDEG=0.;
ANG=ANGDEG/57.3;
VRX=V*cos(1.5708-GAM/57.3+ANG);
VRY=V*sin(1.5708-GAM/57.3+ANG);
ALT=ALTNM/6076.;
S=0.;
SCOUNT=0.;
R0=A+ALT;
R1=V*sin(GAM/57.3);
PSI=0.;
X=(A+ALT)*cos(ANG);
Y=(A+ALT)*sin(ANG);
XFIRST=X;
YFIRST=Y;
X1=VRX;
Y1=VRY;
T=0.;
while ALTNM > -.0001
        R0OLD=R0;
        R1OLD=R1;
        PSIOLD=PSI;
        XOLD=X;
        YOLD=Y;
        X1OLD=X1;
        Y1OLD=Y1;
        STEP=1;
        FLAG=0;
```

```
while STEP <=1
   if FLAG==1
            STEP=2;
            R0=R0+H*R0D;
            R1=R1+H*R1D;
            PSI=PSI+H*PSID;
            X=X+H*XD;
            Y=Y+H*YD;
            X1=X1+H*X1D;
            Y1=Y1+H*Y1D;
            T=T+H;
   end
   PSID=(A+ALT)*V*cos(GAM/57.3)/(R0*R0);
   R1D=-GM/(R0*R0)+R0*PSID*PSID;
   R0D=R1;
   TEMBOT=(X^2+Y^2)^1.5;
   X1D=-GM*X/TEMBOT;
   Y1D=-GM*Y/TEMBOT;
   XD=X1;
   YD=Y1;
   FLAG=1;
end
FLAG=0;
R0=(R0OLD+R0)/2+.5*H*R0D;
R1=(R1OLD+R1)/2+.5*H*R1D;
PSI=(PSIOLD+PSI)/2+.5*H*PSID;
X=(XOLD+X)/2+.5*H*XD;
Y=(YOLD+Y)/2+.5*H*YD;
X1=(X1OLD+X1)/2+.5*H*X1D;
Y1=(Y1OLD+Y1)/2+.5*H*Y1D;
S=S+H;
if S>=9.99999
   S=0.;
   SPOLARNM=A*PSI/6076.;
   ALTPOLARNM=(R0-A)/6076.;
   ALTNM=(sqrt(X^2+Y^2)-A)/6076.;
   R=sqrt(X^2+Y^2);
   RF=sqrt(XFIRST^2+YFIRST^2);
   CBETA=(X*XFIRST+Y*YFIRST)/(R*RF);
   BETA=acos(CBETA);
   DISTNM=A*BETA/6076.;
   count=count+1;
   ArrayT(count)=T;
   ArraySPOLARNM(count)=SPOLARNM;
   ArrayALTPOLARNM(count)=ALTPOLARNM;
   ArrayDISTNM(count)=DISTNM;
```

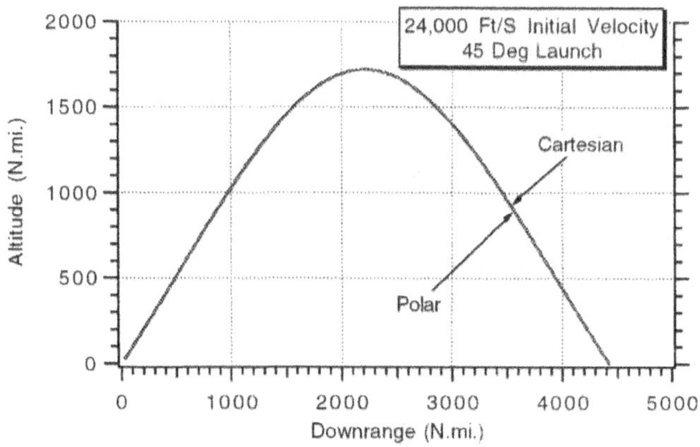

Fig. 11.9 Polar and Earth-centered gravity field equations yield identical trajectories.

```
            ArrayALTNM(count)=ALTNM;
        end
end
figure
plot(ArraySPOLARNM,ArrayALTPOLARNM,ArrayDISTNM,ArrayALTNM),grid
xlabel('Downrange (Nmi)')
ylabel('Altitude (Nmi) ')
clc
output=[ArrayT',ArraySPOLARNM',ArrayALTPOLARNM',ArrayDISTNM',ArrayALTNM'];
save datfil.txt output /ascii
disp 'simulation finished'
```

An experiment was run in which a missile with an initial velocity 24,000ft/s was launched from the surface of the Earth at an angle of 45 deg with respect to the reference. The resultant trajectories for both sets of differential equations appear in Fig. 11.9. We can see that the resultant trajectories are identical for all practical purposes. It is interesting to note that with an initial speed of 24,000 ft/s the impulsive missile traveled nearly 4500 n.mi. and reached an altitude of 1700 n.mi.

CLOSED-FORM SOLUTIONS [4, 5]

For those readers familiar with the literature on astrodynamics, apologies are offered in advance for the text's unconventional notation. Many other authors use *re* or *Re* rather than a for the radius of the Earth and use a rather than a_1 for the semimajor axis of an ellipse. The choice of a for the radius of the Earth is solely historical (it is used in [4]), whereas the use of a_1 for the semimajor axis of an ellipse was to avoid further confusion. In this section we will solve the previously derived differential equations of Newton's law of universal gravitation expressed in polar coordinates. In other words, we seek to find closed-form solutions of the polar differential equations

$$\ddot{r} - r\dot{\theta}^2 + \frac{gm}{r^2} = 0$$

$$r^2\dot{\theta} = (a + alt)V\cos\gamma$$

For convenience let us define constants r_0 and p such that

$$r_0 = a + alt$$

$$p = (a + alt)V\cos\gamma = r_0 V\cos\gamma$$

In addition, we will define an inverse range to be

$$u = 1/r$$

The goal is to convert both polar differential equations to one second-order differential equation in terms of u. First we know from the chain rule that u varies with time according to

$$\frac{\mathrm{d}u}{\mathrm{d}t} = \frac{\mathrm{d}\theta}{\mathrm{d}t}\frac{\mathrm{d}u}{\mathrm{d}\theta} = \dot{\theta}\frac{\mathrm{d}u}{\mathrm{d}\theta} = \frac{p}{r^2}\frac{\mathrm{d}u}{\mathrm{d}\theta}$$

An alternate way of seeing how u changes with respect to time is

$$\frac{\mathrm{d}u}{\mathrm{d}t} = \frac{\mathrm{d}u}{\mathrm{d}r}\frac{\mathrm{d}r}{\mathrm{d}t} = \frac{-1}{r^2}\frac{\mathrm{d}r}{\mathrm{d}t}$$

Equating both expressions yields

$$\frac{\mathrm{d}r}{\mathrm{d}t} = -p\frac{\mathrm{d}u}{\mathrm{d}\theta}$$

Next we define z to be

$$z = \frac{\mathrm{d}r}{\mathrm{d}t}$$

Using the chain rule to see how z changes with respect to time yields

$$\frac{\mathrm{d}z}{\mathrm{d}t} = \frac{\mathrm{d}\theta}{\mathrm{d}t}\frac{\mathrm{d}z}{\mathrm{d}\theta} = \frac{p}{r^2}\frac{\mathrm{d}}{\mathrm{d}\theta}\left[\frac{\mathrm{d}r}{\mathrm{d}t}\right] = \frac{p}{r^2}\frac{\mathrm{d}}{\mathrm{d}\theta}\left[-p\frac{\mathrm{d}u}{\mathrm{d}\theta}\right]$$

Therefore, we can say that

$$\frac{\mathrm{d}z}{\mathrm{d}t} = \frac{\mathrm{d}^2 r}{\mathrm{d}t^2} = \frac{-p^2}{r^2}\frac{\mathrm{d}^2 u}{\mathrm{d}\theta^2}$$

Substitution allows us to rewrite the second-order differential equation in range as

$$\ddot{r} - r\dot{\theta}^2 + \frac{gm}{r^2} = 0 = \frac{-p^2}{r^2}\frac{\mathrm{d}^{2u}}{\mathrm{d}\theta^2} - r\frac{p^2}{r^4} + gm\,u^2$$

Simplification yields

$$\frac{\mathrm{d}^2u}{\mathrm{d}\theta^2} + u = \frac{gm}{p^2}$$

For purposes that will be obvious later we can define a new constant to be

$$\lambda = \frac{r_0 V^2}{gm}$$

We can now summarize the transformed range polar differential equation to be

$$\frac{\mathrm{d}^2u}{\mathrm{d}\theta^2} + u = \frac{1}{\lambda r_0 \cos^2\gamma}$$

The original initial conditions on the polar differential equations were

$$r(0) = r_0$$
$$\dot{r}(0) = V\sin\gamma$$

As we already know that

$$\frac{du}{\mathrm{d}\theta} = -\frac{1}{p}\frac{\mathrm{d}r}{\mathrm{d}t} = \frac{-1}{r_0 V\cos\gamma}\frac{dr}{dt}$$

we can say that

$$\frac{du}{\mathrm{d}\theta}(0) = \frac{-\tan\gamma}{r_0}$$

The other initial condition is simply

$$u(0) = \frac{1}{r(0)} = \frac{1}{r_0}$$

The solution to the preceding second-order differential equation is

$$u = A\sin\theta + B\cos\theta + \frac{1}{\lambda r_0\cos^2\gamma}$$

where A and B can be found from the initial conditions. After some algebra we obtain the complete solution in terms of u as

$$u = \frac{1 - \cos\theta}{\lambda r_0 \cos^2\gamma} + \frac{1}{r_0}\frac{\cos(\theta+\gamma)}{\cos\gamma}$$

However, since

$$u = 1/r$$

the solution in terms of r becomes

$$\frac{r_0}{r} = \frac{1 - \cos\theta}{\lambda\cos^2\gamma} + \frac{\cos(\theta+\gamma)}{\cos\gamma}$$

or, more conveniently,

$$\begin{aligned} r &= \frac{r_0\lambda\cos^2\gamma}{1 - \cos\theta + \lambda\cos\gamma\cos(\theta+\gamma)} \\ &= \frac{r_0\lambda\cos^2\gamma}{1 - \lambda\sin\theta\cos\gamma\sin\gamma - \cos\theta(1 - \lambda\cos^2\gamma)} \end{aligned}$$

Thus, given missile altitude r_0, velocity λ, and flight-path angle γ, we find the missile location r as a function of the central angle θ. The preceding closed-form solution is also the equation of an ellipse in a polar coordinate system. To prove this interesting fact we must first recognize that the equation for an ellipse in polar coordinates is

$$r = \frac{a_1(1 - e^2)}{1 - e\cos(\theta - \omega)} = \frac{a_1(1 - e^2)}{1 - e\sin\theta\sin\omega - e\cos\theta\cos\omega}$$

where a_1 is the semimajor axis, e the eccentricity, and ω the argument of the apogee. The trajectory equation and the equation for an ellipse are equivalent if

$$\begin{aligned} e\sin\omega &= \lambda\cos\gamma\sin\gamma \\ e\cos\omega &= 1 - \lambda\cos^2\gamma \end{aligned}$$

Squaring and adding the preceding equations yields an expression for the eccentricity in terms of λ, or

$$e = [1 + \lambda(\lambda - 2)\cos^2\gamma]^{0.5}$$

The trajectory equation yields a circle if $e = 0$, an ellipse if $0 < e < 1$, a parabola if $e = 1$, and a hyperbola for $e > 1$. If we set the flight-path angle γ to zero, we can see that we get circular motion if $\lambda = 1$, elliptical motion for $0 < \lambda < 2$, parabolic motion for $\lambda = 2$, and hyperbolic motion for $\lambda > 2$. Because we can express the initial velocity in terms of λ as

$$V = \sqrt{\frac{\lambda gm}{r_0}}$$

we can determine the trajectory shape from the magnitude of the velocity!

The Earth-centered trajectory generator was modified so that the initial flight-path angle was zero and the initial velocity expressed in terms of λ according to the preceding velocity equation. In addition, the outputs, rather than being downrange and altitude, were expressed in the natural x, y units (that is, distance from the center of the Earth converted to nautical miles). Listing 11.3 presents the resultant MATLAB orbit generator program. We can see from the listing that the missile is initially at 1000 n.mi. altitude.

A case was run in which λ was set to 1. Figure 11.10 shows that the simulation indicates that the missile trajectory is indeed circular—as theory predicted! Figure 11.11 shows that when λ was set to 1.5 the simulation got an elliptical orbit for the missile—again, as theory predicted! Values of λ between 0 and 2 should yield elliptical orbital motion, with 1 being circular.

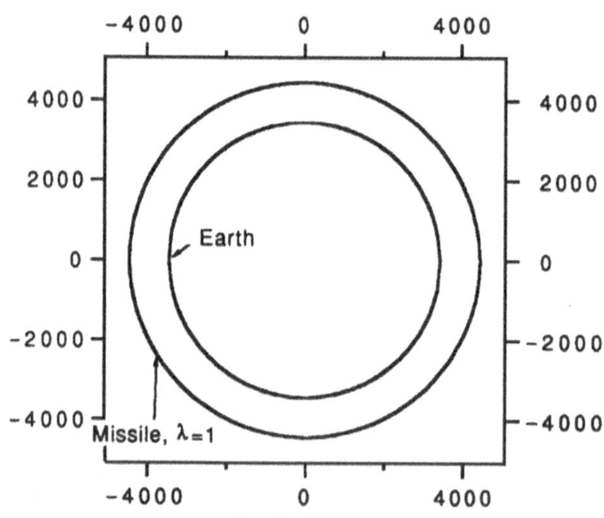

Fig. 11.10 Simulation yields circular orbit when λ is unity.

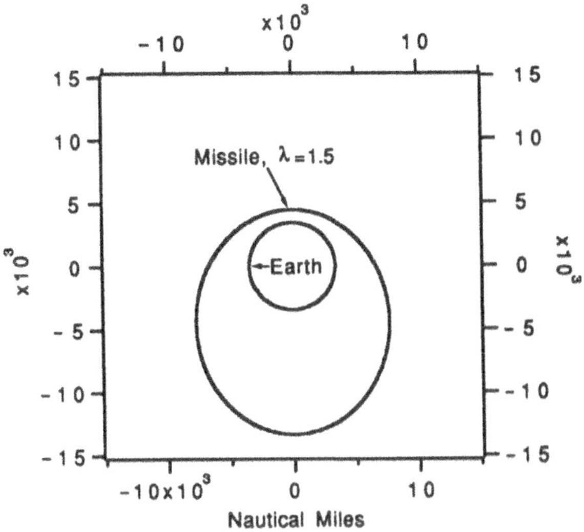

Fig. 11.11 Setting $\lambda = 1.5$ yields elliptical orbit.

LISTING 11.3 MATLAB ORBIT GENERATOR

```
count=0;
H=.01;
A=2.0926e7;
GM=1.4077e16;
GAM=0.;
ALTNM=1000.;
ALT=ALTNM*6076.;
XLAM=1.;
V=sqrt(GM*XLAM/(A+ALT));
ANGDEG=90.;
ANG=ANGDEG/57.3;
VRX=V*cos(1.5708-GAM/57.3+ANG);
VRY=V*sin(1.5708-GAM/57.3+ANG);
S=0.;
SCOUNT=0.;
X=(A+ALT)*cos(ANG);
Y=(A+ALT)*sin(ANG);
XFIRST=X;
YFIRST=Y;
X1=VRX;
Y1=VRY;
T=0.;
TF=30000.;
```

```
while T < TF
        XOLD=X;
        YOLD=Y;
        X1OLD=X1;
        Y1OLD=Y1;
        STEP=1;
        FLAG=0;
        while STEP<=1
           if FLAG==1
                     STEP=2;
                     X=X+H*XD;
                     Y=Y+H*YD;
                     X1=X1+H*X1D;
                     Y1=Y1+H*Y1D;
                     T=T+H;
           end
           TEMBOT=(X^2+Y^2)^1.5;
           X1D=-GM*X/TEMBOT;
           Y1D=-GM*Y/TEMBOT;
           XD=X1;
           YD=Y1;
           FLAG=1;
        end
        FLAG=0;
        X=(XOLD+X)/2+.5*H*XD;
        Y=(YOLD+Y)/2+.5*H*YD;
        X1=(X1OLD+X1)/2+.5*H*X1D;
        Y1=(Y1OLD+Y1)/2+.5*H*Y1D;
        S=S+H;
        if S>=49.99999
           S=0.;
           XNM=X/6076.;
           YNM=Y/6076.;
           count=count+1;
           ArrayT(count)=T;
           ArrayXNM(count)=XNM;
           ArrayYNM(count)=YNM;
        end
end
figure
plot(ArrayXNM,ArrayYNM),grid
xlabel('X (Nmi)')
ylabel('Y (Nmi) ')
clc
output=[ArrayT',ArrayXNM',ArrayYNM'];
save datfil.txt output /ascii
disp 'simulation finished'
```

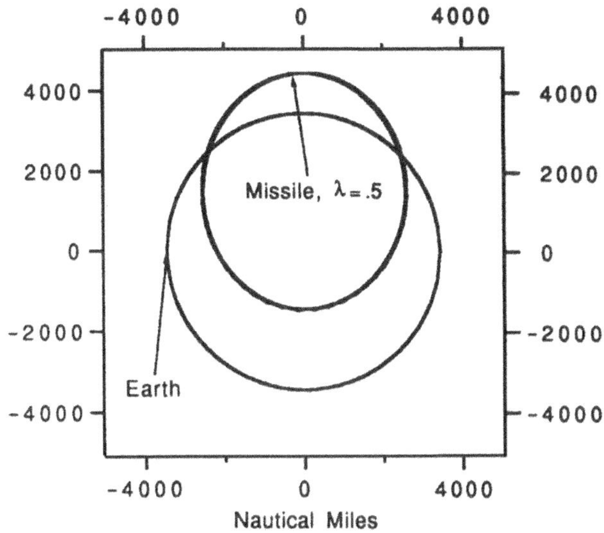

Fig. 11.12 Setting λ too small results in orbit that intersects Earth.

Theory says that if we set $\lambda=0.5$ we should also get an elliptical orbit. However, Fig. 11.12 shows that, although the simulation indicates an elliptical orbit, it is one that intersects the earth! Therefore, values of λ between 0 and 1 yield suborbital motion. Although this type of trajectory is not appropriate for a satellite, it is appropriate for a ballistic missile! Finally Fig. 11.13 shows that when $\lambda = 2$ we have achieved escape velocity and the missile motion is parabolic. This type of orbit does not intersect the Earth.

HIT EQUATION [4, 5]

We have seen that the previously derived trajectory equation is useful in obtaining closed-form solutions for satellite orbits and ballistic missile trajectories. If we specialize in the ballistic missile case, we can also get closed-form solutions from the trajectory equation, which, given an initial missile flight-path angle, altitude, and distance to be traveled (missile hits the Earth at that distance), will define the magnitude of the missile velocity required.

LISTING 11.4 SIMULATION TO DEMONSTRATE VALIDITY OF VELOCITY FORMULA

```
count=0;
H=.01;
A=2.0926e7;
GM=1.4077e16;
GAMDEG=23.;
```

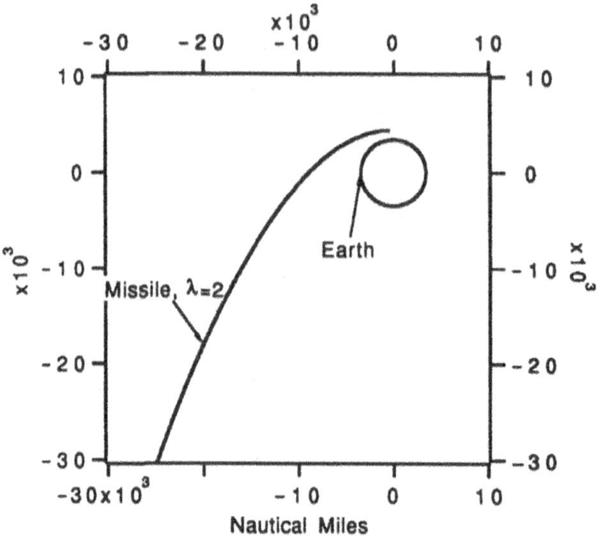

Fig. 11.13 Setting $\lambda = 2$ results in parabolic trajectory for missile.

```
GAM=GAMDEG/57.3;
DISTNM=6000.;
ANGDEG=0.;
ANG=ANGDEG/57.3;
PHI=DISTNM*6076./A;
ALTNM=0.;
ALT=ALTNM*6076.;
R0=A+ALT;
TOP=GM*(1.-cos(PHI));
TEMP=R0*cos(GAM)/A-cos(PHI+GAM);
BOT=R0*cos(GAM)*TEMP;
V=sqrt(TOP/BOT);
VRX=V*cos(1.5708-GAM+ANG);
VRY=V*sin(1.5708-GAM+ANG);
S=0.;
X=(A+ALT)*cos(ANG);
Y=(A+ALT)*sin(ANG);
XFIRST=X;
YFIRST=Y;
X1=VRX;
Y1=VRY;
T=0.;
while ALT >=0.
        XOLD=X;
```

```
YOLD=Y;
X1OLD=X1;
Y1OLD=Y1;
STEP=1;
STEP=1;
FLAG=0;
while STEP <=1
   if FLAG==1
               STEP=2;
               X=X+H*XD;
               Y=Y+H*YD;
               X1=X1+H*X1D;
               Y1=Y1+H*Y1D;
               T=T+H;
   end
   TEMBOT=(X^2+Y^2)^1.5;
   X1D=-GM*X/TEMBOT;
   Y1D=-GM*Y/TEMBOT;
   XD=X1;
   YD=Y1;
   FLAG=1;
end
FLAG=0;
X=(XOLD+X)/2+.5*H*XD;
Y=(YOLD+Y)/2+.5*H*YD;
X1=(X1OLD+X1)/2+.5*H*X1D;
Y1=(Y1OLD+Y1)/2+.5*H*Y1D;
S=S+H;
if S>=9.99999
   S=0.;
   XNM=X/6076.;
   YNM=Y/6076.;
   ALT=sqrt(X^2+Y^2)-A;
   ALTNM=ALT/6076.;
   R=sqrt(X^2+Y^2);
   RF=sqrt(XFIRST^2+YFIRST^2);
   CBETA=(X*XFIRST+Y*YFIRST)/(R*RF);
   BETA=acos(CBETA);
   DISTNM=A*BETA/6076.;
   count=count+1;
   ArrayT(count)=T;
   ArrayXNM(count)=XNM;
   ArrayYNM(count)=YNM;
   ArrayDISTNM(count)=DISTNM;
   ArrayALTNM(count)=ALTNM;
end
```

```
end
figure
plot(ArrayDISTNM,ArrayALTNM),grid
xlabel('Downrange (Nmi)')
ylabel('Altitude (Nmi) ')
clc
output=[ArrayT',ArrayXNM',ArrayYNM',ArrayDISTNM',ArrayALTNM'];
save datfil.txt output /ascii
disp 'simulation finished'
```

If we desire the missile to travel a distance *dist* then the total central angle traveled ϕ is given by (see Fig. 11.3)

$$\phi = dist/a$$

where a is the radius of the Earth. The missile hits the Earth when $r = a$. Therefore, substituting $r = a$ and $\theta = \phi$ into the trajectory equation solution yields

$$\frac{r_0}{a} = \frac{1 - \cos\phi}{\lambda \cos^2 \gamma} + \frac{\cos(\phi + \gamma)}{\cos \gamma}$$

Recognizing that

$$\lambda = \frac{r_0 V^2}{gm}$$

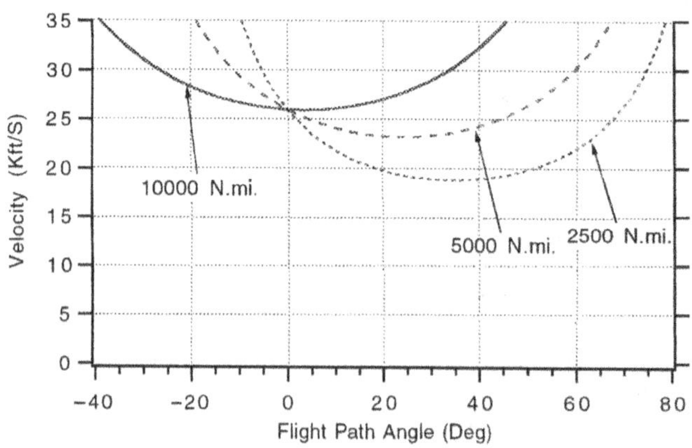

Fig. 11.14 Required velocity depends on range to be traveled and desired flight-path angle.

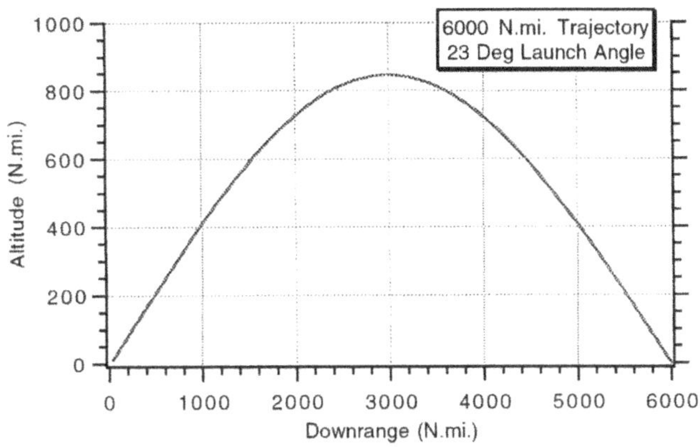

Fig. 11.15 Closed-form solution for velocity is accurate.

we can solve for the velocity. After some algebra we obtain

$$V = \sqrt{\frac{gm(1 - \cos\phi)}{r_0 \cos\gamma[(r_0 \cos\gamma/a) - \cos(\phi + \gamma)]}}$$

This equation tells us the velocity required to hit a target a certain distance away from our launch point, given we want to launch with a certain flight-path angle γ. Figure 11.14 displays the velocity formula in graphic form. We can see that, as expected, longer distances require larger missile velocities. If the initial flight-path angle is too large, the ballistic missile will never hit the Earth because the resultant velocity will exceed the escape velocity ($\lambda = 2$) and the trajectory will not be elliptical.

Listing 11.4 presents a modified ballistic missile simulation using the preceding velocity formula to derive the desired initial velocity given the desired flight-path angle, initial missile altitude, and distance to be covered. From the listing we can see that the missile is launched from the surface of the Earth with an initial flight-path angle of 23 deg. The target is 6000 n.mi. downrange on the surface of the Earth. Although the velocity formula was derived from solutions in the polar coordinate system, the simulation is based in the Cartesian coordinate system.

Figure 11.15 presents simulation results, in the form of an altitude vs downrange plot for the nominal case of Listing 11.4. We can see that the missile indeed travels the desired distance of 6000 n.mi. before hitting the surface of the Earth. The peak altitude for the missile is in excess of 800 n.mi. Figure 11.16 presents the same trajectory information in a way in which the curvature of the Earth is apparent.

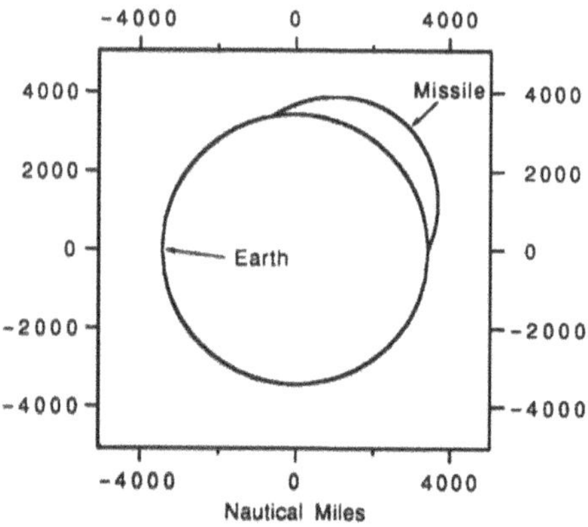

Fig. 11.16 Six thousand nautical miles, 23-deg trajectory.

FLIGHT TIME [4, 5]

We have already seen that, given a distance to be covered and initial flight-path angle, it was possible to derive a formula for the required velocity. Also associated with this velocity is the time to reach the target or time of flight t_F. It is also possible, based on the trajectory equation solution for r, to derive a closed-form solution for the time of flight. From the original gravity field differential equation in polar coordinates, we know that

$$r^2 \frac{d\theta}{dt} = r_0 V \cos \gamma$$

We can cross multiply terms to set up the integrals

$$\int_0^{\phi} r^2 d\theta = \int_0^{t_F} r_0 V \cos \gamma \, dt$$

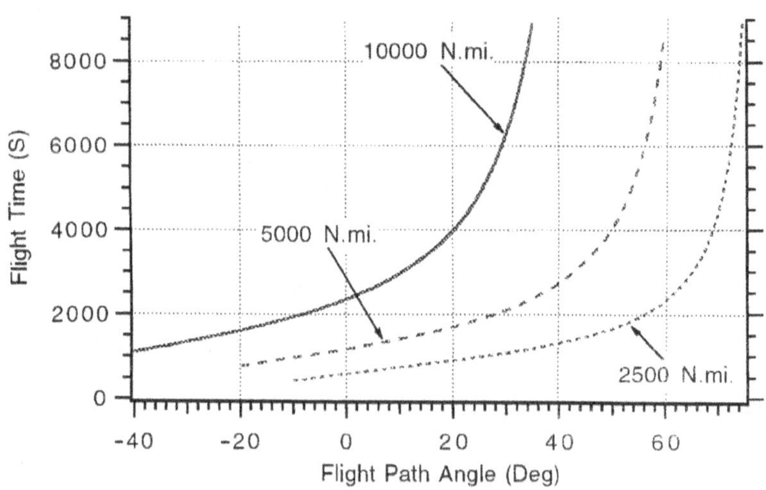

Fig. 11.17 Flight time increases with increasing flight-path angle and increasing distance to be traveled.

Integration of the right-hand side of the equation and substitution of the trajectory solution into the left-hand side yields the integral

$$t_F = \frac{1}{r_0 V \cos\gamma} \int_0^{\phi} \frac{r_0^2 \lambda^2 \cos^4\gamma}{[1 - \cos\theta + \lambda\cos\gamma\cos(\theta+\gamma)]^2}\, d\theta$$

After integration and much algebra, the closed-form solution assuming $\lambda < 2$ for the flight time simplifies to

$$t_F = \frac{r_0}{V\cos\gamma}\left\{\frac{\tan\gamma(1-\cos\phi) + (1-\lambda)\sin\theta}{(2-\lambda)\left[\frac{1-\cos\phi}{\lambda\cos^2\gamma} + \frac{\cos(\gamma+\phi)}{\cos\gamma}\right]} + \frac{2\cos\gamma}{\lambda\left(\frac{2}{\lambda}-1\right)^{1.5}}\tan^{-1}\left(\frac{\sqrt{\frac{2}{\lambda}-1}}{\cos\gamma\cot\frac{\phi}{2} - \sin\gamma}\right)\right\}$$

Figure 11.17 displays the flight time formula in graphic form. We can see that, as expected, it takes longer for a ballistic missile to travel greater distances. In addition, increasing the flight-path angle tends to increase the time of flight. For example, it takes about 1800 s for a ballistic missile to travel 5000 n.mi. when the flight-path angle is 20 deg. Increasing the flight-path angle to 40 deg increases the flight time to nearly 2800 s. We can also see that flight time increases monotonically and smoothly with increasing values of flight-path angle. We shall make use of this interesting observation in Chapter 13.

SUMMARY

This chapter was our first introduction into the strategic world. We saw that the constant-gravity, flat-Earth model used for tactical interceptors was not correct for strategic interceptors. Simulation models based on Newton's law of universal gravitation were derived from first principles. It was shown that an Earth-centered Cartesian system was useful for simulation and a polar model was more useful for analytical work. A closed-form solution was obtained for a ballistic missile's velocity in terms of flight-path angle and distance to be covered, and another expression was derived relating the flight time to the velocity. Simulation results confirmed the closed-form solutions. We shall make much use of these relationships in Chapter 13.

REFERENCES

[1] Kennedy, G. P., *Rockets, Missiles and Spacecraft of the National Air and Space Museum*, Smithsonian Institution Press, Washington DC, 1983.

[2] Bate, R. R., Mueller, D. D., and White, J. E., *Fundamentals of Astrodynamics,* Dover, New York, 1971.

[3] Battin, R. H., *An Introduction to the Mathematics and Methods of Astrodynamics,* AIAA Education Series, New York, 1987.

[4] Wheelon, A. D., "Free Flight of a Ballistic Missile," *ARS Journal,* Vol. 29, Dec. 1959, pp. 915–926.

[5] Regan, F., *Re-Entry Vehicle Dynamics,* AIAA Education Series, New York, 1984.

CHAPTER 12

Boosters

INTRODUCTION

We have seen in Chapter 11 that, in order for a ballistic interceptor to travel long distances or go into orbit, it must attain speeds in excess of 20 kft/s. From the rocket equation we saw in Chapter 10 that, with fuel mass fractions of less than 0.5 (that is, tactical missiles), it was impossible to reach these speeds. In this chapter we shall investigate preliminary booster designs so that speeds required for strategic travel can be achieved.

REVIEW

In Chapter 10 we saw that the change in velocity is related to specific impulse I_{sp} and fuel mass fraction mf according to the rocket equation, or

$$\Delta V = I_{sp} g \, \ell n \frac{1}{1 - mf}$$

Figure 12.1 displays the rocket equation in graphic form. From this figure we can see that fuel mass fractions approaching 0.9 are required if we wish to attain speeds in excess of 20 kft/s for fuel specific impulses of less than 300 s. The figure clearly shows that fuel mass fractions of less than 0.5 (that is, tactical missiles) lead to velocities that are not adequate for a strategic application.

We can think of a strategic interceptor as consisting of two sections: booster and payload. A single-stage booster (we will consider staging later in this chapter) consists of fuel and structure denoted by weights W_p and W_s, respectively, as shown in Fig. 12.2. Initially we will consider that the sole purpose of the single-stage booster is to get the payload up to speed. The payload, denoted by weight W_{pay}, consists of structure, electronics, a divert engine, and fuel. The purpose of the payload for strategic guided interceptors is to acquire the target and maneuver, using divert fuel, to hit the target.

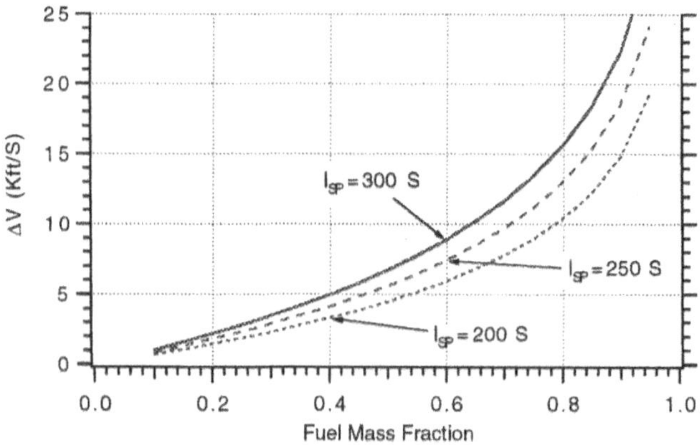

Fig. 12.1 Large fuel mass fractions are required for strategic applications.

If it is desired that the interceptor change its velocity by amount $\boldsymbol{\Delta}V$, then the weight of the structure, fuel, and payload must also follow the rocket equation as

$$W_S + W_P + W_{\text{PAY}} = (W_S + W_{\text{PAY}})\exp\left(\frac{\Delta V}{gI_{\text{sp}}}\right)$$

where I_{sp} denotes the specific impulse of the booster fuel and is measured in seconds. The fuel mass fraction has been defined as the ratio of the propellant weight to the total weight. To simplify computations in this chapter, an approximate fuel mass fraction mf^* is defined as the ratio of the propellant weight to the sum of the propellant weight plus structure or

$$mf^* = \frac{W_P}{W_P + W_S}$$

For small payloads the approximate and actual fuel mass fractions are equivalent. We can express the weight of the booster structure to the propellant weight and fuel approximate mass fraction according to

$$W_S = \frac{W_P(1 - mf^*)}{mf^*}$$

Substitution of the preceding relationship into the rocket equation yields, after some algebra, a formula for the propellant weight in terms of

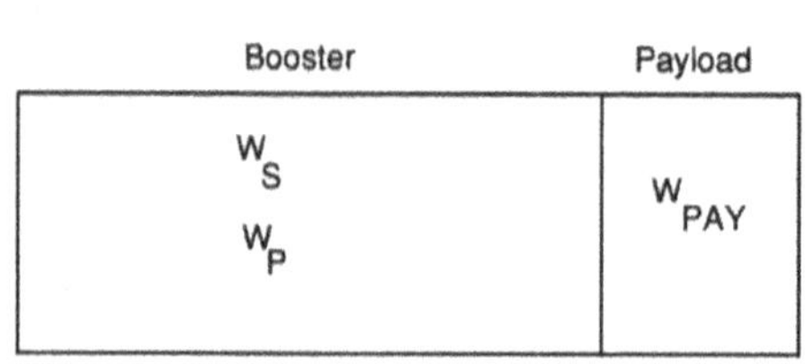

Fig. 12.2 Single-stage strategic interceptor model.

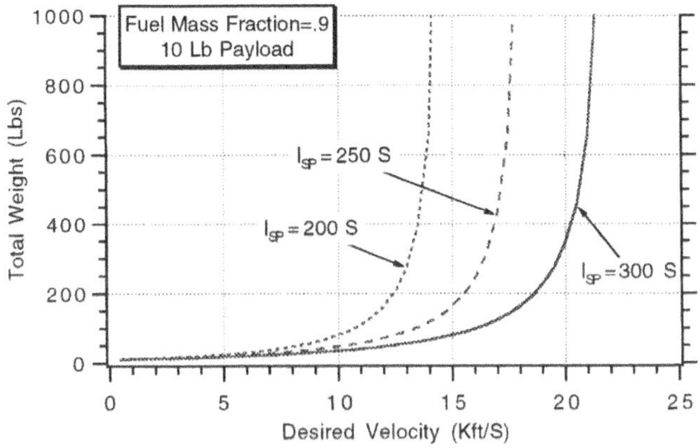

Fig. 12.3 A great deal of weight is required to bring small payloads to strategic speeds.

the payload weight, velocity desired, approximate fuel mass fraction, and specific impulse. This relationship can be expressed as

$$W_P = W_{\mathrm{PAY}}\left[\exp\left(\frac{\Delta V}{gI_{\mathrm{sp}}}\right) - 1\right] \Big/ \left[\frac{1}{mf^*} - \frac{1 - mf^*}{mf^*}\exp\left(\frac{\Delta V}{gI_{\mathrm{sp}}}\right)\right]$$

The total interceptor weight W_{tot} consists of the booster fuel and structure plus the payload, or

$$W_{\mathrm{tot}} = W_S + W_P + W_{\mathrm{PAY}}$$

Based upon the preceding relationships, Fig. 12.3 displays the total weight vs the desired change in velocity for an approximate fuel mass fraction of 0.9 and payload weight of 10 lb. We can see that, for a booster to reach a desired velocity of 20 kft/s from rest (in the absence of atmospheric drag), with a specific impulse of 300 s, more than 150 lb of total weight is required—just for a 10-lb payload! Doubling the payload weight will double the total weight. Decreasing the specific impulse or decreasing the fuel mass fraction both work in the direction of increasing the total weight.

STAGING

We have seen in the previous section that it can take a great deal of total weight to propel small payloads to near-orbital speeds. One way of reducing the total weight for a given approximate fuel mass fraction and specific impulse is to use staging. Figure 12.4 presents a two-stage booster. In this figure, the propellant and structural weights are indicated in each of the stages.

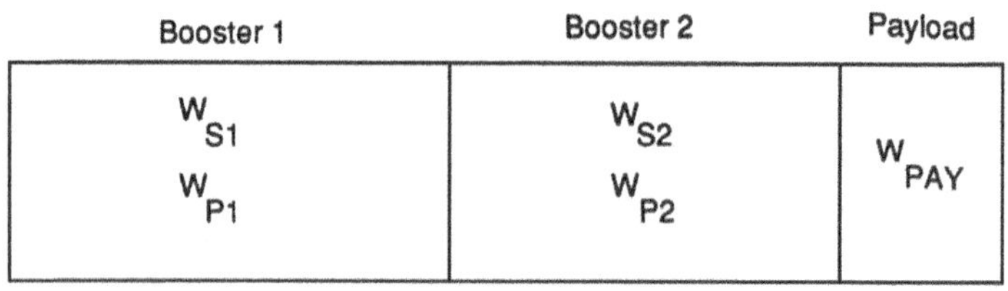

Fig. 12.4 Two-stage booster.

Therefore, the second-stage propellant weight can be expressed as

$$W_{P2} = W_{\text{PAY}}\left[\exp\left(\frac{\Delta V_2}{gI_{\text{sp2}}}\right) - 1\right] \Bigg/ \left[\frac{1}{mf2^*} - \frac{1 - mf2^*}{mf2^*}\exp\left(\frac{\Delta V_2}{gI_{\text{sp2}}}\right)\right]$$

where ΔV_2 is the desired velocity change attributed to the second stage, $mf2^*$ is the second-stage fuel mass fraction, and I_{sp2} is the second-stage specific impulse.

The structural weight of the second stage can then be expressed as

$$W_{S2} = \frac{W_{P2}(1 - mf2^*)}{mf2^*}$$

The weight of the second stage plus payload W_{tot2} is simply

$$W_{\text{tot2}} = W_{P2} + W_{S2} + W_{\text{PAY}}$$

We can now find the propellant weight of the first stage by treating W_{tot2} as an effective payload. The resultant weight is

$$W_{P1} = W_{\text{tot2}}\left[\exp\left(\frac{\Delta V_1}{gI_{\text{sp1}}}\right) - 1\right] \Bigg/ \left[\frac{1}{mf1^*} - \frac{1 - mf1^*}{mf1^*}\exp\left(\frac{\Delta V_1}{gI_{\text{sp1}}}\right)\right]$$

where ΔV_1 is the desired velocity change attributed to the first stage, $mf1^*$ is the first-stage approximate fuel mass fraction, and I_{sp1} is the first-stage specific impulse. The structural weight of the first stage can then be expressed as

$$W_{S1} = \frac{W_{P1}(1 - mf1^*)}{mf1^*}$$

Finally, the total interceptor weight (first stage plus rest) is given by

$$W_{\text{tot}} = W_{P1} + W_{S1} + W_{\text{tot2}}$$

Using the preceding relationships for a two-stage interceptor, the total weight was calculated as a function of desired velocity change for various values of specific impulse. It was assumed that each stage of the interceptor had equal specific impulses and equal approximate fuel mass fractions. In addition, it was

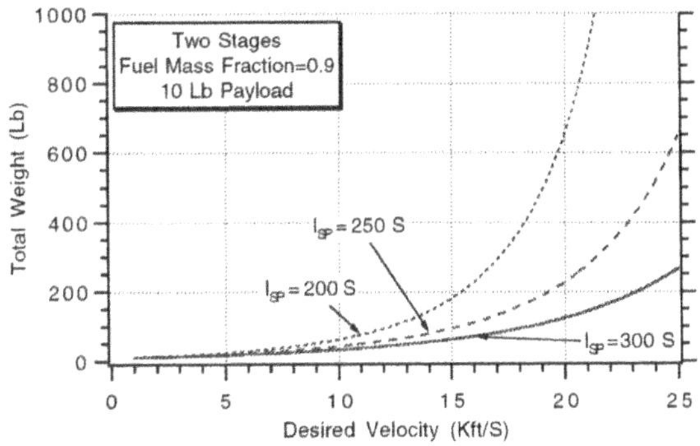

Fig. 12.5 Adding a stage reduces total weight requirements.

also assumed that half of the desired velocity was obtained with the first stage and the second half of the desired velocity was obtained with the second stage. Figure 12.5 shows how the total weight varies. For a desired velocity of 20 kft/s, Fig. 12.5 shows that, for a 300-s specific impulse, approximately 100 lb of total weight are required for a 10-lb payload using a two-stage interceptor. Figure 12.3 shows that for the same case a one-stage interceptor requires more than 150 lb of total weight. Thus, staging appears to be beneficial.

If the approximate fuel mass fraction were unity, the structural weight would be zero. In this case there would be no benefit to staging. In a sense, the unity fuel mass fraction case represents the minimum total weight that can propel a payload to a desired velocity for a given specific impulse. Figure 12.6 presents a comparison of weight requirements for different staging options. In the comparison an approximate fuel mass fraction of 0.9 and specific impulse 250 s were assumed for each of the stages. In addition, it was assumed that each stage contributed an equal fraction to the total desired velocity change. Superimposed on the figure is the infinite stage case (approximate fuel mass fraction equals unity) to represent minimal attainable weight. We can see that three stages get near-optimal

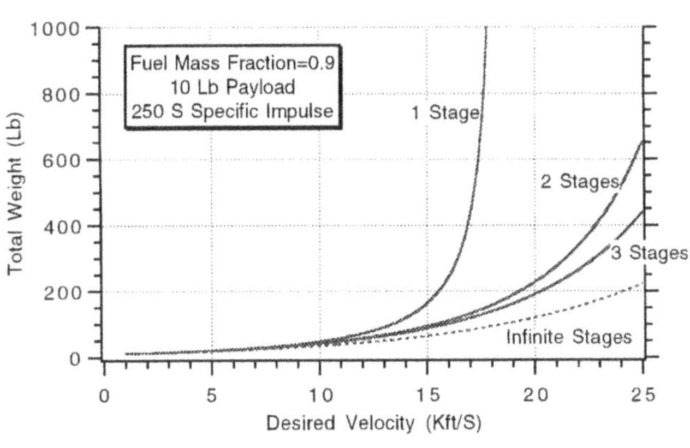

Fig. 12.6 Three stages yield near-minimal weight.

answers for the case in which the approximate fuel mass fraction is 0.9 and specific impulse is 250 s.

BOOSTER NUMERICAL EXAMPLE

We now have enough information so that we can begin, to first order, to model the boost phase of a strategic interceptor. In the previous section we derived formulas so that we could calculate weights based on desired velocity, approximate fuel mass fraction, and specific impulse. The maximum axial acceleration will occur right before staging, since that is where the interceptor weight is a minimum. If the maximum axial acceleration for each stage is given, then we have enough information to find the thrust levels for each of the stages. For example, in a two-stage strategic interceptor, the thrust level during stage 1, T_1, is given by

$$T_1 = a_{\max 1}(W_{\text{tot2}} + W_{S1})$$

where $a_{\max 1}$ is the maximum axial acceleration of the first stage, in units of gravity, and $W_{\text{tot2}} + W_{S1}$ is the weight of the first stage right before staging. The thrust level of the second stage can be found in a similar way and is given by

$$T_2 = a_{\max 2}(W_{\text{pay}} + W_{S2})$$

where $a_{\max 2}$ is the maximum axial acceleration of the second stage, in units of gravity, and $W_{\text{pay}} + W_{S2}$ is the weight of the second stage right before staging. We can find the thrust burn times from specific impulse and thrust information. The first- and second-stage burn times are given by

$$t_{B1} = \frac{I_{\text{sp}1} W_{P1}}{T_1}$$

$$t_{B2} = \frac{I_{\text{sp}2} W_{P2}}{T_2}$$

We now have enough information so that, given sufficient high-level information, we can compute a hypothetical booster's thrust-weight profiles. Listing 12.1 presents a MATLAB program in which thrust-weight information is computed to yield a desired velocity change. The program assumes a two-stage booster with a 100-lb payload. The specific impulse for both stages is the same and is 250 s, and the approximate fuel mass fraction for both stages is also the same and is 0.85. The desired change in velocity is 20,000 ft/s with the first stage contributing one-third of the desired ΔV and the second stage contributing the rest. The maximum axial acceleration in both stages is specified to be 10 g. The program also integrates the computed acceleration to check if the desired velocity is reached.

TABLE 12.1 SIMULATION OUTPUTS

Symbol	Definition	Value
W_{tot}	Total interceptor weight	6169 lb
W_{p1}	First-stage propellant weight	3474 lb
W_{s1}	First-stage structural weight	613 lb
W_{p2}	Second-stage propellant weight	1685 lb
W_{s2}	Second-stage structural weight	297 lb
T_1	Thrust level of first stage	26,950 lb
t_{B1}	Thrust burn time of first stage	32.2 s
T_2	Thrust level of second stage	3973 lb
t_{B2}	Thrust burn time of second stage	106 s

The program was run with the nominal inputs, and the interceptor total weight was computed to be 6169 lb. Table 12.1 summarizes the program's computation of key parameters.

Figure 12.7 presents the information of Table 12.1 in graphic form (but not to scale) as a thrust-weight profile. The sharp weight drops at 32.2 s and 138.2 s represent staging events (structural weight dropped). After the interceptor is finished burning propellant at 138.2 s, the total weight is the payload weight of 100 lb, as can be seen from the figure.

The MATLAB program of Listing 12.1 also had a capability to integrate the one-dimensional equation of motion

$$\dot{V} = \frac{gT}{W}$$

where g is the gravitational acceleration, T is the thrust level, and W the interceptor weight. Values for the instantaneous thrust and weight are obtained

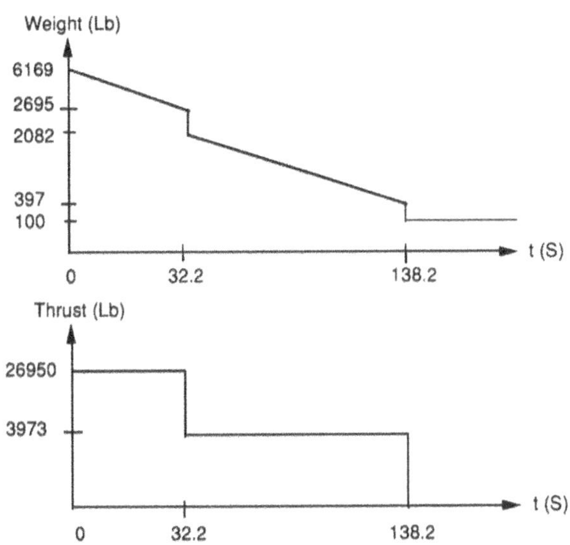

Fig. 12.7 Thrust-weight profiles for nominal case.

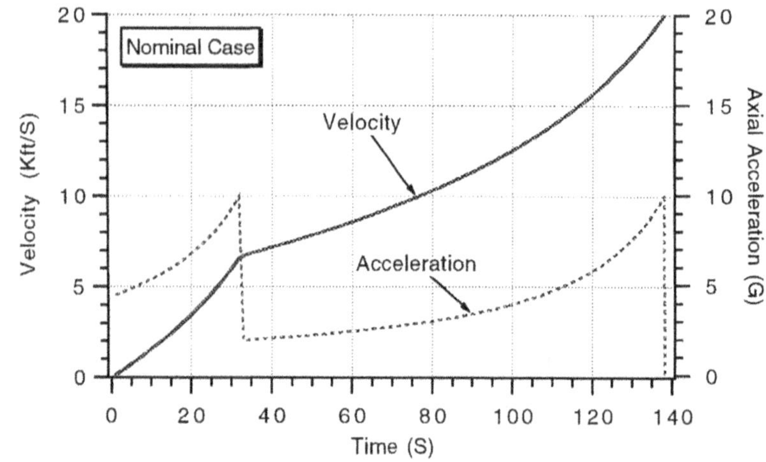

Fig. 12.8 Velocity and acceleration goals met with nominal design.

from Fig. 12.7. Figure 12.8 displays the resultant velocity and acceleration profiles for the nominal case. We can first see that the desired velocity goal of 20 kft/s has been reached by the end of the second-stage burn and that one-third of the velocity was attained at the end of the first-stage burn. We can also see from the acceleration profile that the desired maximum acceleration level of 10 g was also met. However, the axial booster acceleration is not constant and varied between 4 g and 10 g during the first-stage burn and varied between 2 g and 10 g during the second-stage burn.

LISTING 12.1 MATLAB THRUST-WEIGHT COMPUTATIONS

```
count=0;
XISP1=250.;
XISP2=250.;
XMF1=.85;
XMF2=.85;
WPAY=100.;
DELV=20000.;
DELV1=.3333*DELV;
DELV2=.6667*DELV;
AMAX1=10.;
AMAX2=10.;
TOP2=WPAY*(exp(DELV2/(XISP2*32.2))-1.);
BOT2=1/XMF2-((1.-XMF2)/XMF2)*exp(DELV2/(XISP2*32.2));
WP2=TOP2/BOT2;
```

```
WS2=WP2*(1-XMF2)/XMF2;
WTOT2=WP2+WS2+WPAY;
TRST2=AMAX2*(WPAY+WS2);
TB2=XISP2*WP2/TRST2;
TOP1=WTOT2*(exp(DELV1/(XISP1*32.2))-1.);
BOT1=1/XMF1-((1.-XMF1)/XMF1)*exp(DELV1/(XISP1*32.2));
WP1=TOP1/BOT1;
WS1=WP1*(1-XMF1)/XMF1;
WTOT=WP1+WS1+WTOT2;
TRST1=AMAX1*(WTOT2+WS1);
TB1=XISP1*WP1/TRST1;
DELVK=DELV/1000.;
H=.01;
T=0.;
S=0.;
V=0.;
while T <= (TB1+TB2)
      VOLD=V;
      STEP=1;
      FLAG=0;
      while STEP <=1
        if FLAG==1
                STEP=2;
                V=V+H*A;
                T=T+H;
        end
        if T<TB1
                WGT=-WP1*T/TB1+WTOT;
                TRST=TRST1;
        elseif(T<(TB1+TB2))
                WGT=-WP2*T/TB2+WTOT2+WP2*TB1/TB2;
                TRST=TRST2;
        else
                WGT=WPAY;
                TRST=0.;
        end
        A=32.2*TRST/WGT;
        FLAG=1;
      end
      FLAG=0;
      V=(VOLD+V)/2+.5*H*A;
      S=S+H;
      if S > =.99999
        S=0.;
        AG=A/32.2;
        VK=V/1000.;
```

```
            count=count+1;
            ArrayT(count)=T;
            ArrayVK(count)=VK;
            ArrayAG(count)=AG;
        end
end
figure
plot(ArrayT,ArrayVK),grid
xlabel('Time (Sec)')
ylabel('Velocity (Ft/Sec) ')
figure
plot(ArrayT,ArrayAG),grid
xlabel('Time (Sec)')
ylabel('Acceleration (G) ')
clc
output=[ArrayT',ArrayVK',ArrayAG'];
save datfil.txt output /ascii
disp 'simulation finished'
```

GRAVITY TURN [1]

Now that we have a nominal two-stage booster design, we would like to simulate its flight. Booster steering is beyond the scope of this chapter, so we will assume that the booster is launched at an initial flight-path angle γ with respect to the surface of the Earth. For counterclockwise travel Fig. 12.9 indicates the appropriate sign conventions and angle definitions, whereas for clockwise travel Fig. 12.10 is appropriate.

If we attempt to align the thrust vector with the booster velocity vector, we will obtain a gravity turn. The acceleration due to the booster thrusting a_T is given by

$$a_T = \frac{gT}{W}$$

where T is the thrust magnitude in pounds, W the missile weight, and g is 32.2 ft/s^2. The booster velocity V at any time could be found from the velocity components as

$$V = (\dot{x}^2 + \dot{y}^2)^{0.5}$$

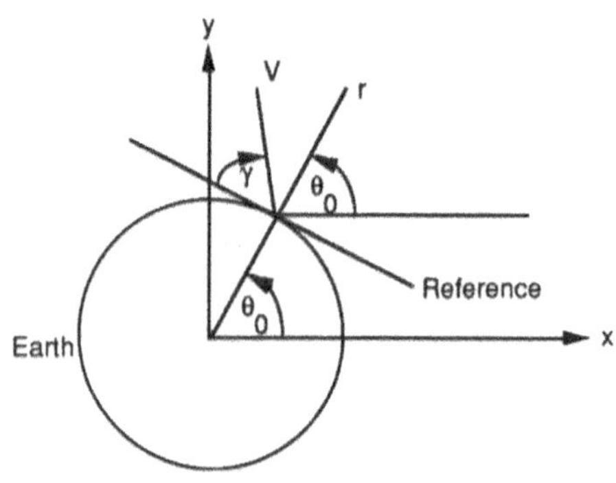

Fig. 12.9 Counterclockwise travel.

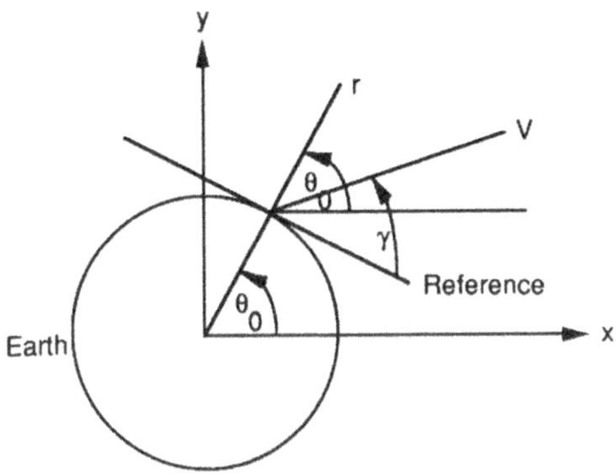

Fig. 12.10 Clockwise travel.

Therefore, during a gravity turn at any time the components of acceleration acting on the booster in our Earth-centered coordinate system are given by

$$\ddot{x} = \frac{-gm\,x}{(x^2+y^2)^{1.5}} + \frac{a_T\dot{x}}{V}$$

$$\ddot{y} = \frac{-gm\,y}{(x^2+y^2)^{1.5}} + \frac{a_T\dot{y}}{V}$$

where the initial conditions on velocity are related to the initial flight-path angle and location. For counterclockwise travel, the velocity initial conditions are

$$\dot{x}(0) = V(0)\cos(\pi/2 - \gamma_0 + \theta_0)$$
$$\dot{y}(0) = V(0)\sin(\pi/2 - \gamma_0 + \theta_0)$$

whereas for clockwise travel the appropriate velocity initial conditions are

$$\dot{x}(0) = V(0)\cos(-\pi/2 + \gamma_0 + \theta_0)$$
$$\dot{y}(0) = V(0)\sin(-\pi/2 + \gamma_0 + \theta_0)$$

The initial components of the booster location are given by

$$x(0) = (a + alt)\cos\theta_0$$
$$y(0) = (a + alt)\sin\theta_0$$

Listing 12.2 presents a MATLAB program that, given some booster design parameters, finds the appropriate thrust-weight profiles and, in addition, flies the booster through a gravity turn. We can see from the listing that the nominal booster design is the default case and that the initial flight-path angle of the booster during the gravity turn is 85 deg. During the trajectory the flight-path angle will start from 85 deg and gradually reduce to smaller values.

Cases were run with the nominal design, and the initial flight-path angle was made a parameter. The resultant trajectories, shown in Fig. 12.11, indicate that large flight-path angles are required just to get a trajectory for a gravity turn! If the flight-path angle is less than 80 deg, the booster will immediately crash into the Earth. As the booster thrusts, the flight-path angle rapidly decreases due to the small booster acceleration (about 4 g at the beginning). Eventually the flight-path angle decreases to the point where the component of the booster acceleration perpendicular to the surface of the Earth is not sufficient to overcome gravity.

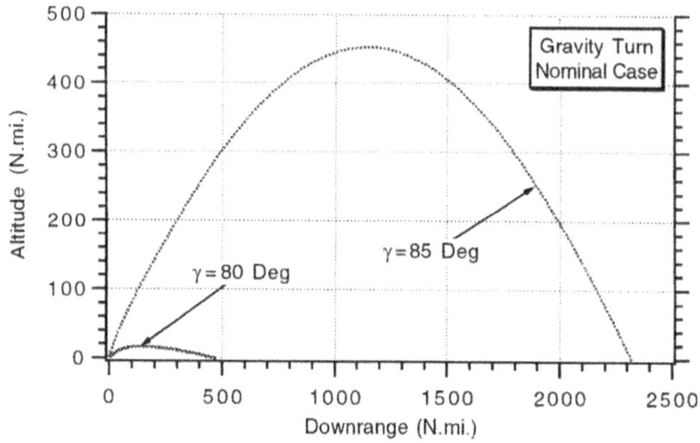

Fig. 12.11 Large flight-path angles are required for initial booster design.

LISTING 12.2 GRAVITY TURN SIMULATION

```
count=0;
LEFT=1;
XISP1=250.;
XISP2=250.;
XMF1=.85;
XMF2=.85;
WPAY=100.;
DELV=20000.;
DELV1=.3333*DELV;
DELV2=.6667*DELV;
AMAX1=10.;
AMAX2=10.;
GAMDEG=85.;
TOP2=WPAY*(exp(DELV2/(XISP2*32.2))-1.);
BOT2=1/XMF2-((1.-XMF2)/XMF2)*exp(DELV2/(XISP2*32.2));
WP2=TOP2/BOT2;
WS2=WP2*(1-XMF2)/XMF2;
WTOT2=WP2+WS2+WPAY;
TRST2=AMAX2*(WPAY+WS2);
TB2=XISP2*WP2/TRST2;
TOP1=WTOT2*(exp(DELV1/(XISP1*32.2))-1.);
BOT1=1/XMF1-((1.-XMF1)/XMF1)*exp(DELV1/(XISP1*32.2));
WP1=TOP1/BOT1;
WS1=WP1*(1-XMF1)/XMF1;
WTOT=WP1+WS1+WTOT2;
TRST1=AMAX1*(WTOT2+WS1);
TB1=XISP1*WP1/TRST1;
```

```
DELVK=DELV/1000.;
H=.01;
T=0.;
S=0.;
A=2.0926e7;
GM=1.4077e16;
ALTNM=0.;
ALT=ALTNM*6076.;
ANGDEG=90.;
ANG=ANGDEG/57.3;
if LEFT==1
        VRX=cos(1.5708-GAMDEG/57.3+ANG);
        VRY=sin (1.5708-GAMDEG/57.3+ANG);
else
        VRX=cos(-1.5708+GAMDEG/57.3+ANG);
        VRY=sin(-1.5708+GAMDEG/57.3+ANG);
end
X=(A+ALT)*cos(ANG);
Y=(A+ALT)*sin(ANG);
ALT=sqrt(X^2+Y^2)-A;
XFIRST=X;
YFIRST=Y;
X1=VRX;
Y1=VRY;
while ~(ALT <0 & T > 10)
        XOLD=X;
        YOLD=Y;
        X1OLD=X1;
        Y1OLD=Y1;
        STEP=1;
        FLAG=0;
        while STEP <=1
           if FLAG==1
                    STEP=2;
                    X=X+H*XD;
                    Y=Y+H*YD;
                    X1=X1+H*X1D;
                    Y1=Y1+H*Y1D;
                    T=T+H;
           end
           if T < TB1
                    WGT=-WP1*T/TB1+WTOT;
                    TRST=TRST1;
           elseif T < (TB1+TB2)
                    WGT=-WP2*T/TB2+WTOT2+WP2*TB1/TB2;
                    TRST=TRST2;
```

```
            else
                    WGT=WPAY;
                    TRST=0.;
            end
            AT=32.2*TRST/WGT;
            VEL=sqrt(X1^2+Y1^2);
            AXT=AT*X1/VEL;
            AYT=AT*Y1/VEL;
            TEMBOT=(X^2+Y^2)^1.5;
            X1D=-GM*X/TEMBOT+AXT;
            Y1D=-GM*Y/TEMBOT+AYT;
            XD=X1;
            YD=Y1;
            FLAG=1;
        end
        FLAG=0;
        X=(XOLD+X)/2+.5*H*XD;
        Y=(YOLD+Y)/2+.5*H*YD;
        X1=(X1OLD+X1)/2+.5*H*X1D;
        Y1=(Y1OLD+Y1)/2+.5*H*Y1D;
        ALT=sqrt(X^2+Y^2)-A;
        S=S+H;
        if S > =9.99999
            S=0.;
            R=sqrt(X^2+Y^2);
            RF=sqrt(XFIRST^2+YFIRST^2);
            CBETA=(X*XFIRST+Y*YFIRST)/(R*RF);
            BETA=acos(CBETA);
            DISTNM=A*BETA/6076.;
            ALTNM=(sqrt(X^2+Y^2)-A)/6076.;
            XNM=X/6076.;
            YNM=Y/6076.;
            count=count+1;
            ArrayT(count)=T;
            ArrayDISTNM(count)=DISTNM;
            ArrayALTNM(count)=ALTNM;
        end
end
figure
plot(ArrayDISTNM,ArrayALTNM),grid
xlabel('Downrange (Nmi)')
ylabel('Altitude (Nmi) ')
clc
output=[ArrayT',ArrayDISTNM',ArrayALTNM'];
save datfil output /ascii
disp 'simulation finished'
```

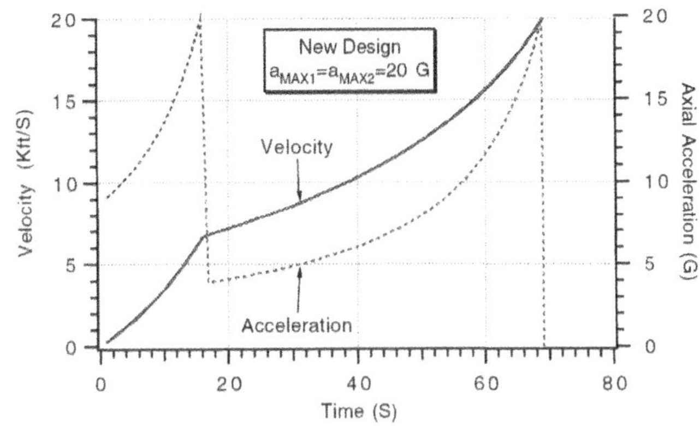

Fig. 12.12 Doubling booster axial acceleration halves burn time.

To remedy the situation so that we could get smaller flight-path angles to yield longer range trajectories, the maximum axial booster acceleration during each stage was increased from 10 *g* to 20 *g*. The resultant velocity and acceleration profiles due to this change appear in Fig. 12.12. We can see that the booster still reaches a velocity of 20 kft/s, but in half the time of the nominal design.

Gravity turns were performed, via the simulation, for the new booster design, and the results for different flight-path angles appear in Fig. 12.13. We can see that the larger axial booster acceleration allowed the booster to experience lower flight-path angles (without crashing into the ground), which increased the booster range. With the nominal design, the maximum range achieved with a flight-path angle of 85 deg was about 2300 n.mi. The new design, which permitted a lower flight-path angle of 65 deg, increased the maximum range about 2600 n.mi.

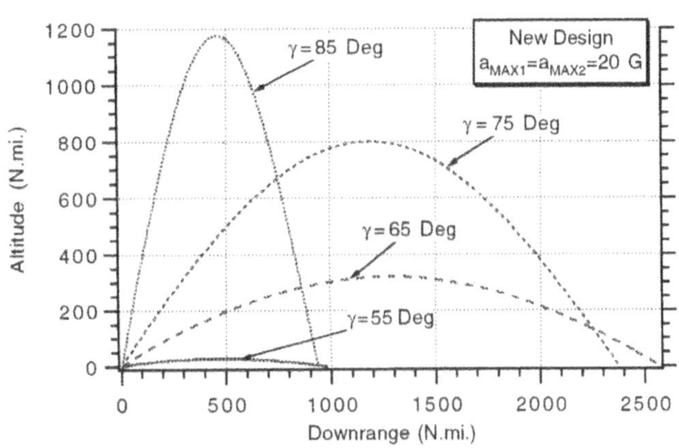

Fig. 12.13 New booster design yields longer flyout ranges.

SUMMARY

In this chapter we have attempted to show that it takes a great deal of booster weight to bring a small payload to near-orbital speeds. High-level formulas were developed and presented so that booster parameters could be specified from fundamental rocket equation relationships. The impact of a key booster parameter on a simple gravity turn trajectory was demonstrated via a simple numerical example.

REFERENCE

[1] Regan, F., *Re-Entry Vehicle Dynamics,* AIAA Education Series, New York, 1984.

CHAPTER 13

Lambert Guidance

INTRODUCTION

A particular problem, known as the problem of Lambert, has intrigued mathematicians for centuries. The solution to this problem is important for navigating spacecraft and for putting strategic missiles on a collision triangle. Elegant numerical solutions exist for the Lambert problem that are based on the known properties of a body in a gravity field [1]. The best of these solutions are numerically very efficient and accurate and, in fact, currently serve as fundamental algorithms in steering both spacecraft and ballistic missiles. Unfortunately, these elegant solutions are extremely difficult to understand because they involve subtle points in conic sections and a detailed understanding of hypergeometric series. In this chapter we shall use an easy to understand but numerically inefficient algorithm for solving Lambert's problem. It will then be shown how to speed up the algorithm by two orders of magnitude using a simple numerical technique. We shall then show how this solution can be used to steer a strategic boosting missile on a collision triangle with a threat.

STATEMENT OF LAMBERT'S PROBLEM

A body in a gravity field satisfies Newton's law of universal gravitation, or

$$\ddot{x} = \frac{-gm\,x}{(x^2 + y^2)^{1.5}}$$

$$\ddot{y} = \frac{-gm\,y}{(x^2 + y^2)^{1.5}}$$

Assume that the initial location of a body in the gravity field is given by

$$x(0) = x_0$$
$$y(0) = y_0$$

and it is desired that t_F seconds later the body be at location

$$x(t_F) = x_F$$
$$y(t_F) = y_F$$

Lambert's problem is to find the initial velocity orientation of the body in the gravity field so that the preceding initial conditions and boundary values are satisfied, or

$$\dot{x}(0) = ?$$
$$\dot{y}(0) = ?$$

SOLUTION TO LAMBERT'S PROBLEM

We showed in Chapter 11 that, given an initial flight-path angle and distance to be traveled, the initial missile velocity required to hit an object on the surface of the Earth is given by

$$V = \sqrt{\frac{gm(1 - \cos\phi)}{r_0 \cos\gamma[(r_0 \cos\gamma/a) - \cos(\phi + \gamma)]}}$$

where ϕ is the central angle separating the initial location of the missile and its intended target, γ the initial flight-path angle of the missile, a the radius of the Earth, and r_0 the initial distance from the center of the Earth to the missile, which can be expressed as

$$r_0 = a + alt$$

where alt is the initial altitude of the missile with respect to the surface of the Earth. Although the velocity equation was derived for hitting an object on the surface of the Earth, it can be made more general. If we desire to hit a target at any location r_F the preceding velocity equation can be modified to

$$V = \sqrt{\frac{gm(1 - \cos\phi)}{r_0 \cos\gamma[(r_0 \cos\gamma/r_F) - \cos(\phi + \gamma)]}}$$

In this new formula r_F is defined as

$$r_F = a + alt_F$$

where alt_F is the altitude of the intended target.

If the velocity vector is oriented for counterclockwise travel as shown in Fig. 13.1, then, given the preceding solution for the total required velocity, we can find the initial conditions on the velocity components in the Earth-centered

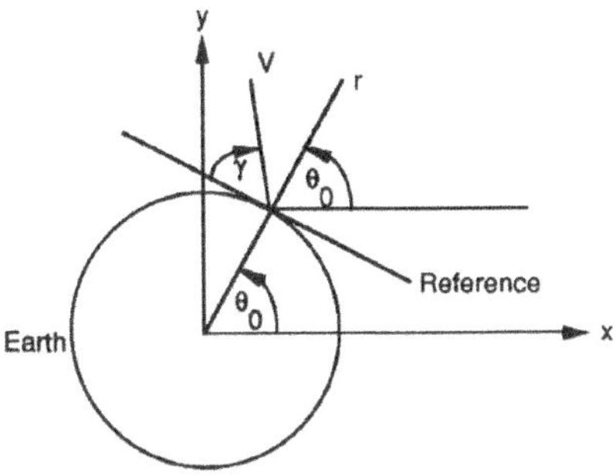

Fig. 13.1 Counterclockwise travel.

system by trigonometry as

$$\dot{x}(0) = V\cos(\pi/2 - \gamma + \theta_0)$$
$$\dot{y}(0) = V\sin(\pi/2 - \gamma + \theta_0)$$

where γ is the orientation of the missile velocity with respect to a reference that is tangent to the Earth and perpendicular to the vector from the center of the Earth to the initial location of the missile. We can see from Fig. 13.1 that θ_0 is the initial angular location of the missile with respect to the x axis of the Earth-centered Cartesian coordinate system.

On the other hand, if the velocity vector is intended to travel clockwise as shown in Fig. 13.2, then the initial conditions on the velocity components in the Earth-centered system can easily be shown to be

$$\dot{x}(0) = V\cos(\gamma - \pi/2 + \theta_0)$$
$$\dot{y}(0) = V\sin(\gamma - \pi/2 + \theta_0)$$

In Chapter 11 we also derived a formula for the time required for the missile to reach its intended target (t_F). The formula, which is valid for elliptical travel ($\lambda < 2$), does not require the target to be on the surface of the Earth and is given by

$$t_F = \frac{r_0}{V\cos\gamma}\left\{\frac{\tan\gamma(1-\cos\phi) + (1-\lambda)\sin\phi}{(2-\lambda)\left[\dfrac{1-\cos\phi}{\lambda\cos^2\gamma} + \dfrac{\cos(\gamma+\phi)}{\cos\gamma}\right]}\right.$$
$$\left. + \frac{2\cos\gamma}{\lambda[(2/\lambda)-1]^{1.5}}\tan^{-1}\left(\frac{\sqrt{(2/\lambda)-1}}{\cos\gamma\cot(\phi/2) - \sin\gamma}\right)\right\}$$

where V is the required velocity to hit the object, and γ was defined in Chapter 11 as

$$\lambda = \frac{r_0 V^2}{gm}$$

and ϕ is the angular distance to be traveled.

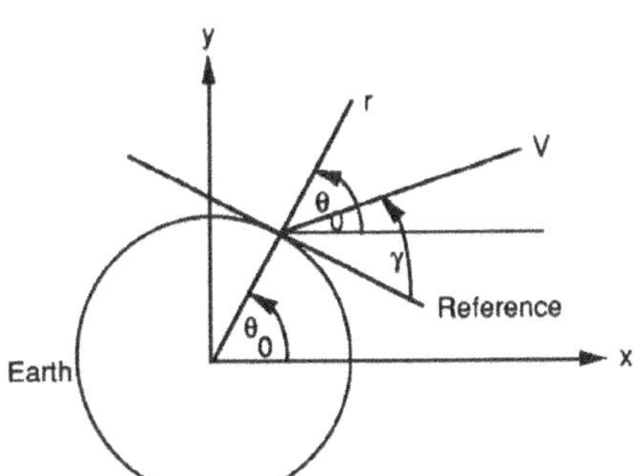

Fig. 13.2 Clockwise travel.

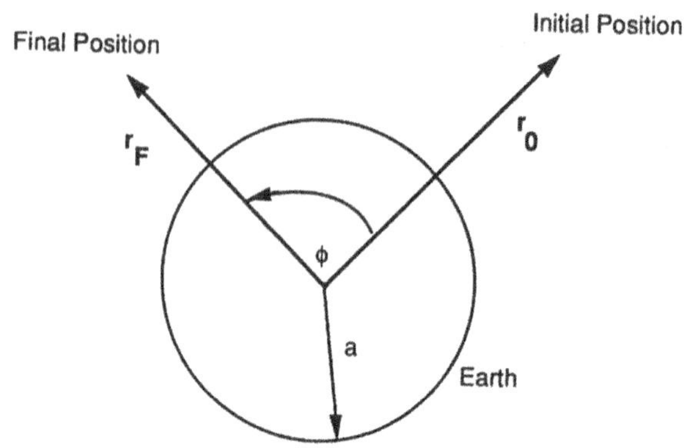

Fig. 13.3 Central angle between initial and final position.

To find the angular distance to be traveled, consider the geometry of Fig. 13.3, in which the initial and final position of an object in a gravity field are shown. In this figure $\boldsymbol{r}_0$ denotes a vector from the center of the Earth to the initial location of the object, and $\boldsymbol{r}_F$ denotes a vector from the center of the Earth to the final location of the object. The angle between the vectors is the central angle ϕ.

The central angle can be found from the definition of the vector dot product, or

$$\phi = \cos^{-1}\frac{\boldsymbol{r}_0 \cdot \boldsymbol{r}_F}{|\boldsymbol{r}_0||\boldsymbol{r}_F|}$$

Believe it or not, we now have sufficient information to numerically solve Lambert's problem!

If we know the initial and final destination of the target, we have just shown that we can find the central angle ϕ. With a central angle, $\boldsymbol{r}_0$, $\boldsymbol{r}_F$, and a flight-path angle γ, sufficient information is available to find the required velocity from our closed-form solution. The resultant velocity can then be used to solve for the flight time from our other closed-form solution. It is important to note that the flight time and velocity obtained are *exact* solutions for the flight-path angle used. Stated mathematically, we can say that given γ, $\boldsymbol{r}_0$, and $\boldsymbol{r}_F$ we can use the following relationships, which are based on exact closed-form solutions:

$$\begin{aligned} \phi &= f(\boldsymbol{r}_0, \boldsymbol{r}_F) \\ V &= f(\boldsymbol{r}_0, \boldsymbol{r}_F, \phi, \gamma) \\ t_F &= f(V, \phi, \gamma) \end{aligned}$$

Recall that in Lambert's problem we are given $\boldsymbol{r}_0$, $\boldsymbol{r}_F$, and $\boldsymbol{t}_F$ and seek to find V and γ. If we use the preceding relationships, we do not know how to choose γ, nor are we guaranteed that a particular value of γ will yield the desired flight time t_F.

We can solve the problem by the method of brute force. That is, we work out all solutions until we find the one that satisfies the constraints of the problem. For example, we start with $\gamma = -90$ deg, solve for the velocity, and then solve for the time of flight. If the flight time is less than the desired flight time, we repeat the procedure with a slightly larger value of γ. We stop the loop when the computed flight time is greater than the desired flight time. If the flight-path angle that satisfies the preceding procedure is negative, we know that the solution must be rejected since it requires the missile to travel through the Earth. This numerical method converges because we saw in Fig. 11.17 that flight time is smooth and monotonically increasing with increasing flight-path angle.

NUMERICAL EXAMPLE

Listing 13.1 presents sample MATLAB code for finding the Lambert solution, based on the procedure developed in the previous section. In the notation of Listing 13.1 we can say that, given an initial angle and altitude for the missile (XLONGMDEG, ALTNMM), an initial angle and altitude for the target (XLONGTDEG, ALTNMT), and a desired flight time (TF), the program iterates on the flight-path angle (GAMDEG) until a solution is found. From the listing we can see that the program consists of two loops. The first loop iterates on the flight-path angle in units of 0.1 deg. When a flight time is found that exceeds the desired flight time, we exit the loop for another loop that increments the flight-path angle (after decreasing the last flight path angle by 0.15 deg) in very fine units of 0.0001 deg. This loop is required to get extremely precise answers. When the desired flight time is achieved, we exit the loop and the routine. The routine, as written, is about 100 times slower than more elegant Lambert routines [1]. We shall show in the next section that by performing a more intelligent search it is possible to find the correct solution to Lambert's problem in a few iterations, thus making this approach very competitive with more elegant Lambert routines. However, the goal in this section is to develop a routine that simply works and is easy to understand.

To demonstrate how the routine works, the nominal case, shown in the listing, was run. In this case the missile is on the surface of the Earth 45 deg away from the target. It is desired to find the velocity orientation of the missile (VRX, VRY) so that the missile will hit the target in exactly 1000 s. Figure 13.4 shows that the solution

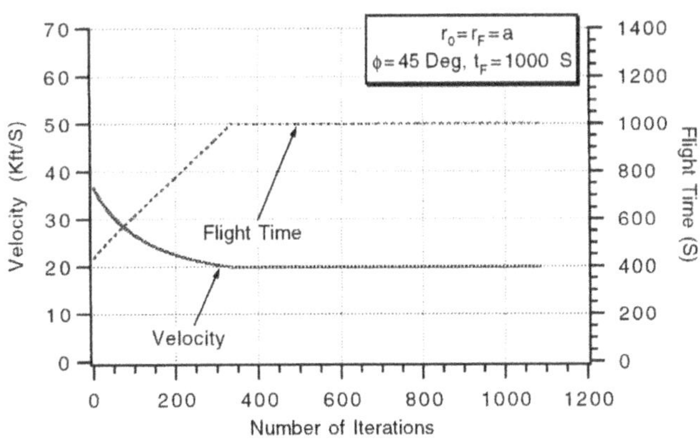

Fig. 13.4 It takes 1084 iterations to get exact solution.

converges to the exact value in 1084 iterations. However, the solution appears to be approximately correct after 335 iterations.

LISTING 13.1 LAMBERT ROUTINE USING BRUTE FORCE APPROACH

```
clear
global a gm count
global ArrayICOUNT ArrayBGAM ArrayVRX ArrayVRY ArrayTF
format long e
XLONGMDEG=45.;
XLONGTDEG=90.;
ALTNMT=0.;
ALTNMM=0.;
TF=1000;
DEGRAD=360./(2.*pi);
a=2.0926e7;
gm=1.4077e16;
ALTT=ALTNMT*6076.;
ALTM=ALTNMM*6076.;
XLONGM=XLONGMDEG/DEGRAD;
XLONGT=XLONGTDEG/DEGRAD;
XM=(a+ALTM)*cos(XLONGM);
YM=(a+ALTM)*sin(XLONGM);
XT=(a+ALTT)*cos(XLONGT);
YT=(a+ALTT)*sin(XLONGT);
[VRXM,VRYM]=olambert(XM,YM,TF,XT,YT,XLONGM,XLONGT)
 output=[ArrayVRX', ArrayVRY', ArrayTF'];
 disp('The final iteration')
 count
VRXM
VRYM
% data to file
 save datfil.txt output /ascii;
% olambert.m
function [vrx,vry]=olambert(xic,yic,tfdes,xf,yf,xlongm,xlongt)
global a gm % In version 3 comment this out!
global count ArrayICOUNT ArrayBGAM ArrayVRX ArrayVRY ArrayTF % for output array (if
reqd)
% Initialise outputs (if reqd)
count=0;
vrx=0;
vry=0;
tf=0;
%A
ric=sqrt(xic^2+yic^2);
```

```
rf=sqrt(xf^2+yf^2);
cphi=(xic*xf+yic*yf)/(ric*rf);
phi=acos(cphi);
r0=ric;
degrad=360./(2.*pi);
% Initialise while loop
SecondTimeThrough=0;
% Program executes this loop twice
while SecondTimeThrough <= 1
     % Initialise for loop
     if SecondTimeThrough == 0
        start=-90;
        step=.1;
        stop=+90;
     else
        start=gamdegnew;
        step=.0001;
        stop=gamdegfin;
     end;
     % Main body of program
     for gamdeg=start:step:stop
        %B
        gam=gamdeg/degrad;
        top=gm*(1-cos(phi));
        temp=r0*cos(gam)/rf-cos(phi+gam);
        bot=r0*cos(gam)*temp;
        if ~(top<0. | bot<0.)
              %C
              v=sqrt(top/bot);
              if xlongt>xlongm
                    vrx=v*cos(pi/2 -gam+xlongm);
                    vry=v*sin(pi/2 -gam+xlongm);
              else
                    vrx=v*cos(-pi/2 +gam+xlongm);
                    vry=v*sin(-pi/2 +gam+xlongm);
              end
              xlam=r0*v*v/gm;
              top1=tan(gam)*(1-cos(phi))+(1-xlam)*sin(phi);
              bot1p=(1-cos(phi))/(xlam*cos(gam)*cos(gam));
              bot1=(2-xlam)*(bot1p+cos(gam+phi)/cos(gam));
              top2=2*cos(gam);
              if ~((2/xlam-1) < 0.)
                    %D
                    bot2=xlam*((2/xlam-1)^1.5);
                    top3=sqrt(2/xlam-1);
                    bot3=cos(gam)/tan(phi/2)-sin(gam);
```

```
                    temp=(top2/bot2)*atan2(top3,bot3);
                    tf=r0*(top1/bot1+temp)/(v*cos(gam));

                    if (tf >tfdes)
                            break % out of the for loop
                    end; % condition #3
                    % output arrays (if reqd)
                    count=count+1;
                    ArrayICOUNT(count)=count;
                    ArrayBGAM(count)=57.3*gam;
                    ArrayVRX(count)=vrx;
                    ArrayVRY(count)=vry;
                    ArrayTF(count)=tf;
               end; % condition #2
          end; % condition #1
     end; % for loop
%E
gamdegnew=gamdeg-.15;
gamdegfin=gamdeg+1.;
SecondTimeThrough=SecondTimeThrough+1;
end % while loop
plot(count,ArrayTF)
```

To investigate the tradeoff between accuracy vs number of iterations required, a simple experiment was conducted. First the second loop of Listing 13.1 was removed from the Lambert subroutine so that the flight-path angle was only incremented in steps of 0.1 deg. Table 13.1 shows that the number of iterations required were reduced from 1084 to 335 and the resultant velocity accuracy (VRX, VRY) appears to be reduced slightly. Actually, the velocities are exact for a 1001-s flight but approximate for a Lambert solution requiring a 1000-s flight. Next, the first loop was modified so that the flight-path angle was incremented in steps of 1 deg (increased from 0.1 deg steps). Table 13.1 shows that the number of iterations was reduced to only 34, but the accuracy loss was more significant if the desired flight time is truly 1000 s. These answers are exact in the sense a hit will result in 1014 s but inaccurate for the Lambert solution requiring exactly 1000 s.

TABLE 13.1 ACCURACY EXPERIMENTS

Condition	V_{RX}	V_{RY}	t_F	Iterations
Nominal	−7696	18,329	1000	1084
Remove second loop	−7668	18,332	1001	335
One-degree increments	−7418	18,360	1014	34

SPEEDING UP LAMBERT ROUTINE

The routine for numerically solving Lambert's problem, presented in the previous section, can be speeded up by more than two orders of magnitude! We have already demonstrated that the brute force search on all possible flight-path angles results in many iterations. We can considerably restrict the brute force search and eliminate many iterations by recalling that the velocity formula was shown in Chapter 11 to be

$$V = \sqrt{\frac{gm(1-\cos\phi)}{r_0\cos\gamma[(r_0\cos\gamma/r_F)-\cos(\phi+\gamma)]}}$$

In this text we are only interested trajectories for ballistic missiles, so we can immediately rule out cases that lead to escape velocity ($\lambda = 2$) or

$$\lambda = 2 = \frac{V^2 r_0}{gm}$$

Substitution of the escape velocity condition into the velocity formula yields

$$2 = \frac{(1-\cos\phi)}{\cos\gamma[(r_0\cos\gamma/r_F)-\cos(\phi+\gamma)]}$$

We can solve the preceding equation for the flight-path angle γ. After much algebra we get two solutions corresponding to the minimum and maximum flight-path angles as

$$\gamma_{\min} = \tan^{-1}\left\{\left[\sin\phi - \sqrt{\frac{2r_0}{r_F}(1-\cos\phi)}\right]\bigg/(1-\cos\phi)\right\}$$

$$\gamma_{\max} = \tan^{-1}\left\{\left[\sin\phi + \sqrt{\frac{2r_0}{r_F}(1-\cos\phi)}\right]\bigg/(1-\cos\phi)\right\}$$

It should not be surprising that there are two solutions for the flight-path angle as we have already observed this phenomenon in Fig. 11.14. We also noticed in Figs. 11.14 and 11.17 that the solution for the velocity and time of flight were smooth, well-behaved functions of the flight-path angle. Based on the nonpathological nature of these solutions and the fact that the flight-path angle is well bounded, we do not have to evaluate each flight-path angle but can instead perform a more efficient search in finding the flight-path angle that corresponds to the desired flight time. For example, we can use an algorithm known as the secant method [3] to perform the search or

$$\gamma_{n+1} = \gamma_n + \frac{(\gamma_n - \gamma_{n-1})(t_{\text{FDES}} - t_{F_n})}{t_{F_n} - t_{F_{n-1}}}$$

We can see from the preceding equation that the new flight-path angle γ_{n+1} is related to previous values γ_n, γ_{n-1}. At each iteration the new computed value of flight-path angle is limited to the minimum and maximum possible values of the flight-path angle derived from the escape velocity condition. The search is terminated when the computed flight time t_{Fn} *is sufficiently close to the desired flight time* t_{TDES}.

Listing 13.2 is identical to the test program of Listing 13.1, except this time the Lambert routine is more efficient. We can see from the new Lambert routine that our initial guess of the flight-path angle is simply the average of the minimum and maximum flight-path angles derived from the escape velocity condition.

LISTING 13.2 MORE EFFICIENT LAMBERT ROUTINE

```
XLONGMDEG=45.;
XLONGTDEG=90.;
ALTNMT=0.;
ALTNMM=0.;
TF=1000.;
PI=3.14159;
DEGRAD=360./(2.*PI);
A=2.0926e7;
GM=1.4077e16;
ALTT=ALTNMT*6076.;
ALTM=ALTNMM*6076.;
XLONGM=XLONGMDEG/DEGRAD;
XLONGT=XLONGTDEG/DEGRAD;
XM=(A+ALTM)*cos(XLONGM);
YM=(A+ALTM)*sin(XLONGM);
XT=(A+ALTT)*cos(XLONGT);
YT=(A+ALTT)*sin(XLONGT);
[VRXM,VRYM]=lambertpz(XM,YM,TF,XT,YT,XLONGM,XLONGT)

% lambertpz.m file
function [vrx,vry]=lambert(xic,yic,tfdes,xf,yf,xlongm,xlongt)
a=2.0926e7;
gm=1.4077e16;
ric=sqrt(xic^2+yic^2);
rf=sqrt(xf^2+yf^2);
cphi=(xic*xf+yic*yf)/(ric*rf);
phi=acos(cphi);
sphi=sin(phi);
r0=ric;
degrad=360./(2.*pi);
icount=0;
```

```
gmin=atan2((sphi-sqrt(2.*r0*(1.-cphi)/rf)),(1-cphi));
gmax=atan2((sphi+sqrt(2.*r0*(1.-cphi)/rf)),(1-cphi));
gam=(gmin+gmax)/2.;
tf=0;
while ~(abs(tfdes-tf)<=(.00000001*tfdes))
        top=gm*(1.-cos(phi));
        temp=r0*cos(gam)/rf-cos(phi+gam);
        bot=r0*cos(gam)*temp;
        v=sqrt(top/bot);
        if  xlongt>xlongm
          vrx=v*cos(pi/2.-gam+xlongm);
          vry=v*sin(pi/2.-gam+xlongm);
        else
          vrx=v*cos(-pi/2.+gam+xlongm);
          vry=v*sin(-pi/2.+gam+xlongm);
        end
        xlam=r0*v*v/gm;
        top1=tan(gam)*(1-cos(phi))+(1-xlam)*sin(phi);
        bot1p=(1-cos(phi))/(xlam*cos(gam)*cos(gam));
        bot1=(2-xlam)*(bot1p+cos(gam+phi)/cos(gam));
        top2=2*cos(gam);
        bot2=xlam*((2/xlam-1)^1.5);
        top3=sqrt(2/xlam-1);
        bot3=cos(gam)/tan(phi/2)-sin(gam);
        temp=(top2/bot2)*atan2(top3,bot3);
        tf=r0*(top1/bot1+temp)/(v*cos(gam));
        icount=icount+1;
        if tf>tfdes
           gmax=gam;
        else
           gmin=gam;
        end
        if icount==1
           xnext=(gmax+gmin)/2.;
        else
           xnext=gam+(gam-gold)*(tfdes-tf)/(tf-told);
           if (xnext>gmax|xnext<gmin)
                      xnext=(gmax+gmin)/2.;
           end
        end
        gold=gam;
        told=tf;
        gam=xnext;
end
```

TABLE 13.2 NUMBER OF ITERATIONS ARE DRAMATICALLY REDUCED

Iteration	Flight-path angle, deg	V_{RX}, ft/s	V_{RY}, ft/s	Flight time, s
1	33.7524947	−3764.57976	18926.02426	1239.37545
2	11.2508376	−12075.71473	18072.66484	813.53185
3	21.1038504	−8103.20444	18287.99279	979.68031
4	22.3088581	−7665.88409	18332.46792	1001.50708
5	22.2256555	−7695.84063	18329.28399	999.98566
6	22.2264396	−7695.55815	18329.31392	999.99999
7	22.2264402	−7695.55795	18329.31394	1000.00000

The nominal case of the previous section was rerun and detailed results for the number of iterations required appear in Table 13.2. We can see that very accurate Lambert solutions are obtained after only four iterations and that after seven iterations we are obtaining a degree of accuracy that is better than obtained with 1084 iterations in the previous section using the brute force approach. *The new Lambert routine is not only more accurate than the one in the previous section but it is also more than two orders of magnitude faster!*

Reference 4 makes extensive tests on this efficient numerical solution to Lambert's problem and shows that it is competitive with the best numerical approaches. In addition, [4] also shows how this efficient solution of Lambert's problem can be extended to parabolic and hyperbolic trajectories.

BOOSTER STEERING

Thus far we have seen that, given that we know where we are and where we want to go and given an arrival time, the Lambert subroutine will tell us the orientation of the velocity vector for an impulsive missile to satisfy the problem. Since we do not have impulsive missiles (missiles that get up to speed immediately), it is desirable to find out if the Lambert subroutine could be of use in enabling a nonimpulsive missile or booster to reach its target. If we neglect the atmosphere, the solution to the problem is quite simple and is known as *Lambert guidance* [1].

Consider the vector diagram shown in Fig. 13.5. All that has to be done at small time increments, while the missile is boosting, is to find the desired velocity from the Lambert subroutine V_{Lambert} and subtract the current missile velocity V_M. The difference in velocities is known as the velocity to be gained ΔV. If the boosting missile thrust vector is aligned with the velocity to be gained vector, then the desired velocity will be obtained in a feedback fashion. When the desired velocity is achieved, the engine is cut off and the missile flies ballistically to the intended target.

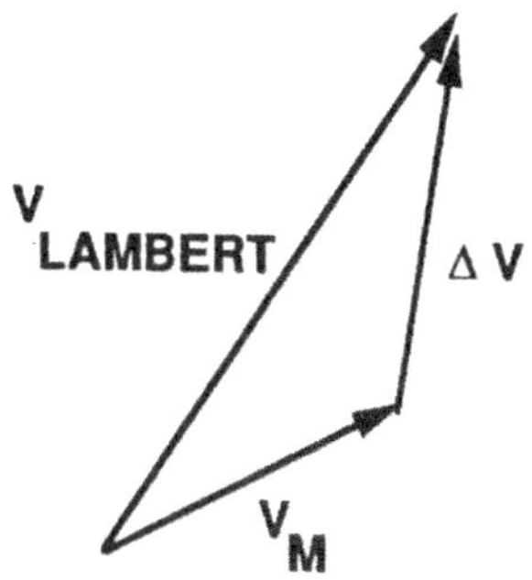

Fig. 13.5 Basis of Lambert guidance.

Mathematically, we are saying that the components of the velocity to be gained are

$$\Delta V_x = V_{\text{Lambertx}} - V_{Mx}$$
$$\Delta V_y = V_{\text{Lamberty}} - V_{My}$$

Therefore, the total velocity to be gained is simply

$$\Delta V = (\Delta V_x^2 + \Delta V_y^2)^{0.5}$$

If the magnitude of the current thrust acceleration is given by a_T, then the direction of the thrust acceleration at each instant of time should be aligned with the velocity to be gained vector, or

$$a_{Tx} = a_T \Delta V_x / \Delta V$$
$$a_{Ty} = a_T \Delta V_y / \Delta V$$

Listing 13.3 presents a MATLAB simulation of a two-stage booster using Lambert guidance during the boost phase. Actually, the scenario is unrealistic because g loading and range safety considerations have been ignored [2], but it is useful for demonstrating how Lambert guidance works. The booster considered in this example has a capability of reaching a velocity of 20,000 ft/s. The booster is assumed to have two stages, each of which has a fuel mass fraction of 0.9 and specific impulse of 300 s. The maximum acceleration in each stage is 20 g. One-third of the speed will be attained in the first stage, and the rest of the speed will be attained in the second stage. Burnout of the second stage will be completed at about 60 s. It is desired that the booster, which is initially at angular location $\theta_0 = 30$ deg (ANGDEG = 30), reach a target at 45 deg (XLONGTDEG = 45) in 500 s (TF = 500).

LISTING 13.3 BOOSTER SIMULATION WITH LAMBERT GUIDANCE

```
count=0;
LEFT=1;
QBOOST=1;
XISP1=300.;
XISP2=300.;
XMF1=.90;
XMF2=.90;
WPAY=100.;
DELV=20000.;
```

```
DELV1=.3333*DELV;
DELV2=.6667*DELV;
AMAX1=20.;
AMAX2=20.;
TOP2=WPAY*(exp(DELV2/(XISP2*32.2))-1.);
BOT2=1/XMF2-((1.-XMF2)/XMF2)*exp(DELV2/(XISP2*32.2));
WP2=TOP2/BOT2;
WS2=WP2*(1-XMF2)/XMF2;
WTOT2=WP2+WS2+WPAY;
TRST2=AMAX2*(WPAY+WS2);
TB2=XISP2*WP2/TRST2;
TOP1=WTOT2*(exp(DELV1/(XISP1*32.2))-1.);
BOT1=1/XMF1-((1.-XMF1)/XMF1)*exp(DELV1/(XISP1*32.2));
WP1=TOP1/BOT1;
WS1=WP1*(1-XMF1)/XMF1;
WTOT=WP1+WS1+WTOT2;
TRST1=AMAX1*(WTOT2+WS1);
TB1=XISP1*WP1/TRST1;
DELVK=DELV/1000.;
H=.01;
T=0.;
S=0.;
A=2.0926e7;
GM=1.4077e16;
ALTNM=0.;
ALT=ALTNM*6076.;
ANGDEG=30.;
ANG=ANGDEG/57.3;
XLONGM=ANG;
X=(A+ALT)*cos(ANG);
Y=(A+ALT)*sin(ANG);
ALT=sqrt(X^2+Y^2)-A;
XFIRST=X;
YFIRST=Y;
X1=0.;
Y1=0.;
AXT=0.;
AYT=0.;
XLONGTDEG=45.;
XLONGT=XLONGTDEG/57.3;
XF=A*cos(XLONGT);
YF=A*sin(XLONGT);
TF=500.;
while ~(ALT<0.&T>10.)
        XOLD=X;
        YOLD=Y;
```

```
X1OLD=X1;
Y1OLD=Y1;
STEP=1;
FLAG=0;
while STEP <=1
   if FLAG==1
            X=X+H*XD;
            Y=Y+H*YD;
            X1=X1+H*X1D;
            Y1=Y1+H*Y1D;
            T=T+H;
            STEP=2;
   end
   if T<TB1
            WGT=-WP1*T/TB1+WTOT;
            TRST=TRST1;
   elseif T < (TB1+TB2)
            WGT=-WP2*T/TB2+WTOT2+WP2*TB1/TB2;
            TRST=TRST2;
   else
            WGT=WPAY;
            TRST=0.;
   end
   AT=32.2*TRST/WGT;
   XD=X1;
   YD=Y1;
   TEMBOT=(X^2+Y^2)^1.5;
   X1D=-GM*X/TEMBOT+AXT;
   Y1D=-GM*Y/TEMBOT+AYT;
   FLAG=1;
end;
FLAG=0;
X=(XOLD+X)/2+.5*H*XD;
Y=(YOLD+Y)/2+.5*H*YD;
X1=(X1OLD+X1)/2+.5*H*X1D;
Y1=(Y1OLD+Y1)/2+.5*H*Y1D;
ALT=sqrt(X^2+Y^2)-A;
if QBOOST==1
   TGOLAM=TF-T;
   XLONGM=atan2(Y,X);
   [VRXM,VRYM]=lambertpz(X,Y,TGOLAM,XF,YF,XLONGM,XLONGT);
   VRX=VRXM;
   VRY=VRYM;
   DELX=VRX-X1;
   DELY=VRY-Y1;
   DEL=sqrt(DELX^2+DELY^2);
```

```
            if (TRST>0. & DEL>500.)
                        AXT=AT*DELX/DEL;
                        AYT=AT*DELY/DEL;
            elseif DEL < 500.
                        TRST=0.;
                        QBOOST=0;
                        AXT=0.;
                        AYT=0.;
                        X1=VRX;
                        Y1=VRY;
                        X1OLD=X1;
                        Y1OLD=Y1;
            else
                        QBOOST=0;
                        AXT=0.;
                        AYT=0.;
            end
        end
        S=S+H;
        if S>=9.99999
            S=0.;
            R=sqrt(X^2+Y^2);
            RF=sqrt(XFIRST^2+YFIRST^2);
            CBETA=(X*XFIRST+Y*YFIRST)/(R*RF);
            BETA=acos(CBETA);
            DISTNM=A*BETA/6076.;
            ALTNM=(sqrt(X^2+Y^2)-A)/6076.;
            count=count+1;
            ArrayT(count)=T;
            ArrayDISTNM(count)=DISTNM;
            ArrayALTNM(count)=ALTNM;
        end
end
figure
plot(ArrayDISTNM,ArrayALTNM),grid
xlabel('Downrange (Nmi)')
ylabel('Altitude (Nmi) ')
clc
output=[ArrayT',ArrayDISTNM',ArrayALTNM'];
save datfil.txt output /ascii
disp 'simulation finished'
%lambertpz.m shown in Listing 13.2
```

The Lambert feedback loop is at the end of the integration routine and is called every integration interval. When the difference between the desired velocity and the attained velocity is less than 500ft/s, the simulation automatically sets the

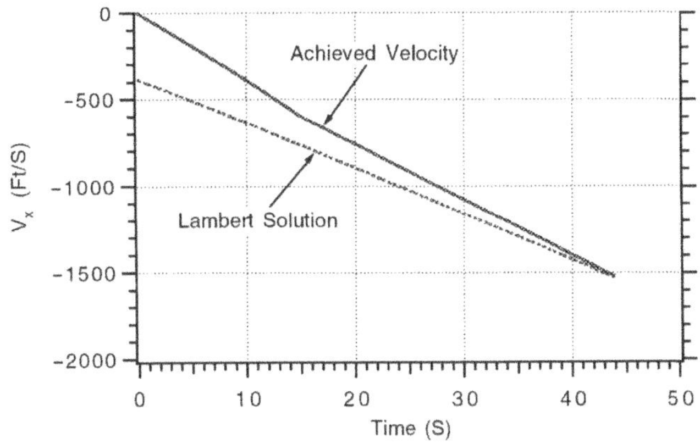

Fig. 13.6 ***X*** **component of achieved velocity reaches Lambert solution.**

actual velocity to the desired velocity to avoid making the integration interval very small in the simulation. At this time the booster cuts off and coasts. The logic in the simulation is self-explanatory.

The nominal case was run where the inputs were previously explained. Figure 13.6 displays the x component of the achieved velocity along with the desired or Lambert velocity. We can see that the two velocities converge at about 45 s. Figure 13.7 presents the y components of the achieved and desired velocities. We can see that this component is much larger than the x component. The discontinuity in the y component at about 15 s is due to staging, and the slight discontinuity near the end of the display is due to setting the achieved velocity to the desired velocity when the velocity to be gained was less than 500 ft/s.

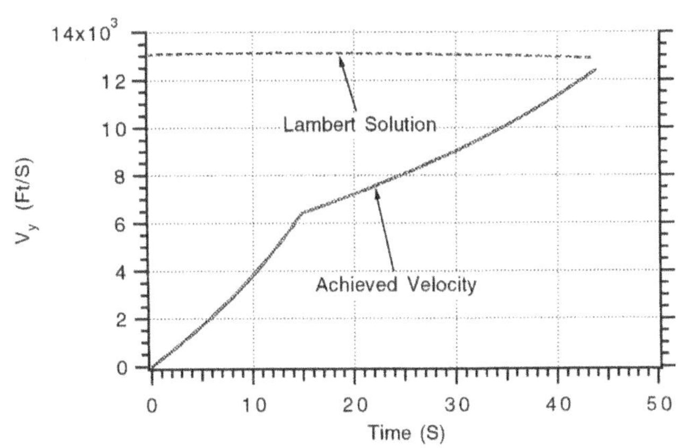

Fig. 13.7 ***Y*** **component of achieved velocity reaches Lambert solution.**

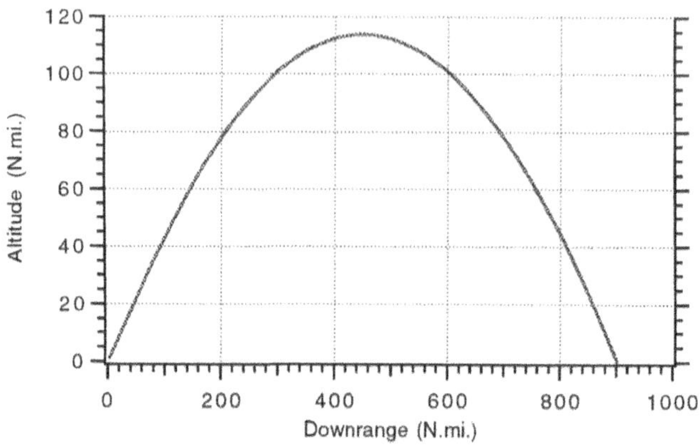

Fig. 13.8 Booster reaches target with Lambert guidance.

Finally, Fig. 13.8 shows the resultant trajectory. The missile reaches the target at exactly 500 s. We can see from the figure that the trajectory is smooth during the boost phase of flight when Lambert guidance is used.

It is interesting to note that the Lambert solution was reached in about 45 s even though the missile was capable of burning fuel for nearly 60 s. Thus, we can see that Lambert guidance can be used to steer a strategic missile with a thrust termination system in the absence of atmospheric effects. The Lambert guidance principal can be used for interceptors that fly ballistically to hit stationary targets. Lambert guidance can also be used for guided interceptors that must hit moving and accelerating targets. In this case, the purpose of Lambert guidance is to place the interceptor on a collision triangle at the end of the boost phase.

GENERAL ENERGY MANAGEMENT (GEM) STEERING [2, 6]

We have seen in the previous section how it was possible to steer a boosting strategic interceptor to a desired intercept point using Lambert guidance. In the example presented in the last section, the thrust had to be terminated before the end of burn in order to achieve the desired Lambert solution. Often there is a restriction, in the absence of a thrust termination system, that all the booster fuel must be consumed. In this case a method other than Lambert guidance must be employed to waste some of the booster's excess energy. A popular energy wasting technique is known as general energy management (GEM) steering.

To explain the concept of energy wasting, consider the simplified geometry of Fig. 13.9. In this figure we have the arc of a circle whose length is denoted V_{cap}. This arc represents the velocity capability of the booster. The radius of the circle forming the arc is denoted r, and the central angle is denoted 2θ. A

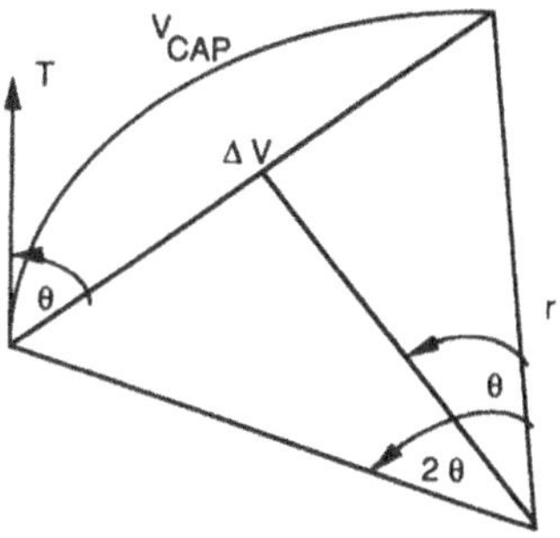

Fig. 13.9 Basic angles in GEM.

chord is drawn connecting both ends of the arc. The chord length represents the velocity to be gained (subtraction of achieved velocity from Lambert solution velocity) and is denoted ΔV. If the thrust vector is drawn tangent to the chord at the beginning of the arc, it is easy to show from geometry that the thrust vector is at an angle of θ with respect to the chord. Finally, a perpendicular is dropped from the chord to the center of the circle. It is also easy to show that the perpendicular bisects the chord and the central angle.

From Fig. 13.9 we can see that the arc length is related to the central angle according to

$$V_{\text{cap}} = 2\theta r$$

Because the perpendicular bisects the central angle, we can also say that

$$\Delta V = 2r \sin\theta$$

Therefore, we can ratio the two velocity expressions, yielding

$$\frac{\Delta V}{V_{\text{cap}}} = \frac{2r\sin\theta}{2r\theta} = \frac{\sin\theta}{\theta}$$

Expanding the sine term into a two-term Taylor series leads to

$$\frac{\Delta V}{V_{\text{cap}}} = \left(\theta - \frac{\theta^3}{6}\right) \Big/ \theta = 1 - \frac{\theta^2}{6}$$

Solving for the angle yields

$$\theta = \sqrt{6\left(1 - \frac{\Delta V}{V_{\text{cap}}}\right)}$$

The formula suggests that if, at each instant of time, we ensure that the thrust vector is at an angle of θ with respect to the velocity to be gained vector, then we can still achieve the Lambert solution at the end of burn and hit the target.

Figure 13.10 shows the proper relationship between the thrust and velocity to be gained vectors relative to the inertial Earth-centered coordinate system. We can see that for counterclockwise travel the components of the thrust

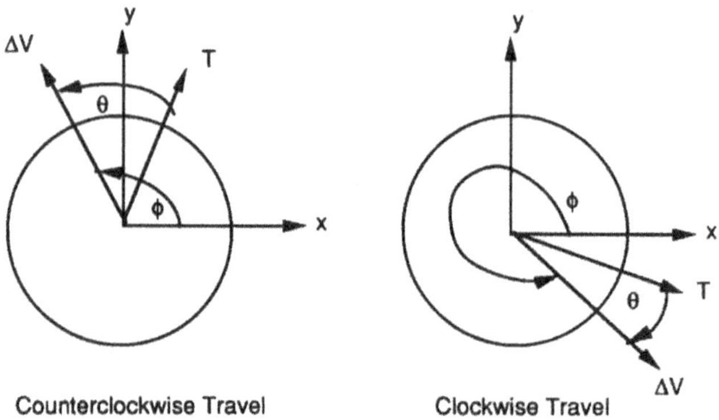

Fig. 13.10 Sign conventions for GEM.

acceleration are given by

$$a_{XT} = a_T \cos(\phi - \theta)$$
$$a_{YT} = a_T \sin(\phi - \theta)$$

where θ is the angle between the thrust vector and the velocity-to-be-gained vector, and ϕ is the angle between the velocity-to-be-gained vector and the x axis. For clockwise travel the thrust acceleration components become

$$a_{XT} = a_T \cos(\phi + \theta)$$
$$a_{YT} = a_T \sin(\phi + \theta)$$

Listing 13.4 presents a simulation of a booster intercepting a ground target using GEM guidance. This simulation and the nominal operating conditions are identical to that of Listing 13.3 except for the GEM logic after the integration routine. We can see from the listing that the axial acceleration capability of the booster is continually being computed according to

$$V_{\text{cap}} = V_{\text{cap}} - Ha_T$$

where H is the integration step size and a_T the instantaneous axial acceleration of the booster. To avoid numerical problems, the GEM logic is terminated when the velocity to be gained drops below 50 ft/s. We can see from the listing that it is still necessary to use the Lambert subroutine in order to implement the GEM guidance technique.

A nominal case was run to see how the GEM guidance logic performed. We can see from Fig. 13.11 that, although the booster burn lasts for nearly 60 s, the angle the thrust vector makes with respect to the velocity to be gained vector

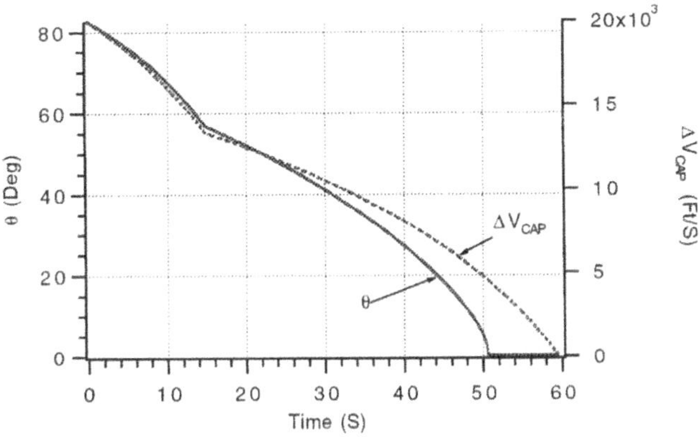

Fig. 13.11 GEM angle reaches steady state quickly.

approaches steady state in slightly over 50 s. Also shown in Fig. 13.11 is a plot of how the velocity capability of the booster diminishes during the burn. The discontinuity in that curve is due to staging.

Figure 13.12 displays the GEM trajectory during the boost phase. It appears from the figure that the booster will never hit the target because it initially appears to be heading in the wrong direction. However, after wasting energy, the GEM-guided booster heads in the right direction. Superimposed on the figure is the Lambert guidance trajectory during boost for the same case. We can see that both trajectories are vastly different during the boost phase.

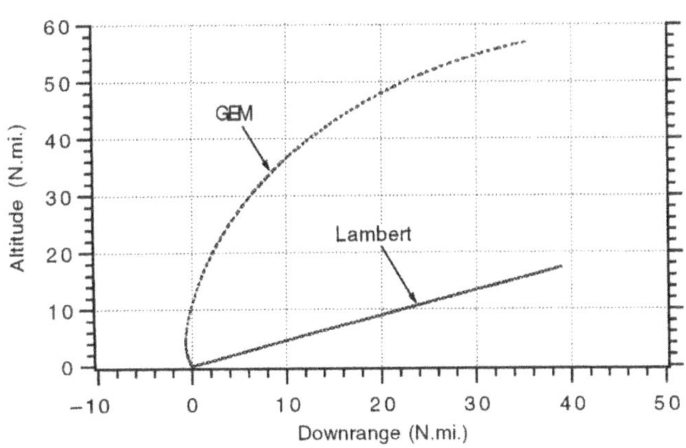

Fig. 13.12 Lambert and GEM trajectories during boost phase are vastly different.

LISTING 13.4 GEM SIMULATION

```
count=0;
LEFT=1;
QBOOST=1;
QZERO=0;
XISP1=300.;
XISP2=300.;
XMF1=.90;
XMF2=.90;
WPAY=100.;
DELV=20000.;
DELV1=.3333*DELV;
DELV2=.6667*DELV;
AMAX1=20.;
AMAX2=20.;
TOP2=WPAY*(exp(DELV2/(XISP2*32.2))-1.);
BOT2=1/XMF2-((1.-XMF2)/XMF2)*exp(DELV2/(XISP2*32.2));
WP2=TOP2/BOT2;
WS2=WP2*(1-XMF2)/XMF2;
WTOT2=WP2+WS2+WPAY;
TRST2=AMAX2*(WPAY+WS2);
TB2=XISP2*WP2/TRST2;
TOP1=WTOT2*(exp(DELV1/(XISP1*32.2))-1.);
BOT1=1/XMF1-((1.-XMF1)/XMF1)*exp(DELV1/(XISP1*32.2));
WP1=TOP1/BOT1;
WS1=WP1*(1-XMF1)/XMF1;
WTOT=WP1+WS1+WTOT2;
TRST1=AMAX1*(WTOT2+WS1);
TB1=XISP1*WP1/TRST1;
DELVK=DELV/1000.;
H=.01;
T=0.;
S=0.;
A=2.0926e7;
GM=1.4077e16;
ALTNM=0.;
ALT=ALTNM*6076.;
ANGDEG=30.;
ANG=ANGDEG/57.3;
XLONGM=ANG;
X=(A+ALT)*cos(ANG);
Y=(A+ALT)*sin(ANG);
ALT=sqrt(X^2+Y^2)-A;
X1=0.;
Y1=0.;
```

```
AXT=0.;
AYT=0.;
XLONGTDEG=45.;
XLONGT=XLONGTDEG/57.3;
XF=A*cos(XLONGT);
YF=A*sin(XLONGT);
XFIRST=XF;;
YFIRST=YF;
R=sqrt(X^2+Y^2);
RF=sqrt(XFIRST^2+YFIRST^2);
CBETA=(X*XFIRST+Y*YFIRST)/(R*RF);
BETA=acos(CBETA);
DISTNM=A*BETA/6076.;
DISTINITNM=DISTNM;
TF=500.;
DVCAP=DELV;
while ~(ALT<0.&T>10.)
      XOLD=X;
      YOLD=Y;
      X1OLD=X1;
      Y1OLD=Y1;
      STEP=1;
      FLAG=0;
      while STEP <=1
        if FLAG==1
                X=X+H*XD;
                Y=Y+H*YD;
                X1=X1+H*X1D;
                Y1=Y1+H*Y1D;
                T=T+H;
                STEP=2;
        end
        if T<TB1
                WGT=-WP1*T/TB1+WTOT;
                TRST=TRST1;
        elseif T<(TB1+TB2)
                WGT=-WP2*T/TB2+WTOT2+WP2*TB1/TB2;
                TRST=TRST2;
        else
                WGT=WPAY;
                TRST=0.;
        end
        AT=32.2*TRST/WGT;
        XD=X1;
        YD=Y1;
        TEMBOT=(X^2+Y^2)^1.5;
```

```
        X1D=-GM*X/TEMBOT+AXT;
        Y1D=-GM*Y/TEMBOT+AYT;
        FLAG=1;
    end;
    FLAG=0;
    X=(XOLD+X)/2+.5*H*XD;
    Y=(YOLD+Y)/2+.5*H*YD;
    X1=(X1OLD+X1)/2+.5*H*X1D;
    Y1=(Y1OLD+Y1)/2+.5*H*Y1D;
    ALT=sqrt(X^2+Y^2)-A;
    TGOLAM=TF-T;
    DVCAP=DVCAP-H*AT;
    if (QBOOST==1 & DVCAP>50.)
        XLONGM=atan2(Y,X);
        [VRXM,VRYM]=lambertpz(X,Y,TGOLAM,XF,YF,XLONGM,XLONGT);
        VRX=VRXM;
        VRY=VRYM;
        DELX=VRX-X1;
        DELY=VRY-Y1;
        DEL=sqrt(DELX^2+DELY^2);
        if (QZERO==0 & DVCAP>DEL)
                THET=sqrt(6.*(1.-DEL/DVCAP));
                DEGTHET=57.3*THET;
        else
                QZERO=1;
        end
        PHI=atan2(DELY,DELX);
        DEGPHI=57.3*PHI;
        if XLONGT>XLONGM
                AXT=AT*cos(PHI-THET);
                AYT=AT*sin(PHI-THET);
        else
                AXT=AT*cos(PHI+THET);
                AYT=AT*sin(PHI+THET);
        end
        R=sqrt(X^2+Y^2);
        RF=sqrt(XFIRST^2+YFIRST^2);
        CBETA=(X*XFIRST+Y*YFIRST)/(R*RF);
        BETA=acos(CBETA);
        DISTNM=A*BETA/6076.;
        DISTNM=DISTINITNM-DISTNM;
        ALTNM=(sqrt(X^2+Y^2)-A)/6076.;
    elseif QBOOST==1
        [VRXM,VRYM]=lambertpz(X,Y,TGOLAM,XF,YF,XLONGM,XLONGT);
        VRX=VRXM;
        VRY=VRYM;
```

```
            TRST=0.;
            QBOOST=0;
            AXT=0.;
            AYT=0.;
            X1=VRX;
            Y1=VRY;
            X1OLD=X1;
            Y1OLD=Y1;
            R=sqrt(X^2+Y^2);
            RF=sqrt(XFIRST^2+YFIRST^2);
            CBETA=(X*XFIRST+Y*YFIRST)/(R*RF);
            BETA=acos(CBETA);
            DISTNM=A*BETA/6076.;
            DISTNM=DISTINITNM-DISTNM;
            ALTNM=(sqrt(X^2+Y^2)-A)/6076.;
        else
            QBOOST=0;
            AXT=0.;
            AYT=0.;
        end
        S=S+H;
        if S>=9.99999
            S=0.;
            R=sqrt(X^2+Y^2);
            RF=sqrt(XFIRST^2+YFIRST^2);
            CBETA=(X*XFIRST+Y*YFIRST)/(R*RF);
            BETA=acos(CBETA);
            DISTNM=A*BETA/6076.;
            DISTNM=DISTINITNM-DISTNM;
            ALTNM=(sqrt(X^2+Y^2)-A)/6076.;
            count=count+1;
            ArrayT(count)=T;
            ArrayDISTNM(count)=DISTNM;
            ArrayALTNM(count)=ALTNM;
        end
end
figure
plot(ArrayDISTNM,ArrayALTNM),grid
xlabel('Downrange (Nmi)')
ylabel('Altitude (Nmi) ')
clc
output=[ArrayT',ArrayDISTNM',ArrayALTNM'];
save datfil.txt output /ascii
disp 'simulation finished'

%lambertpz.m shown in Listing 13.2
```

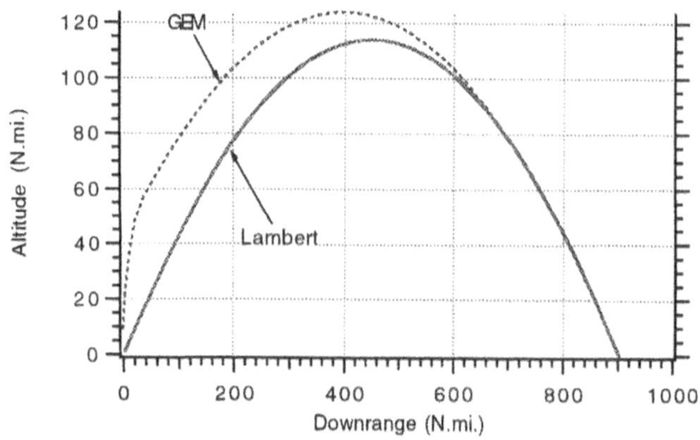

Fig. 13.13 Both Lambert and GEM trajectories hit target at the same time.

Figure 13.13 displays the GEM and Lambert trajectories for the entire flight (boost and coast phases). We can see from the figure that, although both trajectories are vastly different during the boost phase, they eventually converge, and both hit the target at the same time.

SUMMARY

In this chapter the Lambert problem was explained, and a novel numerical technique for solving the problem, based on the closed-form solutions of Chapter 11 was introduced. Two techniques were presented showing how the solution to Lambert's problem was fundamental to steering boosters. Numerical examples were presented illustrating the implementation and effectiveness of the booster steering techniques.

REFERENCES

[1] Battin, R. H., *An Introduction to the Mathematics and Methods of Astrodynamics*, AIAA Education Series, New York, 1987.

[2] Regan, F., *Re-Entry Vehicle Dynamics*, AIAA Education Series, New York, 1984.

[3] Acton, F. S., *Numerical Methods That Work*, Harper and Row, New York, 1970.

[4] Nelson, S. L., and Zarchan, P., "Alternative Approach to the Solution of Lambert's Problem," *Journal of Guidance, Control, and Dynamics*, Vol. 15, July–Aug. 1992, pp. 1003–1009.

[5] Brand, T. J., "A New Approach to Lambert Guidance," Charles Stark Draper Lab., Rept. R-694, Cambridge, MA, June 1971.

CHAPTER 14

Strategic Intercepts

INTRODUCTION

Guidance concepts for tactical homing missiles were introduced, explained, and demonstrated in Chapters 2, 6, and 8. Strategic missiles travel much faster and farther than tactical missiles. We have shown that the coordinate system and gravity models in our tactical simulations had to be modified to handle the new speed and range regimes of strategic missiles. More specifically, we had to shift our coordinate system from the surface of a flat Earth to the center of a round Earth and use a more general formulation for gravitational acceleration (that is, Newton's law of universal gravitation).

Because tactical missiles operate within the atmosphere, they can generate lift by moving control surfaces in order to execute guidance commands. Speed, altitude, and structural considerations limit maximum achievable acceleration levels with tactical missiles. Missile slowdown due to drag limits the tactical missile's range and maneuver capability. Strategic missiles, on the other hand, operate outside the atmosphere and must burn fuel (that is, lateral thrusters) to respond to guidance commands. Achievable engine thrust-to-weight ratios limit maximum strategic lateral acceleration levels. In addition, when the maneuver or divert fuel is exhausted, the strategic missile cannot maneuver at all. Care must be taken to ensure that a strategic missile has sufficient divert fuel so that it can meet system objectives.

Although there are major differences between strategic and tactical missiles, there are also similarities. This chapter will show that tactical guidance laws may be suitable for strategic missiles. Useful design relationships, developed previously in the text for tactical missiles, will be modified and shown to be applicable for strategic missiles as well.

GUIDANCE REVIEW

In Chapter 2 we saw the effectiveness of the proportional navigation guidance law for tactical missiles. A closed-form solution for the required missile acceleration to hit a target, in the presence of heading error, was derived for a zero-lag guidance system. The required missile acceleration was shown to be

$$n_c = \frac{-V_M HE\, N'}{t_F}\left(1 - \frac{t}{t_F}\right)^{N'-2}$$

where V_M is the missile velocity, HE is the angular heading error, N' is the effective navigation ratio, t_F is the flight time, and t is instantaneous time.

With strategic missiles it is often more convenient to talk in terms of prediction error rather than heading error. A prelaunch calculation or prediction must be made of where the target will be at intercept. This location is known as the *predicted intercept point*. If the calculation is imperfect, a prediction error results, and the missile will not be fired on a perfect collision triangle. The prediction error and heading error are related by

$$Pred\ Err = -V_M HE t_F$$

where *Pred Err* is the prediction error in units of feet. Therefore, substitution of the preceding relationship into the closed-form solution indicates that the missile acceleration required by the proportional navigation guidance law to take out an initial prediction error is given by

$$n_c = \frac{Pred\ Err\ N'}{t_F^2}\left(1 - \frac{t}{t_F}\right)^{N'-2}$$

As was mentioned previously, strategic missiles burn fuel to maneuver. The amount of lateral divert or ΔV required is related to the missile acceleration according to

$$\Delta V = \int_0^{t_F} |n_c|\, dt$$

The strategic interceptor ΔV requirements are related to the total interceptor weight by the rocket equation. Increasing a missile's divert requirements can increase the total weight requirements dramatically. We can find a closed-form solution for the required divert to take out a prediction error by substituting the closed-form solution for the missile acceleration into the preceding integral. After some algebra we obtain

$$\Delta V = \frac{Pred\ Err\ N'}{(N'-1)t_F}$$

Thus, increasing the effective navigation ratio or increasing the flight time (or guidance time) will tend to reduce the lateral divert requirements of the interceptor. The preceding formula for the required lateral divert is plotted in Fig. 14.1 for the case in which the prediction error is 100 kft. We can see that, for an effective navigation ratio of 3, it takes more than 10,000 ft/s of divert to remove the error in

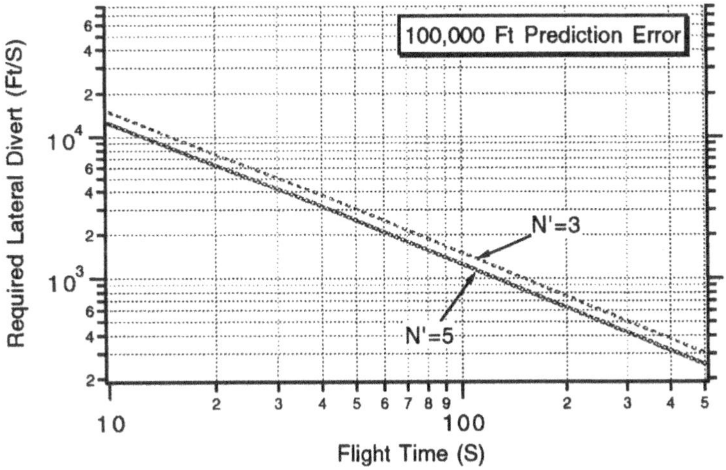

Fig. 14.1 Lateral divert requirements decrease with increasing flight time.

10 s, about 1050 ft/s of divert to remove the error in 100 s, and only about 300 ft/s of divert to remove the error in 500 s. Therefore, larger missile acquisition ranges result in larger guided flight times, which in turn can reduce the lateral divert requirements for a given prediction error. Increasing the effective navigation ratio to 5 only slightly reduces the lateral divert requirements. Doubling the prediction error will double the divert requirements.

BALLISTIC ENGAGEMENT SIMULATION [1]

We can develop a strategic ballistic missile-target engagement simulation by using an Earth-centered coordinate system as shown in Fig. 14.2. In this figure both the missile and target are in a gravity field as described by Newton's law of universal gravitation. The acceleration differential equations acting on a ballistic target were shown in Chapter 11 to be

$$\ddot{x}_T = \frac{-gm\, x_T}{\left(x_T^2 + y_T^2\right)^{1.5}}$$

$$\ddot{y}_T = \frac{-gm\, y_T}{\left(x_T^2 + y_T^2\right)^{1.5}}$$

where gm is the gravitational parameter. These differential equations are in an inertial coordinate system whose origin is at the center of the Earth. Therefore, they can be integrated directly to yield the velocity and position of the target with respect to the center of the Earth. The components of the relative position

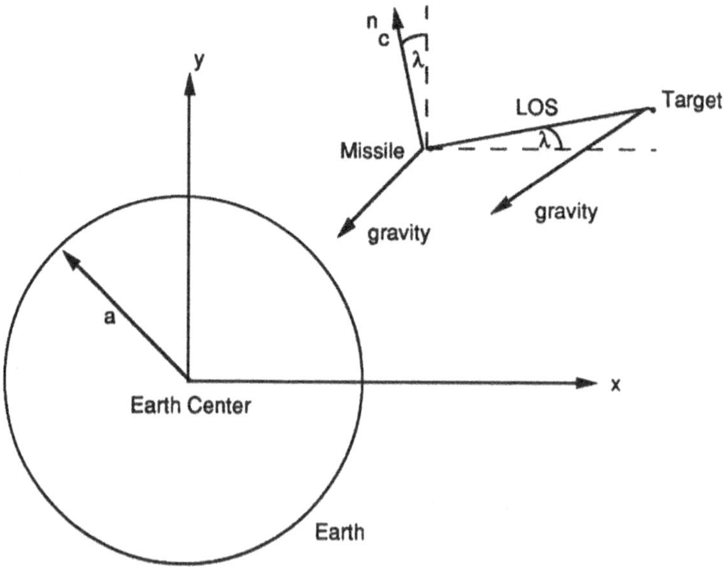

Fig. 14.2 Earth-centered coordinate system and relative engagement geometry.

between the missile and target can be expressed as

$$R_{\text{TM1}} = x_T - x_M$$
$$R_{\text{TM2}} = y_T - y_M$$

and the components of the relative velocity are given by

$$V_{\text{TM1}} = \dot{x}_T - \dot{x}_M$$
$$V_{\text{TM2}} = \dot{y}_T - \dot{y}_M$$

Application of the distance formula shows that the relative separation between the missile and target can be found from

$$R_{\text{TM}} = \left(R_{\text{TM1}}^2 + R_{\text{TM2}}^2\right)^{0.5}$$

The closing velocity, which is defined as the negative rate of change of separation between missile and target, can be obtained by taking the negative derivative of the preceding expression, yielding

$$V_c = \frac{-(R_{\text{TM1}}V_{\text{TM1}} + R_{\text{TM2}}V_{\text{TM2}})}{R_{\text{TM}}}$$

The line-of-sight angle can be found by trigonometry from Fig. 14.2 as

$$\lambda = \tan^{-1}\frac{R_{TM2}}{R_{TM1}}$$

Therefore, the instantaneous value of the line-of-sight rate can be found by taking the derivative of the preceding expression, using the quotient rule, yielding

$$\dot{\lambda} = \frac{R_{TM1}V_{TM2} - R_{TM2}V_{TM1}}{R_{TM}^2}$$

We now have sufficient information to guide a strategic interceptor. The proportional navigation guidance command is proportional to the line-of-sight rate according to

$$n_c = N'V_c\dot{\lambda}$$

where N' is the effective navigation ratio and V_c the closing velocity. This guidance command is perpendicular to the line of sight. From Fig. 14.2 we can see that the components of the guidance command in the Earth-centered coordinate system can be found by trigonometry and are given by

$$a_{XM} = -n_c \sin\lambda$$
$$a_{YM} = n_c \cos\lambda$$

Therefore, the acceleration differential equations describing the missile consist of two parts: the gravitational term and the guidance command term. The components of the missile differential equations in Earth-centered coordinates are

$$\ddot{x}_M = \frac{-gm\, x_M}{(x_M^2 + y_M^2)^{1.5}} + a_{XM}$$
$$\ddot{y}_M = \frac{-gm\, y_M}{(x_M^2 + y_M^2)^{1.5}} + a_{YM}$$

where a_{XM} and a_{YM} have already been defined.

Listing 14.1 presents a MATLAB simulation of an engagement between an impulsive missile and a ballistic target. The simulation, which is based on the differential equations derived in this section, is similar to the tactical engagement simulations presented in the text, except that the coordinate system is Earth-centered.

The program includes a prediction routine to determine where the target will be at the intercept time t_F. Before the main simulation begins, this routine integrates the ballistic target equations forward in time to determine the location of the target at time t_F (that is, predicted intercept point). The Lambert routine determines the velocity components of an impulsive strategic interceptor (VRXM and VRYM) so that it will be on a collision triangle with the ballistic target. In other words, given an initial location, a final location, and an arrival time, the

Lambert routine determines the correct missile velocity components so that it will collide with the target at time t_F. If the predicted intercept point is correct, then no guidance system is required for the strategic interceptor to collide with the target.

We can see from the listing that the guidance equations are virtually identical to those of the two-dimensional tactical simulation of Chapter 2. This is not surprising as the proportional navigation guidance law operates on relative quantities that should be independent of coordinate system. The differential equations for the missile and target and the guidance equations appear before the FLAG=1 statement.

A nominal case was run in which the guidance system was turned off (XNC = 0). The resultant trajectory for the 500-s flight is shown in Fig. 14.3. In this case the missile hit the target. This means that our knowledge of the intercept point was perfect (from the prediction routine) and that the missile was placed on the correct collision triangle (from the Lambert routine). The slight curvature in both missile and target trajectories is due to the fact that both objects are in a gravity field for 500 s.

The same nominal case was rerun, except this time the proportional navigation guidance system was turned on. The resultant commanded acceleration profile, which resulted in a successful intercept, along with the missile lateral divert requirements appear in Fig. 14.4. We can see from the figure that, even though the missile was initially on a collision triangle, the proportional navigation guidance system issued acceleration commands. In this case it appears that proportional navigation is behaving in a counterintuitive way because we could have hit the target without any acceleration commands at all!

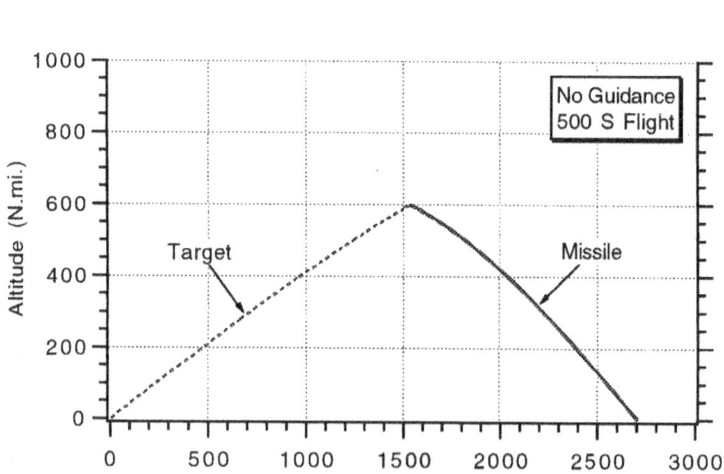

Fig. 14.3 Collision triangle geometry for nominal case.

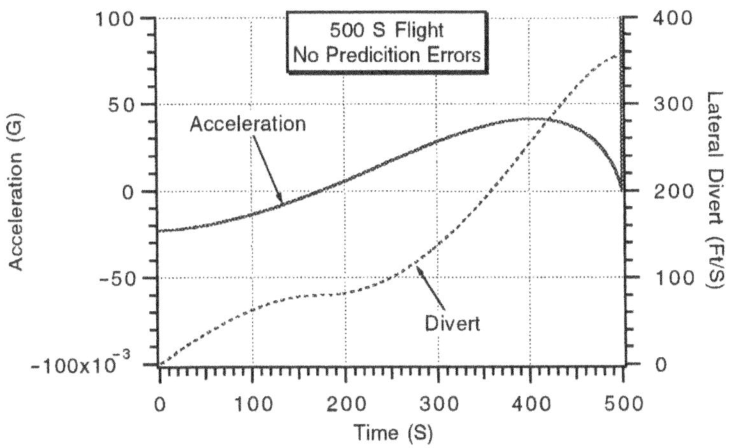

Fig. 14.4 Some divert required with proportional navigation guidance even though missile is on collision triangle.

LISTING 14.1 ENGAGEMENT SIMULATION WITH BALLISTIC TARGET

```
count=0;
XLONGMDEG=45.;
XLONGTDEG=90.;
ALTNMTIC=0.;
ALTNMMIC=0.;
TF=500.;
GAMDEGT=23.;
H=.01;
A=2.0926e7;
GM=1.4077e16;
DEGRAD=360./(2.*pi);
XNP=3.;
PREDERR=0.;
GAMT=GAMDEGT/57.3;
DISTNMT=6000.;
PHIT=DISTNMT*6076./A;
ALTT=ALTNMTIC*6076.;
ALTM=ALTNMMIC*6076.;
R0T=A+ALTT;
TOP=GM*(1.-cos(PHIT));
TEMP=R0T*cos(GAMT)/A-cos(PHIT+GAMT);
BOT=R0T*cos(GAMT)*TEMP;
VT=sqrt(TOP/BOT);
XLONGM=XLONGMDEG/DEGRAD;
```

```
XLONGT=XLONGTDEG/DEGRAD;
if XLONGM>XLONGT
         X1T=VT*cos(pi/2.-GAMT+XLONGT);
         Y1T=VT*sin(pi/2.-GAMT+XLONGT);
else
         X1T=VT*cos(-pi/2.+GAMT+XLONGT);
         Y1T=VT*sin(-pi/2.+GAMT+XLONGT);
end
S=0.;
XLONGM=XLONGMDEG/DEGRAD;
XLONGT=XLONGTDEG/DEGRAD;
XM=(A+ALTM)*cos(XLONGM);
YM=(A+ALTM)*sin(XLONGM);
XT=(A+ALTT)*cos(XLONGT);
YT=(A+ALTT)*sin(XLONGT);
XFIRSTT=XT;
YFIRSTT=YT;
T=0.;
[XTF,YTF]=predictpz(TF,XT,YT,X1T,Y1T);
YTF=YTF+PREDERR;
[VRXM,VRYM]=lambertpz(XM,YM,TF,XTF,YTF,XLONGM,XLONGT);
X1M=VRXM;
Y1M=VRYM;
RTM1=XT-XM;
RTM2=YT-YM;
RTM=sqrt(RTM1^2+RTM2^2);
VTM1=X1T-X1M;
VTM2=Y1T-Y1M;
VC=-(RTM1*VTM1+RTM2*VTM2)/RTM;
DELV=0.;
while VC>=0.
         TGO=RTM/VC;
         if TGO>.1
      H=.01;
         else
            H=.0001;
         end
         XOLDT=XT;
         YOLDT=YT;
         X1OLDT=X1T;
         Y1OLDT=Y1T;
         XOLDM=XM;
         YOLDM=YM;
         X1OLDM=X1M;
         Y1OLDM=Y1M;
         DELVOLD=DELV;
```

```
STEP=1;
FLAG=0;
while STEP <=1
   if FLAG==1
              XT=XT+H*XDT;
              YT=YT+H*YDT;
              X1T=X1T+H*X1DT;
              Y1T=Y1T+H*Y1DT;
              XM=XM+H*XDM;
              YM=YM+H*YDM;
              X1M=X1M+H*X1DM;
              Y1M=Y1M+H*Y1DM;
              DELV=DELV+H*DELVD;
              T=T+H;
              STEP=2;
   end
   TEMBOTT=(XT^2+YT^2)^1.5;
   X1DT=-GM*XT/TEMBOTT;
   Y1DT=-GM*YT/TEMBOTT;
   XDT=X1T;
   YDT=Y1T;
   RTM1=XT-XM;
   RTM2=YT-YM;
   RTM=sqrt(RTM1^2+RTM2^2);
   VTM1=X1T-X1M;
   VTM2=Y1T-Y1M;
   VC=-(RTM1*VTM1+RTM2*VTM2)/RTM;
   TGO=RTM/VC;
   XLAM=atan2(RTM2,RTM1);
   XLAMD=(RTM1*VTM2-RTM2*VTM1)/(RTM*RTM);
   XNC=XNP*VC*XLAMD;
   DELVD=abs(XNC);
   AM1=-XNC*sin(XLAM);
   AM2=XNC*cos(XLAM);
   TEMBOTM=(XM^2+YM^2)^1.5;
   X1DM=-GM*XM/TEMBOTM+AM1;
   Y1DM=-GM*YM/TEMBOTM+AM2;
   XDM=X1M;
   YDM=Y1M;
   FLAG=1;
end;
FLAG=0;
XT=(XOLDT+XT)/2+.5*H*XDT;
YT=(YOLDT+YT)/2+.5*H*YDT;
X1T=(X1OLDT+X1T)/2+.5*H*X1DT;
Y1T=(Y1OLDT+Y1T)/2+.5*H*Y1DT;
```

```
        XM=(XOLDM+XM)/2+.5*H*XDM;
        YM=(YOLDM+YM)/2+.5*H*YDM;
        X1M=(X1OLDM+X1M)/2+.5*H*X1DM;
        Y1M=(Y1OLDM+Y1M)/2+.5*H*Y1DM;
        DELV=(DELVOLD+DELV)/2.+.5*H*DELVD;
        ALTT=sqrt(XT^2+YT^2)-A;
        ALTM=sqrt(XM^2+YM^2)-A;
        S=S+H;
        if S>=.99999
           S=0.;
           ALTNMT=ALTT/6076.;
           R=sqrt(XT^2+YT^2);
           RF=sqrt(XFIRSTT^2+YFIRSTT^2);
           CBETA=(XT*XFIRSTT+YT*YFIRSTT)/(R*RF);
           BETA=acos(CBETA);
           DISTNMT=A*BETA/6076.;
           ALTNMM=ALTM/6076.;
           XNCG=XNC/32.2;
           R=sqrt(XM^2+YM^2);
           RF=sqrt(XFIRSTT^2+YFIRSTT^2);
           CBETA=(XM*XFIRSTT+YM*YFIRSTT)/(R*RF);
           BETA=acos(CBETA);
           DISTNMM=A*BETA/6076.;
           count=count+1;
           ArrayT(count)=T;
           ArrayDISTNMT(count)=DISTNMT;
           ArrayALTNMT(count)=ALTNMT;
           ArrayDISTNMM(count)=DISTNMM;
           ArrayALTNMM(count)=ALTNMM;
           ArrayXNCG(count)=XNCG;
           ArrayDELV(count)=DELV;
        end
end
RTM
DELV
figure
plot(ArrayDISTNMT,ArrayALTNMT,ArrayDISTNMM,ArrayALTNMM),grid
xlabel('Downrange (Nmi)')
ylabel('Altitude (Nmi) ')
figure
plot(ArrayT,ArrayXNCG),grid
xlabel('Time (Sec)')
ylabel('Acceleration (G) ')
figure
plot(ArrayT,ArrayDELV),grid
xlabel('Time (Sec)')
ylabel('Lateral Divert (Ft/Sec) ')
```

```
clc
output=[ArrayT',ArrayDISTNMT',ArrayALTNMT',ArrayDISTNMM',...
        ArrayALTNMM'];
save datfil.txt output /ascii
disp 'simulation finished'

function [xtf,ytf]=predictpz(tf,xdum,ydum,x1dum,y1dum)
h=.01;
a=2.0926e7;
gm=1.4077e16;
t=0.;
x=xdum;
y=ydum;
x1=x1dum;
y1=y1dum;
while t<=(tf-.00001)
      xold=x;
      yold=y;
      x1old=x1;
      y1old=y1;
      step=1;
      flag=0;
      while step <=1
        if flag==1
                x=x+h*xd;
                y=y+h*yd;
                x1=x1+h*x1d;
                y1=y1+h*y1d;
                t=t+h;
                step=2;
        end
        tembot=(x^2+y^2)^1.5;
        x1d=-gm*x/tembot;
        y1d=-gm*y/tembot;
        xd=x1;
        yd=y1;
        flag=1;
      end;
      flag=0;
      x=(xold+x)/2+.5*h*xd;
      y=(yold+y)/2+.5*h*yd;
      x1=(x1old+x1)/2+.5*h*x1d;
      y1=(y1old+y1)/2+.5*h*y1d;
end
xtf=x;
ytf=y;
%lambertpz.m shown in Listing 13.2
```

To understand why guidance commands were required of a missile on a collision triangle, let us review some basics. Consider the case of a constant-velocity missile and constant-velocity target on a collision triangle as shown in Fig. 14.5. If we connect lines between the missile and target at different times during the flight, we have a measure of how the line-of-sight rate changes with time. We can see from Fig. 14.5 that, when both the missile and target are traveling at constant velocities, the line-of-sight lines are parallel. In other words, the *line-of-sight rate is zero!* Since acceleration commands are proportional to the line-of-sight rate in a proportional navigation system, there will be no commands for a constant velocity missile and target on a collision triangle.

Figure 14.6 also shows a missile and target on a collision triangle. However, this time the missile is traveling at a constant velocity while the target velocity is nonconstant. In this case we can see that the line-of-sight lines are not parallel. Thus, we have a line-of-sight rate generated even though the missile and target are on a collision triangle. In this case a proportional navigation guidance system will generate acceleration commands, even though none are required!

In a gravity field, both the missile and target velocities vary with time. Thus, as shown in Fig. 14.6, the line-of-sight rate will not be zero, even though both the missile and target are on a collision triangle. Proportional navigation will waste some fuel in responding to the small line-of-sight rates. In our nominal case about 350 ft/s of lateral divert was required by the missile to intercept the ballistic target.

A guidance system is required as we cannot always be on a collision triangle. There will always be errors in predicting the location of the intercept point. Consider a case for a proportional navigation guidance system with an effective navigation ratio of 3 in which there is a 100-kft prediction error. This means that if we turned off the guidance system we would miss the target by 100 kft. Based on the formula derived at the beginning of this chapter, theory predicts that the required missile lateral divert should be

$$\Delta V = \frac{Pred\ Err\ N'}{(N'-1)t_F} = \frac{100,000 * 3}{2 * 500} = 300\,\text{ft/s}$$

Figure 14.7 shows the resultant commanded acceleration

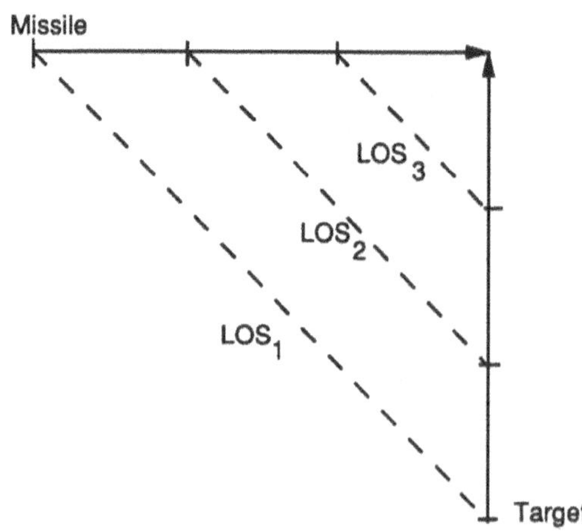

Fig. 14.5 Constant-velocity missile and target on collision triangle.

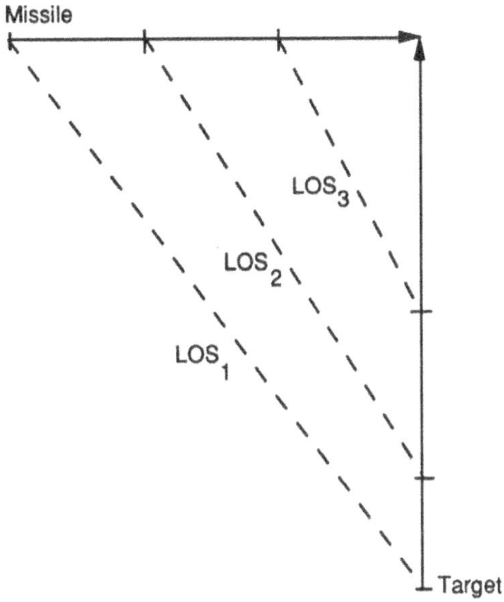

Fig. 14.6 Constant-velocity missile and variable-velocity target on collision triangle.

and lateral divert profiles due to the 100-kft prediction error for a proportional navigation guidance system with an effective navigation ratio of 3. We can see that the missile acceleration requirements are small (less than 0.05 *g*) for the flight. However, even at small acceleration levels, about 480 ft/s of lateral divert was required for a successful intercept. This value is somewhat larger than the theoretically predicted value of 300 ft/s because, as we previously saw, the gravity field also adds to the divert requirements.

Theory tells us that, for a fixed prediction error, the divert requirements will increase if the flight time is decreased. Figure 14.8 presents the engagement geometry for a 100-s flight. Note that in the simulation the flight time was reduced and the initial missile location was moved closer to the target (TF = 100,

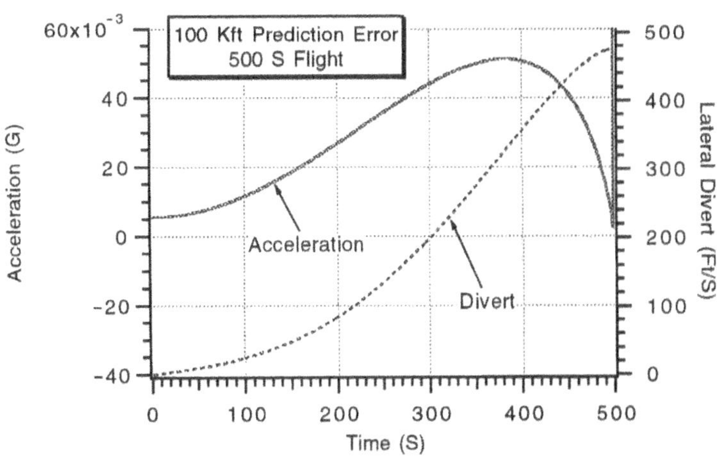

Fig. 14.7 Divert due to prediction error is close to theoretical prediction.

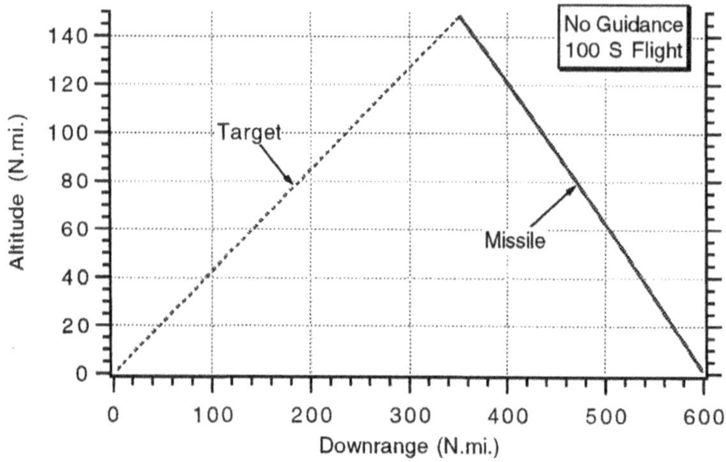

Fig. 14.8 Collision triangle geometry for shorter-range flight.

XLONGMDEG = 80). In this case both the missile and target are initially on a collision triangle.

If we introduce a 100-kft prediction error into the short-range example, theory tells us that the lateral divert requirements should increase substantially. According to the previously presented lateral divert formula, the divert requirements for a 100-s flight should be

$$\Delta V = \frac{Pred\,Err\,N'}{(N'-1)t_F} = \frac{100,000 * 3}{2 * 100} = 1500\,\text{ft/s}$$

Figure 14.9 displays the commanded acceleration and actual divert requirements obtained by running the engagement simulation. We can see from the figure that the required lateral divert required is indeed nearly 1500 ft/s. Thus, we have demonstrated with nonlinear engagement simulation results that the theoretical formula is a useful and accurate indicator of divert requirements for prediction error.

BOOSTING TARGET CONSIDERATIONS [1]

Although a booster does not execute evasive maneuvers, any longitudinal booster acceleration that is perpendicular to the line of sight will appear as a target maneuver to the missile. In Chapter 2 we saw that the closed-form solution for the acceleration required by a missile utilizing proportional navigation guidance was given by

$$n_c|_{PN} = \frac{N'}{N'-2}\left[1-\left(1-\frac{t}{t_F}\right)^{N'-2}\right]n_T$$

where N' is the effective navigation ratio, t is time, t_F the flight time, and n_T the magnitude of the apparent target maneuver. From the definition of lateral divert,

$$\Delta V = \int_0^{t_F} |n_c| \, dt$$

we can derive an expression for the lateral divert required to hit a maneuvering target as

$$\Delta V|_{\text{PN}} = \frac{N'}{N'-1} n_T t_F$$

Thus, the lateral divert due to a maneuvering target increases with increasing flight time and decreases with increasing effective navigation ratio.

We can develop an engagement simulation in which the target is a booster. In Chapter 12 we saw how to model a booster performing a gravity turn. We can express the longitudinal acceleration of the booster as

$$a_T = \frac{32.2T}{W}$$

where T is the booster thrust and W the booster weight. In a gravity turn the thrust and velocity vectors are aligned so that the acceleration differential equations for the booster in a gravity field become

$$\ddot{x}_T = \frac{-gm\, x_T}{(x_T^2 + y_T^2)^{1.5}} + \frac{a_T \dot{x}_T}{V_T}$$

$$\ddot{y}_T = \frac{-gm\, y_T}{(x_T^2 + y_T^2)^{1.5}} + \frac{a_T \dot{y}_T}{V_T}$$

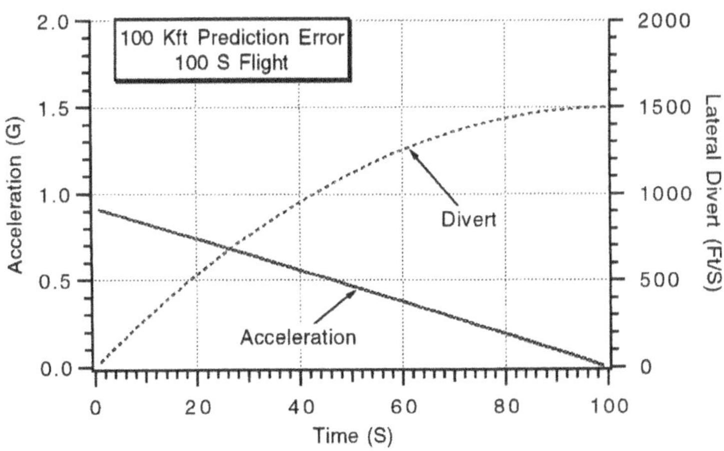

Fig. 14.9 Divert due to prediction error matches theoretical prediction for shorter flight.

where the target velocity V_T is given by

$$V_T = (\dot{x}_T^2 + \dot{y}_T^2)^{0.5}$$

Any component of the booster acceleration perpendicular to the line of sight will appear as an apparent target maneuver to the missile. The component of the booster acceleration perpendicular to the line of sight a_{PLOS} is given by

$$a_{\text{PLOS}} = \ddot{y}_T \cos\lambda - \ddot{x}_T \sin\lambda$$

Listing 14.2 presents a MATLAB listing of an engagement simulation between an impulsive missile and a boosting target. The nominal numbers used to derive the booster characteristics are those of Chapter 12 in which the maximum axial booster acceleration of a two-stage booster is 20 *g*. The total burn time is 69.1 s and the resultant velocity and acceleration profiles can be found in Fig. 12.12. The simulation is identical to that of Listing 14.1 except that the flight time is 50 s (TF = 50), the initial missile location has been moved closer to the target (XLONGMDEG = 85), the target is a booster executing a gravity turn, and a 20-*g* limit has been placed on the commanded acceleration (XNCLIM = 20). Proportional navigation is used against the boosting target but guidance does not begin until 25 s into the flight.

A nominal case was run with the guidance system turned off to ensure that the missile and booster were on a collision triangle. Figure 14.10 shows the missile hitting the target in the nominal 50-s flight in which the booster is accelerating the entire time. The booster acceleration perpendicular to the line of sight

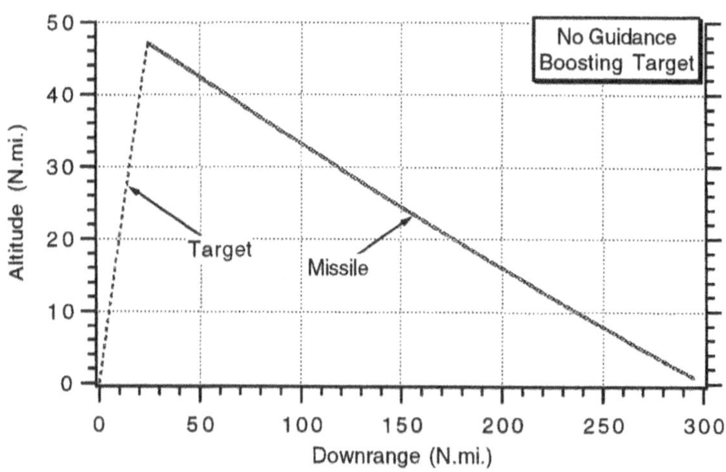

Fig. 14.10 Missile on collision triangle for boosting target.

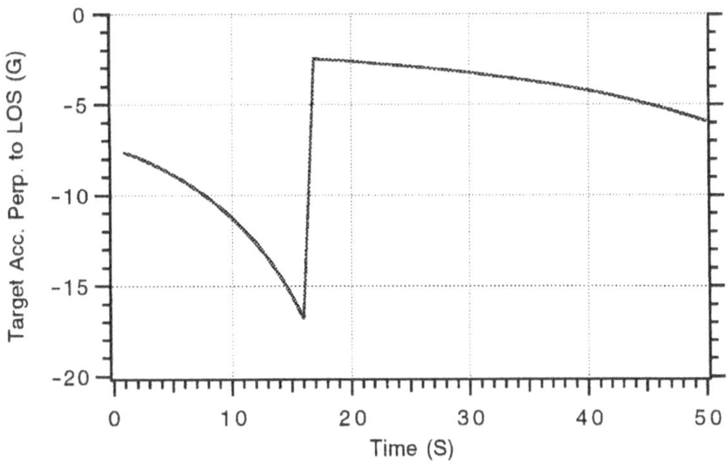

Fig. 14.11 Average target acceleration for last 25 s is approximately 4 *g*.

appears in Fig. 14.11. We can see from the figure that the magnitude of the booster acceleration perpendicular to the line of sight is approximately 4 *g* (129 ft/s^2) on the average for the last 25 s (that is, the time for which proportional navigation is used). This means that the booster appears to the missile to be a target executing a 4-*g* maneuver.

The same nominal case was rerun with the proportional navigation guidance system turned on for the last 25 s. Figure 14.12 displays the missile acceleration

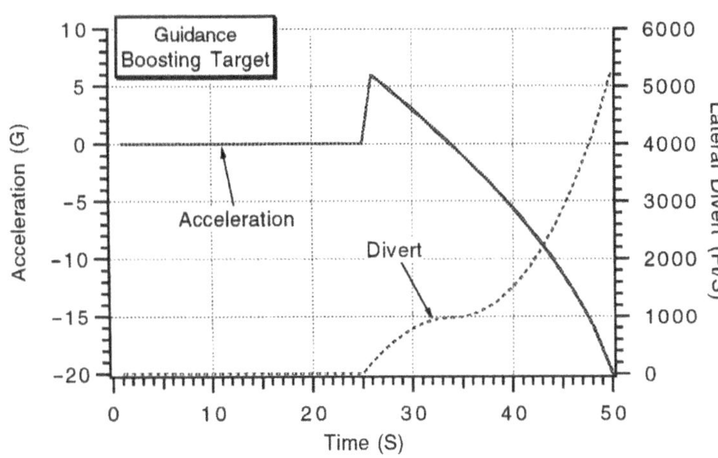

Fig. 14.12 Divert due to apparent maneuver agrees with theory.

along with the resultant lateral divert requirements. We can see that the missile lateral divert requirements for this case are approximately 5300 ft/s.

LISTING 14.2 ENGAGEMENT SIMULATION FOR BOOSTING TARGET

```
count=0;
LEFT=0;
XNCLIM=644.;
A=2.0926e7;
GM=1.4077e16;
DEGRAD=360./(2.*pi);
XNP=3.;
PREDERR=0.;
XISP1=250.;
XISP2=250.;
XMF1=.85;
XMF2=.85;
WPAY=100.;
DELV=20000.;
DELV1=.3333*DELV;
DELV2=.6667*DELV;
AMAX1=20.;
AMAX2=20.;
XKICKDEG=80.;
TOP2=WPAY*(exp(DELV2/(XISP2*32.2))-1.);
BOT2=1/XMF2-((1.-XMF2)/XMF2)*exp(DELV2/(XISP2*32.2));
WP2=TOP2/BOT2;
WS2=WP2*(1-XMF2)/XMF2;
WTOT2=WP2+WS2+WPAY;
TRST2=AMAX2*(WPAY+WS2);
TB2=XISP2*WP2/TRST2;
TOP1=WTOT2*(exp(DELV1/(XISP1*32.2))-1.);
BOT1=1/XMF1-((1.-XMF1)/XMF1)*exp(DELV1/(XISP1*32.2));
WP1=TOP1/BOT1;
WS1=WP1*(1-XMF1)/XMF1;
WTOT=WP1+WS1+WTOT2;
TRST1=AMAX1*(WTOT2+WS1);
TB1=XISP1*WP1/TRST1;
XLONGMDEG=85.;
XLONGTDEG=90.;
ALTNMTIC=0.;
ALTNMMIC=0.;
XKICKDEG=80.;
TF=50.;
ALTT=ALTNMTIC*6076.;
```

```
ALTM=ALTNMMIC*6076.;
S=0.;
XLONGM=XLONGMDEG/DEGRAD;
XLONGT=XLONGTDEG/DEGRAD;
XM=(A+ALTM)*cos(XLONGM);
YM=(A+ALTM)*sin(XLONGM);
XT=(A+ALTT)*cos(XLONGT);
YT=(A+ALTT)*sin(XLONGT);
XFIRSTT=XT;
YFIRSTT=YT;
if LEFT==1
        X1T=cos(pi/2.-XKICKDEG/DEGRAD+XLONGT);
        Y1T=sin(pi/2.-XKICKDEG/DEGRAD+XLONGT);
else
        X1T=cos(-pi/2.+XKICKDEG/DEGRAD+XLONGT);
        Y1T=sin(-pi/2.+XKICKDEG/DEGRAD+XLONGT);
end
XFIRSTT=XT;
YFIRSTT=YT;
T=0.;
[XTF,YTF]=predictb(TF,XT,YT,X1T,Y1T,WP1,WTOT,TB1,TRST1,...
                            TB2,WP2,WTOT2,TRST2,WPAY);
YTF=YTF+PREDERR;
[VRXM,VRYM]=lambertpz(XM,YM,TF,XTF,YTF,XLONGM,XLONGT);
X1M=VRXM;
Y1M=VRYM;
RTM1=XT-XM;
RTM2=YT-YM;
RTM=sqrt(RTM1^2+RTM2^2);
VTM1=X1T-X1M;
VTM2=Y1T-Y1M;
VC=-(RTM1*VTM1+RTM2*VTM2)/RTM;
DELV=0.;
while VC>=0.
        TGO=RTM/VC;
        if TGO>.1
    H=.01;
        else
           H=.0001;
        end
        XOLDT=XT;
        YOLDT=YT;
        X1OLDT=X1T;
        Y1OLDT=Y1T;
        XOLDM=XM;
        YOLDM=YM;
```

```
X1OLDM=X1M;
Y1OLDM=Y1M;
DELVOLD=DELV;
STEP=1;
FLAG=0;
while STEP <=1
   if FLAG==1
            XT=XT+H*XDT;
            YT=YT+H*YDT;
            X1T=X1T+H*X1DT;
            Y1T=Y1T+H*Y1DT;
            XM=XM+H*XDM;
            YM=YM+H*YDM;
            X1M=X1M+H*X1DM;
            Y1M=Y1M+H*Y1DM;
            DELV=DELV+H*DELVD;
            T=T+H;
            STEP=2;
   end
   if T<TB1
            WGT=-WP1*T/TB1+WTOT;
            TRST=TRST1;
   elseif T<(TB1+TB2)
            WGT=-WP2*T/TB2+WTOT2+WP2*TB1/TB2;
            TRST=TRST2;
   else
            WGT=WPAY;
            TRST=0.;
   end
   AT=32.2*TRST/WGT;
   VEL=sqrt(X1T^2+Y1T^2);
   AXT=AT*X1T/VEL;
   AYT=AT*Y1T/VEL;
   TEMBOTT=(XT^2+YT^2)^1.5;
   X1DT=-GM*XT/TEMBOTT+AXT;
   Y1DT=-GM*YT/TEMBOTT+AYT;
   XDT=X1T;
   YDT=Y1T;
   RTM1=XT-XM;
   RTM2=YT-YM;
   RTM=sqrt(RTM1^2+RTM2^2);
   VTM1=X1T-X1M;
   VTM2=Y1T-Y1M;
   VC=-(RTM1*VTM1+RTM2*VTM2)/RTM;
   TGO=RTM/VC;
   XLAM=atan2(RTM2,RTM1);
```

```
    XLAMD=(RTM1*VTM2-RTM2*VTM1)/(RTM*RTM);
    ATPLOS=Y1DT*cos(XLAM)-X1DT*sin(XLAM);
    if T>25.
            XNC=XNP*VC*XLAMD;
    else
            XNC=0.;
    end
    if XNC>XNCLIM
            XNC=XNCLIM;
    end
    if XNC<-XNCLIM
            XNC=-XNCLIM;
    end
    DELVD=abs(XNC);
    AM1=-XNC*sin(XLAM);
    AM2=XNC*cos(XLAM);
    TEMBOTM=(XM^2+YM^2)^1.5;
    X1DM=-GM*XM/TEMBOTM+AM1;
    Y1DM=-GM*YM/TEMBOTM+AM2;
    XDM=X1M;
    YDM=Y1M;
    FLAG=1;
end;
FLAG=0;
XT=(XOLDT+XT)/2+.5*H*XDT;
YT=(YOLDT+YT)/2+.5*H*YDT;
X1T=(X1OLDT+X1T)/2+.5*H*X1DT;
Y1T=(Y1OLDT+Y1T)/2+.5*H*Y1DT;
XM=(XOLDM+XM)/2+.5*H*XDM;
YM=(YOLDM+YM)/2+.5*H*YDM;
X1M=(X1OLDM+X1M)/2+.5*H*X1DM;
Y1M=(Y1OLDM+Y1M)/2+.5*H*Y1DM;
DELV=(DELVOLD+DELV)/2.+.5*H*DELVD;
ALTT=sqrt(XT^2+YT^2)-A;
ALTM=sqrt(XM^2+YM^2)-A;
S=S+H;
if S>=.99999
    S=0.;
    ALTNMT=ALTT/6076.;
    R=sqrt(XT^2+YT^2);
    RF=sqrt(XFIRSTT^2+YFIRSTT^2);
    CBETA=(XT*XFIRSTT+YT*YFIRSTT)/(R*RF);
    BETA=acos(CBETA);
    DISTNMT=A*BETA/6076.;
    ALTNMM=ALTM/6076.;
    XNCG=XNC/32.2;
```

```
            R=sqrt(XM^2+YM^2);
            RF=sqrt(XFIRSTT^2+YFIRSTT^2);
            CBETA=(XM*XFIRSTT+YM*YFIRSTT)/(R*RF);
            BETA=acos(CBETA);
            DISTNMM=A*BETA/6076.;
            ATPLOSG=ATPLOS/32.2;
            count=count+1;
            ArrayT(count)=T;
            ArrayDISTNMT(count)=DISTNMT;
            ArrayALTNMT(count)=ALTNMT;
            ArrayDISTNMM(count)=DISTNMM;
            ArrayALTNMM(count)=ALTNMM;
            ArrayXNCG(count)=XNCG;
            ArrayDELV(count)=DELV;
            ArrayATPLOSG(count)=ATPLOSG;
        end
end
RTM
DELV
figure
plot(ArrayDISTNMT,ArrayALTNMT,ArrayDISTNMM,ArrayALTNMM),grid
xlabel('Downrange (Nmi)')
ylabel('Altitude (Nmi) ')
figure
plot(ArrayT,ArrayATPLOSG),grid
xlabel('Time (Sec)')
ylabel('Acceleration Perp. To LOS (G) ')
figure
plot(ArrayT,ArrayXNCG),grid
xlabel('Time (Sec)')
ylabel('Acceleration (G) ')
figure
plot(ArrayT,ArrayDELV),grid
xlabel('Time (Sec)')
ylabel('Lateral Divert (Ft/Sec) ')
clc
output=[ArrayT',ArrayDISTNMT',ArrayALTNMT',ArrayDISTNMM',...
        ArrayALTNMM',ArrayXNCG',ArrayDELV',ArrayATPLOSG'];
save datfil.txt output /ascii
disp 'simulation finished'

function [xtf,ytf]=predictb(tf,xdum,ydum,x1dum,y1dum,...
                    wp1,wtot,tb1,trst1,tb2,wp2,wtot2,trst2,wpay)
h=.01
a=2.0926e7;
gm=1.4077e16;
```

```
t=0.;
x=xdum;
y=ydum;
x1=x1dum;
y1=y1dum;
while t<=(tf-.00001)
      xold=x;
      yold=y;
      x1old=x1;
      y1old=y1;
      step=1;
      flag=0;
      while step <=1
        if flag==1
                x=x+h*xd;
                y=y+h*yd;
                x1=x1+h*x1d;
                y1=y1+h*y1d;
                t=t+h;
                step=2;
        end
        if t<tb1
                wgt=-wp1*t/tb1+wtot;
                trst=trst1;
        elseif t<(tb1+tb2)
                wgt=-wp2*t/tb2+wtot2+wp2*tb1/tb2;
                trst=trst2;
        else
                wgt=wpay;
                trst=0.;
        end
        at=32.2*trst/wgt;
        vel=sqrt(x1^2+y1^2);
        axt=at*x1/vel;
        ayt=at*y1/vel;
        tembott=(x^2+y^2)^1.5;
        x1d=-gm*x/tembott+axt;
        y1d=-gm*y/tembott+ayt;
        xd=x1;
        yd=y1;
        flag=1;
      end;
      flag=0;
      x=(xold+x)/2+.5*h*xd;
      y=(yold+y)/2+.5*h*yd;
      x1=(x1old+x1)/2+.5*h*x1d;
```

```
        y1=(y1old+y1)/2+.5*h*y1d;
end
        xtf=x;
        ytf=y;
%lambertpz.m shown in Listing 13.2
```

For this case theory says that the divert requirements should be

$$\Delta V|_{\rm PN} = \frac{N'}{N'-1} n_T t_F = \frac{3 * 129 * 25}{2} = 4838\,{\rm ft/s}$$

In other words, theory and simulation are in close agreement.

We saw that with tactical missiles the augmented proportional navigation guidance law reduced the interceptor acceleration requirements. The closed-form solution for the acceleration required to hit a maneuvering target with the augmented proportional navigation guidance law was shown to be given by

$$n_c|_{\rm APN} = 0.5 n_T N' \left[1 - \frac{t}{t_F}\right]^{N'-2}$$

Using the fact that the lateral divert is the integral of the absolute value of the acceleration, it is easy to show that

$$\Delta V|_{\rm APN} = .5 \frac{N'}{N'-1} n_T t_F = .5 \Delta V|_{\rm PN}$$

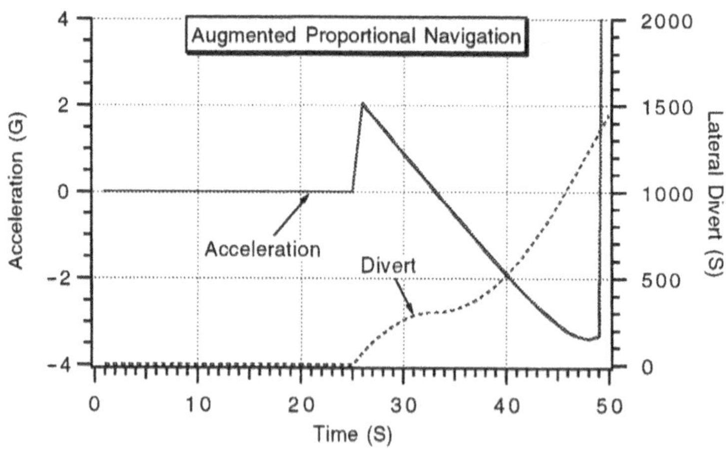

Fig. 14.13 Augmented proportional navigation reduces divert requirements due to boosting target.

In other words, theory says that the divert requirements for an augmented proportional navigation guidance system are half the divert requirements of a proportional navigation guidance system.

To implement augmented proportional navigation guidance in the engagement simulation, it is necessary to modify the guidance command to

$$n_c|_{\text{APN}} = N'V_c\dot{\lambda} + \frac{N'}{2}a_{\text{PLOS}}$$

where a_{PLOS} is the booster acceleration perpendicular to the line of sight. The nominal simulation case was rerun, except this time the augmented proportional navigation guidance law was used. Figure 14.13 shows that the missile lateral divert requirements were dramatically reduced to about 1500 ft/s (down from about 5200 ft/s in the proportional navigation case).

Theory says the divert requirements for the augmented proportional navigation guidance law should be

$$\Delta V|_{\text{APN}} = 0.5\frac{N'}{N'-1}n_T t_F = \frac{0.5 * 3 * 129 * 25}{2} = 2419\,\text{ft/s}$$

which is in approximate agreement with the simulation results.

SUMMARY

In this chapter we have shown that the guidance concepts developed for the tactical world are applicable to the strategic world. In fact, closed-form solutions for the required missile acceleration to hit targets can be converted to lateral divert formulas. Nonlinear engagement simulation results indicate that the divert requirement formulas for prediction error, apparent target acceleration, and guidance law are not only useful but are in fact accurate indicators of strategic interceptor requirements.

REFERENCE

[1] Zarchan, P., "Space Based Interceptor Engagement Simulation and Analysis," Charles Stark Draper Lab., Rept. R-2025, Cambridge, MA, Dec. 1987.

CHAPTER 15

Miscellaneous Topics

INTRODUCTION

In this chapter we shall cover several important topics that have not yet been discussed. First we shall show how lateral divert requirements can be reduced if we add an extra term to the proportional navigation guidance law to account for gravity. Next we will demonstrate that, if complete information on all the target states is available, the ultimate guidance law, predictive guidance, can be used to relax system divert requirements. Finally, a new guidance law, known as pulsed guidance, is developed for those situations in which an interceptor does not have throttleable divert engines.

GRAVITY COMPENSATION

We saw in Chapter 14 that, in the absence of prediction errors, lateral divert fuel was still required to hit a ballistic target. Changes in the missile and target velocities due to gravity caused the line-of-sight to rotate. The proportional navigation guidance law responded to the apparent line-of-sight rate with acceleration commands. If we have knowledge of gravitational acceleration, it seems reasonable that it might be possible to compensate for unnecessary accelerations via the guidance law.

Consider the ballistic missile-target model of Fig. 15.1. In this case the only acceleration acting on the missile and target is gravity. However, since the missile and target are in different locations and since gravitational acceleration is always toward the center of the Earth, the gravitational vectors for the missile and target will have different magnitudes and directions. The missile and target gravitational vectors can be expressed as

$$\boldsymbol{grav}_M = gravx_M\boldsymbol{i} + gravy_M\boldsymbol{j}$$
$$\boldsymbol{grav}_T = gravx_T\boldsymbol{i} + gravy_T\boldsymbol{j}$$

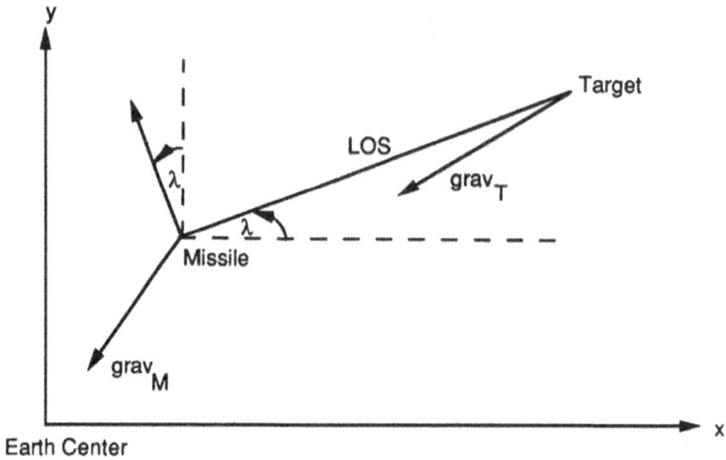

Fig. 15.1 Model for understanding gravity compensation.

where the gravitational components for the missile and target can be found from

$$gravx_M = \frac{-gm\,x_M}{\left(x_M^2 + y_M^2\right)^{1.5}}$$

$$gravy_M = \frac{-gm\,y_M}{\left(x_M^2 + y_M^2\right)^{1.5}}$$

$$gravx_T = \frac{-gm\,x_T}{\left(x_T^2 + y_T^2\right)^{1.5}}$$

$$gravy_T = \frac{-gm\,y_T}{\left(x_T^2 + y_T^2\right)^{1.5}}$$

From Fig. 15.1 we can see that the component of gravity perpendicular to the line of sight for both the missile and target can be found by trigonometry and is given by

$$grav_{\text{MPLOS}} = -gravx_M \sin\lambda + gravy_M \cos\lambda$$

$$grav_{\text{TPLOS}} = -gravx_T \sin\lambda + gravy_T \cos\lambda$$

The gravitational acceleration difference between the target and missile can be treated as an additional term in the zero effort miss. Therefore, we can modify the proportional navigation guidance law to account for gravity. The resultant law, which is similar to augmented proportional navigation, is

$$n_c = N'V_c\dot{\lambda} + \frac{N'}{2}\left(grav_{\text{TPLOS}} - grav_{\text{MPLOS}}\right)$$

The similarity of proportional navigation with gravity compensation and augmented proportional navigation is due to the fact that the gravitational components of the missile and target are treated as an apparent residual target acceleration.

The nominal 500-s ballistic case of Chapter 14 was repeated in which the missile and target are on a collision triangle (see Figs. 14.3 and 14.4). In this case there are zero prediction errors. Figure 15.2 compares the acceleration profiles of proportional navigation, for this case, when gravity compensation is both included and excluded. We can see that in both cases the acceleration levels are low, but the gravity compensation results in even smaller acceleration levels. We can also see from the figure that the lateral divert requirements have been reduced from 356 ft/s to 114 ft/s with gravity compensation.

Although gravity compensation reduced the missile lateral divert requirements for the nominal case, the divert requirements were not very high without the compensation. Gravity compensation should be even more important for longer-range flights since there is more time for a strategic missile to waste fuel. Another case was examined in which the flight time was increased from 500s for the nominal case to 700 s for the new case. Figure 15.3 depicts the engagement geometry for the longer-range case. We can see that the longer flight time results in a more lofted missile trajectory.

Both methods of guidance (with and without gravity compensation) were compared for the new longer flight time case. Figure 15.4, which compares the commanded acceleration profiles, shows that the acceleration requirements are substantially less when gravity compensation is used. The figure indicates that the missile lateral divert requirements were reduced from 1673 ft/s to 508 ft/s

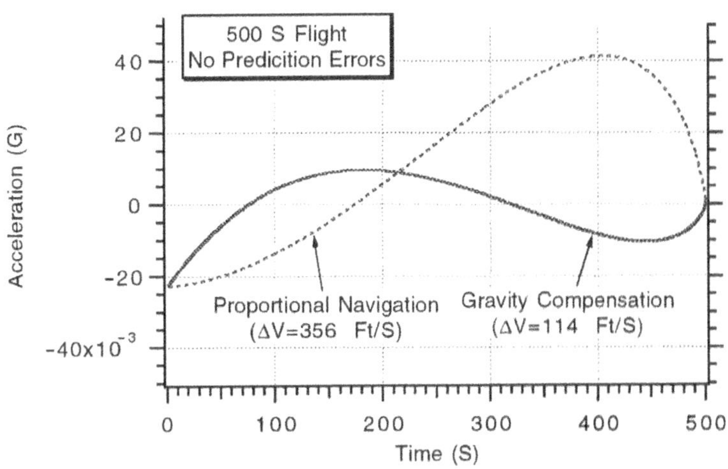

Fig. 15.2 Gravity compensation results in smaller acceleration levels for nominal case.

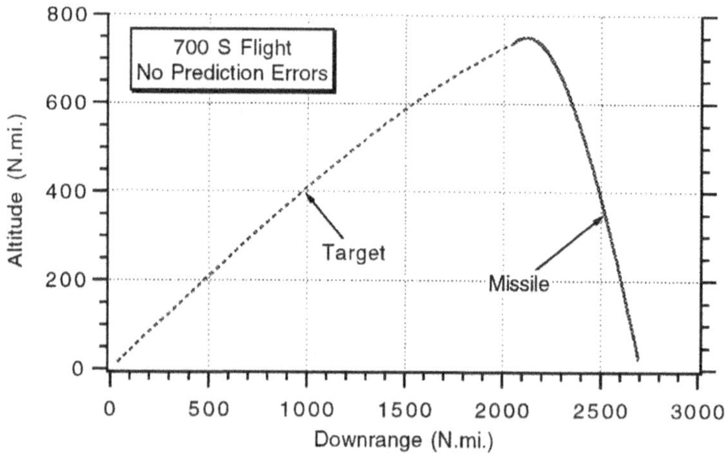

Fig. 15.3 Engagement geometry for longer-range case.

when gravity compensation was used. In this case there would be a dramatic advantage in terms of fuel and weight savings in using gravity compensation.

The gravity compensation guidance law assumes that the required gravitational components of both the missile and target can be measured or estimated precisely. Errors in estimating the gravitational components will of course

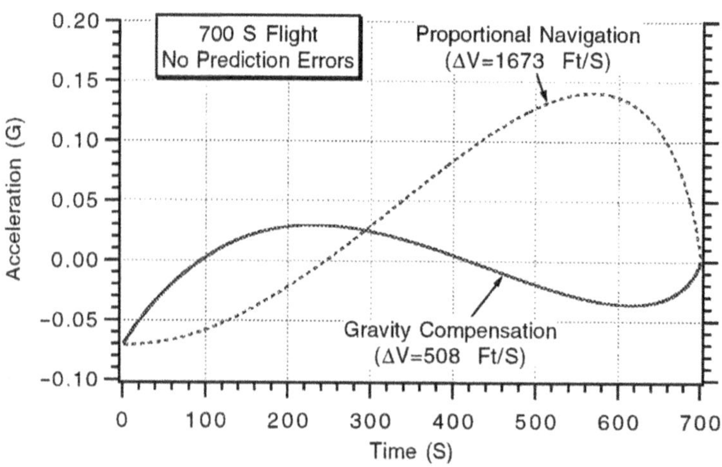

Fig. 15.4 Gravity compensation offers dramatic reduction in divert requirements for long-range case.

degrade the effectiveness of this type of compensation, perhaps to the point where its performance is the same or worse than that of proportional navigation.

PREDICTIVE GUIDANCE

We have seen how interceptor lateral divert requirements can be reduced when extra information, if it exists, is incorporated in the guidance law. If an exact model of the target and missile dynamics were available, one could achieve the best performance with predictive guidance. The principle behind predictive guidance is quite simple. We take our dynamic models of the target and missile and numerically integrate them forward until the desired intercept time. In other words, we are predicting the future location of the missile and target. The difference between the predicted missile and target position at the intercept time is the zero effort miss. If the predicted coordinates of the missile at intercept in the Earth-centered system is given by (x_{MF}, y_{MF}) and the coordinates of the target at intercept are given by (x_{TF}, y_{TF}), then the Earth-centered components of the zero effort miss are given by

$$ZEM_x = x_{TF} - x_{MF}$$
$$ZEM_y = y_{TF} - y_{MF}$$

We can find the component of the zero effort miss perpendicular to the line of sight by trigonometry in Fig. 15.1. The zero effort miss perpendicular to the line of sight is given by

$$ZEM_{\text{PLOS}} = -ZEM_x \sin\lambda + ZEM_y \cos\lambda$$

We saw from our discussion of optimal guidance that the acceleration guidance command should be proportional to the zero effort miss and inversely proportional to the square of time to go until intercept, or

$$n_c = \frac{N' ZEM_{\text{PLOS}}}{t_{\text{go}}^2}$$

Proportional navigation, augmented proportional navigation, and our previously derived optimal guidance law can all be expressed in the preceding form. In these guidance laws we have closed-form expressions for the zero effort miss. In other words, an integration of simple dynamics (assumed to be a polynomial in time) was conducted to get a closed-form expression. In predictive guidance, we ignore closed-form solutions of approximate processes and obtain the exact solution for the zero effort miss at each guidance update by numerical integration. The resultant accuracy of the computed zero effort miss depends on the size of the integration interval. Small integration intervals yield accurate answers but may take too long to be obtained in flight. Of course, the accuracy also depends on the equations used. Having inaccurate models of the target will lead to erroneous predictions of the zero effort miss, and in this case the performance of predictive guidance may be substantially worse than that of proportional navigation.

Listing 15.1 presents an engagement simulation of a ballistic missile and a boosting target. This simulation is identical to the one of Listing 14.2 except that the guidance law has changed from proportional navigation to predictive guidance. We can see from the listing that routine predictg.m is used not only to establish the initial estimate of the intercept point but also is used now at each guidance update to compute the zero effort miss. An examination of the prediction routine shows that the equations of the target and missile have been perfectly modeled. Even the integration step size is an exact match. We can see from the listing that guidance commands, based upon predictive guidance, are calculated at each guidance update throughout the flight.

The nominal 50-s boosting target case of the previous chapter (see Figs. 14.10 and 14.11) was repeated to see the effectiveness of the new guidance law. Figure 15.5 displays the commanded missile acceleration requirements for predictive guidance for the nominal case. We can see that, as expected, predictive guidance requires much less acceleration than either proportional navigation or augmented proportioned navigation. In fact, we can see that predictive guidance virtually requires zero acceleration to intercept the boosting target. The reason for this is that the missile is initially on a collision triangle with the target. Therefore, no commands are really necessary for a successful intercept. However, if for computer throughput reasons the integration step size h in routine predictg.m had to be increased by two orders of magnitude (from .01 s to 1 s), we can see from Fig. 15.5 that the missile acceleration response would be unstable.

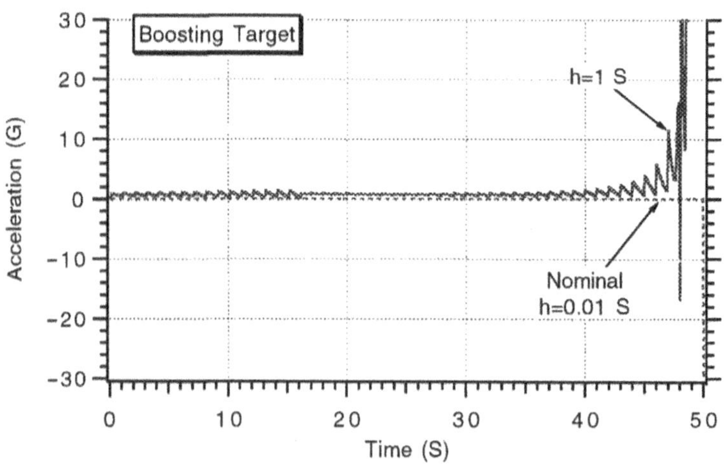

Fig. 15.5 Acceleration requirements for predictive guidance can be very small.

LISTING 15.1 PREDICTIVE GUIDANCE ENGAGEMENT SIMULATION OF BALLISTIC MISSILE AND BOOSTER TARGET

```
count=0;
LEFT=0;
A=2.0926e7;
GM=1.4077e16;
XNP=3.;
AXMGUID=0.;
AYMGUID=0.;
PREDERR=0.;
XISP1=250.;
XISP2=250.;
XMF1=.85;
XMF2=.85;
WPAY=100.;
DELV=20000.;
DELV1=.3333*DELV;
DELV2=.6667*DELV;
AMAX1=20.;
AMAX2=20.;
XKICKDEG=80.;
TOP2=WPAY*(exp(DELV2/(XISP2*32.2))-1.);
BOT2=1/XMF2-((1.-XMF2)/XMF2)*exp(DELV2/(XISP2*32.2));
WP2=TOP2/BOT2;
WS2=WP2*(1-XMF2)/XMF2;
WTOT2=WP2+WS2+WPAY;
TRST2=AMAX2*(WPAY+WS2);
TB2=XISP2*WP2/TRST2;
TOP1=WTOT2*(exp(DELV1/(XISP1*32.2))-1.);
BOT1=1/XMF1-((1.-XMF1)/XMF1)*exp(DELV1/(XISP1*32.2));
WP1=TOP1/BOT1;
WS1=WP1*(1-XMF1)/XMF1;
WTOT=WP1+WS1+WTOT2;
TRST1=AMAX1*(WTOT2+WS1);
TB1=XISP1*WP1/TRST1;
ALTNMT=0.;
ALTNMM=0.;
ALTT=ALTNMT*6076.;
ALTM=ALTNMM*6076.;
DEGRAD=360./(2.*pi);
S=0.;
XLONGMDEG=85.;
XLONGTDEG=90.;
XLONGM=XLONGMDEG/DEGRAD;
```

```
XLONGT=XLONGTDEG/DEGRAD;
XM=(A+ALTM)*cos(XLONGM);
YM=(A+ALTM)*sin(XLONGM);
XT=(A+ALTT)*cos(XLONGT);
YT=(A+ALTT)*sin(XLONGT);
XFIRSTT=XT;
YFIRSTT=YT;
if LEFT==1
        X1T=cos(pi/2.-XKICKDEG/DEGRAD+XLONGT);
        Y1T=sin(pi/2.-XKICKDEG/DEGRAD+XLONGT);
else
        X1T=cos(-pi/2.+XKICKDEG/DEGRAD+XLONGT);
        Y1T=sin(-pi/2.+XKICKDEG/DEGRAD+XLONGT);
end
T=0.;
TF=50.;
TGO=TF-T;
X1M=0.;
Y1M=0.;
[XTF,YTF,ZEM1,ZEM2]=predictg(T,TF,XT,YT,X1T,Y1T,WP1,WTOT,...
           TB1,TRST1,TB2,WP2,WTOT2,TRST2,WPAY,XM,YM,X1M,Y1M,TGO);
YTF=YTF+PREDERR;
[VRXM,VRYM]=lambertpz(XM,YM,TF,XTF,YTF,XLONGM,XLONGT);
X1M=VRXM;
Y1M=VRYM;
RTM1=XT-XM;
RTM2=YT-YM;
RTM=sqrt(RTM1^2+RTM2^2);
VTM1=X1T-X1M;
VTM2=Y1T-Y1M;
VC=-(RTM1*VTM1+RTM2*VTM2)/RTM;
DELV=0.;
XNC=0.;
while VC>=0.
        TGO=RTM/VC;
        if TGO > .1
     H=.01;
        else
           H=.0001;
        end
        XOLDT=XT;
        YOLDT=YT;
        X1OLDT=X1T;
        Y1OLDT=Y1T;
        XOLDM=XM;
        YOLDM=YM;
```

```
X1OLDM=X1M;
Y1OLDM=Y1M;
DELVOLD=DELV;
STEP=1;
FLAG=0;
while STEP <=1
   if FLAG==1
            XT=XT+H*XDT;
            YT=YT+H*YDT;
            X1T=X1T+H*X1DT;
            Y1T=Y1T+H*Y1DT;
            XM=XM+H*XDM;
            YM=YM+H*YDM;
            X1M=X1M+H*X1DM;
            Y1M=Y1M+H*Y1DM;
            DELV=DELV+H*DELVD;
            T=T+H;
            STEP=2;
   end
   if T<TB1
            WGT=-WP1*T/TB1+WTOT;
            TRST=TRST1;
   elseif T<(TB1+TB2)
            WGT=-WP2*T/TB2+WTOT2+WP2*TB1/TB2;
            TRST=TRST2;
   else
            WGT=WPAY;
            TRST=0.;
   end
   AT=32.2*TRST/WGT;
   VEL=sqrt(X1T^2+Y1T^2);
   AXT=AT*X1T/VEL;
   AYT=AT*Y1T/VEL;
   TEMBOTT=(XT^2+YT^2)^1.5;
   X1DT=-GM*XT/TEMBOTT+AXT;
   Y1DT=-GM*YT/TEMBOTT+AYT;
   XDT=X1T;
   YDT=Y1T;
   RTM1=XT-XM;
   RTM2=YT-YM;
   RTM=sqrt(RTM1^2+RTM2^2);
   VTM1=X1T-X1M;
   VTM2=Y1T-Y1M;
   VC=-(RTM1*VTM1+RTM2*VTM2)/RTM;
   TGO=RTM/VC;
   XLAM=atan2(RTM2,RTM1);
```

```
        XLAMD=(RTM1*VTM2-RTM2*VTM1)/(RTM*RTM);
        ATPLOS=AYT*cos(XLAM)-AXT*sin(XLAM);
        AXMGUID=-XNC*sin(XLAM);
        AYMGUID=XNC*cos(XLAM);
        DELVD=abs(XNC);
        TEMBOTM=(XM^2+YM^2)^1.5;
        X1DM=-GM*XM/TEMBOTM+AXMGUID;
        Y1DM=-GM*YM/TEMBOTM+AYMGUID;
        XDM=X1M;
        YDM=Y1M;
        FLAG=1;
    end;
    FLAG=0;
    XT=(XOLDT+XT)/2+.5*H*XDT;
    YT=(YOLDT+YT)/2+.5*H*YDT;
    X1T=(X1OLDT+X1T)/2+.5*H*X1DT;
    Y1T=(Y1OLDT+Y1T)/2+.5*H*Y1DT;
    XM=(XOLDM+XM)/2+.5*H*XDM;
    YM=(YOLDM+YM)/2+.5*H*YDM;
    X1M=(X1OLDM+X1M)/2+.5*H*X1DM;
    Y1M=(Y1OLDM+Y1M)/2+.5*H*Y1DM;
    DELV=(DELVOLD+DELV)/2.+.5*H*DELVD;
    ALTT=sqrt(XT^2+YT^2)-A;
    ALTM=sqrt(XM^2+YM^2)-A;
    S=S+H;
    if S>=.99999
        S=0.;
        [XTF,YTF,ZEM1,ZEM2]=predictg(T,TF,XT,YT,X1T,Y1T,WP1,WTOT,...
                TB1,TRST1,TB2,WP2,WTOT2,TRST2,WPAY,XM,YM,X1M,Y1M,TGO);
        ZEMPLOS=-ZEM1*sin(XLAM)+ZEM2*cos(XLAM);
        XNC=XNP*ZEMPLOS/(TGO*TGO);
        if XNC>966.
                XNC=966.;
        end
       if XNC<-966.
                XNC=-966.;
       end
        ALTNMT=ALTT/6076.;
        R=sqrt(XT^2+YT^2);
        RF=sqrt(XFIRSTT^2+YFIRSTT^2);
        CBETA=(XT*XFIRSTT+YT*YFIRSTT)/(R*RF);
        BETA=acos(CBETA);
        DISTNMT=A*BETA/6076.;
        ALTNMM=ALTM/6076.;
        XNCG=XNC/32.2;
        R=sqrt(XM^2+YM^2);
```

```
            RF=sqrt(XFIRSTT^2+YFIRSTT^2);
            CBETA=(XM*XFIRSTT+YM*YFIRSTT)/(R*RF);
            BETA=acos(CBETA);
            DISTNMM=A*BETA/6076.;
            ATPLOSG=ATPLOS/32.2;
            count=count+1;
            ArrayT(count)=T;
            ArrayDISTNMT(count)=DISTNMT;
            ArrayALTNMT(count)=ALTNMT;
            ArrayDISTNMM(count)=DISTNMM;
            ArrayALTNMM(count)=ALTNMM;
            ArrayXNCG(count)=XNCG;
            ArrayDELV(count)=DELV;
            ArrayATPLOSG(count)=ATPLOSG;
        end
 end
RTM
DELV
figure
plot(ArrayDISTNMT,ArrayALTNMT,ArrayDISTNMM,ArrayALTNMM),grid
xlabel('Downrange (Nmi)')
ylabel('Altitude (Nmi) ')
figure
plot(ArrayT,ArrayATPLOSG),grid
xlabel('Time (Sec)')
ylabel('Acceleration Perp. To LOS (G) ')
figure
plot(ArrayT,ArrayXNCG),grid
xlabel('Time (Sec)')
ylabel('Acceleration (G) ')
figure
plot(ArrayT,ArrayDELV),grid
xlabel('Time (Sec)')
ylabel('Lateral Divert (Ft/Sec) ')
clc
output=[ArrayT',ArrayDISTNMT',ArrayALTNMT',ArrayDISTNMM',...
            ArrayALTNMM',ArrayXNCG',ArrayDELV',ArrayATPLOSG'];
save datfil.txt output /ascii
disp 'simulation finished'
function [xtf,ytf,zem1,zem2]=predictg(tdum,tf,xdum,ydum,x1dum,...
            y1dum,wp1,wtot,tb1,trst1,tb2,wp2,wtot2,trst2,wpay,...
            xmdum,ymdum,x1mdum,y1mdum,tgo)
if tgo>1
  h=.01;
else
        h=tgo;
```

```
end
a=2.0926e7;
gm=1.4077e16;
t=tdum;
x=xdum;
y=ydum;
x1=x1dum;
y1=y1dum;
xm=xmdum;
ym=ymdum;
x1m=x1mdum;
y1m=y1mdum;
while t<=(tf-.00001)
        xold=x;
        yold=y;
        x1old=x1;
        y1old=y1;
        xoldm=xm;
        yoldm=ym;
        x1oldm=x1m;
        y1oldm=y1m;
        step=1;
        flag=0;
        while step <=1
           if flag==1
                        x=x+h*xd;
                        y=y+h*yd;
                        x1=x1+h*x1d;
                        y1=y1+h*y1d;
                        xm=xm+h*xdm;
                        ym=ym+h*ydm;
                        x1m=x1m+h*x1dm;
                        y1m=y1m+h*y1dm;
                        t=t+h;
                        step=2;
           end
           if t<tb1
                        wgt=-wp1*t/tb1+wtot;
                        trst=trst1;
           elseif t<(tb1+tb2)
                        wgt=-wp2*t/tb2+wtot2+wp2*tb1/tb2;
                        trst=trst2;
           else
                        wgt=wpay;
                        trst=0.;
           end
```

```
            at=32.2*trst/wgt;
            vel=sqrt(x1^2+y1^2);
            axt=at*x1/vel;
            ayt=at*y1/vel;
            tembott=(x^2+y^2)^1.5;
            x1d=-gm*x/tembott+axt;
            y1d=-gm*y/tembott+ayt;
            xd=x1;
            yd=y1;
            tembotm=(xm^2+ym^2)^1.5;
            x1dm=-gm*xm/tembotm;
            y1dm=-gm*ym/tembotm;
            xdm=x1m;
            ydm=y1m;
            flag=1;
        end;
        flag=0;
        x=(xold+x)/2+.5*h*xd;
        y=(yold+y)/2+.5*h*yd;
        x1=(x1old+x1)/2+.5*h*x1d;
        y1=(y1old+y1)/2+.5*h*y1d;
        xm=(xoldm+xm)/2+.5*h*xdm;
        ym=(yoldm+ym)/2+.5*h*ydm;
        x1m=(x1oldm+x1m)/2+.5*h*x1dm;
        y1m=(y1oldm+y1m)/2+.5*h*y1dm;
end
        xtf=x;
        ytf=y;
        zem1=x-xm;
        zem2=y-ym;
% lambertpz.m can be found in Listing 13.2
```

The divert requirements for predictive guidance are quite small because the prediction is perfect. In other words, we have a perfect model of the missile and target and a perfect numerical integration technique. The integration is perfect because the same method (second-order Runge–Kutta) and step size are used in the simulation and prediction portion of the program. In a practical application of predictive guidance, a larger integration step size h might have to be used to satisfy flight computer constraints.

Figure 15.6 presents the missile lateral divert requirement profiles when the integration step size h in the prediction routine is 0.01 s (nominal), 0.1 s, and 1 s. We can see that the miss distance and lateral divert increase as the integration step size grows. In fact we can see by comparing Figs. 15.6 and 14.13 that when the integration step size is 1 s the lateral divert requirements are greater with predictive guidance than for augmented proportional navigation. By comparing Fig. 15.6

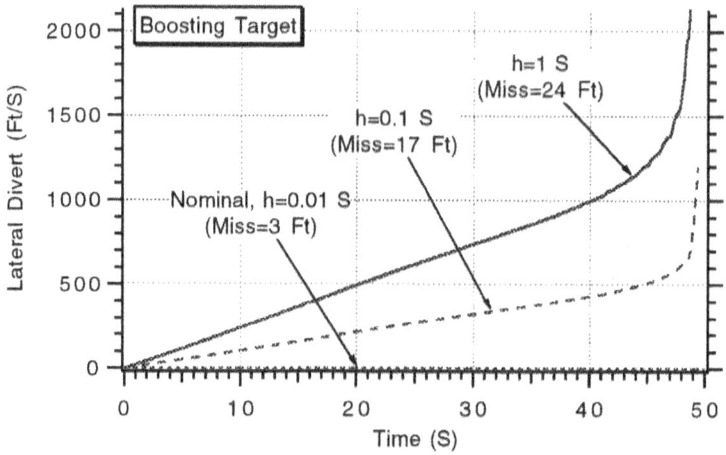

Fig. 15.6 Divert requirements for predictive increase as the integration step size increases.

with Fig. 14.12 we might conclude that the divert requirements for predictive guidance are still smaller than those of proportional navigation, but Fig. 15.6 indicates that predictive guidance misses the target with the larger integration step size.

PULSED GUIDANCE

A strategic interceptor maneuvers with divert engines. Sometimes the engines are effectively throttleable. This means that, by issuing divert commands in opposite directions, any effective acceleration level (within engine thrust-to-weight ratio constraints) can be reached. This is ideal for the implementation of proportional navigation type guidance laws. However, sometimes it is only possible to issue guidance commands of fixed amplitude when the engine is on. In this case we can only influence the duration of the guidance pulse by turning the engine off. In this situation it is not obvious how a guidance law can be implemented or if proportional navigation is appropriate.

Consider the acceleration diagram in Fig. 15.7. Here we have a guidance pulse of magnitude a ft/s^2 lasting for Δt seconds. Implicit in the diagram is the approximation that the acceleration level is constant for the duration of the pulse. Actually, the acceleration level would be increasing for a constant thrust level because of the expenditure of fuel. We wish to derive a guidance law whose output for a given acceleration magnitude a contains information concerning the duration of the guidance pulse Δt.

We have previously seen how proportional navigation is related to the zero effort miss (*ZEM*). In proportional navigation type guidance, we take the entire guided portion of the flight to remove the zero effort miss. If we have a single

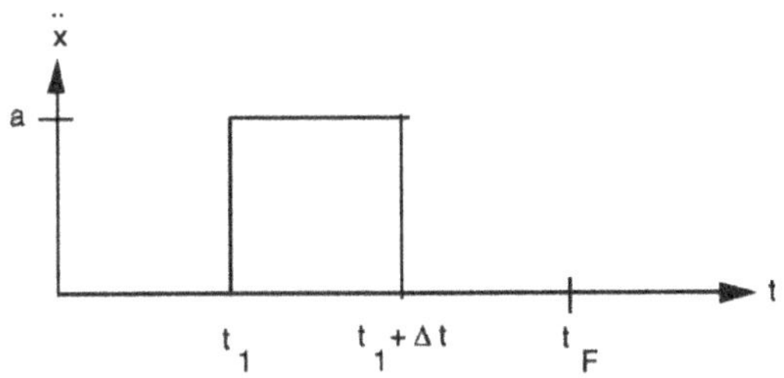

Fig. 15.7 Conceptual diagram for pulsed guidance law.

pulse, as shown in Fig. 15.7, we must remove the zero effort miss in Δt seconds. For a pulse of amplitude a and duration Δt, we can integrate the acceleration twice yielding the velocity profile of Fig. 15.8 and the position profile of Fig. 15.9. Using the information from Fig. 15.9 we can equate the distance traveled due to the acceleration pulse (derived by integrating the acceleration pulse twice) to the zero effort miss or

$$ZEM = .5a\Delta t^2 + a\Delta t(t_{\text{go}} - \Delta t)$$

However, we also know that the proportional navigation guidance law can be expressed as

$$n_c = N'V_c\dot{\lambda} = \frac{N'ZEM}{t_{\text{go}}^2}$$

Therefore, the zero effort miss can be expressed in terms of the line-of-sight rate as

$$ZEM = V_c t_{\text{go}}^2 \dot{\lambda}$$

Equating the expressions for the proportional navigation zero effort miss with the one for the pulse of fixed duration, we get

$$V_c t_{\text{go}}^2 \dot{\lambda} = .5a\Delta t^2 + a\Delta t(t_{\text{go}} - \Delta t)$$

This is a quadratic equation in terms of Δt. Using the quadratic formula and eliminating the unrealistic root, we obtain a closed-form expression for the pulse duration time as

$$\Delta t = t_{\text{go}}\left[1 - \sqrt{1 - \frac{2V_c}{a}\dot{\lambda}}\right]$$

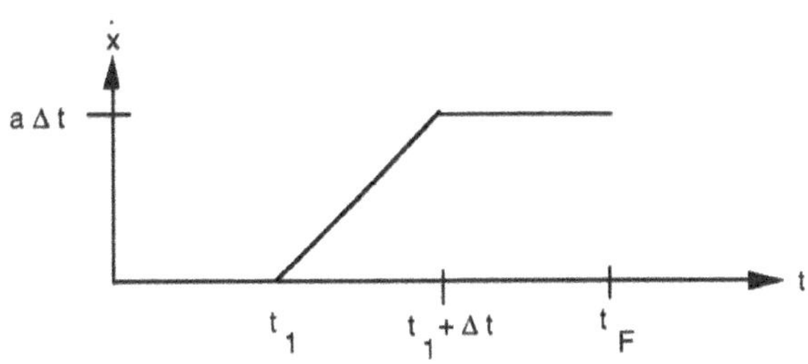

Fig. 15.8 Integrating acceleration pulse once yields velocity.

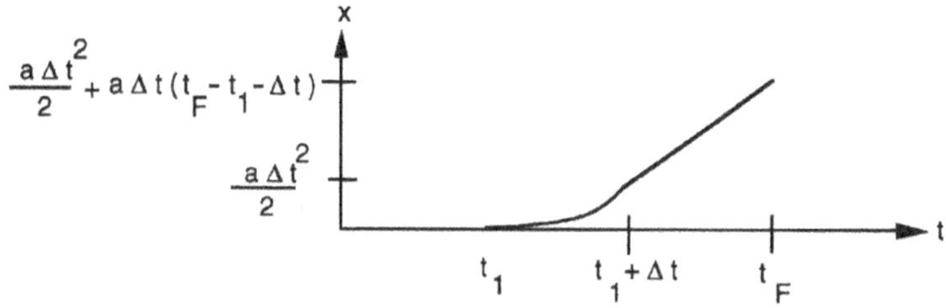

Fig. 15.9 Integrating acceleration pulse twice yields position.

To implement a pulse guidance scheme, we must know in advance the number of pulses to be used. For simplicity, let us assume that the pulses are equally distributed throughout the flight. If it is time for a pulse to commence, we calculate the pulse duration from the preceding formula. An engagement simulation, based upon Listing 14.1, with a pulsed guidance system is presented in Listing 15.2.

A nominal case was run with the pulsed guidance system in which there was 100 kft of prediction error. Figure 15.10 shows the line-of-sight rate profile for a successful intercept in which the missile had 10 guidance pulses. We can see from the plot that the pulsed guidance system is always trying to drive the line-of-sight rate to zero after each pulse is issued. This is not surprising, as the pulsed guidance law was derived with this concept in mind.

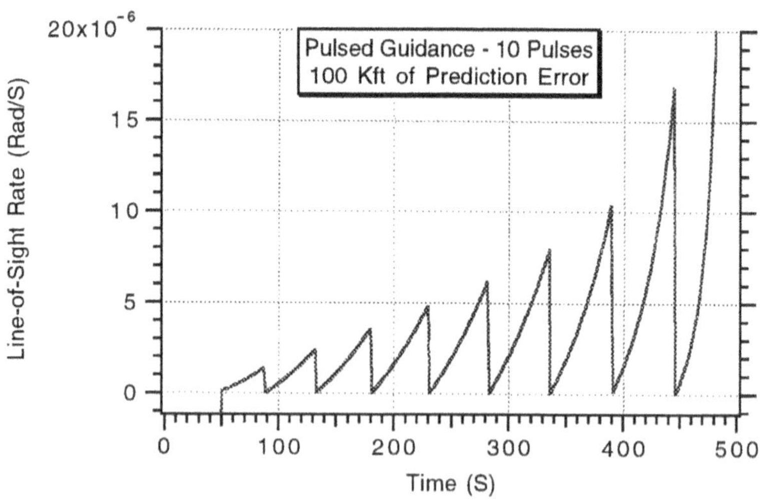

Fig. 15.10 Pulse guidance attempts to drive line-of-sight rate to zero with each pulse.

LISTING 15.2 PULSED GUIDANCE SIMULATION

```
count=0;
XLONGMDEG=45.;
XLONGTDEG=90.;
ALTNMTIC=0.;
ALTNMMIC=0.;
TF=500.;
GAMDEGT=23.;
AMAG=64.4;
PULSES=10.;
PREDERR=-100000.;
PULSE_ON=0;
ACQUIRE=1;
PULSE_NUM=PULSES-1.;
H=.01;
A=2.0926e7;
GM=1.4077e16;
DEGRAD=360./(2.*pi);
XNC=0.;
GAMT=GAMDEGT/57.3;
DISTNMT=6000.;
PHIT=DISTNMT*6076./A;
ALTT=ALTNMTIC*6076.;
ALTM=ALTNMMIC*6076.;
R0T=A+ALTT;
TOP=GM*(1.-cos(PHIT));
TEMP=R0T*cos(GAMT)/A-cos(PHIT+GAMT);
BOT=R0T*cos(GAMT)*TEMP;
VT=sqrt(TOP/BOT);
XLONGM=XLONGMDEG/DEGRAD;
XLONGT=XLONGTDEG/DEGRAD;
if XLONGM>XLONGT
        X1T=VT*cos(pi/2.-GAMT+XLONGT);
        Y1T=VT*sin(pi/2.-GAMT+XLONGT);
else
        X1T=VT*cos(-pi/2.+GAMT+XLONGT);
        Y1T=VT*sin(-pi/2.+GAMT+XLONGT);
end
S=0.;
XLONGM=XLONGMDEG/DEGRAD;
XLONGT=XLONGTDEG/DEGRAD;
XM=(A+ALTM)*cos(XLONGM);
YM=(A+ALTM)*sin(XLONGM);
XT=(A+ALTT)*cos(XLONGT);
YT=(A+ALTT)*sin(XLONGT);
```

```
XFIRSTT=XT;
YFIRSTT=YT;
T=0.;
[XTF,YTF]=predictpz(TF,XT,YT,X1T,Y1T);
YTF=YTF+PREDERR;
[VRXM,VRYM]=lambertpz(XM,YM,TF,XTF,YTF,XLONGM,XLONGT);
X1M=VRXM;
Y1M=VRYM;
RTM1=XT-XM;
RTM2=YT-YM;
RTM=sqrt(RTM1^2+RTM2^2);
VTM1=X1T-X1M;
VTM2=Y1T-Y1M;
VC=-(RTM1*VTM1+RTM2*VTM2)/RTM;
DELV=0.;
while VC>=0.
      TGO=RTM/VC;
      if TGO>.1
   H=.01;
      else
        H=.0001;
      end
      XOLDT=XT;
      YOLDT=YT;
      X1OLDT=X1T;
      Y1OLDT=Y1T;
      XOLDM=XM;
      YOLDM=YM;
      X1OLDM=X1M;
      Y1OLDM=Y1M;
      DELVOLD=DELV;
      STEP=1;
      FLAG=0;
      while STEP <=1
        if FLAG==1
                XT=XT+H*XDT;
                YT=YT+H*YDT;
                X1T=X1T+H*X1DT;
                Y1T=Y1T+H*Y1DT;
                XM=XM+H*XDM;
                YM=YM+H*YDM;
                X1M=X1M+H*X1DM;
                Y1M=Y1M+H*Y1DM;
                DELV=DELV+H*DELVD;
                T=T+H;
                STEP=2;
```

```
        end
        TEMBOTT=(XT^2+YT^2)^1.5;
        X1DT=-GM*XT/TEMBOTT;
        Y1DT=-GM*YT/TEMBOTT;
        XDT=X1T;
        YDT=Y1T;
        RTM1=XT-XM;
        RTM2=YT-YM;
        RTM=sqrt(RTM1^2+RTM2^2);
        VTM1=X1T-X1M;
        VTM2=Y1T-Y1M;
        VC=-(RTM1*VTM1+RTM2*VTM2)/RTM;
        TGO=RTM/VC;
        XLAM=atan2(RTM2,RTM1);
        XLAMD=(RTM1*VTM2-RTM2*VTM1)/(RTM*RTM);
        DELVD=abs(XNC);
        AM1=-XNC*sin(XLAM);
        AM2=XNC*cos(XLAM);
        TEMBOTM=(XM^2+YM^2)^1.5;
        X1DM=-GM*XM/TEMBOTM+AM1;
        Y1DM=-GM*YM/TEMBOTM+AM2;
        XDM=X1M;
        YDM=Y1M;
        FLAG=1;
    end;
    FLAG=0;
    XT=(XOLDT+XT)/2+.5*H*XDT;
    YT=(YOLDT+YT)/2+.5*H*YDT;
    X1T=(X1OLDT+X1T)/2+.5*H*X1DT;
    Y1T=(Y1OLDT+Y1T)/2+.5*H*Y1DT;
    XM=(XOLDM+XM)/2+.5*H*XDM;
    YM=(YOLDM+YM)/2+.5*H*YDM;
    X1M=(X1OLDM+X1M)/2+.5*H*X1DM;
    Y1M=(Y1OLDM+Y1M)/2+.5*H*Y1DM;
    DELV=(DELVOLD+DELV)/2.+.5*H*DELVD;
    ALTT=sqrt(XT^2+YT^2)-A;
    ALTM=sqrt(XM^2+YM^2)-A;
    if PULSE_ON==1
        if T>TOFF
                PULSE_ON=0;
                XNC=0.;
        end
    end
    if PULSE_NUM > 0.
        if TGO<=(TF-0.)*PULSE_NUM/(PULSES-1.)
                PULSE_NUM=PULSE_NUM-1.;
```

```
                        PULSE_ON=1;
                        DISC=1.-2.*VC*abs(XLAMD)/AMAG;
                        if DISC>0.
                                TPULSE=TGO*(1.-sqrt(DISC));
                                if XLAMD>0.
                                        XNC=AMAG;
                                else
                                        XNC=-AMAG;
                                end
                        else
                                TPULSE=0.;
                        end
                        if TGO<TPULSE
                                TOFF=9999999.;
                        else
                                TOFF=T+TPULSE;
                        end
                end
        else
                DISC=1.-2.*VC*abs(XLAMD)/AMAG;
                if DISC>0.
                        TPULSE=TGO*(1.-sqrt(DISC));
                else
                        TPULSE=999999.;
                end
                if TGO<=TPULSE
                        if XLAMD>0.
                                XNC=AMAG;
                        else
                                XNC=-AMAG;
                        end
                        PULSE_ON=1;
                        TOFF=999999.;
                end
        end
        S=S+H;
        if S>=.99999
                S=0.;
                count=count+1;
                ArrayT(count)=T;
                ArrayXNC(count)=XNC;
                ArrayXLAMD(count)=XLAMD;
                ArrayDELV(count)=DELV;
        end
end
RTM
```

```
DELV
figure
plot(ArrayT,ArrayXLAMD),grid
xlabel('Time (Sec)')
ylabel('Line of Sight Rate (Rad/Sec) ')
axis([0 500 0 .00002])
clc
output=[ArrayT',ArrayXNC',ArrayXLAMD',ArrayDELV'];
save datfil.txt output /ascii
disp 'simulation finished'

function [xtf,ytf]=predictpz(tf,xdum,ydum,x1dum,y1dum)
h=.01;
a=2.0926e7;
gm=1.4077e16;
t=0.;
x=xdum;
y=ydum;
x1=x1dum;
y1=y1dum;
while t<=(tf-.00001)
      xold=x;
      yold=y;
      x1old=x1;
      y1old=y1;
      step=1;
      flag=0;
      while step <=1
         if flag==1
                  x=x+h*xd;
                  y=y+h*yd;
                  x1=x1+h*x1d;
                  y1=y1+h*y1d;
                  t=t+h;
                  step=2;
         end
         tembot=(x^2+y^2)^1.5;
         x1d=-gm*x/tembot;
         y1d=-gm*y/tembot;
         xd=x1;
         yd=y1;
         flag=1;
      end;
      flag=0;
      x=(xold+x)/2+.5*h*xd;
      y=(yold+y)/2+.5*h*yd;
```

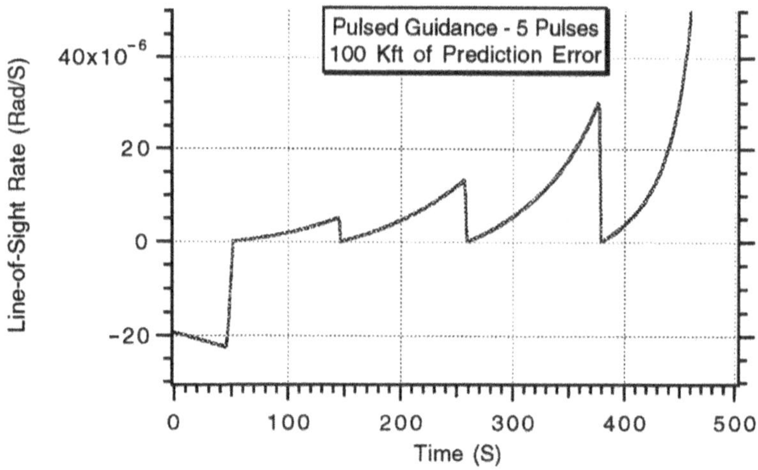

Fig. 15.11 Pulsed guidance works with fewer pulses.

```
        x1=(x1old+x1)/2+.5*h*x1d;
        y1=(y1old+y1)/2+.5*h*y1d;
end
xtf=x;
ytf=y;
% lambertpz.m can be found in Listing 13.2
```

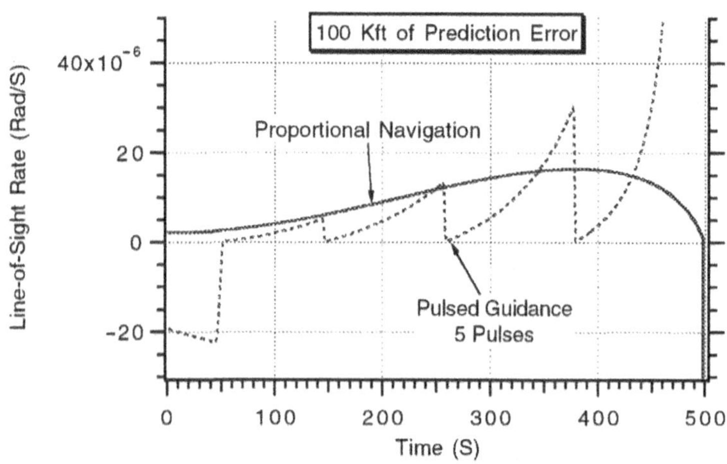

Fig. 15.12 Comparison of proportional navigation and pulsed guidance.

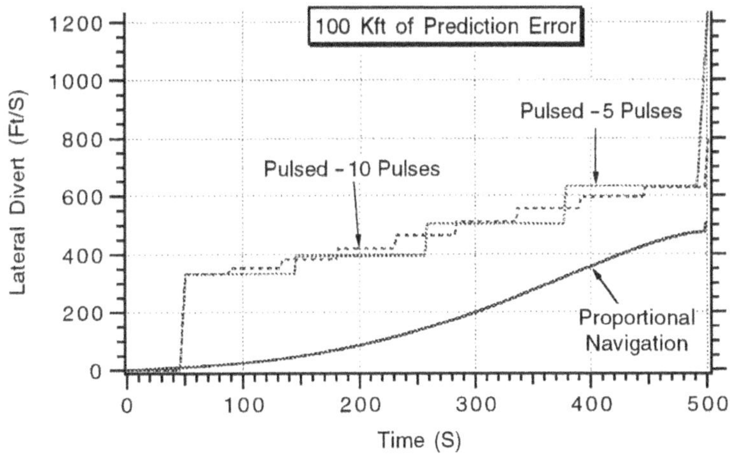

Fig. 15.13 Pulsed guidance does not reduce lateral divert requirements.

Figure 15.11 shows a similar case, except only five guidance pulses are used. We can see that the pulsed guidance law is still attempting to drive the line-of-sight rate to zero after each pulse is issued. However, since there are fewer pulses, the line-of-sight rate builds up to larger values in between pulses.

The five pulse results are overlayed with the proportional navigation results for the identical case and are shown in Figure 15.12. We can see that proportional navigation drives the line-of-sight rate to zero at the end of the flight only.

Although each of the cases studied so far resulted in successful intercepts, the divert requirements for each guidance concept are different. Figure 15.13 displays the required lateral divert profiles for each of the cases. We can see that proportional navigation has the smallest divert requirements. Increasing the number of pulses in a pulsed guidance system does not appear to influence the lateral divert requirements.

We use pulsed guidance when the divert engine characteristics make it infeasible to use proportional navigation. The price paid is somewhat higher for lateral divert requirements.

CHAPTER 16

Ballistic Target Properties

INTRODUCTION [1, 2]

Although surface-to-surface missiles were used as terror weapons during World War II and the Iran–Iraq War, most of the world became familiar with the ballistic missile threat during the 1991 Persian Gulf War. Hundreds of millions of TV viewers will never forget the wail of sirens and images of Scud missiles glowing in the night-time skies over Tel Aviv and Dhahran as they decelerated through the atmosphere toward their civilian targets after having traveled hundreds of miles from their launch sites in Iraq. Viewers, regardless of nationality, were riveted by the drama of the almost nightly duels between the Patriot interceptor and its intended prey—the Scud ballistic target.

In the chapters pertaining to tactical guidance the interceptor's intended target was considered to be an aircraft, whereas in the chapters pertaining to strategic guidance the engagement threat was either considered to be a booster or an exoatmospheric ballistic target. In the next two chapters of the text we will consider the special problems encountered in intercepting an endoatmospheric ballistic target.

BALLISTIC TARGET MODEL

When a ballistic target re-enters the atmosphere after having traveled a long distance, its speed is high and the remaining time to ground impact is relatively short. The small distances traveled by ballistic targets after they re-enter the atmosphere enable us to accurately model these threats using the flat-Earth, constant gravity approximation as was done in modeling tactical interceptors. This simplification is important because it will lead to useful closed-form solutions for ballistic targets.

Figure 16.1 presents the flat-Earth, constant gravity model for the ballistic endoatmospheric threat. In this model, only drag and gravity act on the ballistic target [3]. We can see from Fig. 16.1 that the target has velocity V_T and is initially at re-entry angle γ_T. Note that drag F_{drag} acts in a direction opposite to the velocity

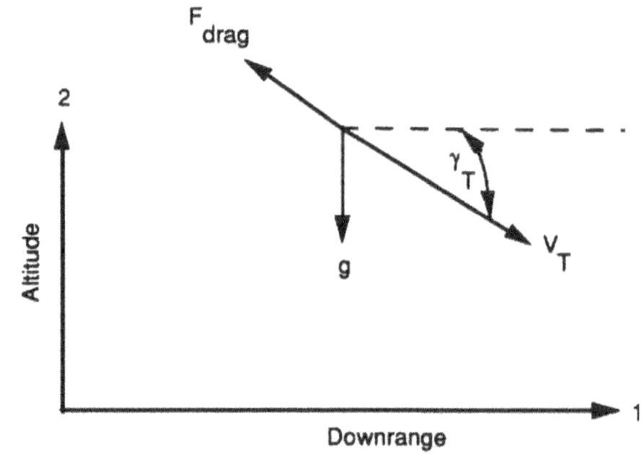

Fig. 16.1 Ballistic target geometry.

vector and gravity g always acts downward in the flat-Earth model. Therefore, if the effect of drag is greater than that of gravity, the target will slow up or decelerate. We will eventually consider the ballistic target as an interceptor threat, so the magnitude of the target deceleration will be of interest to us.

From Fig. 16.1 we can see that the target re-entry angle γ_T can be computed, using trigonometry, from the two inertial components of the target velocity as

$$\gamma_T = \tan^{-1}\frac{-V_{T2}}{V_{T1}}$$

The acceleration components of the ballistic target in the inertial downrange and altitude directions of Fig. 16.1 can either be expressed in terms of the target weight W, reference area S_{ref}, zero lift drag C_{D0}, and gravity g or more simply in terms of the ballistic coefficient β (defined in Chapter 10) according to

$$\frac{\mathrm{d}V_{T1}}{\mathrm{d}t} = \frac{-F_{\text{drag}}}{m}\cos\gamma_T = \frac{-QS_{\text{ref}}C_{D0}g}{W}\cos\gamma_T = \frac{-Qg}{\beta}\cos\gamma_T$$

$$\frac{\mathrm{d}V_{T2}}{\mathrm{d}t} = \frac{-F_{\text{drag}}}{m}\sin\gamma_T - g = \frac{QS_{\text{ref}}C_{D0}g}{W}\sin\gamma_T - g = \frac{Qg}{\beta}\sin\gamma_T - g$$

where Q is the dynamic pressure. Recall that the dynamic pressure has been previously defined as

$$Q = 0.5\rho V_T^2$$

where V_T is the total target velocity, which can be expressed in terms of component velocities as

$$V_T = \sqrt{V_{T1}^2 + V_{T2}^2}$$

and ρ is the air density measured in slug/ft^3 and was shown to be accurately approximated exponentially in Chapter 10 as

$$\rho = 0.0034e^{-R_{T2}/22{,}000}$$

above 30,000 ft and

$$\rho = 0.002378e^{-R_{T2}/30{,}000}$$

below 30,000 ft. The target altitude R_{T2} is measured in feet. Since the acceleration equations are in a fixed or inertial frame, they can be integrated directly to yield velocity and position.

BALLISTIC TARGET EXPERIMENTS

A simulation of a ballistic target, based on the acceleration differential equations of the previous section, appears in Listing 16.1. The ballistic target acceleration differential equations appear before the FLAG=1 statement, and the initial conditions, required for the integration of the differential equations, appear at the beginning of the simulation. We can see that the program is initialized with a target altitude of 100 kft, a target velocity of 6000 ft/s, and a re-entry angle of 45 deg. The nominal ballistic coefficient for the target is 500 lb/ft^2.We can see that the simulation stops when the ballistic threat hits the ground ($R_{T2} < 0$). Every tenth of a second the target location is printed in kft, acceleration in g, and velocity in ft/s. The simulation integration step size of 0.01 s (H = 0.01) is sufficiently small to get accurate answers with the second-order Runge–Kutta numerical integration technique.

LISTING 16.1 BALLISTIC TARGET SIMULATION

```
count=0;
RT1=0.;
RT2=100000.;
VT=6000.;
GAMTDEG=45.;
BETA=500.;
VT1=VT*cos(GAMTDEG/57.3);
VT2=-VT*sin(GAMTDEG/57.3);
T=0.;
H=.01;
S=0.;
while RT2>=0.
        RT1OLD=RT1;
        RT2OLD=RT2;
        VT1OLD=VT1;
        VT2OLD=VT2;
        STEP=1;
        FLAG=0;
        while STEP <=1
```

```
            if FLAG==1
                        RT1=RT1+H*VT1;
                        RT2=RT2+H*VT2;
                        VT1=VT1+H*AT1;
                        VT2=VT2+H*AT2;
                        T=T+H;
                        STEP=2;
            end
            if RT2<=30000.
                        RHO=.002378*exp(-RT2/30000.);
            else
                        RHO=.0034*exp(-RT2/22000.);
            end
            VT=sqrt(VT1^2+VT2^2);
            Q=.5*RHO*VT^2;
            GAMT=atan2(-VT2,VT1);
            AT1=-32.2*Q*cos(GAMT)/BETA;
            AT2=-32.2+32.2*Q*sin(GAMT)/BETA;
            FLAG=1;
        end;
        FLAG=0;
        RT1=.5*(RT1OLD+RT1+H*VT1);
        RT2=.5*(RT2OLD+RT2+H*VT2);
        VT1=.5*(VT1OLD+VT1+H*AT1);
        VT2=.5*(VT2OLD+VT2+H*AT2);
        S=S+H;
        if S>=.09999
            S=0.;
            ATG=sqrt(AT1^2+AT2^2)/32.2;
            RT1K=RT1/1000.;
            RT2K=RT2/1000.;
            VT=sqrt(VT1^2+VT2^2);
            count=count+1;
            ArrayT(count)=T;
            ArrayRT1K(count)=RT1K;
            ArrayRT2K(count)=RT2K;
            ArrayATG(count)=ATG;
            ArrayVT(count)=VT;
        end
end
figure
plot(ArrayRT1K,ArrayRT2K),grid
xlabel('Downrange (Kft)')
ylabel('Altitude (Kft) ')
figure
plot(ArrayRT2K,ArrayATG),grid
```

```
xlabel('Altitude (Kft)')
ylabel('Acceleration (G) ')
figure
plot(ArrayRT2K,ArrayVT),grid
xlabel('Altitude (Kft)')
ylabel('Velocity (Ft/Sec) ')
clc
output=[ArrayT',ArrayRT1K',ArrayRT2K',ArrayATG',ArrayVT'];
save datfil.txt output /ascii
disp 'simulation finished'
```

The nominal case of Listing 16.1 was run, and Fig. 16.2 presents the resultant trajectory of the ballistic target. We can see from the figure that the target trajectory is approximately a straight line (we shall exploit this observation later in this chapter). At the lower altitudes there is slight curvature in the trajectory due to both drag and gravity.

Figure 16.3 displays the deceleration and velocity of the nominal target as a function of altitude. At 100-kft altitude the target has an initial velocity of 6000 ft/s and there is 1 g of acceleration due to gravity (there is too little atmosphere at 100 kft to cause substantial drag). The drag deceleration increases and target velocity decreases as the target descends in altitude. At approximately 40 kft altitude the target deceleration peaks and is nearly 8 g. At this altitude of maximum deceleration the target speed is approximately 63% of its original value ($3800 = 0.63 * 6000$).

The simulation of Listing 16.1 was rerun, and this time the initial target velocity at 100-kft altitude was made a parameter. We can see from the simulation results, displayed in Fig. 16.4, that target deceleration increases as the target

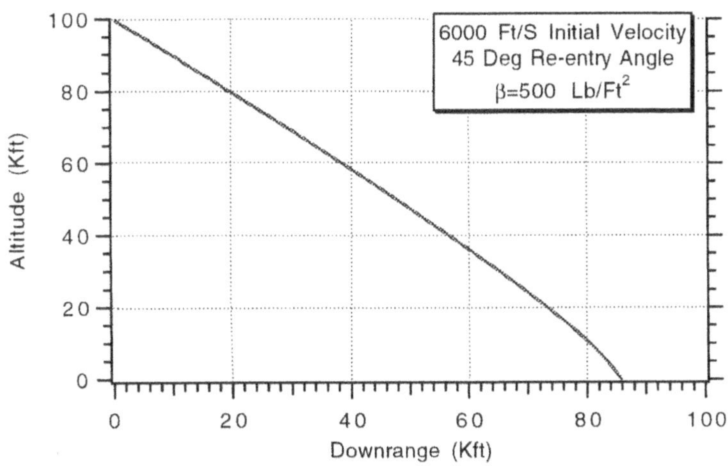

Fig. 16.2 Nominal ballistic target trajectory is approximately a straight line.

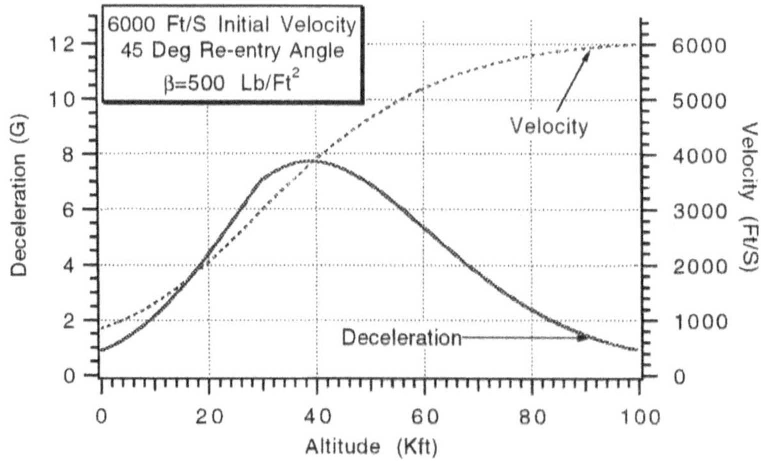

Fig. 16.3 Peak target deceleration occurs at 40 kft.

speed increases. This should not be surprising since the acceleration differential equations tell us that deceleration is proportional to the dynamic pressure (that is, the target velocity squared). Therefore very fast ballistic threats can cause enormous decelerations. Surprisingly, the simulation results in Fig. 16.4 also indicate that the altitude of maximum target deceleration appears to be approximately independent of the target speed!

The simulation of Listing 16.1 was again rerun, and this time the target ballistic coefficient was made a parameter. Simulation results, displayed in Fig. 16.5,

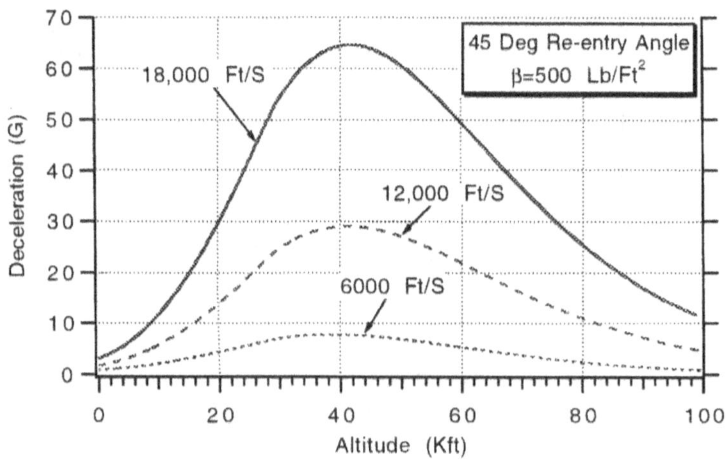

Fig. 16.4 Peak target deceleration increases with increasing target speed.

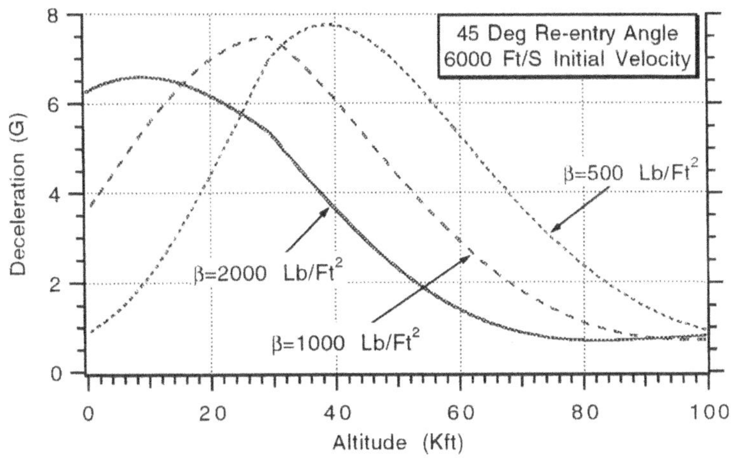

Fig. 16.5 Peak target deceleration is approximately independent of ballistic coefficient.

appear to indicate that the peak target deceleration is approximately independent of ballistic coefficient. This is surprising as there is more drag with lower ballistic coefficients. However, these simulation results also indicate that the altitude at which the peak target deceleration occurs decreases with increasing ballistic coefficient.

Another experiment was conducted using Listing 16.1. This time the initial target velocity and ballistic coefficient were fixed and the re-entry angle was made a parameter. Simulation results, displayed in Fig. 16.6, indicate that the

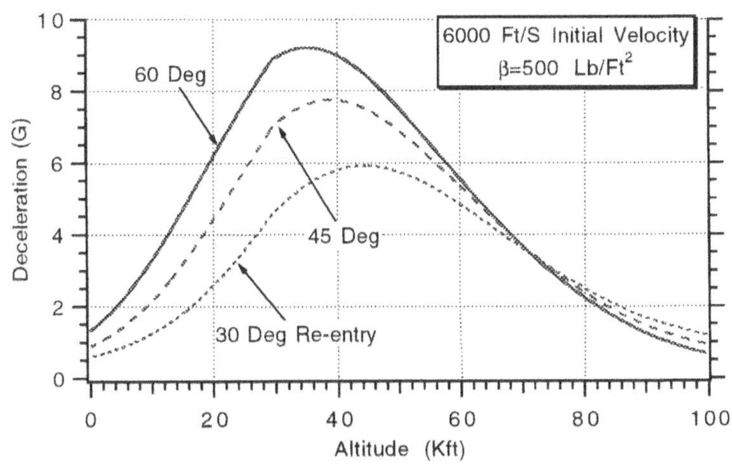

Fig. 16.6 Peak target deceleration increases with increasing re-entry angle.

peak target deceleration increases with increasing re-entry angle. This is reasonable because as the re-entry angle increases the resultant target trajectory tends to become more vertical and more drag is experienced. In addition, Fig. 16.6 shows that the altitude at which the peak target deceleration occurs decreases with increasing re-entry angle.

CLOSED-FORM SOLUTIONS FOR BALLISTIC TARGETS

In this section we will derive some useful closed-form solutions for ballistic targets and compare the theoretical solutions to the simulation results of the previous section. If we neglect gravity, Newton's second law says that we can express the drag force acting on a ballistic target in terms of the dynamic pressure, reference area, and zero lift drag according to

$$F_{\text{drag}} = m\frac{\mathrm{d}V_T}{\mathrm{d}t} = -QS_{\text{ref}}C_{D0}$$

Using definitions of dynamic pressure and ballistic coefficient, we can rewrite the preceding differential equation as

$$\frac{\mathrm{d}V_T}{\mathrm{d}t} = \frac{-\rho g V_T^2}{2\beta}$$

If we assume that the target trajectory is always a straight line (as appeared to be the case in Fig. 16.2), then the re-entry angle γ_T is a constant. Figure 16.1 indicates that for the constant re-entry angle assumption, the altitude component of velocity can be expressed in terms of the total velocity as

$$V_{T2} = \frac{\mathrm{d}R_{T2}}{\mathrm{d}t} = -V_T \sin\gamma_T$$

According to the chain rule we can express the rate of change of the total velocity as

$$\frac{\mathrm{d}V_T}{\mathrm{d}t} = \frac{\mathrm{d}V_T}{\mathrm{d}R_{T2}}\frac{\mathrm{d}R_{T2}}{\mathrm{d}t} = -\frac{\mathrm{d}V_T}{\mathrm{d}R_{T2}}V_T \sin\gamma_T$$

Substitution of the expression for the rate of change of the total velocity into the preceding equation yields

$$\frac{-\rho g V_T^2}{2\beta} = -\frac{\mathrm{d}V_T}{\mathrm{d}R_{T2}}V_T \sin\gamma_T$$

Assuming that the exponential approximation for air density

$$\rho = 0.0034e^{-R_{T2}/22{,}000}$$

applies everywhere (actually another approximation is better below 30,000 ft and is indicated in the previous section), we can rearrange the preceding differential equation so that velocity terms are on one side and altitude terms are on the other side. Rewriting the resultant differential equation with integrals yields

$$\int_{V_{T_{IC}}}^{V_T} \frac{dV_T}{V_T} = \frac{0.0034g}{2\beta \sin \gamma_T} \int_{R_{T2_{IC}}}^{R_{T2}} e^{-R_{T2}/22{,}000}\, dR_{T2}$$

where gravity, the ballistic coefficient, and the re-entry angle have been brought outside the integral because they are considered to be constants. Integrating the preceding expression yields the velocity formula where the target velocity is a function of its initial velocity, re-entry angle, ballistic coefficient, and altitude (or air density) according to

$$V_T = V_{T_{IC}} e^{-22{,}000 g\rho/2\beta \sin \gamma_T}$$

The maximum deceleration experienced by the target will occur at an altitude in which the dynamic pressure is a maximum. Substituting the velocity formula into the definition of dynamic pressure yields

$$Q = 0.5\rho V_T^2 = 0.5\rho V_{T_{IC}}^2 e^{-22{,}000 g\rho/\beta \sin \gamma_T}$$

We can find when the dynamic pressure is a maximum by taking its derivative with respect to the air density and setting the resultant expression to zero or

$$\frac{dQ}{d\rho} = 0 = 0.5 V_{T_{IC}}^2 e^{-22{,}000 g\rho/\beta \sin \gamma_T} - 0.5\rho V_{T_{IC}}^2 \frac{22{,}000g}{\beta \sin \gamma_T} e^{-22{,}000 g\rho/\beta \sin \gamma_T}$$

After some algebra we find that the maximum dynamic pressure condition is

$$\beta \sin \gamma_T = 22{,}000 \rho g$$

The velocity of the target at the maximum dynamic pressure condition can be found by substituting the preceding expression into the velocity formula yielding

$$V_T\big|_{\max Q} = V_{T_{IC}} e^{-22{,}000 g\rho/2\beta \sin \gamma_T} = V_{T_{IC}} e^{-0.5} = 0.606 V_{T_{IC}}$$

In other words, the velocity of the target is always 61% of its initial value when the dynamic pressure is a maximum! This important result was also observed empirically in the simulation results of Fig. 16.3.

To find the altitude at which maximum target deceleration occurs, we must first find the altitude or air density at which the dynamic pressure is greatest. The air density at maximum dynamic pressure can be found from the maximum dynamic pressure condition to be

$$\rho_{\max Q} = \frac{\beta \sin \gamma_T}{22{,}000g}$$

Because the altitude at maximum dynamic pressure is related to the air density at maximum dynamic pressure according to

$$\rho_{\max Q} = 0.0034e^{-R_{T2\max Q}/22{,}000}$$

we can solve for the altitude at this important flight condition. After some algebra we obtain

$$R_{T2_{\max Q}} = 22{,}000\,\ell n \frac{0.0034 * 22{,}000g}{\beta \sin \gamma_T} = 22{,}000\,\ell n \frac{2409}{\beta \sin \gamma_T}$$

where the altitude at which maximum target deceleration occurs is expressed in units of feet. From the preceding relationship we can see that the altitude of maximum target deceleration does not depend on target velocity but only on the ballistic coefficient and re-entry angle! This observation is in agreement with the simulation results of Fig. 16.4.

The closed-form solution for the altitude of maximum target deceleration is displayed as a function of the re-entry angle in Fig. 16.7. Here we can see that the altitude of maximum target deceleration increases with decreasing re-entry angle and decreasing ballistic coefficient. The theoretical results of Fig. 16.7, which neglects gravity and approximates the atmosphere below 30,000 ft, are in excellent agreement with the simulation results of Figs. 16.4–16.6.

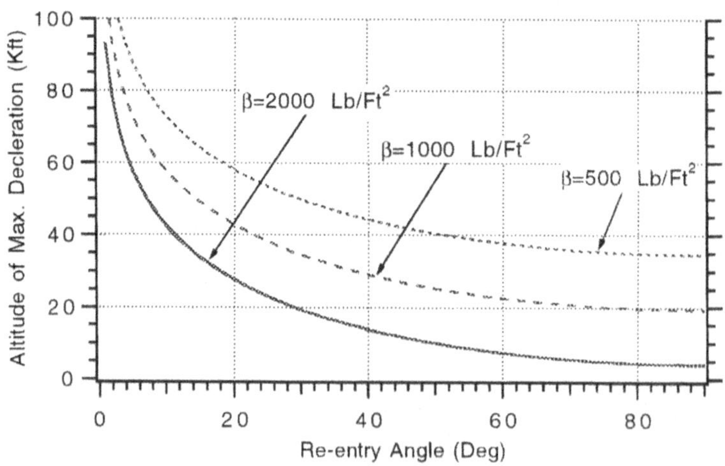

Fig. 16.7 Theory indicates that altitude at which maximum deceleration occurs is independent of velocity.

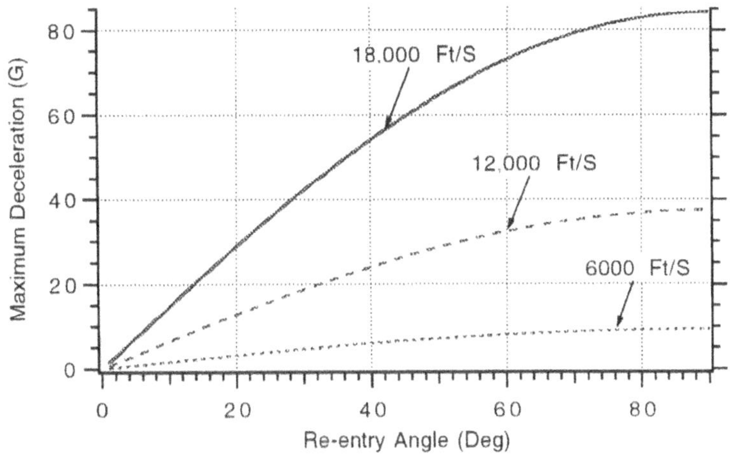

Fig. 16.8 Theory tells us that maximum deceleration is independent of ballistic coefficient.

Since the maximum target deceleration, expressed in units of g, is proportional to the maximum dynamic pressure and inversely proportional to the target ballistic coefficient, we obtain

$$\left.\frac{a}{g}\right|_{\max} = \frac{-Q_{\max}}{\beta} = \frac{-0.5\rho_{\max Q} V_T^2|_{\max Q}}{\beta} = \frac{-0.5\beta \sin\gamma_T}{22{,}000 g\beta} 0.606^2 V_{T_{\mathrm{IC}}}^2$$

where the negative sign indicates target deceleration rather than target acceleration. Simplification of the preceding formula yields

$$\left.\frac{a}{g}\right|_{\max} = -2.6 * 10^{-7} V_{T_{\mathrm{IC}}}^2 \sin\gamma_T$$

Thus we can see that maximum deceleration does not depend on the target ballistic coefficient but only on the velocity and re-entry angle as was also observed empirically in the simulation results displayed in Fig. 16.5! The maximum deceleration formula is displayed as a function of the re-entry angle in Fig. 16.8. We can see that the maximum target deceleration increases with increasing target velocity (actually as the square of target velocity) and increasing re-entry angle. The theoretical results of Fig. 16.8, which neglects gravity and approximates the atmosphere below 30,000 ft, are also in excellent agreement with the simulation results of Figs. 16.4–16.6.

MISSILE AERODYNAMICS

We have just observed the deceleration properties of a ballistic target as a function of its ballistic coefficient, velocity, altitude, and re-entry angle. In this section we want to get an idea of the generic acceleration capability of a pursuing interceptor so that we can better understand ballistic target engagements. To get a first-order estimate of the aerodynamic capability of a missile, we shall treat the interceptor as a cylinder with length L and diameter D.

Basic aerodynamic theory tells us that the lift coefficient C_L for a cylinder is [4]

$$C_L = 2\alpha + \frac{1.5S_{\text{plan}}\alpha^2}{S_{\text{ref}}}$$

where α is the angle of attack or the angle between the missile body and its velocity vector. The planform area S_{plan} and reference area S_{ref} are related to the geometry of a cylinder according to

$$S_{\text{plan}} \approx LD$$

$$S_{\text{ref}} = \frac{\pi D^2}{4}$$

From Chapter 10 we know that the relationship between acceleration and the lift coefficient is given by

$$F = ma = \frac{Wn_L}{g} = QS_{\text{ref}}C_L$$

where W is the missile weight, n_L the lateral missile acceleration, g the acceleration of gravity, and Q the dynamic pressure or

$$Q = 0.5\rho V_M^2$$

The air density ρ can be found from the exponential approximation discussed in both this chapter and Chapter 10. Substitution of the lift coefficient and dynamic pressure into Newton's second law yields the formula for the acceleration capability in units of g of a flying telephone pole (or cylinder) as a function of missile velocity, angle of attack, and altitude (or air density) for a given missile length, diameter, and weight.

$$\frac{n_L}{g} = \frac{QS_{\text{ref}}C_L}{W} = \frac{0.5\rho V_M^2 S_{\text{ref}}}{W}\left[2\alpha + \frac{1.5S_{\text{plan}}\alpha^2}{S_{\text{ref}}}\right]$$

To get a better understanding of the preceding acceleration equation, let us consider a numerical example in which the missile weighs 1000 lb and is 20 ft long and 1 ft in diameter. Figure 16.9 displays the resultant acceleration capability of a cylindrical missile without wings and tails, using the preceding equation, with velocity 3000 ft/s and three different angles of attack, as a function of altitude. We can see that interceptor acceleration capability decreases as altitude increases and as angle of attack decreases. This is why interceptors have a reduced acceleration capability at the higher altitudes and is also why they must operate at higher angles of attack at the higher altitudes in order to maintain the same acceleration capability they had at the lower altitudes. We can see that for this example, the missile

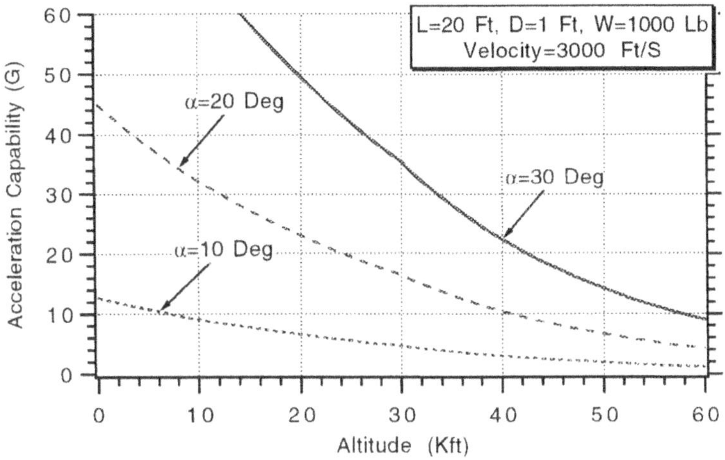

Fig. 16.9 Missile acceleration capability decreases with increasing altitude and decreasing angle of attack.

has only a 7-g capability at 50-kft altitude if its maximum angle of attack is limited to 20 deg. Increasing the maximum angle of attack capability to 30 deg will double the acceleration capability of the interceptor at 50-kft altitude while reducing the maximum angle of attack by 10 deg halves the acceleration capability at that altitude.

Figure 16.10 shows that the interceptor acceleration capability increases as missile velocity increases. Increasing the missile velocity by 1000 ft/s at 50-kft

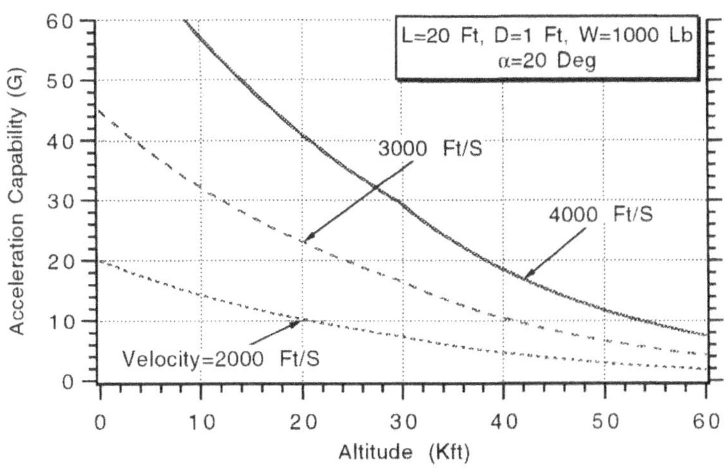

Fig. 16.10 Missile acceleration capability increases with increasing missile velocity.

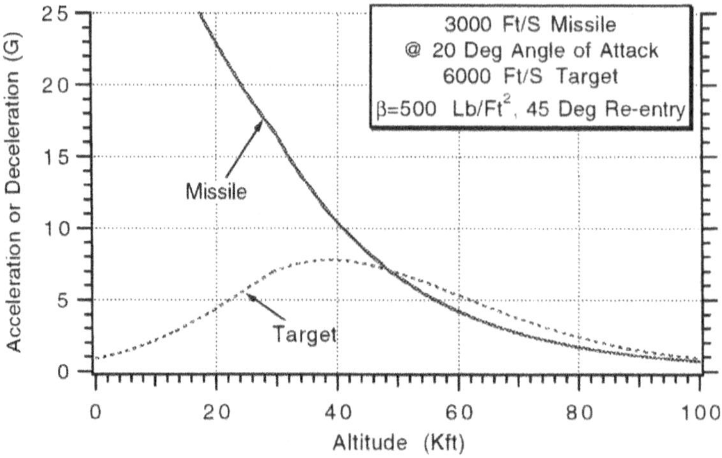

Fig. 16.11 Cylindrical interceptor acceleration capability matches target deceleration characteristics at 50-kft altitude.

altitude nearly doubles the interceptor acceleration capability. Reducing the interceptor velocity by 1000 ft/s approximately halves the acceleration capability of the interceptor at that altitude.

We can compare the lateral acceleration capability of our generic interceptor (or flying telephone pole) to the deceleration levels of the ballistic target. Figure 16.11 compares a 3000-ft/s cylindrical interceptor with a maximum angle-of-attack capability of 20 deg to a 6000-ft/s ballistic target with a 500-lb/ft^2 ballistic coefficient and a 45-deg re-entry angle. We can see that the interceptor acceleration capability increases with decreasing altitude whereas the target deceleration capability increases with decreasing altitude until it peaks at 40 kft. At 50-kft altitude the interceptor acceleration capability and target deceleration characteristics are matched at approximately 7 *g*. From an interceptor acceleration capability point of view, the ideal intercept should take place at very low altitude where the interceptor has enormous capability and a considerable acceleration advantage over the target. In fact, from a missile point of view, the ideal intercept altitude is near sea level where the interceptor acceleration capability is largest and target deceleration capability smallest. However, practical considerations may require the interceptor to engage the ballistic target at much higher altitudes.

INTERCEPTING A BALLISTIC TARGET

In this chapter we have spent considerable time simulating and understanding the properties of ballistic targets. We have seen that fast ballistic targets can go through tremendous deceleration levels as they re-enter the atmosphere. The component of target deceleration perpendicular to the line of sight appears as a target maneuver to a pursuing interceptor. In this section we will investigate some of the difficulties and potential solutions associated with an interceptor trying to engage a ballistic target within the atmosphere.

Listing 16.2 presents the MATLAB source code for an engagement simulation involving a constant speed interceptor and decelerating ballistic target. At the beginning of the simulation, the target is defined as having a 6000-ft/s initial velocity at 200-kft altitude with a re-entry angle of 45 deg and ballistic coefficient of 500 lb/ft^2. It is desired to fire an interceptor immediately and have the intercept take place at 50-kft altitude (RT2DES = 50,000).

Routine initialpz.m is called to predict the location of the target at intercept (RT1F, RT2F) and to compute the time TFDES it will take the target to reach the intercept altitude. Essentially routine initialpz.m is a mini-simulation of the target. Note that we must tell this routine the estimated ballistic coefficient BETEST of the target. If the estimated ballistic coefficient is in error, interceptor launch errors will result. From the outputs of routine initialpz.m the interceptor launch angle GAMMDEG and total velocity VM are computed so that the interceptor will be on a perfect collision course with the target (assuming missile is fired when target is at 200-kft altitude). For simplicity, gravity and drag effects are not included on the interceptor.

The missile and target differential equations appear before the FLAG=1 statement. The ballistic target differential equations are identical to those that have already been modeled in this chapter. The constant speed interceptor differential equations are identical to those found in Chapter 2. Three interceptor guidance options appear. The parameter APN determines which guidance law is used and XNCLIMG determines the interceptor acceleration capability (nominally set to

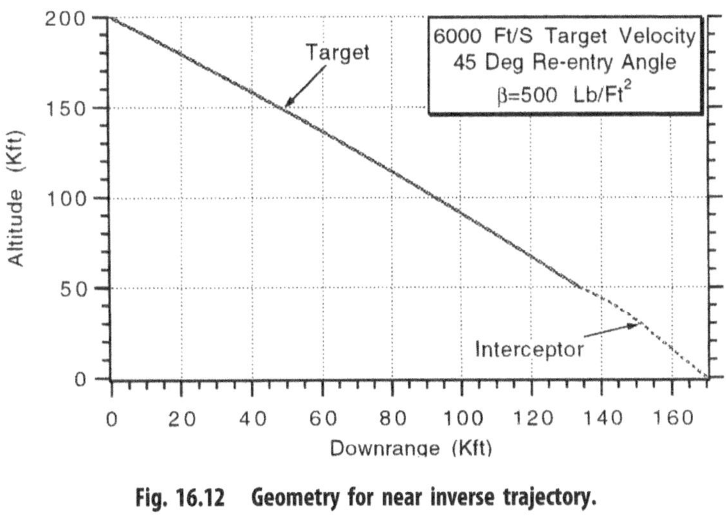

Fig. 16.12 Geometry for near inverse trajectory.

infinity). If APN is 0, we get the proportional navigation guidance law, and if APN is 1, we get augmented proportional navigation where the augmented term includes the component of target acceleration (or deceleration) perpendicular to the line-of-sight ATPLOS.

A nominal case was run in which a proportional navigation interceptor guides on a ballistic target as shown in Fig. 16.12. The geometry is considered near head-on and is also known as an inverse trajectory. The target trajectory is much longer than the missile trajectory since the target is traveling at a much higher velocity.

LISTING 16.2 BALLISTIC TARGET ENGAGEMENT SIMULATION

```
count=0;
APN=0;
XNP=3.;
RT1=0.;
RT2=200000.;
RM1=170000.;
RM2=0.;
VT=6000.;
RT2DES=50000.;
GAMTDEG=45.;
BETA=500.;
BETEST=500.;
XNCLIMG=7.;
XNCLIM=XNCLIMG*32.2;
VT1=VT*cos(GAMTDEG/57.3);
VT2=-VT*sin(GAMTDEG/57.3);
[RT1F,RT2F,TFDES]=initialpz(RT2DES,RT1,RT2,VT1,VT2,BETEST);
RTM1F=RT1F-RM1;
RTM2F=RT2F-RM2;
GAMMDEG=57.3*atan2(RTM2F,RTM1F);
RTMF=sqrt(RTM1F^2+RTM2F^2);
VM=RTMF/TFDES;
VM1=VM*cos(GAMMDEG/57.3);
VM2=VM*sin(GAMMDEG/57.3);
RTM1=RT1-RM1;
RTM2=RT2-RM2;
RTM=sqrt(RTM1^2+RTM2^2);
VTM1=VT1-VM1;
VTM2=VT2-VM2;
VC=-(RTM1*VTM1+RTM2*VTM2)/RTM;
T=0.;
H=.01;
S=0.;
```

```
XNC=0.;
ZEMPLOS=0.;
ZEM1=0.;
ZEM2=0.;
while VC>=0.
    if RTM<1000.
        H=.0002;
    else
        H=.01;
    end
    RT1OLD=RT1;
    RT2OLD=RT2;
    VT1OLD=VT1;
    VT2OLD=VT2;
    RM1OLD=RM1;
    RM2OLD=RM2;
    VM1OLD=VM1;
    VM2OLD=VM2;
    STEP=1;
    FLAG=0;
    while STEP <=1
        if FLAG==1
                RT1=RT1+H*VT1;
                RT2=RT2+H*VT2;
                VT1=VT1+H*AT1;
                VT2=VT2+H*AT2;
                RM1=RM1+H*VM1;
                RM2=RM2+H*VM2;
                VM1=VM1+H*AM1;
                VM2=VM2+H*AM2;
                T=T+H;
                STEP=2;
        end
        if RT2<=30000.
                RHO=.002378*exp(-RT2/30000.);
        else
                RHO=.0034*exp(-RT2/22000.);
        end
        VT=sqrt(VT1^2+VT2^2);
        Q=.5*RHO*VT^2;
        GAMT=atan2(-VT2,VT1);
        AT1=-32.2*Q*cos(GAMT)/BETA;
        AT2=-32.2+32.2*Q*sin(GAMT)/BETA;
        RTM1=RT1-RM1;
        RTM2=RT2-RM2;
        RTM=sqrt(RTM1^2+RTM2^2);
```

```
        VTM1=VT1-VM1;
        VTM2=VT2-VM2;
        VC=-(RTM1*VTM1+RTM2*VTM2)/RTM;
        XLAM=atan2(RTM2,RTM1);
        XLAMD=(RTM1*VTM2-RTM2*VTM1)/(RTM*RTM);
        ATPLOS=-AT1*sin(XLAM)+AT2*cos(XLAM);
        if APN==1
                XNC=XNP*VC*XLAMD+.5*XNP*ATPLOS;
        else
                XNC=XNP*VC*XLAMD;
        end
        if XNC>XNCLIM
                XNC=XNCLIM;
        end
        if XNC<-XNCLIM
                XNC=-XNCLIM;
        end
        AM1=-XNC*sin(XLAM);
        AM2=XNC*cos(XLAM);
        FLAG=1;
    end;
    FLAG=0;
    RT1=.5*(RT1OLD+RT1+H*VT1);
    RT2=.5*(RT2OLD+RT2+H*VT2);
    VT1=.5*(VT1OLD+VT1+H*AT1);
    VT2=.5*(VT2OLD+VT2+H*AT2);
    RM1=.5*(RM1OLD+RM1+H*VM1);
    RM2=.5*(RM2OLD+RM2+H*VM2);
    VM1=.5*(VM1OLD+VM1+H*AM1);
    VM2=.5*(VM2OLD+VM2+H*AM2);
    S=S+H;
    if S>=.09999
        S=0.;
        ATG=sqrt(AT1^2+AT2^2)/32.2;
        RT1K=RT1/1000.;
        RT2K=RT2/1000.;
        RM1K=RM1/1000.;
        RM2K=RM2/1000.;
        XNCG=XNC/32.2;
        ATPLOSG=ATPLOS/32.2;
        count=count+1;
        ArrayT(count)=T;
        ArrayRT1K(count)=RT1K;
        ArrayRT2K(count)=RT2K;
        ArrayRM1K(count)=RM1K;
        ArrayRM2K(count)=RM2K;
```

```
            ArrayATG(count)=ATG;
            ArrayATPLOSG(count)=ATPLOSG;
            ArrayXNCG(count)=XNCG;
        end
 end
 RTM
figure
plot(ArrayRT1K,ArrayRT2K,ArrayRM1K,ArrayRM2K),grid
xlabel('Downrange (Kft)')
ylabel('Altitude (Kft) ')
figure
plot(ArrayT,ArrayATG,ArrayT,ArrayATPLOSG,ArrayT,ArrayXNCG),grid
xlabel('Time (Sec)')
ylabel('Acceleration (G) ')
clc
output=[ArrayT',ArrayRT1K',ArrayRT2K',ArrayRM1K',ArrayRM2K',
                                   ArrayATG',ArrayATPLOSG',ArrayXNCG'];
save datfil.txt output /ascii
disp 'simulation finished'

function [rt1f,rt2f,tfdes]=initialpz(rt2des,rt1ic,rt2ic,...
                                   vt1ic,vt2ic,beta)
rt1=rt1ic;
rt2=rt2ic;
vt1=vt1ic;
vt2=vt2ic;
t=0.;
h=.01;
while rt2>rt2des
      rt1old=rt1;
      rt2old=rt2;
      vt1old=vt1;
      vt2old=vt2;
      step=1;
      flag=0;
      while step <=1
         if flag==1
               rt1=rt1+h*vt1;
               rt2=rt2+h*vt2;
               vt1=vt1+h*at1;
               vt2=vt2+h*at2;
               t=t+h;
               step=2;
         end
         if rt2<=30000.
               rho=.002378*exp(-rt2/30000.);
```

```
            else
                        rho=.0034*exp(-rt2/22000.);
            end
            vt=sqrt(vt1^2+vt2^2);
            q=.5*rho*vt^2;
            gamt=atan2(-vt2,vt1);
            at1=-32.2*q*cos(gamt)/beta;
            at2=-32.2+32.2*q*sin(gamt)/beta;
            flag=1;
        end;
        flag=0;
        rt1=.5*(rt1old+rt1+h*vt1);
        rt2=.5*(rt2old+rt2+h*vt2);
        vt1=.5*(vt1old+vt1+h*at1);
        vt2=.5*(vt2old+vt2+h*at2);
end
rt1f=rt1;
rt2f=rt2;
tfdes=t;
```

Figure 16.13 displays the important accelerations for the nominal case. We can see that the target deceleration is approximately 8 *g* at intercept (or 50 kft). This is in accordance with the ballistic target simulation results of Fig. 16.4 and the theoretical results of Fig. 16.8. We can see that since the engagement geometry is near inverse, there is no target deceleration perpendicular to the line of sight. From a missile point of view, the target does not appear to be maneuvering. Because the

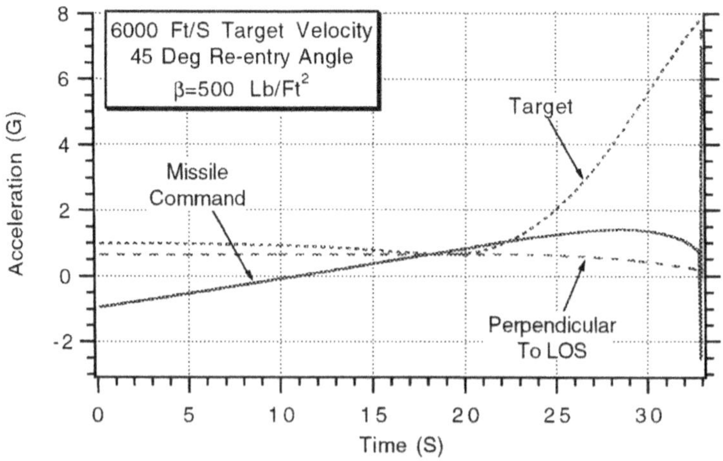

Fig. 16.13 Missile guidance commands are small because very little of target deceleration is perpendicular to line of sight.

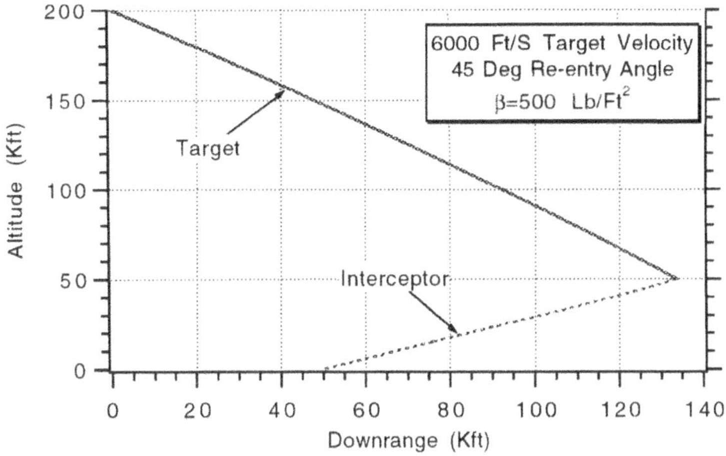

Fig. 16.14 Example of more stressing trajectory.

missile is initially on a collision course and there is no apparent target maneuver, very little acceleration is required by an interceptor using proportional navigation to hit the target.

A more stressing geometry was considered in which the interceptor is initially at 50 kft downrange (RM1 = 50,000). The engagement geometry, shown in Fig. 16.14, is no longer inverse. The intercept still takes place at 50-kft altitude and the missile has a speed of 3000 ft/s in order to be on a collision course. It

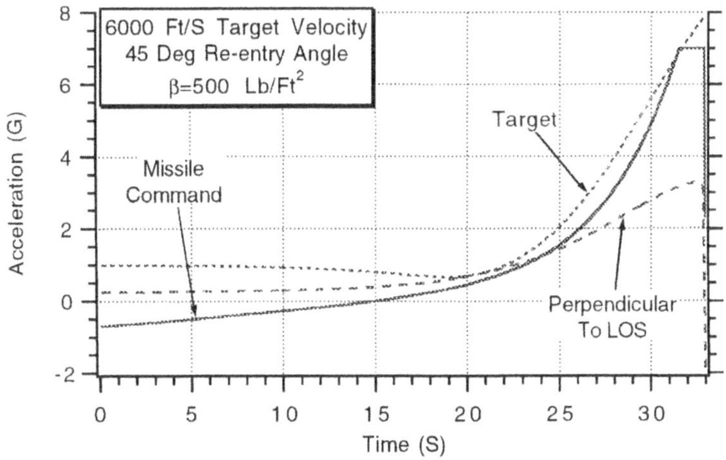

Fig. 16.15 More missile acceleration is required in stressing trajectory because more target deceleration is perpendicular to line of sight.

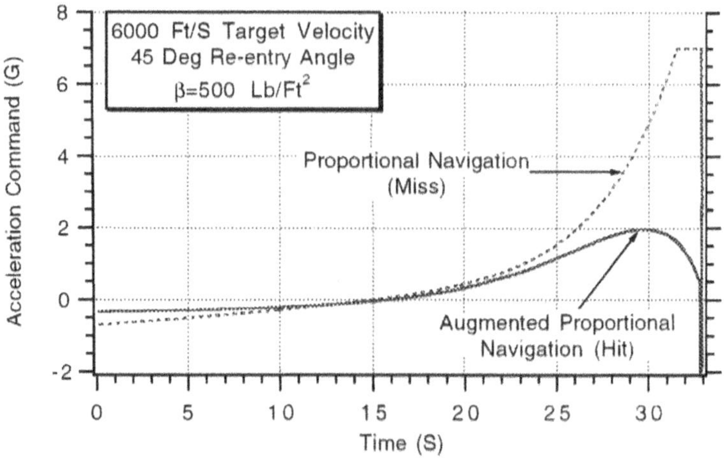

Fig. 16.16 More advanced guidance laws offer significant benefits for stressing trajectory.

is assumed that the interceptor has a maximum angle of attack limit of 20 deg, yielding a maximum acceleration capability of 7 g (XNCLIMG $= 7$) at the intercept altitude based on the results of the previous section for our flying telephone pole ($W = 1000$ lb, $L = 20$ ft, $D = 1$ ft).

Figure 16.15 shows that the target deceleration is unchanged for this new engagement geometry and approaches 8 g near intercept. However, the component of target deceleration perpendicular to the line of sight is much larger than it was for the inverse trajectory case, and the peak value is in excess of 3 g. Therefore it is not surprising that the interceptor requires more than 7 g (missile acceleration limit) to hit the apparent 3-g target maneuver. Acceleration saturation follows causing a large miss distance.

Augmented proportional navigation can be used to relax the acceleration requirements of the interceptor under this stressing engagement geometry. Figure 16.16 shows that the previously unsuccessful intercept can be made successful with this advanced guidance law. However, augmented proportional navigation requires more information than does proportional navigation in order to operate successfully.

SUMMARY

We have seen that ballistic targets can go through enormous decelerations as they re-enter the atmosphere. The magnitude of the target deceleration increases with increasing target speed and increasing target re-entry angle. Any target deceleration that is perpendicular to the line of sight will appear as a target maneuver to the interceptor. It is best for a pursuing interceptor to engage the target on an inverse trajectory where little of the target deceleration is perpendicular to the line of sight [5]. If for practical reasons the target must be engaged under stressing conditions, the interceptor must be sized to have adequate acceleration capability if proportional navigation guidance is used. Advanced guidance laws such as predictive guidance can significantly relax the interceptor acceleration requirements if missile-target range information is available and if the target ballistic coefficient is either known or can be estimated accurately.

REFERENCES

[1] Riezenman, M., "Revising the Script After Patriot," *IEEE Spectrum*, Sept. 1991, pp. 49–52.

[2] Canavan, G., "Strategic Defense in Past and Future Conflicts," *The Journal of Practical Applications In Space*, Vol. 2, No. 3, Spring 1991, pp. 1–42.

[3] Regan, F., *Re-Entry Vehicle Dynamics*, AIAA Education Series, AIAA, New York, 1984.

[4] Jerger, J. J., *System Preliminary Design*, Van Nostrand, Princeton, NJ, 1960.

[5] Lin, C. F., *Modern Navigation, Guidance, and Control Processing*, Prentice-Hall, Englewood Cliffs, NJ, 1991.

CHAPTER 17

Extended Kalman Filtering and Ballistic Coefficient Estimation

INTRODUCTION

Knowledge of the target ballistic coefficient can be used in advanced guidance laws such as predictive guidance to relax the interceptor acceleration requirements in stressing engagement geometries. In addition, knowledge of the target ballistic coefficient is required for fire control due to the importance of accurate intercept point predictions in launching the interceptor on a collision course. Therefore the accurate estimation of the ballistic coefficient of a target re-entering the atmosphere is very important for both guidance and fire control purposes. In this chapter we shall show, in detail, how extended Kalman filtering concepts can be applied to ballistic coefficient estimation.

THEORETICAL EQUATIONS [1]

To apply extended Kalman filtering techniques, it is first necessary to describe the real world by a set of nonlinear differential equations. One standard dynamical model of the system or real world is given by

$$\dot{\boldsymbol{x}} = f(\boldsymbol{x}) + \boldsymbol{w}$$

where $\boldsymbol{x}$ is a vector of the system states, $f(\boldsymbol{x})$ is a nonlinear function of those states, and $\boldsymbol{w}$ is a random zero mean process. The process noise matrix describing the random process $\boldsymbol{w}$ for the preceding model is given by

$$\boldsymbol{Q} = E(\boldsymbol{w}\boldsymbol{w}^T)$$

Finally, the measurement equation, required for the application of an extended Kalman filter, is considered a nonlinear function of the states according to

$$z = \boldsymbol{h}(\boldsymbol{x}) + \boldsymbol{v}$$

where $\boldsymbol{v}$ is a random zero mean process described by the measurement noise matrix $\boldsymbol{R}$, which is defined as

$$\boldsymbol{R} = E(\boldsymbol{v}\boldsymbol{v}^T)$$

For systems in which the measurements are discrete we can rewrite the measurement equation as

$$z_k = \boldsymbol{h}(\boldsymbol{x}_k) + \boldsymbol{v}_k$$

The discrete measurement noise matrix $\boldsymbol{R}_k$ consists of measurement noise source variances. Because the system and measurement equations are nonlinear, a first-order approximation is used in the Ricatti equations for the systems dynamic matrix $\boldsymbol{F}$ and measurement matrix $\boldsymbol{H}$. The matrices are related to the system and measurement equations according to

$$\boldsymbol{F} = \left. \frac{\partial \boldsymbol{f}(\boldsymbol{x})}{\partial \boldsymbol{x}} \right|_{\boldsymbol{x}=\hat{\boldsymbol{x}}}$$

$$\boldsymbol{H} = \left. \frac{\partial \boldsymbol{h}(\boldsymbol{x})}{\partial \boldsymbol{x}} \right|_{\boldsymbol{x}=\hat{\boldsymbol{x}}}$$

The fundamental matrix, also required for the Ricatti equations, is usually approximated by the first two terms of the Taylor series expansion $\exp(\boldsymbol{F}T_s)$ and is given by

$$\Phi_k \approx \boldsymbol{I} + \boldsymbol{F}T_s$$

where T_s is the sampling time and $\boldsymbol{I}$ is the identity matrix. Note that the approximations to the systems dynamics matrix, measurement matrix, and fundamental matrix are time-varying and nonlinear because they depend on the system state estimates. The Ricatti equations, needed for the computation of the Kalman gains, are still given by the matrix difference equations of Chapter 9 and are repeated for convenience as

$$\boldsymbol{M}_k = \Phi_k \boldsymbol{P}_{k-1} \Phi_k^T + \boldsymbol{Q}_k$$

$$\boldsymbol{K}_k = \boldsymbol{M}_k \boldsymbol{H}^T [\boldsymbol{H}\boldsymbol{M}_k \boldsymbol{H}^T + \boldsymbol{R}_k]^{-1}$$

$$\boldsymbol{P}_k = (\boldsymbol{I} - \boldsymbol{K}_k \boldsymbol{H})\boldsymbol{M}_k$$

where $\boldsymbol{P}_k$ is a covariance matrix representing errors in the state estimates before an update and $\boldsymbol{M}_k$ is the covariance matrix representing errors in the state estimates after an update. As Φ_k and $\boldsymbol{H}$ are nonlinear functions of the state estimates, the Kalman gains cannot be computed offline as is possible with a linear Kalman filter. The discrete process noise matrix $\boldsymbol{Q}_k$ can still be found from the continuous process noise matrix $\boldsymbol{Q}$ and the fundamental matrix according to

$$\boldsymbol{Q}_k = \int_0^{T_s} \Phi(\tau) \boldsymbol{Q} \Phi^T(\tau)\, \mathrm{d}\tau$$

If the dynamical model of a linear Kalman filter is matched to the real world, the covariance matrix $\boldsymbol{P}_k$ cannot only be used to calculate Kalman gains but can also provide exact predictions of the errors in the state estimates. The extended Kalman filter offers no such guarantees and in fact the Ricatti equation covariance matrix may indicate excellent performance projections when the filter is performing poorly or is even broken.

The preceding approximations only have to be used in the computation of Kalman gains. The actual extended Kalman filtering equations do not have to use those approximations but instead can be written in terms of the nonlinear measurement equation where the new estimate is the old estimate plus a gain times a residual, or

$$\hat{\boldsymbol{x}}_{k+1} = \hat{\boldsymbol{x}}_k + \boldsymbol{K}_k[z_k - \boldsymbol{h}(\hat{\boldsymbol{x}}_k)]$$

In the preceding equation the residual is the difference between the actual measurement and the nonlinear measurement equation. The new state estimates do not have to be propagated forward from the old estimate with the fundamental matrix but instead can be obtained directly by integrating the actual nonlinear differential equations at each sampling interval. For example, Euler integration can be applied to the nonlinear system differential equations yielding

$$\hat{\dot{\boldsymbol{x}}} = \boldsymbol{f}(\hat{\boldsymbol{x}}_{k-1})$$

$$\hat{\boldsymbol{x}}_k = \hat{\boldsymbol{x}}_{k-1} + \hat{\dot{\boldsymbol{x}}}T_s$$

where state estimates are used instead of the actual states and the sampling time T_s is used as an integration step size.

DIFFERENTIAL EQUATION FOR ONE-DIMENSIONAL BALLISTIC TARGET

To illustrate how extended Kalman filtering concepts can be applied, let us consider the one-dimensional tracking problem originally considered by Gelb [1] and illustrated in Fig. 17.1. In this example a ballistic target is falling on a straight-line path directly toward a surface-based tracking radar. Only drag and gravity act on the ballistic target. This is equivalent to the case in the previous chapter in which the target re-entry angle is 90 deg. In this problem the tracking radar measures the distance from the radar to the target every T_s s. In addition, the tracking radar has the incentive of working well in this application because it is directly in the path of the ballistic target.

We can see from Fig. 17.1 that drag acts upward whereas gravity acts downward. The total acceleration acting on the ballistic target can be expressed in terms of a zero lift drag C_{D0} or a ballistic coefficient β as

$$\frac{\mathrm{d}V_{T2}}{\mathrm{d}t} = \frac{F_{\mathrm{drag}}}{m} - g = \frac{QS_{\mathrm{ref}}C_{D0}g}{W} - g = \frac{Qg}{\beta} - g$$

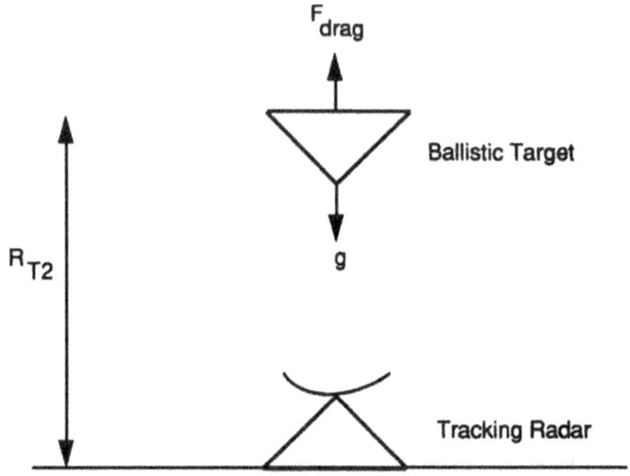

Fig. 17.1 Forces acting on a one-dimensional ballistic target.

where g is the acceleration of gravity and Q is the dynamic pressure. The dynamic pressure can be expressed in terms of the air density ρ and the ballistic target velocity V_{T2} as

$$Q = 0.5\rho V_{T2}^2$$

For purposes of simplicity we can assume that the air density is measured in slug/ft^3 and is exponentially related to altitude R_{T2} measured in feet according to

$$\rho = 0.0034e^{-R_{T2}/22{,}000}$$

and ignore the fact that the coefficients of the exponential approximation change below 30,000 ft. If we assume that the ballistic coefficient of the target is a constant, its derivative must be zero. Therefore the three differential equations that govern the one-dimensional ballistic target can be summarized as

$$\dot{R}_{T2} = V_{T2}$$

$$\dot{V}_{T2} = \frac{0.0034e^{-R_{T2}/22{,}000} g V_{T2}^2}{2\beta} - g$$

$$\dot{\beta} = 0$$

If we want to account for the fact that there may be a large uncertainty in the ballistic coefficient or that it might actually change with time, we could modify the third state equation to be

$$\dot{\beta} = u_s$$

where u_s is white process noise with spectral density Φ_s.

EXTENDED KALMAN FILTER FOR ONE-DIMENSIONAL BALLISTIC TARGET

In the previous section we showed that the differential equations governing the one-dimensional ballistic target could be expressed in terms of position R_{T2}, velocity V_{T2}, and ballistic coefficient β. Therefore a plausible candidate for the system state vector is

$$x = \begin{bmatrix} R_{T2} \\ V_{T2} \\ \beta \end{bmatrix}$$

In the Gelb example the tracking radar measures position directly. Therefore the measurement equation in this example is a linear function of the states or

$$R^*_{T2} = R_{T2} + v_k = \underbrace{[100]}_{H} \begin{bmatrix} R_{T2} \\ V_{T2} \\ \beta \end{bmatrix} + v_k$$

where the uncertainty in the position measurement is simply the scalar variance or

$$\boldsymbol{R}_k = E(\boldsymbol{v}_K \boldsymbol{v}_K^T) = \sigma_k^2$$

The systems dynamics matrix can be obtained from the three differential equations describing the target according to the definition of the theoretical section as

$$\boldsymbol{F} = \left.\frac{\partial \boldsymbol{f}(\boldsymbol{x})}{\partial \boldsymbol{x}}\right|_{x=\hat{x}} = \begin{bmatrix} \dfrac{\partial \dot{R}_{T2}}{\partial R_{T2}} & \dfrac{\partial \dot{R}_{T2}}{\partial V_{T2}} & \dfrac{\partial \dot{R}_{T2}}{\partial \beta} \\ \dfrac{\partial \dot{V}_{T2}}{\partial R_{T2}} & \dfrac{\partial \dot{V}_{T2}}{\partial V_{T2}} & \dfrac{\partial \dot{V}_{T2}}{\partial \beta} \\ \dfrac{\partial \dot{\beta}}{\partial R_{T2}} & \dfrac{\partial \dot{\beta}}{\partial V_{T2}} & \dfrac{\partial \dot{\beta}}{\partial \beta} \end{bmatrix}_{x=\hat{x}}$$

After taking partial derivatives of the three system differential equations we obtain

$$\boldsymbol{F} = \begin{bmatrix} 0 & 1 & 0 \\ \dfrac{-\hat{\rho} g \hat{V}_{T2}^2}{44{,}000\hat{\beta}} & \dfrac{\hat{\rho} g \hat{V}_{T2}}{\hat{\beta}} & \dfrac{-\hat{\rho} g \hat{V}_{T2}^2}{2\hat{\beta}^2} \\ 0 & 0 & 0 \end{bmatrix}$$

where the estimated air density is given by

$$\hat{\rho} = 0.0034 e^{-\hat{R}_{T2}/22{,}000}$$

The fundamental matrix can be obtained from the systems dynamics matrix as

$$\Phi_K \approx \boldsymbol{I} + \boldsymbol{F} T_s = \begin{bmatrix} 1 & T_s & 0 \\ \dfrac{-\hat{\rho} g \hat{V}_{T2}^2 T_s}{44{,}000\hat{\beta}} & 1 + \dfrac{\hat{\rho} g \hat{V}_{T2} T_s}{\hat{\beta}} & \dfrac{-\hat{\rho} g \hat{V}_{T2}^2 T_s}{2\hat{\beta}^2} \\ 0 & 0 & 1 \end{bmatrix}$$

whereas the continuous process noise matrix can be found from

$$\boldsymbol{Q} = \begin{bmatrix} 0 & 0 & 0 \\ 0 & 0 & 0 \\ 0 & 0 & \Phi_s \end{bmatrix}$$

where Φ_s is the spectral density of the process noise. The discrete process noise matrix can be obtained from the continuous process noise matrix according to the relationship

$$\boldsymbol{Q}_k = \int_0^{T_s} \Phi(\tau)\boldsymbol{Q}\Phi^T(\tau)\, \mathrm{d}\tau$$

If we substitute τ for T_s in the previous fundamental matrix approximation, we get

$$\Phi(\tau) = \begin{bmatrix} 0 & \tau & 0 \\ f_{21}\tau & 1 + f_{22}\tau & f_{23}\tau \\ 0 & 0 & 1 \end{bmatrix}$$

where f_{21}, f_{22}, and f_{23} are defined in terms of the state estimates as

$$f_{21} = \frac{-\hat{\rho} g \hat{V}_{T2}^2}{44{,}000\hat{\beta}}$$

$$f_{22} = \frac{\hat{\rho} g \hat{V}_{T2}}{\hat{\beta}}$$

$$f_{23} = \frac{-\hat{\rho} g \hat{V}_{T2}^2}{2\hat{\beta}^2}$$

Assuming that f_{21}, f_{22}, and f_{23} are approximately constant over the sampling interval, we can integrate with respect to τ and obtain the discrete process noise matrix as

$$\boldsymbol{Q}_k = \Phi_s \begin{bmatrix} 0 & 0 & 0 \\ 0 & f_{23}^2 \frac{T_s^3}{3} & f_{23} \frac{T_s^2}{2} \\ 0 & f_{23} \frac{T_s^2}{2} & T_s \end{bmatrix}$$

If we want to neglect process noise in our model, then Φ_s is set to zero.

The measurement noise matrix $\boldsymbol{R}_k$ is a scalar in this three-state system, so there will only be three Kalman gains at each update (that is, K_1, K_2, and K_3). The Kalman gains will also depend on the state estimates because the fundamental matrix depends on the system state estimates. As was mentioned previously, the new extended Kalman filter states will simply be the old states propagated forward by Euler integration, plus a gain times a residual, or

$$\text{Residual} = R^*_{T2} - \hat{R}_{T2_{k-1}} - \overline{\dot{R}_{T2}T_s}$$

$$\hat{R}_{T2_k} = \hat{R}_{T2_{k-1}} + \overline{\dot{R}_{T2}T_s} + K_1 * \text{Residual}$$

$$\hat{V}_{T2_k} = \hat{V}_{T2_{k-1}} + \overline{\dot{V}_{T2}T_s} + K_2 * \text{Residual}$$

$$\hat{\beta}_k = \hat{\beta}_{k-1} + K_3 * \text{Residual}$$

The barred quantities in the preceding set of difference equations represent the derivatives required by Euler integration and are obtained directly from the nonlinear system equations as

$$\overline{\dot{R}_{T2}} = \hat{V}_{T2_{k-1}}$$

$$\overline{\dot{V}_{T2}} = \frac{0.0034e^{-\hat{R}_{T2_{k-1}}/22{,}000} g\hat{V}^2_{T2_{k-1}}}{2\hat{\beta}_{k-1}} - g$$

We now have all of the equations necessary to simulate an extended Kalman filter for the one-dimensional tracking problem.

NUMERICAL EXAMPLE

The same numerical example considered by Gelb [1] is presented here in which a target with ballistic coefficient 500 lb/ft^2 is initially at 100-kft altitude and is traveling downward at a speed of 6000 ft/s. A surface-based radar measures the range from the radar to the target (altitude in this example) every 0.05 s with measurement variance 500 ft^2. The initial estimate of position is 100,025 ft (25-ft error), of velocity is 6150 ft/s (150-ft/s error), and of the ballistic coefficient is 800 lb/ft^2 (300-lb/ft^2 error). Uncertainties in the initial state estimates are also reflected in the initial covariance matrix. The first diagonal element of the initial covariance matrix represents the variance of the error in the initial estimate of position and is taken to be the variance of the measurement noise or 500 ft^2. The second diagonal element of the initial covariance matrix represents the variance of the error in the initial estimate of velocity and is taken to be 20,000 ft^2/s^2 (slightly less than 150^2). The third diagonal element of the initial covariance matrix represents the variance of the error in the initial estimate of ballistic coefficient and is taken to the square of the initial error in estimating the ballistic coefficient or 90,000 lb^2/ft^4 (or 300^2). The off-diagonal elements of the initial covariance matrix are set to zero and it is assumed that there is no process noise.

Listing 17.1 presents the resultant one-dimensional extended Kalman filter for ballistic coefficient estimation derived in the previous section. The initial conditions for the actual ballistic target, filter state estimates, and initial covariance matrix reflect the nominal case. Because Q33 (or Φ_s) is set to zero, there is no process noise. The second-order Runge–Kutta numerical integration technique is used for solving the actual nonlinear differential equations representing the ballistic target with an integration step size of 0.001 s. The exponential approximation for the air density, used in calculating the drag on the actual ballistic target, matches the assumption made in the extended Kalman filter derivation. The actual ballistic coefficient and its estimate are printed every sampling interval. In addition, the actual errors in the estimate of the ballistic coefficient are computed and compared to the square root of the third diagonal element of the covariance matrix. This diagonal element represents the extended Kalman filter's internal prediction of the error in the estimate of the ballistic coefficient.

The nominal case of Listing 17.1 was run, and the estimated and actual ballistic coefficients are displayed versus altitude in Fig. 17.2. At 100-kft altitude (that is, beginning of the estimation process) the initial estimate of the ballistic coefficient is on the high side by 300 lb/ft^2. As the target descends in altitude, the filter's estimate of the ballistic coefficient appears to be continually improving. Below 60-kft altitude, the extended Kalman filter has an excellent estimate of the target's ballistic coefficient.

Figure 17.3 compares single flight results for the actual error in the estimate of the ballistic coefficient (labeled simulation) with the theoretical predictions of the covariance matrix (labeled σ_{THEORY}). Note that the covariance matrix thinks that, as the ballistic target descends in altitude and more measurements are taken, the estimates continually improve (or the error in the estimate of the ballistic coefficient goes to zero). Therefore it appears that the single flight results agree with the covariance matrix predictions in this example.

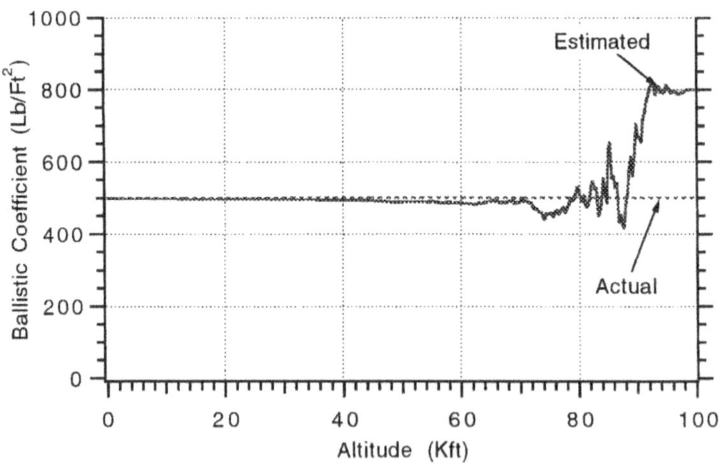

Fig. 17.2 After a while extended Kalman filter is able to estimate ballistic coefficient.

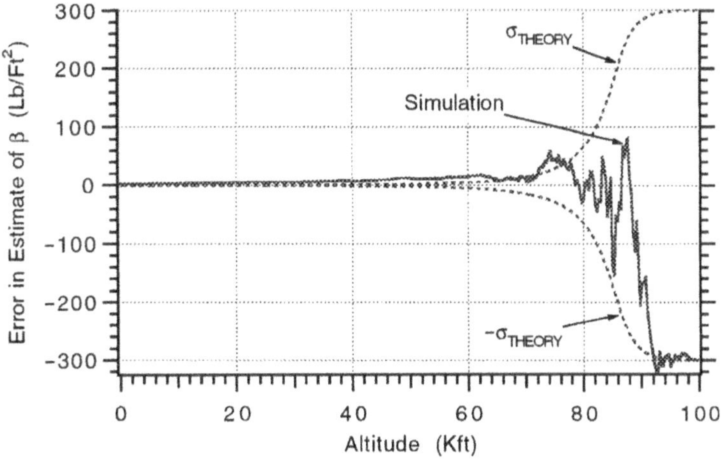

Fig. 17.3 Theory and single flight results appear to agree.

LISTING 17.1 ONE-DIMENSIONAL EXTENDED KALMAN FILTER FOR BALLISTIC COEFFICIENT ESTIMATION

```
clear
count=0;
RT2=100000.;
VT2=-6000.;
BETA=500.;
RT2H=100025.;
VT2H=-6150.;
BETAH=800.;
ORDER=3;
TS=.05;
TF=30.;
Q33=0./TF;
T=0.;
S=0.;
H=.001;
SIGNOISE=sqrt(500.);
PHI=zeros(ORDER,ORDER);
P=zeros(ORDER,ORDER);
Q=zeros(ORDER,ORDER);
IDN=eye(ORDER);
P(1,1)=SIGNOISE*SIGNOISE;
P(2,2)=20000.;
P(3,3)=300.^2;
```

```
HMAT=zeros(1,ORDER);
HMAT(1,1)=1.;
while RT2>=0.
        RT2OLD=RT2;
        VT2OLD=VT2;
        STEP=1;
        FLAG=0;
        while STEP<=1
           if FLAG==1
                    RT2=RT2+H*RT2D;
                    VT2=VT2+H*VT2D;
                    T=T+H;
                    STEP=2;
           end
           RT2D=VT2;
           VT2D=.0034*32.2*VT2*VT2*exp(-RT2/22000.)/(2.*BETA)-32.2;
           FLAG=1;
        end;
        FLAG=0;
        RT2=.5*(RT2OLD+RT2+H*RT2D);
        VT2=.5*(VT2OLD+VT2+H*VT2D);
        S=S+H;
        if S>=(TS-.00001)
           S=0.;
           RHOH=.0034*exp(-RT2H/22000.);
           F21=-32.2*RHOH*VT2H*VT2H/(2.*22000.*BETAH);
           F22=RHOH*32.2*VT2H/BETAH;
           F23=-RHOH*32.2*VT2H*VT2H/(2.*BETAH*BETAH);
           PHI(1,1)=1.;
           PHI(1,2)=TS;
           PHI(2,1)=F21*TS;
           PHI(2,2)=1.+F22*TS;
           PHI(2,3)=F23*TS;
           PHI(3,3)=1.;
           Q(2,2)=F23*F23*Q33*TS*TS*TS/3.;
           Q(2,3)=F23*Q33*TS*TS/2.;
           Q(3,2)=Q(2,3);
           Q(3,3)=Q33*TS;
           M=PHI*P*PHI'+Q;
           HMHT=HMAT*M*HMAT';
           HMHTR=HMHT(1,1)+SIGNOISE*SIGNOISE;
           HMHTRINV=1./HMHTR;
           MHT=M*HMAT';
           for I=1:ORDER
                    GAIN(I,1)=MHT(I,1)*HMHTRINV;
           end
```

```
            P=(IDN-GAIN*HMAT)*M;
            XNOISE=gaussc7(SIGNOISE);
            RT2DB=VT2H;
            VT2DB=.0034*32.2*VT2H*VT2H*exp(-RT2H/22000.)/(2.*BETAH)-32.2;
            RES=RT2+XNOISE-(RT2H+RT2DB*TS);
            RT2H=RT2H+RT2DB*TS+GAIN(1,1)*RES;
            VT2H=VT2H+VT2DB*TS+GAIN(2,1)*RES;
            BETAH=BETAH+GAIN(3,1)*RES;
            ERRY=RT2-RT2H;
            SP11=sqrt(P(1,1));
            ERRV=VT2-VT2H;
            SP22=sqrt(P(2,2));
            ERRBETA=BETA-BETAH;
            SP33=sqrt(P(3,3));
            RT2K=RT2/1000.;
            count=count+1;
            ArrayT(count)=T;
            ArrayRT2(count)=RT2;
            ArrayRT2H(count)=RT2H;
            ArrayRT2K(count)=RT2K;
            ArrayVT2(count)=VT2;
            ArrayVT2H(count)=VT2H;
            ArrayBETA(count)=BETA;
            ArrayBETAH(count)=BETAH;
            ArrayERRBETA(count)=ERRBETA;
            ArraySP33(count)=SP33;
      end
end
figure
plot(ArrayRT2K,ArrayBETA,ArrayRT2K,ArrayBETAH),grid
xlabel('Altitude (Kft)')
ylabel('Ballistic Coefficient (Lb/Ft^2)')
axis([0 100 0 1000])
figure
plot(ArrayRT2K,ArraySP33,ArrayRT2K,-ArraySP33,ArrayRT2K,...
      ArrayERRBETA),grid
xlabel('Altitude (Kft)')
ylabel('Error in Ballistic Coefficient (Lb/Ft^2)')
clc
output=[ArrayT',ArrayRT2K',ArrayRT2',ArrayRT2H',ArrayVT2',...
      ArrayVT2H',ArrayBETA',ArrayBETAH'];
save datfil.txt output /ascii
disp 'simulation finished'
```

We can rerun the nominal case but start at 200-kft altitude rather than 100-kft altitude. Figure 17.4 indicates that both the theoretical and actual errors in the

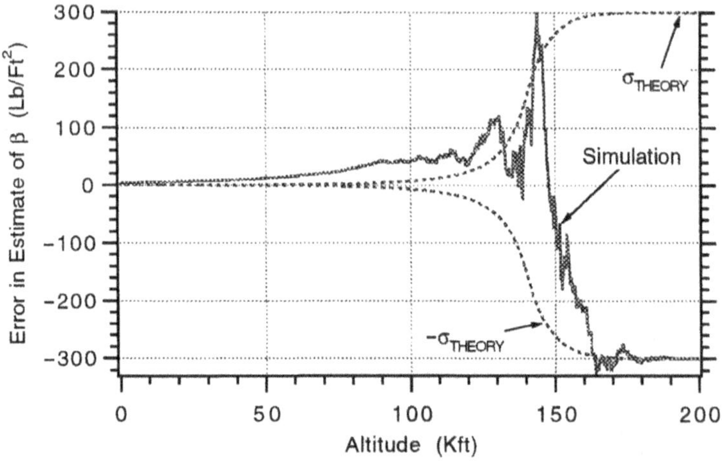

Fig. 17.4 Extended Kalman filter unable to estimate ballistic coefficient above 150 kft.

estimate of the ballistic coefficient do not improve from the initial guess until the ballistic target descends below 150-kft altitude. The lack of estimation capability is due to the absence of drag in the high altitude regime. In other words, at the higher altitudes the ballistic coefficient is not observable from just position measurements. This result can be very important if we must predict the future location of the ballistic target in the atmosphere based on estimates of the ballistic coefficient at very high altitudes.

Another case was run where the ballistic target started at the nominal altitude of 100 kft. However, this time the initial estimate of the ballistic coefficient was 1500 lb/ft^2 (1000-lb/ft^2 error) rather than 800 lb/ft^2 (300-lb/ft^2 error). The third diagonal element of the initial covariance matrix was increased to 1000^2 to reflect the larger initial uncertainty in the ballistic coefficient. Figure 17.5 shows that under these circumstances, the extended Kalman filter is unable to estimate the ballistic coefficient before the ballistic target hits the tracking radar. Unlike a linear Kalman filter, the extended Kalman filter's performance is highly dependent on initial conditions!

Figure 17.6 shows that even though the extended Kalman filter is not able to estimate the ballistic coefficient when we initially severely overestimate the ballistic coefficient by 1000 lb/ft^2, the filter's covariance matrix predictions indicate that the errors in the estimate of the ballistic coefficient are near zero. Apparently this filter does not even realize when it is broken! Therefore we can see that although the covariance matrix is required for Kalman gain computation, its theoretical predictions are not always useful!

In the preceding example the filter is not able to recover when we initially overestimate the ballistic coefficient by a large amount. The filter's lack of robustness is due to a zero process noise matrix ($\boldsymbol{Q}_k = 0$). When the process noise is

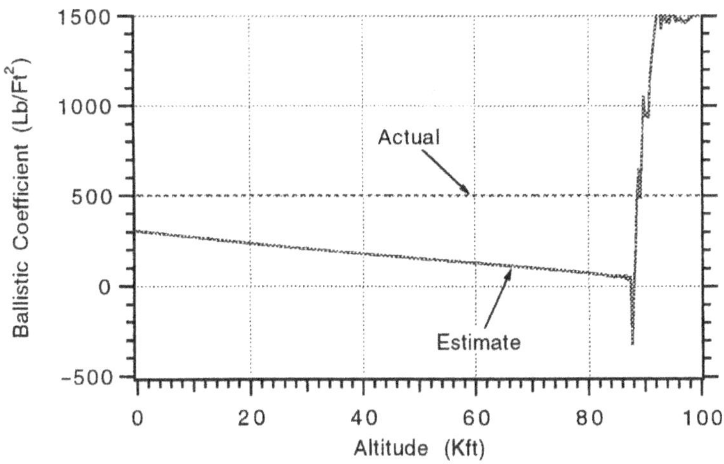

Fig. 17.5 Extended Kalman filter breaks if we severely overestimate ballistic coefficient.

zero, the filter thinks it is very smart (it must have a terrific dynamical model) and eventually stops looking at the measurements. Under these circumstances the filter changes from an extended Kalman filter to an arrogant Kalman filter! Process noise was added to the filter with value $\Phi_s = 1000^2/30$ to indicate large uncertainty in the ballistic coefficient model. Figure 17.7 shows that when the process noise is added, the estimated and actual ballistic coefficients converge fairly quickly.

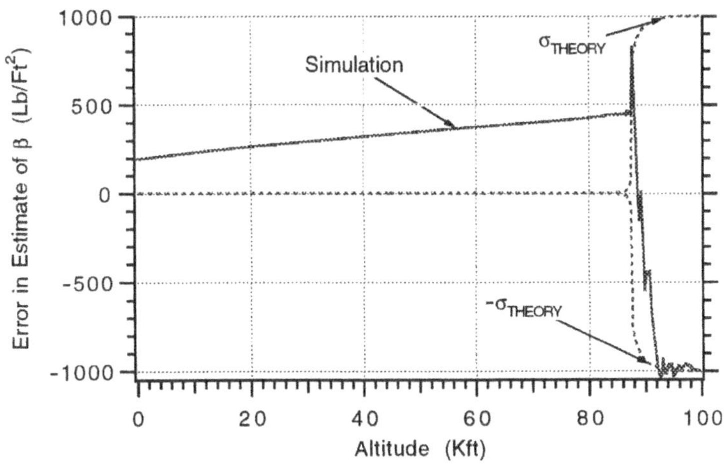

Fig. 17.6 Extended Kalman filter does not even realize it is broken.

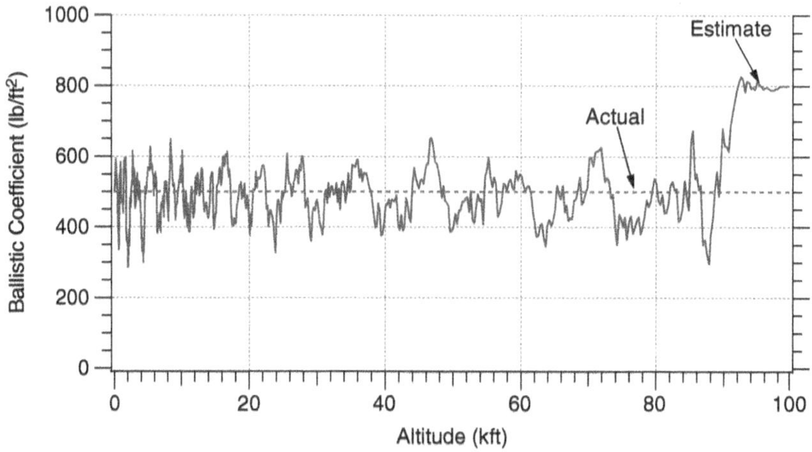

Fig. 17.7 Adding process noise enables extended Kalman filter to recover from overestimating ballistic coefficient.

Figure 17.8 now shows that when realistic process noise is added to reflect large uncertainties, the single flight results and covariance matrix predictions of the error in the estimate of ballistic coefficient are in agreement.

We can now revisit the nominal results of Fig. 17.3 when, without a process noise matrix, the arrogant Kalman filter thought its estimates were continually improving as more measurements were taken. If we rerun the nominal case with $\Phi_s = 300^2/30$ to reflect the fact that we have a smaller uncertainty in our

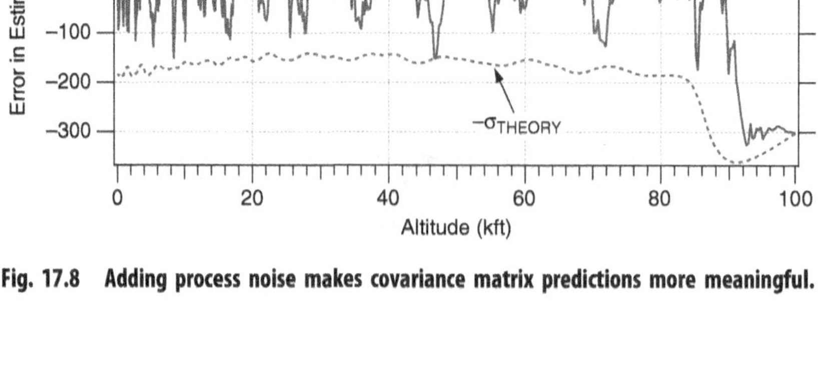

Fig. 17.8 Adding process noise makes covariance matrix predictions more meaningful.

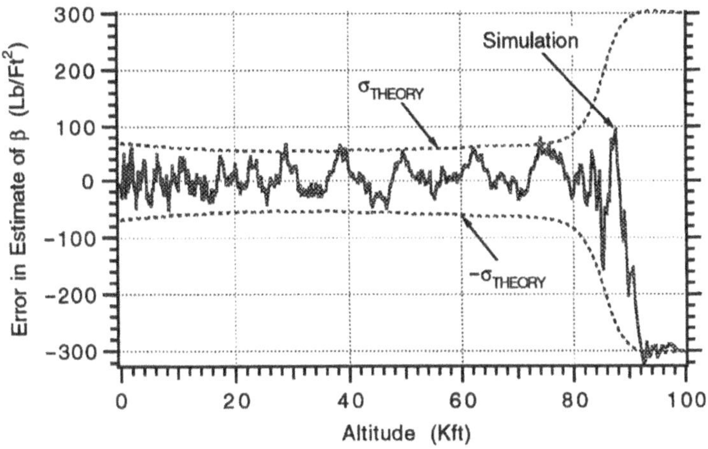

Fig. 17.9 Nominal results are worse when there is process noise but filter is more robust.

knowledge of the ballistic coefficient, we get more sobering results as shown in Fig. 17.9. We can see that after a while the filter's estimate of the ballistic coefficient does not improve. However, our initial uncertainty in our estimate of the ballistic coefficient has been reduced from the initial guess of 300 $\mathrm{lb/ft^2}$ to an estimate with slightly under 100 $\mathrm{lb/ft^2}$ of error. In addition, we now have a filter that is more robust to initialization errors.

SUMMARY

We have seen, using a simplified extended Kalman filter, the difficulties in estimating a target's ballistic coefficient—especially at high altitude where there is very little drag. We have also observed that using zero process noise in the filter gain computations leads to overly optimistic performance projections and makes the filter fragile in the presence of large initialization errors. Adding process noise to the filter's gain computation appears to be the engineering fix when there is large uncertainty. Although adding process noise degrades filter performance under benign conditions, it also enables the filter to perform adequately under more stressing conditions.

REFERENCE

[1] Gelb, A., *Applied Optimal Estimation*, MIT Press, Cambridge, MA, 1974.

CHAPTER 18

Ballistic Target Challenges

INTRODUCTION

In this chapter we shall integrate many of the text's concepts in order to illustrate, from a miss distance point of view, additional reasons why ballistic targets are challenging. First, new miss distance formulas will be derived in order to show how the miss due to noise depends on the closing velocity and guidance system time constant. Next, a new formula will be presented showing how the minimum possible guidance system time constant depends on radome slope, closing velocity, and missile turning rate time constant. For head-on scenarios, numerical examples will be presented showing how low-closing velocity aircraft engagements and high-closing velocity ballistic target engagements yield different miss distances even though the error sources may be the same.

MISS DISTANCE DUE TO NOISE

In Chapter 3 closed-form solutions for various deterministic error sources were derived for a single time constant proportional navigation guidance system. We demonstrated in Chapter 6 that although the miss distances generated with the low-order model of the guidance system were serious underestimates of the actual miss, the closed-form solutions were useful because the miss distance normalization factors did not change for higher order guidance systems. We shall use the same methodology in obtaining miss distance formulas due to noise error sources. First we shall obtain noise miss distance closed-form solutions for the single time constant guidance system and then use the brute force method to extend those solutions for a fifth-order binomial guidance system.

As was done with deterministic error sources in Chapter 3, we shall also use the method of adjoints for finding miss distance formulas due to various noise error sources found in homing guidance systems [1]. The noise sources considered are those usually associated with radar homing missiles. Figure 18.1

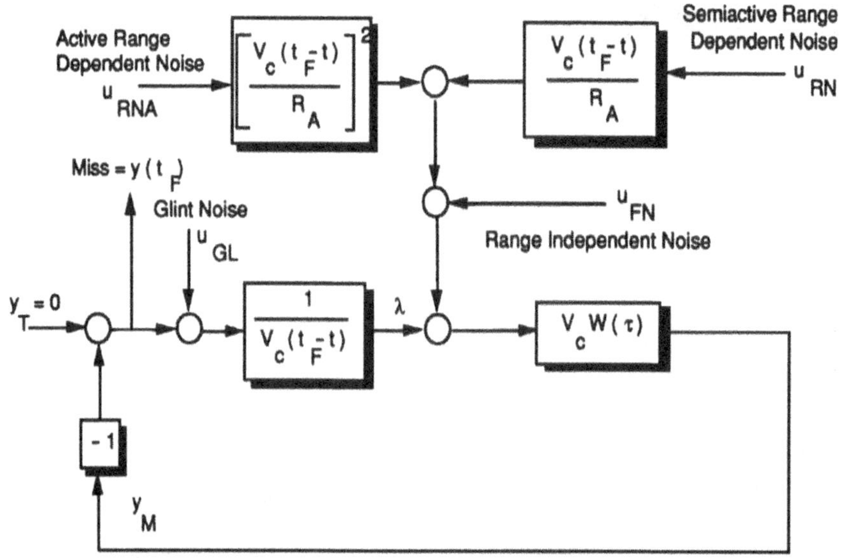

Fig. 18.1 Generalized proportional navigation guidance system with noise sources.

presents a generalized model of the homing loop similar to Fig. 3.16 except this time the error sources are random rather than deterministic. The first error source is glint or scintillation noise and is caused by random fluctuations of the target radar return. The spectral density of this error source is related to the physical dimensions of the target. Strictly speaking, glint should not be modeled as white noise since it may be highly correlated [2]. For semiactive systems, in which the target is illuminated by a transmitter not on the interceptor, range dependent noise is the thermal noise produced in the interceptor radar receiver according to the radar range equation. In this simplified model the spectral density of the noise is defined at a reference range R_A. The noise is proportional to the distance from the missile to the target and goes to zero at intercept [3]. For active systems in which the target is illuminated by a transmitter on the interceptor, range dependent noise is proportional to the square of the distance from the missile to the target [4]. Other noise sources are usually lumped together and termed range independent noise.

The spectral densities for the various white noise error sources are given by

$$\Phi_{FN} = \text{Spectral Density of } u_{FN}$$
$$\Phi_{RN} = \text{Spectral Density of } u_{RN}$$
$$\Phi_{RNA} = \text{Spectral Density of } u_{RNA}$$
$$\Phi_{GL} = \text{Spectral Density of } u_{GL}$$

The generalized homing loop adjoint model, which appears in Fig. 18.2, can be found from Fig. 18.1 by using the rules of adjoints developed in Chapters 3 and 4 and then applying some block diagram manipulation. Note that all white noise inputs of the original system become outputs in the adjoint system by squaring, integrating, and multiplying by the spectral density of each of the white noise error sources. Critical points in the adjoint block diagram have been labeled $H(\tau)$, $g(\tau)$, and $f(\tau)$.

To illustrate how noise miss distance formulas can be derived, let us consider a single time constant guidance system with a navigation ratio of 3. Recall from Chapter 3 that for this case W can be represented in the frequency domain as

$$W(s) = \frac{3}{s(1 + sT)}$$

where T is the guidance system time constant. We showed in Chapter 3 that $1 - H$ could be found in the frequency domain from W according to the relationship

$$1 - H(s) = e^{\int W ds} = \frac{s^3}{(s + 1/T)^3}$$

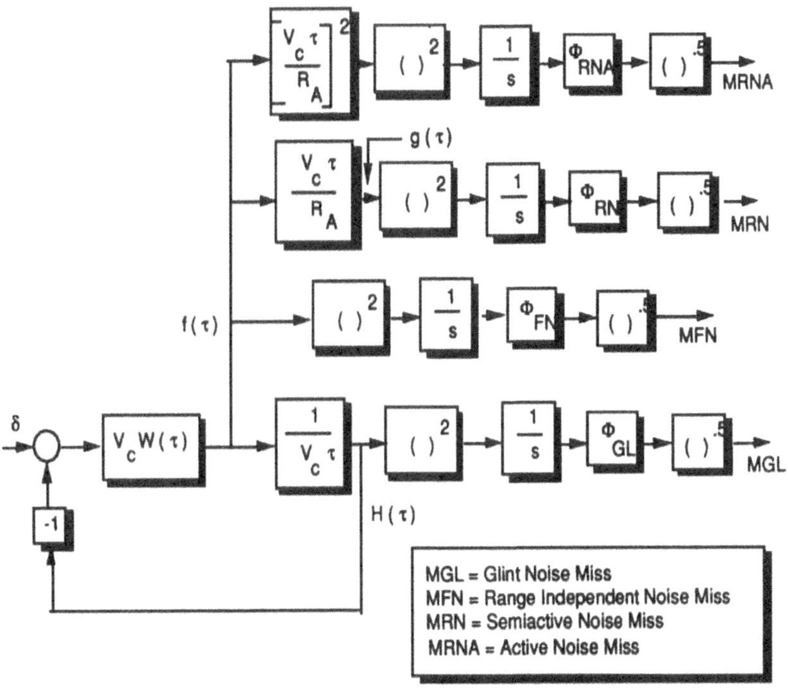

Fig. 18.2 Generalized adjoint homing loop model.

Solving for H by algebraic manipulation of the preceding equation yields

$$H(s) = \frac{1 + 3sT + 3s^2T^2}{(1 + sT)^3}$$

We can convert H from the frequency domain to the adjoint time domain by taking the inverse Laplace transform of the preceding equation yielding

$$H(\tau) = \frac{e^{-\tau/T}}{T}\left[3 - \frac{3\tau}{T} + \frac{\tau^2}{2T^2}\right]$$

where τ can be interpreted as the homing time or time of flight. We can see from Fig. 18.2 that the miss due to glint noise (MGL) can be found by squaring and integrating $H(\tau)$ and then multiplying the result by the square root of the glint noise spectral density. We can simplify matters and integrate from zero to infinity yielding the closed-form solution for the standard deviation of the steady-state miss due to white glint noise as [5]

$$MGL_{N'=3} = \Phi_{GL}^{0.5}\sqrt{\int_0^\infty H^2(\tau)\,d\tau} = 1.44T^{-0.5}\Phi_{GL}^{0.5}$$

From the preceding formula we can see that unlike most other error source results we have studied, if the guidance system time constant is reduced, the miss due to glint noise will increase! We can also see that, unlike deterministic error source results, the miss due to glint noise does not go to zero as the homing time approaches infinity. In other words, there will always be some miss distance due to glint noise, no matter how much homing time we have. Closed-form miss distance formulas for the single time constant guidance system can also be derived, in a similar manner, for different effective navigation ratios. The steady-state standard deviation of the miss due to glint noise for effective navigation ratios of 4 and 5 can be found to be

$$MGL_{N'=4} = 1.71T^{-0.5}\Phi_{GL}^{0.5}$$

$$MGL_{N'=5} = 1.94T^{-0.5}\Phi_{GL}^{0.5}$$

Note that the miss distance normalization factors do not change with different effective navigation ratios. However the miss distance coefficients due to glint noise increase slightly with increasing effective navigation ratio.

To find the steady-state standard deviation of the miss due to range independent noise (MFN), it is first necessary to square and integrate the expression $(1 - H)V_cW$ as can be seen from Fig. 18.2. For a single time constant guidance system with an effective navigation ratio of 3, $(1 - H)V_cW$ in the frequency domain becomes

$$[1 - H(s)]V_c W(s) = \frac{3V_c s^2}{T(s + 1/T)^4}$$

We can convert $(1 - H)V_c W$ to the adjoint time domain by taking the inverse Laplace transform of the preceding equation obtaining

$$f(\tau) = \mathcal{L}^{-1}\{[1 - H(s)]V_c W(s)\} = \frac{3V_c \tau e^{-\tau/T}}{T}\left[1 - \frac{\tau}{T} + \frac{\tau^2}{6T^2}\right]$$

Squaring and integrating the preceding expression from zero to infinity yields the steady-state formula for the standard deviation of the miss distance due to range independent noise as

$$MFN_{N'=3} = \Phi_{FN}^{0.5}\sqrt{\int_0^\infty f^2(\tau)\mathrm{d}\tau} = 0.532 V_c T^{0.5}\Phi_{FN}^{0.5}$$

where V_c is the closing velocity. In this case we can see that there is now a geometry dependence on the miss distance because the miss is proportional to closing velocity. Higher closing velocity engagement scenarios will yield more miss distance due to range independent noise. On the other hand, we can see from the preceding expression that, unlike the glint noise case, reducing the guidance system time constant will decrease the miss due to range independent noise. Closed-form miss distance formulas for range independent noise in a single time constant guidance system can also be derived in a similar manner for different effective navigation ratios. The steady-state standard deviation of the miss due to range independent noise for effective navigation ratios of 4 and 5 can be found to be

$$MFN_{N'=4} = 0.561 V_c T^{0.5}\Phi_{FN}^{0.5}$$

$$MFN_{N'=5} = 0.588 V_c T^{0.5}\Phi_{FN}^{0.5}$$

Here again we can see that there is a slight increase in the miss distance coefficients as the effective navigation ratio increases.

We can find the standard deviation of the steady-state miss due to semiactive range dependent noise (*MRN*) by squaring and integrating $g(\tau)$ as shown in Fig. 18.2. The expression for $g(\tau)$ can be found from $f(\tau)$ as

$$g(\tau) = \frac{V_c \tau f(\tau)}{R_A} = \frac{3V_c^2 \tau^2 e^{-\tau/T}}{R_A T}\left[1 - \frac{\tau}{T} + \frac{\tau^2}{6T^2}\right]$$

Squaring and integrating $g(\tau)$ from zero to infinity yields the standard deviation of the steady-state miss due to semiactive range dependent noise for an effective navigation ratio of 3 as

$$MRN_{N'=3} = \Phi_{RN}^{0.5}\sqrt{\int_0^\infty g^2(\tau)\, d\tau} = \frac{1.06 V_c^2 T^{1.5} \Phi_{RN}^{0.5}}{R_A}$$

where R_A is taken to be a reference range. We can see that the closing velocity and time constant dependence of the miss is much greater for semiactive range dependent noise than it was for range independent noise. Steady-state miss distance formulas for semiactive range dependent noise in a single time constant guidance system can be found for higher effective navigation ratios in a similar way and are

$$MRN_{N'=4} = \frac{1.10 V_c^2 T^{1.5} \Phi_{RN}^{0.5}}{R_A}$$

$$MRN_{N'=5} = \frac{1.15 V_c^2 T^{1.5} \Phi_{RN}^{0.5}}{R_A}$$

Again we can see that the miss distance coefficients increase slightly with increasing effective navigation ratio. For active systems miss distance formulas can also be derived for a single time constant system. The steady-state standard deviation of the miss due to active range dependent noise (*MRNA*) for effective navigation ratios of 3, 4, and 5 are

$$MRNA_{N'=3} = \frac{4.66 V_c^3 T^{2.5} \Phi_{RNA}^{0.5}}{R_A^2}$$

$$MRNA_{N'=4} = \frac{4.68 V_c^3 T^{2.5} \Phi_{RNA}^{0.5}}{R_A^2}$$

$$MRNA_{N'=5} = \frac{4.82 V_c^3 T^{2.5} \Phi_{RNA}^{0.5}}{R_A^2}$$

We can see that the closing velocity and time constant dependence for active range dependent noise is even stronger than it was for semiactive range dependent noise.

FIFTH-ORDER BINOMIAL GUIDANCE SYSTEM MISS DISTANCES

In Chapter 6 we saw that once we had closed-form solutions for a single time constant guidance system, we could obtain solutions for higher order systems by the method of brute force because the miss distance normalization factors remained unchanged with system order (that is, only coefficients change). In this section we shall use the same method to get noise miss distance formulas for higher order systems. The adjoint model for a fifth-order binomial guidance system, first shown in Fig. 6.5, has been modified to include the noise error sources discussed in the previous section and is redrawn in Fig. 18.3. Deterministic error sources are

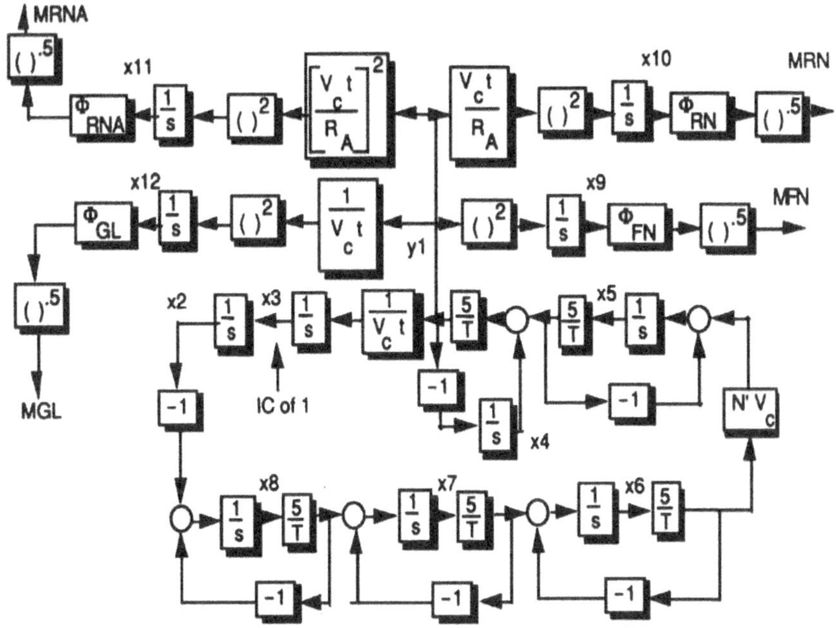

Fig. 18.3 Adjoint of fifth-order binomial guidance system for noise miss distance calculations.

not shown in this diagram. Note that the basic model is unchanged. We have only added new outputs to correspond to the miss due to range independent noise, glint noise, and both semi-active and active range dependent noise. This has been accomplished by squaring and integrating signals proportional to the adjoint variable $y1$ shown in Fig. 18.3.

Adjoint simulation Listing 6.1 has also been modified for noise miss distance calculations and appears in Listing 18.1. The noise spectral densities, closing velocity, guidance system time constant, and reference range have been set to unity so that we can calculate the coefficients for the noise miss distance normalization factors.

Using the method of brute force with the adjoint simulation of Listing 18.1, Table 18.1 was generated for the standard deviation of the steady-state noise missdistances for the fifth-order binomial proportional navigation guidance system under consideration. We can see from Table 18.1 that although the miss distance normalization factors are the same as they were for a single time constant guidance system, the miss distance coefficients are an order of magnitude larger! This means that the miss distances for the higher order system will also be an order of magnitude greater than that of the single time constant guidance system. In addition, we can see that unlike the single time constant guidance

TABLE 18.1 MISS DISTANCE FORMULAS FOR FIFTH-ORDER BINOMIAL GUIDANCE SYSTEM

Error source	Normalization factor	Miss coefficient		
		$N' = 3$	$N' = 4$	$N' = 5$
Range independent noise	$\frac{\sigma_{\text{Miss}}}{V_c T^{0.5} \Phi_{FN}^{0.5}}$	3.04	5.08	8.19
Semiactive range dependent noise	$\frac{\sigma_{\text{Miss}} R_A}{V_c^2 T^{1.5} \Phi_{FN}^{0.5}}$	9.47	18.4	33.7
ACtive range dependent noise	$\frac{\sigma_{\text{Miss}} R_A^2}{V_c^3 T^{2.5} \Phi_{RNA}^{0.5}}$	41.3	89.4	182
Glint noise	$\frac{\sigma_{\text{Miss}} T^{0.5}}{\Phi_{GL}^{0.5}}$	1.68	2.35	3.21

system, the effective navigation ratio has a strong influence on the miss distance for the higher order system. Increasing the effective navigation ratio significantly increases the miss distance due to noise.

LISTING 18.1 ADJOINT OF FIFTH-ORDER BINOMIAL GUIDANCE SYSTEM FOR NOISE MISS DISTANCE CALCULATIONS

```
XNP=3.;
TAU=1.;
TF=10.;
VC=1.;
PHIFN=1;
PHIRN=1;
PHIRNA=1;
PHIGL=1;
RA=1;
T=0.;
S=0.;
TP=T+.00001;
X2=0;
X3=1;
X4=0;
X5=0.;
X6=0.;
X7=0.;
X8=0.;
X9=0.;
X10=0.;
X11=0.;
```

```
X12=0.;
H=.01;
n=0.;
while TP<=(TF-1e-5)
      X2OLD=X2;
      X3OLD=X3;
      X4OLD=X4;
      X5OLD=X5;
      X6OLD=X6;
      X7OLD=X7;
      X8OLD=X8;
      X9OLD=X9;
      X10OLD=X10;
      X11OLD=X11;
      X12OLD=X12;
      STEP=1;
      FLAG=0;
      while STEP<=1
        if FLAG==1
        STEP=2;
        X2=X2+H*X2D;
        X3=X3+H*X3D;
        X4=X4+H*X4D;
        X5=X5+H*X5D;
        X6=X6+H*X6D;
        X7=X7+H*X7D;
        X8=X8+H*X8D;
        X9=X9+H*X9D;
        X10=X10+H*X10D;
                        X11=X11+H*X11D;
        X12=X12+H*X12D;
        TP=TP+H;
        end
        X2D=X3;
        Y1=5.*(5.*X5/TAU+X4)/TAU;
        TGO=TP+.00001;
        X3D=Y1/(VC*TGO);
        X4D=-Y1;
        X5D=-5.*X5/TAU+5.*X6*XNP*VC/TAU;
        X6D=-5.*X6/TAU+5.*X7/TAU;
        X7D=-5.*X7/TAU+5.*X8/TAU;
        X8D=-5.*X8/TAU-X2;
        X9D=Y1^2;
        X10D=(Y1*VC*TGO/RA)^2;
                X11D=(Y1*((VC*TGO/RA)^2))^2;
                X12D=(Y1/(VC*TGO))^2;
```

```
        FLAG=1;
      end
      FLAG=0;
      X2=.5*(X2OLD+X2+H*X2D);
      X3=.5*(X3OLD+X3+H*X3D);
      X4=.5*(X4OLD+X4+H*X4D);
      X5=.5*(X5OLD+X5+H*X5D);
      X6=.5*(X6OLD+X6+H*X6D);
      X7=.5*(X7OLD+X7+H*X7D);
      X8=.5*(X8OLD+X8+H*X8D);
      X9=.5*(X9OLD+X9+H*X9D);
      X10=.5*(X10OLD+X10+H*X10D);
      X11=.5*(X11OLD+X11+H*X11D);
      X12=.5*(X12OLD+X12+H*X12D);
      S=S+H;
      if S>=.0999
        S=0.;
        n=n+1;
        XMFN=sqrt(X9*PHIFN);
        XMRN=sqrt(X10*PHIRN);
        XMRNA=sqrt(X11*PHIRNA);
        XMGL=sqrt(X12*PHIGL);
        ArrayTP(n)=TP;
        ArrayXMFN(n)=XMFN;
        ArrayXMRN(n)=XMRN;
        ArrayXMRNA(n)=XMRNA;
                ArrayXMGL(n)=XMGL;
      end
end
XMFN
XMRN
XMRNA
XMGL
figure
plot(ArrayTP,ArrayXMFN),grid
xlabel('Normalized Flight Time (Sec)')
ylabel('Normalized Fading Noise Miss')
figure
plot(ArrayTP,ArrayXMRN),grid
xlabel('Normalized Flight Time (Sec)')
ylabel('Missile Semiactive Noise Miss')
figure
plot(ArrayTP,ArrayXMRNA),grid
xlabel('Normalized Flight Time (Sec)')
ylabel('Missile Active Noise Miss')
figure
```

```
plot(ArrayTP,ArrayXMGL),grid
xlabel('Normalized Flight Time (Sec)')
ylabel('Missile Glint Noise Miss')
clc
output=[ArrayTP',ArrayXMFN',ArrayXMRN',ArrayXMRNA',ArrayXMGL'];
save datfil.txt output -ascii
disp 'simulation finished'
```

MINIMUM GUIDANCE SYSTEM TIME CONSTANT

We saw in Chapter 6 that the radome aberration effects create an unwanted feedback path in the guidance system, which can cause stability problems. In the presence of radome slope the guidance system transfer function can be derived from Fig. 6.21 and is given by

$$\frac{n_L}{\dot{\lambda}} = N'V_c \Big/ \left\{ \left[1 + \frac{sT}{5}\right]^5 + \frac{N'V_c R}{V_M}(1 + T_\alpha s) \right\}$$

where R is the radome slope, V_M the missile velocity, T_α an aerodynamic parameter known as the turning rate time constant, and T the guidance system time constant. We can see from the preceding equation that if the radome slope is zero, the guidance system transfer function reduces to a fifth-order binomial. For different combinations of guidance system parameters, simulation experiments conducted in Chapter 6 showed that the guidance system can be unstable. One can show mathematically [6] that if the ratio of the turning rate time constant to the guidance system time constant is greater than unity or

$$\frac{T_\alpha}{T} > 1$$

then the guidance system transfer will be stable only if the following inequality is satisfied

$$-0.79 < \frac{N'V_c R T_\alpha}{V_M T} < 2.07$$

If the radome slope is negative, we can find from the preceding inequality that the minimum guidance system time constant to yield a stable guidance system is

$$T_{\min} = \frac{N'V_c R T_\alpha}{0.79 V_M}$$

We can see from the preceding relationship that engagements with larger closing velocities (that is, ballistic targets) or those taking place at high altitudes (that is, larger turning rate time constant) will require a larger guidance system time constant in order to keep the guidance system stable. However, larger guidance system time constants will also tend to increase the miss distance.

MISSILE TURNING RATE TIME CONSTANT [7, 8]

We saw in Chapter 6 that the missile turning rate time constant T_α had a significant interaction with radome effects. The missile turning rate time constant can be defined as the amount of time it takes to turn the missile flight path angle γ through an equivalent angle of attack α or

$$T_\alpha = \frac{\alpha}{\dot{\gamma}} = \frac{\alpha V_M}{n_L}$$

where V_M is missile velocity and n_L is missile acceleration. We showed in Chapter 10 the relationship between the lift coefficient C_L and missile acceleration. Substitution of those relationships into the preceding expression yields

$$T_\alpha = \frac{\alpha V_M W}{gQS_{\text{ref}} C_L}$$

where W is the missile weight, S_{ref} is the missile reference area, g is the acceleration of gravity, and Q is the dynamic pressure. If we assume our missile to be a cylinder or flying telephone pole, we showed in Chapter 16 that the lift coefficient could be expressed as

$$C_L = 2\alpha + \frac{1.5 S_{\text{plan}} \alpha^2}{S_{\text{ref}}}$$

where S_{plan} is the missile planform area. Substitution of the lift coefficient expression into the turning rate time constant formula and expressing the

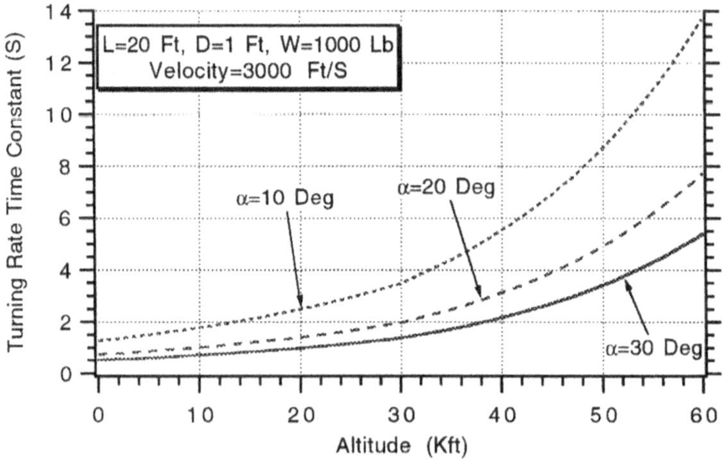

Fig. 18.4 Turning rate time constant increases with increasing altitude and decreasing angle of attack.

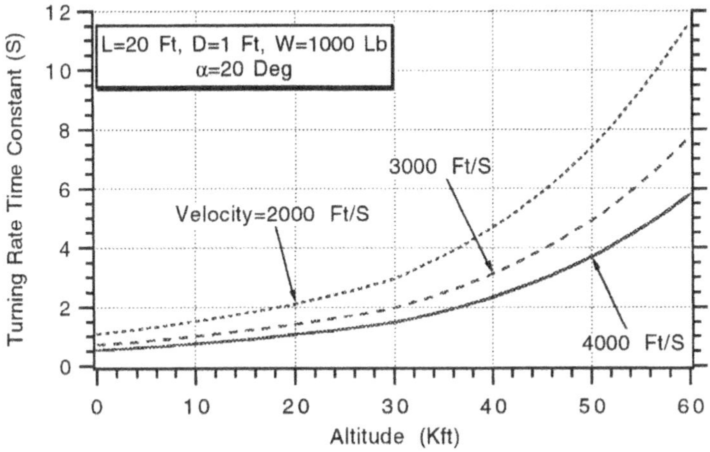

Fig. 18.5 Turning rate time constant increases with decreasing missile velocity.

dynamic pressure in more detail yields

$$T_\alpha = 2W \bigg/ \left\{ g\rho V_M S_{\text{ref}} \left[2 + \frac{1.5 S_{\text{plan}}\alpha}{S_{\text{ref}}} \right] \right\}$$

We can see that the turning rate time constant depends on altitude (or air density ρ), missile velocity, and angle of attack.

Consider the cylindrical missile of Chapter 16 which was 20 ft long, 1 ft in diameter, and weighed 1000 lb. Figure 18.4 shows that for a missile velocity of 3000 ft/s, the turning rate time constant increases with increasing altitude and decreasing angle of attack. At 50-kft altitude and 20-deg angle of attack the turning rate time constant is approximately 5 s.

We can display the turning rate time constant as a function of altitude for different missile velocities as is done in Fig. 18.5. We can see that the missile turning rate time constant increases with decreasing missile velocity. We have already seen in Chapter 6 that larger turning rate time constants exacerbate the radome slope stability problem. In summary, we can say that the radome stability problem will be greatest at the high-altitude, low-missile-velocity portion of the flight envelope.

CHECKING MINIMUM GUIDANCE SYSTEM TIME CONSTANT CONSTRAINTS

To illustrate the increased stability problem caused by ballistic targets, let us compare missile acceleration profiles for both aircraft and ballistic targets at 50-kft altitude. We demonstrated in Chapter 16 that a cylindrical missile (with $W = 1000$ lb, $L = 20$ ft, and $D = 1$ ft) has a 7-g capability for a speed of

3000 ft/s and an angle of attack of 20 deg. In addition, let us assume the radome slope for the interceptor is -0.01. If we consider a head-on case in which an aircraft target is traveling at 1000 ft/s, the resultant closing velocity will be 4000 ft/s. In this chapter we showed that the minimum guidance system time constant permitted for a fifth-order binomial proportional navigation guidance system is

$$T_{\min} = \frac{N' V_c R T_\alpha}{0.79 V_M}$$

From Fig. 18.4 we can see that the turning rate time constant for the cylindrical interceptor is 5 s at this flight condition. Therefore the minimum guidance system time constant at this flight condition becomes

$$T_{\min} = \frac{3 * 4000 * 0.01 * 5}{0.79 * 3000} = 0.25 \text{ s}$$

In other words, the interceptor guidance system time constant must be greater than 0.25 s when engaging this particular aircraft threat.

To test the preceding theoretical limit on the minimum allowable interceptor guidance system time constant against the aircraft threat, our multiple run simulation of a fifth-order binomial guidance system with radome effects (Listing 18.2) was modified to investigate single flights. Listing 18.2 shows that the resultant single flight simulation is set up to monitor the relative separation between missile and target y and the acceleration command n_c. The nominal inputs for the aircraft threat we are considering include the 7-g missile acceleration command limit, 4000-ft/s closing velocity, 5-s turning rate time constant, 3000-ft/s interceptor speed, -0.01 radome slope, and effective navigation ratio of 3. A 1-g target maneuver is considered as the only error source.

LISTING 18.2 SIMULATION OF HOMING LOOP WITH RADOME EFFECTS

```
VC=4000.;
XNT=32.2;
XNCLIMG=7.;
YIC=0.;
VM=3000.;
HEDEG=0.;
TAU=.3;
XNP=3.;
TA=5.;
R=-.01;
TF=10.;
Y=YIC;
YD=-VM*HEDEG/57.3;
```

```
YDIC=YD;
XNL=0.;
ELAMDH=0.;
X4=0.;
X5=0.;
TH=0.;
THH=0.;
T=0.;
H=.01;
S=0.;
XNCLIM=XNCLIMG*32.2;
n=0.;
while T<=(TF-1e-5)
        YOLD=Y;
        YDOLD=YD;
        XNLOLD=XNL;
        ELAMDHOLD=ELAMDH;
        X4OLD=X4;
        X5OLD=X5;
        THOLD=TH;
        THHOLD=THH;
        STEP=1;
        FLAG=0;
        while STEP<=1
           if FLAG==1
           STEP=2;
           Y=Y+H*YD;
           YD=YD+H*YDD;
           XNL=XNL+H*XNLD;
           ELAMDH=ELAMDH+H*ELAMDHD;
           X4=X4+H*X4D;
           X5=X5+H*X5D;
           TH=TH+H*THD;
           THH=THH+H*THHD;
           T=T+H;
           end
           TGO=TF-T+.00001;
           XLAM=Y/(VC*TGO);
           EPS=XLAM-TH-THH+R*THH;
           DD=5.*EPS/TAU;
           ELAMDHD=5.*(DD-ELAMDH)/TAU;
           XNC=XNP*VC*ELAMDH;
           if XNC>XNCLIM
           XNC=XNCLIM;
           end
           if XNC<-XNCLIM
```

```
            XNC=-XNCLIM;
         end
         X4D=5.*(XNC-X4)/TAU;
         X5D=5.*(X4-X5)/TAU;
         XNLD=5.*(X5-XNL)/TAU;
         THD=XNL/VM+TA*XNLD/VM;
         THHD=DD-THD;
         YDD=XNT-XNL;
         FLAG=1;
      end
      FLAG=0;
      Y=.5*(YOLD+Y+H*YD);
      YD=.5*(YDOLD+YD+H*YDD);
      XNL=.5*(XNLOLD+XNL+H*XNLD);
      ELAMDH=.5*(ELAMDHOLD+ELAMDH+H*ELAMDHD);
      X4=.5*(X4OLD+X4+H*X4D);
      X5=.5*(X5OLD+X5+H*X5D);
      TH=.5*(THOLD+TH+H*THD);
      THH=.5*(THHOLD+THH+H*THHD);
      S=S+H;
      if S>=.0999
         S=0.;
         n=n+1;
         ArrayT(n)=T;
         ArrayY(n)=Y;
         ArrayXNCG(n)=XNC/32.2;
      end
end
figure
plot(ArrayT,ArrayXNCG),grid
xlabel('Time (Sec)')
ylabel('Missile Acceleration (G)')
clc
output=[ArrayT',ArrayY',ArrayXNCG'];
save datfil.txt output  -ascii
disp 'simulation finished'
```

The nominal case was run and the guidance system time constant was made a parameter. Figure 18.6 shows that when the guidance system time constant is 0.2 s (less than the minimum permissible time constant) the acceleration command oscillates between $\pm 7\ g$, indicating that the guidance system is indeed unstable. Increasing the guidance system time constant to 0.3 s (more than the minimum permissible time constant) stabilizes the acceleration command. The resultant acceleration profile is a monotonically increasing straight line, as would be expected for the response due to a step target maneuver. Thus the theoretical

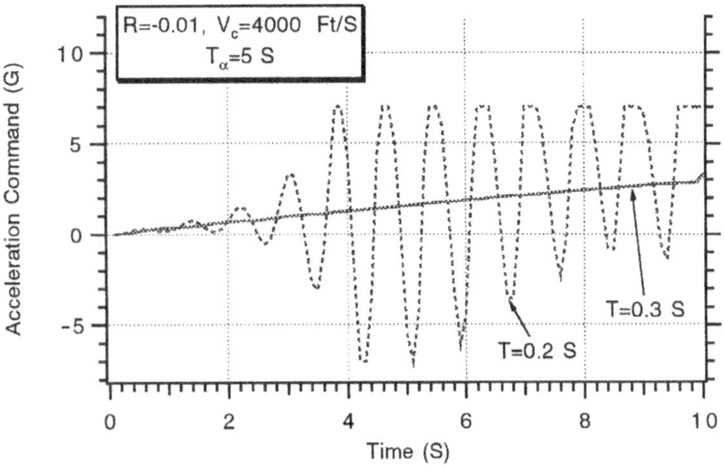

Fig. 18.6 Simulation and theory agree for aircraft threat.

formula for the minimum guidance system time constant and the simulation results of Listing 18.2 appear to be in agreement.

If we fly an inverse trajectory with a 3000 ft/s interceptor against a 6000 ft/s ballistic target, the resultant closing velocity becomes 9000 ft/s. The minimum guidance system time constant at this flight condition reduces to

$$T_{\min} = \frac{3 * 9000 * 0.01 * 5}{0.79 * 3000} = 0.57\ \text{s}$$

which indicates that the minimum permissible interceptor guidance system time constant against this particular ballistic threat must be greater than 0.57 s. In other words, for the case considered, the missile guidance system time constant must be much greater against a ballistic threat than it has to be against an aircraft threat because of the much higher closing velocities.

The simulation of Listing 18.2 was rerun for the case in which the closing velocity was 9000 ft/s and the guidance system time constant was again made a parameter. Figure 18.7 shows that when the guidance system time constant is 0.5 s (less than the minimum permissible time constant) the acceleration command oscillates between $\pm 7\ g$, indicating that the guidance system is unstable. When the guidance system time constant is increased to 0.7 s (more than the minimum permissible time constant), the acceleration command is stable and approximately monotonically increasing as would be expected for the response due to a step target acceleration. We can also see by comparing Figs. 18.6 and 18.7 that the frequency of oscillation depends on the closing velocity.

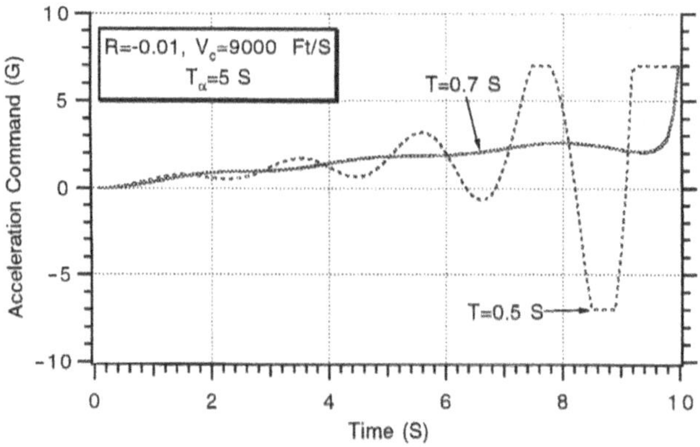

Fig. 18.7 Simulation and theory agree for ballistic threat.

MISS DUE TO NOISE FOR AIRCRAFT AND BALLISTIC TARGETS

In this chapter we have presented formulas for the miss distance due to various noise sources and have also presented stability requirements for the minimum guidance system time constant in a fifth-order binomial proportional navigation guidance system. In this section we shall illustrate, via a numerical example, how the miss distance due to noise increases when the threat changes from an aircraft to a ballistic target.

All of the formulas for the noise miss distances depended on the spectral density Φ of the noise. Often we talk about noise with a standard deviation σ entering a guidance system every T_s seconds. A useful approximation relating the noise spectral density to the standard deviation is given by

$$\Phi = \sigma^2 T_s$$

where Φ is measured in units^2/Hz, σ is measured in units, and T_s is measured in seconds. This relationship is identical to the one we used for simulating white noise in Chapter 4 with the sampling time T_s being replaced by the integration interval.

Let us consider an example in which range independent noise with standard deviation 0.001 rad enters a proportional navigation guidance system ($N' = 3$) every 0.01 s [4, 8]. The miss distance formula for an effective navigation ratio of 3 expressed in terms of the noise standard deviation can be found from Table 18.1 and is given by

$$\sigma_{\text{Miss}}\Bigg|_{\substack{\text{Range}\\\text{Independent}\\\text{Noise}}} = 3.04 V_c T^{0.5} \sigma_{FN} T_s^{0.5}$$

For the aircraft threat with a closing velocity of 4000 ft/s and a minimum guidance time constant of 0.25 s, the minimum miss due to range independent noise is

$$\sigma_{\text{Miss}}\Big|_{\substack{\text{Range}\\\text{Independent}\\\text{Noise}}} = 3.04 * 4000 * 0.25^{0.5} * 0.001 * 0.01^{0.5} = 0.6\ \text{ft}$$

whereas for the ballistic threat with closing velocity of 9000 ft/s and a minimum guidance time constant of 0.57 s, the minimum miss is

$$\sigma_{\text{Miss}}\Big|_{\substack{\text{Range}\\\text{Independent}\\\text{Noise}}} = 3.04 * 9000 * 0.57^{0.5} * 0.001 * 0.01^{0.5} = 2.1\ \text{ft}$$

Although the noise miss is more than three times larger in the ballistic target case than it was in the aircraft threat case, the miss due to range independent noise is negligible.

Let us now consider the influence of semiactive range dependent noise on both targets. The standard deviation of the miss due to semiactive range dependent noise for an effective navigation ratio of 3 can be found from Table 18.1 as

$$\sigma_{\text{Miss}}\Big|_{\substack{\text{Semiactive}\\\text{Range}\\\text{Dependent}\\\text{Noise}}} = \frac{9.47 V_c^2 T^{1.5} \sigma_{RN} T_s^{0.5}}{R_A}$$

We notice from the preceding formula that the dependence on closing velocity and guidance system time constant is more significant than it was in the range dependent noise case. Let us now consider a case in which there is 0.02 rad of semiactive noise at 30,000 ft entering the guidance system every 0.01 s [4, 7]. For the aircraft threat with a closing velocity of 4000 ft/s and a minimum guidance time constant of 0.25 s, the minimum miss due to semiactive range dependent noise is

$$\sigma_{\text{Miss}}\Big|_{\substack{\text{Semiactive}\\\text{Range}\\\text{Dependent}\\\text{Noise}}} = \frac{9.47 * 4000^2 * 0.25^{1.5} * 0.02 * 0.01^{0.5}}{30{,}000} = 1.3\ \text{ft}$$

whereas for the ballistic threat with closing velocity of 9000 ft/s and a minimum guidance time constant of 0.57 s, the minimum miss is

$$\sigma_{\text{Miss}}\Big|_{\substack{\text{Semiactive}\\ \text{Range}\\ \text{Dependent}\\ \text{Noise}}} = \frac{9.47 * 9000^2 * 0.57^{1.5} * 0.02 * 0.01^{0.5}}{30{,}000} = 22\ \text{ft}$$

We can see that the miss due to semiactive noise is very large against ballistic targets because of the strong dependence on closing velocity and guidance system time constant. For this example the ballistic threat miss was nearly 20 times greater than the aircraft threat miss!

Finally, let us now consider the influence of active range dependent noise on both targets. The standard deviation of the miss due to active range dependent noise for an effective navigation ratio of 3 can be found from Table 18.1 as

$$\sigma_{\text{Miss}}\Big|_{\substack{\text{Active}\\ \text{Range}\\ \text{Dependent}\\ \text{Noise}}} = \frac{41.3 V_c^3 T^{2.5} \sigma_{RNA} T_s^{0.5}}{R_A^2}$$

We notice from the preceding formula that the dependence on closing velocity and guidance system time constant is even more significant than it was in the semiactive noise case. Let us now consider a case in which there is 0.02 rad of active noise at 30,000 ft entering the guidance system every 0.01 s [4, 7]. For the aircraft threat with a closing velocity of 4000 ft/s and a minimum guidance time constant of 0.25 s, the minimum miss due to active range dependent noise is

$$\sigma_{\text{Miss}}\Big|_{\substack{\text{Active}\\ \text{Range}\\ \text{Dependent}\\ \text{Noise}}} = \frac{41.3 * 4000^3 * 0.25^{2.5} * 0.02 * 0.01^{0.5}}{30{,}000^2} = 0.18\ \text{ft}$$

whereas for the ballistic threat with closing velocity of 9000 ft/s and a minimum guidance time constant of 0.57 s, the minimum miss is

$$\sigma_{\text{Miss}}\Big|_{\substack{\text{Active}\\ \text{Range}\\ \text{Dependent}\\ \text{Noise}}} = \frac{41.3 * 9000^3 * 0.57^{2.5} * 0.02 * 0.01^{0.5}}{30{,}000^2} = 16.4\ \text{ft}$$

Although the active noise miss is less than the semiactive noise miss it is nearly 100 times larger for the ballistic threat than for the aircraft threat because the miss depends on the cube of closing velocity and the 2.5 power of the guidance system time constant.

SUMMARY

We have shown that from a noise induced miss distance point of view, ballistic targets are more challenging than the aircraft threat. The high-closing velocity of the ballistic target engagement significantly increases the minimum guidance system time constant required for radome slope stability. The large guidance system time constant and high-closing velocity can make ballistic target noise induced miss distances more than an order of magnitude greater than the miss distances experienced against an aircraft threat. To achieve small miss distances against high-closing velocity ballistic targets, methods for reducing the noise and effective radome slope must be found.

REFERENCES

[1] Travers, P., "Interceptor Dynamics," unpublished lecture notes, Raytheon, circa 1971.

[2] Garnell, P., and East, D. J., *Guided Weapon Control System*, Pergamon, Oxford, UK, 1977.

[3] Macfadzean, R. H. M., *Surface-Based Air Defense System Analysis*, Artech House, Norwood, MA, 1992.

[4] Lin, C. F., *Modern Navigation, Guidance, and Control Processing*, Prentice-Hall, Englewood Cliffs, NJ, 1991.

[5] Howe, R. M., "Guidance," *System Engineering Handbook*, eds. R. E. Machol, W. P. Tanner Jr., and S. N. Alexander, McGraw-Hill, New York, 1965, Chap. 19.

[6] Nesline, F. W., and Zarchan, P., "Radome Induced Miss Distance in Aerodynamically Controlled Homing Missiles," *Proceedings of AIAA Guidance and Control Conference*, AIAA, New York, Aug. 1984.

[7] Jerger J. J., System Preliminary Design, Van Nostrand, Princeton, NJ, 1960.

[8] Nesline, F. W., and Zarchan, P., "Miss Distance Dynamics in Homing Missiles," *Proceedings of AIAA Guidance and Control Conference*, AIAA, New York, Aug. 1984.

CHAPTER 19

Multiple Targets

INTRODUCTION AND BACKGROUND

In both tactical and strategic engagements there may be an apparent shift in the target location that can cause unacceptable miss distances. For example, for tactical intercepts, the interceptor may be guiding on the power centroid of two aircraft flying in close formation. When one of the aircraft falls outside the missile seeker beam, the other aircraft will be resolved. In this case it appears to the pursuing interceptor that the target has shifted from the location of the power centroid to the location of the resolved aircraft. In other words, there has been an apparent step change in target position [1]. Similarly, a strategic exoatmospheric missile may be homing on one of two closely spaced objects. After a while discrimination takes place and the interceptor's software may conclude that one object is a decoy while the second object is the real target. In this case too, as far as the missile is concerned, it appears as if the target has changed position (from first object or decoy to second object).

In both preceding examples the target displacement disturbance occurs late in the flight, which is the worst possible time from a missile guidance system point of view. Large miss distances may result because of insufficient remaining homing time. In this chapter we will develop normalized design curves for both a single time constant and a fifth-order binomial proportional navigation guidance system to both illustrate and quantify the multiple target problem. Rules of thumb will be developed relating the necessary ratio of the time left for homing after resolution has taken place to the guidance system time constant and the miss due to the apparent shift in target location.

DEVELOPMENT OF A LINEAR MODEL

To develop an appropriate linear model for the multiple target phenomenon, let us first think of two aircraft flying in close formation being pursued by an interceptor as shown in Fig. 19.1. The first aircraft is at 1200-ft altitude while the

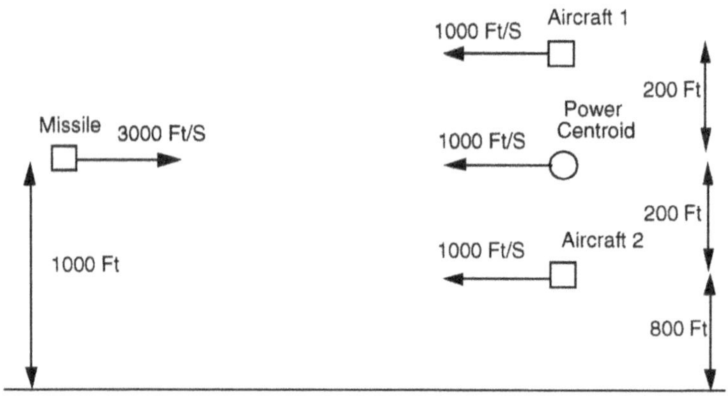

Fig. 19.1 Missile engaging two aircraft flying in formation.

second aircraft is at 800-ft altitude. For simplicity we will assume that the power centroid is located halfway between both aircraft at 1000-ft altitude. In this example both aircraft and the power centroid are traveling at 1000 ft/s. The pursuing missile is at 1000-ft altitude and is moving at 3000 ft/s toward the power centroid. At a certain time to go before intercept with the power centroid, the missile realizes that the true target is aircraft 1 and not the power centroid. At that time it will appear as if the target has changed position (from power centroid to aircraft 1). From a simulation point of view, we only have to model the target the missile sees. Therefore for most of the flight the missile will be guiding on the power centroid and for the rest of the flight the missile will be guiding on aircraft 1. This can easily be modeled as if the target is taking a step displacement at a certain time to go before intercept.

A two-dimensional nonlinear missile-target engagement simulation for a zero-lag guidance system, based on Listing 2.1 and Fig. 19.1, was developed for this problem and appears in Listing 19.1. For simplicity this simulation neglects both gravity and drag. Because the missile is on a collision triangle with the power centroid and there is no target maneuver or heading error, no acceleration commands will be required for most of the engagement. At one second before intercept (THOM = 1) the target jumps 200 ft from the power centroid to aircraft 1 and the missile responds with acceleration commands based on proportional navigation guidance (in this example using a navigation ratio of 3). We can see from Listing 19.1 that in this case there are no lags in the guidance system and so the missile will respond instantaneously, and perhaps unrealistically, to the apparent instantaneous shift in target position.

Figure 19.2 shows the alteration of the missile and target trajectories after the missile makes the guidance switch from the power centroid to aircraft 1. Here we can see that the apparent step change in target position at one second before intercept causes the missile to respond immediately. At the end of the flight, the missile

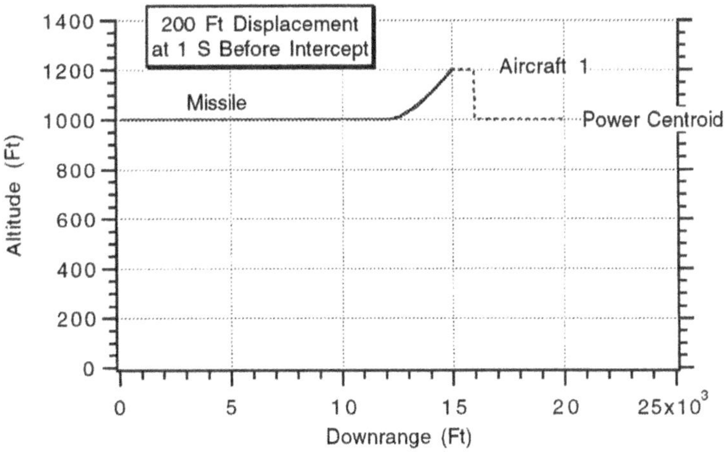

Fig. 19.2 Missile first guides on power centroid and then guides on aircraft 1.

hits aircraft 1. The successful intercept should not be surprising as we can see from Listing 19.1 there was no limit on the amount of acceleration that could be used nor were there any dynamics in the guidance system to cause miss distance.

LISTING 19.1 NONLINEAR ENGAGEMENT SIMULATION WITH STEP IN TARGET DISPLACEMENT

```
n=0;
XNP=3.;
DISPLACE=200.;
THOM=1.;
VM=3000.;
VT=1000.;
RM1=0.;
RM2=1000.;
RT1=20000.;
RT2=1000.;
QSWITCH=0;
VT1=-VT;
VT2=0.;
T=0.;
S=0.;
RTM1=RT1-RM1;
RTM2=RT2-RM2;
RTM=sqrt(RTM1^2+RTM2^2);
XLAM=atan2(RTM2,RTM1);
VM1=VM;
VM2=0.;
```

```
VTM1=VT1-VM1;
VTM2=VT2-VM2;
VC=-(RTM1*VTM1+RTM2*VTM2)/RTM;
TGO=RTM/VC;
H=.005;
while VC>0.
      TGO=RTM/VC;
      if(TGO>.3)
         H=.00005;
      end
      if(TGO<=THOM & QSWITCH==0)
         QSWITCH=1;
         RT2=RT2+DISPLACE;
      end
      RT1OLD=RT1;
      RT2OLD=RT2;
      RM1OLD=RM1;
      RM2OLD=RM2;
      VM1OLD=VM1;
      VM2OLD=VM2;
      STEP=1;
      FLAG=0;
      while STEP<=1
         if FLAG==1
         STEP=2;
         RT1=RT1+H*VT1;
         RT2=RT2+H*VT2;
         RM1=RM1+H*VM1;
         RM2=RM2+H*VM2;
         VM1=VM1+H*AM1;
         VM2=VM2+H*AM2;
         T=T+H;
         end
         RTM1=RT1-RM1;
         RTM2=RT2-RM2;
         RTM=sqrt(RTM1^2+RTM2^2);
         VTM1=VT1-VM1;
         VTM2=VT2-VM2;
         VC=-(RTM1*VTM1+RTM2*VTM2)/RTM;
         XLAM=atan2(RTM2,RTM1);
         XLAMD=(RTM1*VTM2-RTM2*VTM1)/(RTM*RTM);
         XNC=XNP*VC*XLAMD;
         AM1=-XNC*sin(XLAM);
         AM2=XNC*cos(XLAM);
         FLAG=1;
      end
```

```
        FLAG=0;
        RT1=.5*(RT1OLD+RT1+H*VT1);
        RT2=.5*(RT2OLD+RT2+H*VT2);
        RM1=.5*(RM1OLD+RM1+H*VM1);
        RM2=.5*(RM2OLD+RM2+H*VM2);
        VM1=.5*(VM1OLD+VM1+H*AM1);
        VM2=.5*(VM2OLD+VM2+H*AM2);
        S=S+H;
        if S>.049999
           S=0.;
           n=n+1;
           ArrayT(n)=T;
           ArrayRM1(n)=RM1;
           ArrayRM2(n)=RM2;
           ArrayRT1(n)=RT1;
           ArrayRT2(n)=RT2;
           ArrayXNCG(n)=XNC/32.2;
        end
end
RTM
figure
plot(ArrayRM1,ArrayRM2,ArrayRT1,ArrayRT2),grid
xlabel('Downrange (Ft)')
ylabel('Altitude (Ft)')
figure
plot(ArrayT,ArrayXNCG),grid
xlabel('Time (Sec)')
ylabel('Acceleration (G)')
clc
output=[ArrayT',ArrayRM1',ArrayRM2',ArrayRT1',ArrayRT2',ArrayXNCG'];
save datfil.txt output  -ascii
disp 'simulation finished'
```

Figure 19.3 shows that the price paid for the rapid missile trajectory change is that nearly 20 g of missile acceleration is required to make the intercept successful. Apparent step target displacements occurring later in the flight (that is, time to go before intercept is smaller) will cause even more acceleration while apparent step displacements occurring earlier in the flight will place smaller acceleration demands on the interceptor.

We now want to see if we can use our linear analysis tools to understand how an apparent step in target displacement influences guidance system performance. First we must determine if the linearization techniques we used in Chapter 2 are valid for approximating a step in target displacement. Figure 19.4 redraws the simplest possible linearized proportional navigation homing loop of Chapter 2. In this perfect guidance system in which the geometry is linearized, models of the seeker,

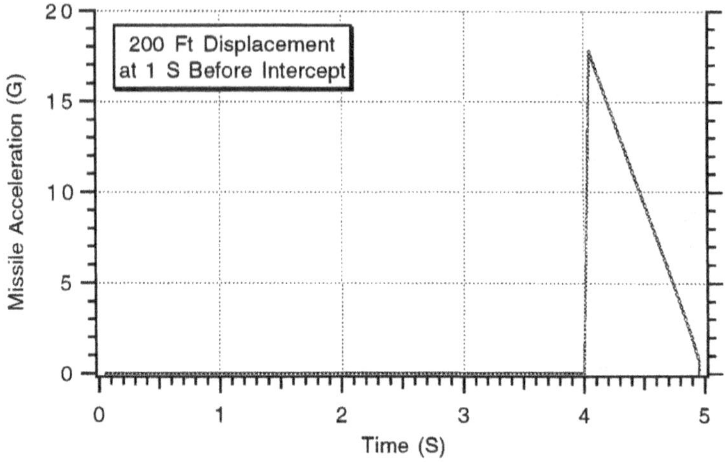

Fig. 19.3 Nearly 20 g of acceleration is required to take out 200 ft of target displacement.

noise filter, guidance, and flight control systems have been considered to be perfect and without dynamics. We are modeling the shift from the power centroid to aircraft 1 as a step in target displacement y_{TIC}. As can be seen from Fig. 19.4, the step target displacement can be modeled as an initial condition on the integrator with output y as $y = y_T - y_M$ and there is no initial y_M.

Following a procedure similar to that of Chapter 2, we can solve the linear time-varying differential equation associated with Fig. 19.4 [2]. From Fig. 19.4

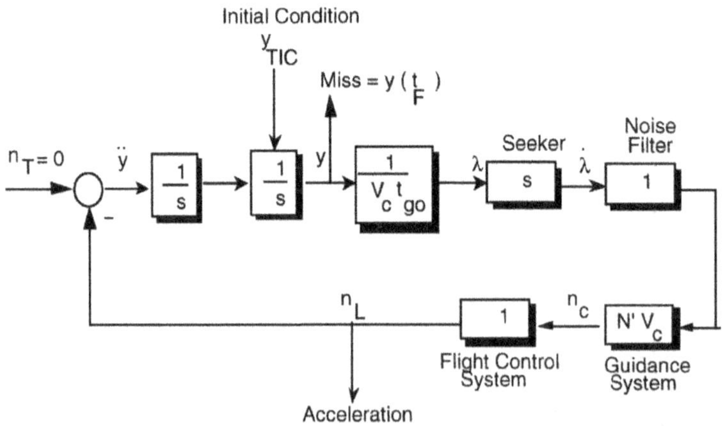

Fig. 19.4 Linear proportional navigation guidance homing loop with step in target displacement.

we can see that the relative acceleration (zero target acceleration minus missile acceleration) can be expressed as

$$\ddot{y} = 0 - n_c = -N'V_c\dot{\lambda}$$

Integrating the preceding differential equation once yields

$$\dot{y} = -N'V_c\lambda + C_1$$

where C_1 is the constant of integration. Substitution of the line-of-sight angle definition from Fig. 19.4 into the preceding expression yields the following linear time-varying first-order differential equation

$$\frac{dy}{dt} + \frac{N'y}{t_F - t} = C_1$$

We have seen in Chapter 2 that a linear first-order differential equation of the form

$$\frac{dy}{dt} + a(t)y = h(t)$$

has the solution

$$y = e^{-\int_0^t a(T)dT} \int_0^t h(n)e^{\int_0^n a(T)dT} dn + C_2 e^{-\int_0^t a(T)dT}$$

Therefore we now have enough information to solve the relative trajectory differential equation exactly. As was mentioned previously, an instantaneous step in target displacement means that the initial condition on the first state y is the value of the displacement or

$$y(0) = y_{TIC}$$

Under these circumstances, after much algebra, we can solve for the closed-form solutions for the relative separation between missile and target y and the missile acceleration n_c due to a step in target displacement. These closed-form solutions are given by

$$y = \frac{[N'(1 - t/t_F) - (1 - t/t_F)^{N'}]y_{TIC}}{N' - 1}$$

and

$$n_c = \frac{N'(1 - t/t_F)^{N'-2}y_{TIC}}{t_F^2}$$

where t is time and t_F is the amount of homing time remaining after the apparent step in target displacement has taken place. From the relative separation expression we can see that the miss distance $y(t_F)$ is always zero! This should not be surprising because we have a zero-lag guidance system and the miss distance for such a system should always be zero provided that the missile has sufficient acceleration capability. In addition, from the acceleration formula we observe that the magnitude of the initial missile acceleration is largest at the beginning and is proportional to the size of the target displacement and inversely proportional to the square of the homing or flight time. Doubling the apparent target displacement will double the initial missile acceleration, whereas doubling the flight time or time remaining after the apparent target displacement occurred will quarter the initial missile acceleration. Of course if the acceleration required by the preceding formula is not available there may be a significant miss distance.

If we overlay the preceding acceleration formula based on a linear model ($Y_{\text{TIC}} = 200$ ft, $N' = 3$, $t_F = 1$ s) with the nonlinear acceleration results of Fig. 19.3, we obtain the overlays of Fig. 19.5. Here we can see that the closed-form solution for the missile acceleration based on the linearized homing loop model approximates the nonlinear results very well. This means that we can use the linear guidance system model to make performance projections when the system disturbance is an apparent step in target displacement.

The closed-form solutions for the missile acceleration response due to a step in target displacement for various effective navigation ratios are displayed in normalized form in Fig. 19.6. We can see that higher effective navigation ratios increase the missile acceleration requirements at the beginning of flight (when apparent step in target displacement occurs) and reduce the acceleration requirements towards the end of the flight. Figure 19.6 indicates that the missile acceleration is always monotonically decreasing as the flight progresses (that is, more acceleration is needed at the beginning of the flight than at the end of the flight). From a

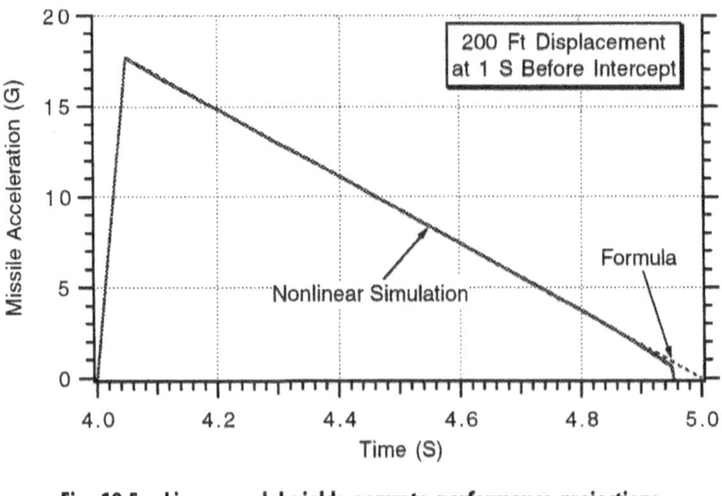

Fig. 19.5 Linear model yields accurate performance projections.

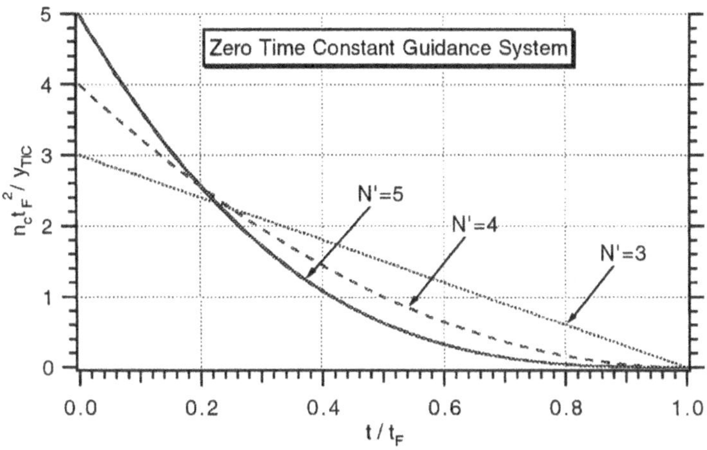

Fig. 19.6 Normalized missile acceleration due to step in target displacement.

system sizing point of view, the designer usually wants to ensure that the acceleration capability of the missile is adequate at the beginning of flight so that saturation can be avoided. For a fixed missile acceleration capability, Fig. 19.6 shows how requirements are placed on the minimum guidance or flight time required after final resolution (or the time remaining after the apparent step in target displacement occurs) and maximum allowable target displacement.

We can illustrate the use of the normalized acceleration curves of Fig. 19.6 and show why these curves are meaningful for the multiple target problem. Consider Fig. 19.7 in which two aircraft flying in formation are being pursued by a missile.

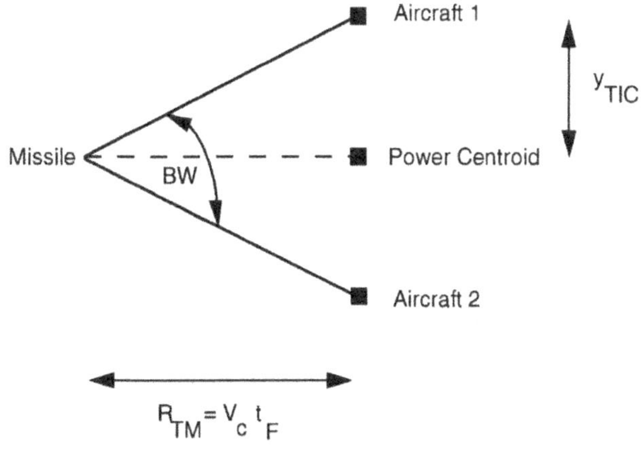

Fig. 19.7 Multiple target geometry.

Initially both aircraft are close enough so that the missile with seeker beamwidth BW homes on the power centroid of the two aircraft. At the point where one of the aircraft falls outside the seeker beam, resolution takes place and it appears to the missile that the aircraft has been instantaneously displaced a distance y_{TIC} (that is, from power centroid to aircraft 1). If the missile and aircraft are traveling at constant speed with closing velocity V_c, the missile will be a distance of $V_c t_F$ from the power centroid at the point of resolution. In this example t_F is the time remaining for guidance after seeker resolution.

From trigonometry we can see that the seeker beamwidth is related to the aircraft displacement according to

$$\tan\frac{BW}{2} = \frac{y_{\mathrm{TIC}}}{V_c t_F}$$

Using the small angle approximation and solving for the effective time remaining for guidance after seeker resolution, we get

$$t_F = \frac{2y_{\mathrm{TIC}}}{V_c BW}$$

If a seeker has a beamwidth of 0.1 rad (nearly 6 deg), and the two aircraft are separated by 400 ft and the closing velocity is 4000 ft/s, the time remaining for guidance after resolution will be 1 s or

$$t_F = \frac{2y_{\mathrm{TIC}}}{V_c BW} = \frac{400}{4000 * 0.1} = 1\ \mathrm{s}$$

Assuming that the missile effective navigation ratio is 3, we can see from either the formula for missile acceleration or the normalized acceleration curves of Fig. 19.6 that the maximum acceleration occurs at the time of resolution and is given by

$$n_{c_{\mathrm{MAX}}} = \frac{3y_{\mathrm{TIC}}}{t_F^2} = \frac{3 * 200}{1^2} = 600\ \mathrm{ft/s^2} = 18.6\,\mathrm{g}$$

which is in agreement with the maximum acceleration indicated by the nonlinear simulation results of Fig. 19.3. In summary, this means that the missile will require nearly 20 g of acceleration to hit the resolved target when the aircraft formation spacing is 400 ft, the missile seeker beamwidth is approximately 6 deg, and the closing velocity is 4000 ft/s.

SINGLE TIME CONSTANT GUIDANCE SYSTEM

If we add a single time constant representation of the guidance system, then Listing 19.1 is modified according to Listing 19.2. However, since the seeker cannot move instantaneously due to an apparent instantaneous change in target displacement, an initial condition is required that says the seeker dish angle D

equals the previous value of the line-of-sight angle λ right before the target appears to change location [3]. The amount of time it takes the seeker dish angle to follow the true target will be determined by the seeker time constant TAU. Statements that have changed from Listing 19.1 are highlighted in bold in Listing 19.2.

LISTING 19.2 NONLINEAR ENGAGEMENT SIMULATION WITH SEEKER DYNAMICS AND STEP IN TARGET DISPLACEMENT

```
n=0;
XNP=3.;
DISPLACE=200.;
THOM=1.;
TAU=1.;
VM=3000.;
VT=1000.;
for THOM = 0.1:0.1:5
    RM1=0.;
    RM2=1000.;
    RT1=20000.;
    RT2=1000.;
    QSWITCH=0;
    VT1=-VT;
    VT2=0.;
    T=0.;
    S=0.;
    RTM1=RT1-RM1;
    RTM2=RT2-RM2;
    RTM=sqrt(RTM1^2+RTM2^2);
    XLAM=atan2(RTM2,RTM1);
    VM1=VM;
    VM2=0.;
    VTM1=VT1-VM1;
    VTM2=VT2-VM2;
    VC=-(RTM1*VTM1+RTM2*VTM2)/RTM;
    TGO=RTM/VC;
    H=.001;
    D=0.;
    while VC>0.
        TGO=RTM/VC;
        if(TGO<.3)
            H=.00001;
        end
        if(TGO<=THOM & QSWITCH==0)
            QSWITCH=1;
```

```
			RT2=RT2+DISPLACE;
			RTM2=RT2-RM2;
			XLAM=atan2(RTM2,RTM1);
			D=XLAM;
end
RT1OLD=RT1;
RT2OLD=RT2;
RM1OLD=RM1;
RM2OLD=RM2;
VM1OLD=VM1;
VM2OLD=VM2;
DOLD=D;
STEP=1;
FLAG=0;
while STEP<=1
if FLAG==1
STEP=2;
RT1=RT1+H*VT1;
RT2=RT2+H*VT2;
RM1=RM1+H*VM1;
RM2=RM2+H*VM2;
VM1=VM1+H*AM1;
VM2=VM2+H*AM2;
			D=D+H*DD;
T=T+H;
end
RTM1=RT1-RM1;
RTM2=RT2-RM2;
RTM=sqrt(RTM1^2+RTM2^2);
VTM1=VT1-VM1;
VTM2=VT2-VM2;
VC=-(RTM1*VTM1+RTM2*VTM2)/RTM;
XLAM=atan2(RTM2,RTM1);
		EPS=XLAM-D;
		DD=(XLAM-D)/TAU;
		XNC=XNP*VC*DD;
AM1=-XNC*sin(XLAM);
AM2=XNC*cos(XLAM);
FLAG=1;
end
FLAG=0;
RT1=.5*(RT1OLD+RT1+H*VT1);
RT2=.5*(RT2OLD+RT2+H*VT2);
RM1=.5*(RM1OLD+RM1+H*VM1);
RM2=.5*(RM2OLD+RM2+H*VM2);
VM1=.5*(VM1OLD+VM1+H*AM1);
```

```
            VM2=.5*(VM2OLD+VM2+H*AM2);
            D=.5*(DOLD+D+H*DD);
        end
        if RTM2>0
            RTMP=RTM;
        else
            RTMP=-RTM;
        end
        n=n+1;
        ArrayTHOM(n)=THOM;
        ArrayRTMP(n)=RTMP;
end
figure
plot(ArrayTHOM,ArrayRTMP),grid
xlabel('Homing Time (s)')
ylabel('Miss (Ft)')
clc
output=[ArrayTHOM',ArrayRTMP'];
save datfil.txt output  -ascii
disp 'simulation finished'
```

The nominal case of Listing 19.2 was run and the miss due to a step in target displacement for different homing times appears in Fig. 19.8. Here we can see that if the homing time after the target displacement is small the miss can be quite large. For long homing times after the target displacement the miss is

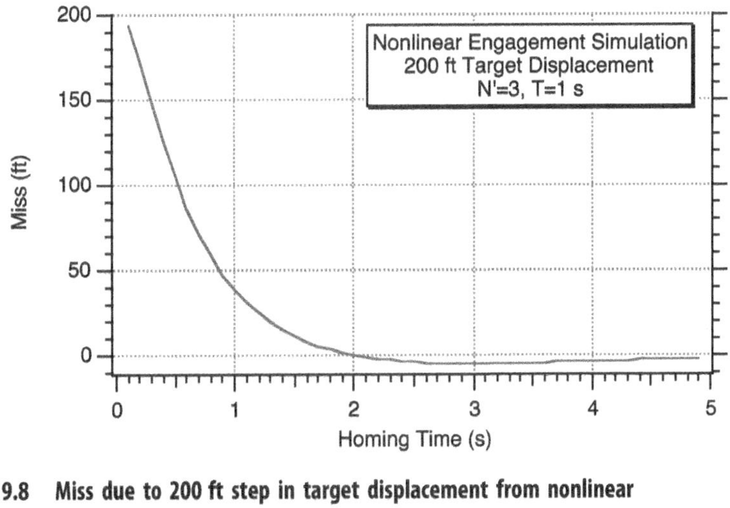

Fig. 19.8 Miss due to 200 ft step in target displacement from nonlinear engagement simulation.

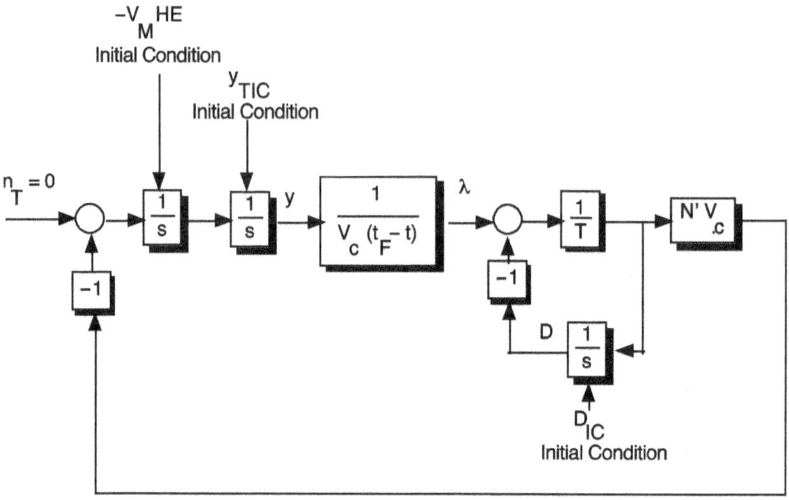

Fig. 19.9 Single-lag guidance system.

approximately zero. A linearized guidance system model, based on the nonlinear single-lag engagement model of Listing 19.2, appears in Fig. 19.9. Here we can see that there appear to be three error sources: the miss due to heading error, the miss due to a pure step in target displacement and the miss due to a seeker dish angle

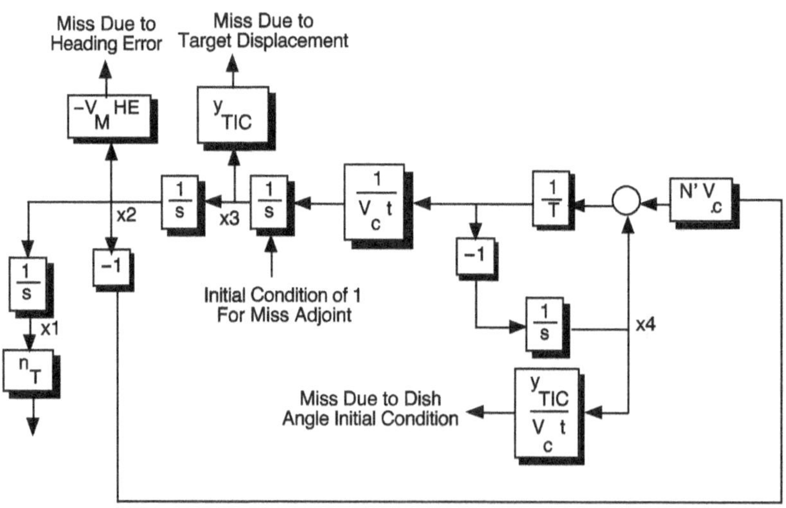

Fig. 19.10 Adjoint of single time constant guidance system with step in target displacement disturbance.

initial condition. In this section we shall show how all three error sources are related.

An adjoint diagram can be constructed from Fig. 19.9 and appears in Fig. 19.10. An adjoint simulation of Fig. 19.10 appears in Listing 19.3. Again, it is important to note that in Listing 19.3 the potential heading error magnitude is related to the target displacement and the missile velocity. The total miss due to the target displacement is made up of two parts—the step in target displacement and the initial condition on the seeker dish angle. We can see from the listing that the heading error miss divided by the homing time is calculated because it is postulated that this term is also equal to the step in target displacement miss.

LISTING 19.3 ADJOINT SIMULATION FOR STEP IN TARGET DISPLACEMENT FOR SINGLE-LAG GUIDANCE SYSTEM

```
n=0;
XNP=3.;
TAU=1.;
TF=5.;
VC=4000.;
DISP=200.;
VM=3000.;
HE=-DISP/VM;
T=0.;
S=0.;
TP=T+.00001;
X1=0;
X2=0;
X3=1;
X4=0;
H=.01;
while ~(TP>(TF-.00001))
      S=S+H;
      X1OLD=X1;
      X2OLD=X2;
      X3OLD=X3;
      X4OLD=X4;
      STEP=1;
      FLAG=0;
      while STEP<=1
        if FLAG==1
        STEP=2;
                X1=X1+H*X1D;
                X2=X2+H*X2D;
```

```
                    X3=X3+H*X3D;
                    X4=X4+H*X4D;
                    TP=TP+H;
            end
            X1D=X2;
            X2D=X3;
            Y1=(X4-XNP*VC*X2)/TAU;
            TGO=TP+.00001;
            X3D=Y1/(VC*TGO);
            X4D=-Y1;
            FLAG=1;
        end
        FLAG=0;
        X1=(X1OLD+X1)/2+.5*H*X1D;
        X2=(X2OLD+X2)/2+.5*H*X2D;
        X3=(X3OLD+X3)/2+.5*H*X3D;
        X4=(X4OLD+X4)/2+.5*H*X4D;
        if S<.09999
            S=0.;
            XMY=DISP*X3;
            XIC=X4*DISP/(VC*TGO);
            TOT=XMY+XIC;
            XMHE=-VM*HE*X2;
            XMHEPZ=XMHE/TP;
            n=n+1;
            ArrayTP(n)=TP;
            ArrayTOT(n)=TOT;
            ArrayXMHEPZ(n)=XMHEPZ;
        end
end
figure
plot(ArrayTP,ArrayTOT,ArrayTP,ArrayXMHEPZ),grid
xlabel('Homing Time (s)')
ylabel('Miss (Ft)')
clc
output=[ArrayTP',ArrayTOT',ArrayXMHEPZ'];
save datfil.txt output  -ascii
disp 'simulation finished'
```

The nominal case of Listing 19.3 was run and the results were overlaid with the nonlinear engagement results of Fig. 19.8. We can see from Fig. 19.11 that the adjoint results are in excellent agreement with the multiple-run nonlinear engagement simulation results of Listing 19.2—indicating that the adjoint was taken correctly. Figure 19.12 shows that if we divide the miss due to heading error by the

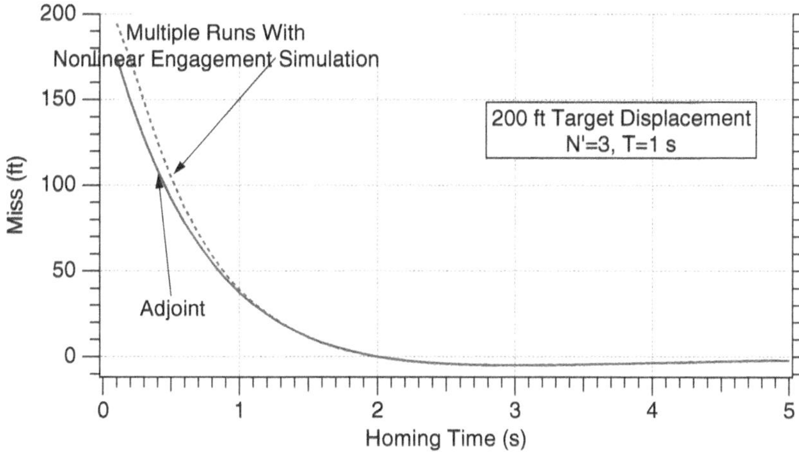

Fig. 19.11 Nonlinear engagement simulation and adjoint results agree.

homing time we also get the same results as the miss due to a step in target displacement [4]. We already have formulas for the miss due to heading error in Chapter 3 for a single time constant system for different effective navigation ratios, so we can easily derive formulas for the miss due to a step in target

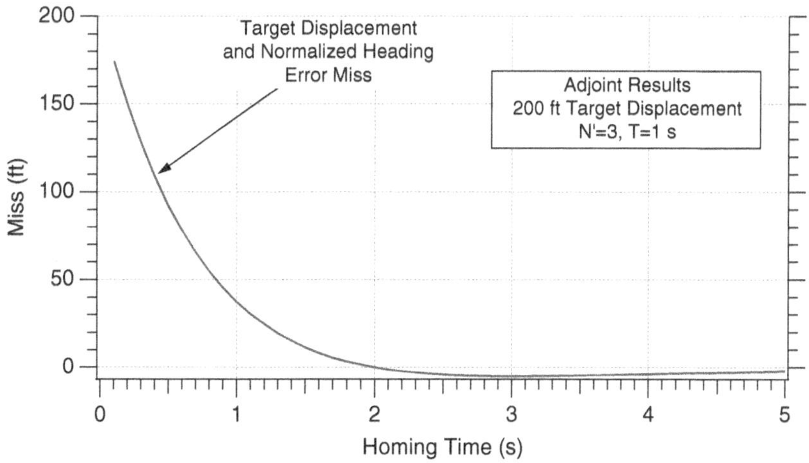

Fig. 19.12 Linear and nonlinear performance projections are identical in single time constant guidance system.

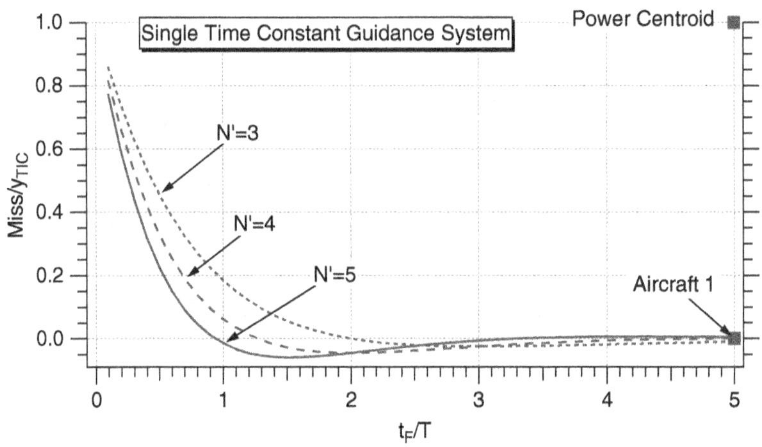

Fig. 19.13 First-order normalized miss for a step in target displacement.

displacement. The resultant miss distance formulas become

$$\left[\frac{Miss}{y_{TIC}}\right]_{N'=3} = e^{-x}(1 - 0.5x)$$

$$\left[\frac{Miss}{y_{TIC}}\right]_{N'=4} = e^{-x}\left(1 - x + \frac{x^2}{6}\right)$$

$$\left[\frac{Miss}{y_{TIC}}\right]_{N'=5} = e^{-x}\left(1 - 1.5x + \frac{x^2}{2} - \frac{x^3}{24}\right)$$

where

$$x = \frac{t_F}{T}$$

If we consider the same example of the previous section where there was no miss distance, the importance of guidance system dynamics and the normalized curves of Fig. 19.13 can be demonstrated. For a 200-ft target displacement (equivalent to 400-ft aircraft separation) 1 s of effective flight time and an overall guidance system time constant of 1 s ($T = 1$), the number of guidance time constants is 1 ($t_F/T = 1/1 = 1$) and the corresponding miss distance for different effective navigation ratios can be computed from Fig. 19.13 as

$$Miss_{N'=3} = 0.18 * 200 = 36\,\text{ft}$$

$$Miss_{N'=4} = 0.061 * 200 = 12\,\text{ft}$$

$$Miss_{N'=5} = -0.015 * 200 = -3\,\text{ft}$$

Thus we can see that there can be significant miss distance due to an apparent step in target displacement when system dynamics are considered. In this example a positive miss distance means that the missile is between aircraft 1 and the power centroid while a negative miss distance means the missile is below aircraft 1 in Fig. 19.13. We can see from Fig. 19.13 that if the missile time constant can be halved or if the seeker beamwidth can be halved then the number of guidance time constants is doubled and the miss will be reduced. The formulas and the curves indicate that large ratios of flight time to guidance system time constant yield small or near zero miss distances whereas small ratios of flight time to guidance system time constant can yield large miss distances. It was pointed out in Chapter 6 that a single-lag representation of the guidance system is not realistic and performance projections based on this model can lead to serious underestimates of the miss distance. We concentrated on the single time constant guidance system in order to derive miss distance formulas. It is postulated that the normalization factors for these formulas will be the same for higher order systems. In the next section we shall consider a higher order model of the guidance system to get a better understanding of the impact of a step in target displacement on system performance.

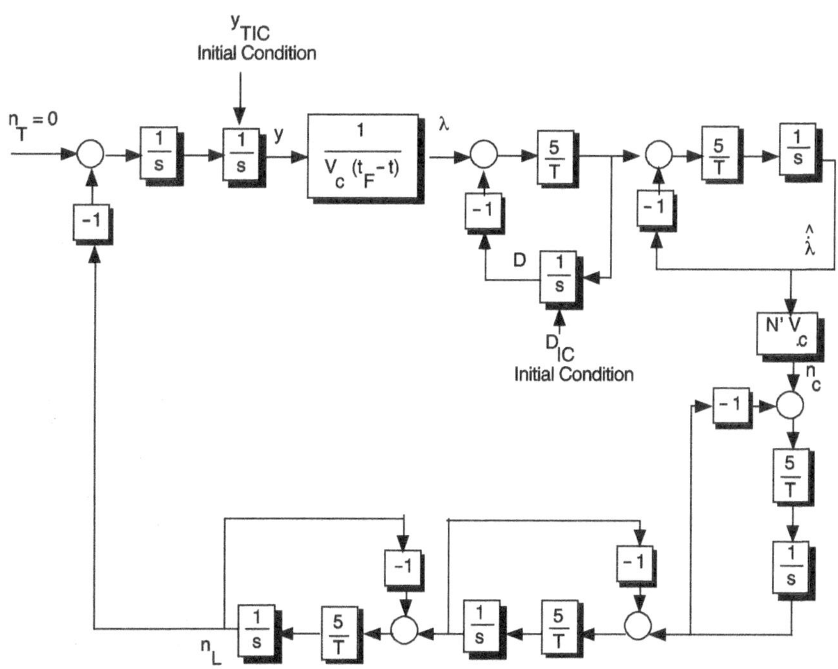

Fig. 19.14 Fifth-order binomial guidance system.

HIGHER-ORDER GUIDANCE SYSTEM DYNAMICS

It was shown in Chapter 6 that a much better and more convenient representation of a missile guidance system transfer function is a canonic fifth-order binomial given by

$$\frac{n_L}{\lambda} = \frac{N' V_c s}{(1 + sT/5)^5}$$

where T is the total guidance system time constant, n_L is the achieved missile acceleration, and λ is the line-of-sight angle. As was pointed out in Chapter 6, one time constant represents the seeker, another represents the noise filter, and the other three time constants represent the flight-control system dynamics (aerodynamics plus autopilot). With this canonic guidance system model the overall guidance system time constant is simply the sum of the five individual time constants or T. For convenience the fifth-order binomial missile homing loop of Chapter 6 is also repeated in Fig. 19.14.

An adjoint block diagram for finding the miss due to a step in target displacement can be constructed from Fig. 19.14 directly or by modifying the adjoint block

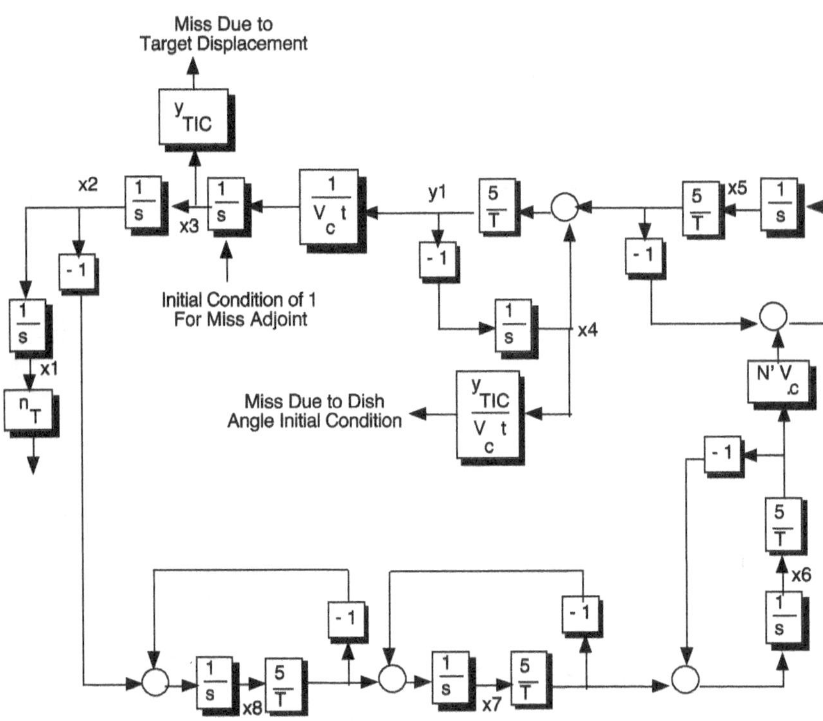

Fig. 19.15 Adjoint of fifth-order binomial guidance system.

diagram of Fig. 6.5. The resultant miss distance adjoint block diagram appears in Fig. 19.15.

An adjoint simulation of the fifth-order binomial guidance system can be constructed directly from Fig. 19.15 and appears in Listing 19.4. This adjoint simulation is nearly identical to that of Listing 6.1 except the states corresponding to the miss due to the ramp and parabolic maneuvers have been removed and the miss due to a step in target displacement TOT has been added.

LISTING 19.4 ADJOINT OF FIFTH-ORDER BINOMIAL GUIDANCE SYSTEM USED TO FIND NORMALIZED MISS DUE TO STEP IN TARGET DISPLACEMENT

```
n=0;
XNP=3.;
TAU=1.;
TF=10.;
DISP=1.;
VC=4000.;
VM=3000.;
HE=-DISP/VM;
T=0.;
S=0.;
TP=T+.00001;
X2=0;
X3=1;
X4=0;
X5=0.;
X6=0.;
X7=0.;
X8=0.;
H=.01;
while ~(TP>(TF-.00001))
      S=S+H;
      X2OLD=X2;
      X3OLD=X3;
      X4OLD=X4;
      X5OLD=X5;
      X6OLD=X6;
      X7OLD=X7;
      X8OLD=X8;
      STEP=1;
      FLAG=0;
      while STEP<=1
        if FLAG==1
                STEP=2;
                X2=X2+H*X2D;
                X3=X3+H*X3D;
```

```
                    X4=X4+H*X4D;
                    X5=X5+H*X5D;
                    X6=X6+H*X6D;
                    X7=X7+H*X7D;
                    X8=X8+H*X8D;
                    TP=TP+H;
            end
            X2D=X3;
            Y1=5.*(5.*X5/TAU+X4)/TAU;
            TGO=TP+.00001;
            X3D=Y1/(VC*TGO);
            X4D=-Y1;
            X5D=-5.*X5/TAU+5.*X6*XNP*VC/TAU;
            X6D=-5.*X6/TAU+5.*X7/TAU;
            X7D=-5.*X7/TAU+5.*X8/TAU;
            X8D=-5.*X8/TAU-X2;
            FLAG=1;
        end
        FLAG=0;
        X2=(X2OLD+X2)/2+.5*H*X2D;
        X3=(X3OLD+X3)/2+.5*H*X3D;
        X4=(X4OLD+X4)/2+.5*H*X4D;
        X5=(X5OLD+X5)/2+.5*H*X5D;
        X6=(X6OLD+X6)/2+.5*H*X6D;
        X7=(X7OLD+X7)/2+.5*H*X7D;
        X8=(X8OLD+X8)/2+.5*H*X8D;
        if S<.09999
            S=0.;
            XMY=DISP*X3;
            XIC=X4*DISP/(VC*TGO);
            TOT=XMY+XIC;
            n=n+1;
            ArrayTP(n)=TP;
            ArrayTOT(n)=TOT;
        end
end
plot(ArrayTP,ArrayTOT),grid
xlabel('Homing Time (s)')
ylabel('Miss (Ft)')
clc
output=[ArrayTP',ArrayTOT'];
save datfil.txt output  -ascii
disp 'simulation finished'
```

Normalized miss distance curves for different effective navigation ratios were generated from Listing 19.4. Because the adjoint technique was used, it was also implicitly assumed that the missile had infinite acceleration capability. As was done in Chapter 6, it was also assumed that the adjoint curves generated with Listing 19.4 had the same normalization factors as those curves for the single time constant guidance system. The hypothesis was checked by running numerous cases with Listing 19.4 using various combinations of guidance system time constant and homing time. The hypothesis was found to be true and the resultant normalized curves are displayed in Fig. 19.16 [2]. Superimposed on Fig. 19.16 is an indication of where the power centroid is in relation to the target we are supposed to be guiding on (aircraft 1 in this case). In other words a normalized miss of unity means we are hitting the power centroid and a normalized miss of zero means we are hitting aircraft 1. By comparing Figs. 19.13 and 19.16 we can conclude that in general the miss distances for the fifth-order guidance system are much larger than the miss for a single time constant guidance system. In addition, Fig. 19.16 also shows that the ratio of the flight time t_F (or time remaining after the apparent step in target displacement has occurred) to the guidance system time constant T must now be greater than eight for there to be negligible miss distance. Recall that for the single time constant system the number of guidance time constants had to be greater than three for the miss distance to be negligible. If the number of guidance time constants is less than eight, it is really a matter of luck as to how large or small the miss distance will be. Luck is involved because in reality the point at which resolution occurs for a specific engagement is random.

If we consider the same example of the previous section, where the miss distances were smaller, the importance of higher order guidance system dynamics can be seen to be even more important. For a 200-ft target displacement (equivalent to 400-ft aircraft separation), 1 s of effective flight time ($t_F = 1$) and an

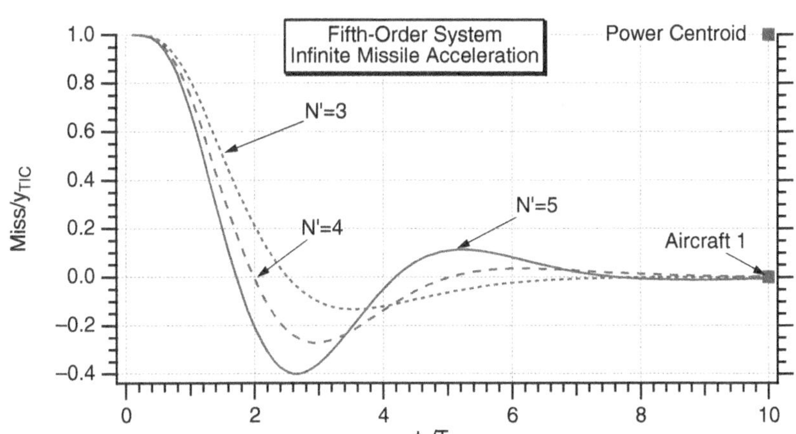

Fig. 19.16 Fifth-order normalized miss for a step in target displacement.

overall guidance system time constant of 1 s ($T = 1$), the number of guidance time constants is 1 ($t_F/T = 1/1 = 1$) and the corresponding miss distance for different effective navigation ratios can be computed from Fig. 19.16 as

$$Miss_{N'=3} = 0.811 * 200 = 162\,\text{ft}$$
$$Miss_{N'=4} = 0.748 * 200 = 150\,\text{ft}$$
$$Miss_{N'=5} = 0.686 * 200 = 137\,\text{ft}$$

Note that these miss distances are more than an order of magnitude greater than those of the single time constant guidance system! Thus we can see that the miss distances can be enormous. If the missile time constant can be halved, or if the seeker beamwidth can be halved, then the number of guidance time constants is doubled to 2 ($t_F/T = 1/0.5 = 2$) thus considerably reducing the miss or

$$Miss_{N'=3} = 0.212 * 200 = 42\,\text{ft}$$
$$Miss_{N'=4} = -0.0073 * 200 = -1.5\,\text{ft}$$
$$Miss_{N'=5} = -0.206 * 200 = -41\,\text{ft}$$

Thus we can see that the ratio of the flight time remaining after resolution has occurred to the guidance system time constant is critical in determining the expected miss distance.

ACCELERATION SATURATION

We have observed in the previous two sections that both the guidance system dynamics and effective navigation ratio play an important role in determining the miss distance due to a step in target displacement. The finite acceleration capability of the interceptor is also important in determining the miss distance. A forward model of the fifth-order binomial guidance with a limit on the acceleration command was constructed from Fig. 19.14 and appears in Listing 19.5. We can see from the code that nominally there is a 200-ft step in target displacement, the guidance system has a 0.2-s time constant, the missile acceleration limit is infinite, and the effective navigation ratio is 3.

LISTING 19.5 FORWARD MODEL FOR FINDING MISS DUE TO STEP IN TARGET DISPLACEMENT FOR FIFTH-ORDER BINOMIAL GUIDANCE SYSTEM IN PRESENCE OF ACCELERATION LIMIT

```
n=0;
VC=4000.;
XNT=0.;
DISPLACE=200.;
VM=3000.;
TAU=.2;
```

```
XNP=3.;
XNCLIM=99999999.;
TF=10.;
for THOM = 0.1:0.1:10
    QSWITCH=0;
    Y=0.;
    YD=0.;
    XNL=0.;
    D=0.;
    ELAMDH=0.;
    X4=0.;
    X5=0.;
    T=0.;
    H=.01;
    S=0.;
    while ~(T>(TF-.0001))
        TGO=TF-T+.00001;
        if TGO<=THOM & QSWITCH==0
            QSWITCH=1;
            Y=Y+DISPLACE;
            XLAM=Y/(VC*TGO);
            D=XLAM;
        end
        YOLD=Y;
        YDOLD=YD;
        XNLOLD=XNL;
        DOLD=D;
        ELAMDHOLD=ELAMDH;
        X4OLD=X4;
        X5OLD=X5;
        STEP=1;
        FLAG=0;
        while STEP<=1
            if FLAG==1
                STEP=2;
                Y=Y+H*YD;
                YD=YD+H*YDD;
                XNL=XNL+H*XNLD;
                ELAMDH=ELAMDH+H*ELAMDHD;
                D=D+H*DD;
                X4=X4+H*X4D;
                X5=X5+H*X5D;
                T=T+H;
            end
            TGO=TF-T+.00001;
            XLAM=Y/(VC*TGO);
```

```
                        DD=5.*(XLAM-D)/TAU;
                        ELAMDHD=5.*(DD-ELAMDH)/TAU;
                        XNC=XNP*VC*ELAMDH;
                        if XNC>XNCLIM
                                XNC=XNCLIM;
                        end
                        if XNC<-XNCLIM
                                XNC=-XNCLIM;
                        end
                        X4D=5.*(XNC-X4)/TAU;
                        X5D=5.*(X4-X5)/TAU;
                        XNLD=5.*(X5-XNL)/TAU;
                        YDD=XNT-XNL;
                        FLAG=1;
                end
                FLAG=0;
                Y=.5*(YOLD+Y+H*YD);
                YD=.5*(YDOLD+YD+H*YDD);
                XNL=.5*(XNLOLD+XNL+H*XNLD);
                D=.5*(DOLD+D+H*DD);
                ELAMDH=.5*(ELAMDHOLD+ELAMDH+H*ELAMDHD);
                X4=.5*(X4OLD+X4+H*X4D);
                X5=.5*(X5OLD+X5+H*X5D);
        end
        n=n+1;
        ArrayTHOM(n)=THOM;
        ArrayY(n)=Y;
end
figure
plot(ArrayTHOM,ArrayY),grid
xlabel('Homing Time (s)')
ylabel('Miss (Ft)')
clc
output=[ArrayTHOM',ArrayY'];
save datfil.txt output  -ascii
disp 'simulation finished'
```

The nominal case of Listing 19.5 was run and compared to adjoint results from Listing 19.4 when the target displacement was 200 ft, the effective navigation ratio was 3, and the guidance system time constant was 0.2 s. Figure 19.17 shows that both the forward and adjoint models yield identical results. At the beginning of this chapter we showed that for a system in which there was zero time constants approximately 20 g of missile acceleration was required to take out the step in target displacement. Figure 19.18 shows how system performance degrades in the presence of an acceleration limit. We can see that for a small guidance

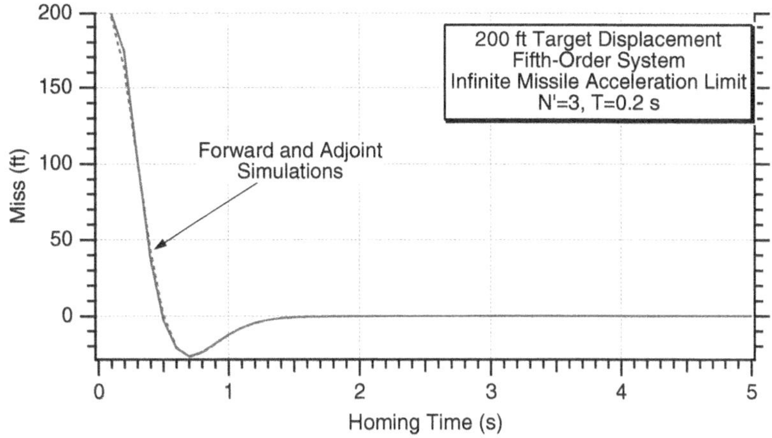

Fig. 19.17 Forward and adjoint models agree for fifth-order guidance systems.

system time constant of 0.2 s the miss degrades significantly when the missile acceleration limit is 20 *g*. More miss distance degradation occurs when the missile acceleration limit is further reduced. However, Fig. 19.19 shows that if the missile guidance system time constant is increased from 0.2 s to 1 s there is less sensitivity to the missile acceleration limit. That is not to say that performance improves as the guidance system time constant is increased. Figure 19.19 simply

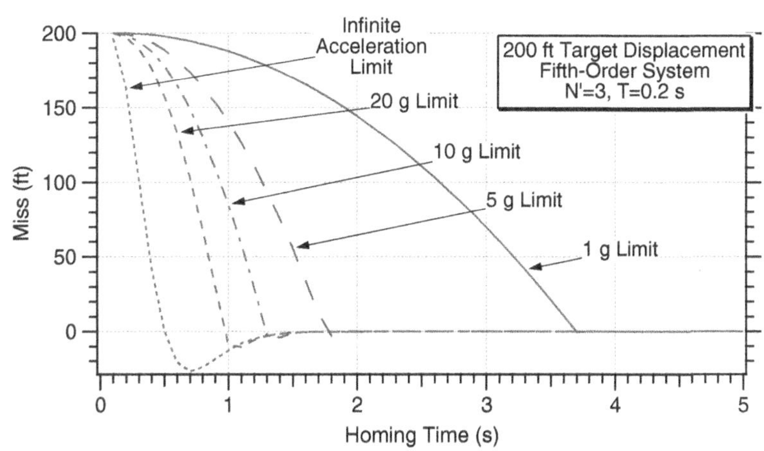

Fig. 19.18 Miss due to saturation and target displacement for large guidance system time constant.

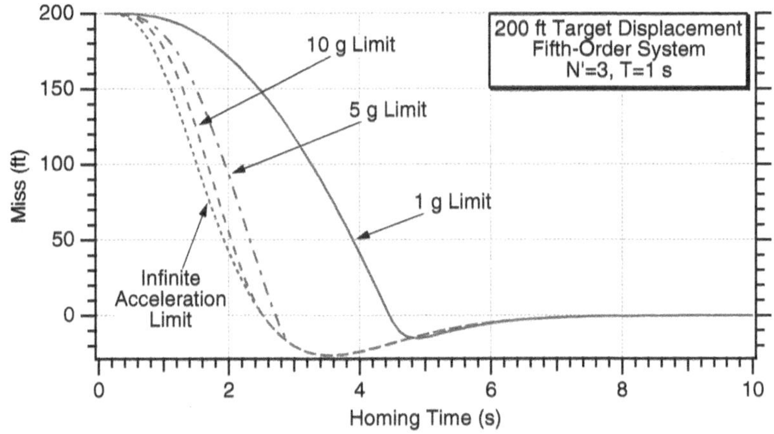

Fig. 19.19 Miss due to saturation and target displacement for large guidance system time constant.

says that when there is poor system performance due to a large guidance system time constant there is less sensitivity to the value of the acceleration limit.

SUMMARY

Normalized miss distance curves were presented showing the designer how to calculate the miss distance due to an apparent step in target displacement. The importance of guidance system dynamics and missile acceleration saturation effects were illustrated with additional design curves and examples.

REFERENCES

[1] Travers, P., "Interceptor Dynamics," unpublished lecture notes, Raytheon Co., circa 1971.

[2] Zarchan, P., "When Bad Things Happen to Good Missiles," *Proceedings of AIAA Guidance, Navigation and Control Conference*, AIAA, Washington, DC, Aug. 1993.

[3] Bucco, D., and Grorecki, R., "On Alternative Formulations for Linearized Miss Distance Analysis," DSTO Technical Report, October 2011.

[4] Gutman, O., and Palmor, Z. J., "Proportional Navigation Against Multiple Targets," 2010 AIAA Guidance, Navigation, and Control Conference, Paper AIAA 2010-8183, Toronto, Canada.

CHAPTER 20

Weaving Targets

INTRODUCTION AND BACKGROUND

We have seen in Chapter 6 that large miss distances could be induced by the target if a maximum acceleration maneuver was initiated at the proper time to go before intercept. It was also shown that the barrel roll or weave maneuver could also generate large miss distances. Because it is well known that tactical ballistic missiles (TBMs) can spiral or weave into resonance (TBM roll rate equals vehicles' natural pitch frequency) as they re-enter the atmosphere due to either mass or configurational asymmetries, the weave maneuver is of particular interest to the guidance system designer [1, 2].

In this chapter we will first study the influence of the target weave maneuver on a single time constant proportional navigation guidance system. Closed-form solutions for the peak steady-state miss distance as a function of the effective navigation ratio, guidance system time constant, weave maneuver amplitude and frequency will be derived [3, 4]. Because we have already shown that the single time constant guidance system seriously underestimates the miss distance, a more realistic, higher-order guidance system will be used to develop normalized miss distance design curves using the normalization factors from the single time constant target maneuver miss distance solutions. The finite acceleration capability of the interceptor also plays an important role in determining system performance. The normalized design curves, which assumed infinite missile acceleration capability, are updated to show how the missile acceleration advantage over the target plays a key role in determining system performance. Finally, methods for improving missile system performance against weaving targets will be explored.

WEAVE MANEUVER IN SINGLE TIME CONSTANT GUIDANCE SYSTEM

Periodic maneuver sequences such as a sinusoidal or weaving target present a challenge for a missile guidance system designer. A planar representation of a weaving target is given by

$$\text{Target Maneuver} = n_T \sin \omega_T t$$

where n_T is the maneuver amplitude, ω_T is the target weave frequency, and t is time. The miss due to a weaving target as a function of flight time can be found using the method of brute force. A nonlinear two-dimensional engagement simulation, based on Listing 2.1, of a missile guiding on a weaving target appears in Listing 20.1. We can see that the listing is based on a single time constant proportional navigation guidance system and that for the nominal case the missile time constant is 1 s, the effective navigation ratio is 3, the target weave frequency is 3 rad/s while the target maneuver amplitude is 193.2 ft/s^2 or 6 g. The target always initiates its maneuver at the beginning of flight in this simulation. The program is set up to run in the brute force mode so that the miss distance results for many flight times can be evaluated. In this nonlinear engagement simulation the initial target downrange position, which is equivalent to the initial missile-target separation, is varied from 500 ft to 40,000 ft in steps of 500 ft, which in the linear world is equivalent to varying the flight time from 0.05 s to 10 s in steps of 0.05 s because the closing velocity is approximately 4000 ft/s. After each run the homing time and miss distance are tabulated. We can also see from the simulation listing that if the target is above the missile at intercept, the miss is considered to be positive whereas if the target is below the missile at intercept the miss is considered to be negative.

LISTING 20.1 NONLINEAR ENGAGEMENT SIMULATION FOR A SINGLE TIME CONSTANT GUIDANCE SYSTEM WITH WEAVING TARGET

```
XNP=3.;
TAU=1.;
XNT=193.2;
W=3.;
n=0;
for RT1IC=500:500:40000
        VM=3000.;
        VT=1000.;
        RM1=0.;
        RM2=0.;
        RT1=RT1IC;
        RT2=0.;
        BETA=0.;
        VT1=-VT*cos(BETA);
        VT2=VT*sin(BETA);
        T=0.;
        S=0.;
        RTM1=RT1-RM1;
```

```
RTM2=RT2-RM2;
RTM=sqrt(RTM1^2+RTM2^2);
XLAM=atan2(RTM2,RTM1);
VM1=VM;
VM2=0.;
VTM1=VT1-VM1;
VTM2=VT2-VM2;
VC=-(RTM1*VTM1+RTM2*VTM2)/RTM;
TGO=RTM/VC;
XLAMH=0.;
H=.01;
while VC > 0.
if(RTM < 1000.)
          H=.0005;
end
  BETAOLD=BETA;
  RT1OLD=RT1;
  RT2OLD=RT2;
  RM1OLD=RM1;
  RM2OLD=RM2;
  VM1OLD=VM1;
  VM2OLD=VM2;
  XLAMHOLD=XLAMH;
  STEP=1;
  FLAG=0;
  while STEP < =1
          if FLAG==1
          STEP=2;
          BETA=BETA+H*BETAD;
          RT1=RT1+H*VT1;
          RT2=RT2+H*VT2;
          RM1=RM1+H*VM1;
          RM2=RM2+H*VM2;
          VM1=VM1+H*AM1;
          VM2=VM2+H*AM2;
          XLAMH=XLAMH+H*XLAMHD;
          T=T+H;
          end
          VT1=-VT*cos(BETA);
          VT2=VT*sin(BETA);
          BETAD=XNT*sin(W*T)/VT;
          RTM1=RT1-RM1;
          RTM2=RT2-RM2;
          RTM=sqrt(RTM1^2+RTM2^2);
          VTM1=VT1-VM1;
          VTM2=VT2-VM2;
```

```
                    VC=-(RTM1*VTM1+RTM2*VTM2)/RTM;
                    XLAM=atan2(RTM2,RTM1);
                    XLAMHD=(XLAM-XLAMH)/TAU;
                    XNC=XNP*VC*XLAMHD;
                    AM1=-XNC*sin(XLAM);
                    AM2=XNC*cos(XLAM);
                    FLAG=1;
        end
        FLAG=0;
        BETA=.5*(BETAOLD+BETA+H*BETAD);
        RT1=.5*(RT1OLD+RT1+H*VT1);
        RT2=.5*(RT2OLD+RT2+H*VT2);
        RM1=.5*(RM1OLD+RM1+H*VM1);
        RM2=.5*(RM2OLD+RM2+H*VM2);
        VM1=.5*(VM1OLD+VM1+H*AM1);
        VM2=.5*(VM2OLD+VM2+H*AM2);
        XLAMH=.5*(XLAMHOLD+XLAMH+H*XLAMHD);
    end
    if RTM2 > 0.
        RTMP=RTM;
    else
        RTMP=-RTM;
    end
    n=n+1;
    ArrayT(n)=T;
    ArrayRTMP(n)=RTMP;
end
figure
plot(ArrayT,ArrayRTMP),grid
xlabel('Flight Time (Sec)')
ylabel('Miss (Ft)')
clc
output=[ArrayT',ArrayRTMP'];
save datfil.txt output  -ascii
disp 'simulation finished'
```

The nominal case of Listing 20.1 was run, and the miss distance results as a function of flight time appear in Fig. 20.1. We can see that unlike the step target maneuver results of Chapters 3 and 6, *the miss distance due to weaving target does not approach zero as the homing time increases!* Depending on the flight time, the miss distance for this example can be as large as 28 ft or as small as zero when the effective navigation ratio is 3. Also note that after an initial transient period the miss is sinusoidal in nature with frequency 3 rad/s, which is identical to the target weave frequency.

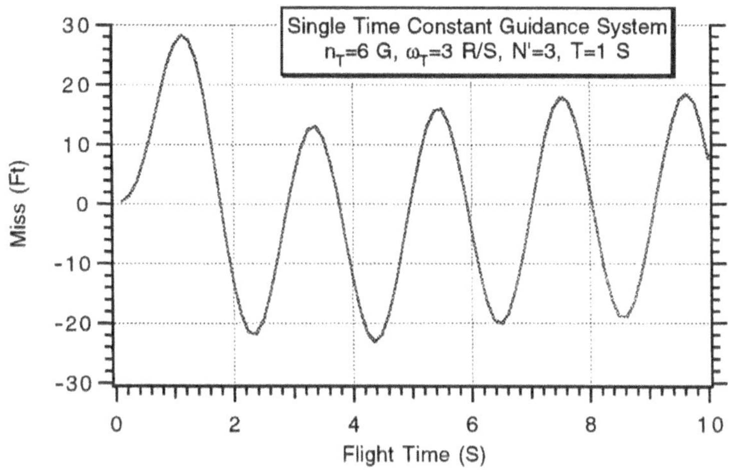

Fig. 20.1 Nonlinear results indicate that weaving target causes miss to oscillate at target weave frequency.

Could these multiple run nonlinear results be generated with one adjoint run? To find out, we first have to assume that the geometry of the engagement can be linearized in a manner similar to that of Chapters 2, 3, and 19. In addition, because adjoint theory requires that all inputs to the original system appear as impulses, we also have to find some way of making the sinusoidal maneuver look like an impulse through a linear shaping network. Fortunately, in Table 1.1 of Chapter 1 we showed that a sinusoidal maneuver could be represented as an impulse through a second-order shaping network since

$$\mathcal{L}[n_T \sin \omega_T t] = \frac{\omega_T n_T}{s^2 + \omega_T^2}$$

Figure 20.2 shows a linearized representation of a single time constant linear proportional navigation guidance system driven by a weave target maneuver (that is, impulsive input through second-order shaping network).

Figure 20.2 is drawn in such a way that the adjoint can be taken by inspection. The new adjoint diagram was constructed from Fig. 20.2 using the adjoint rules of Chapter 3 and appears in Fig. 20.3. In this particular adjoint diagram we are evaluating one disturbance only, the miss due to a weaving target.

An adjoint simulation, based on Fig. 20.3, was constructed and appears in Listing 20.2. Here we can see that the nominal case is set up to be identical to that of the brute force nonlinear engagement simulation of Listing 20.1. However, with the adjoint simulation only one run has to be made to find out how the miss distance varies with flight time.

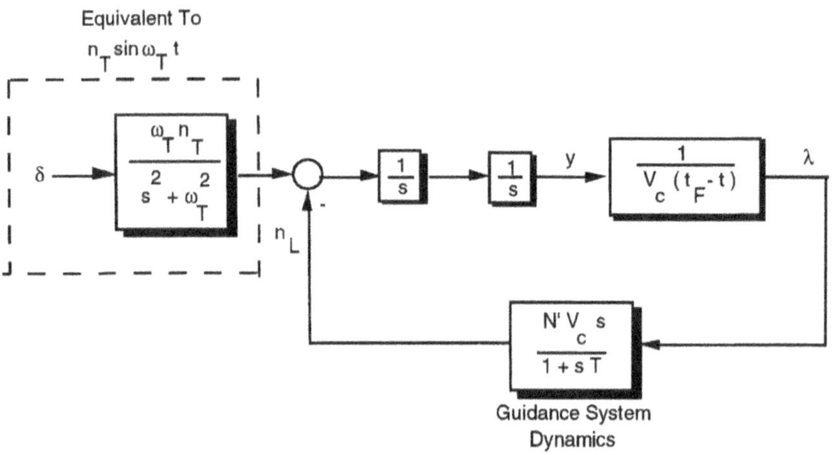

Fig. 20.2 Linearized single time constant guidance system with weave maneuver.

The nominal case of Listing 20.2 was run, and the adjoint miss distance results were plotted alongside the nonlinear, multiple run results of Fig. 20.1. We can see from Fig. 20.4 that the adjoint results are virtually identical to the nonlinear miss distance results thus validating the simplified linear model. In the rest of this chapter, we will use the linearized model of the guidance system to generate performance evaluations and to suggest ways of improving performance.

Another case was run with the adjoint simulation in which the target weave frequency was decreased from 3 rad/s to 1.5 rad/s. We can see from Fig. 20.5 that the miss distance increases significantly with the lower weave frequency

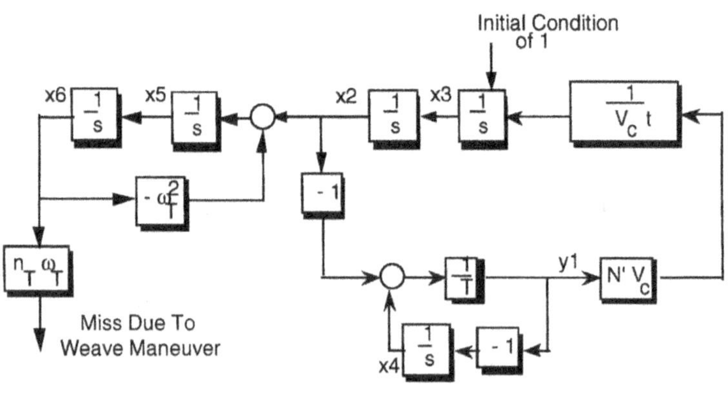

Fig. 20.3 Adjoint of single time constant guidance system with weaving target.

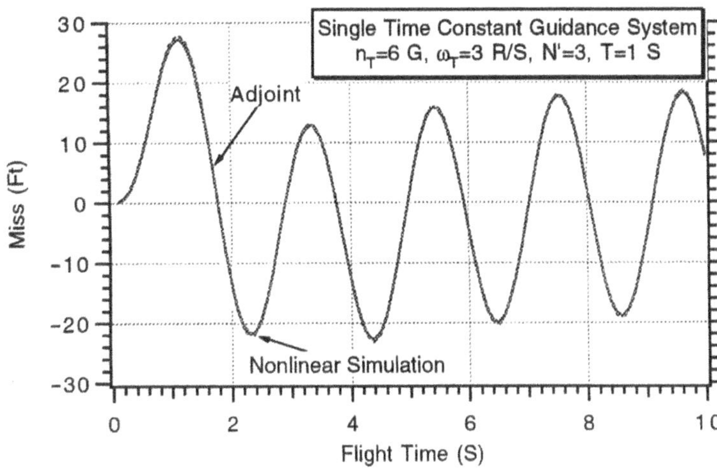

Fig. 20.4 Adjoint simulation results agree with the nonlinear results for the weaving target disturbance.

and the oscillation of the miss with flight time changes to match the new target weave frequency. We now would like to get a deeper understanding of how performance is related to the target weave frequency, guidance system time constant, and effective navigation ratio.

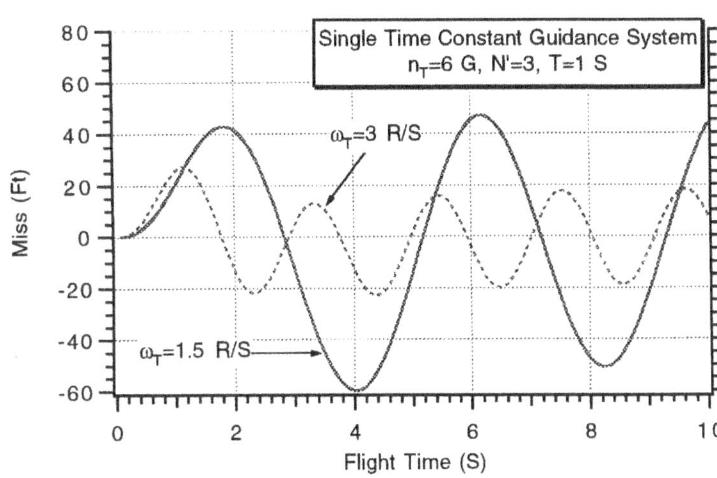

Fig. 20.5 Decreasing weave frequency increases miss distance for this example.

LISTING 20.2 ADJOINT SIMULATION OF SINGLE TIME CONSTANT GUIDANCE SYSTEM AND WEAVING TARGET

```
n=0;
XNT=193.2;
XNP=3.;
TAU=1.;
TF=10.;
VC=4000.;
W=3.;
T=0.;
S=0.;
TP=T+.00001;
X2=0;
X3=1;
X4=0.;
X5=0.;
X6=0.;
H=.01;
while TP < =(TF-1e-5)
      X2OLD=X2;
      X3OLD=X3;
      X4OLD=X4;
      X5OLD=X5;
      X6OLD=X6;
      STEP=1;
      FLAG=0;
      while STEP < =1
        if FLAG==1
        STEP=2;
             X2=X2+H*X2D;
             X3=X3+H*X3D;
             X4=X4+H*X4D;
             X5=X5+H*X5D;
             X6=X6+H*X6D;
             TP=TP+H;
        end
        X2D=X3;
        Y1=(-X2+X4)/TAU;
        TGO=TP+.00001;
        X3D=Y1*XNP/TGO;
        X4D=-Y1;
        X5D=X2-W*W*X6;
        X6D=X5;
        FLAG=1;
      end
      FLAG=0;
```

```
        X2=(X2OLD+X2)/2+.5*H*X2D;
        X3=(X3OLD+X3)/2+.5*H*X3D;
        X4=(X4OLD+X4)/2+.5*H*X4D;
        X5=(X5OLD+X5)/2+.5*H*X5D;
        X6=(X6OLD+X6)/2+.5*H*X6D;
        S=S+H;
        if S > =.09999
           S=0.;
           n=n+1;
                        XMWEAVE=XNT*W*X6;
           ArrayTP(n)=TP;
                        ArrayXMWEAVE(n)=XMWEAVE;
        end
end
figure
plot(ArrayTP,ArrayXMWEAVE),grid
xlabel('Flight Time (S)')
ylabel('Miss (Ft) ')
clc
output=[ArrayTP',ArrayXMWEAVE'];
save datfil.txt output/ascii
disp 'simulation finished'
```

CLOSED-FORM SOLUTIONS FOR MISS DISTANCE

Because we have already shown that the miss due to a weaving target is a sinusoidal function of the flight time, it only makes sense to look at steady-state miss distances in order to quantify system performance. Closed-form solutions for the miss due to a weaving target can be obtained in the steady state (that is, at large flight times when transients die out). Recall that all of the miss distance formulas for a single time constant guidance system were derived in Chapter 3 from the generalized adjoint diagram of Fig. 3.16. Figure 3.16 is redrawn and updated to include the miss due to a weaving target MWEAVE as shown in Fig. 20.6.

In Chapter 3 we found that the miss due to a step target maneuver MNT, expressed in the Laplace transform domain, was given by

$$\mathrm{MNT}(s) = \frac{1 - H(s)}{s^3} * n_T$$

Therefore the miss due to a weaving target can be found by inspection from Fig. 20.6 and can be expressed as

$$\mathrm{MWEAVE}(s) = \frac{1 - H(s)}{s^2} * \frac{n_T \omega_T}{s^2 + \omega_T^2}$$

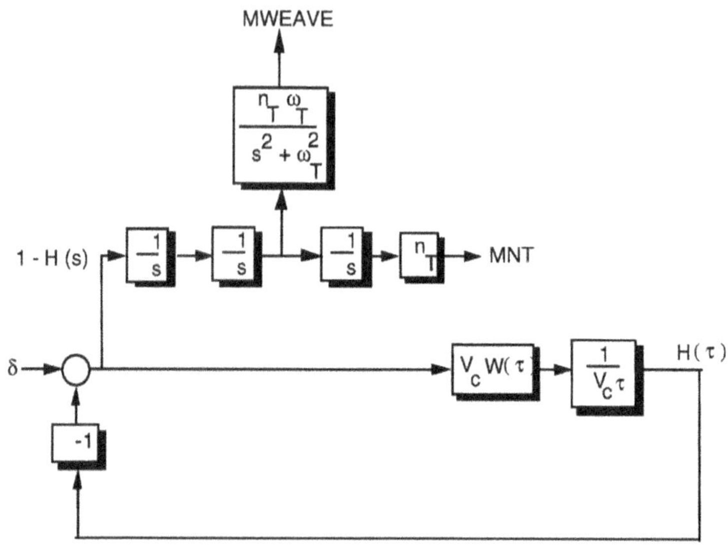

Fig. 20.6 Generalized adjoint diagram showing miss due to weaving target.

where $1 - H(s)$ was shown in Chapter 3 for the single time constant guidance system to be given by

$$1 - H(s) = \left[\frac{s}{s + \frac{1}{T}}\right]^{N'}$$

Therefore the miss due to a weaving target can be expressed as

$$\text{MWEAVE}(s) = \frac{1}{s^2} * \left[\frac{s}{s + \frac{1}{T}}\right]^{N'} * \frac{n_T \omega_T}{s^2 + \omega_T^2}$$

The miss distance in the Laplace transform domain can be evaluated directly by first doing a partial fraction expansion of the terms on the right side of the preceding equation and then taking the inverse Laplace transform to find the miss in the time domain. Some of the terms in the resultant complex expression would be transient in nature while other terms would be sinusoidal. In the steady state the transient terms would go to zero and only the sinusoidal terms would be left.

If we are only interested in the steady-state solution, much work can be saved using a simple technique from electrical engineering. We can rewrite the preceding expression as

$$\frac{\text{Miss}}{\text{weave } n_T}(s) = \frac{32.2}{s^2}\left[\frac{s}{s + \frac{1}{T}}\right]^{N'}$$

where n_T is now the maneuver magnitude in units of g and "weave n_T" reminds us that the target maneuver is sinusoidal. If a linear system has a sine wave input with frequency ω_T in units of rad/s, the output in the steady state will also be a sinusoid of the same frequency but of different magnitude and phase. From basic steady-state electrical engineering circuit analysis techniques it can be shown that the magnitude and phase of the sinusoidal output can be found by replacing s with $j\omega_T$ in the preceding transfer function and then finding the magnitude and phase of the resultant complex transfer function [5]. For example, if the effective navigation ratio is 3, the preceding transfer function becomes

$$\left.\frac{\text{Miss}}{\text{weave } n_T}\right|_{N'=3}(s) = \frac{32.2s}{(s + 1/T)^3}$$

Therefore, the complex weave miss distance transfer function can be derived from the preceding equation by substitution (that is, $s = j\omega_T$) as

$$\left.\frac{\text{Miss}}{\text{weave } n_T}\right|_{N'=3}(j\omega_T) = \frac{32.2j\omega_T}{(j\omega_T + 1/T)^3}$$

The magnitude and phase of this complex transfer function can be written by inspection as

$$\text{Magnitude}|_{N'=3} = \frac{32.2\omega_T}{(\omega_T^2 + 1/T^2)^{1.5}}$$

$$\text{Phase}|_{N'=3} = \frac{\pi}{2} - 3\tan^{-1}\omega_T T$$

Therefore the steady-state miss distance due to a weaving target can be written in the time domain as

$$\left.\frac{\text{Miss}}{\text{weave } n_T}\right|_{\substack{N'=3\\ \text{Steady-State}}} = \text{Magnitude}|_{N'=3}\sin(\omega_T t_F + \text{Phase}|_{N'=3})$$

or

$$\left.\frac{\text{Miss}}{\text{weave } n_T}\right|_{\substack{N'=3\\ \text{Steady-State}}} = \frac{32.2\omega_T}{(\omega_T^2 + \frac{1}{T^2})^{1.5}}\sin\left(\omega_T t_F + \frac{\pi}{2} - 3\tan^{-1}\omega_T T\right)$$

Figure 20.7 presents again the adjoint miss distance results as a function of flight time for the case in which the target weave frequency is 3 rad/s while the missile guidance system time constant is 1 s. Superimposed on the figure is the preceding closed-form solution for the miss distance. We can see that after an initial transient period, the closed-form steady-state miss distance solution and computer generated adjoint results are in excellent agreement thus confirming the validity of the steady-state analysis.

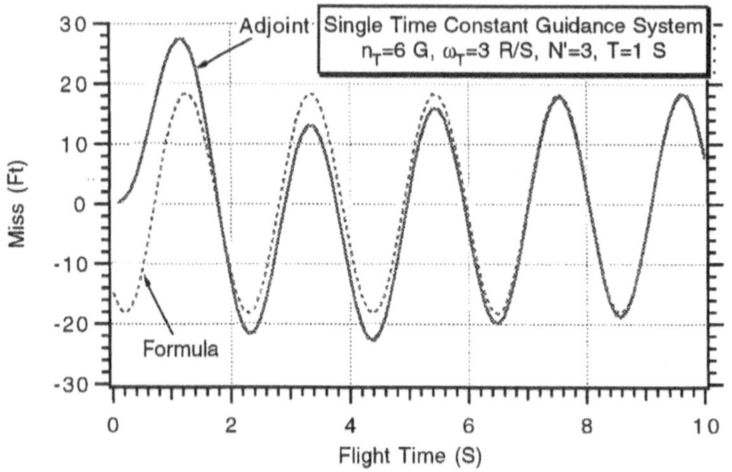

Fig. 20.7 Closed-form miss distance solution agrees with adjoint results.

We have shown mathematically and by simulation that the miss distance due to a weaving target is a sinusoidal function of the flight time. Therefore it is really a matter of luck on how large or small the miss distance will be. Of particular concern to the missile guidance system designer is the maximum or peak value of the sinusoidal miss distance function. The peak value of the miss due to a weave maneuver is simply the magnitude of the steady-state miss distance sinusoid. Therefore the peak miss due to a weave maneuver in a single time constant proportional navigation guidance system with an effective navigation ratio of 3 is given by

$$\left.\frac{\text{Peak Miss}}{\text{weave } n_T}\right|_{N'=3} = \frac{32.2\omega_T}{(\omega_T^2 + 1/T^2)^{1.5}} - \frac{32.2\omega_T T^3}{(1 + \omega_T^2 T^2)^{1.5}}$$

Dividing both sides of the equation by T^2 yields

$$\left.\frac{\text{Peak Miss}}{\text{weave } n_T T^2}\right|_{N'=3} = \frac{32.2\omega_T T}{(1 + \omega_T^2 T^2)^{1.5}}$$

If we let x be the normalized target weave frequency where

$$x = \omega_T T$$

the peak miss distance formula simplifies further to

$$\left.\frac{\text{Peak Miss}}{\text{weave } n_T T^2}\right|_{N'=3} = \frac{32.2\,x}{(1 + x^2)^{1.5}}$$

Similar expressions can be found for the peak miss distance due to a weave maneuver when the effective navigation ratios are 4 and 5 and can be shown to be

$$\left.\frac{\text{Peak Miss}}{\text{weave}\, n_T T^2}\right|_{N'=4} = \frac{32.2\, x^2}{(1+x^2)^2}$$

$$\left.\frac{\text{Peak Miss}}{\text{weave}\, n_T T^2}\right|_{N'=5} = \frac{32.2\, x^3}{(1+x^2)^{2.5}}$$

Figure 20.8 graphically displays the preceding formulas and shows how the steady-state normalized peak miss distance varies with the normalized target maneuver frequency (that is, product of the target weave frequency and the missile guidance system time constant). We can see from Fig. 20.8 that the peak miss distance is close to a maximum when the normalized target maneuver frequency is near unity. Large weave frequencies do not cause much miss distance because very little target displacement is created. On the other hand, small weave frequencies look like step target maneuvers and thus in the steady-state (large flight times) cause very little miss distance. If we were on a collision triangle with the target (with no heading error) and we coasted to the target by turning off the guidance (where $N' = 0$), the peak miss distance would simply be the peak displacement n_T/ω_T^2 caused by the weaving target. Superimposed on Fig. 20.8 is the peak displacement or induced miss distance with no missile guidance (where $N' = 0$) caused by the weaving target. We can see that for the single time constant guidance system, guiding with proportional navigation always yields a smaller miss against a weaving target than coasting without

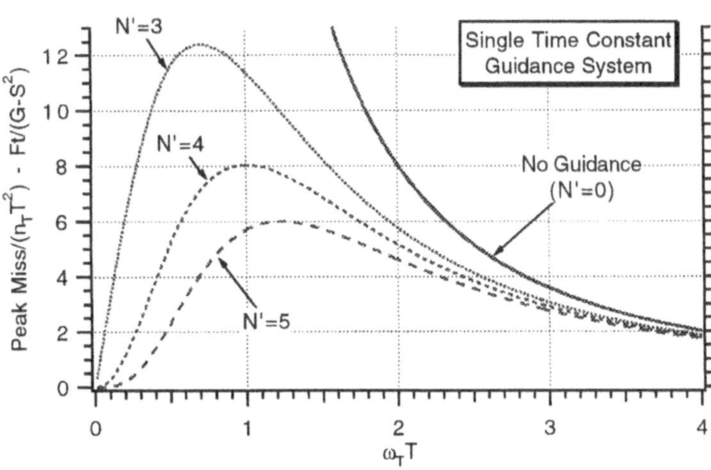

Fig. 20.8 Peak miss distance is maximum when normalized weave frequency is near unity.

guidance. *However, for large values of normalized weave frequency the miss distance with and without guidance is approximately the same!*

To illustrate the use of the normalized miss distance curves of Fig. 20.8, let us consider a numerical example in which there is a 6-*g* weaving target with a weave frequency of 2 rad/s. Assuming that the missile guidance system time constant is 1 s and effective navigation ratio is 3, we first compute the normalized weave frequency as

$$\omega_T T = 2 * 1 = 2$$

which results in a normalized miss of approximately 5.5. Therefore from the ordinate of Fig. 20.8 we can compute the peak steady-state miss distance to be

$$\text{Peak Miss} \approx 5.5 n_T T^2 = 5.5 * 6 * 1^2 = 33\,\text{ft}$$

Reducing the guidance system time constant to 0.5 s changes both the normalized weave frequency and the normalized miss. The new normalized weave frequency is

$$\omega_T T = 2 * 0.5 = 1$$

which results in an increased normalized miss of approximately 11.5. However, the new peak steady-state miss distance is reduced because the guidance system time constant has been halved or

$$\text{Peak Miss} \approx 11.5 n_T T^2 = 11.5 * 6 * 0.5^2 \approx 17\,\text{ft}$$

Keeping the guidance system constant fixed to 0.5 s but increasing the target weave frequency to 4 rad/s increases the normalized weave frequency back to 2 or

$$\omega_T T = 4 * 0.5 = 2$$

which again results in a normalized miss of approximately 5.5. The new peak steady-state miss distance becomes

$$\text{Peak Miss} \approx 5.5 n_T T^2 = 5.5 * 6 * 0.5^2 \approx 8\,\text{ft}$$

Thus we can see that both the guidance system time constant and target weave frequency are important factors in determining the peak steady-state miss distance.

HIGHER-ORDER GUIDANCE SYSTEM DYNAMICS

The single time constant guidance system model, used in the previous section, was useful because it could be used to derive closed-form solutions for the miss distance due to a weave maneuver. The single time constant guidance system miss distance formulas also suggest normalization factors for the miss distance. We have already shown in Chapters 6 and 19 that the disadvantage of the single time constant representation of a missile guidance system is that the miss distance can be seriously underestimated. We have seen that a much better and equally convenient representation of a proportional navigation missile guidance system transfer function is a canonic fifth-order binomial given by

$$\frac{n_L}{\lambda} = \frac{N' V_c s}{(1 + sT/5)^5}$$

where T is the total guidance system time constant, n_L is the achieved missile acceleration, and λ is the line-of-sight angle. As was mentioned in Chapters 6 and 19 for this generic interceptor guidance system model, one time constant represents the seeker, another represents the noise filter, and the three other time constants represent the flight-control system dynamics (aerodynamics plus autopilot). It is easy to show that with this canonic guidance system model, the overall guidance system time constant is simply the sum of the five individual time constants or T. The peak steady-state miss distance due to a weaving target for the fifth-order binomial missile homing loop can either be evaluated using the method of adjoints or the method of brute force. Because the adjoint simulation would have to be extensively modified to figure out when steady-state was reached and special logic would then have to be developed to capture the maximum miss distance, it was considered easier to use the brute force approach. In addition, the brute force approach can easily be extended to the case where there are significant nonlinearities whereas the adjoint method would no longer be valid. Listing 20.3 presents the brute force simulation based on linearized geometry, which we have already shown to be valid for the weaving target case. We can see from Listing 20.3 that acceleration saturation effects can be included by simply reducing the value of the acceleration limit XNCLIM from its near infinite value. The listing shows how the time constant is reduced when the flight times are short to ensure that we are in steady-state. The simulation is set up to generate normalized miss distance curves as a function of the normalized target weave frequency.

Figure 20.9 shows how the steady-state normalized peak miss distance due to a weave maneuver varies with the normalized target weave frequency for the fifth-order binomial guidance system. The curves in this figure are similar in shape to the ones of Fig. 20.8, but as expected, the normalized miss distances are much larger. It is interesting to note the steady-state peak miss distance is still maximum when the normalized target weave frequency is approximately unity. Superimposed on Fig. 20.9 is the zero guidance miss distance or peak displacement n_T/ω_T^2 caused by the weaving target. Surprisingly, we can see that for the fifth-order guidance system, proportional navigation only yields a smaller miss than coasting (where $N' = 0$) when the normalized weave frequency is less than 0.7 (that is, $\omega_T T < 0.7$). *In other words, for normalized weave frequencies greater than 0.7, the weaving target nullifies the effectiveness of a proportional navigation guidance system!*

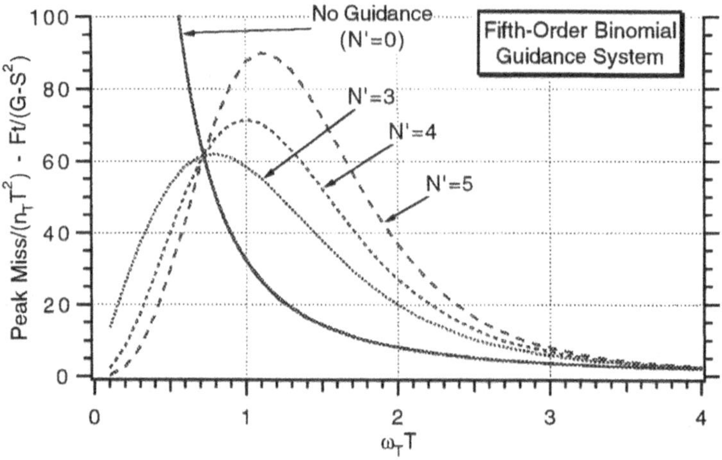

Fig. 20.9 Steady-state peak miss due to weave maneuver is much larger with fifth-order binomial guidance system.

LISTING 20.3 BRUTE FORCE SIMULATION FOR GENERATING NORMALIZED DESIGN CURVES AGAINST WEAVING TARGET

```
n=0;
VC=4000.;
XNT=32.2;
XNP=3.;
XNCLIM=99999.;
for X=.1:.1:4
        if X < .5
            W=1.;
            TAU=X/W;
        else
            W=X;
            TAU=1.;
        end
        XMWEAVEOLD=0.;
        XMWEAVEMAX=0.;
        for TF=.2:.2:20
            PHASE=0.;
            Y=0.;
            YD=0.;
            XNL=0.;
            D=0.;
            ELAMDH=0.;
```

```
X4=0.;
X5=0.;
T=0.;
H=.01;
S=0.;
while T < =(TF-1e-5)
        YOLD=Y;
        YDOLD=YD;
        XNLOLD=XNL;
        DOLD=D;
        ELAMDHOLD=ELAMDH;
        X4OLD=X4;
        X5OLD=X5;
        STEP=1;
        FLAG=0;
while STEP < =1
                if FLAG==1
                        STEP=2;
                        Y=Y+H*YD;
                        YD=YD+H*YDD;
                        XNL=XNL+H*XNLD;
                        ELAMDH=ELAMDH+H*ELAMDHD;
                        D=D+H*DD;
                        X4=X4+H*X4D;
                        X5=X5+H*X5D;
                        T=T+H;
                end
                YTDD=XNT*sin(W*T);
                TGO=TF-T+.00001;
                XLAM=Y/(VC*TGO);
                DD=5.*(XLAM-D)/TAU;
                ELAMDHD=5.*(DD-ELAMDH)/TAU;
                XNC=XNP*VC*ELAMDH;
                if XNC > XNCLIM
                        XNC=XNCLIM;
                end
                if XNC < -XNCLIM
                        XNC=-XNCLIM;
                end
                X4D=5.*(XNC-X4)/TAU;
                X5D=5.*(X4-X5)/TAU;
                XNLD=5.*(X5-XNL)/TAU;
                YDD=YTDD-XNL;
                FLAG=1;
        end
        FLAG=0;
```

```
                Y=.5*(YOLD+Y+H*YD);
                YD=.5*(YDOLD+YD+H*YDD);
                XNL=.5*(XNLOLD+XNL+H*XNLD);
                D=.5*(DOLD+D+H*DD);
                ELAMDH=.5*(ELAMDHOLD+ELAMDH+H*ELAMDHD);
                X4=.5*(X4OLD+X4+H*X4D);
                X5=.5*(X5OLD+X5+H*X5D);
        end
        XMWEAVE=Y;
        if (XMWEAVE > XMWEAVEOLD & XMWEAVE > XMWEAVEMAX & TF > 10.)
                XMWEAVEMAX=XMWEAVE;
        end
        XMWEAVEOLD=XMWEAVE;
    end
    if X < .5
        XMWEAVEMAX=XMWEAVEMAX/TAU^2;
    end
    n=n+1;
    ArrayX(n)=X;
    ArrayXMWEAVEMAX(n)=XMWEAVEMAX;
end
figure
plot(ArrayX,ArrayXMWEAVEMAX),grid
xlabel('X')
ylabel('Normalized Miss')
clc
output=[ArrayX',ArrayXMWEAVEMAX'];
save datfil.txt output  -ascii
disp 'simulation finished'
```

To illustrate the use of the normalized miss distance curves of Fig. 20.9, let us reconsider the numerical example of the previous section in which there is a 6-*g* weaving target with a weave frequency of 2 rad/s. Assuming that the missile guidance system time constant is 1 s and effective navigation ratio is 3, we first compute the normalized weave frequency as

$$\omega_T T = 2 * 1 = 2$$

which results in a normalized miss of approximately 20. Therefore we can compute the peak steady-state miss distance to be

$$\text{Peak Miss} \approx 20\, n_T T^2 = 20 * 6 * 1^2 = 120\,\text{ft}$$

which is four times larger than the miss in a single time constant guidance system (that is, 120 ft vs 33 ft). Reducing the guidance system time constant to 0.5 s changes both the normalized weave frequency and the normalized miss. The new normalized weave frequency is

$$\omega_T T = 2 * 0.5 = 1$$

which results in an increased normalized miss of approximately 60. The new peak steady-state miss distance becomes

$$\text{Peak Miss} \approx 60 n_T T^2 = 60 * 6 * 0.5^2 \approx 90\,\text{ft}$$

which is five times larger than the miss in a single time constant guidance system (that is, 90 ft vs 17 ft). Keeping the guidance system constant fixed to 0.5 s but increasing the weave frequency to 4 rad/s increases the normalized weave frequency back to 2 or

$$\omega_T T = 4 * 0.5 = 2$$

which again results in a normalized miss of approximately 20. The new peak steady-state miss distance becomes

$$\text{Peak Miss} \approx 20 n_T T^2 = 20 * 6 * 0.5^2 \approx 30\,\text{ft}$$

which is approximately four times larger than the miss induced with a single time constant guidance system (that is, 30 ft vs 8 ft). Thus we can see that the higher-order guidance system dynamics of the fifth-order binomial guidance system yield much larger miss distances due to a weaving target than does the single time constant representation of the guidance system.

ACCELERATION SATURATION

We have observed in the preceding two sections that both the guidance system dynamics and effective navigation ratio play an important role in determining the miss distance due to a weaving target. The finite acceleration capability of the interceptor is also important in determining the miss distance. Normalized miss distance curves can also be developed when missile acceleration saturation effects are considered. In this case it is hypothesized that miss distance normalization factors remain unchanged but new curves have to be developed for the nondimensional ratio of the missile to target acceleration advantage or

$$\text{Ratio} = n_{\text{LIM}}/n_T$$

where n_{LIM} is the interceptor acceleration limit.

Using the preceding ratio and the normalization factors for the steady-state peak miss due to a weaving target, we can derive normalized miss distance curves by the method of brute force with Listing 20.3. In other words, we can generate normalized miss distance curves by simulating all of the possibilities. We can then infer performance by making extrapolations from the normalized miss distance curves. Of course, detailed checks have to be made to ensure that the

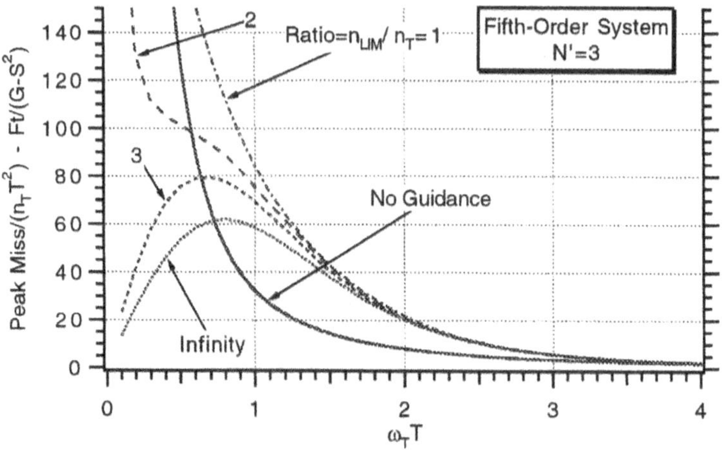

Fig. 20.10 Normalized steady-state peak miss due to weaving target and saturation effects for an effective navigation ratio of 3.

normalization factors are correct. Figures 20.10–20.12 present the normalized steady-state peak miss distances due to a weaving target for effective navigation ratios ranging from 3 to 5 respectively. As expected, we can see that less missile acceleration capability (smaller ratio) means larger miss distances. We can see that at the larger effective navigation ratios (where $N' = 5$), increasing the

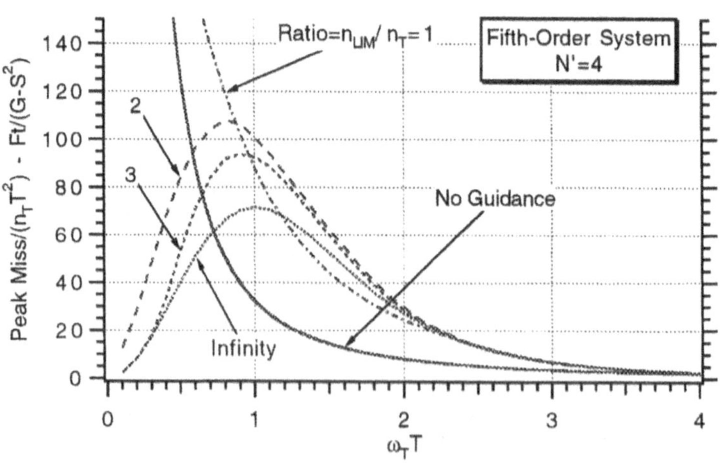

Fig. 20.11 Normalized steady-state peak miss due to weaving target and saturation effects for an effective navigation ratio of 4.

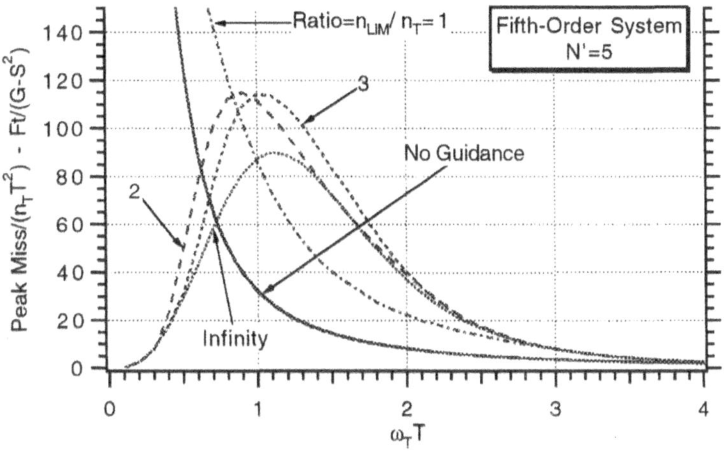

Fig. 20.12 Normalized steady-state peak miss due to weaving target and saturation effects for an effective navigation ratio of 5.

missile acceleration capability may not always reduce the miss ($\omega_T T = 2$). Under these circumstances the weaving target causes proportional navigation to be ineffective. This should not be surprising as we know that when the normalized weave frequency is greater than 0.7, doing nothing or $n_{\text{LIM}}/n_T = 0$ is optimal.

To demonstrate the use of the normalized curves of Figs. 20.10–20.12 let us again consider the same example of the previous section in which there was a 6-*g* target weave maneuver with weave frequency of 2 rad/s and a proportional navigation missile guidance system with overall time constant of 0.5 s and effective navigation ratio of 3 ($n_T = 6$, $\omega_T = 2$, $T = 0.5$, $N' = 3$). In this case the normalized weave frequency is 1 ($\omega_T T = 2^*0.5 = 1$). If the missile acceleration limit is infinite, then the ratio is infinite and we can read from Fig. 20.8 that the steady-state peak miss is 90 ft or

$$\text{Peak Miss}_g = 60\, n_T T^2 = 60 * 6 * 0.5^2 = 90\,\text{ft}$$

Reducing the acceleration limit to 18 *g* reduces the ratio to 3 or

$$\text{Ratio} = n_{\text{LIM}}/n_T = 18/6 = 3$$

For a normalized weave frequency of 1, the new steady-state peak miss increases to 105 ft or

$$\text{Peak Miss}_{18\,g} = 70 n_T T^2 = 70 * 6 * 0.5^2 = 105\,\text{ft}$$

Reducing the acceleration limit further to 12 *g* reduces the ratio to 2 or

$$\text{Ratio} = n_{\text{LIM}}/n_T = 12/6 = 2$$

For a normalized weave frequency of 1, the new steady-state peak miss increases to 113 ft or

$$\text{Peak Miss}_{12g} = 75 n_T T^2 = 75 * 6 * 0.5^2 = 113\,\text{ft}$$

Finally reducing the acceleration limit even further to 6 g reduces the ratio to 12 or

$$\text{Ratio} = n_{\text{LIM}}/n_T = 6/6 = 1$$

For a normalized weave frequency of 1, the new steady-state peak miss increases to 128 ft or

$$\text{Peak Miss}_{6g} = 85\, n_T T^2 = 75 * 6 * 0.5^2 = 128\,\text{ft}$$

In this example, if the missile had no acceleration capability or if the guidance system was turned off, the peak miss would be the maximum value of the weave displacement n_T/ω_T^2 or only 48.4 ft.

If the target weave frequency were increased to 4 rad/s and everything else remained the same, the new normalized weave frequency would be doubled to $2(\omega_T T = 4^* 0.5 = 2)$. In this case we can see from Fig. 20.10 that the miss is independent of the missile-to-target acceleration advantage and that the miss would reduce to 30 ft or

$$\text{Peak Miss}_{\infty\,\sigma, 18\,G, 12\,G, 6\,G} = 20 n_T T^2 = 20 * 6 * 0.5^2 = 30\,\text{ft}$$

Again, turning the guidance system off would make the peak miss equivalent to the maximum value of the weave displacement n_T/ω_T^2 or only 12.1 ft.

REDUCING THE TIME CONSTANT TO IMPROVE PERFORMANCE

In general, the safest and most effective method for improving the performance of a proportional navigation guidance system against the weaving target is to reduce the overall guidance system time constant and to increase the missile-to-target acceleration advantage. In aerodynamically controlled missiles, the major contributor to the guidance system time constant is usually the flight-control system time constant, and the limitation on missile acceleration capability is a function of the maximum angle of attack in which a missile can operate without causing flight catastrophe. The ability to speed up the missile flight-control system and the challenge in increasing the missile's maneuverability depends on advances in flight-control system technology. Radome effects will set a lower limit on how small the missile flight-control system time constant can be made without causing stability problems [6] and flight-control system pitch-yaw-roll cross-coupling will place an upper limit on maximum permissible angle of attack. Although a thorough discussion of the challenges in speeding up a flight-control system and safely achieving high angles of attack are beyond the scope of this text, two numerical examples will be presented in this section showing the benefits to system performance if these goals can be met.

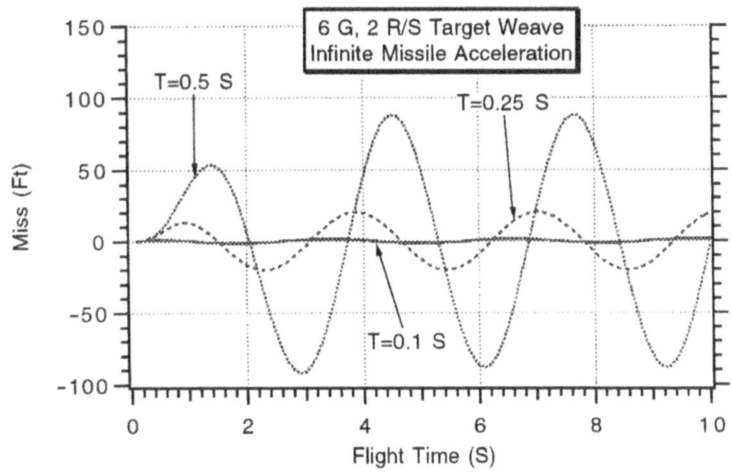

Fig. 20.13 Reducing guidance system time constant dramatically reduces miss.

To illustrate the importance of reducing the guidance system time constant, a nonnormalized, non-steady-state example was chosen in which there was a 6-*g* weaving target with a weave frequency of 2 rad/s. Figure 20.13 shows that the miss distance induced by a weaving target on a fifth-order binomial proportional navigation guidance system dramatically decreases with decreasing guidance system time constant. In fact, when the guidance system time constant is 0.1 s there is virtually no miss due to the weaving target!

If we fix the guidance system time constant at 0.1 s, we can see from Fig. 20.14 that although increasing the target weave frequency increases the miss, the miss is still small. We also could have calculated the maximum peak steady-state miss in this example from the normalized curves of Fig. 20.9. For an effective navigation ratio of 3, the curve of Fig. 20.9 is a maximum when the normalized weave frequency is 0.7. That means for this example the actual target weave frequency is 7 rad/s ($\omega_T\ T = 7^*0.1 = 0.7$). From Fig. 20.9 we can see that the actual maximum peak miss is approximately 4 ft or

$$\text{Peak Miss}|_{N'=3} = 63\, n_T T^2 = 63 * 6 * 0.1 \approx 4\,\text{ft}$$

Thus we can see that a very small miss distance can be achieved against this difficult maneuver if the guidance system time constant can be reduced to 0.1 s. Of course we can also see from Fig. 20.9 that turning the guidance system off would also yield the same miss.

Both previous examples assumed that the missile had infinite acceleration capability. Figure 20.15 shows that when the missile to target acceleration

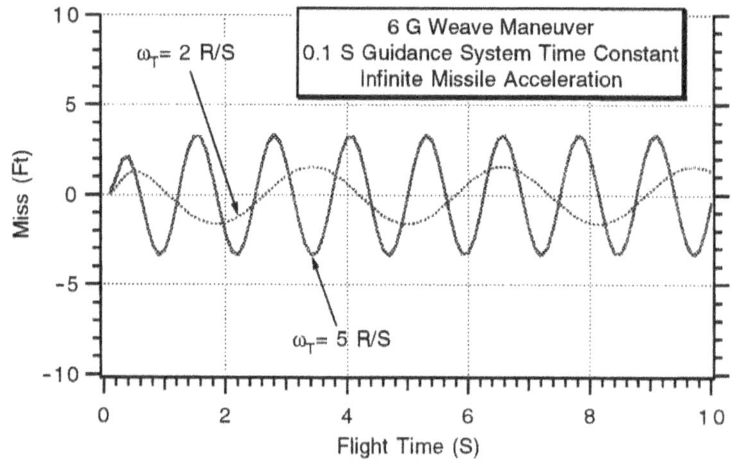

Fig. 20.14 Small guidance time constant yields good performance even when weave frequency increases.

advantage decreases from infinity to only two the miss increases. However, since the guidance system time constant is small the maximum miss distance is not large. Of course, if the missile guidance system were turned off, the miss would only be approximately 4 ft regardless of acceleration limit.

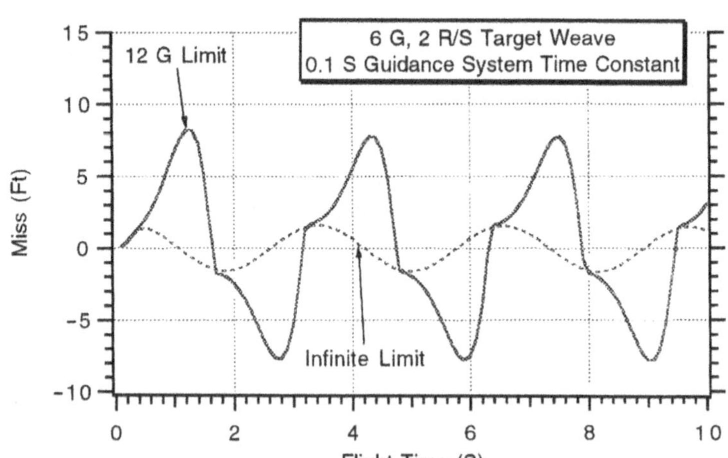

Fig. 20.15 Small miss distances can be achieved even when there is only 2 to 1 acceleration advantage.

ADVANCED GUIDANCE TECHNIQUES TO IMPROVE PERFORMANCE

Traditional guidance laws are a form of proportional navigation (PN) in which the acceleration command is proportional to the measured line-of-sight rate. As we saw in Chapter 2, proportional navigation can also be thought of as a guidance law in which the acceleration command is proportional to the zero effort miss and inversely proportional to the square of the time to go until intercept or

$$n_c = \frac{N'}{t_{\rm go}^2}[y + \dot{y}t_{\rm go}] = N'V_c\dot{\lambda}$$

The zero effort miss can be thought of as a prediction of how much the missile would miss the target by if the target continued to perform as it had done in the past and the missile issued no further acceleration commands (zero effort). We can see from the preceding equation that the zero effort miss term (bracketed quantity) in proportional navigation assumes that the target is not maneuvering. This does not mean that proportional navigation cannot hit a maneuvering target; it just means that this guidance law is not optimal in the sense that it requires the least acceleration when the target is maneuvering.

If it is known that the target is maneuvering in a step-wise fashion, we saw in Chapter 8 that the zero effort miss could be calculated exactly and a new guidance law result, known as augmented proportional navigation (APN). Mathematically this means that the zero effort miss has an acceleration term based on a constant maneuver or

$$n_c = \frac{N'}{t_{\rm go}^2}[y + \dot{y}t_{\rm go} + 0.5t_{\rm go}^2\ddot{y}_T] = N'V_c\dot{\lambda} + 0.5N'\ddot{y}_T$$

Although augmented proportional navigation can hit targets maneuvering in different ways (that is, not step maneuvers), it is only optimal for the step target maneuver in the sense that it requires the least acceleration. From an implementation point of view augmented proportional navigation has one term proportional to the line-of-sight rate and another term proportional to the target acceleration. Therefore when augmented proportional navigation is implemented, a special filter is required to provide an estimate of both the line-of-sight rate and the instantaneous value of the target acceleration.

We can also derive a special guidance law if it is known in advance that the target is weaving [7, 8]. In this case a simple model, similar to the ones of Chapter 8 for guidance law development, is shown in Fig. 20.16. The second-order shaping network shown in Fig. 20.16 represents the weaving or sinusoidal target maneuver (see Chapter 1). As was the case for other guidance laws, we are still trying to derive a guidance law that will yield zero miss distance and at the

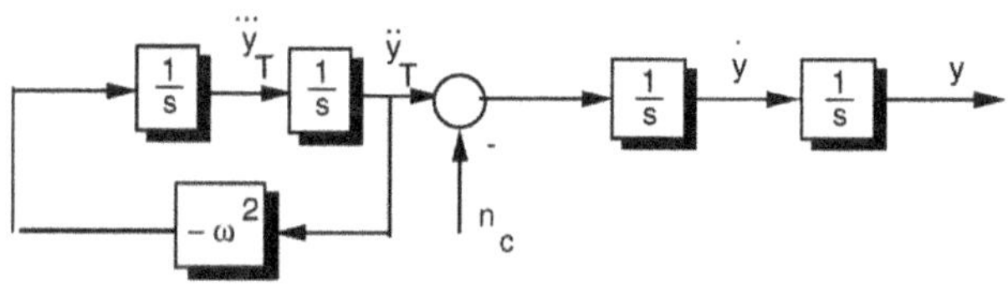

Fig. 20.16 Model for weave guidance law derivation.

same time minimize the integral of the acceleration squared or

$$y(t_F) = 0 \quad \text{subject to minimizing} \int_0^{t_F} n_c^2(t)\mathrm{d}t$$

Figure 20.16 can be expressed in state-space form as

$$\begin{bmatrix} \dot{y} \\ \ddot{y} \\ \dddot{y}_T \\ \ddddot{y}_T \end{bmatrix} = \underbrace{\begin{bmatrix} 0 & 1 & 0 & 0 \\ 0 & 0 & 1 & 0 \\ 0 & 0 & 0 & 1 \\ 0 & 0 & -\omega^2 & 0 \end{bmatrix}}_{F} \begin{bmatrix} y \\ \dot{y} \\ \ddot{y}_T \\ \dddot{y}_T \end{bmatrix} + \underbrace{\begin{bmatrix} 0 \\ -1 \\ 0 \\ 0 \end{bmatrix}}_{G} n_c$$

As was the case in Chapter 8, if a system is expressed in state-space form we can also express the final state of the system at any time according to

$$x(t_F) = \Phi(t_F - t)x(t) + \int_t^{t_F} \Phi(t_F - \lambda)G(\lambda)u(\lambda)\mathrm{d}\lambda$$

where $x(t)$ is the system state vector and $\Phi(t)$ is the fundamental matrix. Because the systems dynamics matrix F in this example is time-invariant, the fundamental matrix can be found directly from F according to

$$\Phi(t) = \mathcal{L}^{-1}[(sI - F)^{-1}]$$

yielding

$$\Phi(t) = \begin{bmatrix} 1 & t & \dfrac{(1 - \cos\omega t)}{\omega^2} & \dfrac{(\omega t - \sin\omega t)}{\omega^3} \\ 0 & 1 & \dfrac{\sin\omega t}{\omega} & \dfrac{(1 - \cos\omega t)}{\omega^2} \\ 0 & 0 & \cos\omega t & \dfrac{\sin\omega t}{\omega} \\ 0 & 0 & -\omega\sin\omega t & \cos\omega t \end{bmatrix}$$

Substitution of the Φ and G matrices into the matrix expression for the final state yields four scalar equations in this example. The first of these scalar equations is given by

$$y(t_F) = y(t) + (t_F - t)\dot{y}(t) + \frac{[1 - \cos\omega(t_F - t)]}{\omega^2}\ddot{y}_T(t) + \frac{[\omega(t_F - t) - \sin\omega(t_F - t)]}{\omega^3}\dddot{y}_T(t) - \int_t^{t_T}(t_F - \lambda)n_c(\lambda)\mathrm{d}\lambda$$

We can use the same shorthand notation of Chapter 8 and define f_1 and h_1 as

$$f_1(t_F - t) = y(t) + (t_F - t)y(t) + \frac{[1 - \cos\omega(t_F - t)]}{\omega^2}\ddot{y}_T(t) + \frac{[\omega(t_F - t) - \sin\omega(t_F - t)]}{\omega^3}\dddot{y}_T(t)$$

and

$$h_1(t_F - \lambda) = t_F - \lambda$$

so that we can say that

$$y(t_F) = f_1 - \int_t^{t_F} h_1(t_F - \lambda)n_c(\lambda)\mathrm{d}\lambda$$

In Chapter 8 we showed via the Schwartz inequality that the general form of the resultant optimal guidance law based on the preceding formulation is given by

$$n_c(\lambda) = kh_1(t_F - \lambda)$$

where

$$k = f_1(t_F - t)\Big/\int_t^{t_F} h_1^2(t_F - \lambda)\mathrm{d}\lambda$$

After some algebra we find that the optimal weave guidance law is given by

$$\begin{aligned} n_c &= \frac{3}{t_{\text{go}}^2}\left[y + \dot{y}t_{\text{go}} + \frac{1 - \cos\omega t_{\text{go}}}{\omega^2}\ddot{y}_T + \frac{\omega t_{\text{go}} - \sin\omega t_{\text{go}}}{\omega^3}\dddot{y}_T\right] \\ &= 3V_c\dot{\lambda} + \frac{3}{t_{\text{go}}^2}\left[\frac{1 - \cos\omega t_{\text{go}}}{\omega^2}\right]\ddot{y}_T + \frac{3}{t_{\text{go}}^2}\left[\frac{\omega t_{\text{go}} - \sin\omega t_{\text{go}}}{\omega^3}\right]\dddot{y}_T \end{aligned}$$

We can see that the weave guidance law is similar to other optimal laws we have derived in Chapter 8 in that guidance commands are still proportional to the zero effort miss and inversely proportional to the square of time to go until intercept. From an implementation point of view, assuming that the target weave frequency can be estimated off line and the time to go until intercept is measured, the weave guidance law consists of three terms: one term proportional to the line-of-sight rate, another term proportional to the target acceleration, and a third term proportional to target jerk.

To better understand the relationship between the new guidance law and its predecessors, let us consider the case in which the target weave frequency approaches zero. One can show using Taylor series approximations that the weave guidance law at zero frequency simplifies to

$$\lim_{\omega \to 0} n_{c_{\text{Weave}}} = \frac{3}{t_{\text{go}}^2}\left[y + \dot{y}t_{\text{go}} + \frac{t_{\text{go}}^2}{2}\ddot{y}_T + \frac{t_{\text{go}}^3}{6}\dddot{y}_T\right]$$

which is simply augmented proportional navigation with an effective navigation ratio of 3 plus an extra term to account for target jerk. The bracketed term can be recognized as the Taylor series expansion for the zero effort miss for constant target jerk.

It is important to note that the new guidance law requires additional information—an estimate of the target weave frequency, target jerk, and the time to go until intercept. A four-state Kalman filter similar to the three-state filter of Chapter 9 can be used to provide estimates of the target acceleration and jerk. The target weave frequency estimate can be derived from either the homing sensors measurements using an extended Kalman filter or from an external sensor (such as ground radar).

Because we have already demonstrated that dynamics within the guidance system will cause miss distance, the preceding guidance law must be modified to account for guidance system lags. With endoatmospheric interceptors, the flight-control system dynamics constitute the bulk of the overall guidance system time constant. If it is known that the target maneuver is sinusoidal in nature, the weave guidance can be modified to compensate for the known dynamics of the interceptor flight-control system. The compensated weave guidance law [7, 8] is very similar to the optimal guidance law derived in Chapter 8 for a single time constant guidance system, which can be expressed as

$$n_{c_{\substack{\text{Weave}\\ \text{Lag}}}} = \frac{N'}{t_{\text{go}}^2}\left[y + \dot{y}t_{\text{go}} + \frac{1 - \cos\omega t_{\text{go}}}{\omega^2}\ddot{y}_T + \frac{\omega t_{\text{go}} - \sin\omega t_{\text{go}}}{\omega^3}\dddot{y}_T - n_L T^2(e^{-x} + x - 1)\right]$$

where x is given by

$$x = \frac{t_{\text{go}}}{T}$$

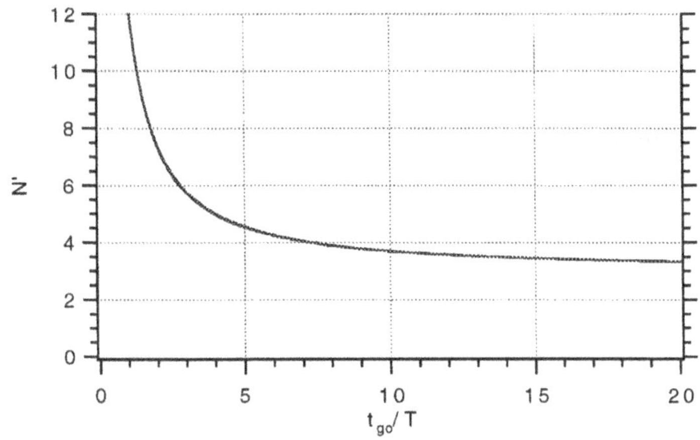

Fig. 20.17 Normalized effective navigation ratio for compensated weave guidance law.

with t_{go} being the time to go until intercept and T being defined as the approximate time constant of the flight control system. The effective navigation ratio in the compensated weave guidance law is now time-varying and is given by

$$N' = \frac{6x^2(e^{-x} - 1 + x)}{2x^3 + 3 + 6x - 6x^2 - 12xe^{-x} - 3e^{-2x}}$$

The effective navigation ratio for the compensated weave guidance law is identical to the effective navigation ratio of the optimal guidance law of Chapter 8 and

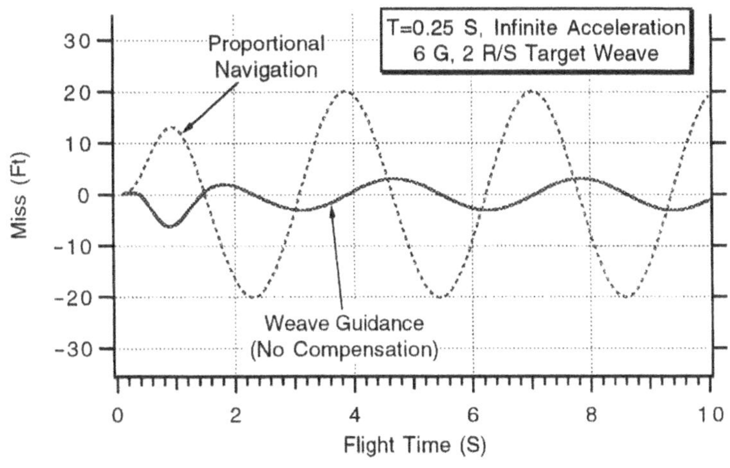

Fig. 20.18 Time lags cause miss—even for uncompensated weave guidance.

is displayed in normalized form in Fig. 20.17. We can see that at the beginning of the flight (long time to go before intercept) the effective navigation ratio is approximately constant and is approaching 3. As we get closer to intercept (small time to go), the effective navigation ratio grows considerably.

To test the effectiveness of the weave and compensated weave guidance laws, the brute force simulation of Listing 20.2 was modified to include various guidance law options (that is, before FLAG=1 statement). We can see from Listing 20.4 that the guidance system under consideration is a fifth-order binomial even though the compensated weave guidance law assumes a single time constant guidance system. In other words, the compensated weave guidance is actually suboptimal in Listing 20.4.

Figure 20.18 shows that proportional navigation can have substantial miss distances against a 6-g, 2 rad/s weaving target in a fifth-order binomial guidance system with a time constant of 0.25 s. We can also see that uncompensated weave guidance (guidance lags are not accounted for) can substantially reduce the miss.

LISTING 20.4 BRUTE FORCE SIMULATION FOR GUIDANCE LAW EVALUATION AGAINST WEAVING TARGET

```
n=0;
VC=4000.;
XNT=193.2;
XNP=3.;
XNCLIM=99999999.;
TAU=.25;
W=2.;
WH=2.;
APN=1;
for TF=.1:.1:10
      Y=0.;
      YD=0.;
      XNL=0.;
      D=0.;
      ELAMDH=0.;
      X4=0.;
      X5=0.;
      T=0.;
      H=.01;
      while T < =(TF-.0001)
         YOLD=Y;
         YDOLD=YD;
         XNLOLD=XNL;
         DOLD=D;
```

```
    ELAMDHOLD=ELAMDH;
    X4OLD=X4;
    X5OLD=X5;
    STEP=1;
    FLAG=0;
while STEP <=1
if FLAG==1
            STEP=2;
                Y=Y+H*YD;
                YD=YD+H*YDD;
                XNL=XNL+H*XNLD;
                ELAMDH=ELAMDH+H*ELAMDHD;
                D=D+H*DD;
                X4=X4+H*X4D;
                X5=X5+H*X5D;
                T=T+H;
            end
            YTDD=XNT*sin(W*T);
            YTDDD=W*XNT*cos(W*T);
            TGO=TF-T+.00001;
            XLAM=Y/(VC*TGO);
            DD=5.*(XLAM-D)/TAU;
            ELAMDHD=5.*(DD-ELAMDH)/TAU;
            if APN==1
                  XNC=XNP*VC*ELAMDH;
            elseif APN==2
                  XP=WH*TGO;
            XNC=XNP*VC*ELAMDH+XNP*YTDD*(1.-cos(XP))/XP^2+...
                  XNP*YTDDD*(XP-sin(XP))/(XP*XP*WH);
            else
                  X=TGO/TAU;
                  XP=WH*TGO;
                  TOP=6.*X*X*(exp(-X)-1.+X);
                  BOT1=2*X*X*X+3.+6.*X-6.*X*X;
                  BOT2=-12.*X*exp(-X)-3.*exp(-2.*X);
                  XNPP=TOP/(.0001+BOT1+BOT2);
    XNC=XNPP*VC*ELAMDH+XNPP*YTDD*(1.cos(XP))/XP^2....
              +XNPP*YTDDD*(XP-sin(XP))/(XP*XP*WH)-...
              XNPP*XNL*TAU*TAU*(exp(-X)+X-1.)/TGO^2;
            end
            if XNC > XNCLIM
                  XNC=XNCLIM;
            end
            if XNC < -XNCLIM
                  XNC=-XNCLIM;
            end
```

```
                    X4D=5.*(XNC-X4)/TAU;
                    X5D=5.*(X4-X5)/TAU;
                    XNLD=5.*(X5-XNL)/TAU;
                    YDD=YTDD-XNL;
                    FLAG=1;
            end
            FLAG=0;
                    Y=.5*(YOLD+Y+H*YD);
                    YD=.5*(YDOLD+YD+H*YDD);
                    XNL=.5*(XNLOLD+XNL+H*XNLD);
                    D=.5*(DOLD+D+H*DD);
                    ELAMDH=.5*(ELAMDHOLD+ELAMDH+H*ELAMDHD);
                    X4=.5*(X4OLD+X4+H*X4D);
                    X5=.5*(X5OLD+X5+H*X5D);
        end
        n=n+1;
        ArrayTF(n)=TF;
        ArrayY(n)=Y;
end
figure
plot(ArrayTF,ArrayY),grid
xlabel('Flight Time (Sec)')
ylabel('Miss (Ft)')
clc
output=[ArrayTF',ArrayY'];
save datfil.txt output -ascii
disp 'simulation finished'
```

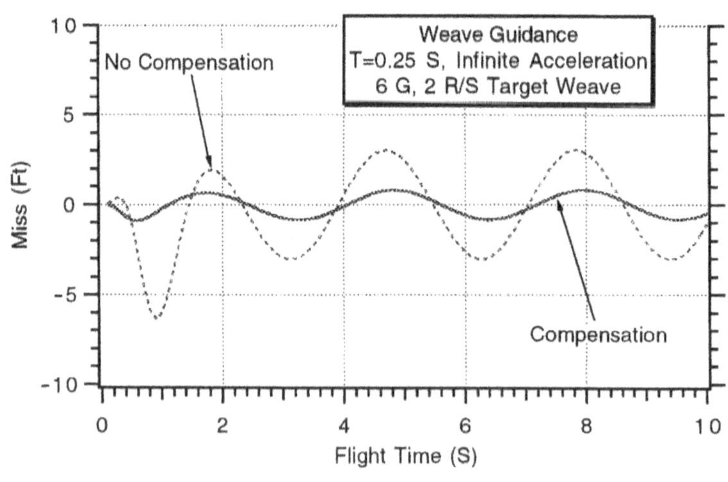

Fig. 20.19 Compensating for guidance system dynamics reduces the miss distance.

Although weave guidance has small miss distances compared to proportional navigation, there is still room for improvement when there are significant guidance system lags. Figure 20.19 shows that when there is imperfect compensation (guidance law optimal for single time constant and there are five time constants in guidance system being tested) the new guidance law reduces the miss distance even more.

SUMMARY

Normalized design curves have been presented showing how a weaving target influences the miss distance of a generic proportional navigation guidance system. This chapter demonstrated how the target weave frequency and amplitude, the missile guidance system time constant, effective navigation ratio, and acceleration capability all play an important role in determining system performance. It was demonstrated that, in general, speeding up a missile guidance system and increasing the missile-to-target acceleration advantage will help reduce the miss distance due to a weaving target. It was also shown how special guidance laws that require more information than proportional navigation can be used to improve system performance.

REFERENCES

[1] Platus, D. H., "Ballistic Re-entry Vehicle Flight Dynamics," *Journal of Guidance, Control, and Dynamics*, Vol. 5, Jan.–Feb. 1982, pp. 4–16.

[2] Chadwick, W. R., and Zarchan, P., "Interception of Spiraling Ballistic Missiles," *Proceedings of American Control Conference*, Seattle, WA, June 1995.

[3] Zarchan, P., "Proportional Navigation and Weaving Targets," *Journal of Guidance, Control, and Dynamics*, Vol. 18, No. 5, 1995, pp. 969–974.

[4] Ohlmeyer, E. J., "Root-Mean-Square Miss Distance of Proportional Navigation Missile Against Sinusoidal Target," *Journal of Guidance, Control, and Dynamics*, Vol. 19, No. 3, 1996, pp. 563–568.

[5] Javid, M., and Brenner, E., *Analysis, Transmission and Filtering of Signals*, McGraw-Hill, New York, 1963.

[6] Nesline, F. W., and Zarchan, P., "Radome Induced Miss Distance in Aerodynamically Controlled Homing Missiles," *Proceedings of AIAA Guidance and Control Conference*, AIAA, New York, Aug. 1984.

[7] Forte, I., and Shinar, J., "Can a Mixed Guidance Strategy Improve Missile Performance," *Journal of Guidance, Control, and Dynamics*, Vol. 11, Jan.–Feb. 1988, pp. 53–59.

[8] Zarchan, P., "The Challenge of Intercepting Spiraling Tactical Ballistic Missiles," *Proceedings of ION Conference*, Cambridge, MA, June 1996.

CHAPTER 21

Representing Missile Airframe with Transfer Functions

INTRODUCTION

So far we have seen that the missile guidance system time constant is extremely important in determining system performance. In aerodynamic missiles the major portion of the total guidance system time constant is governed by the time constant of the flight-control system. In Chapters 6, 19, and 20 for simplicity, we have treated the flight-control system in our fifth-order binomial guidance system as three equal time constants. In this chapter we shall show how the nonlinear force and moment equations determine how the missile airframe responds to control surface deflections. In addition, we shall show one way of linearizing the force and moment equations so that we can find airframe transfer functions. The transfer function representation of the airframe is the necessary first step in the design of the missile autopilot.

In practice both proprietary computer codes and wind tunnel-generated aerodynamic data are used to derive missile airframe transfer functions. However, it is interesting to note that wind tunnel data was not available when the original missiles were designed. Simplified aerodynamic shapes were chosen for the initial designs so that linear theory could be used to derive the necessary airframe transfer functions.

The nonlinear force and moment equations in this chapter are first expressed in terms of physical missile parameters (weight, length, etc.) rather than in terms of wind tunnel-generated functions. Although this is an approximation to reality, it will enable the reader to get a clearer understanding of how the geometry of the missile influences the force and moment equations. Next the force and moment equations are linearized, and it is shown how the airframe can then be represented by various transfer functions. Finally, using numerical examples, this chapter shows that the transfer function representation of the airframe is an excellent approximation to the nonlinear equations for angles of attack less than 20 to 30 deg.

FORCE AND MOMENT EQUATIONS

A typical tail-controlled, aerodynamic winged missile is shown in Fig. 21.1. This type of missile generates lift by moving control surfaces. In this endoatmospheric missile the movable control surface or tail can be deflected about the hinge line through a fin angle δ in order to help the missile develop an angle of attack α.

The fixed surface or wing plus the missile body help the missile develop additional acceleration. The normal force acts through the center of pressure (CP). We can express the normal force equation as

$$F_N = QS_{\text{ref}}C_N$$

where C_N is the normal force coefficient, Q is the dynamic pressure, and S_{ref} is the reference area. The dynamic pressure and reference area are given by

$$Q = 0.5\rho V_M^2$$

$$S_{\text{ref}} = \frac{\pi d^2}{4}$$

where ρ is the air density in units of slug/ft^3, V_M is the missile velocity in units of ft/s, and d is the missile diameter in units of ft.

The total force acting on the missile body consists of component forces on the body, wing tail, and nose as shown in Fig. 21.2. In this simplified diagram all interference effects are neglected, and the total force is simply the sum of the individual forces. Each of the component forces act through their own centers of pressure. The centers of pressure for the body, wing, and nose are denoted X_{CPB}, X_{CPW}, and X_{CPN}, respectively. The force acting on the tail acts through the hinge

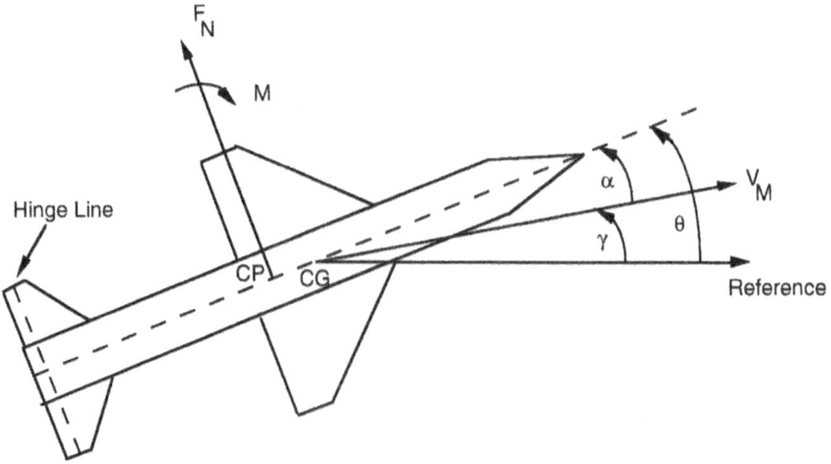

Fig. 21.1 Tail-controlled missile.

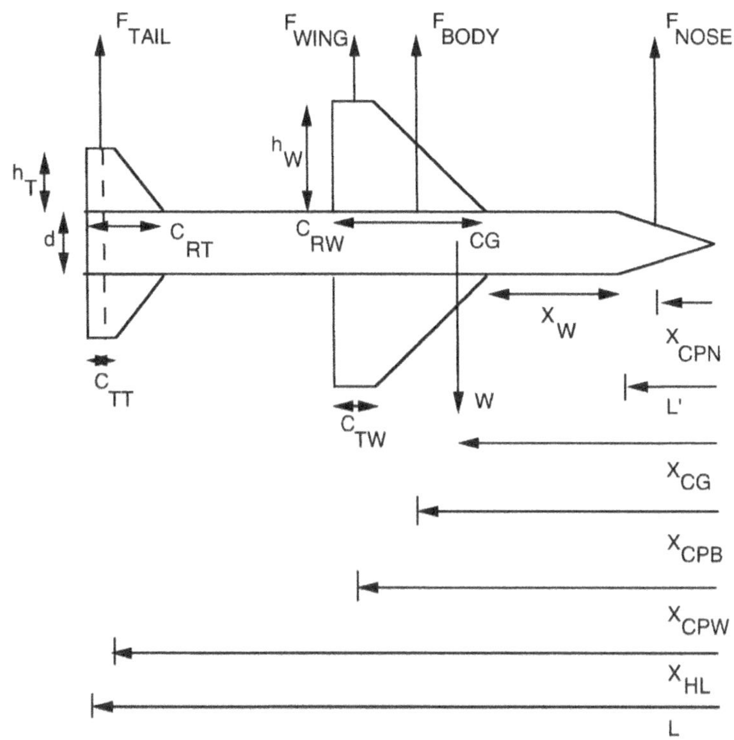

Fig. 21.2 Forces on a tail-controlled missile.

line X_{HL}. Neglecting interference effects, the normal force coefficient can be approximated as [1, 2]

$$C_N = \underbrace{2\alpha}_{\text{Nose}} + \underbrace{\frac{1.5S_{\mathrm{PLAN}}\alpha^2}{S_{\mathrm{ref}}}}_{\text{Body}} + \underbrace{\frac{8S_W\alpha}{\beta S_{\mathrm{ref}}}}_{\text{Wing}} + \underbrace{\frac{8S_T(\alpha+\delta)}{\beta S_{\mathrm{ref}}}}_{\text{Tail}}$$

where α is the angle of attack, δ is the control surface deflection, and S_W, S_T, and S_{PLAN} are per panel wing, tail, and planform areas, respectively. Because the wing and tail are approximated by trapezoids in Fig. 21.2, their panel areas are given by

$$S_W = 0.5h_W(C_{TW} + C_{RW})$$
$$S_T = 0.5h_T(C_{TT} + C_{RT})$$

where the subscript T denotes a tip chord and the subscript R denotes a root chord. For a cylindrical missile body with a parabolic nose (radome), the planform area can be approximated as

$$S_{\mathrm{PLAN}} = (L - L')d + 0.67L'd \approx Ld$$

where L is the missile length and L' is the radome length. The parameter β in the normal force coefficient equation is a normalized speed, for supersonic travel it is

$$\beta = \sqrt{\mathrm{Mach}^2 - 1}$$

The missile Mach number is simply the missile speed divided by the speed of sound. Although the speed of sound is altitude dependent, we shall assume for simplicity that the speed of sound is always 1000 ft/s. Multiplying the force by its moment arm yields the developed moment or

$$M = F_N * \text{Moment Arm}$$

Therefore the moment coefficient can be approximated as

$$C_M = 2\alpha\frac{(X_{\mathrm{CG}} - X_{\mathrm{CPN}})}{d} + \frac{1.5S_{\mathrm{PLAN}}\alpha^2}{S_{\mathrm{ref}}}\frac{(X_{\mathrm{CG}} - X_{\mathrm{CPB}})}{d} + \frac{8S_W\alpha}{\beta S_{\mathrm{ref}}}\frac{(X_{\mathrm{CG}} - X_{\mathrm{CPW}})}{d} + \frac{8S_T(\alpha+\delta)}{\beta S_{\mathrm{ref}}}\frac{(X_{\mathrm{CG}} - X_{\mathrm{HL}})}{d}$$

where X_{CG} is the distance from the nose to the missile center of gravity and X_{CPN}, X_{CPB}, and X_{CPW} are the distances from the nose to the centers of pressure for the nose, body, and wing, respectively. The preceding expression assumes a tail-controlled missile and X_{HL} is the distance from the nose to the missile hinge line. The nose, body, and wing centers of pressure (referenced with respect to the nose) can be analytically approximated because of their geometrical shape and can be shown to be [3]

$$X_{\mathrm{CPN}} = 0.67L'$$

$$X_{\mathrm{CPW}} = L' + X_W + 0.7C_{RW} - 0.2C_{TW}$$

$$X_{\mathrm{CPB}} = \frac{0.67A_NL' + A_B[L' + 0.5(L - L')]}{A_N + A_B}$$

where X_W is the distance from the wing to the radome tangency point as shown in Fig. 21.2. The nose and body areas are given by

$$A_N = 0.67L'd$$

$$A_B = (L - L')d$$

Again for simplicity we will assume that the missile center of gravity is approximately in the center of the missile or

$$X_{\mathrm{CG}} = 0.5L$$

The total moment M on the missile can be expressed in terms of the moment coefficient according to

$$M = QS_{\mathrm{ref}}dC_M$$

We now have enough information to express the normal and angular accelerations acting on the missile in terms of the geometry of the missile configuration. The acceleration normal to the missile body can be expressed in terms of the normal force according to

$$n_B = \frac{F_N g}{W} = \frac{gQS_{\mathrm{ref}}C_M}{W}$$

where W is the missile weight. The angular acceleration acting on the missile can be expressed in terms of the moment according to

$$\ddot{\theta} = \frac{M}{I_{yy}} = \frac{QS_{\mathrm{ref}}dC_M}{I_{yy}}$$

where I_{yy} is the missile moment of inertia. If the missile body is approximated as a cylinder, the formula for the moment of inertia is given by [4]

$$I_{yy} = \frac{W[3(0.5d)^2 + L^2]}{12g} \approx \frac{WL^2}{12g}$$

Finally from Fig. 21.1 we can see that the angle of attack can be expressed in terms of the missile body and flight path angles according to

$$\alpha = \theta - \gamma$$

Taking derivatives of both sides of the equation and recognizing that the flight path rate can also be expressed in terms of the missile acceleration yields

$$\dot{\alpha} = \dot{\theta} - \dot{\gamma} = \dot{\theta} - (n_L/V_M)$$

If we assume that the angle of attack is small, the missile acceleration perpendicular to the velocity n_L is approximately the same as the missile acceleration perpendicular to the body n_B. Therefore the derivative of the angle of attack can be expressed as

$$\dot{\alpha} = \dot{\theta} - \dot{\gamma} = \dot{\theta} - (n_B/V_M)$$

In the next three chapters we shall assume that the acceleration perpendicular to the body is approximately the same as the missile acceleration perpendicular to the velocity vector.

AIRFRAME SIMULATION

We now have enough information to simulate the force and moment equations and thus find out how the missile airframe responds when the tail is deflected. Consider the hypothetical 1000-lb tail-controlled missile, shown in Fig. 21.3, similar to the one first considered by Jerger [2]. In this example both the wing and tail are triangular in shape. The locations of the wing, hinge line, and center of gravity along with all the other airframe dimensions are indicated in Fig. 21.3.

A simulation was written utilizing the preceding nonlinear force and moment equations for the hypothetical missile of Fig. 21.3. The angle of attack rate and missile angular acceleration differential equations, which involve the force and moment equations, are integrated and the simulation appears in Listing 21.1. We can see from the inputs at the beginning of the simulation that the airframe inputs and data of Fig. 21.3 are consistent. The simulation assumes that the speed of sound is always equal to 1000 ft/s. As was mentioned previously, this is an approximation since the speed of sound is altitude dependent and can be as much as 10% different than the number used in the simulation. We can also see that we are using the exponential approximation to the atmosphere that was first introduced in Chapter 10. The airframe differential equations for angle of attack rate and angular body acceleration appear before the FLAG=1 statement.

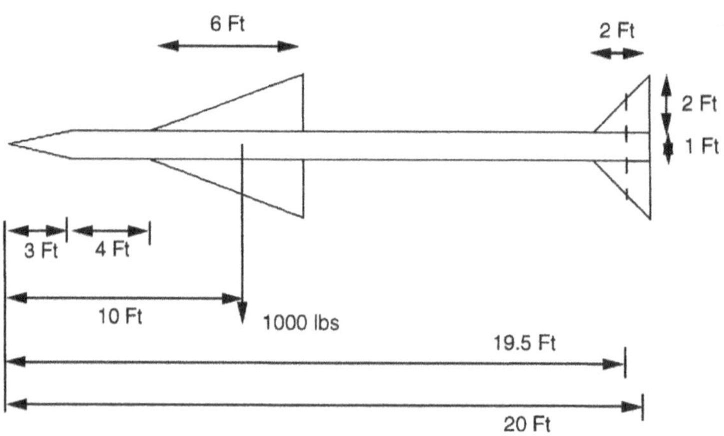

Fig. 21.3 Hypothetical missile.

LISTING 21.1 AIRFRAME SIMULATION

```
n=0;
VM=3000.;
DEL=5./57.3;
ALT=0.;
A=1000.;
DIAM=1.;
FR=3.;
XL=20.;
CTW=0.;
CRW=6.;
HW=2.;
CTT=0.;
CRT=2.;
HT=2.;
XN=4.;
XCG=10.;
XHL=19.5;
WGT=1000.;
if ALT<=30000.
        RHO=.002378*exp(-ALT/30000.);
else
        RHO=.0034*exp(-ALT/22000.);
end
SWING=.5*HW*(CTW+CRW);
STAIL=.5*HT*(CTT+CRT);
SREF=3.1416*DIAM*DIAM/4.;
XLP=FR*DIAM;
SPLAN=(XL-XLP)*DIAM+1.33*XLP*DIAM/2.;
XCPN=2*XLP/3;
AN=.67*XLP*DIAM;
AB=(XL-XLP)*DIAM;
XCPB=(.67*AN*XLP+AB*(XLP+.5*(XL-XLP)))/(AN+AB);
XCPW=XLP+XN+.7*CRW-.2*CTW;
XMACH=VM/A;
XIYY=WGT*(3*((DIAM/2)^2)+XL*XL)/(12*32.2);
TMP1=(XCG-XCPW)/DIAM;
TMP2=(XCG-XHL)/DIAM;
TMP3=(XCG-XCPB)/DIAM;
TMP4=(XCG-XCPN)/DIAM;
B=sqrt(XMACH^2-1);
Q=.5*RHO*VM*VM;
THD=0;
ALF=0;
T=0;
```

```
H=.0025;
S=0.;
while T<1.99999
        THDOLD=THD;
        ALFOLD=ALF;
        STEP=1;
        FLAG=0;
        while STEP<=1
           if FLAG==1
                 STEP=2;
                 THD=THD+H*THDD;
                 ALF=ALF+H*ALFD;
                 T=T+H;
           end
           CN=2*ALF+1.5*SPLAN*ALF*ALF/SREF+8*SWING*ALF/(B*SREF)+...
           8*STAIL*(ALF+DEL)/(B*SREF);
           CM=2*ALF*TMP4+1.5*SPLAN*ALF*ALF*TMP3/SREF+...
                 8*SWING*ALF*TMP1/(B*SREF)...
           +8*STAIL*(ALF+DEL)*TMP2/(B*SREF);
           THDD=Q*SREF*DIAM*CM/XIYY;
           XNL=32.2*Q*SREF*CN/WGT;
           ALFD=THD-XNL/VM;
           FLAG=1;
        end
        FLAG=0;
        THD=.5*(THDOLD+THD+H*THDD);
        ALF=.5*(ALFOLD+ALF+H*ALFD);
        S=S+H;
        if S>=.0099999
           S=0.;
           n=n+1;
           ArrayT(n)=T;
           ArrayXNLG(n)=XNL/32.2;
           ArrayALFDEG(n)=ALF*57.3;
        end
end
figure
plot(ArrayT,ArrayXNLG),grid
xlabel('Time (Sec)')
ylabel('Missile Acceleration (G)')
figure
plot(ArrayT,ArrayALFDEG),grid
xlabel('Time (Sec)')
ylabel('Angle of Attack (Deg)')
clc
output=[ArrayT',ArrayXNLG',ArrayALFDEG'];
```

```
save datfil.txt output -ascii
disp 'simulation finished'
```

A case was run for the hypothetical missile of Fig. 21.3 in which the missile fin was deflected 5 deg when the missile was at sea level traveling at 3000 ft/s. We can see from Fig. 21.4 that a 5-deg fin deflection in the positive direction causes the missile to build up to a negative angle of attack. The angle of attack initially approaches −8 deg, oscillates, and finally approaches a steady-state or trim value of approximately −5 deg after several seconds. The decaying oscillations indicate that the missile airframe has low damping. The transient values of the angle of attack due to the fin deflection are important to the designer because excessive values could cause flight catastrophe.

The buildup in angle of attack enables the missile to accelerate. We can see from Fig. 21.5 that the steady-state acceleration due to a 5-deg fin deflection is approximately 13 g at this flight condition. Again note the oscillatory nature of the achieved acceleration due to a fixed fin deflection.

The simulation was also run when the altitude was increased to 50 kft (ALT= 50000). We can see from Figs. 21.6 and 21.7 that at higher altitudes the missile will pull more angle of attack and have less acceleration available for a given fin deflection. In addition, the airframe natural frequency decreases with increasing altitude. Because the amount of fin travel permitted is limited, the simulation demonstrates that an aerodynamic missile will have less acceleration available at the higher altitudes.

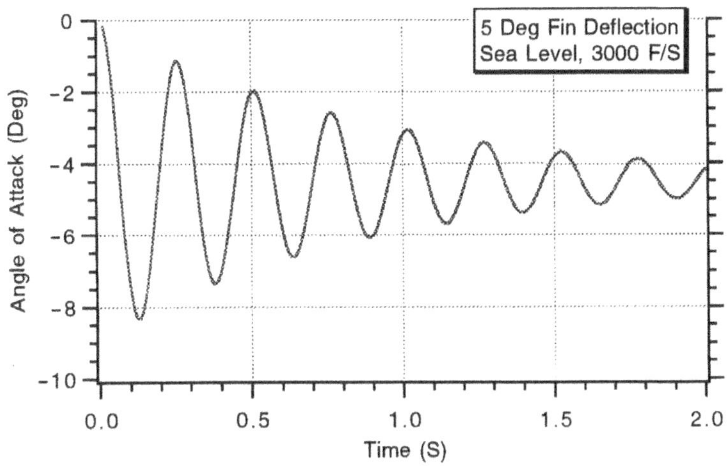

Fig. 21.4 A 5-deg fin deflection results in approximately −5-deg angle of attack at sea level.

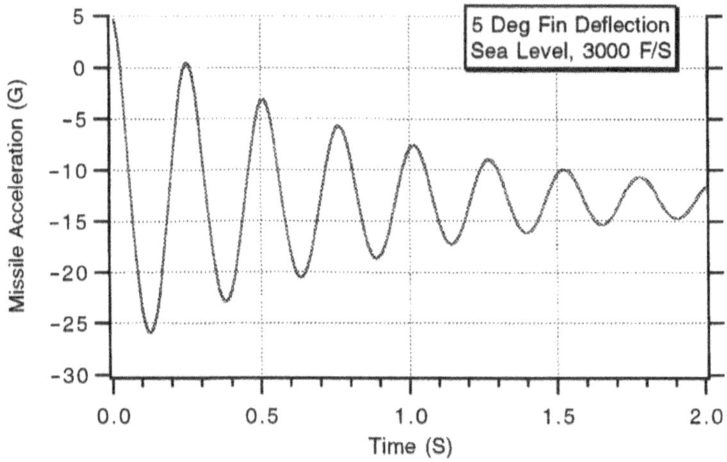

Fig. 21.5 A 5-deg fin deflection results in approximately 13 *g* of acceleration at sea level.

LINEARIZATION OF THE AIRFRAME

An examination of the force and moment coefficients reveals that if we assume that the missile speed and altitude are constant, the equations are mostly linear except for the angle of attack squared term in each of the equations. For the constant speed, constant altitude condition we can linearize by assuming that each

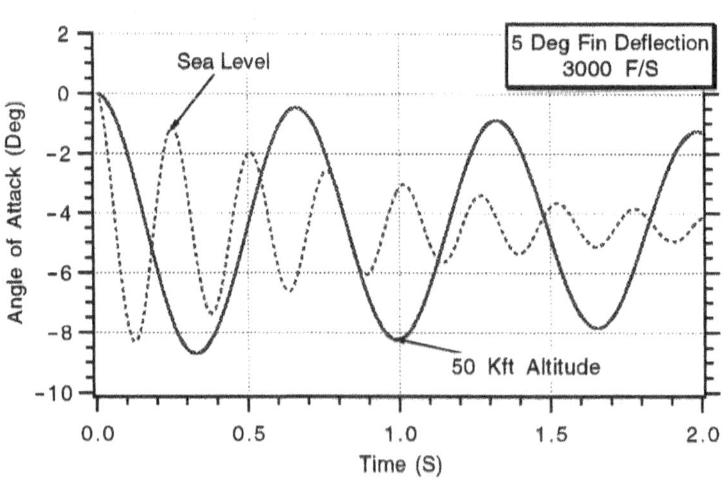

Fig. 21.6 More angle of attack is required at higher altitudes for fixed fin deflection.

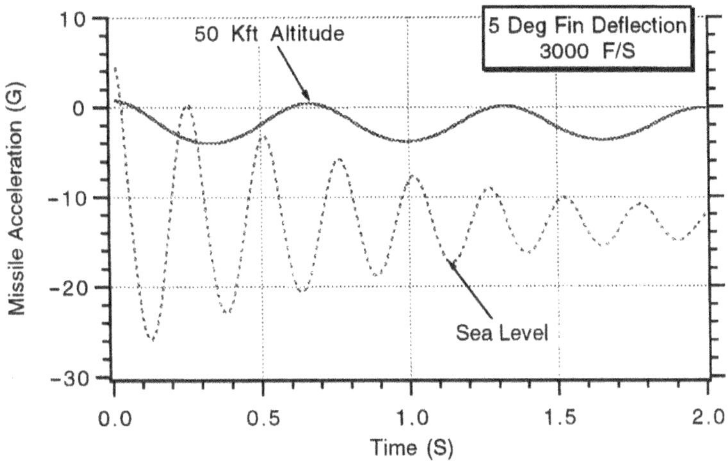

Fig. 21.7 Acceleration capability diminishes at higher altitudes for fixed fin deflection.

equation is linear in angle of attack and fin deflection. This means that the normal force coefficient is approximated as

$$C_N = f(\alpha, \delta) \approx C_{N\alpha}\alpha + C_{N\delta}\delta$$

One method of finding $C_{N\alpha}$ and $C_{N\delta}$ is to simply divide the angle of attack terms in C_N by α to get $C_{N\alpha}$ and then divide the fin deflection term by δ to get $C_{N\delta}$ yielding

$$C_{N\alpha} = 2 + \frac{1.5S_{\text{PLAN}}\alpha}{S_{\text{ref}}} + \frac{8S_W}{\beta S_{\text{ref}}} + \frac{8S_T}{\beta S_{\text{ref}}}$$

$$C_{N\delta} = \frac{8S_T}{\beta S_{\text{ref}}}$$

Note that $C_{N\alpha}$ depends on the angle of attack.

Because we are assuming that the acceleration normal to the body is nearly equal to the missile acceleration perpendicular to the velocity vector (the angle of attack is small), we can express the missile turning rate in terms of $C_{N\alpha}$ and $C_{N\delta}$ or

$$\dot{\gamma} \approx \frac{n_L}{V_M} = \frac{gQS_{\text{ref}}}{WV_M}[C_{N\alpha}\alpha + C_{N\delta}\delta] = -Z_\alpha\alpha - Z_\delta\delta$$

where Z_α and Z_δ are defined as

$$Z_\alpha = \frac{-gQS_{\text{ref}}C_{N\alpha}}{WV_M}$$

$$Z_\delta = \frac{-gQS_{\text{ref}}C_{N\delta}}{WV_M}$$

Therefore we have expressed the missile turning rate or acceleration as a linear function of angle of attack or fin deflection. In a similar way the moment coefficient can be linearized as

$$C_M = f(\alpha, \delta) \approx C_{M\alpha}\alpha + C_{M\delta}\delta$$

As before, we can find $C_{M\alpha}$ and $C_{M\delta}$ by simply dividing the angle of attack terms of C_M first by α and then dividing the fin deflection term by δ yielding

$$C_{M\alpha} = \frac{2(X_{\text{CG}} - X_{\text{CPN}})}{d} + \frac{1.5S_{\text{PLAN}}\alpha}{S_{\text{ref}}}\frac{(X_{\text{CG}} - X_{\text{CPB}})}{d}$$
$$+ \frac{8S_W}{\beta S_{\text{ref}}}\frac{(X_{\text{CG}} - X_{\text{CPW}})}{d} + \frac{8S_T}{\beta S_{\text{ref}}}\frac{(X_{\text{CG}} - X_{\text{HL}})}{d}$$

$$C_{M\delta} = \frac{8S_T}{\beta S_{\text{ref}}}\frac{(X_{\text{CG}} - X_{\text{HL}})}{d}$$

Note that $C_{M\alpha}$ is not a constant for a given speed and altitude but depends on the angle of attack.

We can now express the linearized missile angular acceleration as

$$\ddot{\theta} = \frac{M}{I_{yy}} = \frac{QS_{\text{ref}}d}{I_{yy}}[C_{M\alpha}\alpha + C_{M\delta}\delta] = M_\alpha\alpha + M_\delta\delta$$

where M_α and M_δ are defined as

$$M_\alpha = \frac{QS_{\text{ref}}dC_{M\alpha}}{I_{yy}}$$

$$M_\delta = \frac{QS_{\text{ref}}dC_{M\delta}}{I_{yy}}$$

Because the derivative of the angle of attack is given by

$$\dot{\alpha} = \dot{\theta} - \dot{\gamma}$$

we can say that

$$\dot{\alpha} = \dot{\theta} + Z_\alpha\alpha + Z_\delta\delta$$

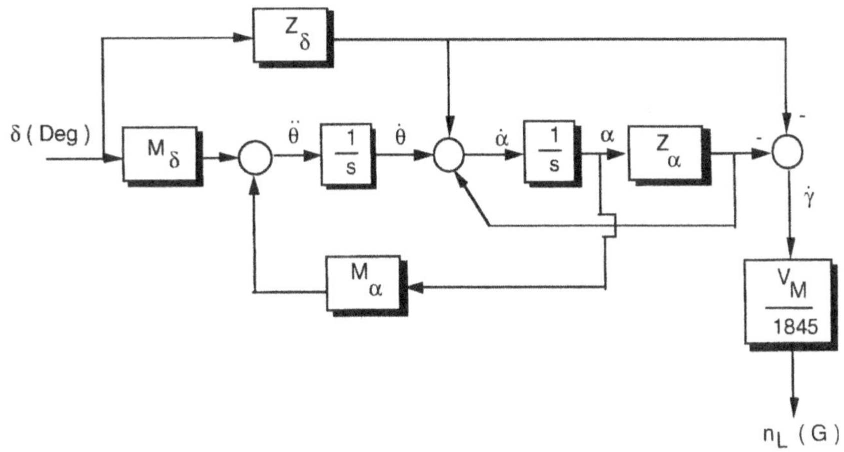

Fig. 21.8 Linearized airframe.

As mentioned previously, M_α and Z_α are not constants in our linearized model but vary with angle of attack. These aerodynamic parameters are usually evaluated at a trim angle of attack. The vehicle is considered to be at trim when the moment is zero ($C_M = 0$). At the trim condition one solves for the angle of attack and uses that value to evaluate M_α and Z_α.

The linearized airframe equations can also be represented in block diagram form as shown in Fig. 21.8. The two integrators shown in the block diagram indicate that the airframe can be considered to be a second-order system. It is important to note that this diagram assumes that the input fin deflection δ is in units of degrees and that the output acceleration n_L is in units of gees. All internal angles and rates are either in units of degrees or degrees per second.

Often it is convenient to have a transfer function representation of the airframe. Strictly speaking, the transfer function is only valid when the missile is at a fixed speed, altitude, and trim angle of attack. After some algebra we can find the transfer function relating the achieved missile acceleration to the fin deflection from Fig. 21.8 as

$$\frac{n_L}{\delta} = \frac{-V_M[M_\alpha Z_\delta - Z_\alpha M_\delta]}{1845 M_\alpha}\left[1 - \frac{Z_\delta s^2}{M_\alpha Z_\delta - Z_\alpha M_\delta}\right] \bigg/ \left(1 + \frac{Z_\alpha}{M_\alpha}s - \frac{s^2}{M_\alpha}\right)$$

The preceding transfer function can be simplified to

$$\frac{n_L}{\delta} = K_1\left(1 - \frac{s^2}{\omega_z^2}\right) \bigg/ \left(1 + \frac{2\zeta_{\mathrm{AF}}}{\omega_{\mathrm{AF}}}s + \frac{s^2}{\omega_{\mathrm{AF}}^2}\right)$$

where

$$K_1 = \frac{-V_M[M_\alpha Z_\delta - Z_\alpha M_\delta]}{1845 M_\alpha}$$

and

$$\omega_z = \frac{M_\alpha Z_\delta - Z_\alpha M_\delta}{Z_\delta}$$

$$\omega_{\text{AF}} = \sqrt{-M_\alpha}$$

$$\zeta_{\text{AF}} = \frac{Z_\alpha \omega_{\text{AF}}}{2M_\alpha}$$

Similarly, the transfer function from missile pitch rate to fin deflection can also be written from Fig. 21.8 as

$$\frac{\dot{\theta}}{\delta} = \frac{-[M_\alpha Z_\delta - Z_\alpha M_\delta]}{M_\alpha}\left[1 + \frac{M_\delta s}{M_\alpha Z_\delta - Z_\alpha M_\delta}\right] \Big/ \left(1 + \frac{Z_\alpha}{M_\alpha}s - \frac{s^2}{M_\alpha}\right)$$

which simplifies to

$$\frac{\dot{\theta}}{\delta} = K_3(1 + T_\alpha s) \Big/ \left(1 + \frac{2\zeta_{\text{AF}}}{\omega_{\text{AF}}}s + \frac{s^2}{\omega_{\text{AF}}^2}\right)$$

where

$$K_3 = \frac{-[M_\alpha Z_\delta - Z_\alpha M_\delta]}{M_\alpha} = \frac{1845 K_1}{V_M}$$

$$T_\alpha = \frac{M_\delta}{M_\alpha Z_\delta - Z_\alpha M_\delta}$$

If an accelerometer is used in a flight-control system to measure the achieved missile acceleration, it will probably not be located at the center of gravity. Its measurement of the actual acceleration will be corrupted by the body angular acceleration according to

$$n_A = n_L + \frac{(X_{\text{CG}} - X_{\text{ACC}})\ddot{\theta}}{1845}$$

where X_{ACC} is the accelerometer location with respect to the nose, X_{CG} is the center of gravity of the missile, and n_A is acceleration measured by the accelerometer. However we shall neglect this effect and assume that the measured and achieved accelerations are identical in order to simplify the ensuing analysis.

NUMERICAL EXAMPLE

To test the accuracy of the airframe linearization, the example presented at the beginning of this chapter was repeated. However, before we begin, we must first have a method of calculating the trim angle of attack for a given fin deflection (fin deflection is 5 deg in this example). The moment coefficient can be written as

$$C_M = 2\alpha\frac{(X_{CG} - X_{CPN})}{d} + \frac{1.5S_{PLAN}\alpha^2}{S_{ref}}\frac{(X_{CG} - X_{CPB})}{d} + \frac{8S_W\alpha}{\beta S_{ref}}\frac{(X_{CG} - X_{CPW})}{d} + \frac{8S_T(\alpha + \delta)}{\beta S_{ref}}\frac{(X_{CG} - X_{HL})}{d}$$

or in shorthand notation as

$$C_M = y_1\alpha + y_2\alpha^2 + y_3\delta$$

where

$$y_1 = \frac{2(X_{CG} - X_{CPN})}{d} + \frac{8S_W}{\beta S_{ref}}\frac{(X_{CG} - X_{CPW})}{d} + \frac{8S_T}{\beta S_{ref}}\frac{(X_{CG} - X_{HL})}{d}$$

$$y_2 = \frac{1.5S_{PLAN}\alpha}{S_{ref}}\frac{(X_{CG} - X_{CPB})}{d}$$

$$y_3 = \frac{8S_T}{\beta S_{ref}}\frac{(X_{CG} - X_{HL})}{d}$$

At trim the moment coefficient is zero. Therefore, for a given fin deflection δ_{NOM} we get the equation for the trim angle of attack α_{TR} to be

$$0 = y_1\alpha_{TR} + y_2\alpha_{TR}^2 + y_3\delta_{NOM}$$

We can use the quadratic formula to solve for the trim angle of attack. After eliminating the unrealistic root we get

$$\alpha_{TR} = \frac{-y_1 - \sqrt{y_1^2 - 4y_2y_3\delta_{NOM}}}{2y_2}$$

and can now evaluate $C_{N\alpha}$ and $C_{M\alpha}$ as

$$C_{N\alpha} = 2 + \frac{1.5_{PLAN}\alpha_{TR}}{S_{ref}} + \frac{8S_W}{\beta S_{ref}} + \frac{8S_T}{\beta S_{ref}}$$

$$C_{M\alpha} = \frac{2(X_{CG} - X_{CPN})}{d} + \frac{1.5S_{PLAN}\alpha_{TR}}{S_{ref}}\frac{(X_{CG} - X_{CPN})}{d} + \frac{8S_W}{\beta S_{ref}}\frac{(X_{CG} - X_{CPW})}{d} + \frac{8S_T}{\beta S_{ref}}\frac{(X_{CG} - X_{HL})}{d}$$

To write a simulation involving the linearized airframe, we must convert the transfer functions to differential equations as was done in Chapter 1. The transfer function relating missile acceleration to fin deflection was already shown to be

$$\frac{n_L}{\delta} = K_1\left(1 - \frac{s^2}{\omega_z^2}\right) \Big/ \left(1 + \frac{2\zeta_{AF}}{\omega_{AF}}s + \frac{s^2}{\omega_{AF}^2}\right)$$

Using the chain rule from calculus we can say that

$$\frac{n_L}{\delta} = \frac{e}{\delta} * \frac{n_L}{e}$$

Therefore, as was done in Chapter 1, we can split the missile acceleration transfer function and get two equivalent transfer functions or

$$\frac{e}{\delta} = 1 \Big/ \left(1 + \frac{2\zeta_{AF}}{\omega_{AF}}s + \frac{s^2}{\omega_{AF}^2}\right) \quad \text{and} \quad \frac{n_L}{e} = K_1\left(1 - \frac{s^2}{\omega_z^2}\right)$$

Cross multiplying the first transfer function and converting Laplace transform notation to the time domain yields

$$e + \frac{2\zeta_{AF}}{\omega_{AF}}\dot{e} + \frac{\ddot{e}}{\omega_{AF}^2} = \delta$$

If we solve the preceding equation for the highest derivative, we get

$$\ddot{e} = \omega_{AF}^2\left(\delta - e - \frac{2\zeta_{AF}}{\omega_{AF}}\dot{e}\right)$$

Repeating the procedure and cross multiplying the second transfer function and converting to the time domain yields the equation for the missile acceleration or

$$n_L = K_1(e - \ddot{e}/\omega_z^2)$$

Similarly recall that the body rate transfer function is given by

$$\frac{\dot{\theta}}{\delta} = K_3(1 + T_\alpha s) \Big/ \left(1 + \frac{2\zeta_{AF}}{\omega_{AF}}s + \frac{s^2}{\omega_{AF}^2}\right)$$

Again we can use the chain rule to split the transfer function as

$$\frac{\dot{\theta}}{\delta} = \frac{e}{\delta} * \frac{\dot{\theta}}{e}$$

The second term on the right-hand side of the preceding equation is simply the numerator of the transfer function or

$$\frac{\dot{\theta}}{e} = K_3(1 + T_\alpha s)$$

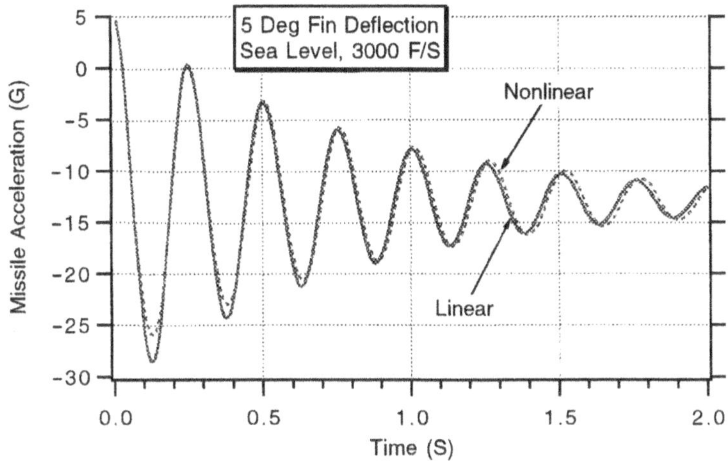

Fig. 21.9 Linear model accurately approximates actual missile acceleration.

Cross multiplying and converting Laplace transforms to the time domain yields the differential equation for the body rate as

$$\dot{\theta} = K_3(e + T_\alpha \dot{e})$$

We now have the two differential equations required to simulate the linearized airframe. A linear simulation of the airframe, based on the definitions of the previous section and the differential equations derived from the airframe transfer functions, appears in Listing 21.2. The linear airframe coefficients are evaluated using the trim value of the angle of attack. The inputs of the linear simulation are identical to those of the nonlinear airframe simulation of Listing 21.1. We can see that the linear differential equations appear before the FLAG=1 statement.

The nominal case was run in which there was a 5-deg fin deflection when the missile was at sea level and traveling at 3000 ft/s. Figure 21.9 shows that the linear airframe acceleration response is a near perfect match to the nonlinear airframe results derived from Listing 21.1. This means that our linearized model is a good approximation to reality.

LISTING 21.2 LINEAR AIRFRAME SIMULATION

```
n=0.;
VM=3000.;
DEL=5./57.3;
ALT=0.;
A=1000.;
DIAM=1.;
```

```
FR=3.;
XL=20.;
CTW=0.;
CRW=6.;
HW=2.;
CTT=0.;
CRT=2.;
HT=2.;
XN=4.;
XCG=10.;
XHL=19.5;
WGT=1000.;
if ALT<=30000.
        RHO=.002378*exp(-ALT/30000.);
else
        RHO=.0034*exp(-ALT/22000.);
end
SWING=.5*HW*(CTW+CRW);
STAIL=.5*HT*(CTT+CRT);
SREF=3.1416*DIAM*DIAM/4.;
XLP=FR*DIAM;
SPLAN=(XL-XLP)*DIAM+1.33*XLP*DIAM/2.;
XCPN=2*XLP/3;
AN=.67*XLP*DIAM;
AB=(XL-XLP)*DIAM;
XCPB=(.67*AN*XLP+AB*(XLP+.5*(XL-XLP)))/(AN+AB);
XCPW=XLP+XN+.7*CRW-.2*CTW;
XMACH=VM/A;
XIYY=WGT*(3*((DIAM/2)^2)+XL*XL)/(12*32.2);
TMP1=(XCG-XCPW)/DIAM;
TMP2=(XCG-XHL)/DIAM;
TMP3=(XCG-XCPB)/DIAM;
TMP4=(XCG-XCPN)/DIAM;
B=sqrt(XMACH^2-1);
Q=.5*RHO*VM*VM;
Y1=2*TMP4+8*SWING*TMP1/(B*SREF)+8*STAIL*TMP2/(B*SREF);
Y2=1.5*SPLAN*TMP3/SREF;
Y3=8*STAIL*TMP2*DEL/(B*SREF);
ALFTR=(-Y1-sqrt(Y1*Y1-4.*Y2*Y3))/(2*Y2);
CNA=2+1.5*SPLAN*ALFTR/SREF+8*SWING/(B*SREF)+8*STAIL/(B*SREF);
CND=8*STAIL/(B*SREF);
CMAP=2*TMP4+1.5*SPLAN*ALFTR*TMP3/SREF+8*SWING*TMP1/(B*SREF);
CMA=CMAP+8*STAIL*TMP2/(B*SREF);
CMD=8*STAIL*TMP2/(B*SREF);
XMA=Q*SREF*DIAM*CMA/XIYY;
XMD=Q*SREF*DIAM*CMD/XIYY;
```

```
ZA=-32.2*Q*SREF*CNA/(WGT*VM);
ZD=-32.2*Q*SREF*CND/(WGT*VM);
WZ=sqrt((XMA*ZD-ZA*XMD)/ZD);
WAF=sqrt(-XMA);
ZAF=.5*WAF*ZA/XMA;
XK1=-VM*(XMA*ZD-XMD*ZA)/(1845*XMA);
XK2=XK1;
TA=XMD/(XMA*ZD-XMD*ZA);
XK3=1845*XK1/VM;
E=0.;
ED=0.;
T=0;
H=.0025;
S=0;
while T<1.99999
      EOLD=E;
      EDOLD=ED;
      STEP=1;
      FLAG=0;
      while STEP<=1
         if FLAG==1
         STEP=2;
            E=E+H*ED;
            ED=ED+H*EDD;
            T=T+H;
         end
         EDD=WAF*WAF*(DEL*57.3-E-2.*ZAF*ED/WAF);
         XNL=XK1*(E-EDD/WZ^2);
         THD=XK3*(E+TA*ED);
     FLAG=1;
      end
      FLAG=0;
      E=.5*(EOLD+E+H*ED);
      ED=.5*(EDOLD+ED+H*EDD);
      S=S+H;
      if S>=.0099999
         S=0.;
         n=n+1;
         ArrayT(n)=T;
         ArrayXNL(n)=XNL;
         ArrayTHD(n)=THD;
      end
end
figure
plot(ArrayT,ArrayXNL),grid
xlabel('Time (Sec)')
```

```
ylabel('Missile Acceleration (G)')
clc
output=[ArrayT',ArrayXNL',ArrayTHD'];
save datfil.txt output  -ascii
disp 'simulation finished'
```

The angle of attack squared terms in the force and moment equations prevent the linear model from being perfect. The linear model is a less accurate representation of reality when the angle of attack is large. To see if the linear model is less accurate when the angle of attack is larger, another case was run in which the fin deflection was increased from 5 deg to 10 deg. We can see from Fig. 21.10 that when the fin deflection is increased the linear approximation to the airframe starts to deteriorate.

Finally, another case was run in which the fin deflection was still 5 deg, but the altitude increased from sea level to 50,000 ft. A 5-deg fin deflection at 50-kft altitude will cause a larger angle of attack than a 5-deg fin deflection at sea level, and so we would expect our linear model to be less accurate because of the angle of attack squared term in the force and moment equations. We can see from Fig. 21.11 that the linear model still approximates reality very well at this high altitude flight condition.

EXPERIMENTS

Using the linear transfer function approach, we can study the effect of flight condition on various important airframe parameters. We have seen that the airframe natural frequency is given by

$$\omega_{\mathrm{AF}} = \sqrt{-M_\alpha}$$

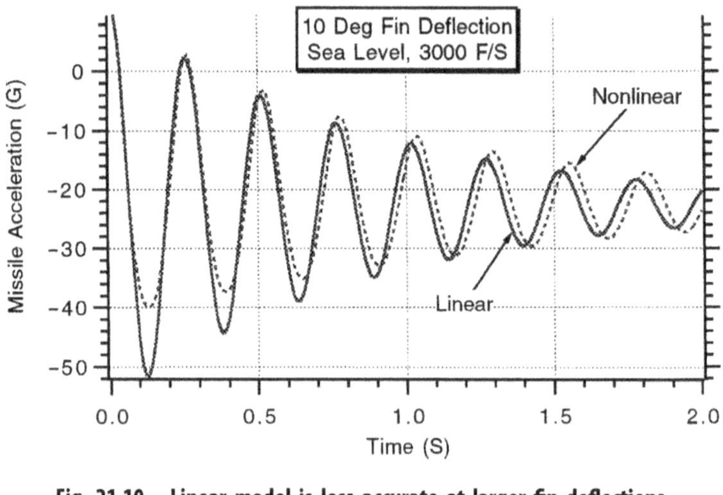

Fig. 21.10 Linear model is less accurate at larger fin deflections.

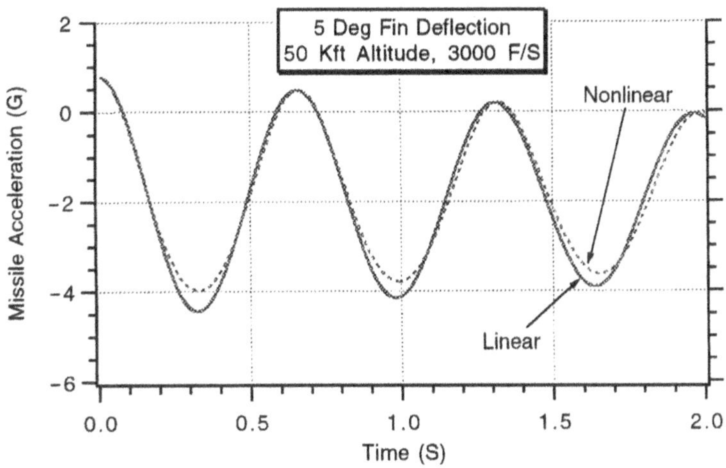

Fig. 21.11 Linear model is reasonable at higher altitudes.

Figure 21.12 shows that for our hypothetical missile the airframe natural frequency decreases with increasing altitude and decreasing speed. For this example the airframe natural frequency varied between 10 rad/s and 30 rad/s. If we think of the airframe time constant being the inverse of the natural frequency, then the time constant variation is between 0.033 s and 0.1 s. In

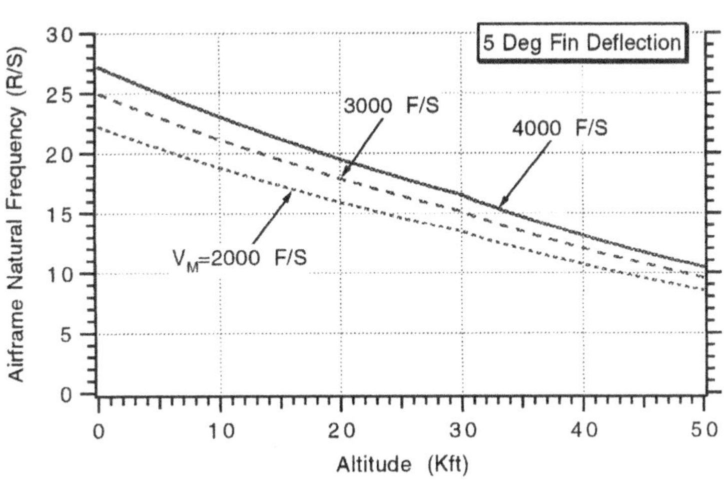

Fig. 21.12 Airframe natural frequency decreases with increasing altitude and decreasing speed.

general the airframe time constant is fast, and an autopilot is usually not required to artificially speed up the airframe response.

We have already shown that the airframe damping is given by

$$\zeta_{AF} = \frac{Z_\alpha \omega_{AF}}{2M_\alpha}$$

Figure 21.13 shows that the airframe damping decreases with increasing altitude and in creasing missile speed. The airframe damping is quite low and in this example varies between 0.02 and 0.065. We shall soon see that this low damping is not satisfactory for overall system performance in a radar homing missile and that a flight-control system is required to artificially increase the low damping of the airframe.

M_α and M_δ are both important airframe parameters and are displayed in Figs. 21.14 and 21.15. A positive M_α indicates that the bare airframe is unstable. There are limits on how negative or positive M_α can be before the design of the flight-control system becomes impossible. During the normal design process it is natural for the baseline airframe to change due to either overly optimistic assumptions concerning weight and size or possibly due to new requirements. In both cases it is important that the flight-control system designer work closely with the aerodynamicist not only when an airframe is being selected but also as it is being modified. For example, large values of M_δ make it difficult to choose actuators that will work with the flight-control system. Therefore it is also important to limit the size of this key aerodynamic parameter.

Finally Fig. 21.16 shows that the missile turning rate time constant T_α increases with increasing altitude and increasing missile speed. At 50 kft altitude

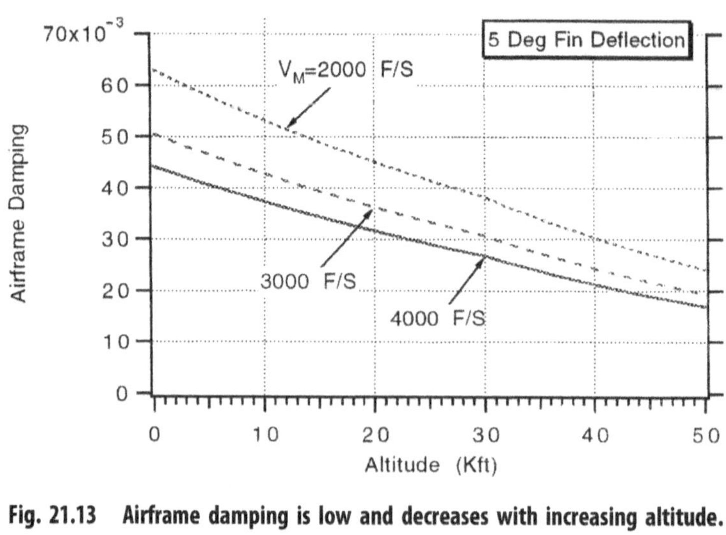

Fig. 21.13 Airframe damping is low and decreases with increasing altitude.

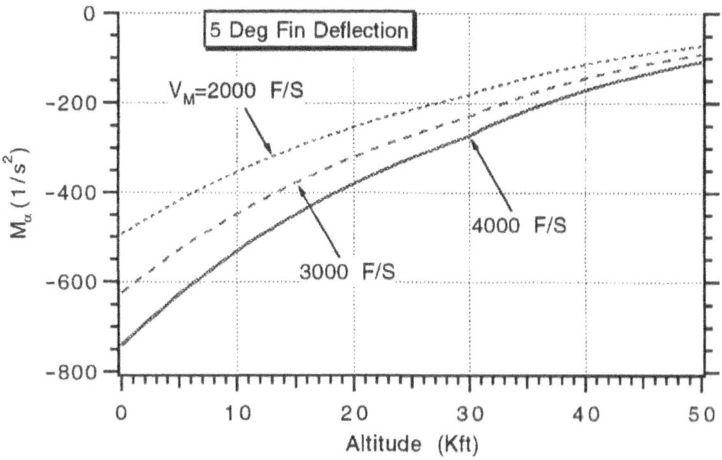

Fig. 21.14 M_α **gets smaller with increasing altitude and decreasing speed.**

we can see that the turning rate time constant is approximately 4 s when the missile is traveling at 3000 ft/s. This is slightly smaller than the value used in Chapter 18. In Chapter 18 the turning rate time constant was calculated based on the nose and body only. We can see that the addition of the wing and tail decreased the turning rate time constant from 5 s to 4 s. Although all missiles have tails, some do not have wings. Wingless missiles will tend to have larger

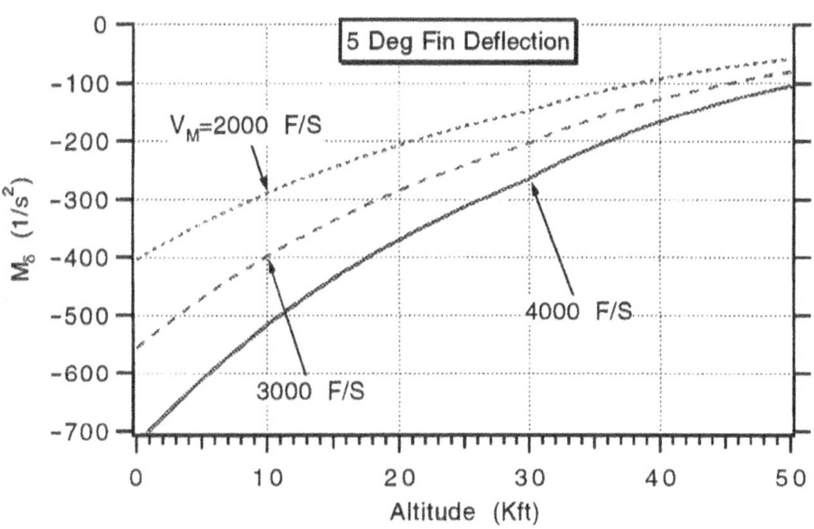

Fig. 21.15 M_δ **gets smaller with increasing altitude and decreasing speed.**

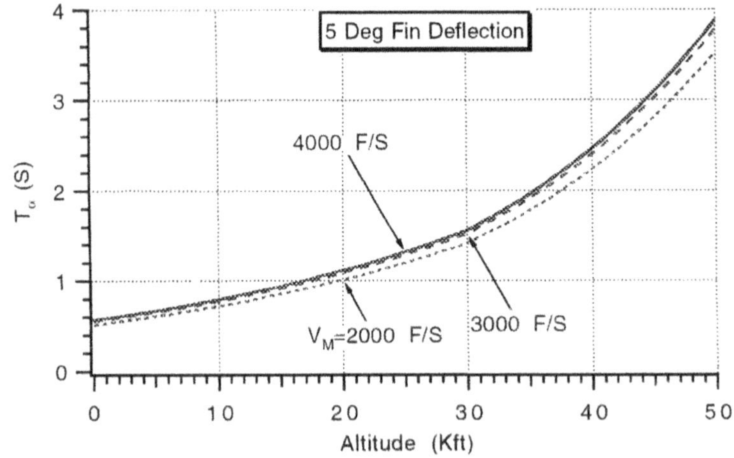

Fig. 21.16 Turning rate time constant increases with increasing altitude and increasing speed.

turning rate time constants. We have shown before that the turning rate time constant is related to the radome stability problem. Large values of turning rate time constant require smaller values of radome slope for a given level of performance.

We have already shown that the airframe zero ω_z can be expressed in terms of the aerodynamic parameters as

$$\omega_z = \frac{M_\alpha Z_\delta - Z_\alpha M_\delta}{Z_\delta}$$

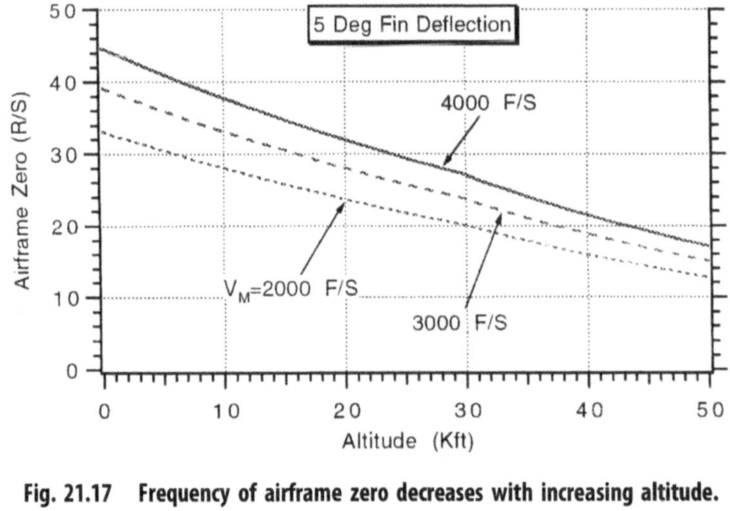

Fig. 21.17 Frequency of airframe zero decreases with increasing altitude.

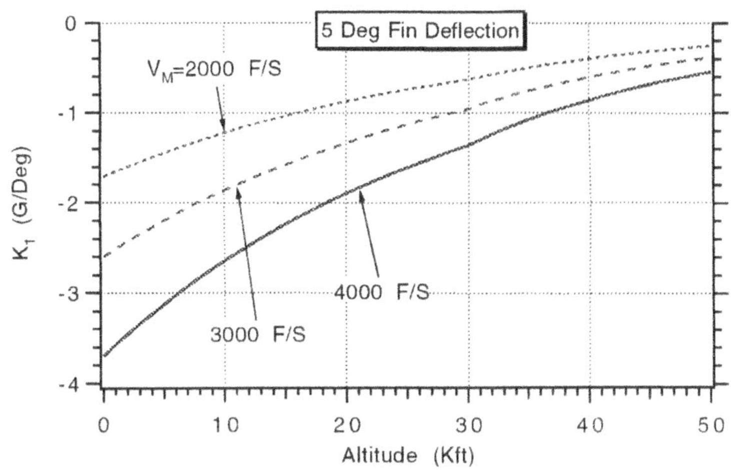

Fig. 21.18 Magnitude of acceleration aerodynamic gain gets smaller as altitude increases.

We can see from Fig. 21.17 that the airframe zero decreases with increasing altitude and decreasing missile velocity. Smaller values of the airframe zero will cause more wrong-way tail effect.

The acceleration aerodynamic gain K_1 is in units of gees per degree and tells how much steady-state acceleration there will be for a given fin deflection. Figure 21.18 shows that the magnitude of the aerodynamic gain gets smaller as the altitude increases and velocity decreases. For example, if the missile is traveling

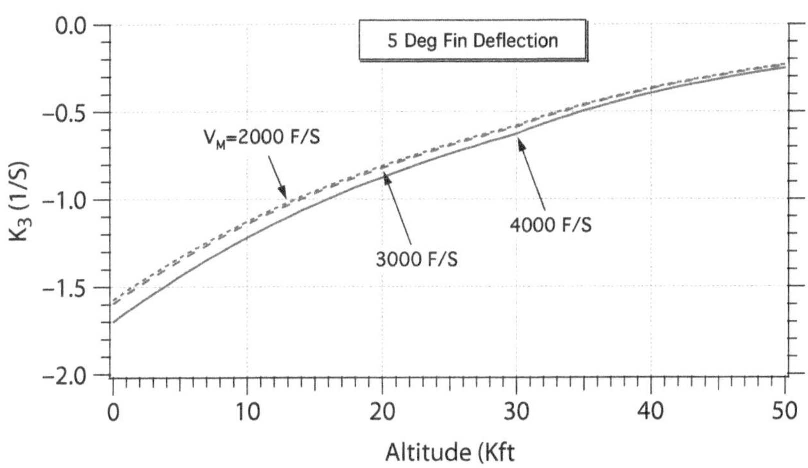

Fig. 21.19 Magnitude of body rate aerodynamic gain is independent of velocity and gets smaller as altitude increases.

at 3000 ft/s the aerodynamic gain is approximately -2.7 at sea level and -0.4 at 50 kft altitude. Therefore a 5-deg fin deflection would result in $-13.5\,g$ at sea level (i.e., $-2.7*5=-13.5$) and only $2\,g$ at 50 kft altitude (i.e., $-0.4*5=-2$). The nonlinear results of Fig. 21.7 confirm these calculations. More fin travel will be required to achieve a given acceleration when the altitude is higher or when the missile velocity is smaller.

The body rate aerodynamic gain K_3 tells how much steady-state body rate there will be for a given fin deflection. Figure 21.19 shows that the magnitude of the aerodynamic gain gets smaller as the altitude increases and velocity decreases. For example, if the missile is traveling at 3000 ft/s the body rate aerodynamic gain is approximately -1.6 s^{-1} at sea level and -0.24 s^{-1} at 50 kft altitude. Therefore a 5-deg fin deflection would result in -8 deg/s at sea level (i.e., $-1.6*5=-8$) and only -1.2 deg/s at 50 kft altitude (i.e., $-0.24*5=-1.2$).

SUMMARY

In this chapter we have seen how the nonlinear missile force and moment equations are related to the geometry of the missile airframe. A simple method for linearizing the force and moment equations was introduced so that transfer functions could be derived for the missile airframe. It was shown that the transfer function approximation to the airframe was a good approximation to the nonlinear force and moment equations at small angles of attack.

REFERENCES

[1] Chin, S. S., *Missile Configuration Design*, McGraw-Hill, New York, 1961.

[2] Jerger, J. J., *System Preliminary Design*, Van Nostrand, Princeton, NJ, 1960.

[3] Giragosian, P., "Aerodynamic Considerations in the Design of a Vertically Launched Advanced Interdiction Missile," *Proceedings of AIAA Atmospheric Flight Mechanics Conference*, AIAA, San Diego, CA, Aug. 9–11, 1982.

[4] Selby, S. M., *Standard Mathematical Tables, Twentieth Edition*, The Chemical Rubber Co., Cleveland, OH, 1972.

CHAPTER 22

Introduction to Flight Control Design

INTRODUCTION

In Chapter 21 we saw how to derive aerodynamic transfer functions and relate them to the missile airframe's force and moment equations. The missile's bare airframe response was shown to be highly oscillatory because of its low damping. The purpose of the flight-control system, shown in Fig. 22.1, is to convert the missile's acceleration command n_c generated by the guidance law to an achieved acceleration n_L. The flight-control system usually must improve the response characteristics of the bare missile airframe to ensure that the achieved acceleration closely follows the commanded acceleration.

As we can see from Fig. 22.1, the missile airframe is just one part of the flight-control system. Mathematically we can think of the airframe as a transfer function whose input is the tail fin deflection δ and whose output is the achieved missile acceleration n_L. The autopilot is another part of the flight-control system and is the mechanism for converting the acceleration command n_c to a fin deflection command δ_c. The actuator then takes the autopilot's electrical output and moves the missile control surfaces (that is, canards, wings, or tails) through the appropriate angular deflection δ in response to the fin deflection command.

In this chapter we shall first see how the flight-control system interacts with the rest of the guidance system and how it influences system performance. Then we shall investigate a simple way in which the principles of feedback can be used in order to improve the flight-control system response so that homing guidance objectives can be met.

OPEN-LOOP FLIGHT-CONTROL SYSTEM

The simplest possible flight-control system is the open-loop system shown in Fig. 22.2. In this diagram, in which the airframe is treated as a transfer function and the actuator dynamics are neglected, the autopilot is simply a gain that attempts to cancel the aerodynamic gain of the airframe. In the steady state the

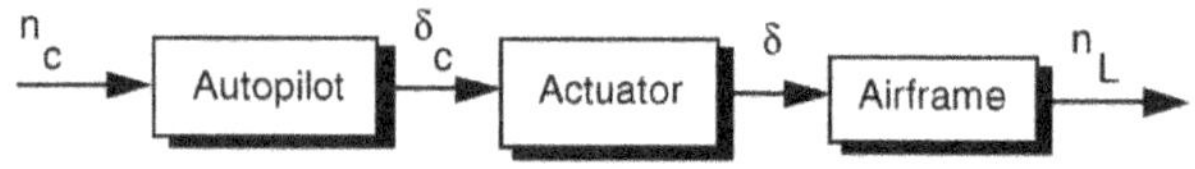

Fig. 22.1 Conceptual block diagram of a flight-control system.

missile-achieved acceleration will match the commanded acceleration provided that the autopilot gain can be changed with flight condition. The open-loop autopilot is the least expensive of all possible autopilots because it does not require a rate gyro or accelerometer.

In Chapter 21 we assumed a certain fin deflection and solved for the trim angle of attack along with the various aeroderivatives. In this chapter we shall specify the desired acceleration and solve for the trim fin deflection and angle of attack.

Recall that the normal force coefficient is given by

$$C_N = 2\alpha + \frac{1.5S_{\text{PLAN}}\alpha^2}{S_{\text{ref}}} + \frac{8S_W\alpha}{\beta S_{\text{ref}}} + \frac{8S_T(\alpha+\delta)}{\beta S_{\text{ref}}}$$

Since

$$F_N = ma = \frac{Wn_L}{g} = QS_{\text{ref}}C_N$$

we can solve for the normal force coefficient at trim according to

$$C_{N\text{TRIM}} = \frac{Wn_{L\text{TRIM}}}{gQS_{\text{ref}}}$$

Using shorthand notation we can now say that at trim we get

$$C_{N\text{TRIM}} = y_1\alpha_{\text{TRIM}} + y_2\alpha_{\text{TRIM}}^2 + y_3\delta_{\text{TRIM}}$$

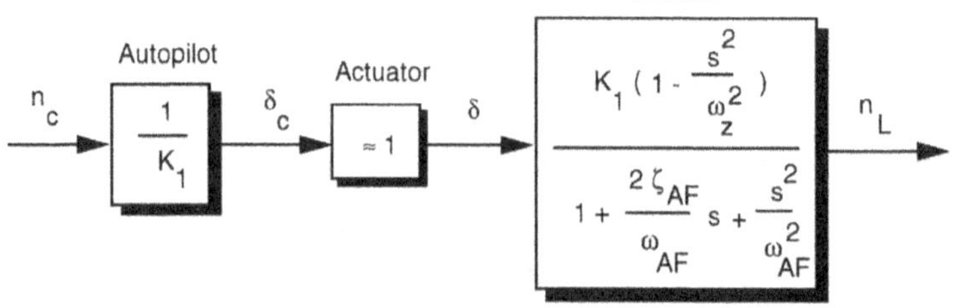

Fig. 22.2 Open-loop flight-control system.

where

$$y_1 = 2 + \frac{8S_W}{\beta S_{\text{ref}}} + \frac{8S_T}{\beta S_{\text{ref}}}$$

$$y_2 = \frac{1.5S_{\text{PLAN}}}{S_{\text{ref}}}$$

$$y_3 = \frac{8S_T}{\beta S_{\text{ref}}}$$

Recall that the moment coefficient equation is given by

$$C_M = 2\alpha \frac{(X_{\text{CG}} - X_{\text{CPN}})}{d} + \frac{1.5S_{\text{PLAN}}\alpha^2}{S_{\text{ref}}} \frac{(X_{\text{CG}} - X_{\text{CPB}})}{d} + \frac{8S_W\alpha}{\beta S_{\text{ref}}} \frac{(X_{\text{CG}} - X_{\text{CPW}})}{d} + \frac{8S_T(\alpha + \delta)}{\beta S_{\text{ref}}} \frac{(X_{\text{CG}} - X_{\text{HL}})}{d}$$

At trim the moment coefficient is zero and we can rewrite the preceding equation in shorthand notation as

$$0 = y_4\alpha_{\text{TRIM}} + y_5\alpha_{\text{TRIM}}^2 + y_6\delta_{\text{TRIM}}$$

where

$$y_4 = \frac{2(X_{\text{CG}} - X_{\text{CPN}})}{d} + \frac{8S_W(X_{\text{CG}} - X_{\text{CPW}})}{\beta S_{\text{ref}} d} + \frac{8S_T(X_{\text{CG}} - X_{\text{HL}})}{\beta S_{\text{ref}} d}$$

$$y_5 = \frac{1.5S_{\text{PLAN}}(X_{\text{CG}} - X_{\text{CPB}})}{S_{\text{ref}} d}$$

$$y_6 = \frac{8S_T(X_{\text{CG}} - X_{\text{HL}})}{\beta S_{\text{ref}} d}$$

We now have two trim equations with two unknowns. The two equations can be reduced to one quadratic equation allowing us to solve for the trim angle of attack as

$$\alpha_{\text{TRIM}} = \frac{-p_3 + \sqrt{p_3^2 + 4p_2C_{N\text{TRIM}}}}{2p_2}$$

where

$$p_2 = y_2 - \frac{y_3 y_5}{y_6}$$

$$p_3 = y_1 - \frac{y_3 y_4}{y_6}$$

Substituting the trim angle of attack into the moment coefficient equation allows us to solve for the trim fin deflection as

$$\delta_{\text{TRIM}} = \frac{-y_4 \alpha_{\text{TRIM}} - y_5 \alpha_{\text{TRIM}}^2}{y_6}$$

We now have enough information to solve for the aerodynamic parameters at any flight condition given a desired acceleration level. Using the notional missile of Chapter 21, Table 22.1 numerically summarizes the values for the various missile transfer function parameters for the case in which the missile is traveling at 3000 ft/s at both sea level and 50-kft altitude and is trying to respond to a 10-*g* acceleration command.

The open-loop flight-control system of Fig. 22.2 was simulated using the linearized aerodynamics. We can see from Listing 22.1 the previously derived equations for the calculation of the trim angle of attack. In addition, we can also see that the open-loop autopilot is simply a gain.

As expected, Fig. 22.3 shows that the flight-control system response due to a 10-*g* step at sea level is oscillatory due to the low damping of the airframe. The frequency of oscillation is also that of the bare airframe and is very high. In other words the flight-control system response is simply the response of the bare airframe. The sole purpose of the flight-control system is to ensure that the achieved acceleration looks like the commanded acceleration in the steady state.

TABLE 22.1 MISSILE TRANSFER FUNCTION PARAMETERS AT TWO FLIGHT CONDITIONS

Airframe parameter	Definition	Sea level	50 kft
ω_{AF}	Airframe natural frequency	25.3 rad/s	10.0 rad/s
ζ_{AF}	Airframe damping	0.058	0.027
ω_z	Airframe zero	43.2 rad/s	18.9 rad/s
K_1	Aerodynamic acceleration gain	−3.07 g/deg	−0.559 g/deg

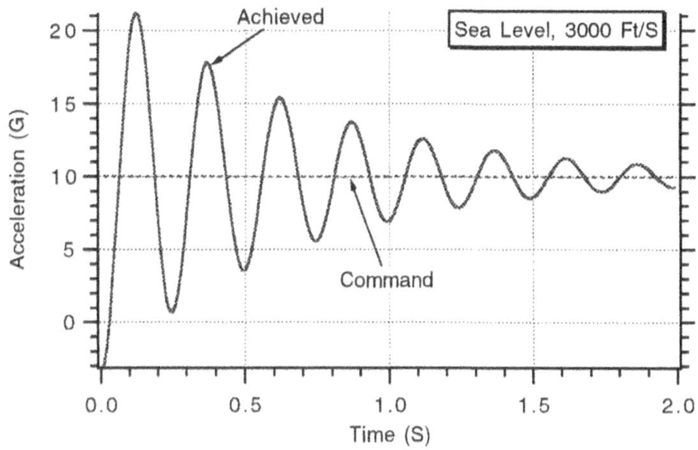

Fig. 22.3 Open-loop flight-control system is lightly damped.

Figure 22.4 shows that increasing the altitude decreases the damping and natural frequency of the open-loop flight-control system response. We can see that the flight-control system response is still that of the bare airframe.

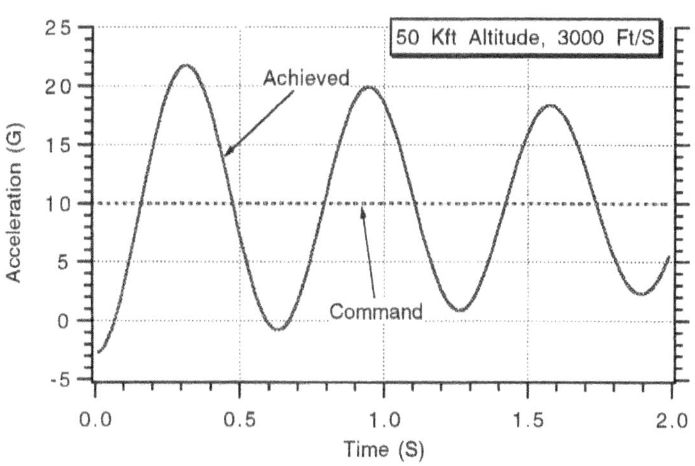

Fig. 22.4 Increasing the attitude decreases both the damping and natural frequency of the open-loop flight-control system.

LISTING 22.1 OPEN-LOOP FLIGHT-CONTROL SYSTEM

```
n=0;
VM=3000.;
XNCG=10.;
ALT=0.;
A=1000.;
DIAM=1.;
FR=3.;
XL=20.;
CTW=0.;
CRW=6.;
HW=2.;
CTT=0.;
CRT=2.;
HT=2.;
XN=4.;
XCG=10.;
XHL=19.5;
WGT=1000.;
if ALT<=30000.
        RHO=.002378*exp(-ALT/30000.);
else
        RHO=.0034*exp(-ALT/22000.);
end
SWING=.5*HW*(CTW+CRW);
STAIL=.5*HT*(CTT+CRT);
SREF=3.1416*DIAM*DIAM/4.;
XLP=FR*DIAM;
SPLAN=(XL-XLP)*DIAM+1.33*XLP*DIAM/2.;
XCPN=2*XLP/3;
AN=.67*XLP*DIAM;
AB=(XL-XLP)*DIAM;
XCPB=(.67*AN*XLP+AB*(XLP+.5*(XL-XLP)))/(AN+AB);
XCPW=XLP+XN+.7*CRW-.2*CTW;
XMACH=VM/A;
XIYY=WGT*(3*((DIAM/2)^2)+XL*XL)/(12*32.2);
TMP1=(XCG-XCPW)/DIAM;
TMP2=(XCG-XHL)/DIAM;
TMP3=(XCG-XCPB)/DIAM;
TMP4=(XCG-XCPN)/DIAM;
B=sqrt(XMACH^2-1);
Q=.5*RHO*VM*VM;
P1=WGT*XNCG/(Q*SREF);
Y1=2.+8*SWING/(B*SREF)+8*STAIL/(B*SREF);
Y2=1.5*SPLAN/SREF;
```

```
Y3=8*STAIL/(B*SREF);
Y4=2*TMP4+8*SWING*TMP1/(B*SREF)+8*STAIL*TMP2/(B*SREF);
Y5=1.5*SPLAN*TMP3/SREF;
Y6=8*STAIL*TMP2/(B*SREF);
P2=Y2-Y3*Y5/Y6;
P3=Y1-Y3*Y4/Y6;
ALFTR=(-P3+sqrt(P3*P3+4.*P2*P1))/(2.*P2);
DELTR=-Y4*ALFTR/Y6-Y5*ALFTR*ALFTR/Y6;
CNA=2+1.5*SPLAN*ALFTR/SREF+8*SWING/(B*SREF)+8*STAIL/(B*SREF);
CND=8*STAIL/(B*SREF);
CMAP=2*TMP4+1.5*SPLAN*ALFTR*TMP3/SREF+8*SWING*TMP1/(B*SREF);
CMA=CMAP+8*STAIL*TMP2/(B*SREF);
CMD=8*STAIL*TMP2/(B*SREF);
XMA=Q*SREF*DIAM*CMA/XIYY;
XMD=Q*SREF*DIAM*CMD/XIYY;
ZA=-32.2*Q*SREF*CNA/(WGT*VM);
ZD=-32.2*Q*SREF*CND/(WGT*VM);
WZ=sqrt((XMA*ZD-ZA*XMD)/ZD);
WAF=sqrt(-XMA);
ZAF=.5*WAF*ZA/XMA;
XK1=-VM*(XMA*ZD-XMD*ZA)/(1845*XMA);
XK2=XK1;
TA=XMD/(XMA*ZD-XMD*ZA);
XK3=1845*XK1/VM;
XKDC=1./XK1;
E=0.;
ED=0.;
T=0;
H=.0001;
S=0;
while T<1.99999
        EOLD=E;
        EDOLD=ED;
        STEP=1;
        FLAG=0;
        while STEP<=1
            if FLAG==1
            STEP=2;
                   E=E+H*ED;
                   ED=ED+H*EDD;
                   T=T+H;
            end
            DEL=XKDC*XNCG;
            EDD=WAF*WAF*(DEL-E-2.*ZAF*ED/WAF);
            XNL=XK1*(E-EDD/WZ^2);
            FLAG=1;
```

```
        end
        FLAG=0;
        E=.5*(EOLD+E+H*ED);
        ED=.5*(EDOLD+ED+H*EDD);
        S=S+H;
        if S>=.0099999
          S=0.;
          n=n+1;
          ArrayT(n)=T;
          ArrayXNL(n)=XNL;
          ArrayXNCG(n)=XNCG;
        end
end
figure
plot(ArrayT,ArrayXNL,ArrayT,ArrayXNCG),grid
xlabel('Time (Sec)')
ylabel('Missile Acceleration (G)')
clc
output=[ArrayT',ArrayXNL',ArrayXNCG'];
save datfil.txt output  -ascii
disp 'simulation finished'
```

GUIDANCE SYSTEM INTERACTIONS

To get a clearer understanding of how the open-loop flight-control system (autopilot plus airframe) interacts with the missile guidance system, let us consider the homing loop block diagram of Fig. 22.5. In this diagram both the seeker and noise filter are represented by single lags while the autopilot is represented simply by a gain. Because actuator dynamics are neglected, the autopilot output is the achieved fin deflection. Airframe transfer functions are present in the block diagram to convert fin deflection to body rate and missile acceleration. The values of the additional parameters for the body rate transfer function along with the guidance system parameters appear in Table 22.2 for the two flight conditions of interest. Because the two airframe transfer functions assume angles are in units of degrees and acceleration is in units of gees, the constants 57.3 and 32.2 appear in the homing loop block diagram to convert quantities to the English system of units. Notice that radome effects are also included in the block diagram.

To investigate how the open-loop flight-control system interacts with the guidance system, an adjoint was constructed based on the model of Fig. 22.5. The rms miss distance for a 10-s flight due to a 1-*g* uniformly distributed target maneuver (see Chapter 4 for more details) was evaluated for different radome slopes when the missile was traveling at 3000 ft/s and the target was traveling at 1000 ft/s at 50-kft altitude. The rms miss distance results are shown in Fig. 22.6. Here we can see that nominally the airframe damping is 0.027 (see Table 22.1) and that the rms

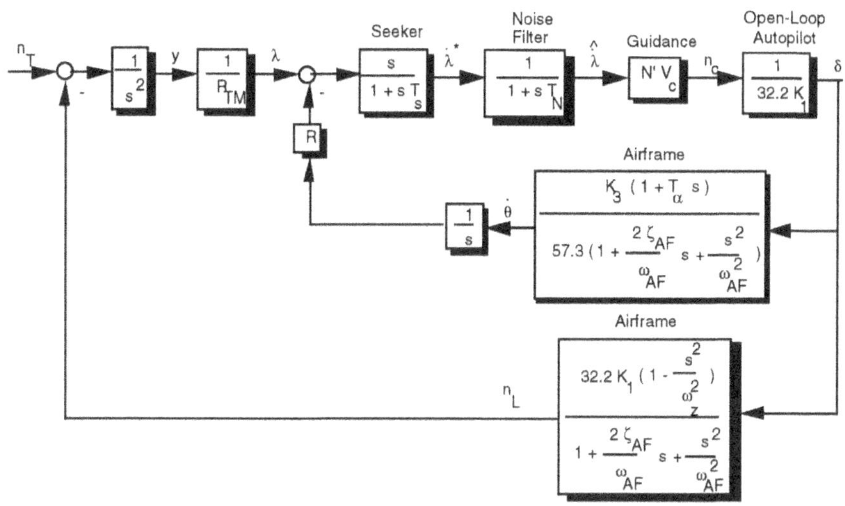

Fig. 22.5 Homing loop with open-loop flight-control system.

miss distance can be enormous if the radome slope is more negative than −0.005 or more positive than 0.04. However, if somehow the airframe damping could be increased to 0.7 we can also see from Fig. 22.6 that the system would be less sensitive to radome slope because the rms miss distance would only start to increase if the radome slope became more negative than −0.03 or more positive than 0.06. In many radar homing applications the low damping of the airframe must be increased artificially so that a wider range of radome slopes can be tolerated. Therefore in these applications the low cost open-loop flight-control system would not be suitable.

TABLE 22.2 ADDITIONAL LINEARIZED AERODYNAMICS AT TWO FLIGHT CONDITIONS

Parameter	Definition	Sea level	50 kft
T_α	Turning rate time constant	0.457 s	2.40 s
K_3	Aerodynamic body rate gain	−1.89 1/s	−0.344 1/s
T_s	Seeker time constant	0.1 s	0.1 s
T_N	Noise filter time constant	0.1 s	0.1 s
V_c	Closing velocity	4000 ft/s	4000 ft/s
N'	Effective navigation ratio	3	3
T_F	Flight time	10 s	10 s

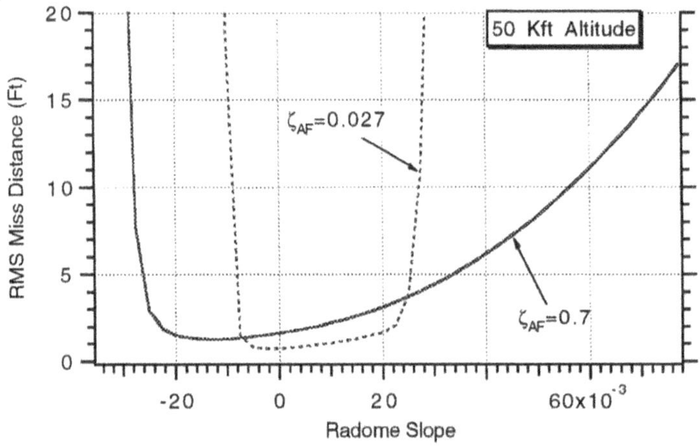

Fig. 22.6 Law airframe damping increases sensitivity to radome.

RATE GYRO FLIGHT-CONTROL SYSTEM

The rate gyro flight-control system artificially increases the low damping of the open-loop flight-control system by the use of a rate gyro sensor and the principles of feedback. In this flight-control system the rate gyro measures the missile body rate and feeds back this information to develop an error signal. The rate gyro flight-control system shown in Fig. 22.7 has two autopilot gains. The gain K_R

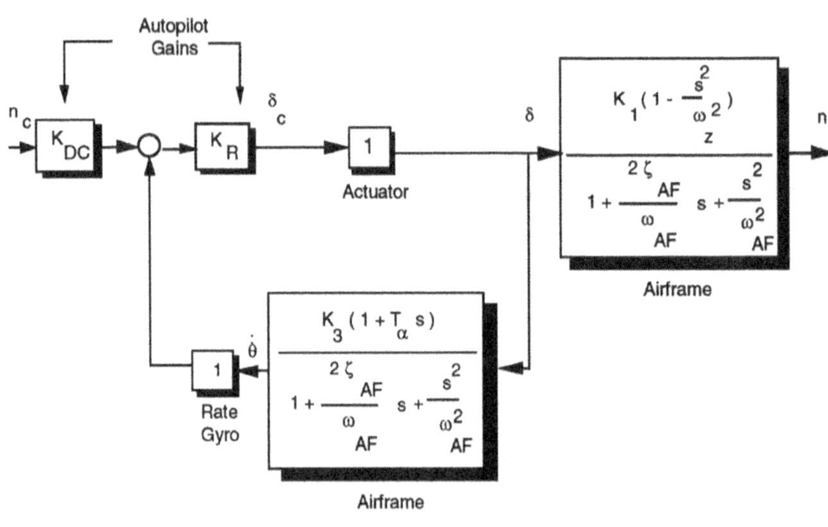

Fig. 22.7 Rate gyro flight-control system.

will determine the response of the flight-control system while the gain K_{DC} controls the steady-state value of the response.

The transfer function from the achieved to the commanded acceleration can be obtained from Fig. 22.7. After some algebraic manipulations we obtain

$$\frac{n_L}{n_c} = \frac{K_{DC}K_1K_R}{1 - K_RK_3}\left\{\left[1 - \frac{s^2}{\omega_z^2}\right] \Big/ \left[1 + \frac{(2\zeta_{\mathrm{AF}}/\omega_{\mathrm{AF}}) - K_RK_3T_\alpha}{1 - K_RK_3}s + \frac{s^2}{\omega_{\mathrm{AF}}^2(1 - K_RK_3)}\right]\right\}$$

To get the achieved acceleration to match the commanded acceleration in the steady state, we can see from the preceding equation that the gain K_{DC} must be set to

$$K_{DC} = \frac{1 - K_RK_3}{K_1K_R}$$

Because the denominator of the flight-control system is a quadratic, we can find the equivalent natural frequency ω and damping ζ of the rate gyro flight-control system by equating

$$1 + \frac{2\zeta}{\omega}s + \frac{s^2}{\omega^2} = 1 + \frac{(2\zeta_{\mathrm{AF}}/\omega_{\mathrm{AF}}) - K_RK_3T_\alpha}{1 - K_RK_3}s + \frac{s^2}{\omega_{\mathrm{AF}}^2(1 - K_RK_3)}$$

Solving for the frequency and damping of the rate gyro flight-control system yields

$$\omega = \omega_{\mathrm{AF}}\sqrt{1 - K_RK_3}$$

$$\zeta = \frac{\omega}{2}\left[\frac{(2\zeta_{\mathrm{AF}}/\omega_{\mathrm{AF}}) - K_RK_3T_\alpha}{1 - K_RK_3}\right]$$

Therefore we can see that the rate gyro autopilot gain K_R influences both the flight-control system frequency and damping.

Figures 22.8 and 22.9 show that increasing the autopilot gain K_R increases the damping of the flight-control system above that of the bare airframe both at low and high altitudes. At sea level an autopilot gain of approximately 0.1 is required to increase the damping of the rate gyro flight-control system to unity whereas at 50-kft altitude a gain of approximately 0.25 is required to get the same damping. In other words the autopilot gain must be changed with flight condition to ensure adequate damping in order to desensitize the guidance system to radome slope effects. Changing the autopilot gain or gains with flight condition is known as *gain scheduling*. It is important to note that for the rate gyro autopilot the autopilot gain does not cause the natural frequency of the flight-control system to be significantly different from the airframe's natural frequency.

Listing 22.2 presents the simulation of the rate gyro flight-control system. In this simulation parameters are first expressed in terms of the geometry of the airframe and then the linearized aerodynamic parameters are derived. Note

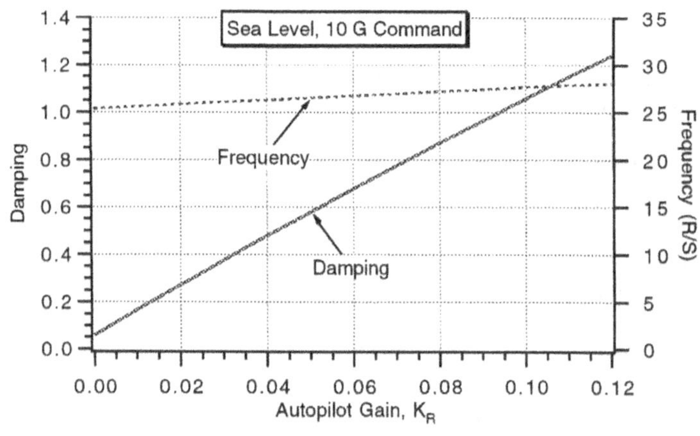

Fig. 22.8 Autopilot gain determines damping of flight-control system at sea level.

that the simulation integration interval is very small ($h = 0.0001$ s). If the integration interval were 10 times larger, the system would appear to go unstable when the autopilot gain is large ($K_R = 5$). However this instability is due simply to numerical integration and can be corrected by choosing a smaller integration interval. In this simplified model of the world, the autopilot gain can be made arbitrarily large without causing an instability in the guidance system. The specific equations for the rate gyro flight-control system appear before the FLAG = 1 statement.

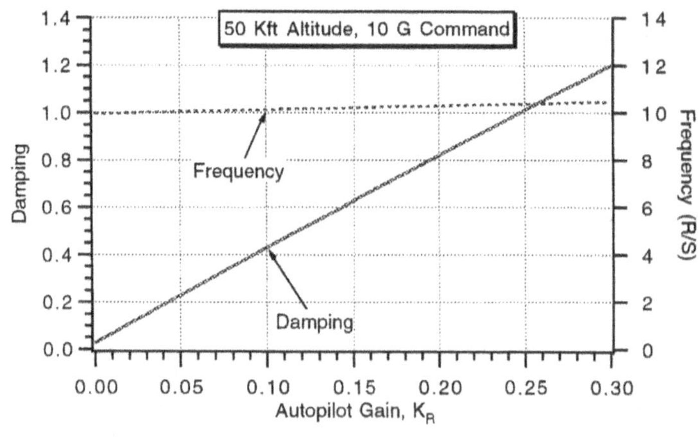

Fig. 22.9 Autopilot gain also determines damping of flight-control system at 50-kft altitude.

LISTING 22.2 RATE GYRO FLIGHT-CONTROL SYSTEM IN PRESENCE OF LINEAR AIRFRAME

```
n=0;
VM=3000.;
XNCG=10.;
ALT=0.;
A=1000.;
DIAM=1.;
FR=3.;
XL=20.;
CTW=0.;
CRW=6.;
HW=2.;
CTT=0.;
CRT=2.;
HT=2.;
XN=4.;
XCG=10.;
XHL=19.5;
WGT=1000.;
XKR=.1;
if ALT<=30000.
        RHO=.002378*exp(-ALT/30000.);
else
        RHO=.0034*exp(-ALT/22000.);
end
SWING=.5*HW*(CTW+CRW);
STAIL=.5*HT*(CTT+CRT);
SREF=3.1416*DIAM*DIAM/4.;
XLP=FR*DIAM;
SPLAN=(XL-XLP)*DIAM+1.33*XLP*DIAM/2.;
XCPN=2*XLP/3;
AN=.67*XLP*DIAM;
AB=(XL-XLP)*DIAM;
XCPB=(.67*AN*XLP+AB*(XLP+.5*(XL-XLP)))/(AN+AB);
XCPW=XLP+XN+.7*CRW-.2*CTW;
XMACH=VM/A;
XIYY=WGT*(3*((DIAM/2)^2)+XL*XL)/(12*32.2);
TMP1=(XCG-XCPW)/DIAM;
TMP2=(XCG-XHL)/DIAM;
TMP3=(XCG-XCPB)/DIAM;
TMP4=(XCG-XCPN)/DIAM;
B=sqrt(XMACH^2-1);
Q=.5*RHO*VM*VM;
P1=WGT*XNCG/(Q*SREF);
Y1=2.+8*SWING/(B*SREF)+8*STAIL/(B*SREF);
```

```
Y2=1.5*SPLAN/SREF;
Y3=8*STAIL/(B*SREF);
Y4=2*TMP4+8*SWING*TMP1/(B*SREF)+8*STAIL*TMP2/(B*SREF);
Y5=1.5*SPLAN*TMP3/SREF;
Y6=8*STAIL*TMP2/(B*SREF);
P2=Y2-Y3*Y5/Y6;
P3=Y1-Y3*Y4/Y6;
ALFTR=(-P3+sqrt(P3*P3+4.*P2*P1))/(2.*P2);
DELTR=-Y4*ALFTR/Y6-Y5*ALFTR*ALFTR/Y6;
CNA=2+1.5*SPLAN*ALFTR/SREF+8*SWING/(B*SREF)+8*STAIL/(B*SREF);
CND=8*STAIL/(B*SREF);
CMAP=2*TMP4+1.5*SPLAN*ALFTR*TMP3/SREF+8*SWING*TMP1/(B*SREF);
CMA=CMAP+8*STAIL*TMP2/(B*SREF);
CMD=8*STAIL*TMP2/(B*SREF);
XMA=Q*SREF*DIAM*CMA/XIYY;
XMD=Q*SREF*DIAM*CMD/XIYY;
ZA=-32.2*Q*SREF*CNA/(WGT*VM);
ZD=-32.2*Q*SREF*CND/(WGT*VM);
WZ=sqrt((XMA*ZD-ZA*XMD)/ZD);
WAF=sqrt(-XMA);
ZAF=.5*WAF*ZA/XMA;
XK1=-VM*(XMA*ZD-XMD*ZA)/(1845*XMA);
XK2=XK1;
TA=XMD/(XMA*ZD-XMD*ZA);
XK3=1845*XK1/VM;
XKDC=(1.-XKR*XK3)/(XK1*XKR);
E=0.;
ED=0.;
T=0;
H=.0001;
S=0;
while T<1.99999
        EOLD=E;
        EDOLD=ED;
        STEP=1;
        FLAG=0;
        while STEP<=1
           if FLAG==1
        STEP=2;
              E=E+H*ED;
              ED=ED+H*EDD;
              T=T+H;
           end
           THD=XK3*(E+TA*ED);
           DEL=XKR*(XKDC*XNCG+THD);
           EDD=WAF*WAF*(DEL-E-2.*ZAF*ED/WAF);
```

```
            XNL=XK1*(E-EDD/WZ^2);
            FLAG=1;
        end
        FLAG=0;
        E=.5*(EOLD+E+H*ED);
        ED=.5*(EDOLD+ED+H*EDD);
        S=S+H;
        if S>=.0099999
            S=0.;
            n=n+1;
            ArrayT(n)=T;
            ArrayXNL(n)=XNL;
            ArrayXNCG(n)=XNCG;
        end
end
figure
plot(ArrayT,ArrayXNL,ArrayT,ArrayXNCG),grid
xlabel('Time (Sec)')
ylabel('Missile Acceleration (G)')
clc
output=[ArrayT',ArrayXNL',ArrayXNCG'];
save datfil.txt output -ascii
disp 'simulation finished'
```

The simulation of the linear rate gyro flight-control system of Listing 22.2 was run for the cases in which the missile was traveling at 3000 ft/s at both sea level and 50-kft altitude. An autopilot gain of 0.1 (XKR = 0.1) was chosen at sea level and a gain of 0.25 (XKR = 0.25) was chosen at 50 kft in order to ensure that the flight-control system damping was approximately unity. We can see from Fig. 22.10 that although the desired damping is achieved at both flight conditions the response is more sluggish at the higher altitude. We shall see in Chapter 23 that a more advanced flight-control system will be required to control both the damping and the time constant at the same time.

To make the simulation of the flight-control system more realistic, a simplified model of the actuator was used. Nominally the second-order actuator, shown in Fig. 22.11, has a natural frequency of 150 rad/s and a damping of 0.7 ($\omega_{ACT} =$ 150 rad/s, $\zeta_{ACT} = 0.7$).

Figure 22.12 shows then when the dynamics of the actuator are included in the simulation of the rate gyro flight-control system the step response only changes slightly. It appears that the inclusion of the actuator dynamics has very little effect on the flight-control system response.

Experiments were conducted to see how large the autopilot gain could be made when actuator dynamics were included. If we increase the autopilot gain to 0.4, we can see from Fig. 22.13 that the flight-control system response goes unstable. Because there are 24 peaks in 1 s, the frequency of oscillation is 24 Hz

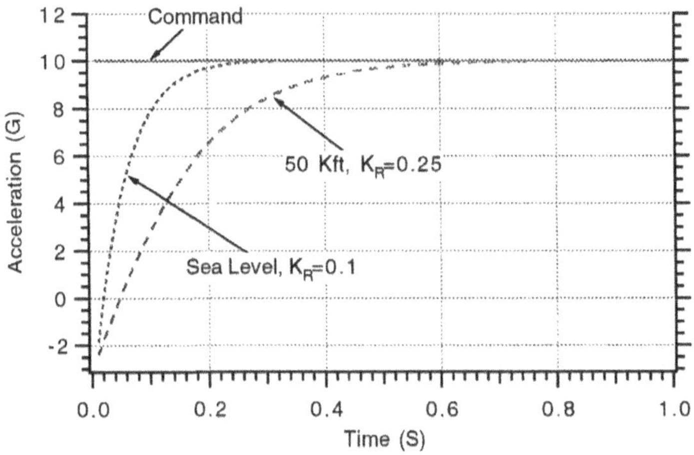

Fig. 22.10 Linear rate gyro flight-control system response is well damped when autopilot gain is a function of flight condition.

or 150 rad/s. Therefore we can see empirically that actuator dynamics place an upper limit on the achievable autopilot gain with the rate gyro flight-control system. When the actuator dynamics were neglected, we could have made the autopilot gain arbitrarily large.

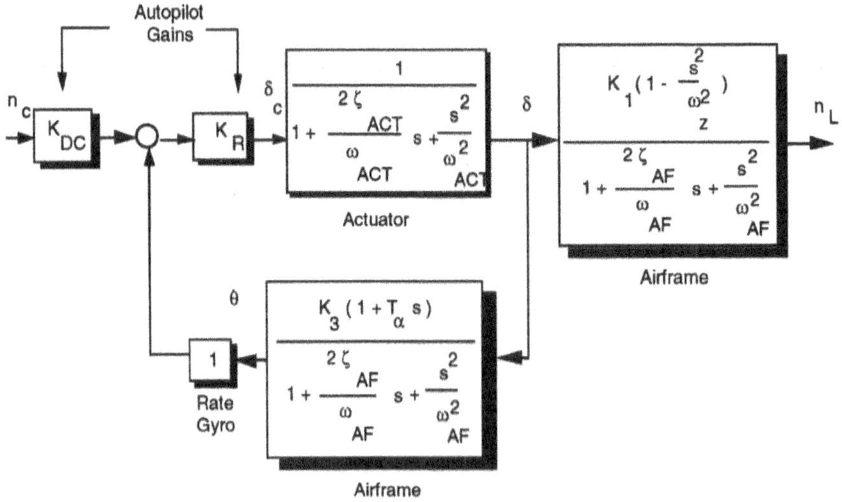

Fig. 22.11 Rate gyro flight-control system with actuator dynamics.

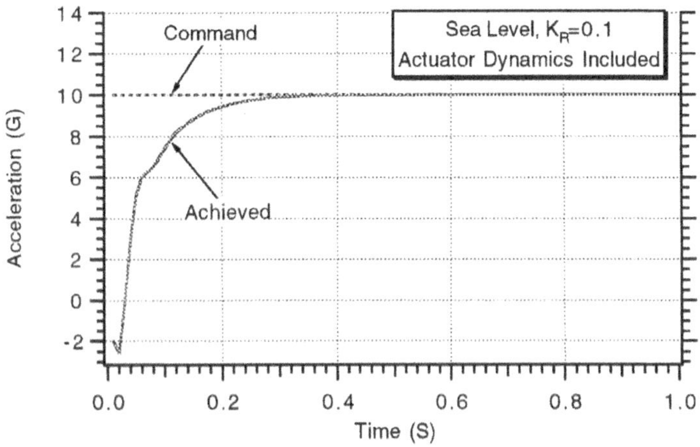

Fig. 22.12 Inclusion of actuator dynamics appears to have very little effect on flight-control system response.

OPEN-LOOP TRANSFER FUNCTION

We have seen through out this text that valuable information is available from the time domain simulation of the system differential equations. However, additional information is also available from the system's open-loop transfer function, which is in the frequency domain. The concept of the open-loop transfer function is the basis of classical feedback control systems analysis [1]. Both relative stability and robustness can be determined from an analysis of the magnitude and phase of the

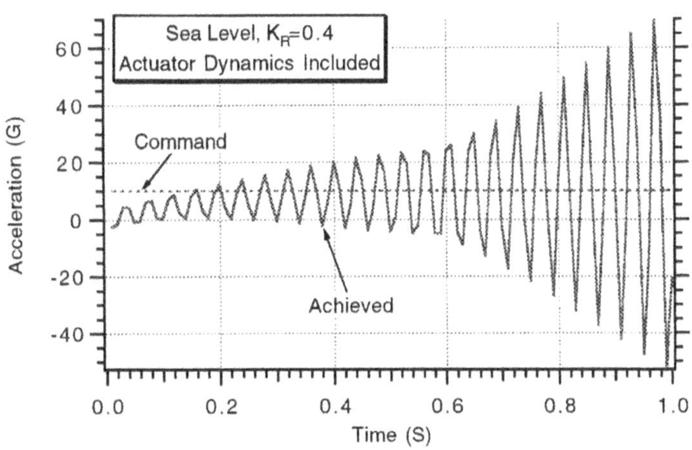

Fig. 22.13 Actuator dynamics have a destabilizing effect when autopilot gain is large.

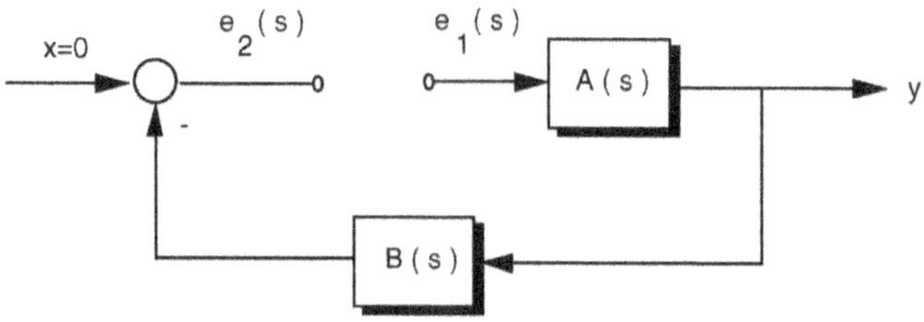

Fig. 22.14 Sample open-loop system.

open-loop frequency response, and, even more importantly, the designer can determine from it what changes to make in order to achieve design goals. The open-loop transfer function is the transfer function around the loop when the loop is broken at a point. Although the loop can be broken anywhere, it is usually broken in series with some parameter whose value the designer can control to achieve a desired characteristic. For example, we can break the loop of a single-loop feedback control system at the error signal as shown in Fig. 22.14. In this case the open-loop transfer is defined as

$$HG(s) = -\frac{e_2(s)}{e_1(s)} = A(s)B(s)$$

To fully understand open-loop concepts, it is first required to understand the mechanics of finding the magnitude and phase of an open-loop transfer function. This can be done by replacing the complex frequency s in the transfer function with

$$s = j\omega$$

where

$$j = \sqrt{-1}$$

Usually the magnitude of the open-loop transfer function is expressed in dB where

$$\mathrm{dB} = 20\log_{10}(\mathrm{Magnitude})$$

and the phase is expressed in degrees.

With the open-loop transfer function other quantities are also important. For example, the gain margin (gm) is the value of additional gain required at the loop break (assuming the phase remains constant) to cause instability while the phase margin φ_{pm} is the amount of phase loss required at the loop break (assuming that the gain remains constant) to cause instability. In addition to these margins, crossover frequencies are also of interest. The gain crossover frequency ω_{CR} is the frequency at which the open-loop magnitude is unity or zero dB, while the phase crossover frequency ω_{180} is the frequency at which the open-loop phase is -180 deg. Both these crossover frequencies indicate the frequency of the ensuing oscillation in the time domain should the system go unstable due to either an increase in gain or decrease in phase.

To demonstrate the utility of the open-loop transfer function, let us revisit the rate gyro flight-control system of Fig. 22.11. Figure 22.15 shows the same system, except this time the loop is broken at the input to the actuator. The loop is broken here because the designer can control the autopilot gain K_R. From the definition of open-loop transfer function, we can express $HG(s)$ as

$$HG(s) = -K_3K_R(1 + T_\alpha s) \Bigg/ \left\{ \left[1 + \frac{2\zeta_{\mathrm{ACT}}}{\omega_{\mathrm{ACT}}} s + \frac{s^2}{\omega_{\mathrm{ACT}}^2} \right] \left[1 + \frac{2\zeta_{\mathrm{AF}}}{\omega_{\mathrm{AF}}} s + \frac{s^2}{\omega_{\mathrm{AF}}^2} \right] \right\}$$

By going to the complex frequency domain, we can rewrite the open-loop transfer function as

$$HG(j\omega) = -K_3K_R(1 + j\omega T_\alpha) \Bigg/ \left\{ \left[1 + \frac{\omega^2}{\omega_{\mathrm{ACT}}^2} + \frac{j2\zeta_{\mathrm{ACT}}\omega}{\omega_{\mathrm{ACT}}} \right] \left[1 + \frac{\omega^2}{\omega_{\mathrm{AF}}^2} + \frac{j2\zeta_{\mathrm{AF}}\omega}{\omega_{\mathrm{AF}}} \right] \right\}$$

where care has been taken in the preceding equation to separate the real and imaginary parts. The magnitude and phase of the open-loop transfer function can now be expressed as

$$|HG(j\omega)| = -K_RK_3$$

$$\times \sqrt{1 + \omega^2 T_\alpha^2 \Bigg/ \left\{ \left[\left[1 - \frac{\omega^2}{\omega_{\mathrm{ACT}}^2} \right]^2 + \left[\frac{2\zeta_{\mathrm{ACT}}\omega}{\omega_{\mathrm{ACT}}} \right]^2 \right] \left[\left[1 - \frac{\omega^2}{\omega_{\mathrm{AF}}^2} \right]^2 + \left[\frac{2\zeta_{\mathrm{AF}}\omega}{\omega_{\mathrm{AF}}} \right]^2 \right] \right\}}$$

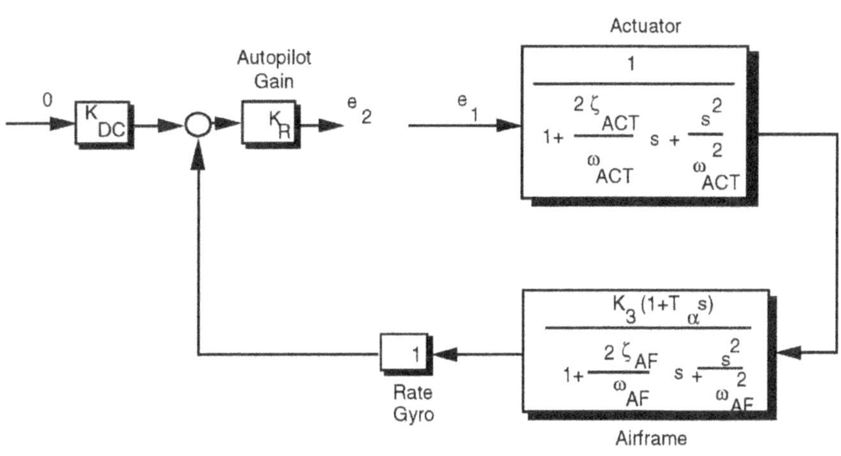

Fig. 22.15 Open-loop model of rate gyro flight-control system.

$$\angle HG(j\omega) = \tan^{-1}\omega T_\alpha - \tan^{-1}\left[\frac{2\zeta_{ACT}\omega}{\omega_{ACT}} \Big/ \left(1 + \frac{\omega^2}{\omega_{ACT}^2}\right)\right] - \tan^{-1}\left[\frac{2\zeta_{AF}\omega}{\omega_{AF}} \Big/ \left(1 - \frac{\omega^2}{\omega_{AF}^2}\right)\right]$$

LISTING 22.3 OPEN-LOOP BODE RESPONSE FOR RATE GYRO FLIGHT-CONTROL SYSTEM

```
count=0;
ZACT=.7;
WACT=150.;
K3=-1.89;
TA=.457;
ZAF=.058;
WAF=25.3;
KR=.1;
for I=2:160
	W=10^(.025*I-1);
	XMAG1=sqrt(1+(W*TA)^2);
	XMAG2=sqrt((1-(W/WAF)^2)^2+(2*ZAF*W/WAF)^2);
	XMAG3=sqrt((1-(W/WACT)^2)^2+(2*ZACT*W/WACT)^2);
	GAIN=20*log10(-K3*KR*XMAG1/(XMAG2*XMAG3));
	PHASE1=57.3*atan2(W*TA,1.);
	PHASE2=57.3*atan2(2*ZAF*W/WAF,1-(W/WAF)^2);
	PHASE3=57.3*atan2(2*ZACT*W/WACT,1-(W/WACT)^2);
	PHASE=PHASE1-PHASE2-PHASE3;
	count=count+1;
	ArrayW(count)=W;
	ArrayGAIN(count)=GAIN;
	ArrayPHASE(count)=PHASE;
end
figure
semilogx(ArrayW,ArrayGAIN),grid
xlabel('Frequency (Rad/Sec)')
ylabel('Gain (Db)')
axis([.1 1000 -60 40])
figure
semilogx(ArrayW,ArrayPHASE),grid
xlabel('Frequency (Rad/Sec)')
ylabel('Phase (Deg)')
axis([.1 1000 -400 100])
clc
output=[ArrayW',ArrayGAIN',ArrayPHASE'];
save datfil.txt output /ascii
disp 'simulation finished'
```

Therefore the open-loop gain (magnitude) and phase can be expressed in conventional units as

$$\text{Gain} = 20\log_{10}|HG(j\omega)| \quad (\text{dB})$$
$$\text{Phase} = 57.3\angle HG(j\omega) \quad (\text{deg})$$

Designers have found several useful ways of displaying open-loop data. One of these ways is a Bode plot in which the magnitude, expressed in dB, and phase, expressed in degrees, are displayed versus frequency on a logarithmic scale [2]. The preceding equations for the magnitude and phase of the rate gyro flight-control system were programmed in order to generate a Bode plot and the resultant program appears in Listing 22.3. Note that in this program we are incrementally updating the frequency logarithmically and then solving for the magnitude and phase. Unlike most other programs in this text, this program runs extremely rapidly because numerical integration is not involved.

Figure 22.16 presents the resultant Bode plot, using the data generated by the MATLAB program. Here we can see that the gain (or magnitude) peaks due to the low airframe damping ($\zeta_{AF} = 0.058$) and then is quickly attenuated due to the dynamics of the actuator. At the gain crossover frequency (that is, the frequency at which gain is zero dB) the phase is -125 deg. Because the phase margin represents the phase departure from -180 deg, the phase margin is 55 deg ($180 - 125 = 55$). At the phase crossover frequency (i.e., frequency at which phase is -180 deg) the gain is -11.5 dB. Because the gain margin represents the gain departure from 0 dB, the gain margin is 11.5 dB. The various margins and crossover frequencies have important practical interpretations. For example, if the system phase is decreased by the phase margin the system will go unstable and oscillate at the gain crossover frequency. If the system gain is increased by the gain margin, the system will go unstable and

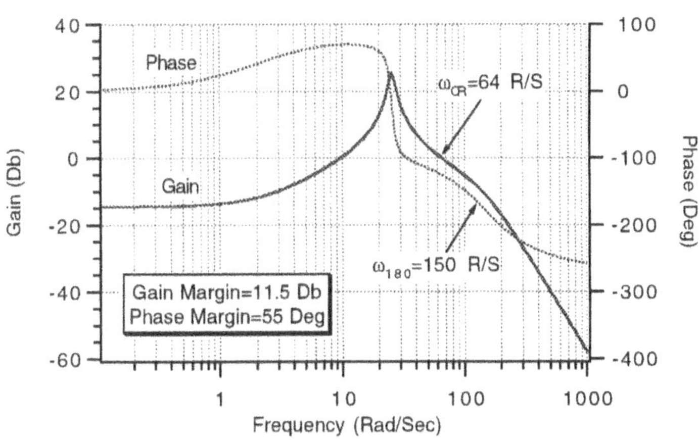

Fig. 22.16 Bode plot for rate gyro flight-control system.

oscillate at the phase crossover frequency. We can see from Fig. 22.16 that the gain and phase crossover frequencies are 64 rad/s and 150 rad/s respectively.

TIME DOMAIN VERIFICATION OF OPEN-LOOP RESULTS [3]

The open-loop analysis of the previous section indicated that the system gain margin was 11.5 dB and the phase crossover frequency was 150 rad/s. This means that if the gain K_R was increased by 11.5 dB the system would go unstable and oscillate at 150 rad/s. A gain increase of 11.5 dB means that K_R must increase from 0.1 to 0.376 to destabilize the system. In other words,

$$20\log_{10}\frac{K_{\text{UNSTABLE}}}{0.1} = 11.5$$

$$\log_{10}\frac{K_{\text{UNSTABLE}}}{0.1} = 0.575$$

$$\frac{K_{\text{UNSTABLE}}}{0.1} = 10^{0.575} = 3.76$$

$$K_{\text{UNSTABLE}} = 0.376$$

We have already seen from Fig. 22.13 that when the autopilot gain was increased to 0.4 the rate gyro flight-control system indeed oscillated at 150 rad/s. This means that the time domain and frequency domain results are in total agreement.

We can also illustrate the concept of phase margin by first observing that an ideal delay can be represented by the transfer function

$$\text{DELAY} = e^{-sT}$$

Converting this representation to the complex frequency domain yields

$$\text{DELAY}(j\omega) = e^{-j\omega T} = \cos\omega T - j\sin\omega T$$

The magnitude and phase of the ideal delay is therefore

$$|\text{DELAY}(j\omega)| = (\cos^2\omega T + \sin^2\omega T)^{\frac{1}{2}} = 1$$

$$\angle\text{DELAY}(j\omega) = \tan^{-1}\left[\frac{\sin\omega T}{\cos\omega T}\right] = -\omega T$$

In summary, an ideal delay can be represented in the frequency domain as a transfer function with unity magnitude and pure phase loss. The phase loss at

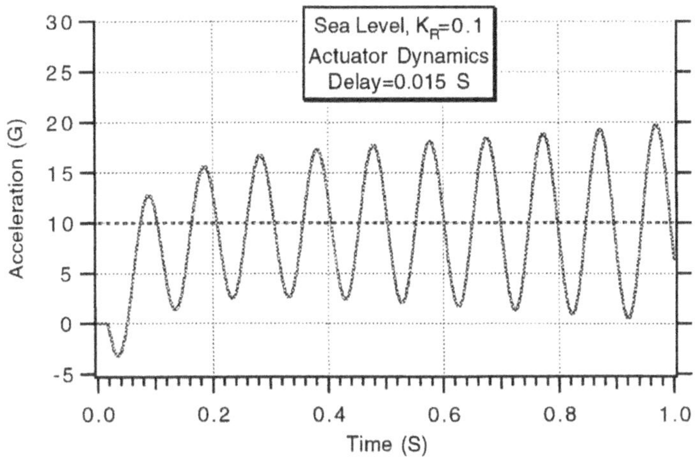

Fig. 22.17 Rate gyro flight-control system goes unstable when delay of 0.015 s is inserted.

64 rad/s (open-loop gain crossover frequency ω_{CR}) in units of rad can be obtained from the preceding equation as

$$\text{DELAY PHASE LOSS} = -64T$$

Therefore a delay of 0.015 s in the time domain corresponds to a phase loss of 55 deg ($64 * 0.015 = 0.96$ rad = 55 deg = phase margin). Because the phase margin of the open-loop system (with the loop broken at K_R) is 55 deg, this means that if a pure delay of 0.015 s were inserted in series with K_R, the system would go unstable and oscillate at a frequency of 64 rad/s (that is, at the open-loop gain crossover frequency). The rate gyro flight-control time domain simulation of Listing 22.2 was modified to include a pure time delay of 0.015 s, and the new simulation appears in Listing 22.4. Again note that a very small integration step size is used to avoid numerical difficulties.

The system step response is shown in Fig. 22.17. Here we can see that the system does go unstable when a delay of 0.015 s is inserted before the actuator. Because there are 10 peaks in the response, the frequency of oscillation is 10 Hz or 62.8 rad/s, which is approximately the gain crossover frequency. Thus again we can see that the time and frequency domain results are in total agreement.

LISTING 22.4 RATE GYRO FLIGHT-CONTROL SYSTEM WITH PURE DELAY

```
count=0;
Z=zeros(size(1:20002));
DELAY=.015;
VM=3000;
```

```
XNCG=10;
ALT=0;
A=1000;
DIAM=1;
FR=3;
XL=20;
CTW=0;
CRW=6;
HW=2;
CTT=0;
CRT=2;
HT=2;
XN=4;
XCG=10;
XHL=19.5;
WGT=1000;
XKR=.1;
WACT=150;
ZACT=.7;
if (ALT < 30000.)
        RHO=.002378*exp(-ALT/30000.);
else
        RHO=.0034*exp(-ALT/22000.);
end
SWING=.5*HW*(CTW+CRW);
STAIL=.5*HT*(CTT+CRT);
SREF=3.1416*DIAM*DIAM/4;
XLP=FR*DIAM;
SPLAN=(XL-XLP)*DIAM+1.33*XLP*DIAM/2;
XCPN=2*XLP/3;
AN=.67*XLP*DIAM;
AB=(XL-XLP)*DIAM;
XCPB=(.67*AN*XLP+AB*(XLP+.5*(XL-XLP)))/(AN+AB);
XCPW=XLP+XN+.7*CRW-.2*CTW;
XMACH=VM/A;
XIYY=WGT*(3*((DIAM/2)^2)+XL*XL)/(12*32.2);
TMP1=(XCG-XCPW)/DIAM;
TMP2=(XCG-XHL)/DIAM;
TMP3=(XCG-XCPB)/DIAM;
TMP4=(XCG-XCPN)/DIAM;
B=sqrt(XMACH^2-1);
Q=.5*RHO*VM*VM;
P1=WGT*XNCG/(Q*SREF);
Y1=2.+8*SWING/(B*SREF)+8*STAIL/(B*SREF);
Y2=1.5*SPLAN/SREF;
Y3=8*STAIL/(B*SREF);
```

```
Y4=2*TMP4+8*SWING*TMP1/(B*SREF)+8*STAIL*TMP2/(B*SREF);
Y5=1.5*SPLAN*TMP3/SREF;
Y6=8*STAIL*TMP2/(B*SREF);
P2=Y2-Y3*Y5/Y6;
P3=Y1-Y3*Y4/Y6;
ALFTR=(-P3+sqrt(P3*P3+4.*P2*P1))/(2.*P2);
DELTR=-Y4*ALFTR/Y6-Y5*ALFTR*ALFTR/Y6;
CNA=2+1.5*SPLAN*ALFTR/SREF+8*SWING/(B*SREF)+8*STAIL/(B*SREF);
CND=8*STAIL/(B*SREF);
CMAP=2*TMP4+1.5*SPLAN*ALFTR*TMP3/SREF+8*SWING*TMP1/(B*SREF);
CMA=CMAP+8*STAIL*TMP2/(B*SREF);
CMD=8*STAIL*TMP2/(B*SREF);
XMA=Q*SREF*DIAM*CMA/XIYY;
XMD=Q*SREF*DIAM*CMD/XIYY;
ZA=-32.2*Q*SREF*CNA/(WGT*VM);
ZD=-32.2*Q*SREF*CND/(WGT*VM);
WZ=sqrt((XMA*ZD-ZA*XMD)/ZD);
WAF=sqrt(-XMA);
ZAF=.5*WAF*ZA/XMA;
XK1=-VM*(XMA*ZD-XMD*ZA)/(1845*XMA);
XK2=XK1;
TA=XMD/(XMA*ZD-XMD*ZA);
XK3=1845*XK1/VM;
XKDC=(1.-XKR*XK3)/(XK1*XKR);
E=0;
ED=0;
DELD=0;
DEL=0;
THD=0;
DELC=0;
DELCP=0;
T=0;
H=.0001;
Z(1)=0;
I=1;
DINT=floor(DELAY/H); %ROUND(X) rounds the elements of X to the nearest integers.
S=0;
while ~(T > .99999)
        S=S+H;
        EOLD=E;
        EDOLD=ED;
        DELOLD=DEL;
        DELDOLD=DELD;
        STEP=1;
        FLAG=0;
        while STEP <=1
```

```
            if FLAG==1
                E=E+H*ED;
                ED=ED+H*EDD;
                DEL=DEL+H*DELD;
                DELD=DELD+H*DELDD;
                T=T+H;
                STEP=2;
            end
            DELCP=XKR*(XKDC*XNCG+THD);
            DELDD=WACT*WACT*(DELC-DEL-2.*ZACT*DELD/WACT);
            EDD=WAF*WAF*(DEL-E-2.*ZAF*ED/WAF);
            XNL=XK1*(E-EDD/WZ^2);
            THD=XK3*(E+TA*ED);
            FLAG=1;
        end
        FLAG=0;
        E=.5*(EOLD+E+H*ED);
        ED=.5*(EDOLD+ED+H*EDD);
        DEL=.5*(DELOLD+DEL+H*DELD);
        DELD=.5*(DELDOLD+DELD+H*DELDD);
        Z(I+1)=DELCP;
        if ((I+1) < DINT)
            DELC=Z(1);
        else
            DELC=Z(I+2-DINT); % I Have changed the code here from FORTRAN!
        end
        I=I+1;
        if S > .00099999
            S=0;
            count=count+1;
            ArrayT(count)=T;
            ArrayXNL(count)=XNL;
            ArrayXNCG(count)=XNCG;
        end
end
figure
plot(ArrayT,ArrayXNL,ArrayT,ArrayXNCG),grid
xlabel('Time (S)')
ylabel('Acceleration (G)')
title('Fig 22.17: Rate gyro flight control system ')
clc
output=[ArrayT', ArrayXNL', ArrayXNCG'];
save datfil.txt output /ascii
disp 'simulation finished'
```

We have just seen that there is a relationship between the time and frequency domains. The gain margin of the open-loop response tells us how much the gain of the flight-control system can be increased by in order for the system to go unstable. When the system oscillates in the time domain due to the increased gain, the frequency of oscillation will be the phase crossover frequency of the open-loop response. The phase margin of the open-loop response tells us the amount of phase loss (or pure time delay) that the system can tolerate before it goes unstable. When the system oscillates in the time domain due to the phase loss (that is, time delay), the frequency of oscillation will be the open-loop gain crossover frequency.

SIMPLIFIED EXPRESSION FOR OPEN-LOOP CROSSOVER FREQUENCY

We have seen that if the autopilot gain K_R is made too large the rate gyro flight-control system can go unstable. To place realistic bounds on the autopilot gain, it is necessary to go back to the frequency domain. If we neglect actuator dynamics, the open-loop transfer function of the rate gyro flight-control system can be written by inspection of Fig. 22.7 as

$$HG = -K_R K_3(1 + T_\alpha s) \Big/ \left(1 + \frac{2\zeta_{AF}}{\omega_{AF}} s + \frac{s^2}{\omega_{AF}^2}\right)$$

The magnitude of the open-loop transfer function can be found from the preceding expression. Recall that the open loop gain crossover frequency occurs when the magnitude of the open-loop transfer function is unity. Therefore if we assume that the open-loop crossover frequency is beyond the airframe dynamics we can say that

$$1 \approx \frac{-K_R K_3 T_\alpha \omega_{CR} \omega_{AF}^2}{\omega_{CR}^2}$$

Solving for the crossover frequency yields

$$\omega_{CR} \approx -K_R K_3 T_\alpha \omega_{AF}^2 = K_R K_3 T_\alpha M_\alpha$$

Therefore we can see that open-loop crossover frequency is proportional to the autopilot gain. Increasing the autopilot gain will increase the crossover frequency. However it is important to note that since we have assumed that the crossover frequency is far beyond the airframe dynamics the preceding expression is approximate. For our 3000 ft/s missile at sea level, the preceding formula indicates that the approximate crossover frequency is 55.2 rad/s or

$$\omega_{CR} = 0.1 * 1.89 * 0.457 * 25.3^2 = 55.2\,\text{rad/s}$$

From our open-loop Bode response we already know that the actual crossover frequency is 60 rad/s. Therefore in this case the approximate expression for the

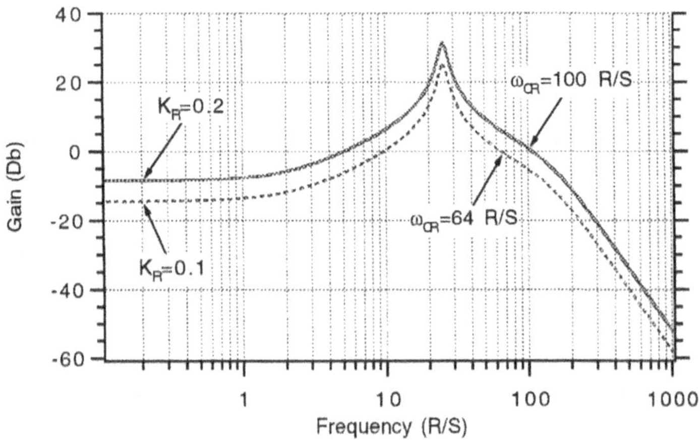

Fig. 22.18 Doubling autopilot gain approximately doubles crossover frequency.

open-loop crossover frequency is in error by slightly less than 10%. The technique for deriving crossover frequency will prove to be very useful when we deal with the three-loop autopilot in the next chapter.

We can simplify the expression for the open-loop crossover frequency even further by recalling that

$$K_1 = \frac{V_M(M_\alpha Z_\delta - Z_\alpha M_\delta)}{1845\,M_\alpha} \approx \frac{V_M Z_\delta}{1845}$$

$$K_3 = \frac{1845\,K_1}{V_M} \approx Z_\delta$$

$$T_\alpha = \frac{M_\delta}{(M_\alpha Z_\delta - Z_\alpha M_\delta)} \approx \frac{M_\delta}{M_\alpha Z_\delta}$$

Therefore substitution yields an even simpler expression for the crossover frequency

$$\omega_{\mathrm{CR}} \approx K_R K_3 T_\alpha M_\alpha = -K_R M_\delta$$

Therefore at a given flight condition the autopilot gain and linearized aerodynamic parameter M_δ determine the open-loop crossover frequency. According to the preceding equation, doubling the autopilot gain from 0.1 to 0.2 should double the crossover frequency from 52.6 rad/s to 105 rad/s. If the open-loop frequency response program of Listing 22.3 is run with actuator dynamics (neglected in approximate analysis), we can see from Fig. 22.18 that the actual crossover frequency increases from 60 rad/s to 103 rad/s when the autopilot gain increases from 0.1 to 0.2. Thus we can see that at the higher crossover frequency, simulation results are more in agreement with the approximate analysis. The agreement is better because the crossover frequency is now far beyond the airframe dynamics. Later on we shall see that the crossover frequency is chosen to be no more than one third of the bandwidth of the actuator to ensure a well-behaved flight-control system response.

SUMMARY

In this chapter we have seen how the flight-control system interacts with the guidance system. The open-loop flight-control system has the dynamics of the bare airframe. This type of flight-control system is usually not acceptable in radar homing applications because of low damping. The rate gyro flight-control system improves the system damping by using a sensor and feedback. We have also seen that the autopilot gain in a rate gyro flight-control system also can be used to control the open-loop crossover frequency.

REFERENCES

[1] Del Toro, V., and Parker, S., *Principles of Control Systems Engineering*, McGraw-Hill, New York, 1960.

[2] D'Azzo, J. J., and Houpis, C. H., *Feedback Control System Analysis and Synthesis*, McGraw-Hill, New York, 1960.

[3] Zarchan, P., "Micro Based Technology—A New Tool for Missile Guidance System Design and Visualization," *AGARD Lecture Series 173*, Oct. 1990, pp 8.1–8.16.

CHAPTER 23

Three-Loop Autopilot

INTRODUCTION

In this chapter we shall see that by adding an accelerometer to the flight-control system we can independently select the system damping, time constant, and open-loop crossover frequency. Controlling the system damping will ensure that the guidance system is not overly sensitive to radome slope effects at the high altitudes. Selecting the system time constant means that we can have adequate performance against maneuvering targets. Finally, controlling the open-loop crossover frequency means that we will have a robust design that is not overly sensitive to unmodeled high frequency dynamics.

THREE-LOOP AUTOPILOT CONFIGURATION

The flight-control system with the three-loop autopilot appears in Fig. 23.1. In this system the rate gyro feeds body rate information into the autopilot while the accelerometer feeds back achieved acceleration information. For simplicity, it has been assumed that the accelerometer location is at the missile center of gravity so that the acceleration sensed is the true acceleration. The three autopilot gains K_A, ω_I, and K_R must be chosen to satisfy some designer-chosen criteria and the gain K_{DC} is computed from the other gains so that the achieved acceleration will match the commanded acceleration. An interesting discussion of the initial design considerations of the three-loop autopilot can be found in [1].

An example of a particularly useful methodology [2–6] in gain selection is to choose the open-loop crossover frequency so that many stability problems can be avoided. In addition, the dominant flight-control system time constant can be selected so that rapid speed of response can be achieved in order to hit maneuvering targets. Finally, adequate damping can also be chosen by the designer to alleviate potential radome coupling problems.

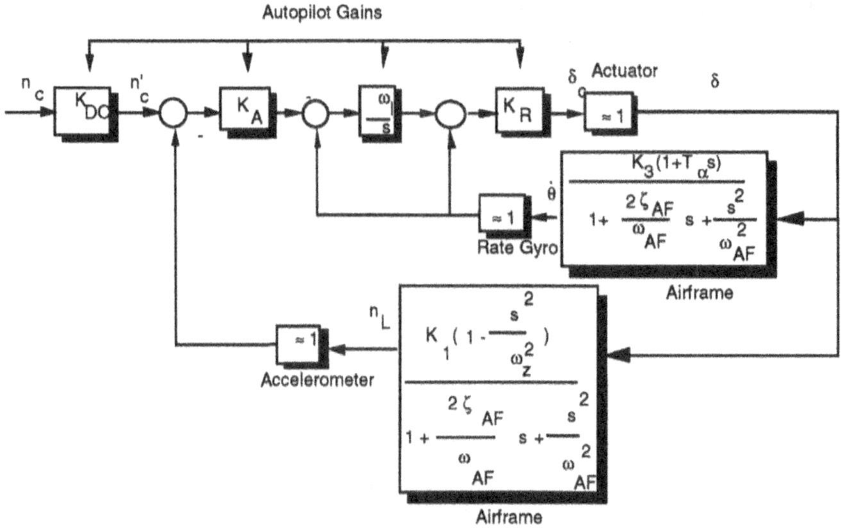

Fig. 23.1 Flight-control system with three-loop autopilot.

OPEN-LOOP ANALYSIS

Because the open-loop crossover frequency has no meaning in the time domain, we must first shift to the frequency domain to see how the autopilot gains influence the crossover frequency. Figure 23.2 shows the three-loop autopilot with the loop broken right before the actuator as was done in Chapter 22. By inspection of Fig. 23.2, we can write an expression for the open- loop transfer function $HG(s)$ as

$$HG(s) = -\frac{y}{x} = -K_R\left[G_3 + \frac{G_3\omega_I}{s} + \frac{G_1K_A\omega_I}{s}\right]$$

where G_1 and G_3 are shorthand notation for the airframe transfer functions that were derived in Chapter 21 and are given by

$$G_1 = \frac{n_L}{\delta} = K_1\left(1 - \frac{s^2}{\omega_z^2}\right) \bigg/ \left(1 + \frac{2\zeta_{AF}}{\omega_{AF}}s + \frac{s^2}{\omega_{AF}^2}\right)$$

$$G_3 = \frac{\dot{\theta}}{\delta} = K_3(1 + T_\alpha s) \bigg/ \left(1 + \frac{2\zeta_{AF}}{\omega_{AF}}s + \frac{s^2}{\omega_{AF}^2}\right)$$

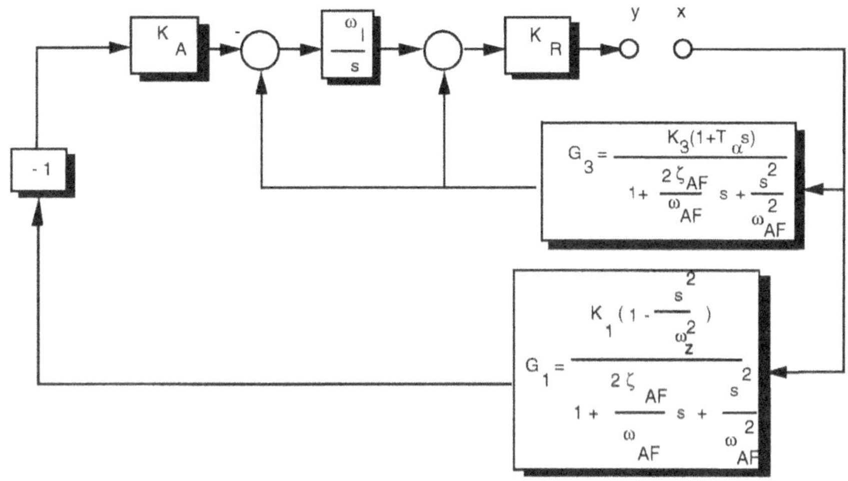

Fig. 23.2 Open-loop representation of three-loop autopilot.

After much algebra the open-loop transfer function can be rewritten as

$$HG = -K_R\omega_1 K_A\left(\frac{K_3}{K_A}+K_1\right)\bigg/\left[s\left(1+\frac{2\zeta_{\mathrm{AF}}}{\omega_{\mathrm{AF}}}s+\frac{s^2}{\omega_{\mathrm{AF}}^2}\right)\right]$$

$$\times\left[1+\frac{s(K_3+\omega_I K_3 T_\alpha)}{\omega_I K_3+K_A\omega_I K_1}+\frac{s^2\left[K_3T_\alpha-\left(K_A\omega_I K_1/\omega_z^2\right)\right]}{\omega_I K_3+K_A\omega_I K_1}\right]$$

which simplifies to

$$HG = -K_0\left(1+\frac{2\zeta_0}{\omega_0}s+\frac{s^2}{\omega_0^2}\right)\bigg/\left[s\left(1+\frac{2\zeta_{\mathrm{AF}}}{\omega_{\mathrm{AF}}}s+\frac{s^2}{\omega_{\mathrm{AF}}^2}\right)\right]$$

where the gain K_0 is given by

$$K_0 = K_R\omega_I K_A\left(\frac{K_3}{K_A}+K_1\right)$$

and the numerator coefficients can be expressed in terms of the autopilot gains and aerodynamic parameters according to

$$\frac{2\zeta_0}{\omega_0}=\frac{(K_3+\omega_I K_3T_\alpha)}{\omega_I K_3+K_A\omega_I K_1}$$

$$\frac{1}{\omega_0^2}=\frac{\left[K_3T_\alpha-\left(K_A\omega_I K_1/\omega_z^2\right)\right]}{\omega_I K_3+K_A\omega_I K_1}$$

If we define

$$K_c = \frac{K_3}{K_A K_1}$$

then

$$\frac{2\zeta_0}{\omega_0} = \frac{(K_c/\omega_I) + K_c T_\alpha}{1 + K_c}$$

$$\frac{1}{\omega_0^2} = \frac{-(1/\omega_z^2) + (K_c T_\alpha/\omega_I)}{1 + K_c}$$

If we define the intermediate gain K by

$$K = K_R \omega_I K_A$$

We can say that

$$K_0 = K_R \omega_I K_A[(K_3/K_A) + K_1] = K(K_1 K_c + K_1) = K K_1(1 + K_c)$$

If we assume that the crossover frequency is beyond the airframe dynamics, we can set the magnitude of the open-loop crossover frequency to unity as we did in the previous chapter in order to obtain

$$1 \approx \frac{-(K_0 \omega_{\mathrm{CR}}^2/\omega_0^2)}{(\omega_{\mathrm{CR}} \omega_{\mathrm{CR}}^2/\omega_{\mathrm{AF}}^2)} = \frac{-K_0 \omega_{\mathrm{AF}}^2}{\omega_{\mathrm{CR}} \omega_0^2}$$

Solving for the open-loop crossover frequency yields

$$\omega_{\mathrm{CR}} = \frac{-K_0 \omega_{\mathrm{AF}}^2}{\omega_0^2}$$

From the preceding equation we can see that the open-loop crossover frequency is a function of both the aerodynamics and the autopilot gains.

CLOSED-LOOP ANALYSIS

Now we can go back to the time domain to complete the autopilot design. By inspection of Fig. 23.1, we can write an expression for the relationship between the output control surface deflection and the input acceleration command as

$$\frac{\delta}{n_c'} = \frac{-K_A K_R \omega_I/s}{1 - K_R G_3 - (K_R \omega_I G_3/s) - (K_R \omega_I K_A G_1/s)}$$

Rewriting the preceding expression in terms of the open-loop transfer function yields

$$\frac{\delta}{n_c'} = \frac{-K_A K_R \omega_I / s}{1 + HG}$$

The relationship between the flight-control system output acceleration and the input command can then be obtained from the chain rule as

$$\frac{n_L}{n_c'} = \frac{\delta}{n_c'} * \frac{n_L}{\delta} = \left[\frac{-K_A K_R \omega_I / s}{1 + HG}\right] \times \left[K_1\left(1 - \frac{s^2}{\omega_z^2}\right) \Big/ \left(1 + \frac{2\zeta_{AF}}{\omega_{AF}} s + \frac{s^2}{\omega_{AF}^2}\right)\right]$$

After much algebra one can show that the flight-control system transfer function becomes

$$\frac{n_L}{n_c'} = \frac{K_A K_R \omega_I K_1 \left(1 - \dfrac{s^2}{\omega_z^2}\right) \Big/ K_0}{1 + s\left[\dfrac{2\zeta_0}{\omega_0} - \dfrac{1}{K_0}\right] + s^2\left[\dfrac{1}{\omega_0^2} - \dfrac{2\zeta_{AF}}{\omega_{AF} K_0}\right] - \dfrac{s^3}{\omega_{AF}^2 K_0}}$$

We are not interested in controlling the airframe zeros (that is, the numerator in flight-control system transfer function) but would like to have the preceding closed-loop transfer function to have the form

$$\frac{n_L}{n_c'} = \frac{K_A K_R \omega_I K_1 \left(1 - \dfrac{s^2}{\omega_z^2}\right) \Big/ K_0}{(1 + \tau s)\left[1 + \dfrac{2\zeta s}{\omega} + \dfrac{s^2}{\omega^2}\right]}$$

where we have made the denominator of the flight-control system transfer function a real pole times a quadratic. The two preceding flight-control system transfer functions are equivalent if the denominators are the same or

$$\frac{2\zeta}{\omega} + \tau = \frac{2\zeta_0}{\omega_0} - \frac{1}{K_0}$$

$$\frac{1}{\omega^2} + \frac{2\zeta\tau}{\omega} = \frac{1}{\omega_0^2} - \frac{2\zeta_{AF}}{\omega_{AF} K_0}$$

$$\frac{\tau}{\omega^2} = -\frac{1}{\omega_{AF}^2 K_0}$$

In addition, we have already derived an expression for the open-loop crossover frequency to be

$$\omega_{CR} = \frac{-K_0\omega_{AF}^2}{\omega_0^2}$$

The preceding four equations have four unknowns. At a given flight condition we know the aerodynamics or ω_{AF} and ζ_{AF}. If we specify the desired time constant τ, damping ζ, and open-loop crossover frequency ω_{CR} of the flight-control system, the remaining four unknowns are ω, ζ_0, ω_0, and K_0. Because we have four equations, there is sufficient information to solve for these four unknowns.

If we solve the fourth equation for K_0 and substitute it into the second equation, we get

$$\frac{1}{\omega^2} + \frac{2\zeta\tau}{\omega} = \frac{1}{\omega_0^2} - \frac{2\zeta_{AF}}{\omega_{AF}K_0} = \frac{1}{\omega_0^2} + \frac{2\zeta_{AF}\omega_{AF}^2}{\omega_{AF}\omega_{CR}\omega_0^2} = \frac{1}{\omega_0^2}\left(1 + \frac{2\zeta_{AF}\omega_{AF}}{\omega_{CR}}\right)$$

We can also substitute K_0 into the third equation yielding

$$\frac{\tau}{\omega^2} = -\frac{1}{\omega_{AF}^2 K_0} = \frac{\omega_{AF}^2}{\omega_{AF}^2\omega_{CR}\omega_0^2} = \frac{1}{\omega_{CR}\omega_0^2}$$

Substituting this result into the preceding equation yields

$$\frac{1}{\omega^2} + \frac{2\zeta\tau}{\omega} = \frac{\omega_{CR}\tau}{\omega^2}\left(1 + \frac{2\zeta_{AF}\omega_{AF}}{\omega_{CR}}\right)$$

Note that all of the terms in the preceding equation are known except for ω. We can solve the preceding equation for ω yielding

$$\omega = \left[\tau\omega_{CR}\left(1 + \frac{2\zeta_{AF}\omega_{AF}}{\omega_{CR}}\right) - 1\right] \Big/ (2\zeta\tau)$$

Because we already know that

$$\frac{\tau}{\omega^2} = \frac{1}{\omega_{CR}\omega_0^2}$$

we can solve for ω_0 in terms of known quantities as

$$\omega_0 = \frac{\omega}{\sqrt{\tau\omega_{CR}}}$$

Substituting the solutions for ω, ω_0, and K_0 into the first of the four equations with four unknowns allows us to solve for ζ_0 as

$$\zeta_0 = .5\omega_0\left(\frac{2\zeta}{\omega} + \tau + \frac{1}{K_0}\right) = .5\omega_0\left[\frac{2\zeta}{\omega} + \tau - \frac{\omega_{AF}^2}{\omega_{CR}\omega_0^2}\right]$$

Recalling that

$$\frac{2\zeta_0}{\omega_0} = \left[\left(\frac{K_c}{\omega_I} + K_c T_\alpha\right) \Big/ (1 + K_c)\right] = a_1$$

$$\frac{1}{\omega_0^2} = -\left[\left(\frac{1}{\omega_z^2} + \frac{K_c T_\alpha}{\omega_I}\right) \Big/ (1 - K_c)\right] = a_2$$

we would like to solve for K_c and eliminate ω_I. Therefore

$$a_1(1 + K_c) = \frac{K_c}{\omega_I} + K_c T_\alpha$$

$$a_2(1 + K_c) = \frac{T_\alpha K_c}{\omega_I} + \frac{1}{\omega_z^2}$$

Multiplying the first of the preceding two equations by T_a on both sides and then subtracting it from the second equation yields

$$a_2(1 + K_c) - a_1 T_\alpha (1 + K_c) = -(1/\omega_z^2) - T_\alpha^2 K_c$$

Therefore

$$K_c(a_2 - a_1 T_\alpha + T_\alpha^2) = -(1/\omega_z^2) - a_2 + a_1 T_\alpha$$

Solving for K_c yields

$$K_c = \frac{-(1/\omega_z^2) - a_2 + a_1 T_\alpha}{a_2 - a_1 T_\alpha + T_\alpha^2}$$

Substitution of the expressions for a_1 and a_2 into the preceding equation yields after some algebra

$$K_c = \frac{-(\omega_0^2/\omega_z^2) - 1 + 2\zeta_0 \omega_0 T_\alpha}{1 - 2\zeta_0 \omega_0 T_\alpha + \omega_0^2 T_\alpha^2}$$

Because we already know that

$$K_c = \frac{K_3}{K_A K_1}$$

we can solve for the autopilot gain K_A yielding

$$K_A = \frac{K_3}{K_c K_1}$$

Because we already know that

$$\frac{1}{\omega_0^2}=\left[\left(\frac{-1}{\omega_z^2}+\frac{K_cT_\alpha}{\omega_I}\right)\Big/(1+K_c)\right]$$

we can invert the preceding equation and solve for the autopilot gain ω_I or

$$\omega_I=\frac{T_\alpha K_c\omega_0^2}{1+K_c+\omega_0^2/\omega_z^2}$$

Recall that

$$\frac{\tau}{\omega^2}=-\frac{1}{\omega_{AF}^2K_0}$$

We can now solve for K_0 yielding

$$K_0=-\frac{\omega^2}{\tau\omega_{AF}^2}$$

Recall that

$$K_0=KK_1(1+K_c)$$

We can now invert the preceding expression in order to solve for K or

$$K=\frac{K_0}{K_1(1+K_c)}$$

Because

$$K=K_R\omega_IK_A$$

we can solve for the final autopilot gain K_R yielding

$$K_R=\frac{K}{K_A\omega_I}$$

Finally, in order to get unity flight-control system gain we set the gain of the closed-loop transfer function to unity or

$$\frac{K_{DC}K_AK_R\omega_IK_1}{K_0}=1$$

and get

$$K_{DC}=\frac{K_0}{K_AK_R\omega_IK_1}=\frac{K_0}{KK_1}=\frac{KK_1(1+K_c)}{KK_1}=1+K_c$$

which simplifies to

$$K_{DC} = 1 + \frac{1}{K_A V_M}$$

We now have enough information to simulate the three-loop flight-control system with the autopilot gain algorithm we just derived. Listing 23.1 presents a time domain step response simulation of the three-loop autopilot in the presence of the linear airframe dynamics. We can see from the listing that the required aerodynamic parameters are derived from the geometry of the airframe. For a given flight condition these airframe parameters are used to both describe the airframe and to determine the autopilot gains using the gain algorithm we just derived. Although the original gain algorithm derivation neglected the dynamics of the actuator, these dynamics are included in the step response simulation to test the robustness of the autopilot gains. The actuator is modeled as a second-order transfer function or

$$\frac{\delta}{\delta_c} = 1 \Bigg/ \left(1 + \frac{2\zeta_{\mathrm{ACT}}}{\omega_{\mathrm{ACT}}} s + \frac{s^2}{\omega_{\mathrm{ACT}}^2}\right)$$

with a natural frequency ω_{ACT} of 150 rad/s and damping ζ_{ACT} of 0.7.

Using Listing 23.1 for the flight condition in which the missile was at sea level and traveling at 3000 ft/s, a 10-*g* command was issued to the autopilot. We can see from Listing 23.1 that the nominal design goals given the autopilot gain algorithm were to achieve a time constant of 0.3 s, an open-loop crossover frequency of 50 rad/s, and a damping of 0.7. The three autopilot gains for these requirements at this flight condition turn out to be $K_A = 1.15$ deg/g-s, $\omega_I = 12.9$ rad/s, and $K_R = 0.0928$ s. We can see from Fig. 23.3 that the overall time constant of the

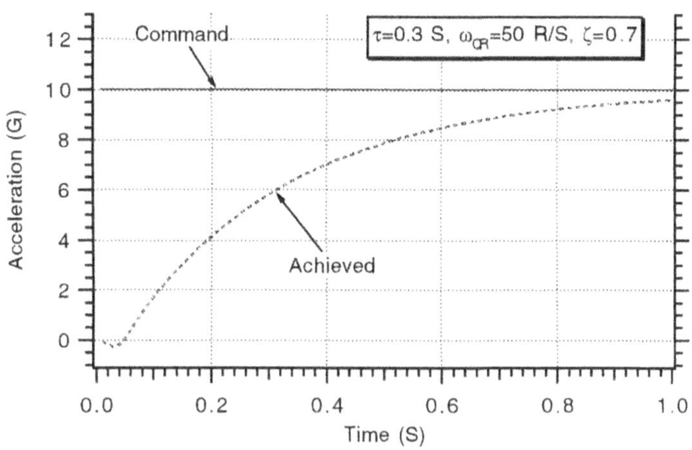

Fig. 23.3 Autopilot gain algorithm allows us to select time constant.

flight-control system is slightly in excess of 0.3 s. This is not in disagreement with theory since the overall time constant of the third-order flight-control system can be approximated as

$$\tau_{\mathrm{TOT}} = \tau + \frac{2\zeta}{\omega}$$

Because the damping is 0.7 and the natural frequency is approximately 35 rad/s, the overall or total time constant is 0.34 s, which is consistent with Fig. 23.3. In other words the response reaches 63% of the steady-state value in 0.34 s.

LISTING 23.1 THREE-LOOP AUTOPILOT STEP RESPONSE SIMULATION

```
count=0;
FR=3.;
DIAM=1.;
XL=20.;
CTW=0.;
CRW=6.;
HW=2.;
CTT=0.;
CRT=2.;
HT=2.;
XN=4.;
XCG=10.;
XHL=19.5;
WACT=150.;
ZACT=.7;
TF=1.;
VM=3000.;
XNCG=10.;
WCR=50.;
ZETA=.7;
TAU=.3;
ALT=0.;
A=1000.;
if ALT<=30000.
        RHO=.002378*exp(-ALT/30000.);
else
        RHO=.0034*exp(-ALT/22000.);
end
WGT=1000.;
XNLLIN=0.;
XACC=XCG;
SWING=.5*HW*(CTW+CRW);
STAIL=.5*HT*(CTT+CRT);
```

```
SREF=3.1416*DIAM*DIAM/4.;
XLP=FR*DIAM;
SPLAN=(XL-XLP)*DIAM+1.33*XLP*DIAM/2.;
XCPN=2*XLP/3;
AN=.67*XLP*DIAM;
AB=(XL-XLP)*DIAM;
XCPB=(.67*AN*XLP+AB*(XLP+.5*(XL-XLP)))/(AN+AB);
XCPW=XLP+XN+.7*CRW-.2*CTW;
XMACH=VM/A;
XIYY=WGT*(3*((DIAM/2)^2)+XL*XL)/(12*32.2);
TMP1=(XCG-XCPW)/DIAM;
TMP2=(XCG-XHL)/DIAM;
TMP3=(XCG-XCPB)/DIAM;
TMP4=(XCG-XCPN)/DIAM;
B=sqrt(XMACH^2-1);
Q=.5*RHO*VM*VM;
P1=WGT*XNCG/(Q*SREF);
Y1=2+8*SWING/(B*SREF)+8*STAIL/(B*SREF);
Y2=1.5*SPLAN/SREF;
Y3=8*STAIL/(B*SREF);
Y4=2*TMP4+8*SWING*TMP1/(B*SREF)+8*STAIL*TMP2/(B*SREF);
Y5=1.5*SPLAN*TMP3/SREF;
Y6=8*STAIL*TMP2/(B*SREF);
P2=Y2-Y3*Y5/Y6;
P3=Y1-Y3*Y4/Y6;
ALFTR=(-P3+sqrt(P3*P3+4.*P2*P1))/(2.*P2);
DELTR=-Y4*ALFTR/Y6-Y5*ALFTR*ALFTR/Y6;
CNA=2+1.5*SPLAN*ALFTR/SREF+8*SWING/(B*SREF)+8*STAIL/(B*SREF);
CND=8*STAIL/(B*SREF);
CMAP=2*TMP4+1.5*SPLAN*ALFTR*TMP3/SREF+8*SWING*TMP1/(B*SREF);
CMA=CMAP+8*STAIL*TMP2/(B*SREF);
CMD=8*STAIL*TMP2/(B*SREF);
XMA=Q*SREF*DIAM*CMA/XIYY;
XMD=Q*SREF*DIAM*CMD/XIYY;
ZA=-32.2*Q*SREF*CNA/(WGT*VM);;
ZD=-32.2*Q*SREF*CND/(WGT*VM);
WZ=sqrt((XMA*ZD-ZA*XMD)/ZD);
WAF=sqrt(-XMA);
ZAF=.5*WAF*ZA/XMA;
XK1=-VM*(XMA*ZD-XMD*ZA)/(1845*XMA);
XK2=XK1;
TA=XMD/(XMA*ZD-XMD*ZA);
XK3=1845*XK1/VM;
W=(TAU*WCR*(1+2.*ZAF*WAF/WCR)-1)/(2*ZETA*TAU);
W0=W/sqrt(TAU*WCR);
Z0=.5*W0*(2*ZETA/W+TAU-WAF^2/(W0*W0*WCR));
```

```
XKC=(-W0^2/WZ^2-1.+2.*Z0*W0*TA)/(1.-2.*Z0*W0*TA+W0*W0*TA*TA);
XKA=XK3/(XK1*XKC);;
XK0=-W*W/(TAU*WAF*WAF);
XK=XK0/(XK1*(1+XKC));
WI=XKC*TA*W0*W0/(1+XKC+W0^2/WZ^2);
XKR=XK/(XKA*WI);
XKDC=1.+1845./(XKA*VM);
E=0.;
ED=0.;
DELD=0.;
DEL=0.;
X=0.;
T=0;
H=.0001;
S=0;
while T<=(TF-.00001)
	EOLD=E;
	EDOLD=ED;
	DELOLD=DEL;
	DELDOLD=DELD;
	XOLD=X;
	STEP=1;
	FLAG=0;
	while STEP<=1
		if FLAG==1
			STEP=2;
			E=E+H*ED;
			ED=ED+H*EDD;
			DEL=DEL+H*DELD;
			DELD=DELD+H*DELDD;
			X=X+H*XD;
			T=T+H;
		end
		THD=XK3*(E+TA*ED);
		DELC=XKR*(X+THD);
		DELDD=WACT*WACT*(DELC-DEL-2.*ZACT*DELD/WACT);
		EDD=WAF*WAF*(DEL-E-2.*ZAF*ED/WAF);
		XNL=XK1*(E-EDD/WZ^2);
		XD=WI*(THD+XKA*(XNL-XNCG*XKDC));
		FLAG=1;
	end
	FLAG=0;
	E=.5*(EOLD+E+H*ED);
	ED=.5*(EDOLD+ED+H*EDD);
	DEL=.5*(DELOLD+DEL+H*DELD);
	DELD=.5*(DELDOLD+DELD+H*DELDD);
	X=.5*(XOLD+X+H*XD);
```

```
        S=S+H;
        if S>.0099999
                S=0.;
                count=count+1;
                ArrayT(count)=T;
                ArrayXNL(count)=XNL;
                ArrayXNCG(count)=XNCG;
        end
end
figure
plot(ArrayT,ArrayXNL,ArrayT,ArrayXNCG),grid
xlabel('Time (Sec)')
ylabel('Acceleration (G) ')
clc
output=[ArrayT',ArrayXNL',ArrayXNCG'];
save datfil.txt output /ascii
disp 'simulation finished'
```

The open-loop transfer function for the three-loop flight-control system neglecting actuator dynamics has already been derived as

$$HG_{\substack{NO\\ACTUATOR}} = -K_0\left(1+\frac{2\zeta_0}{\omega_0}s+\frac{s^2}{\omega_0^2}\right)\Bigg/\left[s\left(1+\frac{2\zeta_{AF}}{\omega_{AF}}s+\frac{s^2}{\omega_{AF}^2}\right)\right]$$

If the dynamics of the actuator are now included, it is easy to show that the open-loop transfer function is simply a multiplication of the previous open-loop transfer function with the actuator transfer function or

$$HG_{ACTUATOR} = \frac{-K_0\left(1+\dfrac{2\zeta_0}{\omega_0}s+\dfrac{s^2}{\omega_0^2}\right)}{s\left(1+\dfrac{2\zeta_{AF}}{\omega_{AF}}s+\dfrac{s^2}{\omega_{AF}^2}\right)\left(1+\dfrac{2\zeta_{ACT}}{\omega_{ACT}}s+\dfrac{s^2}{\omega_{ACT}^2}\right)}$$

Therefore the magnitude and phase of the open-loop transfer function can be written by inspection of the previous expression as

$$|HG_{ACTUATOR}| = \frac{-K_0}{\omega}\sqrt{\frac{\left(1-\dfrac{\omega^2}{\omega_0^2}\right)^2+\left(\dfrac{2\zeta_0\omega}{\omega_0}\right)^2}{\left[\left(1-\dfrac{\omega^2}{\omega_{AF}^2}\right)^2+\left(\dfrac{2\zeta_{AF}\omega}{\omega_{AF}}\right)^2\right]\left[\left(1-\dfrac{\omega^2}{\omega_{ACT}^2}\right)^2+\left(\dfrac{2\zeta_{ACT}\omega}{\omega_{ACT}}\right)^2\right]}}$$

$$\phi = -90+57.3\left[\tan^{-1}\left[\frac{\dfrac{2\zeta_0\omega}{\omega_0}}{1-\dfrac{\omega^2}{\omega_0^2}}\right]-\tan^{-1}\left[\frac{\dfrac{2\zeta_{AF}\omega}{\omega_{AF}}}{1-\dfrac{\omega^2}{\omega_{AF}^2}}\right]-\tan^{-1}\left[\frac{\dfrac{2\zeta_{ACT}\omega}{\omega_{ACT}}}{1-\dfrac{\omega^2}{\omega_{ACT}^2}}\right]\right]$$

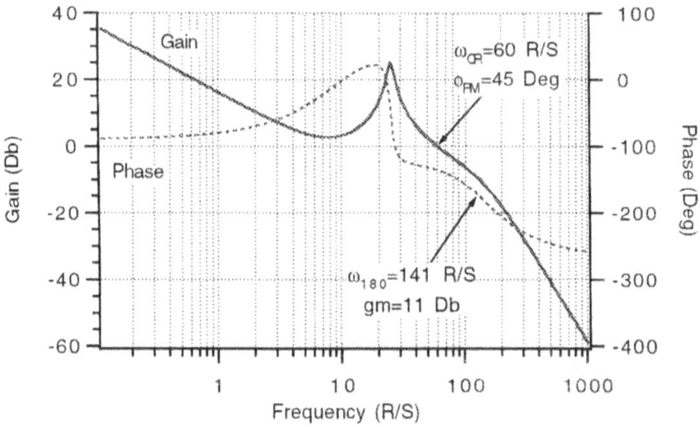

Fig. 23.4 Desired crossover frequency is achieved—even in presence of actuator dynamics.

Listing 23.2 is a program that evaluates the magnitude and phase of the open-loop transfer function (the preceding two equations) for different frequencies. We can see from the listing that the magnitude is expressed in units of dB whereas the phase is expressed in units of deg. The program runs quickly because numerical integration is not involved.

The output of Listing 23.2 is a Bode plot as shown in Fig. 23.4. We can see that the achieved gain crossover frequency of 60 rad/s is close to the desired goal of 50 rad/s. We did not meet the exact design goal because in the derivation of the formula for the open-loop gain crossover frequency it was assumed that the

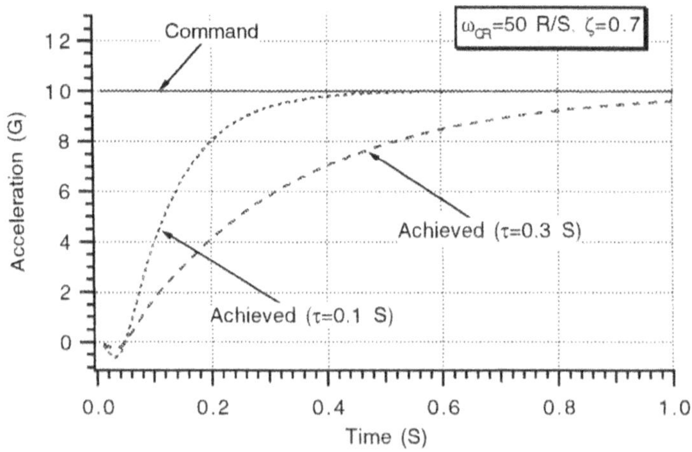

Fig. 23.5 Time constant control can be achieved with autopilot gain algorithm.

crossover frequency was far beyond the dynamics of the airframe, which is not quite true. Figure 23.4 also indicates that the flight-control system stability margins are adequate since the gain margin (gm) is 11 dB and the phase margin (ϕ_{PM}) is 45 deg.

Figure 23.5 indicates that the autopilot gain algorithm is successful in controlling the flight-control system time constant. When the desired time constant is reduced to 0.1 s, the new gains become $K_A = 3.72$ deg/g-s, $\omega_I = 11.2$ rad/s, and $K_R = 0.098$ s. Recall that when the desired autopilot time constant was 0.3 s the autopilot gains were $K_A = 1.15$ deg/g-s, $\omega_I = 12.9$ rad/s, and $K_R = 0.0928$ s.

LISTING 23.2 OPEN-LOOP RESPONSE OF THREE-LOOP AUTOPILOT

```
count=0;
FR=3.;
DIAM=1.;
XL=20.;
CTW=0.;
CRW=6.;
HW=2.;
CTT=0.;
CRT=2.;
HT=2.;
XN=4.;
XCG=10.;
XHL=19.5;
WACT=150.;
ZACT=.7;
VM=3000.;
XNCG=10.;
WCR=50.;
ZETA=.7;
TAU=.3;
ALT=0.;
A=1000.;
if ALT<=30000.
        RHO=.002378*exp(-ALT/30000.);
else
        RHO=.0034*exp(-ALT/22000.);
end
WGT=1000.;
XNLLIN=0.;
XACC=XCG;
SWING=.5*HW*(CTW+CRW);
STAIL=.5*HT*(CTT+CRT);
SREF=3.1416*DIAM*DIAM/4.;
```

```
XLP=FR*DIAM;
SPLAN=(XL-XLP)*DIAM+1.33*XLP*DIAM/2.;
XCPN=2*XLP/3;
AN=.67*XLP*DIAM;
AB=(XL-XLP)*DIAM;
XCPB=(.67*AN*XLP+AB*(XLP+.5*(XL-XLP)))/(AN+AB);
XCPW=XLP+XN+.7*CRW-.2*CTW;
XMACH=VM/A;
XIYY=WGT*(3*((DIAM/2)^2)+XL*XL)/(12*32.2);
TMP1=(XCG-XCPW)/DIAM;
TMP2=(XCG-XHL)/DIAM;
TMP3=(XCG-XCPB)/DIAM;
TMP4=(XCG-XCPN)/DIAM;
B=sqrt(XMACH^2-1);
Q=.5*RHO*VM*VM;
P1=WGT*XNCG/(Q*SREF);
Y1=2+8*SWING/(B*SREF)+8*STAIL/(B*SREF);
Y2=1.5*SPLAN/SREF;
Y3=8*STAIL/(B*SREF);
Y4=2*TMP4+8*SWING*TMP1/(B*SREF)+8*STAIL*TMP2/(B*SREF);
Y5=1.5*SPLAN*TMP3/SREF;
Y6=8*STAIL*TMP2/(B*SREF);
P2=Y2-Y3*Y5/Y6;
P3=Y1-Y3*Y4/Y6;
ALFTR=(-P3+sqrt(P3*P3+4.*P2*P1))/(2.*P2);
DELTR=-Y4*ALFTR/Y6-Y5*ALFTR*ALFTR/Y6;
CNA=2+1.5*SPLAN*ALFTR/SREF+8*SWING/(B*SREF)+8*STAIL/(B*SREF);
CND=8*STAIL/(B*SREF);
CMAP=2*TMP4+1.5*SPLAN*ALFTR*TMP3/SREF+8*SWING*TMP1/(B*SREF);
CMA=CMAP+8*STAIL*TMP2/(B*SREF);
CMD=8*STAIL*TMP2/(B*SREF);
XMA=Q*SREF*DIAM*CMA/XIYY;
XMD=Q*SREF*DIAM*CMD/XIYY;
ZA=-32.2*Q*SREF*CNA/(WGT*VM);
ZD=-32.2*Q*SREF*CND/(WGT*VM);
WZ=sqrt((XMA*ZD-ZA*XMD)/ZD);
WAF=sqrt(-XMA);
ZAF=.5*WAF*ZA/XMA;
XK1=-VM*(XMA*ZD-XMD*ZA)/(1845*XMA);
XK2=XK1;
TA=XMD/(XMA*ZD-XMD*ZA);
XK3=1845*XK1/VM;
W=(TAU*WCR*(1+2.*ZAF*WAF/WCR)-1)/(2*ZETA*TAU);
W0=W/sqrt(TAU*WCR);
Z0=.5*W0*(2*ZETA/W+TAU-WAF^2/(W0*W0*WCR));
XKC=(-W0^2/WZ^2-1.+2.*Z0*W0*TA)/(1.-2.*Z0*W0*TA+W0*W0*TA*TA);
```

```
XKA=XK3/(XK1*XKC);
XK0=-W*W/(TAU*WAF*WAF);
XK=XK0/(XK1*(1+XKC));
WI=XKC*TA*W0*W0/(1+XKC+W0^2/WZ^2);
XKR=XK/(XKA*WI);
XKDC=1.+1845./(XKA*VM);
for I=2:160
        W=10^(.025*I-1);
        XMAGTOP=-XK0*sqrt((1.-(W/W0)^2)^2+(2.*Z0*W/W0)^2);
        XMAGBOT=W*sqrt((1.-(W/WAF)^2)^2+(2.*ZAF*W/WAF)^2);
        XMAG=XMAGTOP/XMAGBOT;
        XMAGACT=1./sqrt((1.-W*W/(WACT*WACT))^2+(2.*ZACT*W/WACT)^2);
        PHASETOP=atan2(2.*Z0*W/W0,1.-(W/W0)^2);
        PHASEBOT=atan2(2.*ZAF*W/WAF,1.-(W/WAF)^2);
        PHASEACT=atan2(2.*ZACT*W/WACT,1.-W*W/(WACT*WACT));
        GAIN=20.*log10(XMAG*XMAGACT);
        PHASE=-90.+57.3*(PHASETOP-PHASEBOT-PHASEACT);
        count=count+1;
        ArrayW(count)=W;
        ArrayGAIN(count)=GAIN;
        ArrayPHASE(count)=PHASE;
end
figure
semilogx(ArrayW,ArrayGAIN),grid
xlabel('Frequency (Rad/Sec)')
ylabel('Gain (Db)')
axis([.1 1000 -60 40])
figure
semilogx(ArrayW,ArrayPHASE),grid
xlabel('Frequency (Rad/Sec)')
ylabel('Phase (Deg)')
axis([.1 1000 -400 100])
clc
output=[ArrayW',ArrayGAIN',ArrayPHASE'];
save datfil.txt output /ascii
disp 'simulation finished'
```

To check if the open-loop crossover frequency remains the same when one reduces the desired flight-control system time constant from 0.3 s to 0.1 s, it is necessary to go back to the frequency domain. Figure 23.6 shows that although the open-loop gain changes at some frequencies when the time constant is reduced, the gain crossover frequency remains unchanged thus showing that the open-loop gain crossover frequency is independent of the flight-control system time constant.

We can see from Fig. 23.7 that if the desired time constant in the flight-control algorithm is reset to 0.3 s and if we increase the desired crossover frequency from

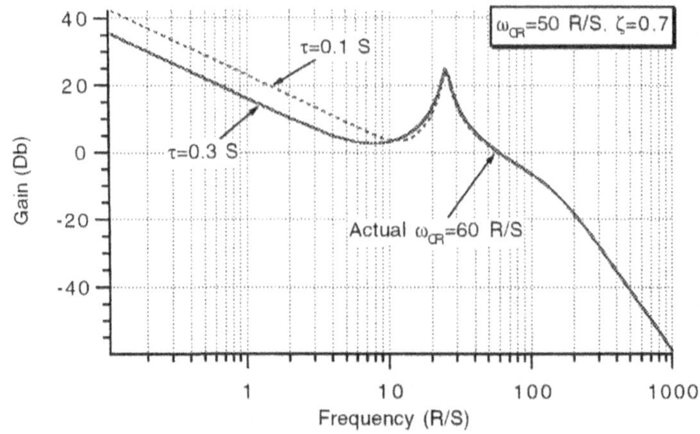

Fig. 23.6 Actual crossover frequency is independent of time constant.

50 rad/s to 100 rad/s the achieved crossover frequency is 104 rad/s, which is close to the desired goal. In this case the open-loop crossover frequency is far beyond the airframe dynamics thus justifying the approximation made in the derivation of the formula for the crossover frequency. However we can also see from Fig. 23.7 that the inclusion of the actuator dynamics tends to reduce the stability margins when the crossover frequency is increased. In this case the gain margin has been reduced from 11 dB to 1.5 dB while the phase margin has been reduced from 45 deg to 5 deg.

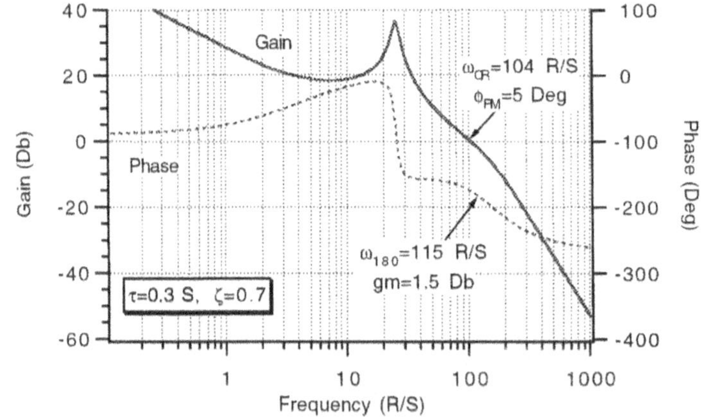

Fig. 23.7 Crossover frequency can be controlled but higher crossover frequency reduces stability margins.

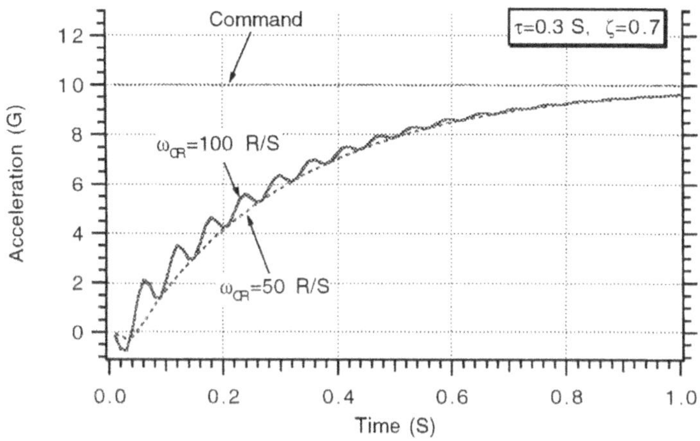

Fig. 23.8 High crossover frequencies are dangerous when actuator dynamics are considered.

Figure 23.8 confirms from a time domain point of view that the higher crossover frequency flight-control system (which has lower stability margins) has an oscillatory response. In practice the selection of the open-loop crossover frequency is limited by the dynamics of the actuator.

Figure 23.9 shows that when the actuator natural frequency is increased to 300 rad/s the step response stabilizes and is well behaved when the open-loop crossover frequency is 100 rad/s. In practice the maximum open-loop crossover frequency should be approximately one third the bandwidth of the actuator.

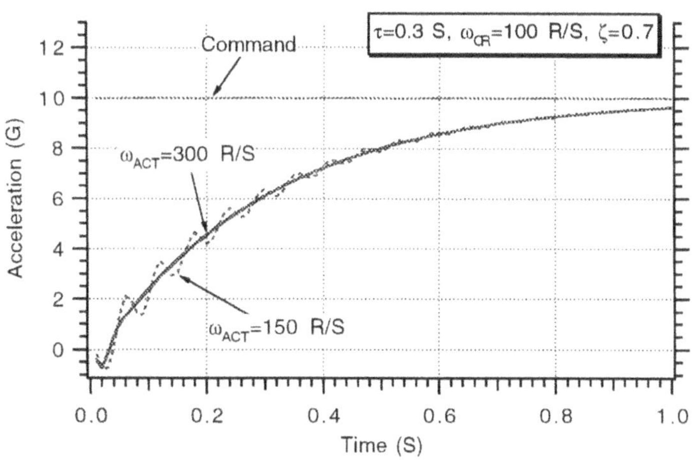

Fig. 23.9 Increasing actuator bandwidth allows operation at high crossover frequencies.

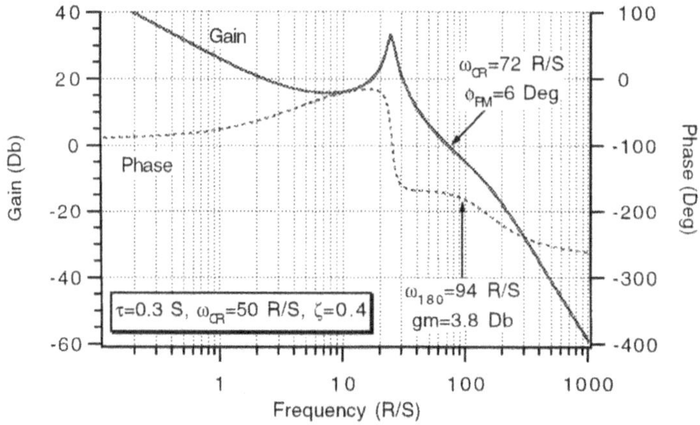

Fig. 23.10 Decreasing damping decreases stability margins.

If the flight-control system damping is reduced from 0.7, system performance will suffer. Figure 23.10 shows that for the nominal case in which it is desired that the flight-control system time constant be 0.3 s and the open-loop crossover frequency be 50 rad/s, reducing the damping to 0.4 reduces the stability margins. In this case the phase margin reduces from 43 deg to 6 deg while the gain margin reduces from 11 dB to 3.8 dB.

Finally, Fig. 23.11 confirms from a time domain point of view that decreasing the system damping reduces the stability margins. We can see that when the

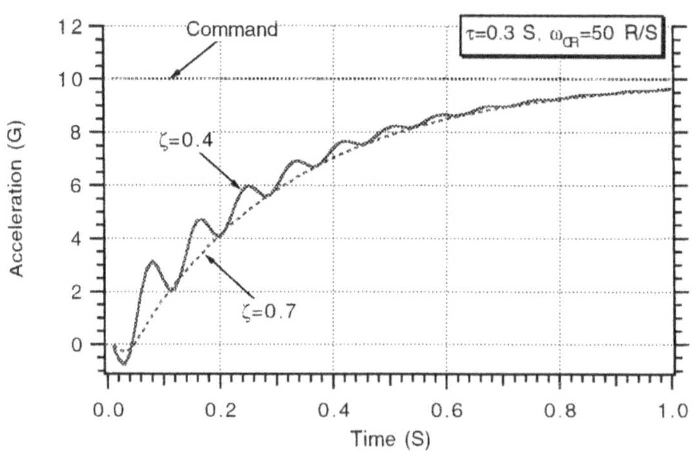

Fig. 23.11 Damping can be controlled, but low damping is dangerous when actuator dynamics are considered.

damping is decreased from 0.7 to 0.4 the system step response becomes oscillatory.

EXPERIMENTS WITH FLIGHT CONDITION

The autopilot gain algorithm was used on the hypothetical missile airframe of Chapter 21 for the flight condition in which there was a 10-*g* step acceleration command when the missile was traveling at 3000 ft/s. Three-loop autopilot gains were determined at sea level, 30-kft altitude, and 50-kft altitude. The gain algorithm was told that the desired flight-control system time constant was 0.3 s, the damping was 0.7, and the open-loop crossover frequency was 50 rad/s. The resultant autopilot gains for each of the three flight conditions are summarized in Table 23.1. We can see from the table that the autopilot gain K_A increases by a factor of 4 in going from sea level to 50-kft altitude. The gain K_R increases by a factor of nearly 10 whereas the gain ω_I remains approximately constant.

It is apparent from Fig. 23.12 that the autopilot gains enable the flight-control system to maintain the same step response for the three different flight conditions. The only difference between each of the step responses is that the wrong-way effect increases as the altitude increases. The autopilot gain algorithm does not attempt to control the airframe zeroes and therefore the wrong-way effect will get worse as the altitude increases.

To understand why the wrong-way tail effect gets worse at the higher altitudes, it is necessary to examine the flight-control system transfer functions in more detail. For example, the flight-control system transfer function from commanded to achieved acceleration is given by

$$\frac{n_L}{n_c} = \left(1 - \frac{s^2}{\omega_z^2}\right) \Big/ (1 + s\tau)\left[1 + \frac{2\zeta s}{\omega} + \frac{s^2}{\omega^2}\right]$$

where the negative sign in the numerator of the preceding equation indicates that the airframe zero is in the right-half plane. We recall from Chapter 21 that the airframe zero ω_z is related to the airframe parameters according to

$$\omega_z = \sqrt{\frac{M_\alpha Z_\delta - M_\delta Z_\alpha}{Z_\delta}}$$

TABLE 23.1 AUTOPILOT GAINS VARY WITH FLIGHT CONDITION

Altitude	K_A	K_R	ω_I
0 kft	1.15 deg/g	0.0928 s	12.9 rad/s
30 kft	1.98 deg/g	0.264 s	19.3 rad/s
50 kft	4.16 deg/g	0.726 s	19.7 rad/s

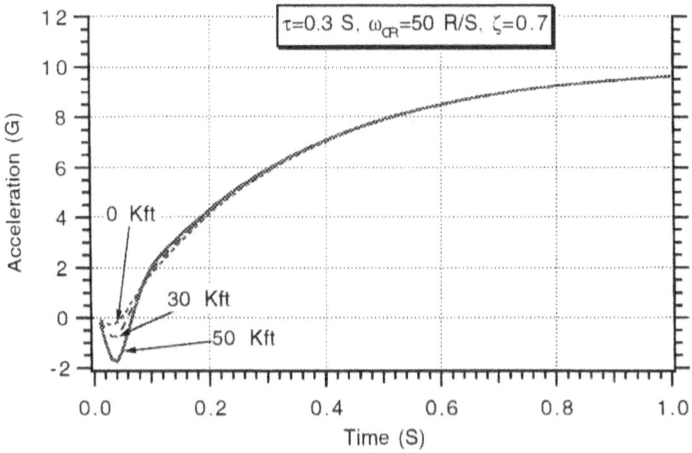

Fig. 23.12 With exception of wrong-way effect, acceleration response of autopilot is independent of flight condition.

For the three different flight conditions the linearized airframe parameters were calculated and are summarized in Table 23.2. Using the preceding linearized aerodynamic parameters, the airframe zero can be evaluated at the three different altitudes yielding

$$\omega_z|_{0\,\mathrm{kft}} = \sqrt{\frac{642 * 0.65 - 555 * 2.94}{-0.65}} = 43.2\,\mathrm{rad/s}$$

$$\omega_z|_{30\,\mathrm{kft}} = \sqrt{\frac{240 * 0.239 - 204 * 1.17}{-0.239}} = 27.6\,\mathrm{rad/s}$$

$$\omega_z|_{50\,\mathrm{kft}} = \sqrt{\frac{99.1 * 0.0957 - 81.7 * 0.533}{-0.0957}} = 18.8\,\mathrm{rad/s}$$

We can see from the preceding calculations that the right-half plane airframe zero decreases as the altitude increases thus causing more wrong-way effect.

For the same 10-g step acceleration input command, Fig. 23.13 shows how the missile fin rate response changes with time. We can see that the maximum fin rate increases with increasing altitude. At sea level only a few deg/s of fin rate are required whereas at 50 kft altitude nearly 600 deg/s are required to ensure the well-behaved step response of Fig. 23.12. In practice, the actuator must be sized to handle the expected maximum fin rate or else saturation occurs and flight catastrophe can result. It is apparent from Fig. 23.13 that the requirements for actuator fin rate sizing will take place at the highest expected altitude in the flight envelope.

TABLE 23.2 LINEARIZED AERODYNAMIC PARAMETERS FOR DIFFERENT FLIGHT CONDITIONS

Altitude	M_α	M_δ	Z_α	Z_δ
0 kft	$-642\ s^{-2}$	$-555\ s^{-2}$	$-2.94\ s^{-1}$	$-0.65\ s^{-1}$
30 kft	$-240\ s^{-2}$	$-204\ s^{-2}$	$-1.17\ s^{-1}$	$-0.239\ s^{-1}$
50 kft	$-99.1\ s^{-2}$	$-81.7\ s^{-2}$	$-0.533\ s^{-1}$	$-0.0957\ s^{-1}$

Using a transfer function type of analysis, we can try and understand why more fin rate is required at the higher altitudes. We have shown in Chapter 21 that the airframe transfer function from fin rate to achieved acceleration is given by

$$\frac{n_L}{\delta} = \left[K_1\left(1 - \frac{s^2}{\omega_z^2}\right) \bigg/ \left(1 + \frac{2\zeta_{AF}s}{\omega_{AF}} + \frac{s^2}{\omega_{AF}^2}\right)\right]$$

Using the chain rule from calculus, we can find the transfer function of fin rate due to commanded acceleration by combining the previous airframe transfer function with the flight-control system transfer function yielding

$$\frac{\dot{\delta}}{n_c} = \frac{s\delta}{n_L} * \frac{n_L}{n_c} = s\left[1 + \frac{2\zeta_{AF}s}{\omega_{AF}} + \frac{s^2}{\omega_{AF}^2}\right] \bigg/ \left\{K_1(1 + s\tau)\left[1 + \frac{2\zeta s}{\omega} + \frac{s^2}{\omega^2}\right]\right\}$$

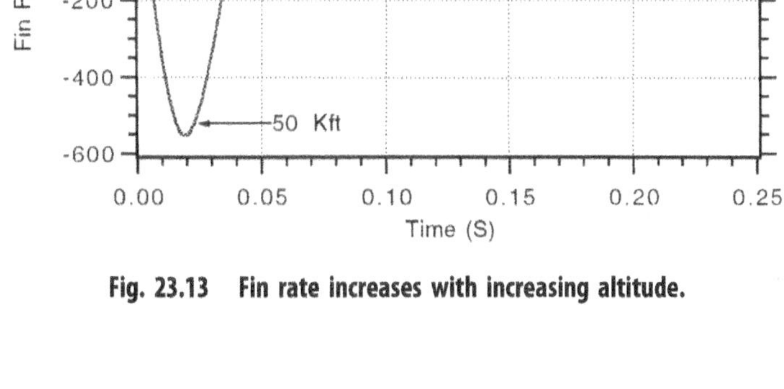

Fig. 23.13 Fin rate increases with increasing altitude.

We can see from the preceding transfer function that the magnitude of the fin rate is inversely proportional to the aerodynamic gain K_1. We have shown in Chapter 21 that the aerodynamic gain is given by

$$K_1 = \frac{-V_M(M_\alpha Z_\delta - Z_\alpha M_\delta)}{1845 M_\alpha}$$

Evaluating the aerodynamic gain at the three different flight conditions using the values of Table 23.2 for the various aerodynamic parameters reveals that the magnitude of the aerodynamic gain decreases as the altitude increases or

$$K_1|_{0\,\text{kft}} = \frac{-3000(642 * 0.65 - 2.94 * 555)}{-1845 * 642} = -3.07\,\text{g/deg}$$

$$K_1|_{30\,\text{kft}} = \frac{-3000(240 * 0.239 - 1.17 * 204)}{-1845 * 240} = -1.23\,\text{g/deg}$$

$$K_1|_{50\,\text{kft}} = \frac{-3000(99.1 * 0.0957 - 0.533 * 81.7)}{-1845 * 99.1} = -0.558\,\text{g/deg}$$

Because the fin rate transfer function is inversely proportional to the aerodynamic gain, the fin rate magnitude will increase with increasing altitude as was already demonstrated in Fig. 23.13.

GUIDANCE SYSTEM ANALYSIS

The guidance system with the flight-control system modeled explicitly appears in Fig. 23.14. In this model the airframe is represented by the second-order transfer function derived in Chapter 21, and the three-loop autopilot is modeled in conjunction with a second-order actuator model. In this homing loop model we can see that there are four random error sources that will cause miss distance: white semiactive noise, white glint noise, white range independent noise, and a uniformly distributed target maneuver. The airframe parameters and the autopilot gains in the flight-control system are a function of the flight condition. In this model we have assumed that the missile speed and altitude are constant so that the aerodynamic parameters and flight-control gains will be constant for a given engagement. The various homing loop parameters used in the study of this section appear in Table 23.3.

The adjoint of the homing loop was taken and appears in Fig. 23.15. Shown in the adjoint diagram are the miss distance outputs due to the four random error sources. The values of the error sources used in the ensuing study appear

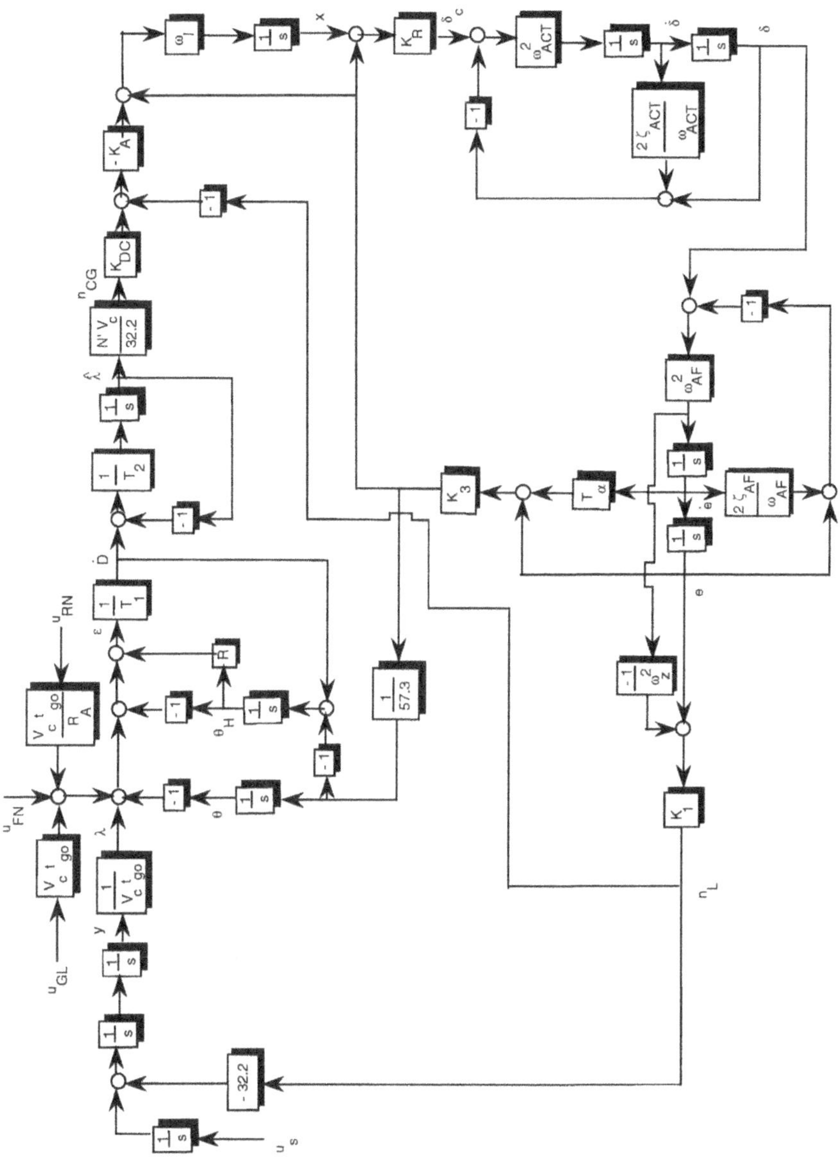

Fig. 23.14 Homing loop with flight-control system.

TABLE 23.3 NOMINAL GUIDANCE SYSTEM PARAMETERS

Name	Definition	Value
V_c	Closing velocity	4000 ft/s
T_1	Seeker track loop time constant	0.1 s
T_2	Noise filter time constant	0.15 s
N'	Effective navigation ratio	3
ω_{ACT}	Actuator natural frequency	150 rad/s
ζ_{ACT}	Actuator damping	0.7

in Table 23.4. The spectral densities for each of the error sources can be calculated from the preceding table according to

$$\Phi_{RN} = 2T_{RN}\sigma_{RN}^2 = 2*0.01*0.01^2 = 2*10^{-6}\frac{\text{rad}^2}{\text{hz}}$$

$$\Phi_{GL} = 2T_{GL}\sigma_{GL}^2 = 2*0.1*10^2 = 20\frac{\text{ft}^2}{\text{hz}}$$

$$\Phi_{FN} = 2T_{FN}\sigma_{FN}^2 = 2*0.01*0.002^2 = 8*10^{-8}\frac{\text{rad}^2}{\text{hz}}$$

$$\Phi_{n_T} = \frac{n_T^2}{t_F} = \frac{64.4^2}{5} = 829\frac{\text{ft/s}^2}{\text{hz}}$$

The adjoint simulation, derived from Fig. 23.15, appears in Listing 23.3. We can see from the listing that the adjoint can yield miss distance results (MISS = 1, TINT = 0), fin rate results at a particular time to go (MISS = 2, TINT = t_{go}), or commanded acceleration results at a particular time to go (MISS = 3, TINT = t_{go}) by simply changing the initial conditions on the adjoint differential equations. We can also see from Listing 23.3 that the aerodynamic parameters and autopilot gains are computed automatically for a given flight condition using the formulas derived in Chapter 22 and this chapter.

Running the adjoint simulation to calculate miss distances for the three different flight conditions yields the miss distance error budget of Table 23.5. Here we can see that the total rms miss distance is independent of flight condition because the autopilot gain algorithm has maintained the same flight-control system response at the three different flight conditions. We can also see from Table 23.5, that for the values of the error sources selected, the major contributors to the miss distance are target maneuver and glint noise. In this particular case, improving the sensor (such as reducing

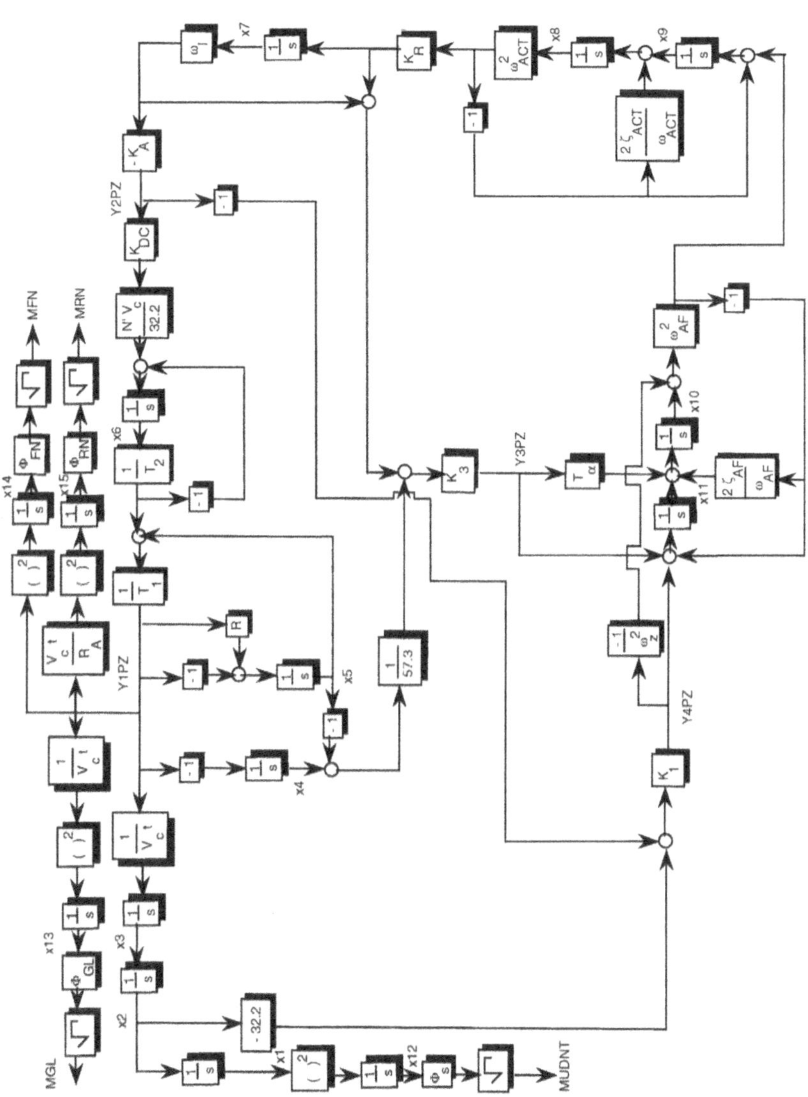

Fig. 23.15 Adjoint model of homing loop.

TABLE 23.4 NOMINAL ERROR SOURCES

Error source	Parameter values
Semiactive noise	.01 rad @ 30 kft in 0.01 s
Glint noise	10 ft in 0.1 s
Range independent noise	0.002 rad in 0.01 s
Random target maneuver	2 *g* over 5 s

TABLE 23.5 MISS DISTANCE PERFORMANCE APPEARS TO BE INDEPENDENT OF ALTITUDE

Error source	0 kft	30 kft	50 kft
Random maneuver	12.4 ft	12.7 ft	13.2 ft
Range independent noise	2.01 ft	2.06 ft	2.17 ft
Semiactive noise	2.27 ft	2.33 ft	2.44 ft
Glint noise	9.03 ft	9.21 ft	9.61 ft
RMS miss	15.6 ft	16.0 ft	16.6 ft

range independent noise and semiactive noise) will not significantly reduce the total rms miss distance.

Miss distance results were generated as a function of the radome slope for each of the three different flight conditions. We can see from Fig. 23.16 that at

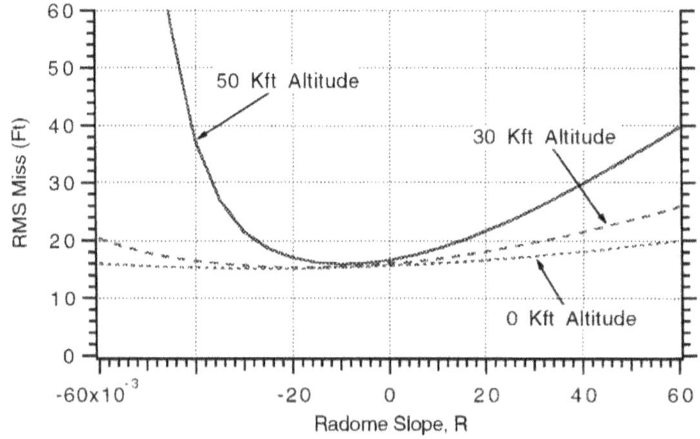

Fig. 23.16 High-altitude performance is more sensitive to radome slope.

the lower altitudes there is very little miss distance variation with radome slope. However, at 50-kft altitude we can see that the miss distance can be very large due to either large negative or positive slopes. The specification on the allowable radome slope swing will be set at the highest operating altitude [6].

LISTING 23.3 ADJOINT SIMULATION OF HOMING LOOP WITH DETAILED FLIGHT-CONTROL SYSTEM

```
count=0;
FR=3.;
DIAM=1.;
XL=20.;
CTW=0.;
CRW=6.;
HW=2.;
CTT=0.;
CRT=2.;
HT=2.;
XN=4.;
XCG=10.;
XHL=19.5;
WACT=150.;
ZACT=.7;
TF=5.;
VM=3000.;
XNCG=10.;
WCR=50.;
ZETA=.7;
TAU=.3;
ALT=0.;
XNT=64.4;
XNP=3.;
VC=4000.;
T1=.1;
T2=.15;
PHIRN=.000002;
PHIGL=20.;
PHIFN=.00000008;
RA=30000.;
R=0.;
MISS=1;
TINT=0.;
A=1000.;
if ALT<=30000.
        RHO=.002378*exp(-ALT/30000.);
```

```
else
        RHO=.0034*exp(-ALT/22000.);
end
WGT=1000.;
XNLLIN=0.;
XACC=XCG;
SWING=.5*HW*(CTW+CRW);
STAIL=.5*HT*(CTT+CRT);
SREF=3.1416*DIAM*DIAM/4.;
XLP=FR*DIAM;
SPLAN=(XL-XLP)*DIAM+1.33*XLP*DIAM/2.;
XCPN=2*XLP/3;
AN=.67*XLP*DIAM;
AB=(XL-XLP)*DIAM;
XCPB=(.67*AN*XLP+AB*(XLP+.5*(XL-XLP)))/(AN+AB);
XCPW=XLP+XN+.7*CRW-.2*CTW;
XMACH=VM/A;
XIYY=WGT*(3*((DIAM/2)^2)+XL*XL)/(12*32.2);
TMP1=(XCG-XCPW)/DIAM;
TMP2=(XCG-XHL)/DIAM;
TMP3=(XCG-XCPB)/DIAM;
TMP4=(XCG-XCPN)/DIAM;
B=sqrt(XMACH^2-1);
Q=.5*RHO*VM*VM;
P1=WGT*XNCG/(Q*SREF);
Y1=2+8*SWING/(B*SREF)+8*STAIL/(B*SREF);
Y2=1.5*SPLAN/SREF;
Y3=8*STAIL/(B*SREF);
Y4=2*TMP4+8*SWING*TMP1/(B*SREF)+8*STAIL*TMP2/(B*SREF);
Y5=1.5*SPLAN*TMP3/SREF;
Y6=8*STAIL*TMP2/(B*SREF);
P2=Y2-Y3*Y5/Y6;
P3=Y1-Y3*Y4/Y6;
ALFTR=(-P3+sqrt(P3*P3+4.*P2*P1))/(2.*P2);
DELTR=-Y4*ALFTR/Y6-Y5*ALFTR*ALFTR/Y6;
CNA=2+1.5*SPLAN*ALFTR/SREF+8*SWING/(B*SREF)+8*STAIL/(B*SREF);
CND=8*STAIL/(B*SREF);
CMAP=2*TMP4+1.5*SPLAN*ALFTR*TMP3/SREF+8*SWING*TMP1/(B*SREF);
CMA=CMAP+8*STAIL*TMP2/(B*SREF);
CMD=8*STAIL*TMP2/(B*SREF);
XMA=Q*SREF*DIAM*CMA/XIYY;
XMD=Q*SREF*DIAM*CMD/XIYY;
ZA=-32.2*Q*SREF*CNA/(WGT*VM);
ZD=-32.2*Q*SREF*CND/(WGT*VM);
WZ=sqrt((XMA*ZD-ZA*XMD)/ZD);
WAF=sqrt(-XMA);
```

```
ZAF=.5*WAF*ZA/XMA;
XK1=-VM*(XMA*ZD-XMD*ZA)/(1845*XMA);
XK2=XK1;
TA=XMD/(XMA*ZD-XMD*ZA);
XK3=1845*XK1/VM;
W=(TAU*WCR*(1+2.*ZAF*WAF/WCR)-1)/(2*ZETA*TAU);
W0=W/sqrt(TAU*WCR);
Z0=.5*W0*(2*ZETA/W+TAU-WAF^2/(W0*W0*WCR));
XKC=(-W0^2/WZ^2-1.+2.*Z0*W0*TA)/(1.-2.*Z0*W0*TA+W0*W0*TA*TA);
XKA=XK3/(XK1*XKC);
XK0=-W*W/(TAU*WAF*WAF);
XK=XK0/(XK1*(1+XKC));
WI=XKC*TA*W0*W0/(1+XKC+W0^2/WZ^2);
XKR=XK/(XKA*WI);
XKDC=1.+1845./(XKA*VM);
for R=-.06:.01:.06
        X1=0.;
        X2=0.;
        X3=0.;
        X4=0.;
        X5=0.;
        X6=0.;
        X7=0.;
        X8=0.;
        X9=0.;
        X10=0.;
        X11=0.;
        X12=0.;
        X13=0.;
        X14=0.;
        X15=0.;
        if MISS==1
                X3=1.;
        elseif MISS==2
                X8=1.;
        elseif MISS==3
                X6=XNP*VC/32.2;
        end
        T=0;
        H=.0001;
        S=0;
        TP=T+.00001+TINT;
        while TP<=(TF - 1e-5)
                X1OLD=X1;
                X2OLD=X2;
                X3OLD=X3;
```

```
X4OLD=X4;
X5OLD=X5;
X6OLD=X6;
X7OLD=X7;
X8OLD=X8;
X9OLD=X9;
X10OLD=X10;
X11OLD=X11;
X12OLD=X12;
X13OLD=X13;
X14OLD=X14;
X15OLD=X15;
STEP=1;
FLAG=0;
while STEP<=1
        if FLAG==1
                X1=X1+H*X1D;
                X2=X2+H*X2D;
                X3=X3+H*X3D;
                X4=X4+H*X4D;
                X5=X5+H*X5D;
                X6=X6+H*X6D;
                X7=X7+H*X7D;
                X8=X8+H*X8D;
                X9=X9+H*X9D;
                X10=X10+H*X10D;
                X11=X11+H*X11D;
                X12=X12+H*X12D;
                X13=X13+H*X13D;
                X14=X14+H*X14D;
                X15=X15+H*X15D;
                TP=TP+H;
                STEP=2;
        end
        TGO=TP+.00001;
        X1D=X2;
        X2D=X3;
        Y1PZ=(X6/T2+X5)/T1;
        X3D=Y1PZ/(VC*TGO);
        X4D=-Y1PZ;
        X5D=-Y1PZ+R*Y1PZ;
        Y2PZ=-XKA*WI*X7;
        X6D=-X6/T2+XNP*VC*XKDC*Y2PZ/32.2;
        X7D=XKR*WACT*WACT*X8;
        X8D=X9-2.*ZACT*WACT*X8;
        Y4PZ=XK1*(-32.2*X2-Y2PZ);
```

```
                    X9D=-WACT*WACT*X8+X10*WAF*WAF-WAF*WAF*Y4PZ/WZ^2;
                    Y3PZ=XK3*(X7D+WI*X7+(X4-X5)/57.3);
                    X10D=-2.*ZAF*WAF*(X10-Y4PZ/WZ^2)+X11+TA*Y3PZ;
                    X11D=-WAF*WAF*(X10-Y4PZ/WZ^2)+Y4PZ+Y3PZ;
                    X12D=X1*X1;
                    X13D=(Y1PZ/(VC*TGO))^2;
                    X14D=(Y1PZ*VC*TGO/RA)^2;
                    X15D=Y1PZ^2;
                    FLAG=1;
               end;
               FLAG=0;
               X1=.5*(X1OLD+X1+H*X1D);
               X2=.5*(X2OLD+X2+H*X2D);
               X3=.5*(X3OLD+X3+H*X3D);
               X4=.5*(X4OLD+X4+H*X4D);
               X5=.5*(X5OLD+X5+H*X5D);
               X6=.5*(X6OLD+X6+H*X6D);
               X7=.5*(X7OLD+X7+H*X7D);
               X8=.5*(X8OLD+X8+H*X8D);
               X9=.5*(X9OLD+X9+H*X9D);
               X10=.5*(X10OLD+X10+H*X10D);
               X11=.5*(X11OLD+X11+H*X11D);
               X12=.5*(X12OLD+X12+H*X12D);
               X13=.5*(X13OLD+X13+H*X13D);
               X14=.5*(X14OLD+X14+H*X14D);
               X15=.5*(X15OLD+X15+H*X15D);
          end
          XMFN=sqrt(X15*PHIFN);
          XMRN=sqrt(X14*PHIRN);
          XMGL=sqrt(X13*PHIGL);
          XMUDNT=XNT*sqrt(X12/TGO);
          RMS=sqrt(XMFN^2+XMRN^2+XMGL^2+XMUDNT^2);
          count=count+1;
          ArrayR(count)=R;
          ArrayXMFN(count)=XMFN;
          ArrayXMRN(count)=XMRN;
          ArrayXMGL(count)=XMGL;
          ArrayXMUDNT(count)=XMUDNT;
          ArrayRMS(count)=RMS;
end
figure
plot(ArrayR,ArrayXMFN,ArrayR,ArrayXMRN,ArrayR,ArrayXMGL,ArrayR,...
   ArrayXMUDNT,ArrayR,ArrayRMS),grid
xlabel('Radome Slope')
ylabel('Standard Deviation of Miss (Ft) ')
clc
```

```
output=[ArrayR',ArrayXMFN',ArrayXMRN',ArrayXMGL',ArrayXMUDNT',...
        ArrayRMS'];
save datfil.txt output /ascii
disp 'simulation finished'
```

We can examine the miss distance results at high altitude in more detail. Figure 23.17 shows the miss distance error budget at 50-kft altitude as a function of radome slope. For large negative slopes all the noise error sources cause the miss to be big whereas for large positive slopes the dominant contributor to the miss is random target maneuver. If the maximum rms miss that can be tolerated at 50-kft altitude is 25 ft, then the radome slope swing would have to be confined to range from $R = -0.03$ to $R = 0.03$ for a total radome swing of 0.06.

Increasing the flight-control system time constant at high altitude is sometimes effective in reducing system sensitivity to radome slope. Figure 23.18 shows that when the flight-control system time constant is increased from 0.3 s to 0.5 s the guidance system performance appears to be less sensitive to radome slope. However a closer examination of Fig. 23.18 reveals that if the allowable rms miss is 25 ft then the maximum negative slope that could be tolerated is -0.05 and the maximum positive slope that could be tolerated is 0.01 for a total radome swing of 0.06, which is the same radome swing as before. Soon we will find another reason why increasing the time constant may be beneficial at high altitudes.

We can also run the adjoint program to generate fin rate results at a specific time to go before intercept. Therefore a number of adjoint runs will yield fin rate results for all times to go. These results can be inverted for a particular flight time to yield rms fin rate as a function of time. Figure 23.19 displays rms fin rate as a function of time, obtained from the adjoint program of Listing 23.3, for a 5-s flight

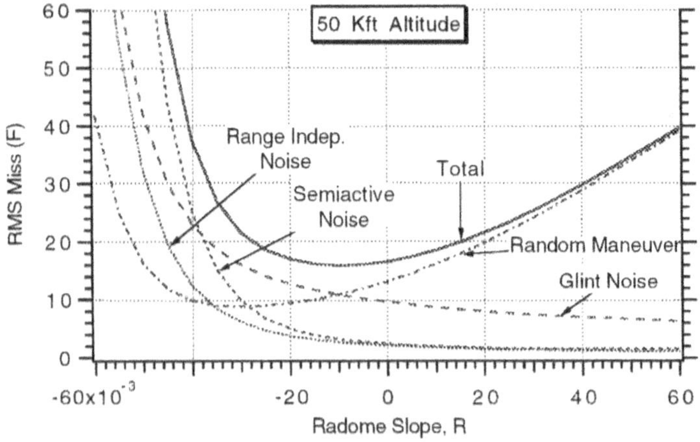

Fig. 23.17 Miss distance error budget at 50-kft altitude.

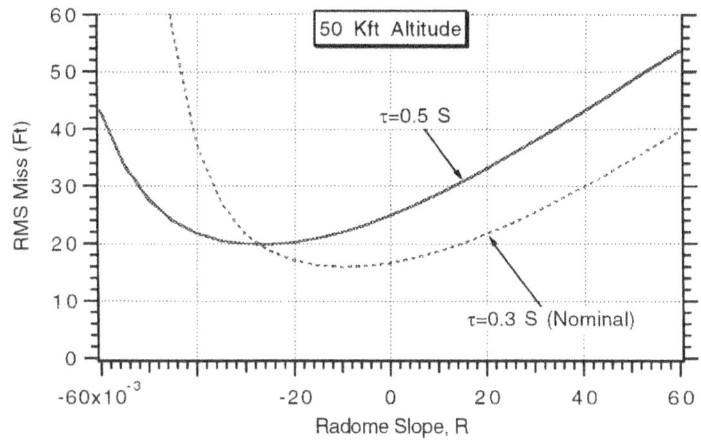

Fig. 23.18 Sometimes increasing flight-control system time constant reduces sensitivity to radome slope.

assuming zero radome slope for the three different flight conditions. We can see that the largest fin rates occur at the higher altitudes and at the end of the flight. Figure 23.19 indicates that if intercepts are to be supported at 50-kft altitude for this example, the actuator must be able to support rms fin rates in excess of 500 deg/s.

The fin rate results can be examined in more detail at 50-kft altitude. Figure 23.20 presents the adjoint fin rate error budget. We can see that at the

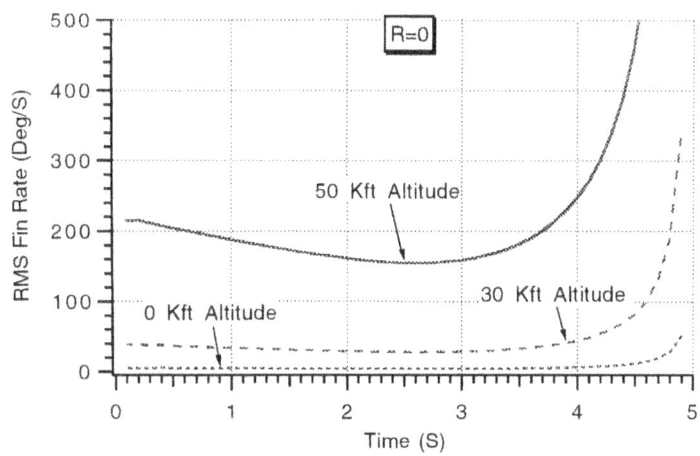

Fig. 23.19 Fin rate increases with increasing altitude.

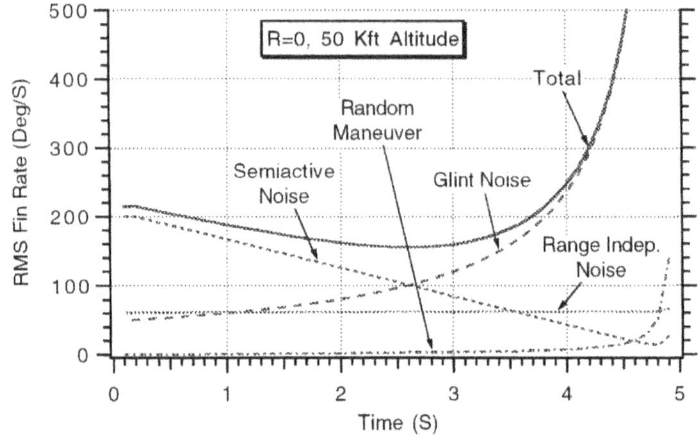

Fig. 23.20 Fin rate error budget at 50 kft altitude.

beginning of flight the large fin rates are due to semiactive noise. An improved sensor will relax these fin rate requirements. However at the end of the flight the large fin rates are mainly due to glint noise. Because the glint noise is usually a function of the target and not the sensor, it would be difficult to relax the fin rate requirement at this altitude.

Figure 23.21 shows that the fin rate can be significantly reduced by increasing the flight-control system time constant from 0.3 s to 0.5 s. Therefore increasing the flight-control system time constant at the higher altitudes might be considered to be a viable option for a given level of actuator performance.

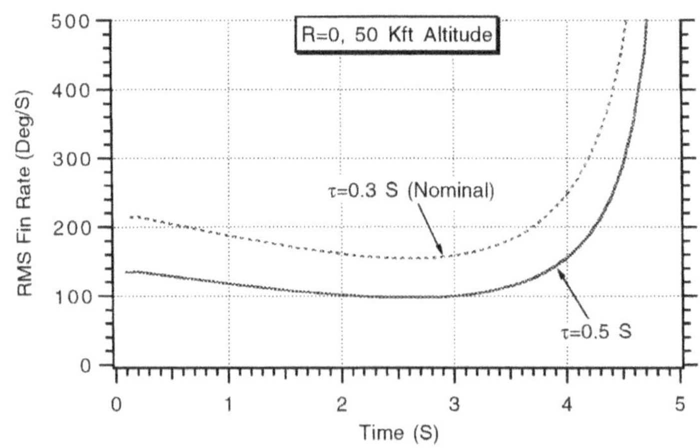

Fig. 23.21 Fin rate can be reduced by increasing flight-control system time constant.

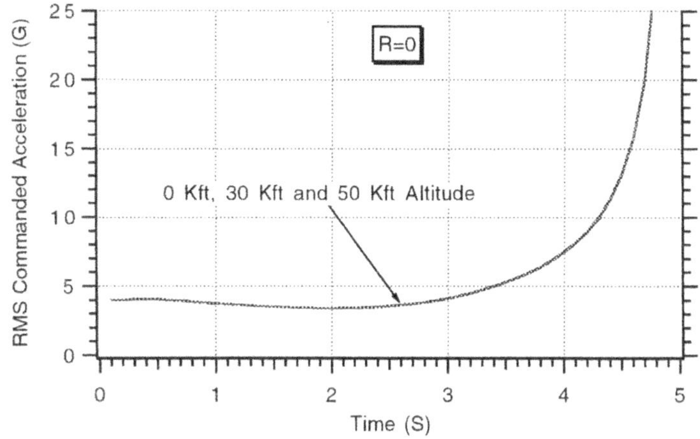

Fig. 23.22 Commanded acceleration profile is independent of altitude.

The adjoint program was run again to yield commanded acceleration information. We can see from Fig. 23.22 that the rms commanded acceleration is independent of flight condition. For this example 25 g of rms missile acceleration capability would be required to prevent the missile from acceleration saturation. Avoiding saturation will validate the rms miss distance results of the previous figures. If saturation occurs, the rms miss distance will be considerably higher.

A typical rms commanded acceleration error budget, obtained from the adjoint simulation of Listing 23.3, appears in Fig. 23.23. Here we can see that near the end of the flight most of the rms commanded acceleration is due to

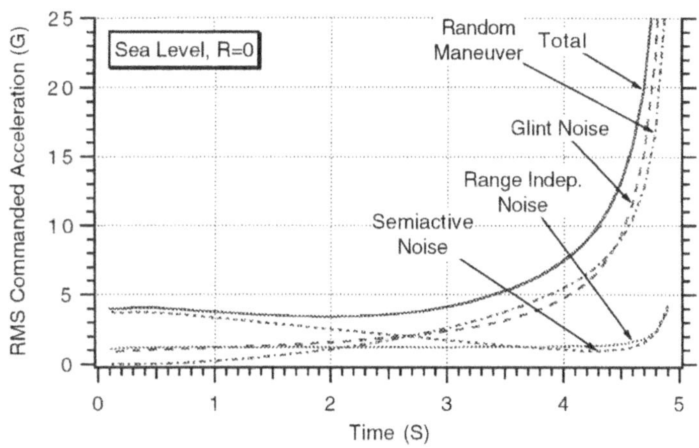

Fig. 23.23 Acceleration error budget at sea level.

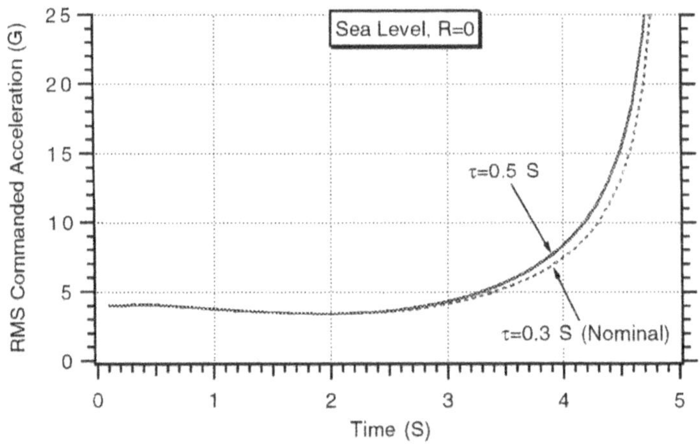

Fig. 23.24 Increasing flight-control system time constant does not reduce acceleration requirements.

both random target maneuver and glint noise. Therefore an improved sensor will not relax the acceleration requirements because only the acceleration due to semi-active and range independent noise will be reduced. The guidance system must be sized to handle the commanded acceleration at the end of the flight due to the random target maneuver and glint noise.

Finally Fig. 23.24 shows that increasing the flight-control system time constant does not reduce the rms commanded acceleration requirements. Therefore increasing the time constant is mainly an option for reducing the fin rate requirements and possibly also to desensitize the system to radome effects but not for reducing the acceleration requirements.

SUMMARY

In this chapter we have shown how the three-loop autopilot and associated gain selection algorithm can be used to independently specify the time constant, damping, and crossover frequency. Typical performance studies were conducted using the method of adjoints to show how miss distance results could be generated, actuator requirements set, and missile acceleration requirements derived.

REFERENCES

[1] Fossier, M. W., "The Development of Radar Homing Missiles," *Journal of Guidance, Control, and Dynamics*, Vol. 7, Nov.–Dec. 1984, ppp. 641–651.

[2] Stallard, D. V., "An Approach to Autopilot Design for Homing Interceptor Missiles," *Proceedings of the 1991AIAA Guidance and Control Conference*, AIAA, Washington, DC, 1991.

[3] Nesline, F. W., and Nabbefeld, N. C., "Design of Digital Autopilots for Homing Missiles," *Proceedings of AGARD Flight Mechanics Panel Symposium*, London, May 1979.

[4] Wells, B. H., "Tactical Missile Structural Testing and Model Verification for Autopilot Design," *Proceedings of the 1991 AIAA Guidance and Control Conference*, AIAA, Washington, DC, 1991.

[5] Gratt, H. J., and Zarchan, P., "A Practical Approach to Augmenting Missile Autopilot Design (The 3-Loop Autopilot on Steroids)," *Proceedings of AIAA/BMDO Interceptor Technology Conference*, San Diego, CA, July 1994.

[6] Horton, M. P., "A Study of Autopilots for the Adaptive Control of Tactical Guided Missiles," MSc Thesis, Univ. of Bath, Bath, UK, 1992.

CHAPTER 24

Trajectory Shaping Guidance

INTRODUCTION

In all of the guidance work done thus far, the goal has been to hit the target using the least amount of energy. In some applications, in addition to hitting the target, it may also be desirable to shape the missile trajectory near impact. For example, in antitank or antiballistic missile applications we may want to have the missile approach the target at certain strike angles to improve lethality. In this chapter we will show how the guidance problem can be reformulated so that a new guidance law can be developed that both hits the target using minimum energy and, in addition, travels on the desired trajectory. It will be shown that the new guidance law is actually the same one used to land the Apollo spacecraft on the moon. We will then evaluate the trajectory shaping guidance law and see how it performs in a more realistic nonlinear environment.

PROBLEM SETUP

Before we derive the new guidance law, we must first express mathematically what we desire to do. Let us first revisit our homing loop model for a zero-time constant guidance system as shown in Fig. 24.1.

As was the case in Chapter 8 for deriving augmented proportional navigation, we are still assuming a constant target maneuver, which means that the derivative of n_T must be zero. Therefore we can express the model of Fig. 24.1 in matrix form as

$$\begin{bmatrix} \dot{y} \\ \ddot{y} \\ \dot{n}_T \end{bmatrix} = \begin{bmatrix} 0 & 1 & 0 \\ 0 & 0 & 1 \\ 0 & 0 & 0 \end{bmatrix} \begin{bmatrix} y \\ \dot{y} \\ n_T \end{bmatrix} + \begin{bmatrix} 0 \\ -1 \\ 0 \end{bmatrix} n_c$$

This is the same form as the state space matrix differential equation

$$\dot{\boldsymbol{x}} = \boldsymbol{F}\boldsymbol{x} + \boldsymbol{G}\boldsymbol{u}$$

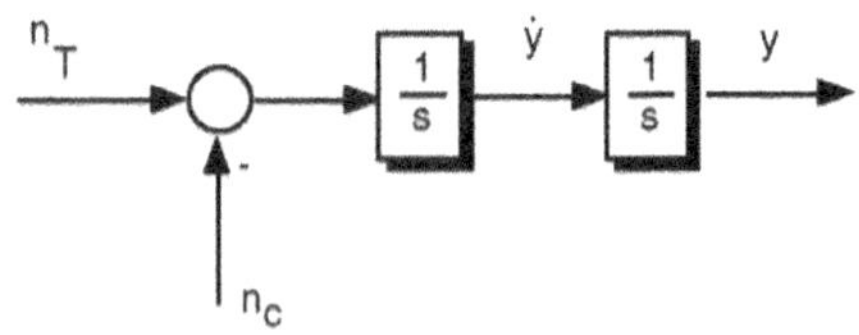

Fig. 24.1 Zero-time constant homing loop model for guidance law development.

As with our other guidance problems, we still desire to minimize the integral of the commanded acceleration squared. In addition, we want the miss distance to be zero. We will soon see that selecting the relative velocity at the end of flight to be some specified value is the same as shaping the missile trajectory. Mathematically we have just stated that our goals are

$$y(t_F) = 0 \quad \text{and} \quad \dot{y}(t_F) = \dot{y}_F \ \text{ subject to minimizing } \int_0^{t_F} n_c^2(t)\,dt$$

Recall from Chapter 8 that the general solution to the state space differential equation describing the homing loop model is given by the vector relationship

$$\boldsymbol{x}(t_F) = \boldsymbol{\Phi}(t_F - t)\boldsymbol{x}(t) + \int_t^{t_F} \boldsymbol{\Phi}(t_F - \lambda)\,\boldsymbol{G}(\lambda)\boldsymbol{u}(\lambda)\,\mathrm{d}\lambda$$

where Φ is the fundamental matrix. By comparing the state space equation to our matrix equation representing the homing loop model of Fig. 24.1 we can see that $\boldsymbol{F}$, $\boldsymbol{G}$, and $\boldsymbol{u}$ are given by

$$\boldsymbol{F} = \begin{bmatrix} 0 & 1 & 0 \\ 0 & 0 & 1 \\ 0 & 0 & 0 \end{bmatrix}$$

$$\boldsymbol{G} = \begin{bmatrix} 0 \\ -1 \\ 0 \end{bmatrix}$$

$$\boldsymbol{u} = n_c$$

We have already shown in Chapter 8 that for the systems dynamics matrix $\boldsymbol{F}$ under consideration the continuous fundamental matrix is given by

$$\boldsymbol{\Phi}(t) = \begin{bmatrix} 1 & t & .5t^2 \\ 0 & 1 & t \\ 0 & 0 & 1 \end{bmatrix}$$

Substitution of the preceding matrices into the general solution of the state space equation yields

$$\begin{bmatrix} y(t_F) \\ \dot{y}(t_F) \\ n_T(t_F) \end{bmatrix} = \begin{bmatrix} 1 & t_F - t & .5(t_F - t)^2 \\ 0 & 1 & t_F - t \\ 0 & 0 & 1 \end{bmatrix} \begin{bmatrix} y(t) \\ \dot{y}(t) \\ n_T(t) \end{bmatrix} + \int_t^{t_F} \begin{bmatrix} 1 & t_F - \lambda & .5(t_F - \lambda)^2 \\ 0 & 1 & t_F - \lambda \\ 0 & 0 & 1 \end{bmatrix} \begin{bmatrix} 0 \\ -1 \\ 0 \end{bmatrix} n_c(\lambda)\,\mathrm{d}\lambda$$

We can multiply out the preceding matrix equation and end up with three scalar equations. Because we are only interested in controlling the relative position and velocity at the end of flight we can ignore the last scalar equation of the three equations and obtain

$$y(t_F) = y(t) + (t_F - t)\dot{y}(t) + .5(t_F - t)^2 n_T(t) - \int_t^{t_F} (t_F - \lambda) n_c(\lambda)\,\mathrm{d}\lambda$$

$$\dot{y}(t_F) = \dot{y}(t) + (t_F - t) n_T(t) - \int_t^{t_F} n_c(\lambda)\,\mathrm{d}\lambda$$

We will see in the next section how the Schwartz inequality can be used on the preceding two scalar equations to obtain the trajectory shaping guidance law.

USING THE SCHWARTZ INEQUALITY FOR TRAJECTORY SHAPING GUIDANCE

We will first simplify the guidance problem by using shorthand notation to represent the two scalar equations that were just derived in the preceding section. Let us first define

$$\begin{aligned} f_1 &= y(t) + (t_F - t)\dot{y}(t) + .5(t_F - t)^2 n_T(t) \\ f_2^* &= \dot{y}(t) + (t_F - t) n_T(t) \\ h_1(\lambda) &= t_F - \lambda \\ h_2(\lambda) &= 1 \end{aligned}$$

We can now rewrite the two scalar equations at the end of the preceding section in shorthand notation as

$$y(t_F) = f_1 - \int_t^{t_F} h_1(\lambda) n_c(\lambda)\,\mathrm{d}\lambda$$

$$\dot{y}(t_F) = f_2^* - \int_t^{t_F} h_2(\lambda) n_c(\lambda)\,\mathrm{d}\lambda$$

Recall that we want the miss distance to be zero $[y(t_F) = 0]$ and we also want to specify the value of the relative velocity at the terminal time. Therefore, the two preceding equations simplify to

$$f_1 = \int_t^{t_F} h_1(\lambda) n_c(\lambda)\, d\lambda$$

$$f_2^* - \dot{y}(t_F) = \int_t^{t_F} h_2(\lambda) n_c(\lambda)\, d\lambda = f_2$$

Let us now combine the two preceding scalar equations into one equation by introducing a new variable δ or

$$f_1 - \delta f_2 = \int_t^{t_F} [h_1(\lambda) - \delta h_2(\lambda)] n_c(\lambda)\, d\lambda$$

If we apply the Schwartz inequality to the preceding expression we obtain

$$(f_1 - \delta f_2)^2 \le \int_t^{t_F} [h_1(\lambda) - \delta h_2(\lambda)]^2\, d\lambda \int_t^{t_F} n_c^2(\lambda)\, d\lambda$$

Inverting the preceding equation and solving for the integral of the square of the commanded acceleration yields

$$\int_t^{t_F} n_c^2(\lambda)\, d\lambda \ge \frac{(f_1 - \delta f_2)^2}{\int_t^{t_F} [h_1(\lambda) - \delta h_2(\lambda)]^2\, d\lambda}$$

The integral of the square of the acceleration will be minimized when the equality sign of the preceding inequality holds. According to the Schwartz inequality, the equality sign holds when

$$n_c(\lambda) = \mathrm{K}[h_1(\lambda) - \delta h_2(\lambda)]$$

where K is a constant. We will soon use the preceding equation. When the equality sign holds, the integral of the square of the commanded acceleration can be expanded to

$$z = \int_t^{t_F} n_c^2(\lambda)\, d\lambda = \frac{(f_1 - \delta f_2)^2}{\int_t^{t_F} [h_1(\lambda) - \delta h_2(\lambda)]^2\, d\lambda}$$

$$= \frac{(f_1 - \delta f_2)^2}{\int_t^{t_F} h_1^2(\lambda)\, d\lambda - 2\delta \int_t^{t_F} h_1(\lambda) h_2(\lambda)\, d\lambda + \delta^2 \int_t^{t_F} h_2^2(\lambda)\, d\lambda}$$

In the work that follows we will be performing many algebraic manipulations. To simplify our task, we will again use shorthand notation. We can define

$$\|h_1^2\| = \int_t^{t_F} h_1^2(\lambda)\,\mathrm{d}\lambda$$
$$\|h_2^2\| = \int_t^{t_F} h_2^2(\lambda)\,\mathrm{d}\lambda$$
$$\|h_1 h_2\| = \int_t^{t_F} h_1(\lambda)h_2(\lambda)\,\mathrm{d}\lambda$$

Therefore, our expression for the integral of the square of the commanded acceleration simplifies to

$$z = \frac{(f_1 - \delta f_2)^2}{\|h_1^2\| - 2\delta\|h_1 h_2\| + \delta^2\|h_2^2\|}$$

Now we have to choose a value for the variable δ. The best value of δ is one that minimizes the preceding expression. We know from calculus that the value of δ that minimizes the preceding expression can be found by taking the derivative of the preceding expression with respect to δ and setting the result to zero. Using the quotient rule from calculus to take the derivative yields

$$\frac{dz}{d\delta}=0$$
$$=\frac{[\|h_1^2\|-2\delta\|h_1h_2\|+\delta^2\|h_2^2\|]2(f_1-\delta f_2)(-f_2)-(f_1-\delta f_2)^2[-2\|h_1h_2\|+2\delta\|h_2^2\|]}{[\|h_1^2\|-2\delta\|h_1h_2\|+\delta^2\|h_2^2\|]^2}$$

We can now solve the preceding expression for δ. After some algebra, we obtain

$$\delta = \frac{f_1\|h_1h_2\| - f_2\|h_1^2\|}{f_1\|h_2^2\| - f_2\|h_1h_2\|}$$

Because we know that

$$f_1 = \int_t^{t_F} h_1(\lambda)n_c(\lambda)\,\mathrm{d}\lambda$$

and

$$n_c(\lambda) = \mathrm{K}[h_1(\lambda) - \delta h_2(\lambda)]$$

Substitution yields

$$f_1 = \int_t^{t_F} h_1(\lambda)\mathrm{K}[h_1(\lambda) - \delta h_2(\lambda)]\,\mathrm{d}\lambda$$

If we solve for the constant K we obtain

$$\mathrm{K} = \frac{f_1}{\int_t^{t_F} h_1(\lambda)[h_1(\lambda) - \delta h_2(\lambda)]\,\mathrm{d}\lambda} = \frac{f_1}{\|h_1^2\| - \delta\|h_1h_2\|}$$

Therefore, the optimal acceleration command can be rewritten as

$$n_c(\lambda) = \mathrm{K}[h_1(\lambda) - \delta h_2(\lambda)] = \frac{f_1[h_1(\lambda) - \delta h_2(\lambda)]}{\|h_1^2\| - \delta\|h_1h_2\|}$$

Substitution of the optimal value of δ in the preceding expression yields

$$n_c(\lambda) = \frac{f_1\left[h_1(\lambda) - h_2(\lambda)\left[\dfrac{f_1\|h_1h_2\| - f_2\|h_1^2\|}{f_1\|h_2^2\| - f_2\|h_1h_2\|}\right]\right]}{\|h_1^2\| - \|h_1h_2\|\left[\dfrac{f_1\|h_1h_2\| - f_2\|h_1^2\|}{f_1\|h_2^2\| - f_2\|h_1h_2\|}\right]}$$

After some algebra, we obtain

$$n_c(\lambda) = \frac{f_1h_1(\lambda)\|h_2^2\| - \|h_1h_2\|[f_2h_1(\lambda) + f_1h_2(\lambda)] + f_2h_2(\lambda)\|h_1^2\|}{\|h_1^2\|\|h_2^2\| - \|h_1h_2\|^2}$$

or in the time domain

$$n_c(t) = \frac{f_1h_1(t)\|h_2^2\| - \|h_1h_2\|[f_2h_1(t) + f_1h_2(t)] + f_2h_2(t)\|h_1^2\|}{\|h_1^2\|\|h_2^2\| - \|h_1h_2\|^2}$$

Now we have enough information to evaluate the numerator and denominator of the preceding expression to find the new guidance law. Recall that for this problem

$$f_1 = y(t) + (t_F - t)\dot{y}(t) + .5(t_F - t)^2 n_T(t) = y + t_{\mathrm{go}}\dot{y} + .5t_{\mathrm{go}}^2 n_T$$

$$f_2 = f_2^* - \dot{y}(t_F) = \dot{y}(t) + (t_F - t)n_T(t) - \dot{y}(t_F) = \dot{y} + t_{\mathrm{go}}n_T - \dot{y}(t_F)$$

$$h_1(t) = t_F - t = t_{\mathrm{go}}$$

$$h_2(t) = 1$$

Therefore, we can easily solve the necessary integrals as

$$\|h_1^2\| = \int_t^{t_F} h_1^2(\lambda)\,\mathrm{d}\lambda = \int_t^{t_F} (t_F - \lambda)^2\,\mathrm{d}\lambda = \frac{t_{\mathrm{go}}^3}{3}$$

$$\|h_2^2\| = \int_t^{t_F} h_2^2(\lambda)\,\mathrm{d}\lambda = \int_t^{t_F} \mathrm{d}\lambda = t_{\mathrm{go}}$$

$$\|h_1h_2\| = \int_t^{t_F} h_1(\lambda)h_2(\lambda)\,\mathrm{d}\lambda = \int_t^{t_F} (t_F - \lambda)\,\mathrm{d}\lambda = \frac{t_{\mathrm{go}}^2}{2}$$

where the time to go until intercept is given by

$$t_{\mathrm{go}} = t_F - t$$

Substituting the preceding expressions into the formula for the acceleration command yields

$$n_c(t) = \left\{ (y + t_{go}\dot{y} + .5t_{go}^2 n_T)t_{go}t_{go} - .5t_{go}^2 \left[(\dot{y} + t_{go}n_T - \dot{y}_F)t_{go} \right.\right.$$

$$\left.\left. + (y + t_{go}\dot{y} + .5t_{go}^2 n_T)(1) \right] + (\dot{y} + t_{go}n_T - \dot{y}_F)(1)\frac{t_{go}^3}{3} \right\} \Big/ \left[\frac{t_{go}^3}{3} t_{go} - \left(\frac{t_{go}^2}{2} \right)^2 \right]$$

After some algebra, we see that the new trajectory shaping guidance law simplifies to

$$n_c(t) = \frac{6y + 4\dot{y}t_{go} + n_T t_{go}^2 + 2\dot{y}(t_F)t_{go}}{t_{go}^2}$$

The guidance law that landed the Apollo spacecraft on the moon in 1969 used the preceding guidance law. In the Apollo case, there was no target acceleration and the relative velocity at intercept was chosen to be zero (that is, this special case is also known as a rendezvous). Therefore, the Apollo guidance law is simply [1–4]

$$n_c(t)_{\text{Apollo}} = \frac{6y + 4\dot{y}t_{go}}{t_{go}^2}$$

ALTERNATE FORM OF TRAJECTORY SHAPING GUIDANCE LAW

We can rewrite the trajectory shaping guidance law of the preceding section as

$$n_c(t) = \frac{4y + 4\dot{y}t_{go} + n_T t_{go}^2 + 2y + 2\dot{y}(t_F)t_{go}}{t_{go}^2} = \frac{4(y + \dot{y}t_{go})}{t_{go}^2} + \frac{2\left[y + \dot{y}(t_F)t_{go}\right] + n_T t_{go}^2}{t_{go}^2}$$

Recall that the formula for the line of sight angle is given by

$$\lambda = \frac{y}{R_{TM}} = \frac{y}{V_c t_{go}} = \frac{y}{V_c(t_F - t)}$$

Therefore, the line of sight rate can be found by differentiating the preceding expression using the quotient rule from calculus. After some algebra, we obtain

$$\dot{\lambda} = \frac{y + \dot{y}t_{go}}{V_c t_{go}^2}$$

Therefore, the trajectory shaping guidance law simplifies to

$$n_c(t) = 4V_c\dot{\lambda} + \frac{2[\lambda V_c + \dot{y}(t_F)]}{t_{go}} + n_T$$

To express the guidance law in terms of a final angle rather than a final relative velocity, we can invert the expression for the line of sight rate and solve for the relative velocity or

$$\dot{y} = \frac{\dot{\lambda} V_c t_{\text{go}}^2 - y}{t_{\text{go}}} = \frac{\dot{\lambda} V_c t_{\text{go}}^2 - \dot{\lambda} V_c t_{\text{go}}}{t_{\text{go}}} = \dot{\lambda} V_c t_{\text{go}}^2 - \lambda V_c$$

We can evaluate the preceding expression at intercept. At the end of the flight, time to go is zero and the final line of sight angle is λ_F. Therefore, at the end of the flight we can see that the relative velocity is simply

$$\dot{y}(t_F) = -\lambda(t_F)V_c = -\lambda_F V_c$$

and the trajectory shaping guidance law simplifies to

$$n_c(t) = 4V_c\dot{\lambda} + \frac{2V_c[\lambda - \lambda_F]}{t_{go}} + n_T$$

Thus, we can see that we can think of the trajectory shaping guidance law as one that minimizes the integral of the square of the commanded acceleration, makes the miss zero, and drives the final line of sight angle to the designer-chosen value λ_F. The trajectory shaping guidance law appears to be a form of augmented proportional navigation (with an effective navigation ratio of 4 and a different multiplier for the target acceleration term) plus an extra term that is proportional to the difference between the true line of sight angle and the desired line of sight angle at the end of the flight.

TESTING TRAJECTORY SHAPING GUIDANCE IN THE LINEAR WORLD

Now that the trajectory shaping guidance law has been derived, it is important to first test the new guidance law in the linear world to see if it works as anticipated. In addition, we would like to compare trajectory shaping guidance with proportional navigation in terms of both accuracy and acceleration requirements. Figure 24.2 presents a block diagram of a zero-lag homing loop to be used in evaluating both proportional navigation and trajectory shaping guidance. We can see from the homing loop that the two sources of error considered are target maneuver, n_T, and heading error, HE. We can see from Fig. 24.2 that the two measures of performance will be the miss distance, $y(t_F)$, and the final line of sight angle, λ_F.

The homing loop model of Fig. 24.2 was programmed, and the resultant engagement simulation appears in Listing 24.1. We can see from the listing that the parameter PN determines the type of guidance law to be used. If PN=1 then proportional navigation is used, whereas if PN=0 the trajectory shaping guidance

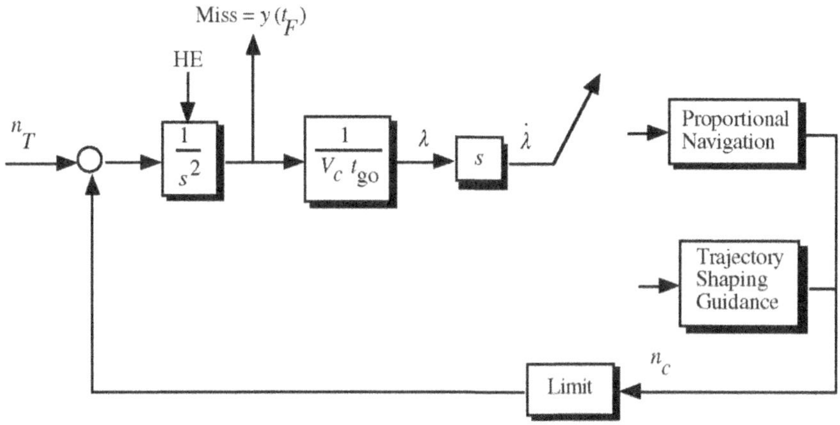

Fig. 24.2 Homing loop model for guidance law comparison.

law is used. When trajectory shaping guidance is used the final specified line of sight angle is denoted XLAMFDEG and is in units of degrees. The trajectory shaping guidance law assumes that time-to-go information and line of sight angle and rate information are available. It is important to note that proportional navigation does not require time-to-go information.

The nominal case of Listing 24.1 was run in which there was a 10-s flight and −20 deg of heading error. When the trajectory shaping guidance law is used it is specified that the final line of sight angle should be −30 deg. We can see from Fig. 24.3 that both guidance laws enable the missile to take out the heading

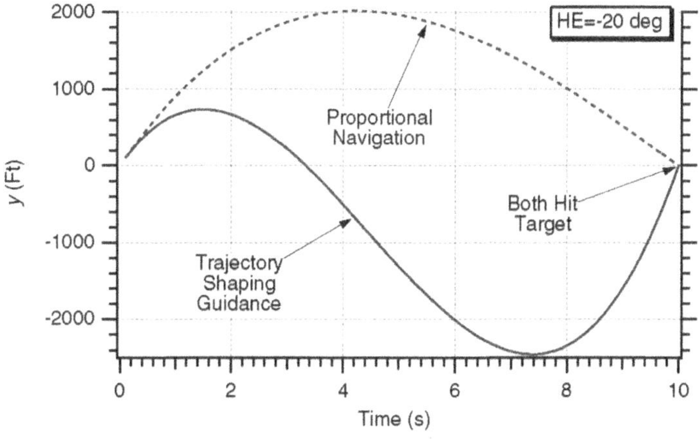

Fig. 24.3 Both guidance laws enable missile to take out the heading error disturbance and hit the target using different relative trajectories.

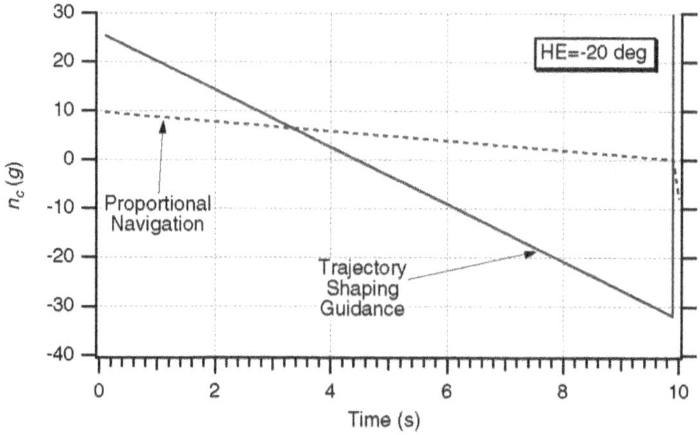

Fig. 24.4 Trajectory shaping guidance requires more acceleration than proportional navigation to take out heading error disturbance.

error and hit the target because $y(10)=0$ in both cases. However, we can see that the relative trajectories are totally different for both guidance laws.

The acceleration requirements for both guidance laws are displayed in Fig. 24.4. We can see that significantly more commanded acceleration is required for trajectory shaping guidance to take out 20 deg of heading error (25 g at the beginning of the flight and approximately $-30\,g$ near the end of the flight) than is required with proportional navigation (10 g at the beginning of the flight and 0 g near the end of the flight).

LISTING 24.1 LINEAR ENGAGEMENT SIMULATION FOR GUIDANCE LAW COMPARISON

```
n=0;
XNT=0.;
HEDEG=-20.;
XNCLIM=999999.;
PN=0;
XLAMFDEG=-30.;
VC=4000.;
VM=3000.;
TF=10.;
XNP=3.;
XLAMF=XLAMFDEG/57.3;
Y=0.;
YD=-VM*HEDEG/57.3;
T=0.;
```

```
H=.001;
S=0.;
while T<=(TF-.0001)
        YOLD=Y;
        YDOLD=YD;
        STEP=1;
        FLAG=0;
        while STEP <=1
          if FLAG==1
        STEP=2;
                Y=Y+H*YD;
                YD=YD+H*YDD;
                T=T+H;
          end
          TGO=TF-T+.00001;
          XLAM=Y/(VC*TGO);
          XLAMD=(Y+YD*TGO)/(VC*TGO*TGO);
          if PN==1
                XNC=XNP*VC*XLAMD;
          else
                XNC=4.*VC*XLAMD+XNT+2.*VC*(XLAM-XLAMF)/TGO;
          end
          if XNC>XNCLIM
                 XNC=XNCLIM;
          end
          if XNC<-XNCLIM
                XNC=-XNCLIM;
          end
          YDD=XNT-XNC;
          FLAG=1;
        end
        FLAG=0;
        Y=.5*(YOLD+Y+H*YD);
        YD=.5*(YDOLD+YD+H*YDD);
        S=S+H;
        if S>=.09999
          S=0.;
          n=n+1;
          XLAMDEG=XLAM*57.3;
          XNCG=XNC/32.2;
          ArrayT(n)=T;
          ArrayY(n)=Y;
          ArrayXNCG(n)=XNCG;
          ArrayXLAMDEG(n)=XLAMDEG;
        end
```

```
end
figure
plot(ArrayT,ArrayY),grid
title('Relative Trajectory')
xlabel('Time (Sec) ')
ylabel('Y (Ft)')
figure
plot(ArrayT,ArrayXNCG),grid
title('Commanded Acceleration')
xlabel('Time (Sec) ')
ylabel('XNC (G)')
axis([0 10 -40 30])
figure
plot(ArrayT,ArrayXLAMDEG),grid
title('Line-of-Sight Angle')
xlabel('Time (Sec) ')
ylabel('XLAM (Deg)')
axis([0 10 -30 10])
clc
output=[ArrayT',ArrayY',ArrayXNCG',ArrayXLAMDEG'];
save datfil.txt output /ascii
disp '*** Simulation Complete'
```

Finally, we can see from Fig. 24.5 that with trajectory shaping guidance the line of sight angle matches the design goal of -30 deg at the end of the flight. With proportional navigation the final line of sight angle is not controlled and it is really a matter of luck on what that angle will be (that is, approximately 8 deg

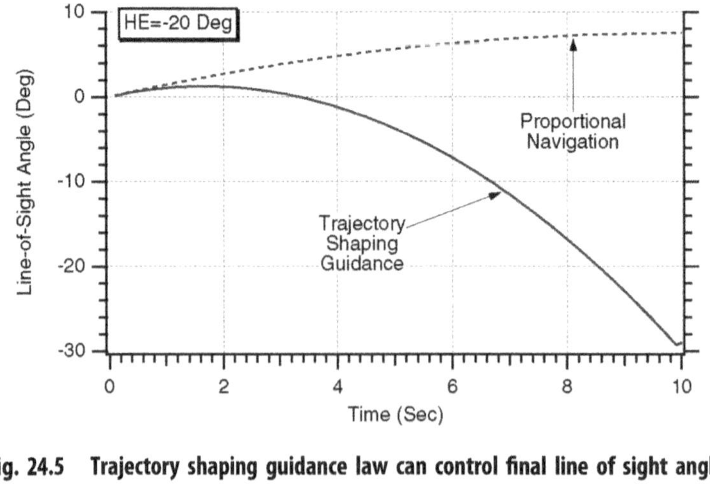

Fig. 24.5 Trajectory shaping guidance law can control final line of sight angle.

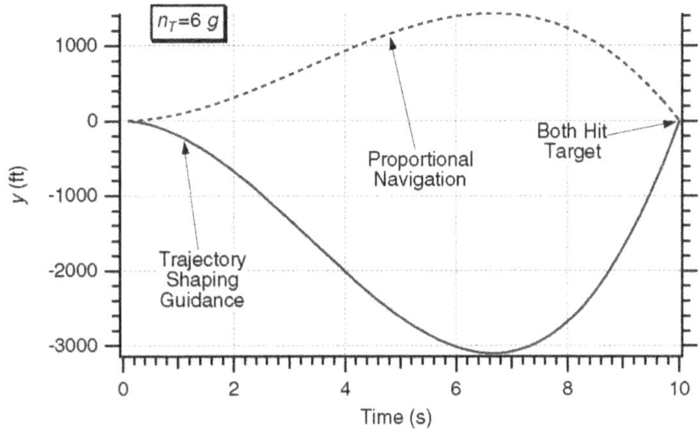

Fig. 24.6 Trajectory shaping guidance law can also hit maneuvering target.

in this example). Thus, we can say that simulation results indicate that trajectory shaping guidance appears to be working correctly against the heading error disturbance.

Next, both guidance laws were compared in terms of their response to a 6 g target maneuver. Again, we can see from Fig. 24.6 that both guidance laws enable the missile to hit the maneuvering target because $y(10)=0$ in both cases. As was the case before, both guidance laws result in relative trajectories that are significantly different.

Again, we can see from Fig. 24.7 that trajectory shaping guidance requires more acceleration than proportional navigation to hit the maneuvering target. For the case of the 6 g maneuvering target and desired final line of sight angle of -30 deg, trajectory shaping guidance required 20 g at the beginning of flight and $-20\,g$ at the end of flight. On the other hand, proportional navigation, which did not reach the final line of sight angle goal, required 0 g at the beginning of flight and nearly 20 g at the end of flight.

From Fig. 24.8 we can see that with trajectory shaping guidance we achieved the goal of the line of sight angle becoming -30 deg at the end of the flight. We can also see that for this example proportional navigation ended up with a final line of sight angle of 14 deg. Thus, we can conclude that simulation results indicate that trajectory shaping guidance appears to be working correctly against the target maneuver disturbance.

Other cases were run with the trajectory shaping guidance law for the case in which there was -20 deg of heading error and the final line of sight angle is made a parameter. Figure 24.9 shows that the acceleration requirements for the trajectory shaping guidance law are dependent on the final line of sight angle. Figure 24.10 shows that the various design goals for the final line of sight angle

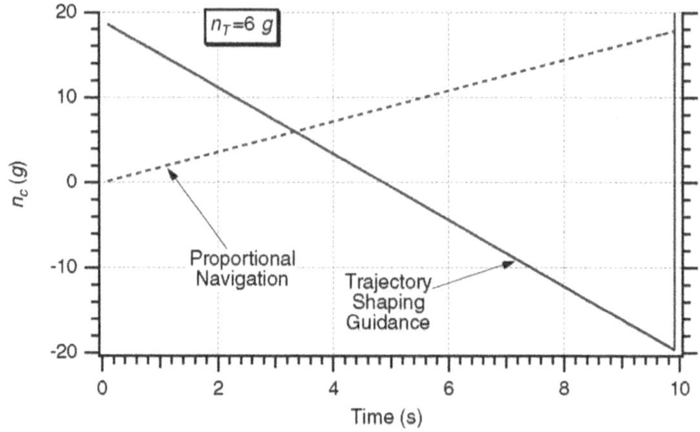

Fig. 24.7 Trajectory shaping guidance law requires more acceleration than proportional navigation against maneuvering target.

are all met with trajectory shaping guidance in the presence of the heading error disturbance, provided adequate acceleration is available.

Finally, even more cases were run with the trajectory shaping guidance law for the situation in which there was a 6 *g* maneuvering target and the final line of sight angle was made a parameter. Figure 24.11 shows that the acceleration requirements for the trajectory shaping guidance law are again dependent on the final line of sight angle. Figure 24.12 shows that the various design goals for the final

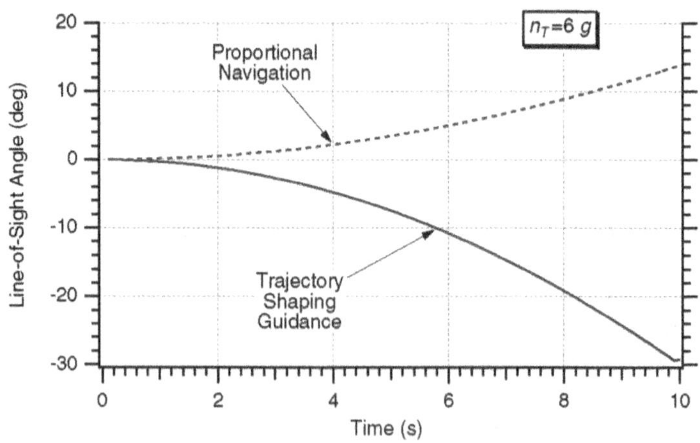

Fig. 24.8 Trajectory shaping guidance law can still control final line of sight angle—even in presence of maneuvering target.

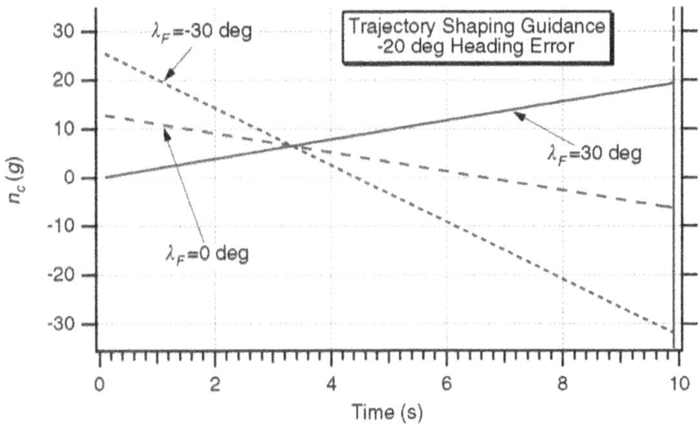

Fig. 24.9 Acceleration requirements depend on final line of sight angle specification when disturbance is heading error.

line of sight angle are all met with trajectory shaping guidance in the presence of the target maneuver disturbance.

CLOSED-FORM SOLUTIONS

With the proportional and augmented proportional navigation guidance laws, we were able to derive closed-form solutions for the missile acceleration due to a step

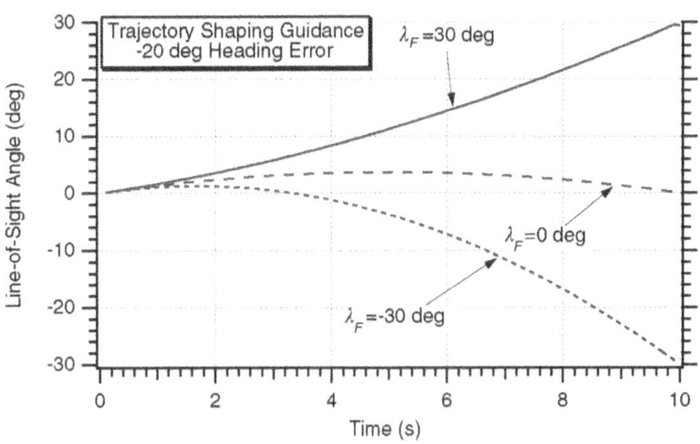

Fig. 24.10 Final line of sight angle goals are met in presence of heading error with trajectory shaping guidance.

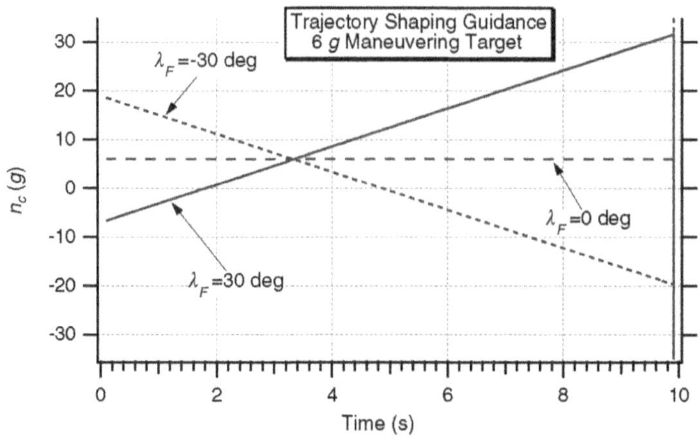

Fig. 24.11 Acceleration requirements still depend on final line of sight angle specification when disturbance is target maneuver.

in target maneuver, heading error, and a step in target displacement for a zero-time constant missile guidance system. The solutions were obtained by solving a linear, first-order, time-varying differential equation as was demonstrated in Chapters 2 and 19. Let us see if we can use the same techniques to derive acceleration formulas for the trajectory shaping guidance law.

Consider the case in which the only disturbance to the guidance system is target maneuver. From Fig. 24.2 we can see that the relative acceleration is

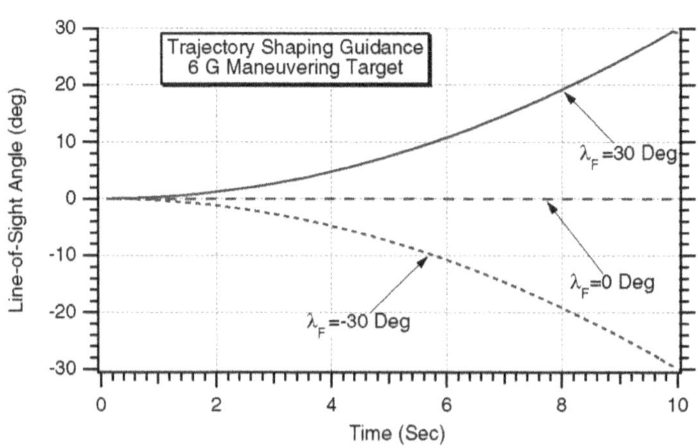

Fig. 24.12 Final line of sight angle goals are met in presence of target maneuver with trajectory shaping guidance.

simply target acceleration minus missile acceleration, or

$$\ddot{y} = n_T - n_c$$

Substituting the original expression for the trajectory shaping guidance law into the preceding equation yields

$$\ddot{y} = n_T - n_c(t) = n_T - \left[\frac{6y + 4\dot{y}t_{\text{go}} + n_T t_{\text{go}}^2 + 2\dot{y}(t_F)t_{\text{go}}}{t_{\text{go}}^2}\right]$$

$$= \frac{-6y - 4\dot{y}t_{\text{go}} - 2\dot{y}(t_F)t_{\text{go}}}{t_{\text{go}}^2}$$

or more simply

$$\ddot{y} + \frac{4\dot{y}}{t_F - t} + \frac{6y}{(t_F - t)^2} = \frac{-2\dot{y}(t_F)}{t_F - t}$$

with initial conditions

$$y(0) = 0 \quad \text{and} \quad \dot{y}(0) = 0$$

Recall that in Chapters 2 and 19 we obtained closed-form solutions by solving a first-order linear differential equation with time-varying coefficients. However, now we now have a second-order linear differential equation with time-varying coefficients. The solution to such an equation is extremely difficult at best. Let us see if we can take another, less conventional, approach to the problem.

Recall that the general solution to the state space equation at the final time

$$\dot{\boldsymbol{x}} = \boldsymbol{Fx} + \boldsymbol{Gu}$$

was given by

$$\mathbf{x}(t_F) = \Phi(t_F - t)\mathbf{x}(t) + \int_t^{t_F} \Phi(t_F - \lambda)\mathbf{G}(\lambda)\mathbf{u}(\lambda)\,\mathrm{d}\lambda$$

We can also say that the general solution to the state space equation can be expressed in terms of the initial conditions at time zero as

$$\mathbf{x}(t_F) = \Phi(t_F)\mathbf{x}(0) + \int_0^{t_F} \Phi(t_F - \lambda)\mathbf{G}(\lambda)\mathbf{u}(\lambda)\,\mathrm{d}\lambda$$

Therefore, for the homing-loop problem under consideration we have

$$\begin{bmatrix} y(t_F) \\ \dot{y}(t_F) \\ n_T(t_F) \end{bmatrix} = \begin{bmatrix} 1 & t_F & .5t_F^2 \\ 0 & 1 & t_F \\ 0 & 0 & 1 \end{bmatrix} \begin{bmatrix} y(0) \\ \dot{y}(0) \\ n_T(0) \end{bmatrix} + \int_0^{t_F} \begin{bmatrix} 1 & t_F - \lambda & .5(t_F - \lambda)^2 \\ 0 & 1 & t_F - \lambda \\ 0 & 0 & 1 \end{bmatrix} \begin{bmatrix} 0 \\ -1 \\ 0 \end{bmatrix} n_c(\lambda)\, \mathrm{d}\lambda$$

Multiplying out the preceding matrix equation and leaving out the third scalar equation yields

$$y(t_F) = y(0) + t_F\dot{y}(0) + .5t_F^2 n_T(0) - \int_0^{t_F} (t_F - \lambda) n_c(\lambda)\, \mathrm{d}\lambda$$

$$\dot{y}(t_F) = \dot{y}(0) + t_F n_T(0) - \int_0^{t_F} n_c(\lambda)\, \mathrm{d}\lambda$$

We still want to minimize the integral of the square of the commanded acceleration subject to the miss being zero and the relative velocity at the end of flight being specified or

$$y(t_F) = 0 \quad \text{and} \quad \dot{y}(t_F) = \dot{y}_F \quad \text{subject to minimizing} \int_0^{t_F} n_c^2(t)\, \mathrm{d}t$$

We get the same solution for the acceleration command as before or

$$n_c(t) = \frac{f_1 h_1(t)\|h_2^2\| - \|h_1 h_2\|[f_2 h_1(t) + f_1 h_2(t)] + f_2 h_2(t)\|h_1^2\|}{\|h_1^2\|\|h_2^2\| - \|h_1 h_2\|^2}$$

except that this time the shorthand notation definitions have changed to

$$\begin{aligned} f_1 &= y(0) + t_F\dot{y}(0) + .5t_F^2 n_T(0) \\ f_2 &= \dot{y}(0) + t_F n_T(0) - \dot{y}(t_F) \\ h_1(t) &= t_F - t = t_{go} \\ h_2(t) &= 1 \end{aligned}$$

$$\|h_1^2\| = \int_0^{t_F} h_1^2(t)\, \mathrm{d}t = \int_0^{t_F} (t_F - t)^2\, \mathrm{d}\lambda = \frac{t_F^3}{3}$$

$$\|h_2^2\| = \int_0^{t_F} h_2^2(t)\, \mathrm{d}t = \int_0^{t_F} \mathrm{d}t = t_F$$

$$\|h_1 h_2\| = \int_0^{t_F} h_1(t) h_2(t)\, \mathrm{d}t = \int_0^{t_F} (t_F - t)\, \mathrm{d}t = \frac{t_F^2}{2}$$

Substitution of the preceding definitions into the formula for the acceleration command yields

$$n_c(t) = \Bigg\{ \big[y(0) + t_F\dot{y}(0) + .5t_F^2 n_T(0)\big] t_{go} t_F - .5t_F^2\big[\,[\dot{y}(0) + t_F n_T(0) - \dot{y}(t_F)]t_{go}$$

$$+[y(0) + t_F\dot{y}(0) + .5t_F^2 n_T(0)](1)\big] + [\dot{y}(0) + t_F n_T(0) - \dot{y}(t_F)](1)\frac{t_F^3}{3}\Bigg\} \Bigg/$$

$$\left[\frac{t_F^3}{3}t_F - \left[\frac{t_F^2}{2}\right]^2\right]$$

After much algebra, we obtain

$$n_c(t) = \frac{12y(0)\left(t_{go} - \frac{t_F}{2}\right) + t_F\dot{y}(0)\left(6t_{go} - 2t_F\right) + t_F^3 n_T(0) + t_F\dot{y}(t_F)\left(6t_{go} - 4t_F\right)}{t_F^3}$$

The preceding expression is the closed-form solution for the total missile acceleration due to the various initial conditions or error sources when using the trajectory shaping guidance law. Therefore, the missile acceleration due to an initial condition in relative velocity can be written by inspection from the preceding formula as

$$n_c(t)\big|\dot{y}_0 = \frac{t_F\dot{y}(0)\left(6t_{go} - 2t_F\right)}{t_F^3} = \frac{2\dot{y}(0)}{t_F}\left[2 - \frac{3t}{t_F}\right]$$

Because the initial relative velocity and heading error are related by

$$\dot{y}(0) = -V_M\text{HE}$$

we can say that the acceleration due to heading error is given by

$$n_c(t)\big|_{\text{HE}} = \frac{-2V_M\text{HE}}{t_F}\left[2 - \frac{3t}{t_F}\right]$$

Therefore, the acceleration required to take out the heading error is proportional to the amount of heading error and inversely proportional to the amount of homing time. More heading error and less homing time both work in the direction of increasing the missile acceleration requirements.

From the general closed-form acceleration formula, we can see that the acceleration due to a target maneuver is given by

$$n_c(t)\big|_{n_T} = \frac{t_F^3 n_T(0)}{t_F^3} = n_T(0)$$

We can see from the preceding expression that in this case the missile is simply matching the target acceleration. Therefore, as expected, larger target maneuvers will require more acceleration capability from the missile.

Finally, we can see that the acceleration due to shaping the trajectory to match a desired final relative velocity can also be written by inspection of the total acceleration formula as

$$n_c(t)\big|_{\dot{y}_F} = \frac{t_F\dot{y}(t_F)(6t_{go} - 4t_F)}{t_F^3} = \frac{2\dot{y}(t_F)}{t_F}\left[1 - \frac{3t}{t_F}\right]$$

Recall that the final relative velocity can also be expressed in terms of the final line of sight angle as

$$\dot{y}(t_F) = -V_c\lambda(t_F)$$

Therefore, the acceleration due to shaping the final line of sight angle can be rewritten as

$$n_c(t)|_{\lambda_F} = \frac{-2V_c\lambda(t_F)}{t_F}.\left[1 - \frac{3t}{t_F}\right]$$

The acceleration requirements are proportional to the amount of shaping we want to do and inversely proportional to the amount of homing time. Larger desired final line of sight angles (that is, more shaping) will require more missile acceleration.

To check the formulas derived in this section, cases were run with the linear engagement simulation of Listing 24.1. First, a case was run with trajectory shaping guidance in which the desired final line of sight angle was zero and there was a 6 g target maneuver for a 10-s flight. Recall that the formula for the commanded missile acceleration due to a maneuvering target is given by

$$n_c(t)|n_T = n_T(0)$$

We can see from Fig. 24.13 that the simulation results of Listing 24.1 (namely, XNT $= 193.2$, HEDEG $= 0$, PN $= 0$, XLAMFDEG $= 0$) and the preceding formula are in exact agreement, thus demonstrating that the acceleration due to target maneuver formula is correct.

Listing 24.1 was again run with trajectory shaping guidance in which the desired final line of sight was zero and there was a -20 deg heading error for a 10-s flight. Recall that the formula for the commanded missile acceleration due to heading error is given by

$$n_c(t)|\text{HE} = \frac{-2V_M\text{HE}}{t_F}\left[2 - \frac{3t}{t_F}\right]$$

We can see from Fig. 24.14 that the simulation results of Listing 24.1 (namely, XNT $= 0$, HEDEG $= -20$, PN $= 0$, XLAMFDEG $= 0$) and the preceding formula are in exact agreement, thus demonstrating that the acceleration due to heading error formula is also correct.

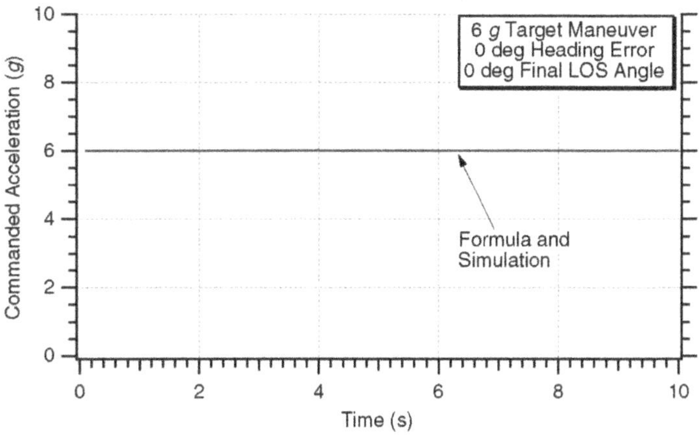

Fig. 24.13 Formula for acceleration due to target maneuver is accurate.

Finally, Listing 24.1 was run again with trajectory shaping guidance in which the desired final line of sight angle is set to −30 deg. In this case there is no heading error or target maneuver. However, the flight time is still 10-s. Recall that the formula for the commanded missile acceleration due to specifying the final line of sight angle is given by

$$n_c(t)|\lambda_F = \frac{-2V_c\lambda(t_F)}{t_F}\left[1 - \frac{3t}{t_F}\right]$$

We can see from Fig. 24.15 that again the simulation results of Listing 24.1 (namely, XNT = 0, HEDEG = 0, PN=0, XLAMFDEG = −30) and the preceding

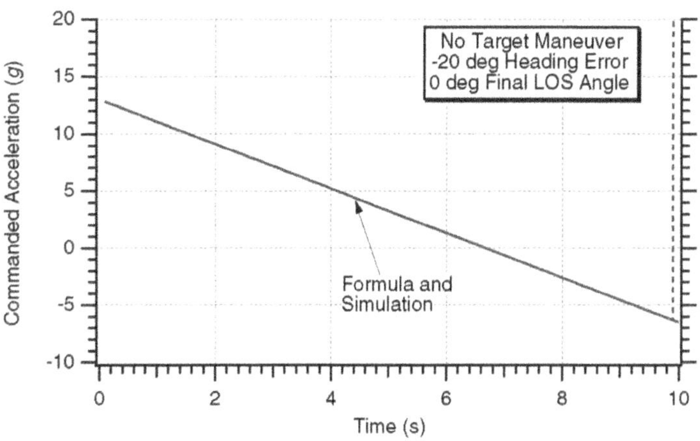

Fig. 24.14 Formula for acceleration heading error is accurate.

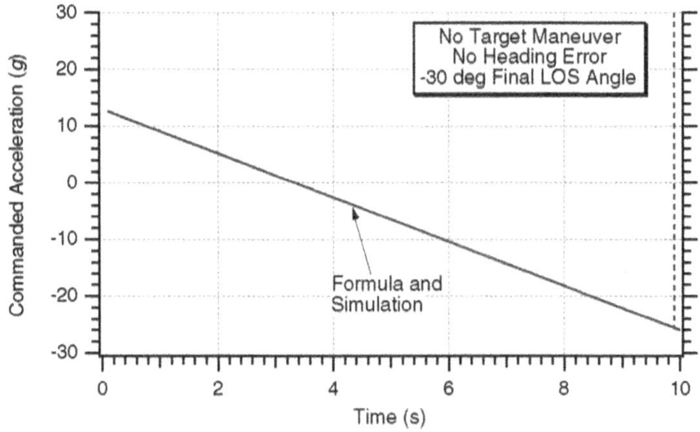

Fig. 24.15 Formula for acceleration because of controlling final line of sight angle is accurate.

formula are in exact agreement, thus demonstrating that the acceleration due to specifying the final line of sight angle formula is also correct.

NONLINEAR RESULTS

The trajectory shaping guidance law was really derived for operation in the world in which the geometry was linear. This implies small angle approximations. It is now of interest to see how the new guidance law works in the two-dimensional world in which the equations of motion are nonlinear. Listing 24.2 is a slight modification of the original zero-time constant nonlinear missile-target engagement simulation of Listing 2.1. An option has been included so that the target can either be stationary (target flight path rate has been modified so there is no division by zero) or moving. It is important to note that the trajectory shaping guidance law requires knowledge of the target acceleration. In the nonlinear engagement simulation, the target acceleration perpendicular to the line of sight is used in the guidance law. The components of the target acceleration in the downrange and altitude direction can be expressed in terms of the target flight path angle as

$$n_{T1} = n_T \sin\beta$$
$$n_{T2} = n_T \cos\beta$$

Therefore, the target acceleration that appears perpendicular to the line of sight can be obtained from trigonometry and can be expressed as

$$n_{T_{\text{PLOS}}} = -n_{T1}\sin\lambda + n_{T2}\cos\lambda$$

Now the trajectory shaping guidance law for the nonlinear world can be written as

$$n_c(t) = 4V_c\dot{\lambda} + \frac{2V_c[\lambda - \lambda_F]}{t_{go}} + n_{T_{PLOS}}$$

where the direction of the commanded acceleration is perpendicular to the line of sight. The new nonlinear engagement simulation appears in Listing 24.2. We can see that the simulation can also be run using proportional navigation by simply setting APN = 0.

The nominal case of Listing 24.2 was run for the example in which the target is considered to be stationary (namely, VT = 0) and is located 30-kft downrange from the missile (namely, RT1IC = 30000). The missile is traveling at 3000 ft/s and is initially at 10-kft altitude (namely, VM = 3000, RM1IC = 10000). The geometry is such that the missile is on a collision path with the target (that is, zero heading error). We can see from Fig. 24.16 that when proportional navigation is used the missile essentially travels in a straight line to the target because it is already on a collision triangle with the target. However, in this application we would like to hit the target vertically for lethality reasons (such as an antitank application). This means that for the trajectory shaping guidance law we would like the final line of sight angle to be −90 deg (namely, XLAMFDEG = −90). We can see from Fig. 24.16 that it indeed appears that the trajectory shaping guidance law is enabling the missile to hit the target near vertically. In addition, we can see that the trajectory shaping guidance law trajectory is entirely different than the proportional navigation guidance trajectory. Of course, trajectory shaping guidance requires time-to-go information, whereas proportional navigation does not.

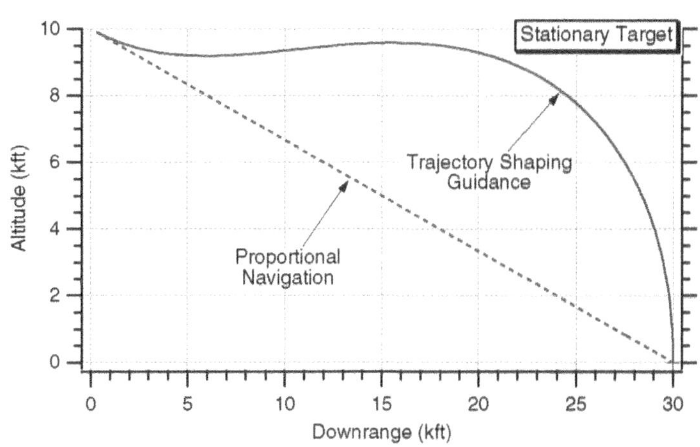

Fig. 24.16 Trajectory shaping also works in the nonlinear world.

LISTING 24.2 NONLINEAR ENGAGEMENT SIMULATION TO TEST TRAJECTORY SHAPING GUIDANCE LAW

```
n=0;
XNTG=0.;
HEDEG=0.;
XNP=3.;
RM1IC=0.;
RM2IC=10000.;
RT1IC=30000.;
RT2IC=0.;
VM=3000.;
VT=0.;
XNCLIMG=9999999.;
APN=1;
XLAMFDEG=-90.;
H=.0001;
XNCLIM=32.2*XNCLIMG;
XLAMF=XLAMFDEG/57.3;
XNT=32.2*XNTG;
RM1=RM1IC;
RM2=RM2IC;
RT1=RT1IC;
RT2=RT2IC;
BETA=0.;
VT1=-VT*cos(BETA);
VT2=VT*sin(BETA);
HE=HEDEG/57.3;
T=0.;
S=0.;
RTM1=RT1-RM1;
RTM2=RT2-RM2;
RTM=sqrt(RTM1^2+RTM2^2);
XLAM=atan2(RTM2,RTM1);
XLEAD=asin(VT*sin(BETA+XLAM)/VM);
THET=XLAM+XLEAD;
VM1=VM*cos(THET+HE);
VM2=VM*sin(THET+HE);
VTM1=VT1-VM1;
VTM2=VT2-VM2;
VC=-(RTM1*VTM1+RTM2*VTM2)/RTM;
while VC >= 0
        if RTM < 1000
          H=.00001;
        else
```

```
    H=.0001;
  end
  BETAOLD=BETA;
  RT1OLD=RT1;
  RT2OLD=RT2;
  RM1OLD=RM1;
  RM2OLD=RM2;
  VM1OLD=VM1;
  VM2OLD=VM2;
  STEP=1;
  FLAG=0;
  while STEP <=1
   if FLAG==1
  STEP=2;
         BETA=BETA+H*BETAD;
         RT1=RT1+H*VT1;
         RT2=RT2+H*VT2;
         RM1=RM1+H*VM1;
         RM2=RM2+H*VM2;
         VM1=VM1+H*AM1;
         VM2=VM2+H*AM2;
         T=T+H;
   end
   RTM1=RT1-RM1;
   RTM2=RT2-RM2;
   RTM=sqrt(RTM1^2+RTM2^2);
   VTM1=VT1-VM1;
   VTM2=VT2-VM2;
   VC=-(RTM1*VTM1+RTM2*VTM2)/RTM;
   XLAM=atan2(RTM2,RTM1);
   XLAMD=(RTM1*VTM2-RTM2*VTM1)/(RTM*RTM);
   TGO=RTM/VC;
   if APN==0
          XNC=XNP*VC*XLAMD;
   else
          XNT1=XNT*sin(BETA);
          XNT2=XNT*cos(BETA);
          XNTPLOS=-XNT1*sin(XLAM)+XNT2*cos(XLAM);
          XNC=4.*VC*XLAMD+XNTPLOS+2.*VC*(XLAM-XLAMF)/TGO;
   end
   if XNC>XNCLIM
          XNC=XNCLIM;
   end
   if XNC<-XNCLIM
          XNC=-XNCLIM;
   end
```

```
			AM1=-XNC*sin(XLAM);
			AM2=XNC*cos(XLAM);
			VT1=-VT*cos(BETA);
			VT2=VT*sin(BETA);
			if VT==0.
				BETAD=0.;
			else
				BETAD=XNT/VT;
			end
			FLAG=1;
		end
		FLAG=0;
		BETA=.5*(BETAOLD+BETA+H*BETAD);
		RT1=.5*(RT1OLD+RT1+H*VT1);
		RT2=.5*(RT2OLD+RT2+H*VT2);
		RM1=.5*(RM1OLD+RM1+H*VM1);
		RM2=.5*(RM2OLD+RM2+H*VM2);
		VM1=.5*(VM1OLD+VM1+H*AM1);
		VM2=.5*(VM2OLD+VM2+H*AM2);
		S=S+H;
		if S>=.09999
			S=0.;
			n=n+1;
			RT1K=RT1/1000.;
			RT2K=RT2/1000.;
			RM1K=RM1/1000.;
			RM2K=RM2/1000.;
			XLAMDEG=XLAM*57.3;
			XNCG=XNC/32.2;
			ArrayT(n)=T;
			ArrayRT1K(n)=RT1K;
			ArrayRT2K(n)=RT2K;
			ArrayRM1K(n)=RM1K;
			ArrayRM2K(n)=RM2K;
			ArrayXNCG(n)=XNCG;
			ArrayXLAMDEG(n)=XLAMDEG;
		end
end
RTM
figure
plot(ArrayRT1K,ArrayRT2K,ArrayRM1K,ArrayRM2K),grid
title('Engagement Geometry')
xlabel('Downrange (Kft) ')
ylabel('Altitude (Kft)')
figure
plot(ArrayT,ArrayXNCG),grid
```

```
title('Commanded Acceleration')
xlabel('Time (Sec) ')
ylabel('XNC (G)')
axis([0 14 -20 25])
figure
plot(ArrayT,ArrayXLAMDEG),grid
title('Line-of-Sight Angle')
xlabel('Time (Sec) ')
ylabel('XLAM (Deg)')
axis([0 14 -100 0])
clc
output=[ArrayT',ArrayRT1K',ArrayRT2K',ArrayRM1K',...
        ArrayRM2K',ArrayXNCG',ArrayXLAMDEG'];
save datfil.txt output /ascii
disp '*** Simulation Complete'
```

Figure 24.17 shows that the price paid for shaping the trajectory is that considerable acceleration is required by the missile to hit the target. Unlike proportional navigation, which does not require any acceleration to hit the target in this scenario because it is already on a collision triangle, trajectory shaping guidance requires more than 20 g of acceleration at the beginning of the flight and nearly $-10\ g$ at the end of the flight.

Finally, we can see from Fig. 24.18 that the line of sight angle for trajectory shaping guidance achieved the design goal by reaching -90 deg at the end of the flight. Figure 24.18 also shows that proportional navigation, which does not shape the trajectory, ended up with a final line of sight angle of -18.4 deg. We

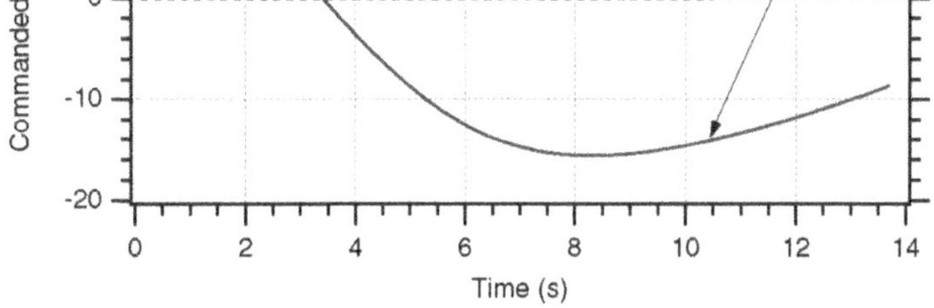

Fig. 24.17 A great deal of acceleration may be required to shape trajectory to get a final line of sight angle of −90 deg.

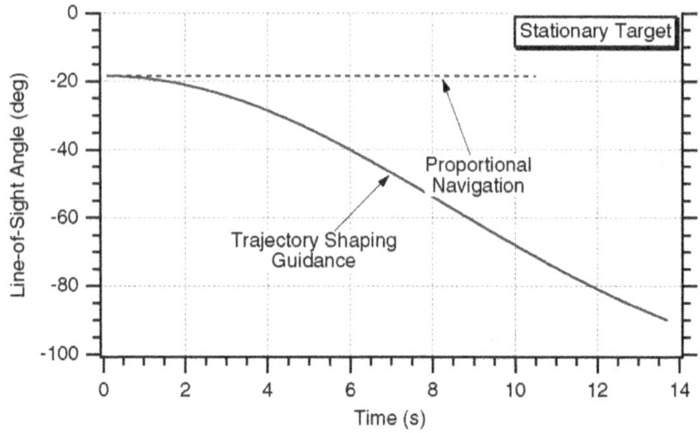

Fig. 24.18 Trajectory shaping enables line of sight angle to reach its goal.

can also see from Fig. 24.18 that because proportional navigation enabled the missile to fly directly to the target, the flight time was nearly 4 s shorter than when the trajectory shaping guidance law was used.

It is of considerable interest to see if the formulas we derived in the previous section for the commanded acceleration are useful in predicting the nonlinear results. Recall that in the previous section we showed that the acceleration in units of $\mathrm{ft/s^2}$ required to turn the missile through and angle of λ_F in units of radians is given by

$$n_c(t)|_{\lambda_F} = \frac{-2V_c\lambda_F}{t_F}\left[1 - \frac{3t}{t_F}\right]$$

where V_c is the closing velocity in ft/s, t_F is the amount of flight or guidance time, and t is instantaneous time both in units of seconds. For the problem depicted in Fig. 24.16, the final line of sight angle that would have been achieved without trajectory shaping guidance is −18.4 deg (see Fig. 24.18). Therefore, trajectory shaping guidance is attempting to change the angle from −18.4 deg to −90 deg, or a change of −71.6 deg. In addition, we can tell from Fig. 24.18 that the amount of guidance time for trajectory shaping guidance is 13.6 s.

Figure 24.19 indicates that the linear formula for missile acceleration is not very accurate in this example for predicting the nonlinear commanded missile acceleration. It is hypothesized that perhaps the angular change in the line of sight angle is too great for linear theory to hold. Another case was run with the nonlinear simulation in which the desired final line of sight angle was −30 deg (a change of only 11.6 deg from −18.4 deg that could be obtained with proportional navigation). Because there is less trajectory shaping, the flight time

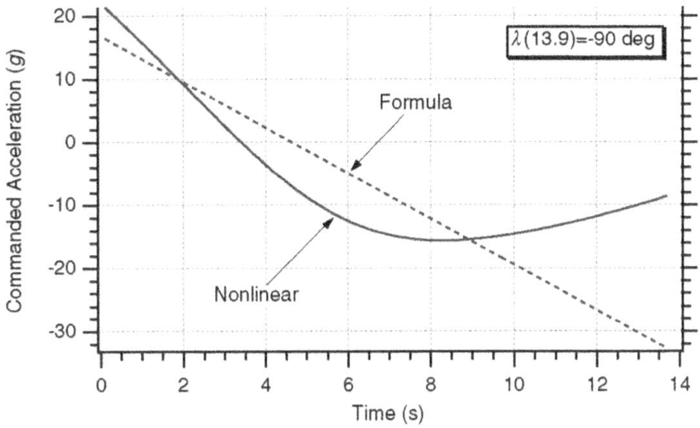

Fig. 24.19 Linear formula is not a great match to nonlinear results.

reduces to 10.6 s. We can see from Fig. 24.20 that the formula now matches the nonlinear results quite accurately.

Another, more stressing case was considered in which the target was both moving and maneuvering. In this example, the missile had a −20 deg heading error, while the target was executing a 6 *g* maneuver. Two cases were considered—one in which the desired final line of sight angle was −30 deg and the other in which the desired final line of sight angle was 30 deg. Figure 24.21 shows that when trajectory shaping guidance was used intercepts were achieved in both cases. It is too difficult to tell from Fig. 24.21 if the final line of sight

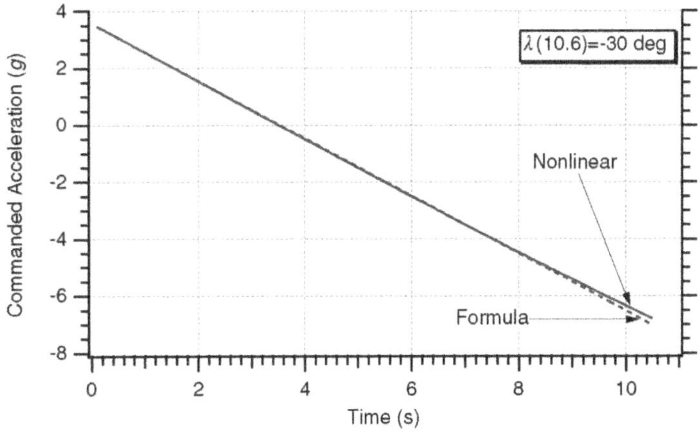

Fig. 24.20 Linear formula is nearly perfect when angular turn is smaller.

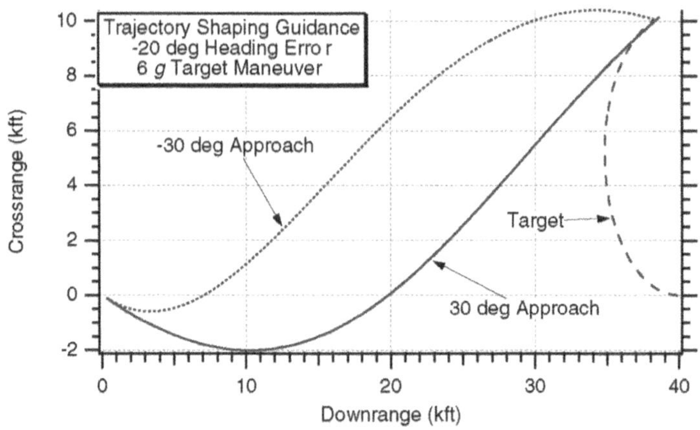

Fig. 24.21 Trajectory shaping guidance works against maneuvering target for different approach angles.

angle design goals have actually been met. Figure 24.22 displays the commanded acceleration profiles that were required in both cases for successful intercepts. We can see that for the −30 deg intercept the maximum positive acceleration was 30 *g* while the maximum negative acceleration was −10 *g*. We can also see that for the 30-deg intercept the maximum positive acceleration was only 10 *g* while the maximum negative acceleration was −5 *g*. Finally, we can see from Fig. 24.23 that the design goals for the final line of sight angles for both cases were met.

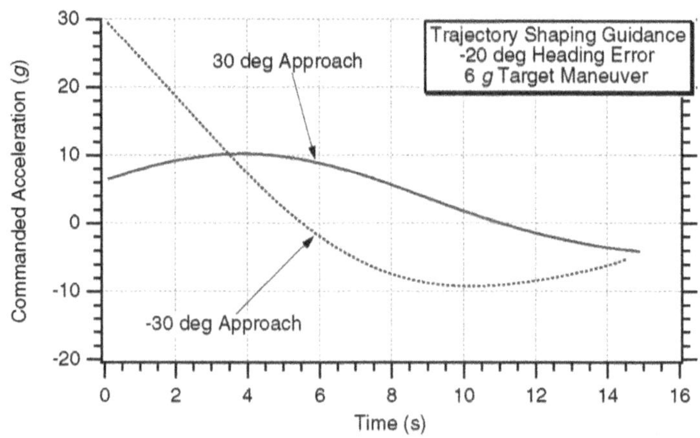

Fig. 24.22 Acceleration requirements are larger with maneuvering target than with stationary target.

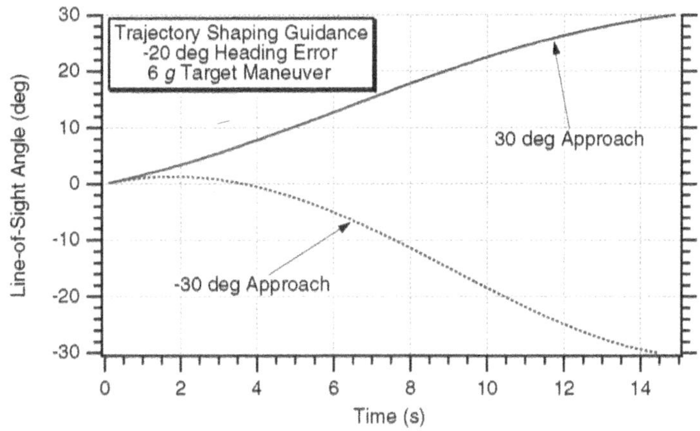

Fig. 24.23 Trajectory shaping guidance meets design goals against maneuvering target.

Again, it is of considerable interest to see if the formulas we derived in the previous section for the commanded acceleration are useful in predicting the nonlinear results. Recall that the total acceleration in units of ft/s^2 required to turn the missile through an angle of λ_F in units of radians in the presence of target maneuver n_T in units of ft/s^2 and heading error HE in units of radians is given by

$$n_c(t)|_{\text{Total}} = \frac{-2V_c\lambda_F}{t_F}\left[1 - \frac{3t}{t_F}\right] + n_T + \frac{-2V_M\text{HE}}{t_F}\left[2 - \frac{3t}{t_F}\right]$$

For the case of interest the closing velocity is approximately 4000 ft/s while the flight time turns out to be 14.5 s. We can see from Fig. 24.24 that the match between the formula and simulation results is not very accurate.

It is hypothesized that the reason for the inaccurate comparison of Fig. 24.24 is because of the highly maneuvering target. As the target maneuvers, the portion of the maneuver perpendicular to the line of sight diminishes, making the formula less accurate. To test the hypothesis another case was run in which the maneuver level decreased to 3 g. The heading error remained at -20 deg, and the desired final line of sight angle remained at -30 deg. For this case, the time of flight reduced to 11.4 s. We can see from Fig. 24.25 that now the formula is an excellent approximation to the nonlinear results.

Thus, we can see that the trajectory shaping guidance law also works in the nonlinear world. We have demonstrated that under many circumstances we also have formulas that can be used to predict or explain the resultant acceleration requirements on the missile when trajectory shaping guidance is used.

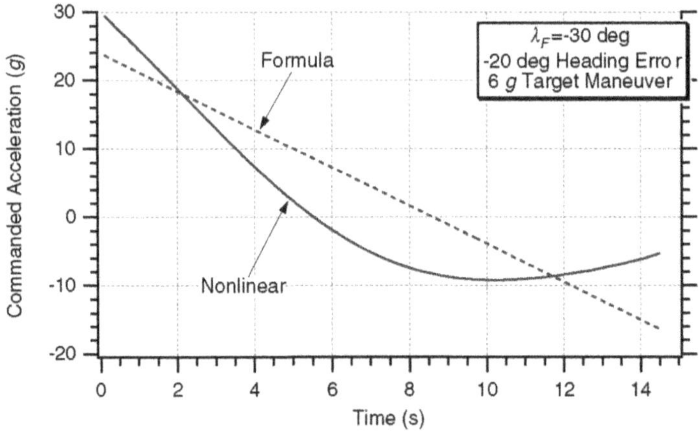

Fig. 24.24 Linear formula not very accurate for highly maneuvering target.

SUMMARY

In this chapter the trajectory shaping guidance law has been derived. It was demonstrated that with this new guidance law we could not only hit the target but could also control the final line of sight angle. The price paid for the trajectory shaping was that more acceleration was required to hit the target. Formulas were also derived that could be used to predict the missile acceleration requirements for the new guidance law under a variety of circumstances. It was demonstrated that the formulas were also an accurate indicator of performance in the nonlinear world.

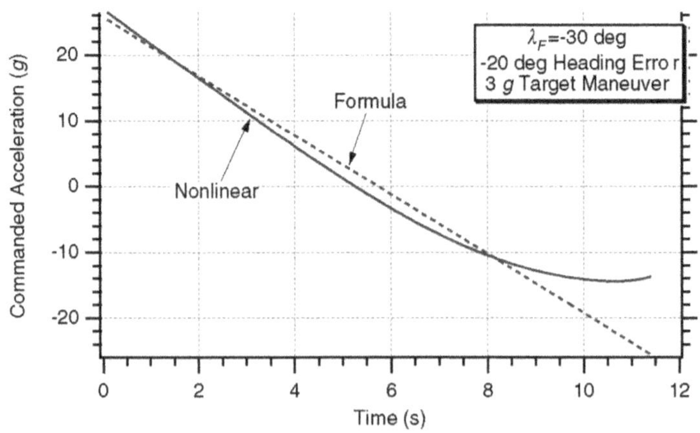

Fig. 24.25 Linear formula is much more accurate when target maneuver level is lower.

REFERENCES

[1] Battin, R., *An Introduction to the Mathematics and Methods of Astrodynamics*, AIAA, New York, 1987, p. 561.

[2] Lin, C. F., *Modern Navigation, Guidance, and Control Processing*, Prentice Hall, New Jersey, 1991, p. 605.

[3] Ben-Asher, J., and Yaesh, I., *Advances in Missile Guidance Theory*, AIAA, Reston, VA, 1998, pp. 25–88.

[4] Bryson, A. E., and Ho, Y. C., *Applied Optimal Control*, Chapter 5, Hemisphere, New York, 1975.

CHAPTER 25

Filtering and Weaving Targets

INTRODUCTION

So far we have shown that we could considerably improve our performance against weaving targets if we could either use a special purpose guidance law or somehow achieve a smaller guidance system time constant. If we choose to use the guidance law that is optimal against weaving targets, we then have to estimate the target acceleration, target jerk, and target weave frequency.

In this chapter we will explore the various filtering options that can be used against a weaving target by using a step-by-step approach. First we shall see how our original linear three-state Kalman filter from Chapter 9 is able to function in the presence of a weaving target. Although with this filter we cannot use the weave guidance law that was derived in Chapter 20, we can use either proportional navigation, augmented proportional navigation, or optimal guidance. Next we will assume that the target weave frequency is known (that is, estimated or derived using other sensors or phenomenology) and proceed to derive an optimal linear four-state weave Kalman filter that estimates both target acceleration and jerk. This filter can be used with either the weave guidance law or compensated weave guidance law that were both derived in Chapter 20. Finally, we will assume that the target weave frequency is not known in advance but must also be estimated. In this case an extended five-state Kalman filter that estimates the relative position, relative velocity, target acceleration, jerk, and weave frequency will be derived. All three Kalman filters and appropriate guidance laws will be compared in terms of both performance and robustness.

REVIEW OF ORIGINAL THREE-STATE LINEAR KALMAN FILTER

The original three-state linear Kalman filter from Chapter 9 was derived based on the homing loop model of Fig. 25.1. Recall that in this guidance system model we measured noisy relative position y^* and were attempting to estimate relative position, relative velocity, and target acceleration. As was the case in Chapter 9, the

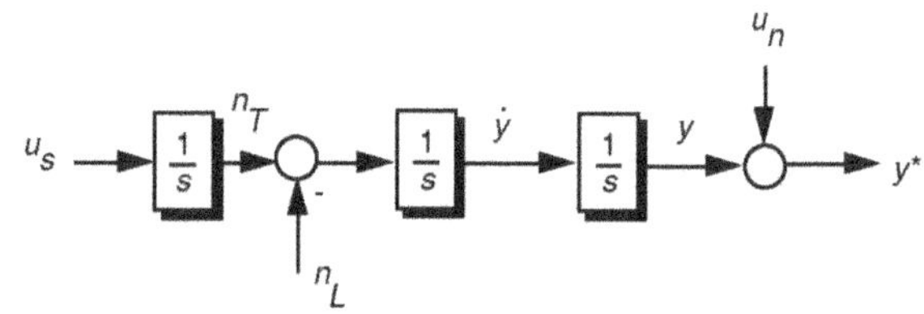

Fig. 25.1 Homing loop model for Kalman filter development.

achieved missile acceleration n_L was assumed to be known, and the target acceleration was considered to be modeled as white noise through an integrator. It is important to note that we have already shown in Chapter 4 that the shaping filter equivalent of a target maneuver with constant amplitude and random starting time is mathematically equivalent (that is, in terms of second-order statistics) to white noise through an integrator.

According to the results of Chapter 4, the spectral density of the white noise source u_s depicted in Fig. 25.1 was shown to be

$$\Phi_s = \frac{n_{\text{TMAX}}^2}{t_F}$$

where n_{TMAX} is the assumed maximum target maneuver level magnitude and t_F is the flight time. The model of Fig. 25.1 can be expressed in state space form as

$$\begin{bmatrix} \dot{y} \\ \ddot{y} \\ \dot{n}_T \end{bmatrix} = \begin{bmatrix} 0 & 1 & 0 \\ 0 & 0 & 1 \\ 0 & 0 & 0 \end{bmatrix} \begin{bmatrix} y \\ \dot{y} \\ n_T \end{bmatrix} + \begin{bmatrix} 0 \\ -1 \\ 0 \end{bmatrix} n_L + \begin{bmatrix} 0 \\ 0 \\ u_s \end{bmatrix}$$

Because the systems dynamics matrix of the preceding equation is given by

$$\mathbf{F} = \begin{bmatrix} 0 & 1 & 0 \\ 0 & 0 & 1 \\ 0 & 0 & 0 \end{bmatrix}$$

the continuous fundamental matrix can easily be derived (as was the case in Chapter 9) as

$$\Phi(t) = \begin{bmatrix} 1 & t & .5t^2 \\ 0 & 1 & t \\ 0 & 0 & 1 \end{bmatrix}$$

By replacing t with the sampling time T_s we can obtain the discrete form of the fundamental matrix as

$$\boldsymbol{\Phi}_{\mathbf{k}} = \begin{bmatrix} 1 & T_s & .5T_s^2 \\ 0 & 1 & T_s \\ 0 & 0 & 1 \end{bmatrix}$$

The discrete measurement equation can be written by inspection of Fig. 25.1 as

$$y_k^* = \begin{bmatrix} 1 & 0 & 0 \end{bmatrix} \begin{bmatrix} y_k \\ \dot{y}_k \\ n_{T_k} \end{bmatrix} + v_k$$

Therefore, the discrete measurement matrix can be written by inspection of the preceding equation as

$$\mathbf{H}_{\mathbf{k}} = \begin{bmatrix} 1 & 0 & 0 \end{bmatrix}$$

The continuous control matrix $\mathbf{G}$ can also be written by inspection of the original state space equation as

$$\mathbf{G} = \begin{bmatrix} 0 \\ -1 \\ 0 \end{bmatrix}$$

Therefore the discrete control matrix $\mathbf{G}_{\mathbf{k}}$ becomes

$$\mathbf{G}_{\mathbf{k}} = \int_0^{T_s} \boldsymbol{\Phi}(\tau)\mathbf{G}(\tau)\mathrm{d}\tau = \int_0^{T_s} \begin{bmatrix} 1 & \tau & .5\tau^2 \\ 0 & 1 & \tau \\ 0 & 0 & 1 \end{bmatrix} \begin{bmatrix} 0 \\ -1 \\ 0 \end{bmatrix} \mathrm{d}\tau = \begin{bmatrix} -.5T_s^2 \\ -T_s \\ 0 \end{bmatrix}$$

Recall the discrete Kalman filtering equation is given by

$$\hat{\mathbf{x}}_{\mathbf{k}} = \boldsymbol{\Phi}_{\mathbf{k}}\hat{\mathbf{x}}_{\mathbf{k}-1} + \mathbf{G}_{\mathbf{k}}\mathbf{u}_{\mathbf{k}-1} + \mathbf{K}_{\mathbf{k}}(\mathbf{z}_{\mathbf{k}} - \mathbf{H}\boldsymbol{\Phi}_{\mathbf{k}}\hat{\mathbf{x}}_{\mathbf{k}-1} - \mathbf{H}\mathbf{G}_{\mathbf{k}}\mathbf{u}_{\mathbf{k}-1})$$

Substitution of the appropriate matrices into the preceding matrix difference equation yields

$$\begin{bmatrix} \hat{y}_k \\ \hat{\dot{y}}_k \\ \hat{n}_{T_k} \end{bmatrix} = \begin{bmatrix} 1 & T_s & .5T_s^2 \\ 0 & 1 & T_s \\ 0 & 0 & 1 \end{bmatrix} \begin{bmatrix} \hat{y}_{k-1} \\ \hat{\dot{y}}_{k-1} \\ \hat{n}_{T_{k-1}} \end{bmatrix} + \begin{bmatrix} -.5T_s^2 \\ -T_s \\ 0 \end{bmatrix} n_{L_{k-1}} + \begin{bmatrix} K_{1_k} \\ K_{2_k} \\ K_{3_k} \end{bmatrix}$$

$$\times \left(y_k^* - \begin{bmatrix} 1 & 0 & 0 \end{bmatrix} \begin{bmatrix} 1 & T_s & .5T_s^2 \\ 0 & 1 & T_s \\ 0 & 0 & 1 \end{bmatrix} \begin{bmatrix} \hat{y}_{k-1} \\ \hat{\dot{y}}_{k-1} \\ \hat{n}_{T_{k-1}} \end{bmatrix} - \begin{bmatrix} 1 & 0 & 0 \end{bmatrix} \begin{bmatrix} -.5T_s^2 \\ -T_s \\ 0 \end{bmatrix} n_{L_{k-1}} \right)$$

We can multiply out the terms of the preceding matrix equation to yield the linear three-state Kalman filter scalar equations as

$$\mathrm{RES}_k = y_k^* - \hat{y}_{k-1} - T_s\hat{\dot{y}}_{k-1} - .5T_s^2(n_{T_{k-1}} - n_{L_{k-1}})$$

$$\hat{y}_k = \hat{y}_{k-1} + T_s\hat{\dot{y}}_{k-1} + .5T_s^2(\hat{n}_{T_{k-1}} - n_{L_{k-1}}) + K_{1_k}\ \mathrm{RES}_k$$

$$\hat{\dot{y}}_k = \hat{\dot{y}}_{k-1} + T_s(\hat{n}_{T_{k-1}} - n_{L_{k-1}}) + K_{2_k}\ \mathrm{RES}_k$$

$$\hat{n}_{T_k} = \hat{n}_{T_{k-1}} + K_{3_k}\ \mathrm{RES}_k$$

A simulation of the linear three-state linear Kalman filter as part of a missile guidance system is a modified form of Listing 9.2 and appears in Listing 25.1. The simulation now has a single time constant representation of the flight control system plus a weaving target rather than a constant target maneuver. We can see from the listing that there is 3-*g* weaving target with a weave frequency of 2 rad/s. Nominally there is still 1 mr of measurement noise, but the closing velocity has been increased to 9000 ft/s to reflect a ballistic target engagement. We can see from the listing that the guidance law options for this filter are either proportional navigation, augmented proportional navigation, or optimal guidance (namely, APN = 0, 1 or 2, respectively). Because there is a single time constant representation of the flight control system, the achieved missile acceleration rather than the commanded acceleration enters the filtering equations.

LISTING 25.1 ORIGINAL THREE-STATE LINEAR KALMAN FILTER AND WEAVING TARGET

```
n=0;
TAU=.5;
APN=0;
VC=9000.;
XNT=96.6;
XNTREAL=96.6;
XNTMAX=96.6;
```

```
W=2.;
YIC=0.;
VM=3000.;
HEDEG=0.;
HEDEGFIL=20.;
XNP=3.;
SIGRIN=.001;
TS=.01;
TF=10.;
Y=YIC;
YD=-VM*HEDEG/57.3;
YDIC=YD;
TS2=TS*TS;
TS3=TS2*TS;
TS4=TS3*TS;
TS5=TS4*TS;
PHIN=XNTMAX*XNTMAX/TF;
RTM=VC*TF;
SIGNOISE=SIGRIN;
SIGPOS=RTM*SIGNOISE;
SIGN2=SIGPOS^2;
P11=SIGN2;
P12=0.;
P13=0.;
P22=(VM*HEDEGFIL/57.3)^2;
P23=0.;
P33=XNTMAX*XNTMAX;
T=0.;
H=.001;
S=0.;
YH=0.;
YDH=0.;
XNTH=0.;
XNC=0.;
XNL=0.;
while T<=TF
        YOLD=Y;
        YDOLD=YD;
        XNLOLD=XNL;
        STEP=1;
        FLAG=0;
        while STEP <=1
                if FLAG==1
        STEP=2;
                        Y=Y+H*YD;
                        YD=YD+H*YDD;
```

```
                    XNL=XNL+H*XNLD;
                    T=T+H;
            end
            XNT=XNTREAL*sin(W*T);
            TGO=TF-T+.00001;
            RTM=VC*TGO;
            XLAM=Y/(VC*TGO);
            XLAMD=(RTM*YD+Y*VC)/(RTM^2);
            XNLD=(XNC-XNL)/TAU;
            YDD=XNT-XNL;
            FLAG=1;
    end
    FLAG=0;
    Y=.5*(YOLD+Y+H*YD);
    YD=.5*(YDOLD+YD+H*YDD);
    XNL=.5*(XNLOLD+XNL+H*XNLD);
    S=S+H;
    if S>=(TS-.0001)
            S=0.;
            TGO=TF-T+.000001;
            RTM=VC*TGO;
            SIGNOISE=SIGRIN;
            SIGPOS=RTM*SIGNOISE;
            SIGN2=SIGPOS^2;
            M11=P11+TS*P12+.5*TS2*P13+TS*(P12+TS*P22+.5*TS2*P23);
            M11=M11+.5*TS2*(P13+TS*P23+.5*TS2*P33)+TS5*PHIN/20.;
            M12=P12+TS*P22+.5*TS2*P23+TS*(P13+TS*P23+.5*TS2*P33)...
                    +TS4*PHIN/8.;
            M13=P13+TS*P23+.5*TS2*P33+PHIN*TS3/6.;
            M22=P22+TS*P23+TS*(P23+TS*P33)+PHIN*TS3/3.;
            M23=P23+TS*P33+.5*TS2*PHIN;
            M33=P33+PHIN*TS;
            K1=M11/(M11+SIGN2);
            K2=M12/(M11+SIGN2);
            K3=M13/(M11+SIGN2);
            P11=(1.-K1)*M11;
            P12=(1.-K1)*M12;
            P13=(1.-K1)*M13;
            P22=-K2*M12+M22;
            P23=-K2*M13+M23;
            P33=-K3*M13+M33;
            XLAMNOISE=SIGNOISE*randn;
            YSTAR=RTM*(XLAM+XLAMNOISE);
            RES=YSTAR-YH-TS*YDH-.5*TS*TS*(XNTH-XNL);
            YH=K1*RES+YH+TS*YDH+.5*TS*TS*(XNTH-XNL);
            YDH=K2*RES+YDH+TS*(XNTH-XNL);
```

```
            XNTH=K3*RES+XNTH;
            XLAMDH=(YH+YDH*TGO)/(VC*TGO*TGO);
            if APN==0
                XNC=XNP*(YH+YDH*TGO)/(TGO*TGO);
            elseif APN==1
                XNC=XNP*(YH+YDH*TGO+.5*XNTH*TGO*TGO)/(TGO*TGO);
            else
                XS=TGO/TAU;
                TOP=6.*XS*XS*(exp(-XS)-1.+XS);
                BOT1=2*XS*XS*XS+3.+6.*XS-6.*XS*XS;
                BOT2=-12.*XS*exp(-XS)-3.*exp(-2.*XS);
                XNPP=TOP/(.0001+BOT1+BOT2);
                C1=XNPP/(TGO*TGO);
                C2=XNPP/TGO;
                C3=.5*XNPP;
                C4=-XNPP*(exp(-XS)+XS-1.)/(XS*XS);
                XNC=C1*YH+C2*YDH+C3*XNTH+C4*XNL;
            end
            n=n+1;
            XNTG=XNT/32.2;
            XNTHG=XNTH/32.2;
            ArrayT(n)=T;
            ArrayXNTG(n)=XNTG;
            ArrayXNTHG(n)=XNTHG;
            ArrayY(n)=Y;
            ArrayYSTAR(n)=YSTAR;
        end
end
figure
plot(ArrayT,ArrayXNTG,ArrayT,ArrayXNTHG),grid
title('Acceleration Estimate')
xlabel('Time (Sec) ')
ylabel('Acceleration (G)')
figure
plot(ArrayT,ArrayY,ArrayT,ArrayYSTAR),grid
title('Measurement and Signal')
xlabel('Time (Sec) ')
ylabel('Y (Ft)')
clc
output=[ArrayT',ArrayXNTG',ArrayXNTHG',ArrayY',ArrayYSTAR'];
save datfil.txt output /ascii
disp '*** Simulation Complete'
```

Before we see how well the filter can estimate the weaving target maneuver, it is important to see how much effective noise there is on the filter measurement in this high closing velocity engagement. Figure 25.2 shows that although there is

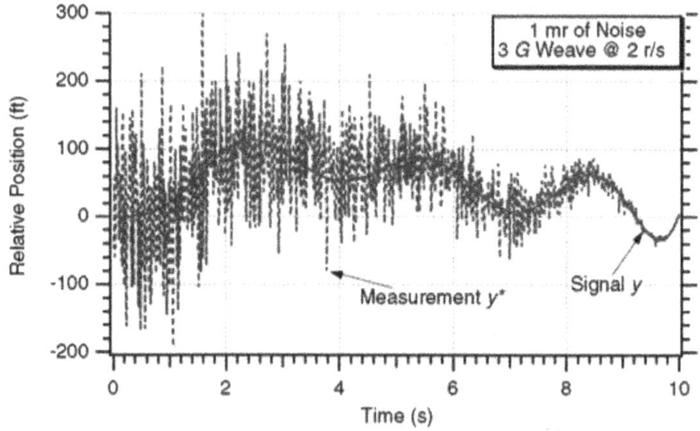

Fig. 25.2 With high closing velocity 1 mr of measurement noise translates into a great deal of noise on relative position.

only 1 mr of measurement noise on the line of sight angle, the high closing velocity causes there to be significant noise on the effective measured relative position y^*. Essentially, the filter will have to take two derivatives of this noisy measurement to estimate target acceleration.

The nominal case of Listing 25.1 was run and Fig. 25.3 shows that the three-state linear Kalman filter's estimate of target acceleration is not very

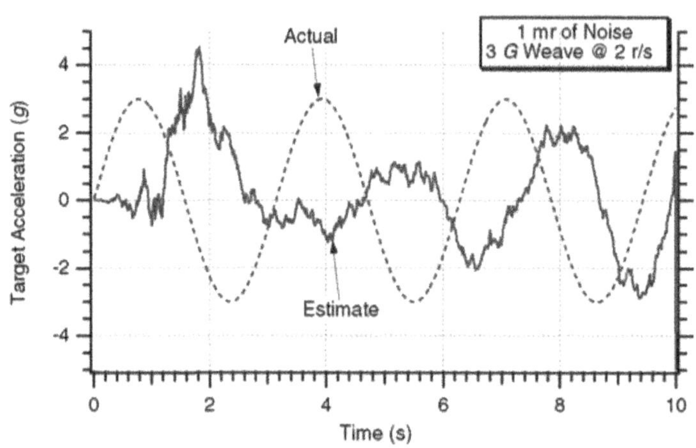

Fig. 25.3 With 1 mr of measurement noise the three-state Kalman filter has difficulty in estimating the sinusoidal nature of the weave maneuver.

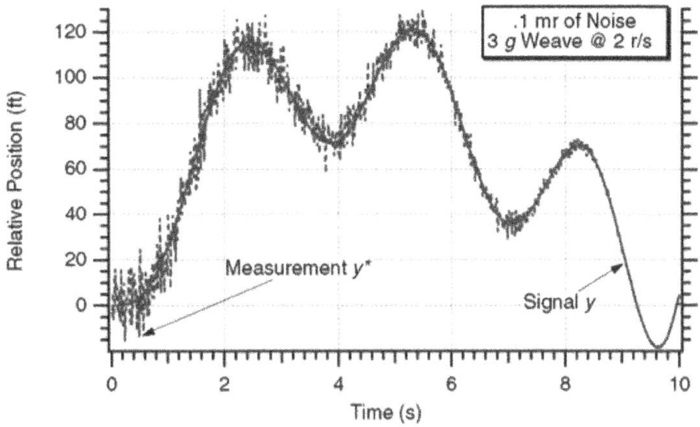

Fig. 25.4 **Reducing measurement noise by order of magnitude is beneficial.**

good. The effective high-noise environment prevents the filter from accurately estimating the sinusoidal motion of the target maneuver. Near the end of the flight, where the effective measurement noise on relative position is diminished, the filter estimate improves but lags the actual target maneuver. It is important to note that the Kalman filter is really optimized for a constant target maneuver and is therefore suboptimal in this example because it is mismatched to the real world. However, the presence of process noise in the filter enables the Kalman filter to track all types of target maneuvers. Process noise lets the filter know that its model of the real world may be in error.

Listing 25.1 was modified so that the measurement noise was decreased by an order of magnitude to .1 mr (namely, SIGRIN = .0001). We can see from Fig. 25.4 that the effective measurement of relative position now more closely resembles the actual relative position. As a consequence of the reduced measurement noise we can see from Fig. 25.5 that the filter's estimate of the target maneuver now better approximates the sinusoidal nature of the maneuver. The estimate of the target maneuver is nearly perfect, except there is approximately a half-second lag between the actual maneuver and the estimate.

Thus, we can conclude that the original three-state linear Kalman filter of Chapter 9 can track a weaving target quite effectively if the measurement noise can be made small.

FOUR-STATE WEAVE KALMAN FILTER

If we had *a priori* information that the target maneuver was sinusoidal in nature, one would think that a better Kalman filter could be designed. To design a Kalman filter optimized to estimate the states of a weaving target, we must first express the

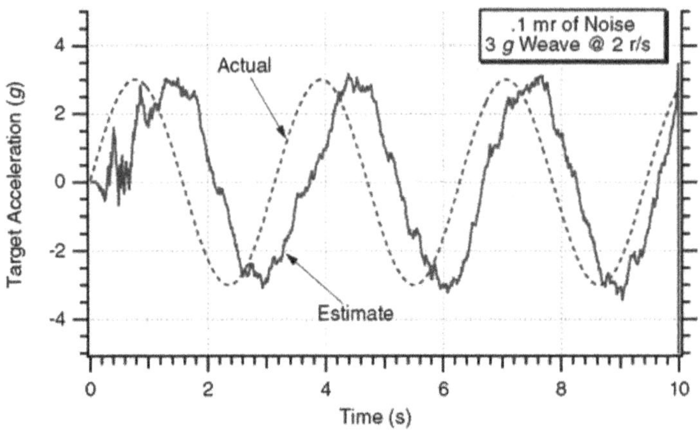

Fig. 25.5 Reducing measurement noise by order of magnitude improves target acceleration estimate.

sinusoidal target motion in some statistical fashion. First recall from Chapter 1 that the Laplace transform of a sinusoidal signal is given by

$$£(\sin\ \omega t) = \frac{\omega}{s^2 + \omega^2}$$

Therefore, if we assume that the target maneuver is sinusoidal in shape and that the starting time is still uniformly distributed over the flight time, we get the model of Fig. 25.6. Here the input to the sinusoidal transfer function is white noise u_s with spectral density

$$\Phi_s = \frac{n_{\text{TMAX}}^2}{t_F}$$

where again n_{TMAX} is the peak of the sinusoidal maneuver and t_F the flight time. It was shown in Chapter 4 that mathematically this is the shaping filter equivalent

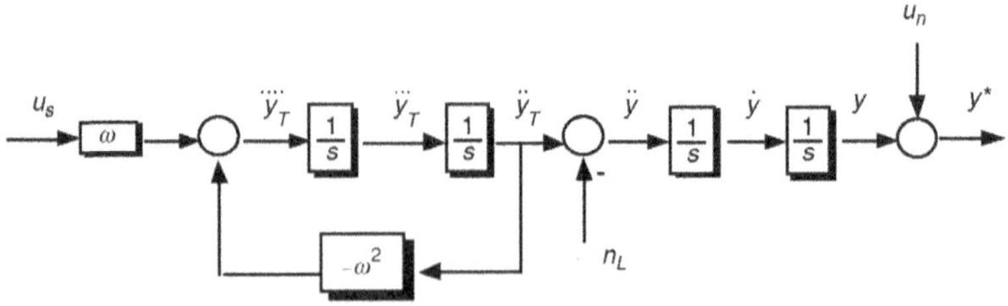

Fig. 25.6 Homing loop model for Kalman filter to be designed for sinusoidal target maneuver.

of a target maneuver with sinusoidal amplitude but random starting time (where the starting time is uniformly distributed over the flight time).

In this homing system model we also effectively measure noisy relative position y^*. If the range from the interceptor to the target is known, it is easy to show that measuring relative position is equivalent to measuring the line of sight angle. The linear four-state weave Kalman filter will estimate relative position, relative velocity, target acceleration, and target jerk. The homing loop model of Fig. 25.2 assumes that the achieved missile acceleration n_L and the target weave frequency ω are both known and do not have to be estimated.

The model of Fig. 25.6 can be expressed in state space form as

$$\begin{bmatrix} \dot{y} \\ \ddot{y} \\ \dddot{y}_T \\ \ddddot{y}_T \end{bmatrix} = \begin{bmatrix} 0 & 1 & 0 & 0 \\ 0 & 0 & 1 & 0 \\ 0 & 0 & 0 & 1 \\ 0 & 0 & -\omega^2 & 0 \end{bmatrix} \begin{bmatrix} y \\ \dot{y} \\ \ddot{y}_T \\ \dddot{y}_T \end{bmatrix} + \begin{bmatrix} 0 \\ -1 \\ 0 \\ 0 \end{bmatrix} n_L + \begin{bmatrix} 0 \\ 0 \\ 0 \\ \omega u_s \end{bmatrix}$$

The systems dynamics matrix of the preceding equation can be written by inspection and is given by

$$\mathbf{F} = \begin{bmatrix} 0 & 1 & 0 & 0 \\ 0 & 0 & 1 & 0 \\ 0 & 0 & 0 & 1 \\ 0 & 0 & -\omega^2 & 0 \end{bmatrix}$$

The fundamental matrix can be derived from the systems dynamics matrix according to

$$\Phi(\mathbf{t}) = \pounds^{-1}\left[(\mathbf{sI} - \mathbf{F})^{-1}\right]$$

Therefore, the fundamental matrix in the Laplace transform domain can be expressed as

$$\Phi(s) = (\mathbf{sI} - \mathbf{F})^{-1} = \begin{bmatrix} s & -1 & 0 & 0 \\ 0 & s & -1 & 0 \\ 0 & 0 & s & -1 \\ 0 & 0 & s + \omega^2 & 0 \end{bmatrix}^{-1}$$

From the preceding equation we can see that first we must take the inverse of a four-by-four matrix and then take its inverse Laplace transform to find the fundamental matrix in the time domain. After considerable algebra, the continuous fundamental matrix turns out to be

$$\Phi(t) = \begin{bmatrix} 1 & t & \dfrac{1-\cos\omega t}{\omega^2} & \dfrac{\omega t - \sin\omega t}{\omega^3} \\ 0 & 1 & \dfrac{\sin\omega t}{\omega} & \dfrac{1-\cos\omega t}{\omega^2} \\ 0 & 0 & \cos\omega t & \dfrac{\sin\omega t}{\omega} \\ 0 & 0 & -\omega\sin\omega t & \cos\omega t \end{bmatrix}$$

By replacing time t with the sampling time T_s we obtain the discrete form of the fundamental matrix as

$$\Phi_k = \begin{bmatrix} 1 & T_s & \dfrac{1-\cos x}{\omega^2} & \dfrac{x - \sin x}{\omega^3} \\ 0 & 1 & \dfrac{\sin x}{\omega} & \dfrac{1-\cos x}{\omega^2} \\ 0 & 0 & \cos x & \dfrac{\sin x}{\omega} \\ 0 & 0 & -\omega\sin x & \cos x \end{bmatrix}$$

where

$$x = \omega T_s$$

The discrete measurement equation can be written by inspection of Fig. 25.6 as

$$y_k^* = \begin{bmatrix} 1 & 0 & 0 & 0 \end{bmatrix} \begin{bmatrix} y_k \\ \dot{y}_k \\ \ddot{y}_{T_k} \\ \dddot{y}_{T_k} \end{bmatrix} + v_k$$

which means that the discrete measurement matrix is given by

$$\mathbf{H_k} = \begin{bmatrix} 1 & 0 & 0 & 0 \end{bmatrix}$$

The continuous control matrix $\mathbf{G}$ can be written by inspection of the original state space equation as

$$\mathbf{G} = \begin{bmatrix} 0 \\ -1 \\ 0 \\ 0 \end{bmatrix}$$

After some algebra the discrete control matrix $\mathbf{G_k}$ becomes

$$\mathbf{G_k} = \int_0^{T_s} \Phi(\tau)\mathbf{G}(\tau)d\tau = \begin{bmatrix} -.5T_S^2 \\ -T_S \\ 0 \\ 0 \end{bmatrix}$$

Finally, the continuous process noise matrix can be written from the system state space equation by inspection as

$$\mathbf{Q} = \mathbf{E}\left\{ \begin{bmatrix} 0 & 0 & 0 & \omega u_s \end{bmatrix} \begin{bmatrix} 0 \\ 0 \\ 0 \\ \omega u_s \end{bmatrix} \right\} = \begin{bmatrix} 0 & 0 & 0 & 0 \\ 0 & 0 & 0 & 0 \\ 0 & 0 & 0 & 0 \\ 0 & 0 & 0 & \omega^2\Phi_s \end{bmatrix}$$

After some algebra the discrete process noise matrix can be derived from the continuous process noise matrix according to

$$\mathbf{Q_k} = \int_0^{T_s} \Phi(\tau)\mathbf{Q}\Phi^T(\tau)d\tau = \begin{bmatrix} Q_{11} & Q_{12} & Q_{13} & Q_{14} \\ Q_{12} & Q_{22} & Q_{23} & Q_{24} \\ Q_{13} & Q_{23} & Q_{33} & Q_{34} \\ Q_{14} & Q_{24} & Q_{34} & Q_{44} \end{bmatrix}$$

where

$$Q_{11} = \frac{\Phi_s}{\omega^5}\left[.333x^3 - 2\sin x + 2x\cos x + .5x - .25\sin 2x\right]$$

$$Q_{12} = \frac{\Phi_s}{\omega^4}\left[.5x^2 - x\sin x + .5\sin^2 x\right]$$

$$Q_{13} = \frac{\Phi_s}{\omega^3}\left[\sin x - x\cos x - .5x + .25\sin 2x\right]$$

$$Q_{14} = \frac{\Phi_s}{\omega^2}\left[\cos x + x\sin x - .5\sin^2 x - 1\right]$$

$$Q_{22} = \frac{\Phi_s}{\omega^3}\left[1.5x - 2\sin x + .25\sin 2x\right]$$

$$Q_{23} = \frac{\Phi_s}{\omega^2}\left[1 - \cos x - .5\sin^2 x\right]$$

$$Q_{24} = \frac{\Phi_s}{\omega}\left[\sin x - .5x - .25\sin 2x\right]$$

$$Q_{33} = \frac{\Phi_s}{\omega}\left[.5x - .25\sin 2x\right]$$

$$Q_{34} = .5\Phi_s \sin^2 x$$

$$Q_{44} = \omega\Phi_s\left[.5x + .25\sin 2x\right]$$

Recall that in the preceding set of expressions the process noise and normalized weave frequency have been defined as

$$\Phi_s = \frac{\omega^2 n_{TMAX}^2}{t_F}$$

$$x = \omega T_s$$

Recall that the discrete Kalman filtering equation is given by

$$\hat{\mathbf{x}}_{\mathbf{k}} = \mathbf{\Phi}_{\mathbf{k}}\hat{\mathbf{x}}_{\mathbf{k-1}} + \mathbf{G}_{\mathbf{k}}\mathbf{u}_{\mathbf{k-1}} + \mathbf{K}_{\mathbf{k}}(\mathbf{z}_{\mathbf{k}} - \mathbf{H}\mathbf{\Phi}_{\mathbf{k}}\hat{\mathbf{x}}_{\mathbf{k-1}} - \mathbf{H}\mathbf{G}_{\mathbf{k}}\mathbf{u}_{\mathbf{k-1}})$$

Substitution of the appropriate matrices into the preceding matrix difference equation yields

$$\begin{bmatrix} \hat{y}_k \\ \hat{\dot{y}}_k \\ \hat{\ddot{y}}_{T_k} \\ \hat{\dddot{y}}_{T_k} \end{bmatrix} = \begin{bmatrix} 1 & T_s & \dfrac{1-\cos x}{\omega^2} & \dfrac{x-\sin x}{\omega^3} \\ 0 & 1 & \dfrac{\sin x}{\omega} & \dfrac{1-\cos x}{\omega^2} \\ 0 & 0 & \cos x & \dfrac{\sin x}{\omega} \\ 0 & 0 & -\omega\sin x & \cos x \end{bmatrix} \begin{bmatrix} \hat{y}_{k-1} \\ \hat{\dot{y}}_{k-1} \\ \hat{\ddot{y}}_{T_{k-1}} \\ \hat{\dddot{y}}_{T_{k-1}} \end{bmatrix} + \begin{bmatrix} -.5T_s^2 \\ -T_s \\ 0 \\ 0 \end{bmatrix} n_{L_{k-1}} + \begin{bmatrix} K_1 \\ K_2 \\ K_3 \\ K_4 \end{bmatrix}$$

$$\times \left[y_k^* - [1 \quad 0 \quad 0 \quad 0] \begin{bmatrix} 1 & T_s & \dfrac{1-\cos x}{\omega^2} & \dfrac{1-\sin x}{\omega^3} \\ 0 & 1 & \dfrac{\sin x}{\omega} & \dfrac{1-\cos x}{\omega^2} \\ 0 & 0 & \cos x & \dfrac{\sin x}{\omega} \\ 0 & 0 & -\omega\sin x & \cos x \end{bmatrix} \begin{bmatrix} \hat{y}_{k-1} \\ \hat{\dot{y}}_{k-1} \\ \hat{\ddot{y}}_{T_{k-1}} \\ \hat{\dddot{y}}_{T_{k-1}} \end{bmatrix} \right.$$

$$\left. -[1 \quad 0 \quad 0 \quad 0] \begin{bmatrix} -.5T_s^2 \\ -T_s \\ 0 \\ 0 \end{bmatrix} n_{L_{k-1}} \right]$$

We can multiply out the terms of the preceding matrix equation to yield the Kalman filter scalar equations

$$\mathrm{RES}_k = y_k^* - \hat{y}_{k-1} - T_s\hat{\dot{y}}_{k-1} - \frac{(1-\cos x)}{\omega^2}\hat{\ddot{y}}_{T_{k-1}} - \frac{(x-\sin x)}{\omega^3}\hat{\dddot{y}}_{T_{k-1}} + .5T_s^2 n_{L_{k-1}}$$

$$\hat{y}_k = \hat{y}_{k-1} + T_s\hat{\dot{y}}_{k-1} + \frac{(1-\cos x)}{\omega^2}\hat{\ddot{y}}_{T_{k-1}} - \frac{(x-\sin x)}{\omega^3}\hat{\dddot{y}}_{T_{k-1}} - .5T_s^2 n_{L_{k-1}} + K_{1_k}\mathrm{RES}_k$$

$$\hat{\dot{y}}_k = \hat{\dot{y}}_{k-1} + \frac{(\sin x)}{\omega}\hat{\ddot{y}}_{T_{k-1}} + \frac{(1-\cos x)}{\omega^2}\hat{\dddot{y}}_{T_{k-1}} - T_s n_{L_{k-1}} + K_{2_k}\mathrm{RES}_k$$

$$\hat{\dot{y}}_{T_k} = \cos x \hat{\dot{y}}_{T_{k-1}} + \frac{\sin x}{\omega} \hat{\ddot{y}}_{T_{k-1}} + K_{3_k} \text{RES}_k$$

$$\hat{\ddot{y}}_{T_k} = -\omega \sin x \hat{\dot{y}}_{T_{k-1}} + \cos x \hat{\ddot{y}}_{T_{k-1}} + K_{4_k} \text{RES}_k$$

Both the Riccati equations and Kalman filtering equations for the linear four-state weave Kalman filter were programmed as part of the homing loop, and the resultant linearized missile-target engagement simulation appears in Listing 25.2. We can see that the simulation has a single time constant representation of the flight control system plus a 3 *g* weaving target with a weave frequency of 2 rad/s. Nominally there is 1 mr of measurement noise on the line of sight angle and the closing velocity is 9000 ft/s to reflect a ballistic target engagement. We can see from Listing 25.2 that the guidance law options for this filter are either proportional navigation, augmented proportional navigation, optimal guidance, weave guidance, or compensated weave guidance (namely, APN = 0, 1, 2, 3, or 4, respectively).

The nominal case of Listing 25.2 was run, and Fig. 25.7 shows that the linear four-state weave Kalman filter's estimate of target acceleration is much better than the general purpose linear three-state Kalman filter when the measurement noise is 1 mr (see Fig. 25.3). Compared with the previous section, when the amount of measurement noise is large, it is now easier to see from the state estimates of the weave Kalman filter that the target maneuver is indeed sinusoidal. In addition, the weave Kalman filter also yields estimates of the target jerk. We can see from Fig. 25.8 that the weave Kalman filter provides fairly good estimates of the target jerk.

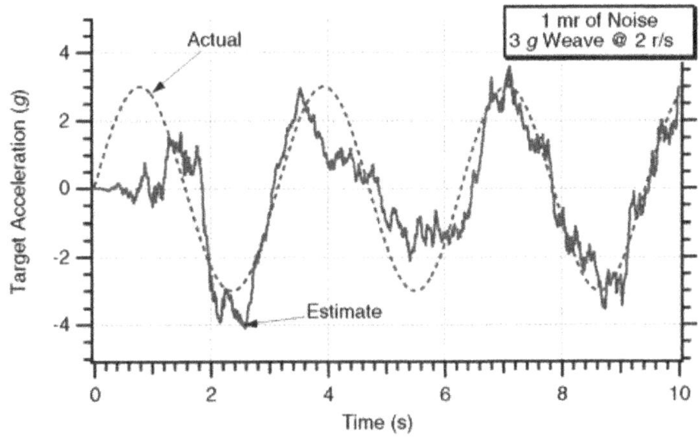

Fig. 25.7 Four-state weave Kalman filter yields better estimates than three-state Kalman filter when measurement noise is large.

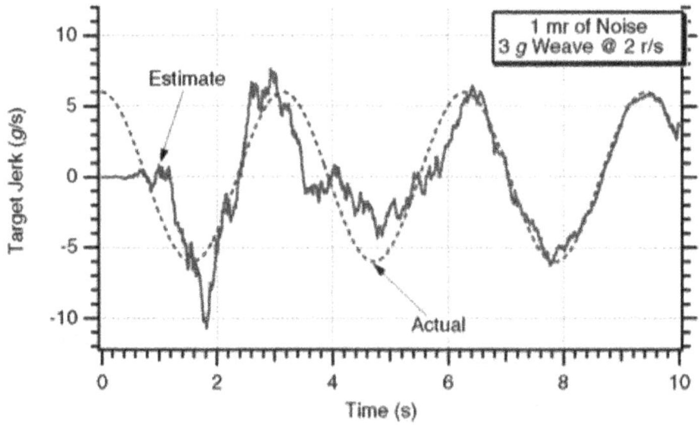

Fig. 25.8 Weave Kalman filter also provides a fairly good estimate of target jerk when measurement noise is large.

To see if the four-state weave Kalman filter is truly working properly, it is necessary to examine the errors in the state estimates. Figures 25.9 and 25.10 show that the errors in the estimates of target acceleration and jerk appear to lie within the theoretical bounds (that is, $\pm$ square root of P_{33} and P_{44}, respectively) approximately 68% of the time, indicating that the filter is working properly.

Figures 25.11 and 25.12 indicate that when the measurement noise on the line of sight angle is reduced by an order of magnitude to .1 mr, the four-state weave

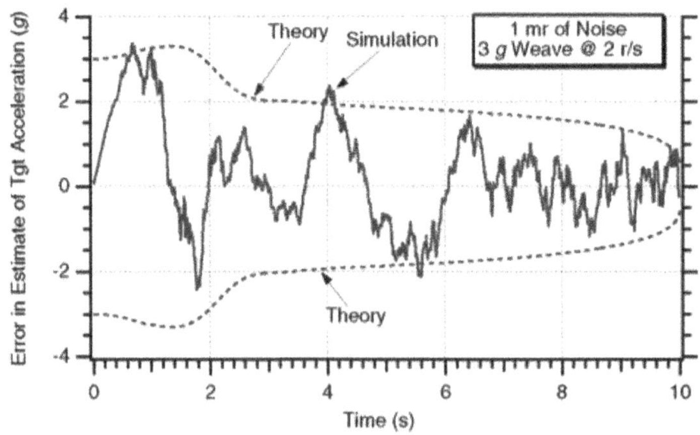

Fig. 25.9 Weave Kalman filter estimation errors for target acceleration are within theoretical bounds.

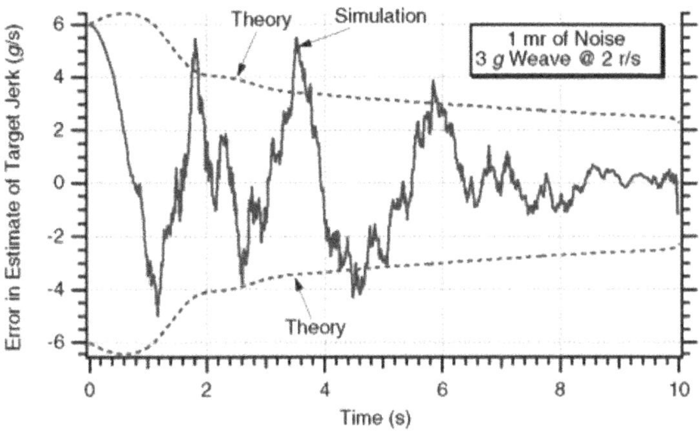

Fig. 25.10 Weave Kalman filter estimation errors for target jerk are within theoretical bounds.

Kalman filter estimates of target acceleration and jerk improve significantly. In fact, we can see that the estimates of these states are nearly perfect in the low-noise environment.

Thus, we can conclude that if the target is maneuvering in a sinusoidal fashion, and we have knowledge of the target wave frequency, superior estimates of the target acceleration and jerk can be obtained with the linear four-state weave Kalman filter.

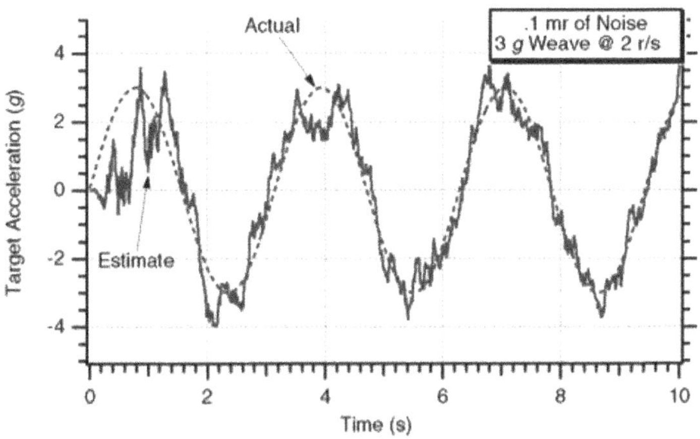

Fig. 25.11 Reducing measurement noise improves four-state weave Kalman filter's target acceleration estimate.

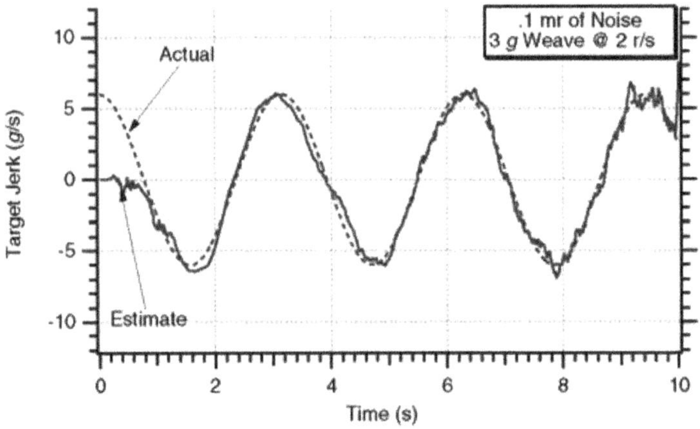

Fig. 25.12 Reducing measurement noise improves weave Kalman filter's target jerk estimate.

LISTING 25.2 WEAVE KALMAN FILTER AND WEAVING TARGET

```
n=0;
TAU=.5;
APN=0;
ORDER=4;
MVR=1;
VC=9000.;
W=2.;
WREAL=2.;
WH=W;
XNT=96.6;
XNTREAL=96.6;
TS=.01;
YIC=0.;
VM=3000.;
HEDEG=0.;
HEDEGFIL=20.;
XNP=3.;
SIGRIN=.001;
SIGGL=0.;
RA=21000.;
SRN=0.;
TF=10.;
QPERFECT=0;
PHASE=0./57.3;
```

```
X=WH*TS;
Y=YIC;
YD=-VM*HEDEG/57.3;
PHIS=WH*WH*XNT*XNT/TF;
RTM=VC*TF;
SIGNOISE=sqrt(SIGRIN^2+(SIGGL/RTM)^2+(SRN*RTM*RTM/(RA*RA))^2);
SIGPOS=RTM*SIGNOISE;
SIGN2=SIGPOS^2;
PHI=zeros(ORDER);
P=zeros(ORDER);
Q=zeros(ORDER);
IDNP=eye(ORDER);
PHI(1,1)=1;
PHI(1,2)=TS;
PHI(1,3)=(1-cos(X))/(WH*WH);
PHI(1,4)=(X-sin(X))/(WH*WH*WH);
PHI(2,2)=1;
PHI(2,3)=sin(X)/WH;
PHI(2,4)=(1-cos(X))/(WH*WH);
PHI(3,3)=cos(X);
PHI(3,4)=sin(X)/WH;
PHI(4,3)=-WH*sin(X);
PHI(4,4)=cos(X);
Q(1,1)=PHIS*(.333*X^3-2*sin(X)+2*X*cos(X)+.5*X-.25*sin(2*X))/(WH^5);
Q(1,2)=PHIS*(.5*X*X-X*sin(X)+.5*sin(X)*sin(X))/(WH^4);
Q(2,1)=Q(1,2);
Q(1,3)=PHIS*(sin(X)-X*cos(X)-.5*X+.25*sin(2*X))/(WH^3);
Q(3,1)=Q(1,3);
Q(1,4)=PHIS*(cos(X)+X*sin(X)-.5*sin(X)*sin(X)-1)/(WH*WH);
Q(4,1)=Q(1,4);
Q(2,2)=PHIS*(1.5*X-2*sin(X)+.25*sin(2*X))/(WH^3);
Q(2,3)=PHIS*(1-cos(X)-.5*sin(X)*sin(X))/(WH*WH);
Q(3,2)=Q(2,3);
Q(2,4)=PHIS*(sin(X)-.5*X-.25*sin(2*X))/WH;
Q(4,2)=Q(2,4);
Q(3,3)=PHIS*(.5*X-.25*sin(2*X))/WH;
Q(3,4)=.5*PHIS*sin(X)*sin(X);
Q(4,3)=Q(3,4);
Q(4,4)=WH*PHIS*(.5*X+.25*sin(2*X));
P(1,1)=SIGN2;
P(2,2)=(VM*HEDEGFIL/57.3)^2;
P(3,3)=XNT*XNT;
P(4,4)=WH*WH*XNT*XNT;
HMAT=[1 0 0 0];
HT=HMAT';
PHIT=PHI';
```

```
T=0.;
H=.001;
S=0.;
XNC=0.;
XNL=0.;
XLAM=Y/RTM;
if MVR==0
        YTDD=XNTREAL;
        YTDDD=0.;
else
        YTDD=XNTREAL*sin(WREAL*T);
        YTDDD=XNTREAL*WREAL*cos(WREAL*T);
end
if QPERFECT==1
        YH=Y;
        YDH=YD;
        YTDDH=YTDD;
        YTDDDH=YTDDD;
else
        YH=0.;
        YDH=0.;
        YTDDH=0.;
        YTDDDH=0.;
end
while T<=(TF-.0001)
        YOLD=Y;
        YDOLD=YD;
        XNLOLD=XNL;
        STEP=1;
        FLAG=0;
        while STEP <=1
                if FLAG==1
        STEP=2;
                        Y=Y+H*YD;
                        YD=YD+H*YDD;
                        XNL=XNL+H*XNLD;
                        T=T+H;
                end
                TGO=TF-T+.000001;
                RTM=VC*TGO;
                XLAM=Y/(VC*TGO);
                if MVR==0
                        YTDD=XNTREAL;
                else
                        YTDD=XNTREAL*sin(WREAL*T);
                end
```

```
            XNLD=(XNC-XNL)/TAU;
            YDD=YTDD-XNL;
            FLAG=1;
      end
      FLAG=0;
      Y=.5*(YOLD+Y+H*YD);
      YD=.5*(YDOLD+YD+H*YDD);
      XNL=.5*(XNLOLD+XNL+H*XNLD);
      S=S+H;
      if S>=(TS-.00001)
            S=0.;
            TGO=TF-T+.000001;
            RTM=VC*TGO;
            SIGNOISE=sqrt(SIGRIN^2+(SIGGL/RTM)^2+(SRN*RTM*RTM/(RA*RA))^2);
            SIGPOS=RTM*SIGNOISE;
            SIGN2=SIGPOS^2;
            RMAT=[SIGN2];
            PHIP=PHI*P;
            PHIPPHIT=PHIP*PHIT;
            M=PHIPPHIT+Q;
            HM=HMAT*M;
            HMHT=HM*HT;
            HMHTR=HMHT+RMAT;
            HMHTRINV=inv(HMHTR);
            MHT=M*HT;
            GAIN=MHT*HMHTRINV;
            KH=GAIN*HMAT;
            IKH=IDNP-KH;
            P=IKH*M;
            if MVR==0
                  YTDD=XNTREAL;
                  YTDDD=0.;
            else
                  YTDD=XNTREAL*sin(WREAL*T);
                  YTDDD=XNTREAL*WREAL*cos(WREAL*T);
            end
            XLAMNOISE=SIGNOISE*randn;
            YSTAR=RTM*(XLAM+XLAMNOISE);
            RES=YSTAR-YH-TS*YDH-(1-cos(X))*YTDDH/(WH*WH)-(X-sin(X))...
                        *YTDDDH/(WH*WH*WH)+.5*TS*TS*XNL;
            YH=YH+TS*YDH+(1-cos(X))*YTDDH/(WH*WH)+(X-sin(X))...
                        *YTDDDH/(WH*WH*WH)+GAIN(1,1)*RES-.5*TS*TS*XNL;
            YDH=YDH+sin(X)*YTDDH/WH+(1-cos(X))*YTDDDH/(WH*WH)...
                        +GAIN(2,1)*RES-TS*XNL;
            YTDDHNEW=cos(X)*YTDDH+sin(X)*YTDDDH/WH+GAIN(3,1)*RES;
            YTDDDH=-WH*sin(X)*YTDDH+cos(X)*YTDDDH+GAIN(4,1)*RES;
```

```
YTDDH=YTDDHNEW;
if APN==0
        XNC=XNP*(YH+YDH*TGO)/(TGO*TGO);
elseif APN==1
        XNC=XNP*(YH+YDH*TGO+.5*YTDDH*TGO*TGO)/(TGO*TGO);
elseif APN==2
        XS=TGO/TAU;
        TOP=6.*XS*XS*(exp(-XS)-1.+XS);
        BOT1=2*XS*XS*XS+3.+6.*XS-6.*XS*XS;
        BOT2=-12.*XS*exp(-XS)-3.*exp(-2.*XS);
        XNPP=TOP/(.0001+BOT1+BOT2);
        C1=XNPP/(TGO*TGO);
        C2=XNPP/TGO;
        C3=.5*XNPP;
        C4=-XNPP*(exp(-XS)+XS-1.)/(XS*XS);
        XNC=C1*YH+C2*YDH+C3*YTDDH+C4*XNL;
elseif APN==3
        XP=WH*TGO;
        XNC=XNP*(YH+YDH*TGO)/(TGO*TGO)+XNP*YTDDH*...
        (1.-cos(XP))/XP^2+XNP*YTDDDH*(XP-sin(XP))/(XP*XP*WH);
else
        XS=TGO/TAU;
        TOP=6.*XS*XS*(exp(-XS)-1.+XS);
        BOT1=2*XS*XS*XS+3.+6.*XS-6.*XS*XS;
        BOT2=-12.*XS*exp(-XS)-3.*exp(-2.*XS);
        XNPP=TOP/(.0001+BOT1+BOT2);
        C1=XNPP/(TGO*TGO);
        C2=XNPP/TGO;
        C3=XNPP*(1.-cos(WH*TGO))/(WH*WH*TGO*TGO);
        C4=-XNPP*(exp(-XS)+XS-1.)/(XS*XS);
        C5=XNPP*(WH*TGO-sin(WH*TGO))/(WH*WH*WH*TGO*TGO);
        XNC=C1*YH+C2*YDH+C3*YTDDH+C4*XNL+C5*YTDDDH;
end
YTDDG=YTDD/32.2;
YTDDHG=YTDDH/32.2;
ERRY=Y-YH;
SP11=sqrt(P(1,1));
SP11P=-SP11;
ERRYD=YD-YDH;
SP22=sqrt(P(2,2));
SP22P=-SP22;
ERRYTDDG=(YTDD-YTDDH)/32.2;
SP33G=sqrt(P(3,3))/32.2;
SP33GN=-SP33G;
ERRYTDDDG=(YTDDD-YTDDDH)/32.2;
SP44G=sqrt(P(4,4))/32.2;
```

```
                SP44GN=-SP44G;
                YTDDG=YTDD/32.2;
                YTDDHG=YTDDH/32.2;
                YTDDDG=YTDDD/32.2;
                YTDDDHG=YTDDDH/32.2;
                n=n+1;
                ArrayT(n)=T;
                ArrayYTDDG(n)=YTDDG;
                ArrayYTDDHG(n)=YTDDHG;
                ArrayYTDDDG(n)=YTDDDG;
                ArrayYTDDDHG(n)=YTDDDHG;
                ArrayERRYTDDG(n)=ERRYTDDG;
                ArraySP33G(n)=SP33G;
                ArraySP33GN(n)=SP33GN;
                ArrayERRYTDDDG(n)=ERRYTDDDG;
                ArraySP44G(n)=SP44G;
                ArraySP44GN(n)=SP44GN;
        end
end
figure
plot(ArrayT,ArrayYTDDG,ArrayT,ArrayYTDDHG),grid
xlabel('Time (Sec)')
ylabel('Acceleration and Estimate (G)')
figure
plot(ArrayT,ArrayYTDDDG,ArrayT,ArrayYTDDDHG),grid
xlabel('Time (Sec)')
ylabel('Jerk and Estimate (G/S)')
figure
plot(ArrayT,ArrayERRYTDDG,ArrayT,ArraySP33G,ArrayT,ArraySP33GN)...
        ,grid
xlabel('Time (Sec)')
ylabel('Error in Estimate of Acceleration (G)')
figure
plot(ArrayT,ArrayERRYTDDDG,ArrayT,ArraySP44G,ArrayT,ArraySP44GN)...
        ,grid
xlabel('Time (Sec)')
ylabel('Error in Estimate of Jerk (G/S)')
clc
output=[ArrayT',ArrayYTDDG',ArrayYTDDHG',ArrayYTDDDG',...
                ArrayYTDDDHG'];
save datfil.txt output  -ascii
output=[ArrayT',ArrayERRYTDDG',ArraySP33G',ArraySP33GN'...
            ,ArrayERRYTDDDG',ArraySP44G',ArraySP44GN'];
save covfil.txt output  -ascii
disp 'simulation finished'
```

MISS DISTANCE ANALYSIS

To see how the various filtering and guidance law options perform in terms of the RMS miss distance in the presence of a weaving target, the homing loop model of Fig. 25.13 is considered. With this guidance system model the Kalman filter can either be the general-purpose, linear, three-state filter or the special-purpose, four-state, weave filter. When the linear, three-state Kalman filter is used, the possible guidance laws that can be used are either proportional navigation or optimal guidance. Recall that these two guidance laws can be expressed as

$$n_{c_{PN}} = \frac{N'}{t_{go}^2}(y + \dot{y}t_{go})$$

$$n_{c_{\text{Optimal}}} = \frac{N'}{t_{go}^2}\left[y + \dot{y}t_{go} + \frac{t_{go}^2}{2}\ddot{y}_T - n_L T^2(e^{-x} + x - 1)\right]$$

With proportional navigation, the effective navigation ratio is usually chosen to be a constant in the range of 3 to 5. With optimal guidance the effective navigation ratio is not constant but can be computed from

$$N' = \frac{6x^2(e^{-x} - 1 + x)}{2x^3 + 3 + 6x - 6x^2 - 12xe^{-x} - 3e^{-2x}}$$

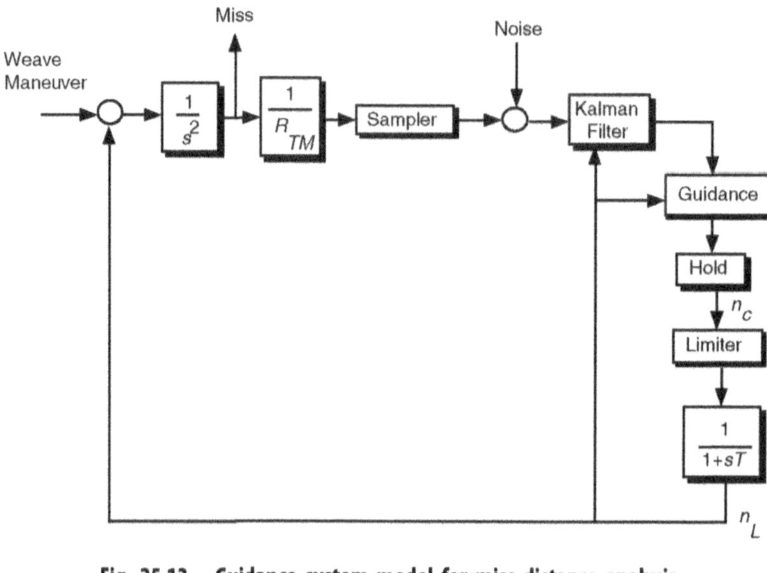

Fig. 25.13 Guidance system model for miss distance analysis.

We showed in Chapter 8 that when the missile is very far away from the target (that is, t_{go} is large) the effective navigation ratio approaches 3. As the missile gets closer to the target, the navigation ratio increases to a larger number. Both guidance laws make use of the state estimates from the linear three-state Kalman filter (that is, estimates of relative position, relative velocity, and target acceleration). The compensated weave guidance law can only be used with the four-state weave Kalman filter given by

$$n_{c_{\text{Weave Lag}}} = \frac{N'}{t_{go}^2}\left[y + \dot{y}t_{go} + \frac{1-\cos\omega t_{go}}{\omega^2}\ddot{y}_T + \frac{\omega t_{go} - \sin\omega t_{go}}{\omega^3}\dddot{y}_T \right.$$
$$\left. - n_L T^2(e^{-x} + x - 1)\right]$$

where the effective navigation ratio is the same as it was with the optimal guidance law. Again, this guidance law makes use of the state estimates from the linear four-state weave Kalman filter (that is, estimates of relative position, relative velocity, target acceleration, and target jerk). It is important to note that when this guidance law is used in conjunction with the weave Kalman filter it is assumed that the target weave frequency is known.

The parameters used for the guidance system analysis appear in Table 25.1. We can see from the table that the weaving target has a 3 g maneuver amplitude and a weave frequency of 2 r/s. The flight control system time constant is set at .5 s. Notice the high closing velocity in Table 25.1 is representative of a ballistic target engagement.

Experiments were run with the different guidance system configurations. Twenty-five Monte Carlo sets were run for flight times ranging from .5 s to 10 s in steps of .5 s. We can see from Fig. 25.14 that for the case in which there is a infinite missile acceleration capability and 1 mr of measurement noise, both the three-state Kalman filter using optimal guidance and the four-state weave Kalman filter using the compensated weave guidance law yield approximately

TABLE 25.1 NOMINAL VALUES FOR EXPERIMENT

Parameter	Value
Autopilot time constant	.5 s
Missile velocity	3000 ft/s
Closing velocity	9000 ft/s
Target acceleration level	3 g
Target weave frequency	2 r/s

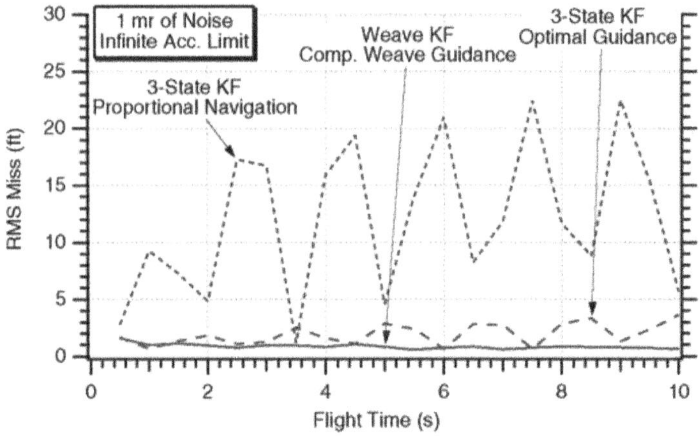

Fig. 25.14 Both optimal guidance and compensated weave guidance have similar performance.

the same results. Both guidance laws yield significantly smaller RMS miss distances than proportional navigation.

The previous case assumed that the missile had infinite acceleration capability. If the acceleration limit is set to 10 g (more than a 3 to 1 advantage over the target), we can see from Fig. 25.15 that the RMS miss distance performance of optimal guidance deteriorates significantly. In this particular case we can see that the

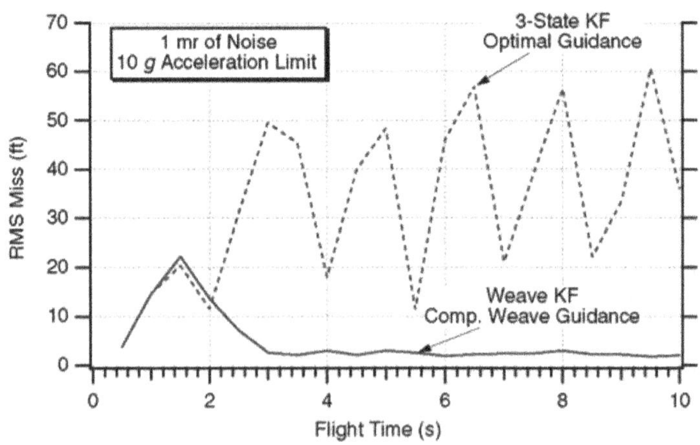

Fig. 25.15 Weave guidance can be superior to optimal guidance when acceleration saturation effects are considered.

compensated weave guidance law in combination with the four-state weave Kalman filter yields dramatic performance advantages. Thus, we can conclude that it is only advantageous to use compensated weave guidance rather than optimal guidance if the missile has a small missile to target acceleration advantage and knows the target weave frequency.

If we compare Figs. 25.16 and 25.15, we can see that reducing the measurement noise by an order of magnitude to .1 mr improves the performance of both optimal guidance and compensated weave guidance. However, because in this example there is still a small missile to target acceleration advantage, it is still better to use compensated weave guidance if the target weave frequency is known.

Recall that the four-state weave Kalman filter and compensated weave guidance law both required knowledge of the target weave frequency. So far it has been assumed that the target weave frequency has been known perfectly. Errors in the knowledge of the target weave frequency will degrade both the performance of the four-state weave Kalman filter and compensated weave guidance law. Figure 25.17 shows that when the estimated weave frequency is either twice as large or half as small as the actual target weave frequency, significant performance degradation may occur. By comparing Figs. 25.17 and 25.16 we can see that when the target weave frequency is in error we can do just as well and sometimes better by using optimal guidance and the three-state Kalman filter because this combination does not require knowledge of the target weave frequency. Thus, we can see that if the target weave frequency is not known in advance it must somehow be estimated in real time if we wish to derive the benefits of the compensated weave guidance law.

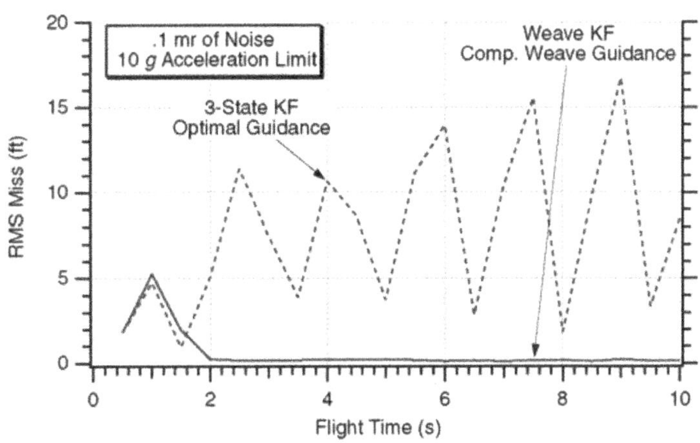

Fig. 25.16 Reducing measurement noise improves performance of both systems but compensated weave guidance is still better.

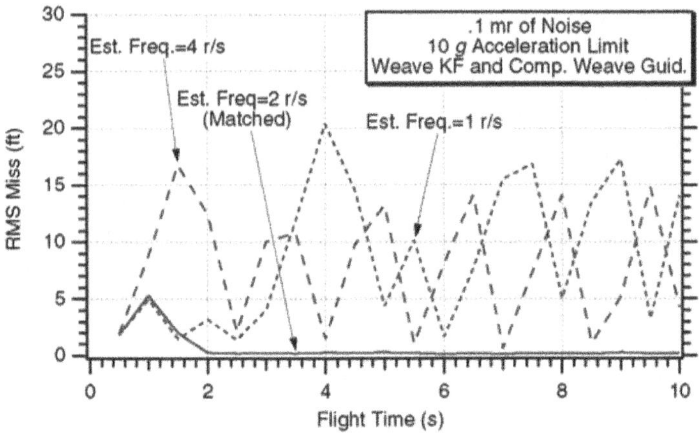

Fig. 25.17 Compensated weave guidance performance can be worse than optimal guidance if estimated weave frequency is in error.

EXTENDED KALMAN FILTER [1]

To build a Kalman filter that can estimate the target weave frequency, it is first necessary to write the state equations representing our model of the real world. From Fig. 25.6 we can say that the equations for the homing loop with a weave target maneuver are still given by

$$\begin{aligned}\dot{y} &= \dot{y}\\ \ddot{y} &= \ddot{y}_T - n_L\\ \dddot{y}_T &= \dddot{y}_T\\ \ddddot{y}_T &= -\omega^2\ddot{y}_T + \omega u_{s1}\end{aligned}$$

We now need an additional equation that says something about the target weave frequency. If the target weave frequency is constant its derivative must be zero. However, for protection we can say that the derivative of the frequency is simply white noise or

$$\dot{\omega} = u_{s2}$$

In the preceding five scalar differential equations, u_{s1} and u_{s2} are white process noise sources. Uncertainty in when the target starts to maneuver is reflected in u_{s1}, while uncertainty in the fact that the weave frequency may not be a constant is reflected in u_{s2}. In the previous section we used for the spectral density of the first white process noise

$$\Phi_{s1} = \frac{\omega_{\text{EXP}}^2 n_{\text{TMAX}}^2}{t_F}$$

where ω_{EXP} can be interpreted as the maximum expected target weave frequency, $n_{T\text{MAX}}$ is the maximum target maneuver level, and t_F is the amount of homing time. For now we will simply treat the spectral density of the second white process noise Φ_{s2} as a fudge factor whose value will be determined by experiment.

Because the target weave frequency is a state, the preceding differential equations describing our model of the real world are nonlinear and the resultant filter will be an extended Kalman filter rather than a linear Kalman filter. However, the measurement equation for this model is still linear and turns out to be

$$y^* = \begin{bmatrix} 1 & 0 & 0 & 0 & 0 \end{bmatrix} \begin{bmatrix} y \\ \dot{y} \\ \ddot{y}_T \\ \dddot{y}_T \\ \omega \end{bmatrix} + v$$

The systems dynamics matrix can be determined from the system state equations as a matrix of partial derivatives given by

$$\mathbf{F} = \begin{bmatrix} \frac{\partial \dot{y}}{\partial y} & \frac{\partial \dot{y}}{\partial \dot{y}} & \frac{\partial \dot{y}}{\partial \ddot{y}_T} & \frac{\partial \dot{y}}{\partial \dddot{y}_T} & \frac{\partial \dot{y}}{\partial \omega} \\ \frac{\partial \ddot{y}}{\partial y} & \frac{\partial \ddot{y}}{\partial \dot{y}} & \frac{\partial \ddot{y}}{\partial \ddot{y}_T} & \frac{\partial \ddot{y}}{\partial \dddot{y}_T} & \frac{\partial \ddot{y}}{\partial \omega} \\ \frac{\partial \dddot{y}_T}{\partial y} & \frac{\partial \dddot{y}_T}{\partial \dot{y}} & \frac{\partial \dddot{y}_T}{\partial \ddot{y}_T} & \frac{\partial \dddot{y}_T}{\partial \dddot{y}_T} & \frac{\partial \dddot{y}_T}{\partial \omega} \\ \frac{\partial \ddddot{y}_T}{\partial y} & \frac{\partial \ddddot{y}_T}{\partial \dot{y}} & \frac{\partial \ddddot{y}_T}{\partial \ddot{y}_T} & \frac{\partial \ddddot{y}_T}{\partial \dddot{y}_T} & \frac{\partial \ddddot{y}_T}{\partial \omega} \\ \frac{\partial \dot{\omega}}{\partial y} & \frac{\partial \dot{\omega}}{\partial y} & \frac{\partial \dot{\omega}}{\partial \ddot{y}_T} & \frac{\partial \dot{\omega}}{\partial \dddot{y}_T} & \frac{\partial \dot{\omega}}{\partial \omega} \end{bmatrix}$$

where the partial derivatives are evaluated at the current state estimates. After taking the appropriate partial derivatives, the systems dynamics matrix turns out to be

$$\mathbf{F} = \frac{\partial f(\mathbf{x})}{\partial \mathbf{x}} = \begin{bmatrix} 0 & 1 & 0 & 0 & 0 \\ 0 & 0 & 1 & 0 & 0 \\ 0 & 0 & 0 & 1 & 0 \\ 0 & 0 & -\hat{\omega}^2 & 0 & -2\hat{\omega}\hat{\ddot{y}}_T \\ 0 & 0 & 0 & 0 & 0 \end{bmatrix}$$

We will use a two-term Taylor series expansion to obtain the fundamental matrix. The number of terms used in the series approach is not critical because the fundamental matrix will only be used in the Riccati equations [2]. The approximate fundamental matrix turns out to be

$$\Phi_k \approx I + \boldsymbol{F}T_s = \begin{bmatrix} 1 & T_s & 0 & 0 & 0 \\ 0 & 1 & T_s & 0 & 0 \\ 0 & 0 & 1 & T_s & 0 \\ 0 & 0 & -\hat{\omega}^2 T_s & 1 & -2\hat{\omega}\hat{\dot{y}}_T T_s \\ 0 & 0 & 0 & 0 & 1 \end{bmatrix}$$

From our model of the real world, the continuous process noise matrix can be found from

$$\mathbf{Q} = E(ww^T) = E\left[\begin{bmatrix} 0 \\ 0 \\ 0 \\ u_{s1} \\ u_{s2} \end{bmatrix} \begin{bmatrix} 0 & 0 & 0 & u_{s1} & u_{s2} \end{bmatrix}\right] = \begin{bmatrix} 0 & 0 & 0 & 0 & 0 \\ 0 & 0 & 0 & 0 & 0 \\ 0 & 0 & 0 & 0 & 0 \\ 0 & 0 & 0 & \Phi_{s1} & 0 \\ 0 & 0 & 0 & 0 & \Phi_{s2} \end{bmatrix}$$

where the spectral densities Φ_{s1} and Φ_{s2} have been previously defined. The discrete process noise matrix can be obtained from the continuous process noise matrix according to

$$Q_k = \int_0^{T_s} \Phi(\tau)\mathbf{Q}\Phi^T(\tau)\mathrm{d}t$$

After some algebra we obtain

$$Q_k = \begin{bmatrix} 0 & 0 & 0 & 0 & 0 \\ 0 & 0 & 0 & 0 & 0 \\ 0 & 0 & \Phi_{s1}\dfrac{T_s^3}{3} & \Phi_{s1}\dfrac{T_s^2}{2} & 0 \\ 0 & 0 & \Phi_{s1}\dfrac{T_s^2}{2} & \Phi_{s1}T_s + 4\dfrac{T_s^3}{3}\hat{\omega}^2\hat{\dot{y}}_T\Phi_{s2} & -\hat{\omega}T_s^2\hat{\dot{y}}_T\Phi_{s2} \\ 0 & 0 & 0 & -\hat{\omega}T_s^2\hat{\dot{y}}_T\Phi_{s2} & T_s\Phi_{s2} \end{bmatrix}$$

Note that the elements of the discrete process noise matrix are also evaluated at the current state estimates. Finally, the equations for the extended Kalman filter are simply

$$\hat{y}_k = \bar{y}_k + K_{1_k}(y_k^* - \bar{y}_k)$$

$$\hat{\dot{y}}_k = \bar{\dot{y}}_k + K_{2_k}(y_k^* - \bar{y}_k)$$

$$\hat{\dot{y}}_{T_k} = \bar{\dot{y}}_{T_k} + K_{3_k}(y_k^* - \bar{y}_k)$$

$$\hat{\ddot{y}}_{T_k} = \bar{\ddot{y}}_{T_k} + K_{4_k}(y_k^* - \bar{y}_k)$$

$$\hat{\omega}_k = \hat{\omega}_{k-1} + K_{5_k}(x_k^* - \bar{x}_k)$$

where the barred quantities represent projections of the previous state estimates to the current time. Normally, the barred quantities would be obtained by multiplying the previous state estimates by the fundamental matrix to project the states ahead one sampling interval. However, because the fundamental matrix in this example is not exact (because it was obtained by using a two-term Taylor series approximation to a linearized systems dynamics matrix), it is better to use brute force to obtain the necessary projections of all the states. In this case we actually numerically integrate the nonlinear equations of motion forward one sampling interval. Euler integration is used with an integration step size that is much smaller than the sampling interval to accurately numerically integrate the state equations forward.

Listing 25.3 presents a simulation of the extended Kalman filter as part of the homing loop. We can see that routine PROJECT is used to propagate ahead the state estimates one sampling interval. It is important to note that the extended Kalman filter's estimate of the target weave frequency is intentionally initialized wrong to -1 r/s (that is, WHIC $= -1$) rather than to 2 r/s (W $= 2$). The incorrect initialization is used to ensure that the extended Kalman filter is robust to initialization errors.

The nominal case of Listing 25.3 was run in which there was 1 mr of measurement noise and the second process noise spectral density Φ_{s2} was set to zero. At first, it appears from Fig. 25.18 that the extended Kalman filter is unable to estimate the target weave frequency. However, a closer examination of Fig. 25.18 reveals that the estimated magnitude of the target weave frequency in the steady state is nearly correct (that is, 2.5 r/s rather than 2 r/s), but the sign is wrong. If the filter was initialized with a positive frequency, the sign would

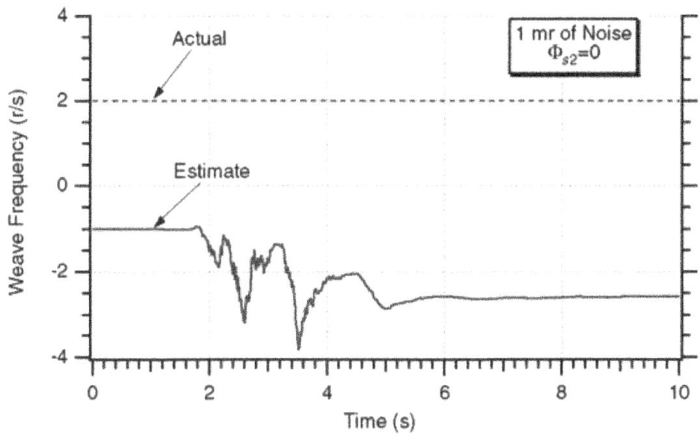

Fig. 25.18 Extended Kalman filter is able to estimate target weave frequency magnitude but not the sign.

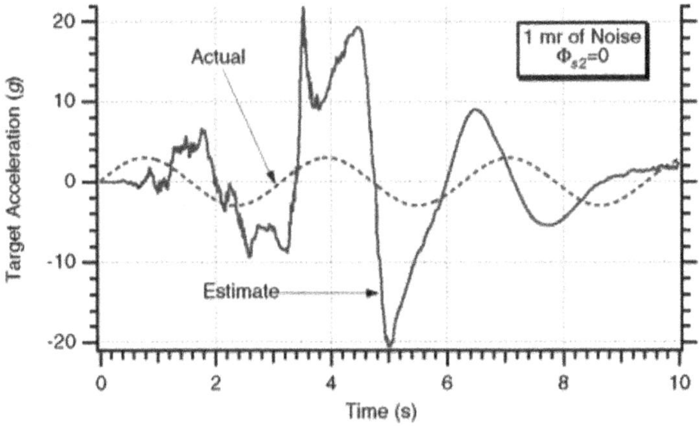

Fig. 25.19 Except for the end of flight, large amount of measurement noise causes bad estimates of target acceleration.

have been correct. Figures 25.19 and 25.20 reveal that the large amount of measurement noise causes the estimate of the target acceleration and jerk to be fairly bad, except at the end of the flight.

We can see from Fig. 25.21 that reducing the measurement noise by an order of magnitude to .1 mr improves the estimate of the magnitude of the target weave frequency. However, we are still unable to estimate the sign of the target weave frequency. Figures 25.22 and 25.23 now show that the estimates of target

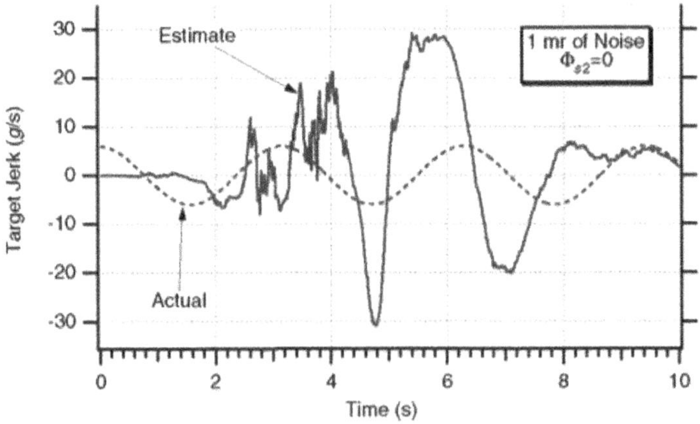

Fig. 25.20 Except for the end of flight, large amount of measurement noise causes bad estimate of target jerk.

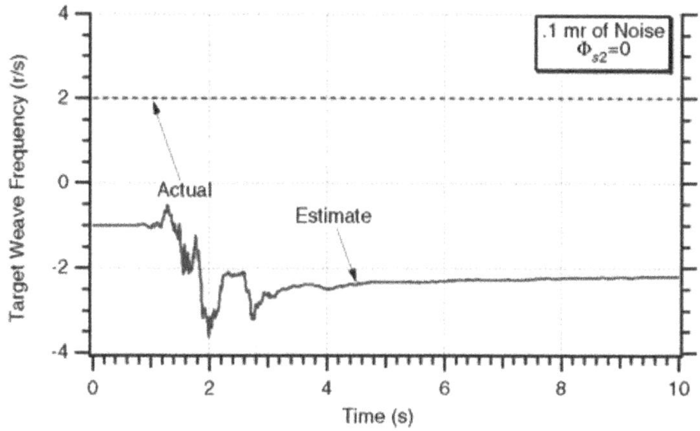

Fig. 25.21 Reducing the measurement noise improves the extended Kalman filter's estimate of the magnitude of the target weave frequency.

acceleration and jerk are very good for most of the flight with the reduced measurement noise, even though the sign of the target weave frequency is in error. However, these estimates are not quite as good as those estimates that were obtained with the four-state linear weave Kalman filter when the target weave frequency was known precisely (see Figs. 25.11 and 25.12 for comparison).

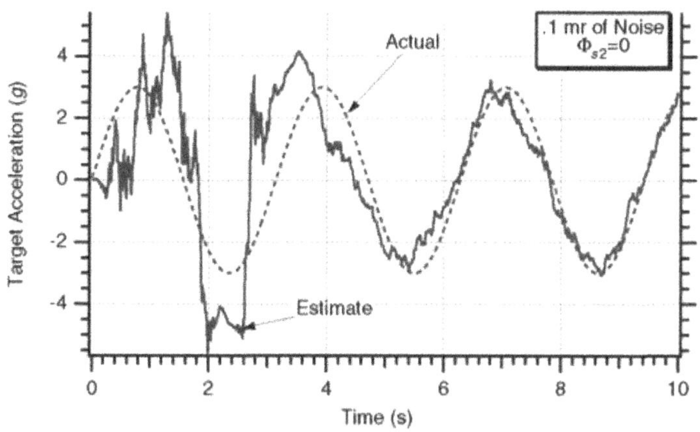

Fig. 25.22 Extended Kalman filter's estimate of target acceleration is fairly good after a transient period when measurement noise is reduced by an order of magnitude.

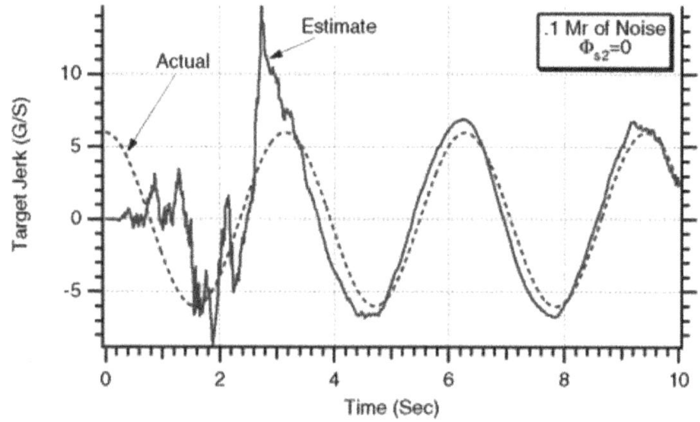

Fig. 25.23 Extended Kalman filter's estimate of target jerk is also fairly good after a transient period when measurement noise is reduced an order of magnitude.

Figure 25.24 shows that if we add a small amount of process noise (namely, $\Phi_{s2} = .1$) to the frequency state, our estimate of the target weave frequency magnitude improves. However, if we add too much process noise (namely, $\Phi_{s2} = 1$) to the frequency state, Fig. 25.25 shows that our estimate of the target weave frequency can actually diverge. Therefore, for safety reasons we will simply keep the second process noise source at zero (namely, $\Phi_{s2} = 0$) for future experiments.

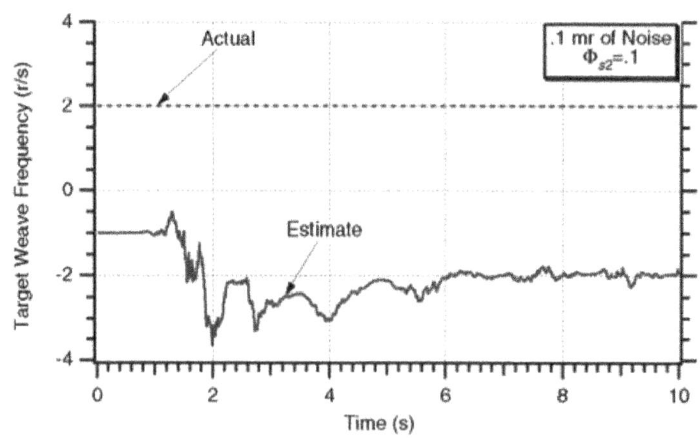

Fig. 25.24 Adding small amount of process noise to frequency state slightly improves frequency estimate.

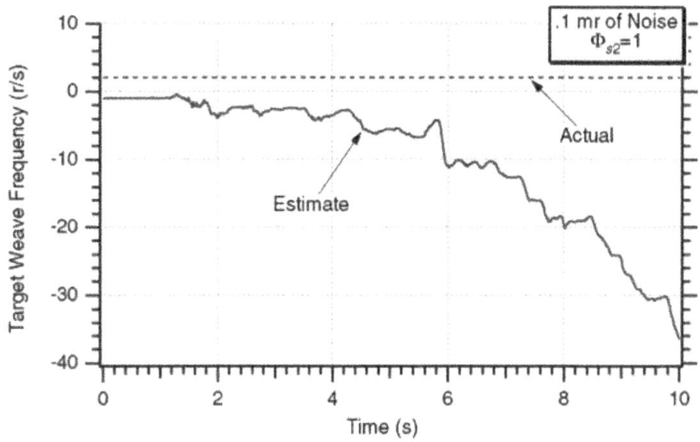

Fig. 25.25 Adding too much process noise to frequency state causes frequency estimate to diverge.

LISTING 25.3 EXTENDED KALMAN FILTER

```
n=0;
PHIS2=0.;
XNT=96.6;
W=2.;
PHASEDEG=0.;
SIGRIN=.0001;
SIGGL=0.;
SRN=0.;
RA=21000.;
WHIC=-1.;
TS=.01;
TF=10.;
PHIS1=W*W*XNT*XNT/TF;
QPERFECT=0;
VC=9000.;
XNP=3.;
XNCLIM=9999999.;
APN=4;
TAU=.5;
HEDEG=0.;
VM=3000.;
QEKF=0;
PHASE=PHASEDEG/57.3;
ORDER=5;
```

```
TGO=TF;
T=0.;
X=W*T;
S=0.;
Y=0.;
YD=-XNT/W-VM*HEDEG/57.3;
YTDD=XNT*sin(W*T);
YTDDD=XNT*W*cos(W*T);
XNC=0.;
XNL=0.;
H=.001;
HP=.001;
TS2=TS*TS;
TS3=TS2*TS;
TS4=TS3*TS;
TS5=TS4*TS;
TS6=TS5*TS;
TS7=TS6*TS;
WH=WHIC;
if QPERFECT==1
        YH=Y;
        YDH=YD;
        YTDDH=YTDD;
        YTDDDH=YTDDD;
        WH=W;
else
        YH=0.;
        YDH=0.;
        YTDDH=0.;
        YTDDDH=0.;
end
PHI=zeros(ORDER);
P=zeros(ORDER);
Q=zeros(ORDER);
IDNP=eye(ORDER);
RTM=VC*TF;
SIGNOISE=sqrt(SIGRIN^2+(SIGGL/RTM)^2+(SRN*RTM*RTM/(RA*RA))^2);
YNOISE=SIGNOISE*RTM;
P(1,1)=YNOISE*YNOISE;
P(2,2)=(VM*20./57.3)^2;
P(3,3)=XNT*XNT;
P(4,4)=(W*XNT)^2;
P(5,5)=W^2;
HMAT=[1 0 0 0 0];
HT=HMAT';
while T<=(TF-.0001)
```

```
YOLD=Y;
YDOLD=YD;
XNLOLD=XNL;
STEP=1;
FLAG=0;
while STEP <=1
        if FLAG==1
                STEP=2;
                Y=Y+H*YD;
                YD=YD+H*YDD;
                XNL=XNL+H*XNLD;
                T=T+H;
        end
        YTDD=XNT*sin(W*T);
        TGO=TF-T+.00001;
        XNLD=(XNC-XNL)/TAU;
        YDD=YTDD-XNL;
        FLAG=1;
end
FLAG=0;
Y=.5*(YOLD+Y+H*YD);
YD=.5*(YDOLD+YD+H*YDD);
XNL=.5*(XNLOLD+XNL+H*XNLD);
S=S+H;
if S>=(TS-.00001)
        S=0.;
        YTDD=XNT*sin(W*T);
        YTDDD=XNT*W*cos(W*T);
        PHI(1,1)=1.;
        PHI(1,2)=TS;
        PHI(2,2)=1.;
        PHI(2,3)=TS;
        PHI(3,3)=1.;
        PHI(3,4)=TS;
        PHI(4,3)=-WH*WH*TS;
        PHI(4,4)=1.;
        PHI(4,5)=-2.*WH*YTDDH*TS;
        PHI(5,5)=1.;
        Q(3,3)=PHIS1*TS*TS*TS/3.;
        Q(3,4)=PHIS1*TS*TS/2.;
        Q(4,3)=Q(3,4);
        Q(4,4)=4.*WH*WH*YTDDH*YTDDH*PHIS2*TS*TS*TS/3.+PHIS1*TS;
        Q(4,5)=-WH*YTDDH*TS*TS*PHIS2;
        Q(5,4)=Q(4,5);
        Q(5,5)=PHIS2*TS;
        PHIT=PHI';
```

```
PHIP=PHI*P;
PHIPPHIT=PHIP*PHIT;
M=PHIPPHIT+Q;
HM=HMAT*M;
HMHT=HM*HT;
RTM=VC*TGO;
SIGNOISE=sqrt(SIGRIN^2+(SIGGL/RTM)^2+(SRN*RTM*RTM/...
        (RA*RA))^2);
YNOISE=SIGNOISE*RTM;
RMAT=[YNOISE^2];
HMHTR=HMHT+RMAT;
HMHTRINV=inv(HMHTR);
MHT=M*HT;
GAIN=MHT*HMHTRINV;
KH=GAIN*HMAT;
IKH=IDNP-KH;
P=IKH*M;
RTM=VC*TGO;
XLAM=Y/RTM;
XNOISE=SIGNOISE*randn;
XLAMS=XLAM+XNOISE;
[YB,YDB,YTDDB,YTDDDB]=PROJECT(T,TS,YH,YDH,YTDDH,YTDDDH,...
        HP,XNL,WH);
RES=RTM*XLAMS-YB;
YH=YB+GAIN(1,1)*RES;
YDH=YDB+GAIN(2,1)*RES;
YTDDH=YTDDB+GAIN(3,1)*RES;
YTDDDH=YTDDDB+GAIN(4,1)*RES;
WH=WH+GAIN(5,1)*RES;
if APN==0
        XNC=XNP*(YH+YDH*TGO)/(TGO*TGO);
elseif APN==1
        XNC=XNP*(YH+YDH*TGO+.5*YTDDH*TGO*TGO)/(TGO*TGO);
elseif APN==2
        XS=TGO/TAU;
        TOP=6.*XS*XS*(exp(-XS)-1.+XS);
        BOT1=2*XS*XS*XS+3.+6.*XS-6.*XS*XS;
        BOT2=-12.*XS*exp(-XS)-3.*exp(-2.*XS);
        XNPP=TOP/(.0001+BOT1+BOT2);
        C1=XNPP/(TGO*TGO);
        C2=XNPP/TGO;
        C3=.5*XNPP;
        C4=-XNPP*(exp(-XS)+XS-1.)/(XS*XS);
        XNC=C1*YH+C2*YDH+C3*YTDDH+C4*XNL;
elseif APN==3
        XP=WH*TGO;
```

```
        XNC=XNP*(YH+YDH*TGO)/(TGO*TGO)+XNP*YTDDH*...
        (1.-cos(XP))/XP^2+XNP*YTDDDH*(XP-sin(XP))/(XP*XP*WH);
else
        XS=TGO/TAU;
        TOP=6.*XS*XS*(exp(-XS)-1.+XS);
        BOT1=2*XS*XS*XS+3.+6.*XS-6.*XS*XS;
        BOT2=-12.*XS*exp(-XS)-3.*exp(-2.*XS);
        XNPP=TOP/(.0001+BOT1+BOT2);
        C1=XNPP/(TGO*TGO);
        C2=XNPP/TGO;
        C3=XNPP*(1.-cos(WH*TGO))/(WH*WH*TGO*TGO);
        C4=-XNPP*(exp(-XS)+XS-1.)/(XS*XS);
        C5=XNPP*(WH*TGO-sin(WH*TGO))/(WH*WH*WH*TGO*TGO);
        XNC=C1*YH+C2*YDH+C3*YTDDH+C4*XNL+C5*YTDDDH;
end
if XNC>XNCLIM
        XNC=XNCLIM;
end
if XNC<-XNCLIM
        XNC=-XNCLIM;
end
ERRYTDD=YTDD-YTDDH;
ERRYTDDG=ERRYTDD/32.2;
ERRYTDDD=YTDDD-YTDDDH;
ERRYTDDDG=ERRYTDDD/32.2;
ERRW=W-WH;
SP44=sqrt(P(4,4));
SP44P=-SP44;
SP33=sqrt(P(3,3));
SP33P=-SP33;
SP33G=SP33/32.2;
SP33PG=SP33P/32.2;
SP44G=SP44/32.2;
SP44PG=SP44P/32.2;
SP55=sqrt(P(5,5));
SP55P=-SP55;
YTDDG=YTDD/32.2;
YTDDHG=YTDDH/32.2;
YTDDDG=YTDDD/32.2;
YTDDDHG=YTDDDH/32.2;
XNCG=XNC/32.2;
n=n+1;
ArrayT(n)=T;
ArrayYTDDG(n)=YTDDG;
ArrayYTDDHG(n)=YTDDHG;
ArrayYTDDDG(n)=YTDDDG;
```

```
                ArrayYTDDDHG(n)=YTDDDHG;
                ArrayW(n)=W;
                ArrayWH(n)=WH;
                ArrayERRYTDDG(n)=ERRYTDDG;
                ArraySP33G(n)=SP33G;
                ArraySP33PG(n)=SP33PG;
                ArrayERRYTDDDG(n)=ERRYTDDDG;
                ArraySP44G(n)=SP44G;
                ArraySP44PG(n)=SP44PG;
                ArrayERRW(n)=ERRW;
                ArraySP55(n)=SP55;
                ArraySP55P(n)=SP55P;
        end
end
figure
plot(ArrayT,ArrayYTDDG,ArrayT,ArrayYTDDHG),grid
xlabel('Time (Sec)')
ylabel('Acceleration and Estimate (G)')
figure
plot(ArrayT,ArrayYTDDDG,ArrayT,ArrayYTDDDHG),grid
xlabel('Time (Sec)')
ylabel('Jerk and Estimate (G/S)')
figure
plot(ArrayT,ArrayW,ArrayT,ArrayWH),grid
xlabel('Time (Sec)')
ylabel('Frequency and Estimate (G/S)')
figure
plot(ArrayT,ArrayERRYTDDG,ArrayT,ArraySP33G,ArrayT,...
        ArraySP33PG),grid
xlabel('Time (Sec)')
ylabel('Error in Estimate of Acceleration (G)')
figure
plot(ArrayT,ArrayERRYTDDDG,ArrayT,ArraySP44G,ArrayT,...
        ArraySP44PG),grid
xlabel('Time (Sec)')
ylabel('Error in Estimate of Jerk (G/S)')
clc
output=[ArrayT',ArrayYTDDG',ArrayYTDDHG',ArrayYTDDDG',...
                ArrayYTDDDHG',ArrayW',ArrayWH'];
save datfil.txt output  -ascii
output=[ArrayT',ArrayERRYTDDG',ArraySP33G',ArraySP33PG',...
            ArrayERRYTDDDG',ArraySP44G',ArraySP44PG'];
save covfil.txt output  -ascii
disp 'simulation finished'

function[YB,YDB,YTDDB,YTDDDB]=PROJECT(TP,TS,YPH,YDPH,...
                YTDDPH,YTDDDPH,HP,XNLP,WPH)
```

```
T=0.;
Y=YPH;
YD=YDPH;
YTDD=YTDDPH;
YTDDD=YTDDDPH;
W=WPH;
XNL=XNLP;
H=HP;
while T<=(TS-.0001)
        YTDDDD=-W*W*YTDD;
        YTDDD=YTDDD+H*YTDDDD;
        YTDD=YTDD+H*YTDDD;
        YDD=YTDD-XNL;
        YD=YD+H*YDD;
        Y=Y+H*YD;
        T=T+H;
end
YB=Y;
YDB=YD;
YTDDB=YTDD;
YTDDDB=YTDDD;
```

To demonstrate the robustness of the five-state extended Kalman filter, another experiment was conducted. This time a case was considered in which the target maneuvered, but not sinusoidally. Instead, the target performed a constant 3-g maneuver. Figure 25.26 shows that in this case the extended Kalman filter correctly estimated, after an initial transient period, that the target weave frequency was zero. Figure 25.27 shows that after 4 s the filter is able to estimate that the level of the constant target maneuver is 3 g. Finally, Fig. 25.28 shows that the estimate of the target jerk is zero. This should be the case when the target maneuver is constant. Thus, we can conclude that the five-state extended Kalman filter is indeed robust.

Miss distance experiments were conducted with the three-state linear Kalman filter and the five-state extended Kalman filter when both were part of the homing loop. The guidance law options for the linear Kalman filter were proportional navigation and optimal guidance. The extended Kalman filter was always used in conjunction with the compensated weave guidance law. A close examination of this guidance law reveals that the sign of the target weave frequency does not influence the guidance command. Therefore, the inability of the five-state extended Kalman filter to correctly determine the sign of the target weave frequency should not be important for guidance purposes.

Figure 25.29 shows that when there are no constraints on the missile acceleration and there is 1 mr of measurement noise, both the three-state linear Kalman filter with optimal guidance and five-state extended Kalman filter with compensated weave guidance yield superior miss distance performance to that of a proportional navigation guidance system. Both optimal guidance and compensated

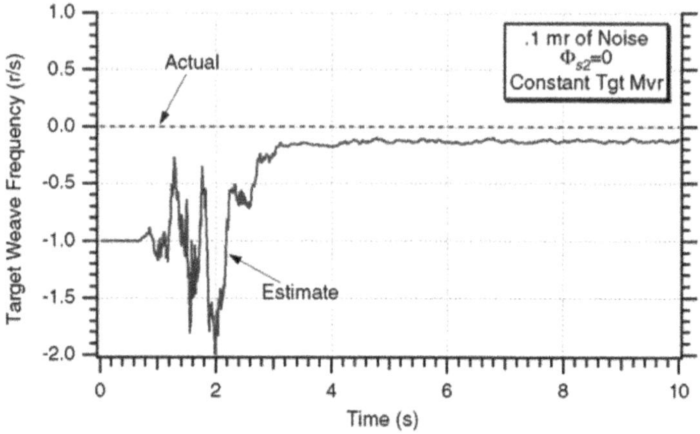

Fig. 25.26 Filter correctly estimates weave frequency of zero in presence of constant target maneuver.

weave guidance yield similar performance as measured by the RMS miss distance. However, we can see from Fig. 25.30 that if there is a 10-g missile acceleration limit, optimal guidance degrades severely and the best performance is obtained by the five-state extended Kalman filter with compensated weave guidance. In this case, the performance of optimal guidance and proportional navigation are

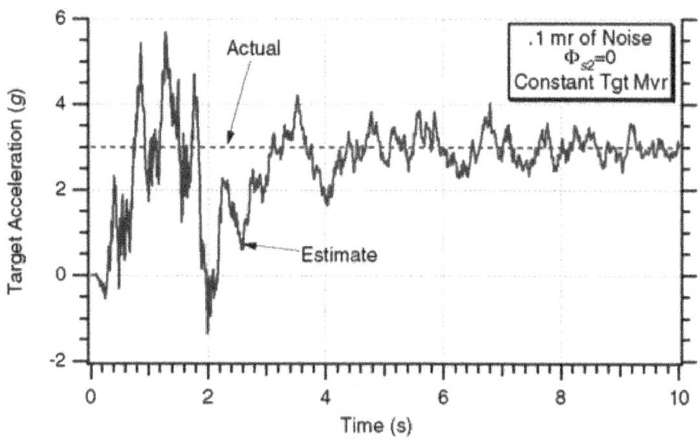

Fig. 25.27 After initial transient period extended Kalman filter is able to estimate constant target maneuver.

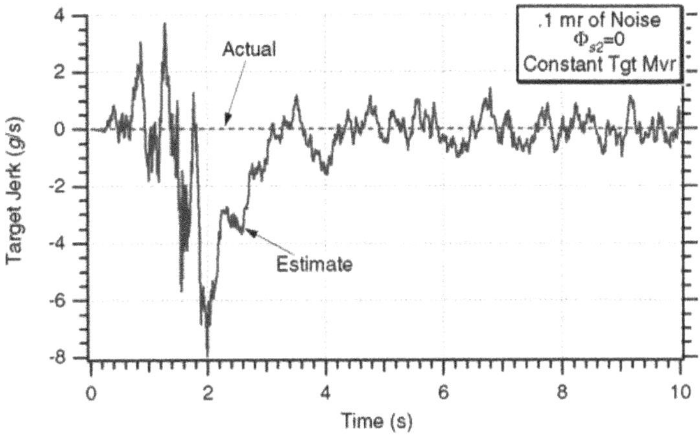

Fig. 25.28 Extended Kalman filter is able to correctly estimate that constant target maneuver has no jerk term.

similar and both yield much larger RMS miss distances when compared with the compensated weave guidance law.

Reducing the measurement noise by an order of magnitude to .1 mr improves the performance of all the guidance systems. When there is no constraint on the available missile acceleration Fig. 25.31 again shows that optimal guidance and compensated weave guidance yield much better performance than a proportional navigation guidance system. We can also see from Fig. 25.31 that the

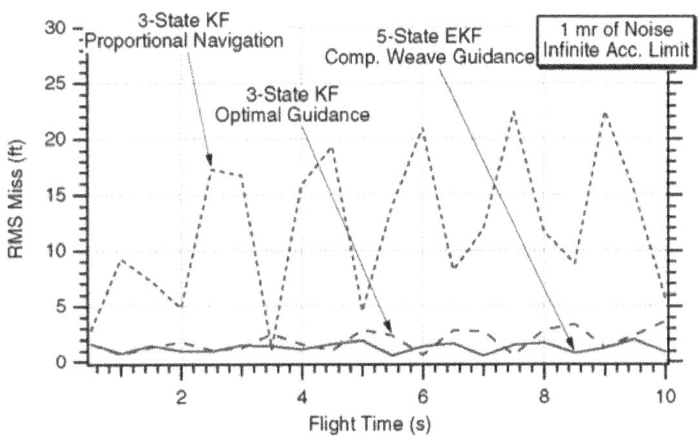

Fig. 25.29 Both compensate weave guidance and optimal guidance yield similar performance when there are no missile acceleration constraints.

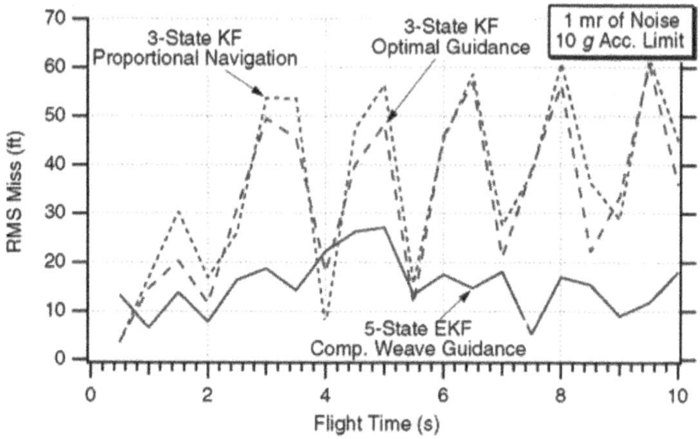

Fig. 25.30 With missile acceleration constraints, five-state extended Kalman filter with compensated weave guidance yields the best results.

performances of both optimal guidance and compensated weave guidance are virtually identical. However, we can see from Fig. 25.32 that the performance of the optimal guidance system degrades when there is a 10-g missile acceleration limit, and the performance of the extended Kalman filter with the compensated weave guidance law is much better. The trends are identical to the case in which the measurement noise was an order of magnitude larger.

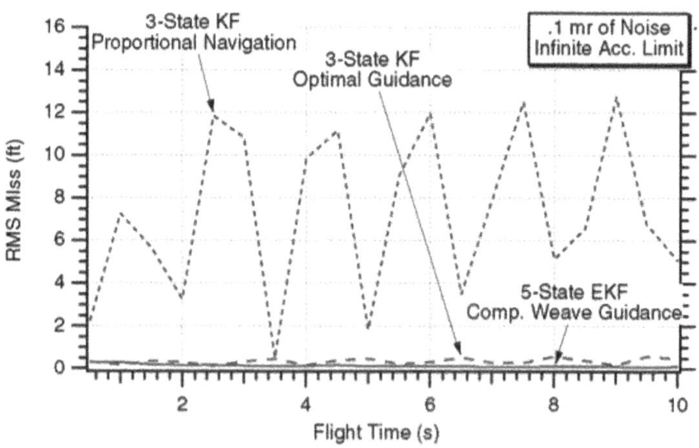

Fig. 25.31 With reduced measurement noise, both compensated weave guidance and optimal guidance still yield similar performance when there are no missile acceleration constraints.

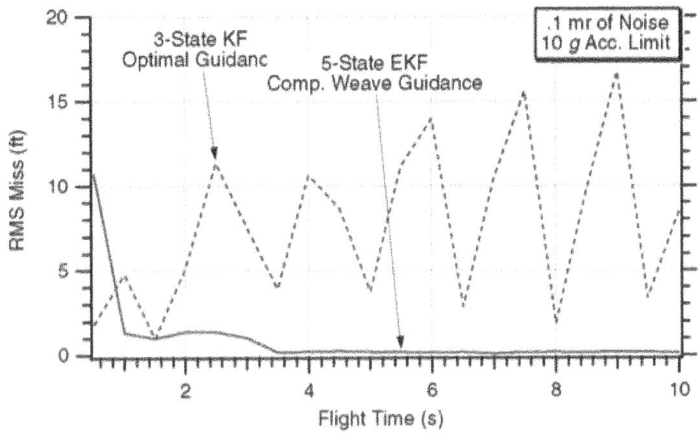

Fig. 25.32 With missile acceleration constraints, five-state extended Kalman filter with compensated weave guidance yields the best results when measurement noise is reduced.

SUMMARY

This chapter shows various filtering options that can be used in conjunction with advance guidance laws to improve the performance of missile guidance system against weaving targets. A conventional linear three-state Kalman filter can be used in conjunction with the optimal guidance law to yield significant performance improvements compared with proportional navigation. Similar performance improvements can also be obtained with a four-state weave Kalman filter in conjunction with the compensated weave guidance law if the target weave frequency is known. This filtering and guidance approach will perform better than optimal guidance when there is a low missile-to-target acceleration advantage. If the target weave frequency is not known it can be estimated with a five-state extended Kalman filter. Similar performance improvements can be obtained with this nonlinear filter and guidance law approach.

REFERENCES

[1] Zarchan, P., "Tracking and Intercepting Spiraling Ballistic Missiles," *Proceedings of IEEE PLANS Conference*, San Diego, CA, March 2000.

[2] Zarchan, P., and Musoff, H., *Fundamentals of Kalman Filtering: A Practical Approach*, Progress in Astronautics and Aeronautics, AIAA, Reston, VA, 2000.

CHAPTER 26

Alternative Approaches to Guidance Law Development

INTRODUCTION

So far this text has made use of the Schwartz inequality to derive guidance laws analytically. The Schwartz inequality was used because it was the simplest technique known to the author for the derivation of missile guidance laws. In this chapter we will investigate two alternative ways of developing guidance laws. These techniques can either be used to check analytically derived guidance laws or can be used to derive more advanced guidance laws when the Schwartz inequality technique becomes too cumbersome. The first alternative approach presented is based on optimal control theory and is the most general. The optimal control approach consists of the numerical integration of the nonlinear matrix Ricatti differential equation to solve for the guidance law control gains. In addition, a totally different approach for the numerical derivation of guidance laws is also presented in this chapter. This innovative technique is numerically much faster than the optimal control method and is especially convenient when the dynamics of the flight control system are too complex for any analytical approach. In both preceding approaches numerical examples are presented showing how the results of the new techniques compare to the analytical expressions for the previously derived guidance laws. Finally, a new guidance law will be developed in this chapter for the case in which there is a significant right-half-plane zero in the flight control system of a tail-controlled missile. It will be shown that the new guidance law, which is developed numerically, offers improved system performance at very high altitudes.

OPTIMAL CONTROL

Guidance laws do not have to be derived analytically as has been done so far in this text. For example, optimal control theory [1, 2] can be used to numerically derive guidance laws. With optimal control theory a performance index is set up, and a nonlinear matrix Ricatti differential equation needs to be solved in order to obtain the control gains of the resultant guidance law. At the very least this numerical approach can also be useful as an independent check of the accuracy of the analytically derived guidance laws that were presented earlier in this text. In addition, the numerical approach is useful in situations in which the analytical approach is either impossible or too complex (that is, very high order model of flight control system dynamics) to apply. To demonstrate the utility of the optimal control approach, we will first state the theoretical equations to be solved and then derive two guidance laws numerically for which we already have closed-form solutions.

Optimal control theory states that guidance laws can be derived either analytically or numerically if our model of the real world can be expressed in state-space form as

$$\dot{\mathbf{x}} = \mathbf{F}\mathbf{x} + \mathbf{G}\mathbf{u}$$

where the control $\mathbf{u}$ (guidance command) is linearly related to the states $\boldsymbol{x}$ according to

$$\mathbf{u} = -\mathbf{C}\mathbf{x}$$

In the preceding equation $\mathbf{C}$ is a set of control gains. The expression involving $\mathbf{C}$ is another way of writing a missile guidance law. With optimal control techniques we are trying to minimize a performance index. Previously when we derived guidance laws in this text, we simply tried to make the miss distance zero subject to minimizing the integral of the acceleration squared or

$$y(t_F) = 0 \quad \text{subject to minimizing} \quad \int_0^{t_F} n_c^2(t)\,\mathrm{d}t$$

With standard optimal control techniques we cannot quite minimize the preceding performance index. However a nearly equivalent performance index J can be expressed in matrix form and is given by

$$J = \mathbf{x}^T(t_F)\mathbf{S}_F\mathbf{x}(t_F) + \int_0^{t_F} \mathbf{u}^T\mathbf{B}\mathbf{u}\,\mathrm{d}t$$

where $\mathbf{x}(t_F)$ is the state vector at the final time. The miss $y(t_F)$ is one element of that state vector. By judiciously choosing $\mathbf{S}_F$, we can make the first term of J equal to the miss distance squared $[y^2(t_F)]$. In addition, by making $\mathbf{B}$ a small scalar we can come close to our desire of making the miss zero subject to minimizing the integral of the acceleration squared.

Optimal control theory shows that minimizing J for our state space model of the real world yields a guidance law in which the control gains are linearly related to the states. Obtaining those control gains involves solving the nonlinear matrix Riccati differential equation

$$\dot{\mathbf{S}} = -\mathbf{F}^T\mathbf{S} - \mathbf{S}\mathbf{F} + \mathbf{C}^T\mathbf{B}\mathbf{C}$$

with boundary value

$$\mathbf{S}(t_F) = \mathbf{S}_F$$

The control gains for the optimal guidance law are related to the solution of the preceding differential equation for **S** and can be obtained from

$$\mathbf{C} = \mathbf{B}^{-1}\mathbf{G}^T\mathbf{S}$$

There can be significant numerical difficulty in solving the matrix Riccati differential equation numerically because it is a boundary-value problem. In fact, in this text we have not yet solved a boundary-value problem numerically. However, this difficulty can be easily avoided by simply changing variables in order to convert the nonlinear matrix differential equation with a boundary value to one with an initial condition. Recall that from the overdot notation we know that

$$\dot{\mathbf{S}} = \frac{\mathrm{d}\mathbf{S}}{\mathrm{d}t}$$

Therefore if we define

$$\tau = t_F - t$$

we can say that

$$\frac{\mathrm{d}\mathbf{S}}{\mathrm{d}t} = -\frac{\mathrm{d}\mathbf{S}}{\mathrm{d}\tau}$$

Therefore by changing variables the nonlinear matrix differential equation becomes

$$\dot{\mathbf{S}} = \mathbf{F}^T\mathbf{S} + \mathbf{S}\mathbf{F} - \mathbf{C}^T\mathbf{B}\mathbf{C}$$

with initial value

$$\mathbf{S}(0) = \mathbf{S}_F$$

Because we have changed variables, **S** is now a function of τ or time to go rather than t or time. The control gains for the optimal guidance law are still obtained from

$$\mathbf{C} = \mathbf{B}^{-1}\mathbf{G}^T\mathbf{S}$$

but again the gains are now functions of time to go rather than a function of time. Double-precision arithmetic plus a very small integration step size are often required for the successful integration of the Riccati equations because of their numerical fragility.

The preceding optimal control equations are only valid for driving the relative position and/or relative velocity or some other quantities to zero at the end of the flight. The technique cannot be used to drive a quantity, such as the relative velocity, to a specified value at the end of the flight such as was done in using the Schwartz inequality to derive the trajectory-shaping guidance law of Chapter 24.

USING OPTIMAL CONTROL TO DERIVE GUIDANCE LAW FOR SINGLE-LAG FLIGHT CONTROL SYSTEM

In Chapter 8 we have already derived an optimal guidance law when the missile flight control system was modeled as a single lag. For the single-lag flight control system the overall model for guidance law development was previously displayed in Fig. 8.13 and is repeated here for convenience in Fig. 26.1. Recall that in this model the relative acceleration is simply the difference between target acceleration n_T and the achieved missile acceleration n_L. Integrating the relative acceleration twice yields relative position y. At the end of the flight, the relative position is the miss distance $y(t_F)$.

As was shown in Chapter 8, the model of Fig. 26.1 can be put in state-space form as shown next. Here we are making the assumption that the target maneuver is constant and therefore its derivative is zero.

$$\begin{bmatrix} \dot{y} \\ \ddot{y} \\ \dot{n}_T \\ \dot{n}_L \end{bmatrix} = \begin{bmatrix} 0 & 1 & 0 & 0 \\ 0 & 0 & 1 & -1 \\ 0 & 0 & 0 & 0 \\ 0 & 0 & 0 & -\frac{1}{T} \end{bmatrix} \begin{bmatrix} y \\ \dot{y} \\ n_T \\ n_L \end{bmatrix} + \begin{bmatrix} 0 \\ 0 \\ 0 \\ \frac{1}{T} \end{bmatrix} n_c$$

Because the state-space equation is given by

$$\dot{\mathbf{x}} = \mathbf{F}\mathbf{x} + \mathbf{G}u$$

the appropriate state-space matrices for the single-lag flight control system can be written by inspection as

$$\mathbf{F} = \begin{bmatrix} 0 & 1 & 0 & 0 \\ 0 & 0 & 1 & -1 \\ 0 & 0 & 0 & 0 \\ 0 & 0 & 0 & -\frac{1}{T} \end{bmatrix}$$

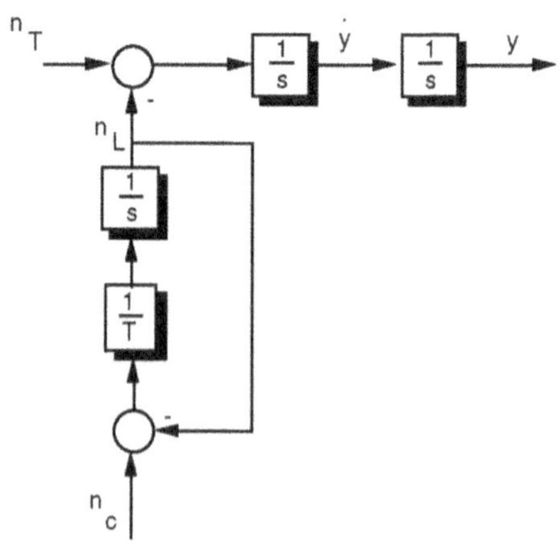

Fig. 26.1 Single-lag flight control system model for guidance law development.

$$\mathbf{G} = \begin{bmatrix} 0 \\ 0 \\ 0 \\ -\frac{1}{T} \end{bmatrix}$$

$$\mathbf{u} = n_c$$

As was mentioned earlier at the beginning of this section, the performance index that is usually chosen for guidance problems is one in which we make the miss zero subject to minimizing the integral of the square of acceleration. A convenient way of expressing this desire is shown next for the performance index J. Here the parameter γ is used as a fudge factor in the derivation of the guidance law (meaning that γ numerically prevents the guidance gains from becoming infinite when the time to go approaches zero). Using smaller values of γ will cause us to pay more attention to the miss (that is, make the miss smaller). The performance index for our optimal control problem is given by

$$J = y^2(t_F) + \gamma \int_0^{t_F} n_c^2 \, \mathrm{d}t$$

Because the generalized performance index for optimal control problems is given by

$$J = \mathbf{x}^T(t_F)\mathbf{S}_F\mathbf{x}(t_F) + \int_0^{t_F} \mathbf{u}^T\mathbf{B}\mathbf{u} \, \mathrm{d}t$$

the matrices in the preceding performance index expression can also be written by inspection as

$$\mathbf{S}_F = \begin{bmatrix} 1 & 0 & 0 & 0 \\ 0 & 0 & 0 & 0 \\ 0 & 0 & 0 & 0 \\ 0 & 0 & 0 & 0 \end{bmatrix}$$

$$\mathbf{B} = \gamma$$

We now have all of the information we need to integrate the matrix Ricatti equations and solve for the control gains. In this example the Ricatti equations can also be solved in closed form [3, 4]; however, we shall simply use second-order Runge–Kutta numerical integration to obtain the Ricatti equation solution.

Analytically, we already know from Chapter 8 that for the single-lag flight control system the closed-form solution for the optimal guidance law is given by

$$n_c = \frac{N'}{t_{\text{go}}^2}[y + \dot{y}t_{\text{go}} + 0.5n_T t_{\text{go}}^2 - n_L T^2(e^{-x} + x - 1)]$$

In the preceding equation the effective navigation ratio can be expressed as

$$N' = \frac{6x^2(e^{-x} - 1 + x)}{2x^3 + 3 + 6x - 6x^2 - 12xe^{-x} - 3e^{-2x}}$$

where T is the flight control system time constant and x is defined as

$$x = \frac{t_{go}}{T}$$

Alternatively we can also present the optimal guidance law in terms of control gains or

$$n_c = C_1 y + C_2 \dot{y} + C_3 n_T + C_4 n_L$$

where C_1 through C_4 are the control gains. Equating the preceding expression with the optimal guidance law yields the equations for each of the control gains as

$$C_1 = \frac{N'}{t_{go}^2}$$

$$C_2 = \frac{N'}{t_{go}}$$

$$C_3 = 0.5N'$$

$$C_4 = \frac{-T^2(e^{-x} + x - 1)N'}{t_{go}^2} = \frac{-N'(e^{-x} + x - 1)}{x^2}$$

The preceding equations will be considered the exact solution for the control gains for our guidance problem.

We can numerically integrate the nonlinear matrix Ricatti differential equation and solve for the control gains and then compare the numerical results with the preceding closed-form solutions. Listing 26.1 uses the matrices of this section to numerically integrate the nonlinear matrix differential equation with the appropriate initial conditions. The nominal parameters for Listing 26.1 are a flight-control-system time constant of 1 s, a flight time of 10 s, and a performance index weighting of 0.00001. Second-order Runge–Kutta integration is used to numerically integrate the Ricatti equation. After some experimentation an integration step size of 0.0001 s was chosen because it yielded accurate answers and still allowed the program to run quickly. Note that the denominator in the closed-form solution for the effective navigation ratio goes to zero as time to go approaches zero. To avoid that singularity, 0.0001 was added to the denominator as can be seen from Listing 26.1.

LISTING 26.1 INTEGRATING RICATTI EQUATIONS TO GET OPTIMAL CONTROL GAINS FOR SINGLE-LAG FLIGHT CONTROL SYSTEM

```
clear
TAU=1;
GAM=.00001;
TF=10;
F=zeros([4,4]);
S=zeros([4,4]);
count=0;
G(1,1)=0;
G(2,1)=0;
G(3,1)=0;
G(4,1)=1./TAU;
F(1,2)=1;
F(2,3)=1;
F(2,4)=-1;
F(4,4)=-1./TAU;
S(1,1)=1;
T=0;
H=.0001;
S1=0;
while ~(T >= (TF-.0001))
        S1=S1+H;
        SOLD=S;
        STEP=1;
        FLAG=0;
        while STEP<=1
                if FLAG==1
        STEP=2;
        HSD=H*SD;
        S=S+HSD;
                        T=T+H;
                end
                SF=S*F;
                GT=G';
                GTS=GT*S;
                GAMINV=1./GAM;
                C=GAMINV*GTS;
                SFT=SF';
                CT=C';
                CTC=CT*C;
                CTBC=GAM*CTC;
                SFSFT=SF+SFT;
                SD=SFSFT-CTBC;
        FLAG=1;
```

```
		end
		FLAG=0;
		H2=.5*H;
		HSDP=H2*SD;
		SS=SOLD+S;
		SSP=.5*SS;
		S=SSP+HSDP;
		if S1>=.009999
			S1=0;
			C1=-C(1,1);
			C2=-C(1,2);
			C3=-C(1,3);
			C4=-C(1,4);
			NP=C2*T;
			XS=T/TAU;
			W1=1./TAU;
			TOP=6.*XS*XS*(exp(-XS)-1.+XS);
			BOT1=2*XS*XS*XS+3.+6.*XS-6.*XS*XS;
			BOT2=-12.*XS*exp(-XS)-3.*exp(-2.*XS);
			XNPP=TOP/(BOT1+BOT2+.0001);
			C1TH=XNPP/(T*T);
			C2TH=XNPP/T;
			C3TH=.5*XNPP;
			C4TH=-XNPP*(exp(-XS)+XS-1.)/(XS*XS);
			count=count+1;
			ArrayT(count)=T;
			ArrayC1(count)=C1;
			ArrayC1TH(count)=C1TH;
			ArrayC2(count)=C2;
			ArrayC2TH(count)=C2TH;
			ArrayC3(count)=C3;
			ArrayC3TH(count)=C3TH;
			ArrayC4(count)=C4;
			ArrayC4TH(count)=C4TH;
			ArrayNP(count)=NP;
			ArrayXNPP(count)=XNPP;
		end
end
output=[ArrayT',ArrayC1',ArrayC1TH',ArrayC2',ArrayC2TH', ...
		ArrayC3',ArrayC3TH',ArrayC4',ArrayC4TH',ArrayNP',ArrayXNPP'];
save datfil.txt output /ascii
disp 'simulation finished'
clc
figure
semilogy(ArrayT,ArrayC1,ArrayT,ArrayC1TH),grid
xlabel('Time (s) ')
```

```
ylabel('C1')
axis([0 10 .01 1000])
figure
semilogy(ArrayT,ArrayC2,ArrayT,ArrayC2TH),grid
xlabel('Time (s) ')
ylabel('C2')
axis([0 10 .1 200])
figure
semilogy(ArrayT,ArrayC3,ArrayT,ArrayC3TH),grid
xlabel('Time (s) ')
ylabel('C3')
axis([0 10 1 20])
figure
semilogy(ArrayT,-ArrayC4,ArrayT,-ArrayC4TH),grid
xlabel('Time (s) ')
ylabel('-C4')
axis([0 10 .1 20])
figure
semilogy(ArrayT,ArrayNP,ArrayT,ArrayXNPP),grid
xlabel('Time (s) ')
ylabel('Effective Navigation Ratio')
axis([0 10 .1 40])
```

The nominal case of Listing 26.1 was run, and the control gain outputs are displayed along with the analytical solution ("Formula") in Figs. 26.2–26.5. We can see that the numerical and analytical results are in near perfect agreement, thus verifying the accuracy of the numerical integration in the optimal control approach. Figure 26.6 compares the effective navigation ratios for the optimal

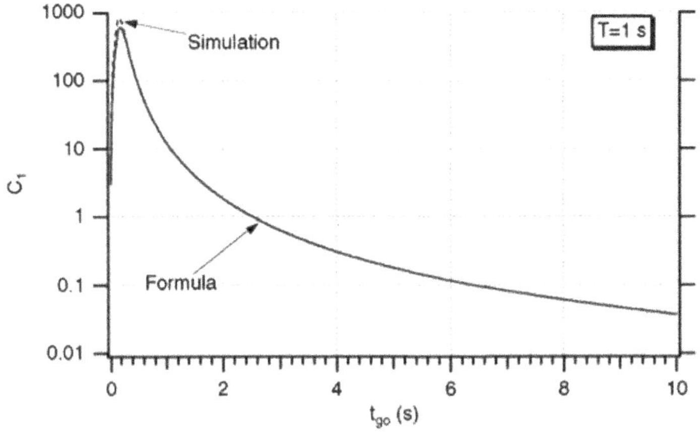

Fig. 26.2 Formula and simulation results agree for first control gain.

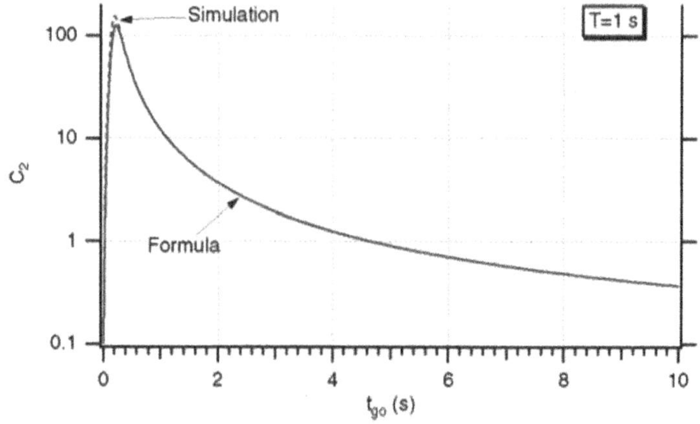

Fig. 26.3 Formula and simulation results agree for second control gain.

guidance law from both the analytical and numerical approaches and shows they are equivalent. Thus we can now feel confident about using Listing 26.1, with minor modifications for the matrices, to generate other guidance laws using the optimal control technique.

DERIVING GUIDANCE LAW FOR WEAVING TARGET USING OPTIMAL CONTROL

As another example of the utility of the optimal control approach for deriving guidance laws, let us consider an example involving a weaving target. In this case we shall consider the flight control system to be perfect (that is, zero time constant). We have already shown in Chapter 20 that a weave maneuver could

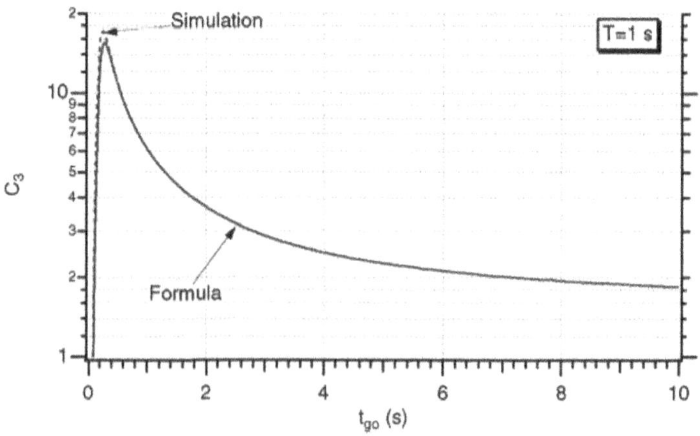

Fig. 26.4 Formula and simulation results agree for third control gain.

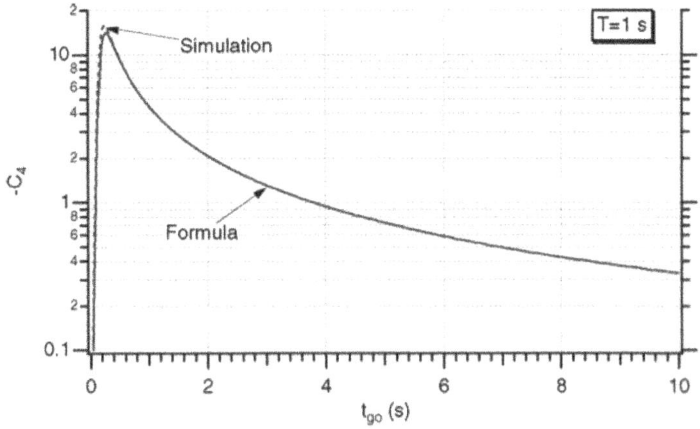

Fig. 26.5 Formula and simulation results agree for fourth control gain.

be represented analytically by a shaping network. The overall model for guidance law development was presented in Fig. 20.16 for the weaving target and is repeated in Fig. 26.7 for convenience.

As was shown in Chapter 20, the shaping network to the left of the summing junction in Fig. 26.7 represents the sinusoidal maneuver. The model of Fig. 26.7 can be put in state space form as shown here:

$$\begin{bmatrix} \dot{y} \\ \ddot{y} \\ \dot{n}_T \\ \ddot{n}_T \end{bmatrix} = \begin{bmatrix} 0 & 1 & 0 & 0 \\ 0 & 0 & 1 & 0 \\ 0 & 0 & 0 & 1 \\ 0 & 0 & -\omega^2 & 0 \end{bmatrix} \begin{bmatrix} y \\ \dot{y} \\ n_T \\ \dot{n}_T \end{bmatrix} + \begin{bmatrix} 0 \\ -1 \\ 0 \\ 0 \end{bmatrix} n_c$$

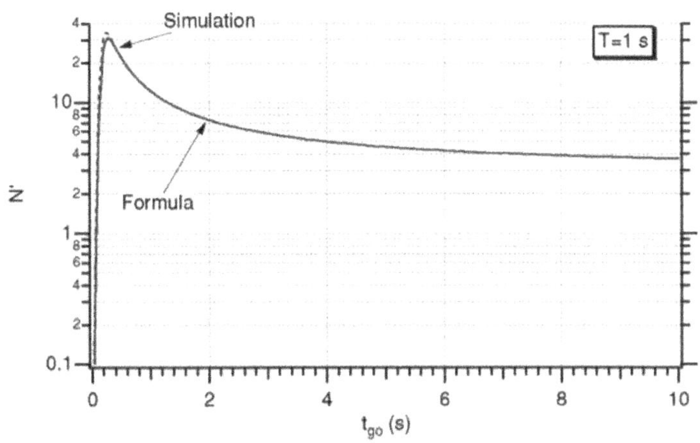

Fig. 26.6 Formula and simulation results agree for effective navigation ratio.

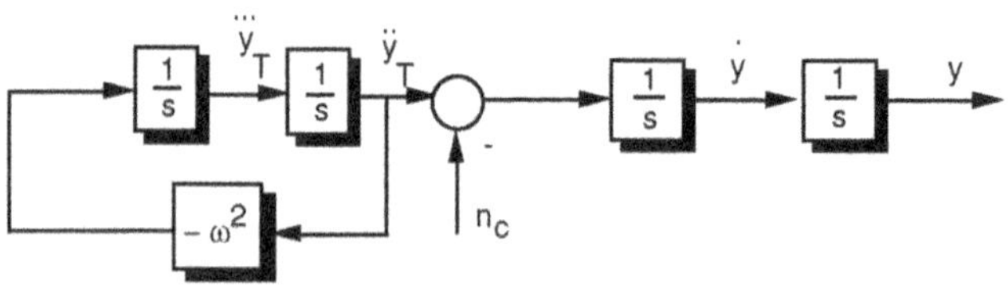

Fig. 26.7 Model for weave guidance law derivation.

As was the case in the preceding section, we will still assume that the performance index to be minimized is given by

$$J = y^2(t_F) + \gamma \int_0^{t_F} n_c^2 \, dt$$

Now that we have formulated the guidance problem, the appropriate matrices required by the matrix Ricatti differential equation can be written by inspection as

$$\mathbf{F} = \begin{bmatrix} 0 & 1 & 0 & 0 \\ 0 & 0 & 1 & 0 \\ 0 & 0 & 0 & 1 \\ 0 & 0 & -\omega^2 & 0 \end{bmatrix}$$

$$\mathbf{G} = \begin{bmatrix} 0 \\ -1 \\ 0 \\ 0 \end{bmatrix}$$

$$\mathbf{S}_F = \begin{bmatrix} 1 & 0 & 0 & 0 \\ 0 & 0 & 0 & 0 \\ 0 & 0 & 0 & 0 \\ 0 & 0 & 0 & 0 \end{bmatrix}$$

$$\mathbf{u} = n_c \quad \text{and} \quad \mathbf{B} = \gamma$$

We have already shown in Chapter 20 that for the weaving target the closed-form solution for the optimal guidance law, known as weave guidance, is given by

$$n_c = \frac{3}{t_{\text{go}}^2}\left[y + \dot{y}t_{\text{go}} + \left(\frac{1 - \cos \omega t_{\text{go}}}{\omega^2}\right)n_T + \left(\frac{\omega t_{\text{go}} - \sin \omega t_{\text{go}}}{\omega^3}\right)\dot{n}_T\right]$$

Note that in the preceding equation for the weave guidance law the effective navigation ratio is three.

As was done in the preceding section, we can also express the guidance law in terms of control gains or

$$n_c = C_1 y + C_2 \dot{y} + C_3 n_T + C_4 \dot{n}_T$$

By comparing the preceding equation with the weave guidance law, we can easily see that the control gains for the weave guidance law are given by

$$C_1 = \frac{3}{t_{go}^2}$$

$$C_2 = \frac{3}{t_{go}}$$

$$C_3 = \frac{3}{t_{go}^2}\left(\frac{1 - \cos \omega t_{go}}{\omega^2}\right)$$

$$C_4 = \frac{3}{t_{go}^2}\left(\frac{\omega t_{go} - \sin \omega t_{go}}{\omega^3}\right)$$

The preceding equations can be viewed as the exact solution for the control gains for the weaving guidance law.

We can slightly modify Listing 26.1 to solve for the control gains in the weaving target problem. Listing 26.2 shows that we still numerically integrate the nonlinear matrix Ricatti equation with a small integration step size and solve for the control gains. A comparison is made between the results obtained by numerical integration and the preceding closed-form solutions. Statements that have changed from Listing 26.1 in order to find the weave guidance law are highlighted in bold in Listing 26.2. The nominal case considered by Listing 26.2 had a target weave frequency of 1 r/s, a flight time of 10 s, and a performance index weighting of 0.00001. Again, second-order Runge–Kutta integration was used to numerically integrate the matrix Ricatti differential equation, and an integration step size of 0.0001 s was still used to get accurate answers.

LISTING 26.2 INTEGRATING MATRIX RICATTI EQUATION TO GET OPTIMAL CONTROL GAINS FOR WEAVING TARGET

```
clear
W=1;
GAM=.00001;
TF=10;
F=zeros([4,4]);
S=zeros([4,4]);
count=0;
G(1,1)=0;
G(2,1)=-1.;
```

```
G(3,1)=0;
G(4,1)=0.;
F(1,2)=1;
F(2,3)=1;
F(3,4)=1.;
F(4,3)=-W*W;
S(1,1)=1;
T=0;
H=.0001;
S1=0;
while ~(T >= (TF-.0001))
        S1=S1+H;
        SOLD=S;
        STEP=1;
        FLAG=0;
        while STEP<=1
                if FLAG==1
        STEP=2;
        HSD=H*SD;
        S=S+HSD;
                T=T+H;
                end
                SF=S*F;
                GT=G';
                GTS=GT*S;
                GAMINV=1./GAM;
                C=GAMINV*GTS;
                SFT=SF';
                CT=C';
                CTC=CT*C;
                CTBC=GAM*CTC;
                SFSFT=SF+SFT;
                SD=SFSFT-CTBC;
                FLAG=1;
        end
        FLAG=0;
        H2=.5*H;
        HSDP=H2*SD;
        SS=SOLD+S;
        SSP=.5*SS;
        S=SSP+HSDP;
        if S1>=.009999
                S1=0;
                C1=-C(1,1);
                C2=-C(1,2);
```

```
                C3=-C(1,3);
                C4=-C(1,4);
                NP=C2*T;
                XNPP=3.;
                C1TH=XNPP/(T*T);
                C2TH=XNPP/T;
                C3TH=XNPP*((1.-cos(W*T))/W^2)/T^2;
                C4TH=XNPP*((W*T-sin(W*T))/W^3)/T^2;
                count=count+1;
                ArrayT(count)=T;
                ArrayC1(count)=C1;
                ArrayC1TH(count)=C1TH;
                ArrayC2(count)=C2;
                ArrayC2TH(count)=C2TH;
                ArrayC3(count)=C3;
                ArrayC3TH(count)=C3TH;
                ArrayC4(count)=C4;
                ArrayC4TH(count)=C4TH;
                ArrayNP(count)=NP;
                ArrayXNPP(count)=XNPP;
        end
end
output=[ArrayT',ArrayC1',ArrayC1TH',ArrayC2',ArrayC2TH',...
        ArrayC3',ArrayC3TH',ArrayC4',ArrayC4TH',ArrayNP',ArrayXNPP'];
save datfil.txt output /ascii
disp 'simulation finished'
clc
figure
plot(ArrayT,ArrayC3,ArrayT,ArrayC3TH),grid
xlabel('Time (s) ')
ylabel('C3')
axis([0 10 0 1.5])
figure
plot(ArrayT,ArrayC4,ArrayT,ArrayC4TH),grid
xlabel('Time (s) ')
ylabel('C4')
axis([0 10 0 1])
```

The nominal case of Listing 26.2 was run, and the control gain outputs for C_3 and C_4 are displayed along with the analytical solution ("Formula") in Figs. 26.8 and 26.9. We can see that again the numerical and analytical results are in near perfect agreement, thus again verifying the accuracy of the numerical integration. Therefore we have again demonstrated that the optimal control technique is a viable alternative approach for numerically deriving missile guidance laws.

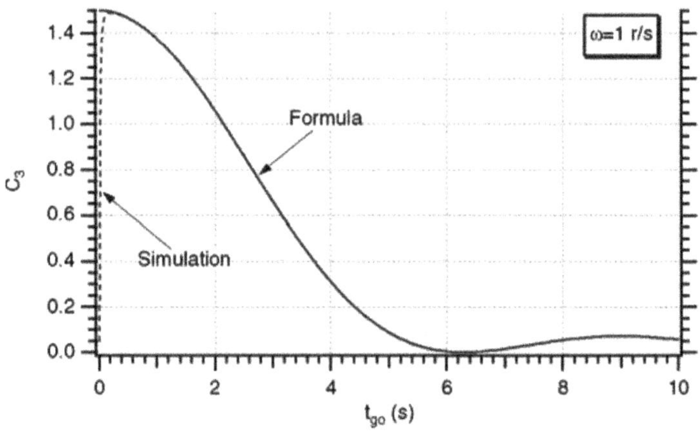

Fig. 26.8 Formula and simulation results agree for third control gain.

GUIDANCE PORTION DUE TO MANEUVERING TARGETS

We have seen in this text that all of the guidance laws can be expressed as a navigation ratio time the zero effort miss and divided by the square of time to go until intercept. In addition, we have noted that the first two terms in the guidance law relate to proportional navigation and the other terms relate to target maneuver and the dynamics of the flight control system. Therefore we can say that in general all guidance laws can be expressed as

$$n_c = \frac{N'}{t_{\text{go}}^2}\text{ZEM} = \frac{N'}{t_{\text{go}}^2}[y + \dot{y}t_{\text{go}} + \text{ZEM}_{\text{TGT}} + \text{ZEM}_{\text{FCS}}]$$

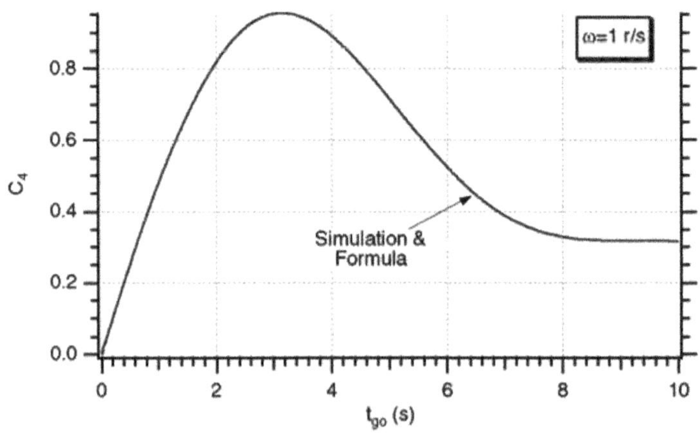

Fig. 26.9 Formula and simulation results agree for fourth control gain.

where ZEM_{TGT} is the zero effort miss due to the target maneuver and ZEM_{FCS} is the zero effort miss due to the flight control system. In this section we shall present an alternative analytical technique for finding the zero effort miss due to target maneuver and demonstrate that the new technique yields answers that we know to be true.

Reference 5 shows that in general the zero effort miss caused by the target maneuver is given by

$$\text{ZEM}_{\text{TGT}} = -\int_t^{t_F} (\alpha - t_F) n_T(\alpha)\, \mathrm{d}\alpha$$

where α is a dummy variable, n_T is the target maneuver (which does not have to be a constant), and t_F is the time of flight. Let us consider two cases to which we know the answers in order to demonstrate that the preceding relationship is correct.

First let us consider the example of a constant target maneuver. In this case the closed-form solution for the zero effort miss becomes

$$\underset{\text{TGT}}{\text{ZEM}_{\text{CONST}}} = -n_T \int_t^{t_F} (\alpha - t_F)\, \mathrm{d}\alpha = -n_T \left[\int_t^{t_F} \alpha\, \mathrm{d}\alpha - t_F \int_t^{t_F} \mathrm{d}\alpha \right]$$

Integration of the preceding expression yields

$$\underset{\text{TGT}}{\text{ZEM}_{\text{CONST}}} = -n_T \left[\frac{\alpha^2}{2} \bigg|_t^{t_F} - t_F \alpha \bigg|_t^{t_F} \right] = -n_T \left[\frac{t_F^2}{2} - \frac{t^2}{2} - t_F^2 + t_F t \right]$$

which simplifies to

$$\underset{\text{TGT}}{\text{ZEM}_{\text{CONST}}} = \frac{n_T (t_F - t)^2}{2} = \frac{n_T t_{\text{go}}^2}{2}$$

where

$$t_{\text{go}} = t_F - t$$

We recognize the zero effort miss caused by constant target maneuver as the term we used in the augmented proportional navigation guidance law, thus verifying the theoretical expression presented at the beginning of this section.

Let us now consider another example in which the target maneuver is not a constant but is a sinusoid with weave frequency ω and arbitrary phase angle ϕ. In this case we can express the time-varying target acceleration as

$$n_T(t) = a_T \sin(\omega t + \phi)$$

where a_T is the amplitude of the sinusoidal maneuver. By taking the derivative of the preceding expression, we can also find the target jerk as

$$\dot{n}_T(t) = a_T \omega \cos(\omega t + \phi)$$

Substitution of the expression for the sinusoidal target maneuver in the formula for the zero effort miss yields

$$\mathrm{ZEM}_{\mathrm{WEAVE}} = -\int_t^{t_F} (\alpha - t_F) n_T(\alpha)\, \mathrm{d}\alpha = -a_T \int_t^{t_F} (\alpha - t_F) \sin(\omega\alpha + \phi)\, \mathrm{d}\alpha$$

Let us recall from integral tables that

$$\int \sin(\omega\alpha + \phi)\, \mathrm{d}\alpha = -\frac{1}{\omega}\cos(\omega\alpha + \phi)$$

$$\int \alpha \sin(\omega\alpha + \phi)\, \mathrm{d}\alpha = -\frac{\alpha}{\omega}\cos(\omega\alpha + \phi) + \frac{1}{\omega^2}\sin(\omega\alpha + \phi)$$

Therefore the zero effort miss becomes

$$\mathrm{ZEM}_{\mathrm{WEAVE}} = -a_T \left\{ \left[-\frac{\alpha}{\omega}\cos(\omega\alpha + \phi) + \frac{1}{\omega^2}\sin(\omega\alpha + \phi) \right]_t^{t_F} + \left[\frac{t_F}{\omega}\cos(\omega\alpha + \phi) \right]_t^{t_F} \right\}$$

After evaluation of the limits and some simplification, we obtain

$$\mathrm{ZEM}_{\mathrm{WEAVE}} = -a_T \left\{ \left(\frac{t}{\omega} - \frac{t_F}{\omega} \right) \cos(\omega t + \phi) + \frac{1}{\omega^2} [\sin(\omega t_F + \phi) - \sin(\omega t + \phi)] \right\}$$

We can expand

$$\sin(\omega t_F + \phi) = \sin[\omega(t_F - t + t) + \phi] = \sin[\omega(t_{\mathrm{go}} + t) + \phi] = \sin(\omega t_{\mathrm{go}} + \omega t + \phi)$$

and recognize that

$$\sin(\omega t_{\mathrm{go}} + \omega t + \phi) = \sin \omega t_{\mathrm{go}} \cos(\omega t_{\mathrm{go}} + \phi) + \cos \omega t_{\mathrm{go}} \sin(\omega t + \phi)$$

The expression for the zero effort miss caused by the weave maneuver then becomes

$$\mathrm{ZEM}_{\mathrm{WEAVE}} = -a_T \left\{ \frac{-t_{\mathrm{go}}}{\omega}\cos(\omega t + \phi) + \frac{1}{\omega^2} [\sin \omega t_{\mathrm{go}} \cos(\omega t + \phi) + \cos \omega t_{\mathrm{go}} \sin(\omega t + \phi) - \sin(\omega t + \phi)] \right\}$$

Simplification of the preceding expression yields

$$\text{ZEM}_{\text{WEAVE}} = -a_T\left[\left(\frac{-\omega t_{\text{go}} + \sin\omega t_{\text{go}}}{\omega^2}\right)\cos(\omega t + \phi) + \sin(\omega t + \phi)\left(\frac{\cos\omega t_{\text{go}} - 1}{\omega^2}\right)\right]$$

Substitution of the expressions for target acceleration and jerk yields

$$\text{ZEM}_{\text{WEAVE}} = n_T\left(\frac{1 - \cos\omega t_{\text{go}}}{\omega^2}\right) + \dot{n}_T\left(\frac{\omega t_{\text{go}} - \sin\omega t_{\text{go}}}{\omega^3}\right)$$

The preceding expression is identical to the portion of the zero effort miss related to target acceleration and jerk in the weave guidance law. Thus we have again verified the theoretical formula presented at the beginning of this section.

ALTERNATIVE NUMERICAL APPROACH AS A RESULT OF FLIGHT CONTROL SYSTEM DYNAMICS

For completeness an alternative and efficient numerical method for calculating the optimal guidance law as a result of the dynamics of the flight control system is presented in this section. It is based upon the work of Rusnak and Meir [6]. Consider a flight control system transfer function $H(s)$, where $H(s) = n_L(s)/n_C(s)$. Reference 6 shows that the optimal control gains can be calculated by first computing the quantity $\Lambda(t_{\text{go}})$

$$\Lambda(t_{\text{go}}) = L^{-1}\left[\frac{1}{s^2}\cdot\frac{n_L(s)}{n_C(s)}\right]_{t_{\text{go}}} \bigg/ \left[\gamma + \int_0^{t_{\text{go}}}\left\{L^{-1}\left[\frac{1}{s^2}\cdot\frac{n_L(s)}{n_C(s)}\right]_{t_{\text{go}}}\right\}^2 \text{d}\tau\right] = \frac{N}{D}$$

where L^{-1} is the inverse Laplace transform. The symbols N and D have been used as shorthand for numerator and denominator, respectively. Therefore the inverse Laplace transform of the item in the brackets of the preceding equation represents the time response of the flight control system to a ramp t acceleration command $n_c(t)$ (the Laplace transform of t is $1/s^2$). The guidance law control gains for the first three states (assuming constant target acceleration) are computed from $\Lambda(t_{\text{go}})$ as

$$C_1(t_{\text{go}}) = \Lambda(t_{\text{go}})$$
$$C_2(t_{\text{go}}) = \Lambda(t_{\text{go}})t_{\text{go}}$$
$$C_3(t_{\text{go}}) = 0.5\Lambda(t_{\text{go}})t_{\text{go}}^2$$

Reference 6 also states that the control gains for the flight control system states $i =$ 1 through N can be expressed as

$$C_{3+i}(t_{go}) = -\Lambda(t_{go})L^{-1}\left[\frac{1}{s^2}\cdot\frac{n_L(s)}{p_i(0)}\right]_{t_{go}}$$

where p_1 through p_N are the states of the flight-control-system transfer function. Therefore the quantity in brackets of the preceding equation represents the flight control systems response to a ramp input t on the derivative of the ith state of the flight control system. Note that the actual effective navigation ratio is related to $\Lambda(t_{go})$ according to

$$N'(t_{go}) = \Lambda(t_{go})t_{go}^2$$

The best way of illustrating this innovative technique is to work an example for the single time constant flight control system as was done in Ref. 7. Figure 26.10 is the block diagram equivalent of the preceding equations for the effective navigation ratio and control gains. The resultant differential equations resulting from this block diagram are simple and well behaved numerically.

Recall that for the preceding system the optimal guidance law can be expressed in terms of the gains as

$$n_c = C_1 y + C_2 \dot{y} + C_3 n_T + C_4 n_L$$

Figure 26.10 was simulated, and the effective navigation ratio and C_4 gain were compared to the analytical results derived earlier. Listing 26.3 is a simulation of

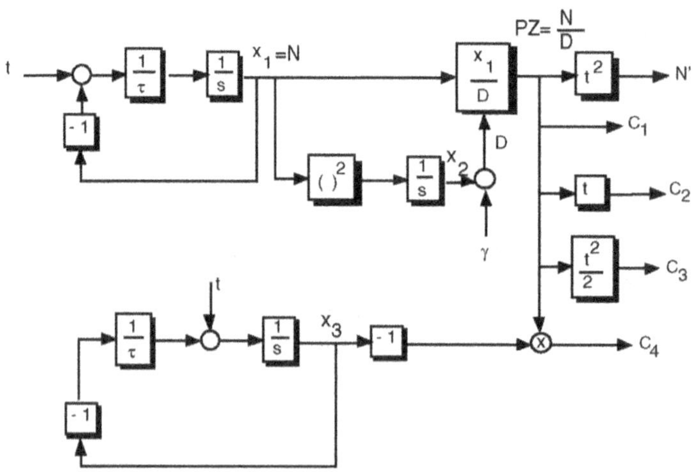

Fig. 26.10 Reference 5 method for finding classic optimal guidance law for single time constant flight control system.

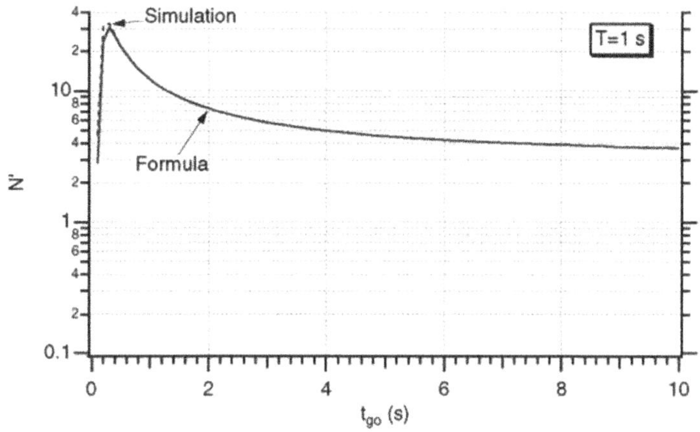

Fig. 26.11 Rusnak and Meir technique is equivalent to Riccati equation method for single-lag flight control system.

Fig. 26.10 using second-order Runge-Kutta integration with an integration step size of 0.01 s. This technique is not as numerically fragile as the optimal control method. Therefore this simulation will run approximately 100 times faster than the simulation using the optimal control approach because of the much larger integration step size. Listing 26.3 also presents the formulas for the effective navigation ratio and C_4 gain of the actual optimal guidance law.

The nominal case of Listing 26.3 was run. Figures 26.11 and 26.12 indicate that there is excellent agreement for the effective navigation ratio and fourth control gains using this new technique when compared to the closed-form solutions. Thus we have confidence that no programming errors were made in Listing 26.3 and that a large integration interval yielded the correct answers.

LISTING 26.3 NUMERICAL APPROACH TO GUIDANCE-LAW DEVELOPMENT FOR SINGLE-LAG FLIGHT CONTROL SYSTEM USING RUSNAK AND MEIR TECHNIQUE

```
clear
count=0;
TAU=1.;
GAM=.00001;
T=0.;
H=.01;
S=0.;
X1=0.;
X2=0.;
```

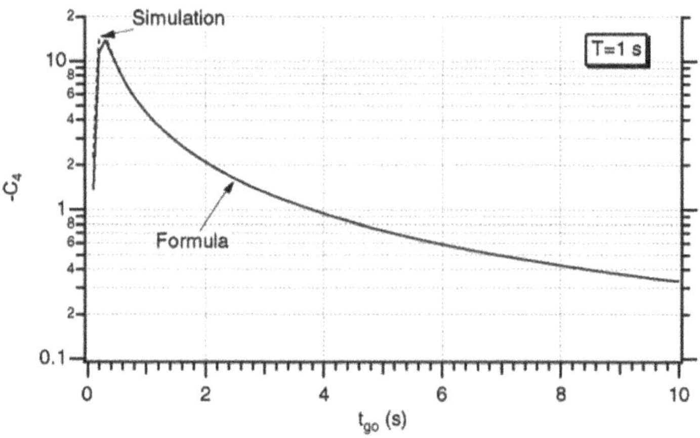

Fig. 26.12 Rusnak and Meir technique yields accurate fourth control gain for single-lag flight control system.

```
X3=0.;
while ~(T >= (10.-.0001))
        S=S+H;
        X1OLD=X1;
        X2OLD=X2;
        X3OLD=X3;
        STEP=1;
        FLAG=0;
        while STEP<=1
                if FLAG==1
        STEP=2;
                        X1=X1+H*X1D;
                        X2=X2+H*X2D;
                        X3=X3+H*X3D;
                        T=T+H;
                end
                X1D=(T-X1)/TAU;
                X2D=X1^2;
                D=X2+GAM;
                X3D=-X3/TAU+T;
                FLAG=1;
        end
        FLAG=0;
        X1=.5*(X1OLD+X1+H*X1D);
        X2=.5*(X2OLD+X2+H*X2D);
        X3=.5*(X3OLD+X3+H*X3D);
        if S>=.09999
```

```
            S=0.;
            PZ=X1/D;
            XNP=PZ*T*T;
            C1=PZ;
            C2=PZ*T;
            C3=.5*PZ*T*T;
            C4=-X3*PZ;
            XS=T/TAU;
            TOP=6.*XS*XS*(exp(-XS)-1.+XS);
            BOT1=2*XS*XS*XS+3.+6.*XS-6.*XS*XS;
            BOT2=-12.*XS*exp(-XS)-3.*exp(-2.*XS);
            XNPTH=TOP/(BOT1+BOT2+.0001);
            C4TH=-XNPTH*(exp(-XS)+XS-1.)/(XS*XS);
            count=count+1;
            ArrayT(count)=T;
            ArrayC4(count)=C4;
            ArrayC4TH(count)=C4TH;
            ArrayXNP(count)=XNP;
            ArrayXNPTH(count)=XNPTH;
    end
end
output=[ArrayT',ArrayC4',ArrayC4TH',ArrayXNP',ArrayXNPTH'];
save datfil.txt output /ascii
disp 'simulation finished'
clc
figure
semilogy(ArrayT,ArrayXNP,ArrayT,ArrayXNPTH),grid
xlabel('Time (s) ')
ylabel('NP')
axis([0 10 .1 40])
figure
semilogy(ArrayT,-ArrayC4,ArrayT,-ArrayC4TH),grid
xlabel('Time (s) ')
ylabel('-C4')
axis([0 10 .1 20])
```

DERIVING NEW GUIDANCE LAW FOR CUBIC FLIGHT CONTROL SYSTEM

So far in this chapter we have used the new techniques to derive guidance laws for which we already had the closed-form solution. This was done to give us confidence that the alternative techniques actually worked. In the next two sections we shall use both new techniques to numerically develop a new guidance law that would not be possible to derive using the Schwartz inequality.

When more accurate representations of the missile flight control system are used, closed-form solutions for the resultant guidance law either become impossible to derive or unwieldy because of the flight control system complexity. However guidance laws do not have to be derived in closed form for them to be useful. The more advanced guidance laws can be developed numerically, and the resultant control gains can be stored in a flight computer for guidance-law implementation [7]. In fact, this approach is similar to the way autopilot gains are stored as a function of flight condition in the missile flight computer [8].

We saw in Chapter 23 that a more realistic model for the flight control system of a tail-controlled missile (compared to the single-lag representation) is given by the transfer function

$$\frac{n_L}{n_c} = \frac{\left(1 - \frac{s^2}{\omega_z^2}\right)}{\left[(1 + s\tau)\left(1 + \frac{2\zeta}{\omega}s + \frac{s^2}{\omega^2}\right)\right]}$$

In this transfer function we have modeled the "wrong-way" tail effect with left- and right-half-plane zeros. The denominator of the flight control system consists of a dominant real pole followed by a higher-frequency quadratic. To use optimal control techniques to derive a new guidance law, we must first place the preceding transfer function in state-space form. The first step is to multiply out the denominator of the flight control system transfer function or

$$\frac{n_L}{n_c} = \left(1 - \frac{s^2}{\omega_z^2}\right) \Big/ \left[1 + \left(\frac{2\zeta}{\omega} + \tau\right)s + \left(\frac{2\zeta\tau}{\omega} + \frac{1}{\omega^2}\right)s^2 + \frac{\tau}{\omega^2}s^3\right]$$

We would like to split the transfer function so that we can start to write differential equations in state-space format. This can be accomplished by using the chain rule from calculus, as was done in Chapter 1, or

$$\frac{n_L}{n_c} = \frac{e}{n_c} * \frac{n_L}{e}$$

Using the preceding equation yields the two transfer functions:

$$\frac{e}{n_c} = 1 \Big/ \left[1 + \left(\frac{2\zeta}{\omega} + \tau\right)s + \left(\frac{2\zeta\tau}{\omega} + \frac{1}{\omega^2}\right)s^2 + \frac{\tau}{\omega^2}s^3\right]$$

$$\frac{n_L}{e} = 1 - \frac{s^2}{\omega_z^2}$$

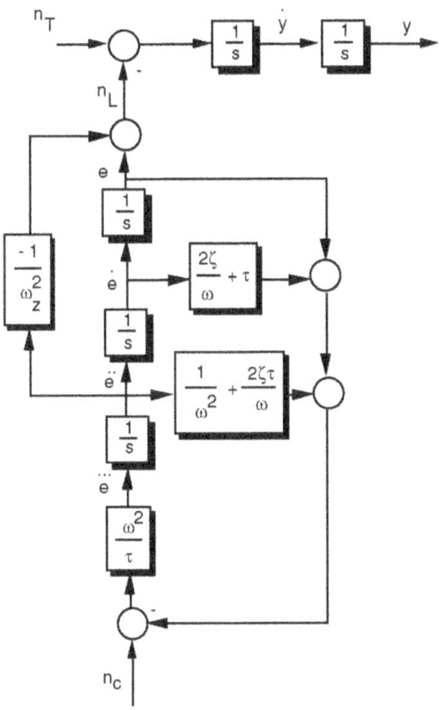

Fig. 26.13 Cubic flight control system model for guidance-law development.

Cross multiplying and converting to the time domain yields the two differential equations describing the cubic flight control system as

$$\dddot{e} = \frac{\omega^2}{\tau}\left[n_c - e - \left(\frac{2\zeta}{\omega} + \tau\right)\dot{e} - \left(\frac{2\zeta\tau}{\omega} + \frac{1}{\omega^2}\right)\ddot{e}\right]$$

$$n_L = e - \frac{\ddot{e}}{\omega_z^2}$$

Using the preceding two differential equations, we can express the new model for guidance-law development in block diagram form as shown in Fig. 26.13.

Figure 26.13 can be converted to state-space form by inspection. Thus the state-space equation for the homing loop with the cubic flight control system is given by

$$\begin{bmatrix} \dot{y} \\ \ddot{y} \\ \dot{n}_T \\ \dot{e} \\ \ddot{e} \\ \dddot{e} \end{bmatrix} = \begin{bmatrix} 0 & 1 & 0 & 0 & 0 & 0 \\ 0 & 0 & 1 & -1 & 0 & \frac{1}{\omega_z^2} \\ 0 & 0 & 0 & 0 & 0 & 0 \\ 0 & 0 & 0 & 0 & 1 & 0 \\ 0 & 0 & 0 & 0 & 0 & 1 \\ 0 & 0 & 0 & -\frac{\omega^2}{\tau} & -\frac{\omega^2}{\tau}\left(\frac{2\zeta}{\omega} + \tau\right) & -\frac{\omega^2}{\tau}\left(\frac{1}{\omega^2} + \frac{2\zeta\tau}{\omega}\right) \end{bmatrix} \begin{bmatrix} y \\ \dot{y} \\ n_T \\ e \\ \dot{e} \\ \ddot{e} \end{bmatrix} + \begin{bmatrix} 0 \\ 0 \\ 0 \\ 0 \\ 0 \\ \frac{\omega^2}{\tau} \end{bmatrix} n_c$$

Therefore the systems dynamics and control matrices are given by

$$\mathbf{F} = \begin{bmatrix} 0 & 1 & 0 & 0 & 0 & 0 \\ 0 & 0 & 1 & -1 & 0 & \frac{1}{\omega_z^2} \\ 0 & 0 & 0 & 0 & 0 & 0 \\ 0 & 0 & 0 & 0 & 1 & 0 \\ 0 & 0 & 0 & 0 & 0 & 1 \\ 0 & 0 & 0 & -\frac{\omega^2}{\tau} & -\frac{\omega^2}{\tau}\left(\frac{2\zeta}{\omega} + \tau\right) & -\frac{\omega^2}{\tau}\left(\frac{1}{\omega^2} + \frac{2\zeta\tau}{\omega}\right) \end{bmatrix}$$

$$\mathbf{G} = \begin{bmatrix} 0 \\ 0 \\ 0 \\ 0 \\ 0 \\ \frac{\omega^2}{\tau} \end{bmatrix}$$

Because we are still trying to minimize the performance index

$$J = y^2(t_F) + \gamma \int_0^{t_F} n_c^2 \mathrm{d}t$$

we can say that

$$\mathbf{S}_F = \begin{bmatrix} 1 & 0 & 0 & 0 & 0 & 0 \\ 0 & 0 & 0 & 0 & 0 & 0 \\ 0 & 0 & 0 & 0 & 0 & 0 \\ 0 & 0 & 0 & 0 & 0 & 0 \\ 0 & 0 & 0 & 0 & 0 & 0 \\ 0 & 0 & 0 & 0 & 0 & 0 \end{bmatrix}$$

$$\mathbf{u} = n_c \quad \text{and} \quad \mathbf{B} = \gamma$$

To see what the new guidance law looks like, let us consider a numerical example. The flight-control-system transfer function used to test the new guidance law is considered to be

$$\frac{n_L}{n_c} = \left(1 - \frac{s^2}{5^2}\right) \Big/ \left[(1+s)\left(1 + \frac{2^*0.7}{20}s + \frac{s^2}{20^2}\right)\right]$$

This means that the right-half-plane zero is at the low frequency of 5 r/s. The dominant time constant is 1 s, and the damping and natural frequency of the flight control system are 0.7 and 20 r/s, respectively. This type of transfer function might be representative of a tail-controlled missile, such as the one of Fig. 21.3, flying at very high altitude and low velocity [7]. Because our model of the real world has six states, the new guidance law will have six control gains. Listing 26.4 is our optimal control program with the new matrices and changes in code (from Listings 26.1 and 26.2) highlighted in bold.

The nominal case of Listing 26.4 was run. The time constant of the cubic flight control system is identical to that of the single-lag flight control system previously considered in Listing 26.1. To compare the new guidance law to the guidance law that was optimal for the single-lag flight control system, it is easiest to compare the effective navigation ratios. Figure 26.14 shows that both effective navigation ratios

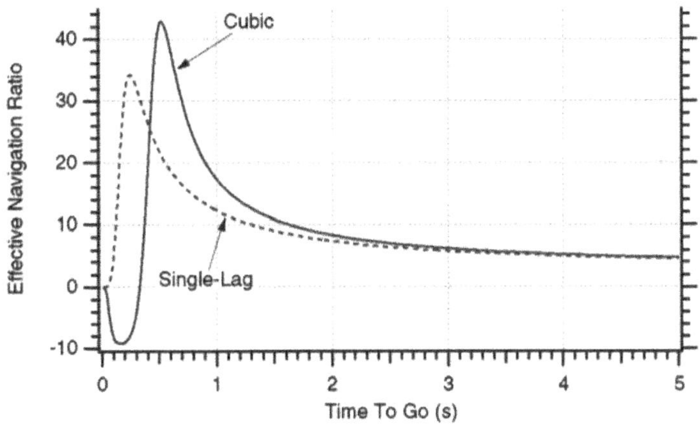

Fig. 26.14 New guidance-law effective navigation ratio goes negative near intercept.

approach 3 when the time to go before intercept is very large and that both navigation ratios get large near intercept and go to zero at intercept. However, the new effective navigation ratio also goes negative near intercept in an attempt to compensate for the right-half-plane zero of the flight-control-system transfer function.

The resultant guidance law is given by

$$n_c = C_1 y + C_2 \dot{y} + C_3 n_T + C_4 e + C_5 \dot{e} + C_6 \ddot{e}$$

where the y, $\dot{y}$, and n_T are obtained from a Kalman filter and e, $\dot{e}$, and $\ddot{e}$ can be derived by reconstructing states within the flight control system. As was already mentioned, there is no need to compute all six control gains in the missile in real time. These gains can be computed as part of the missile design and stored in the onboard computer memory as part of the overall data input. During a missile intercept, the control gains would be determined by table look-up. At each guidance update, the values of time to go, ω_z, ω, τ, and ζ, are the inputs to a table look-up algorithm. The outputs are the six control gains C_1 through C_6. The autopilot is usually designed to have a particular response as a function of Mach and altitude, so that ω_z, ω, τ, and ζ can be looked up based on Mach and altitude. Alternatively and more simply, the control gains can be stored and looked up directly as a function of Mach, altitude, and time to go.

LISTING 26.4 INTEGRATING RICATTI EQUATIONS TO GET OPTIMAL CONTROL GAINS FOR CUBIC FLIGHT CONTROL SYSTEM

```
clear
TAU=1.;
WZ=5.;
```

```
W=20.;
Z=.7;
TF=10.;
GAM=.00001;
F=zeros([6,6]);
S=zeros([6,6]);
count=0;
G(1,1)=0.;
G(2,1)=0.;
G(3,1)=0.;
G(4,1)=0.;
G(5,1)=0.;
G(6,1)=W*W/TAU;
F(1,2)=1;
F(2,3)=1;
F(2,4)=-1.;
F(2,6)=1./WZ^2;
F(4,5)=1.;
F(5,6)=1.;
F(6,4)=-W*W/TAU;
F(6,5)=-W*W*(2.*Z/W+TAU)/TAU;
F(6,6)=-W*W*(1./W^2+2.*Z*TAU/W)/TAU;
S(1,1)=1;
T=0;
H=.0001;
S1=0;
while ~(T >= (TF-.0001))
        S1=S1+H;
        SOLD=S;
        STEP=1;
        FLAG=0;
        while STEP<=1
                if FLAG==1
        STEP=2;
        HSD=H*SD;
        S=S+HSD;
                        T=T+H;
                end
                SF=S*F;
                GT=G';
                GTS=GT*S;
                GAMINV=1./GAM;
                C=GAMINV*GTS;
                SFT=SF';
                CT=C';
                CTC=CT*C;
                CTBC=GAM*CTC;
```

```
          SFSFT=SF+SFT;
          SD=SFSFT-CTBC;
          FLAG=1;
     end
     FLAG=0;
     H2=.5*H;
     HSDP=H2*SD;
     SS=SOLD+S;
     SSP=.5*SS;
     S=SSP+HSDP;
     if S1>=.009999
          S1=0;
          C1=-C(1,1);
          C2=-C(1,2);
          C3=-C(1,3);
          C4=-C(1,4);
          C5=-C(1,5);
          C6=-C(1,6);
          NP=C2*T;
          count=count+1;
          ArrayT(count)=T;
          ArrayC4(count)=C4;
          ArrayC5(count)=C5;
          ArrayC6(count)=C6;
          ArrayNP(count)=NP;
     end
end
output=[ArrayT',ArrayC4',ArrayC5',ArrayC6',ArrayNP'];
save datfil.txt output /ascii
disp 'simulation finished'
clc
figure
plot(ArrayT,ArrayNP),grid
xlabel('Time (s) ')
ylabel('NP')
axis([0 10 -10 50])
```

ALTERNATIVE APPROACH TO CUBIC FLIGHT-CONTROL-SYSTEM GUIDANCE LAW

For completeness the alternative method was used for the derivation of the guidance law for the cubic flight control system. Recall that the transfer function of the cubic flight control system was given by

$$\frac{n_L}{n_c} = \left(1 - \frac{s^2}{\omega_z^2}\right) \Big/ \left[(1 + s\tau)\left(1 + \frac{2\zeta}{\omega}s + \frac{s^2}{\omega^2}\right)\right]$$

Applying the methodology of Ref. 6 yields the block diagram of Fig. 26.15. The first three control gains can be found from the solution for the effective navigation ratio as

$$C_1 = \frac{N'}{t_{go}^2}$$

$$C_2 = \frac{N'}{t_{go}}$$

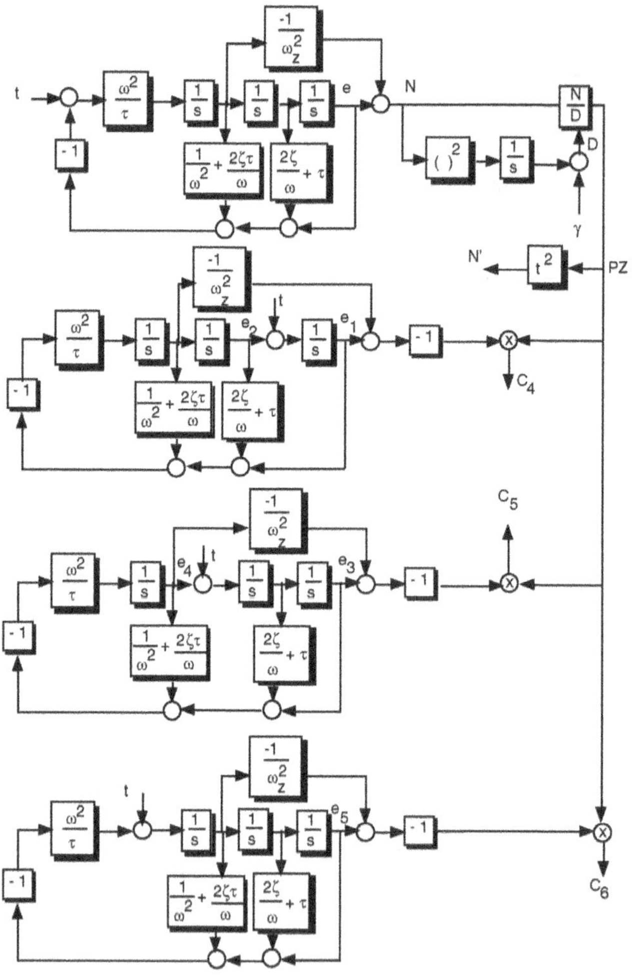

Fig. 26.15 Reference 6 method for finding control gains for cubic autopilot.

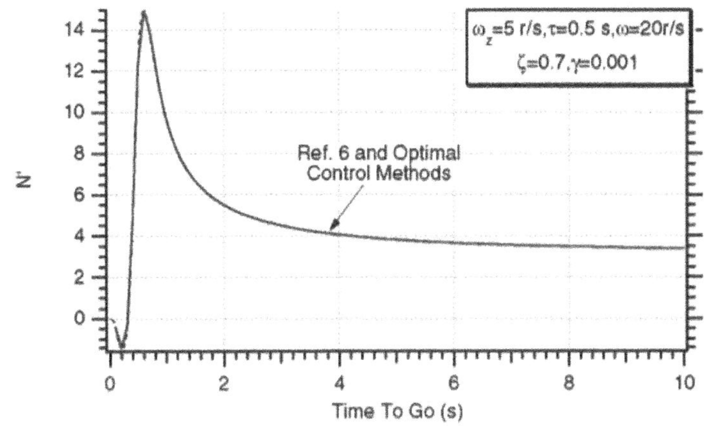

Fig. 26.16 Method of Ref. 6 is equivalent to optimal control method for cubic autopilot.

$$C_3 = 0.5N'$$

where t in the block diagram represents t_{go}.

Listing 26.5 simply programs Fig. 26.15. The gains C_1 through C_6 are generated by numerically solving the differential equations of Fig. 26.15, and the program is set up to run a case for the autopilot transfer function

$$\frac{n_L}{n_c} = \left(1 - \frac{s^2}{5^2}\right) \bigg/ \left[(1+s)\left(1 + \frac{2^*0.7}{20}s + \frac{s^2}{20^2}\right)\right]$$

Note that the integration step size H is again 100 times larger than the step size required for the optimal control technique.

The nominal case of Listing 26.5 was run, and the resultant effective navigation ratio is displayed in Fig. 26.16. This figure shows that the effective navigation ratio generated by the method of Ref. 6 is completely equivalent to that generated by the optimal control method. Thus we have further confirmation that the alternative method for generating guidance gains can either be used as a substitute for the more traditional method or can be used as an independent check in control gain selection.

LISTING 26.5 ALTERNATIVE APPROACH FOR OBTAINING CONTROL GAINS FOR CUBIC FLIGHT-CONTROL-SYSTEM GUIDANCE LAW

```
clear
count=0;
TAU=1.;
```

```
GAM=.00001;
WZ=5.;
W=20.;
Z=.7;
T=0.;
H=.01;
S=0.;
X=0.;
E=0.;
ED=0.;
EDD=0.;
E1=0.;
E2=0.;
E2D=0.;
E3=0.;
E3D=0.;
E4=0.;
E4D=0.;
E5=0.;
E5D=0.;
E5DD=0.;
while ~(T >= (10.-.0001))
        S=S+H;
        XOLD=X;
        EOLD=E;
        EDOLD=ED;
        EDDOLD=EDD;
        E1OLD=E1;
        E2OLD=E2;
        E2DOLD=E2D;
        E3OLD=E3;
        E3DOLD=E3D;
        E4OLD=E4;
        E5OLD=E5;
        E5DOLD=E5D;
        E5DDOLD=E5DD;
        STEP=1;
        FLAG=0;
        while STEP<=1
                if FLAG==1
        STEP=2;
                        X=X+H*XD;
                        E=E+H*ED;
                        ED=ED+H*EDD;
                        EDD=EDD+H*EDDD;
                        E1=E1+H*E1D;
```

```
                E2=E2+H*E2D;
                E2D=E2D+H*E2DD;
                E3=E3+H*E3D;
                E3D=E3D+H*E3DD;
                E4=E4+H*E4D;
                E5=E5+H*E5D;
                E5D=E5D+H*E5DD;
                E5DD=E5DD+H*E5DDD;
                T=T+H;
        end
        EDDD=W*W*(T-(1./W^2+2.*Z*TAU/W)*EDD-(2.*Z/W+TAU)*ED-E)/TAU;
        XN=E-EDD/WZ^2;
        XD=XN^2;
        D=X+GAM;
        PZ=XN/D;
        XNP=T*T*PZ;
        E2DD=W*W*(-(1./W^2+2.*Z*TAU/W)*E2D-(2.*Z/W+TAU)*E2-E1)/TAU;
        E1D=E2+T;
        C4=-(E1-E2D/WZ^2)*PZ;
        E4D=W*W*(-(1./W^2+2.*Z*TAU/W)*E4-(2.*Z/W+TAU)*E3D-E3)/TAU;
        E3DD=E4+T;
        C5=-(E3-E4/WZ^2)*PZ;
        E5DDD=T-W*W*((1./W^2+2.*Z*TAU/W)*E5DD+(2.*Z/W+TAU)*E5D+E5)
        /TAU;C6=PZ*(-E5+E5DD/WZ^2);
        FLAG=1;
end
FLAG=0;
X=.5*(XOLD+X+H*XD);
E=.5*(EOLD+E+H*ED);
ED=.5*(EDOLD+ED+H*EDD);
EDD=.5*(EDDOLD+EDD+H*EDDD);
E1=.5*(E1OLD+E1+H*E1D);
E2=.5*(E2OLD+E2+H*E2D);
E2D=.5*(E2DOLD+E2D+H*E2DD);
E3=.5*(E3OLD+E3+H*E3D);
E3D=.5*(E3DOLD+E3D+H*E3DD);
E4=.5*(E4OLD+E4+H*E4D);
E5=.5*(E5OLD+E5+H*E5D);
E5D=.5*(E5DOLD+E5D+H*E5DD);
E5DD=.5*(E5DDOLD+E5DD+H*E5DDD);
if S>=.09999
        S=0.;
        count=count+1;
        ArrayT(count)=T;
        ArrayC4(count)=C4;
        ArrayC5(count)=C5;
```

```
            ArrayC6(count)=C6;
            ArrayXNP(count)=XNP;
    end
end
output=[ArrayT',ArrayC4',ArrayC5',ArrayC6',ArrayXNP'];
save datfil.txt output /ascii
disp 'simulation finished'
clc
figure
plot(ArrayT,ArrayXNP),grid
xlabel('Time (s) ')
ylabel('NP')
axis([0 10 -10 50])
```

PERFORMANCE COMPARISON OF GUIDANCE LAWS IN PRESENCE OF CUBIC FLIGHT CONTROL SYSTEM

In the last two sections we have numerically derived a guidance law that is optimal when the flight-control-system transfer function is given by

$$\frac{n_L}{n_c} = \left(1 - \frac{s^2}{5^2}\right) \Big/ \left[(1+s)\left(1 + \frac{2^*0.7}{20}s + \frac{s^2}{20^2}\right)\right]$$

Figure 26.17 indicates that the step response of the preceding cubic transfer function is totally different than the step response for the single-lag representation when the dominant time constant is 1 s. The single-lag inaccuracy in the step response is mainly caused by the low-frequency numerator zero of 5 r/s.

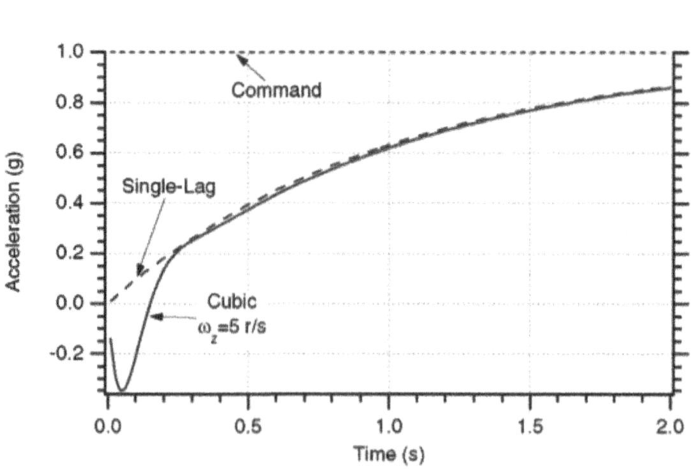

Fig. 26.17 Single-lag flight control system response does not match cubic response when numerator zero is 5 r/s.

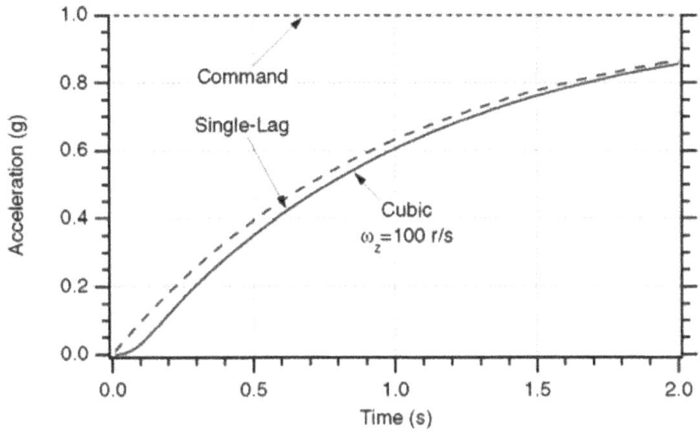

Fig. 26.18 Single-lag flight-control-system response matches cubic response when numerator zero is 100 r/s.

Figure 26.18 demonstrates than when the numerator zero is increased to 100 r/s the single-lag and cubic representations of the flight control system are virtually identical.

We are now ready to see if there is a performance benefit in using the numerical solution for the cubic guidance law. As was mentioned earlier, there is no need to compute all six control gains in the missile in real time. These gains can be computed as part of the missile design and stored in the onboard computer memory as part of the overall data input as was suggested in Ref. 7. To illustrate part of that concept, Listing 26.6 computes the six control gains for the cubic guidance law for a given set of flight-control-system parameters in routine generatepains.m using the optimal control approach. The resultant gains, which are a function of the time-to-go until intercept, are stored in memory and called when needed. For academic purposes and to get to the heart of the matter, the simulation is set up so that all states are known perfectly (that is, no filtering is required), and the only disturbance is a 0.5-g target maneuver. The program evaluates the miss distances for a variety of flight times. The real flight control system in the simulation is identical to the cubic transfer function. By setting the parameter APN, we can choose from a variety of guidance laws (0 = proportional navigation, 1 = augmented proportional navigation, 2 = optimal guidance for single-lag flight control system, and 3 = optimal guidance for cubic flight control system).

LISTING 26.6 SIMULATION FOR EVALUATING DIFFERENT GUIDANCE LAWS

```
count=0;
TAU=1.;
W=20.;
```

```
Z=.7;
WZ=5.;
APN=3;
XNT=16.1;
TS=.01;
GAM=.00001;
XLIM=9999999.;
TFMAX=10.;
XNP=3.;
VC=4000.;
VM=3000.;
if APN==3
        [C1,C2,C3,C4,C5,C6]=GENERATEGAINS(TAU,W,Z,WZ,GAM,TFMAX,TS);
end
for TF=.1:.1:10.
        E=0.;
        ED=0.;
        EDD=0.;
        T=0;
        H=.0001;
        S=0.;
        Y=0.;
        YD=0.;
        XNC=0.;
        RTM=VC*TF;
        while T< (TF-.00001)
                S=S+H;
                EOLD=E;
                EDOLD=ED;
                EDDOLD=EDD;
                YOLD=Y;
                YDOLD=YD;
                STEP=1;
                FLAG=0;
                while STEP<=1
                        if FLAG==1
                STEP=2;
                E=E+H*ED;
                                ED=ED+H*EDD;
                                EDD=EDD+H*EDDD;
                                Y=Y+H*YD;
                                YD=YD+H*YDD;
                                T=T+H;
                        end
                        TGO=TF-T+.0001;
                        RTM=VC*TGO;
```

```
            XLAM=Y/RTM;
            XNCG=XNC/32.2;
            EDDD=W*W*(XNC-E-(2.*Z/W+TAU)*ED-(2.*Z*TAU/W+1./W^2)...
                        *EDD)/TAU;
            XNL=E-EDD/WZ^2;
            YDD=XNT-XNL;
            FLAG=1;
    end
    FLAG=0;
    E=.5*(EOLD+E+H*ED);
    ED=.5*(EDOLD+ED+H*EDD);
    EDD=.5*(EDDOLD+EDD+H*EDDD);
    Y=.5*(YOLD+Y+H*YD);
    YD=.5*(YDOLD+YD+H*YDD);
    if S>=(TS-.0001)
            S=0.;
            if APN==0
                    XNC=XNP*(Y+YD*TGO)/(TGO*TGO);
                    XNPP=XNP;
            elseif APN==1
                    XNC=XNP*(Y+YD*TGO+.5*XNT*TGO*TGO)/(TGO*TGO);
                    XNPP=XNP;
            elseif APN==2
                    XS=TGO/TAU;
                    TOP=6.*XS*XS*(exp(-XS)-1.+XS);
                    BOT1=2*XS*XS*XS+3.+6.*XS-6.*XS*XS;
                    BOT2=-12.*XS*exp(-XS)-3.*exp(-2.*XS);
                    XNPP=TOP/(.0001+BOT1+BOT2);
                    C1P=XNPP/(TGO*TGO);
                    C2P=XNPP/TGO;
                    C3P=.5*XNPP;
                    C4P=-XNPP*(exp(-XS)+XS-1.)/(XS*XS);
                    XNC=C1P*Y+C2P*YD+C3P*XNT+C4P*XNL;
            else
                    JJ=fix(TGO/TS)+1;
                    XNC=C1(JJ)*Y+C2(JJ)*YD+C3(JJ)*XNT+C4(JJ)*E...
                            +C5(JJ)*ED+C6(JJ)*EDD;
                    XNPP=C2(JJ)*TGO;
            end
            if XNC>XLIM
                    XNC=XLIM;
            elseif XNC<-XLIM
                    XNC=-XLIM;
            end
            XNCG=XNC/32.2;
    end
```

```
        end
        count=count+1;
        ArrayTF(count)=TF;
        ArrayY(count)=Y;
end
figure
plot(ArrayTF,ArrayY),grid
xlabel('Flight Time (s)')
ylabel('Miss (ft) ')
clc
output=[ArrayTF',ArrayY'];
save datfil.txt output /ascii

function[C1,C2,C3,C4,C5,C6]=GENERATEGAINS(TAU,W,Z,WZ,GAM,TF,TS)
F=zeros([6,6]);
S=zeros([6,6]);
G(1,1)=0.;
G(2,1)=0.;
G(3,1)=0.;
G(4,1)=0.;
G(5,1)=0.;
G(6,1)=W*W/TAU;
F(1,2)=1;
F(2,3)=1;
F(2,4)=-1.;
F(2,6)=1./WZ^2;
F(4,5)=1.;
F(5,6)=1.;
F(6,4)=-W*W/TAU;
F(6,5)=-W*W*(2.*Z/W+TAU)/TAU;
F(6,6)=-W*W*(1./W^2+2.*Z*TAU/W)/TAU;
S(1,1)=1          ;
T=0;
H=.0001;
S1=0;
ICOUNT=1;
while T< (TF-.0001)
        S1=S1+H;
        SOLD=S;
        STEP=1;
        FLAG=0;
        while STEP<=1
                if FLAG==1
                        STEP=2;
                        HSD=H*SD;
                        S=S+HSD;
```

```
                T=T+H;
            end
            SF=S*F;
            GT=G';
            GTS=GT*S;
            GAMINV=1./GAM;
            C=GAMINV*GTS;
            SFT=SF';
            CT=C';
            CTC=CT*C;
            CTBC=GAM*CTC;
            SFSFT=SF+SFT;
            SD=SFSFT-CTBC;
            FLAG=1;
        end
        FLAG=0;
        H2=.5*H;
        HSDP=H2*SD;
        SS=SOLD+S;
        SSP=.5*SS;
        S=SSP+HSDP;
        if S1>=(TS-.0001)
            S1=0;
            C1(ICOUNT)=-C(1,1);
            C2(ICOUNT)=-C(1,2);
            C3(ICOUNT)=-C(1,3);
            C4(ICOUNT)=-C(1,4);
            C5(ICOUNT)=-C(1,5);
            C6(ICOUNT)=-C(1,6);
            ICOUNT=ICOUNT+1;
        end
end
```

A case was first run with Listing 26.6 in which the numerator zero of the flight control system was set to 100 r/s (WZ = 100). We know from Fig. 26.18 that in this case the single-lag representation of the flight control system is excellent. Therefore it comes as no surprise that the performance of the optimal guidance law, as shown in Fig. 26.19, is very good and far superior to the performance of proportional navigation.

Another case was run with Listing 26.6 in which the numerator zero of the flight control system was reduced to 10 r/s (WZ = 10). We can see from Fig. 26.20 that the performance of the single-lag optimal guidance law deteriorates significantly for flight times less than 25. On the other hand the cubic guidance law yields nearly perfect performance in that the miss distance is always zero. The decrease of the right-half-plane zero does not significantly influence the

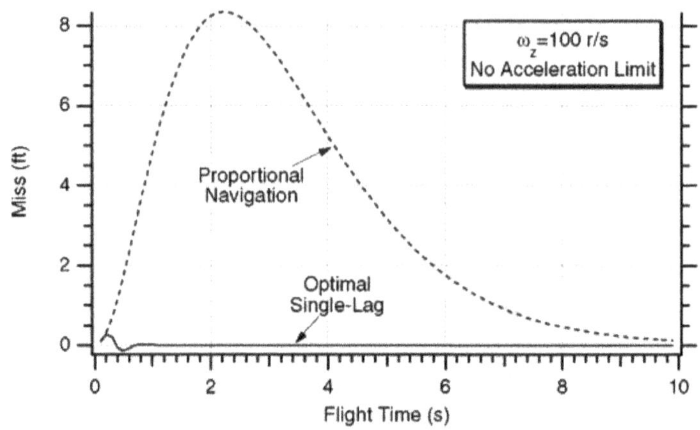

Fig. 26.19 Optimal guidance law for single-lag flight control system works well when numerator zero is 100 r/s.

proportional-navigation results. However the proportional-navigation results are generally worse than that of both other guidance laws.

Finally a case was run with Listing 26.6 in which the numerator zero of the flight control system was further reduced to 5 r/s (WZ = 5). We know from Fig. 26.17 that in this case the single-lag representation of the flight control system is terrible because it totally misses the significant wrong-way tail effect. Therefore it comes as no surprise that the performance of the optimal single-lag guidance law, as shown in Fig. 26.21, is terrible and in fact is much worse than

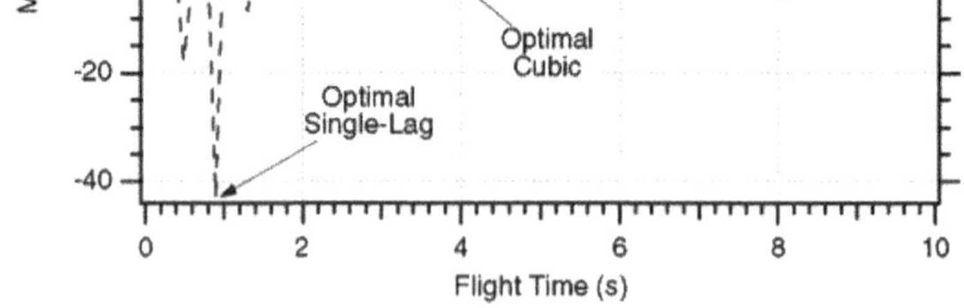

Fig. 26.20 Optimal guidance law for single-lag deteriorates when numerator zero is 10 r/s while optimal guidance law for cubic works perfectly.

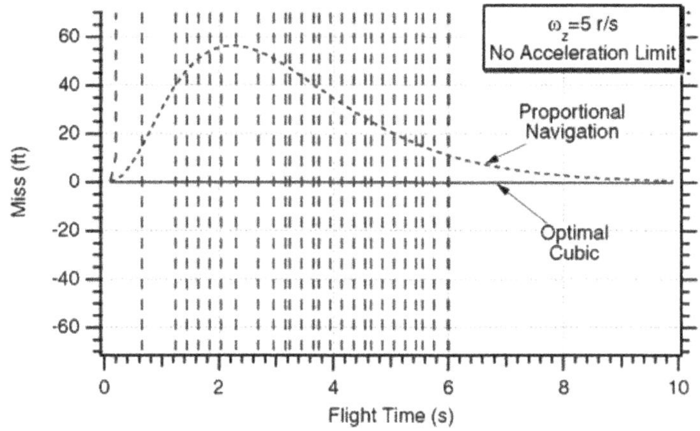

Fig. 26.21 Optimal guidance law for single lag goes unstable when numerator zero is 5 r/s while optimal guidance law for cubic works perfectly.

proportional navigation. The decrease in the right-half-plane zero to 5 r/s now significantly affects the performance of proportional navigation. Again we can see that the cubic guidance law yields nearly perfect performance.

SUMMARY

In this chapter we have presented two new techniques for deriving missile guidance laws numerically. It was shown that both techniques can accurately numerically derive guidance laws that have previously been derived analytically. In addition, a new guidance law for a complex missile flight control system was derived using the new techniques. It was shown that the new guidance law can yield significant performance benefits for a tail-controlled missile traveling at low speed and high altitude.

REFERENCES

[1] Bryson, A. E., and Ho, Y. C., *Applied Optimal Control*, Blaisdell, Waltham, MA, 1969.

[2] Gelb, A., *Applied Optimal Estimation*, MIT Press, Cambridge, MA, 1974, pp. 356–361.

[3] Cottrell, R. G., "Optimal Intercept Guidance for Short-Range Tactical Missiles," *AIAA Journal*, Vol. 9, No. 7, 1971, pp. 1414, 1415.

[4] Nesline, F. W., and Zarchan, P., "A New Look at Classical Versus Modern Homing Guidance," *Journal of Guidance and Control*, Vol. 4, No. 1, 1981, pp. 78–85.

[5] Asher, R. B., and Matuszewski, J. P., "Optimal Guidance with Maneuvering Targets," *Journal of Spacecraft and Rockets*, Vol. 11, No. 3, 1974, pp. 204–206.

[6] Rusnak, I., and Meir, L., *Journal of Guidance, Control, and Dynamics*, Vol. 14, No. 5, 1991, pp. 1056–1058.

[7] Zarchan, P., Greenberg, E., and Alpert, J., "Improving the High Altitude Performance of Tail-Controlled Endoatmospheric Missiles," AIAA Paper 2002-4770, Aug. 2002.

[8] Wells, B. H., "Tactical Missile Structural Testing and Model Verification for Autopilot Design," *Proceedings of the 1991 AIAA Guidance and Control Conference*, AIAA, Washington, DC, 1991.

CHAPTER 27

Filter Bank Approach to Weaving Target Problem

INTRODUCTION

We showed in Chapter 20 that interceptor performance against a weaving target could be improved considerably by use of a special-purpose compensated weave guidance law. Essentially the compensated weave guidance law makes an internal prediction of where the target will be at intercept and also compensates for the missile's own flight-control-system dynamics. This information enables the missile to guide directly to the predicted intercept point using a minimum of energy. If the weave frequency of the target is known in advance, it was shown in Chapter 25 that when the measurement noise was low a four-state linear weave Kalman filter could be designed to yield excellent estimates of the necessary states to implement the compensated weave guidance law. If, on the other hand, the target weave frequency was not known, we also showed in Chapter 25 that the target weave frequency could be estimated under certain circumstances, with a five-state extended Kalman filter. Because all extended Kalman filters are sensitive to large initialization errors, this type of filter will only work if it is initialized with a target frequency estimate that is close to the actual target weave frequency. In this chapter we shall show how a bank of linear four-state weave Kalman filters, each one tuned to a different target weave frequency, can be used to estimate the target weave frequency and improve system performance when the measurement noise is low. It will be demonstrated that the filter bank approach is more robust than an extended Kalman filter when there is a large uncertainty in the target weave frequency.

REVIEW OF FIVE-STATE EXTENDED-KALMAN-FILTER PERFORMANCE

The guidance system used in this chapter for the weaving target problem is identical to the one used in Chapter 25 and is repeated in Fig. 27.1 for convenience. Note that the two sources of error in the guidance system are measurement noise and a sinusoidal target maneuver. The Kalman filter, shown in the homing

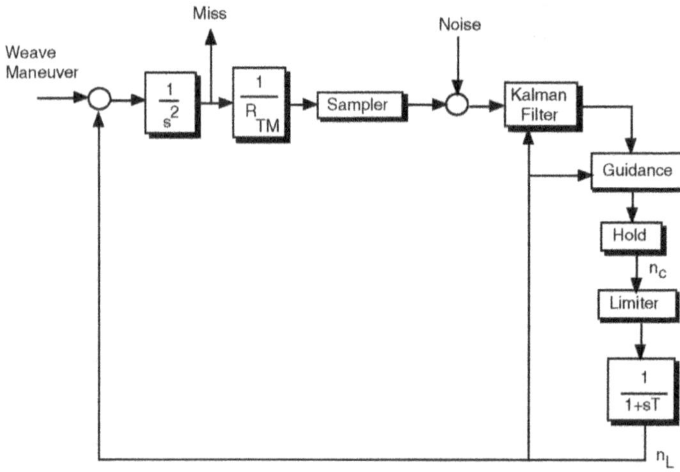

Fig. 27.1 Guidance system model for miss distance analysis.

loop of Fig. 27.1, can either be a four-state linear weave Kalman filter, a five-state extended Kalman filter (both of which were fully described in Chapter 25), or a bank of linear weave four-state Kalman filters (which will be fully described later in this chapter). The parameters to be used in the analysis of the Fig. 27.1 guidance system are the same ones used in Chapter 25 and are repeated in Table 27.1 for convenience.

It is well known that, in general, the extended Kalman filter is sensitive to initialization errors [1]. This means that if the five-state extended Kalman filter of Chapter 25 is not initialized with a target weave frequency estimate that is fairly close to the actual target weave frequency the performance of the extended Kalman filter can degrade considerably. For example, Listing 25.3 was run for the

TABLE 27.1 NOMINAL SYSTEM INPUTS FOR VARIOUS STUDIES

Parameter	Definition	Value
n_T	Target acceleration	3 g
n_{LIM}	Missile acceleration limit	10 g
σ_{Noise}	Seeker measurement noise	0.1 mr
V_c	Closing velocity	9000 ft/s
T	Flight-control-system time constant	0.5 s
ω	Target weave frequency	Varies from 1 to 10 r/s
t_F	Flight time	10 s
T_s	Sampling time	0.01 s

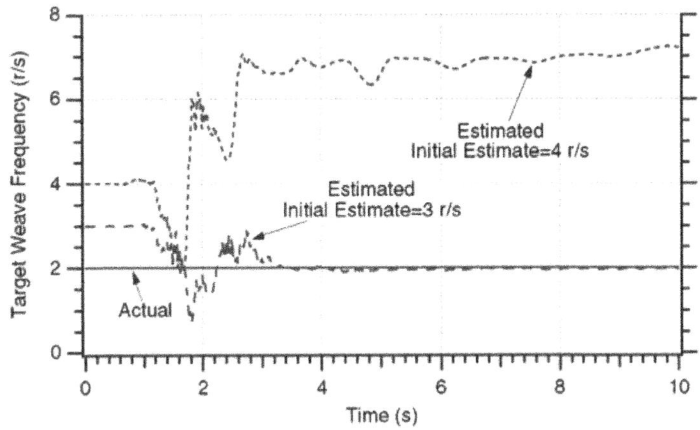

Fig. 27.2 Extended Kalman filter is unable to estimate target weave frequency if initial frequency estimate is in error by 2 r/s.

case in which the actual target weave frequency was 2 r/s and the initial estimate of the target weave frequency WHIC was varied. In Fig. 27.2 we can see that if the initial estimate of the target weave frequency is 3 r/s (in error by 1 r/s) the extended Kalman filter is able to estimate the target weave frequency after 3 s. However, if the initial estimate of the target weave frequency is 4 r/s (in error by 2 r/s), the filter's estimate of the target weave frequency diverges from the truth and is in considerable error.

Thus it appears that we need some other way of estimating the target weave frequency when we have no idea of what the target weave frequency might be.

REVIEW OF FOUR-STATE LINEAR WEAVE KALMAN-FILTER PERFORMANCE

The four-state linear weave Kalman filter was derived in Chapter 25. The states of that filter are relative position y, relative velocity $\dot{y}$, target acceleration $\ddot{y}_T$, and target jerk $\dddot{y}_T$. With this linear Kalman filter it is assumed that the weave frequency of the target ω is known and does not have to be estimated. Chapter 25 showed that the scalar equations for the four-state linear weave Kalman filter are given by

$$\mathrm{RES}_k = y_k^* - \hat{y}_{k-1} - T_s\hat{\dot{y}}_{k-1} - \left(\frac{1-\cos x}{\omega^2}\right)\hat{\ddot{y}}_{T_{k-1}} - \left(\frac{x-\sin x}{\omega^3}\right)\hat{\dddot{y}}_{T_{k-1}} + 0.5T_s^2 n_{L_{k-1}}$$

$$\hat{y}_k = \hat{y}_{k-1} + T_s\hat{\dot{y}}_{k-1} + \left(\frac{1-\cos x}{\omega^2}\right)\hat{\ddot{y}}_{T_{k-1}} + \left(\frac{x-\sin x}{\omega^3}\right)\hat{\dddot{y}}_{T_{k-1}} - 0.5T_s^2 n_{L_{k-1}} + K_{1_k}\mathrm{RES}_k$$

$$\hat{y}_k = \hat{y}_{k-1} + \left(\frac{\sin x}{\omega}\right)\hat{\dot{y}}_{T_{k-1}} + \left(\frac{1 - \cos x}{\omega^2}\right)\hat{\ddot{y}}_{T_{k-1}} - T_s n_{L_{k-1}} + K_{2_k}\text{RES}_k$$

$$\hat{\dot{y}}_{T_k} = \cos x\hat{\dot{y}}_{T_{k-1}} + \left(\frac{\sin x}{\omega}\right)\hat{\ddot{y}}_{T_{k-1}} + K_{3_k}\text{RES}_k$$

$$\hat{\ddot{y}}_{T_k} = -\omega \sin x\hat{\dot{y}}_{T_{k-1}} + \cos x\hat{\ddot{y}}_{T_{k-1}} + K_{4_k}\text{RES}_k$$

where

$$x = \omega T_s$$

T_s is the sampling time or time between measurements, and ω is the target weave frequency, which is assumed to be known. The Kalman gains (K_1, K_2, K_3, and K_4) are obtained from the matrix Ricatti equations. Details concerning the fundamental, measurement, process noise, and measurement noise matrices can also be found in Chapter 25. The filter is optimal if the real target maneuver is a sinusoid, the target weave frequency is known, and the compensated weave guidance law is used.

The compensated weave guidance law, which is optimal in the sense that it requires the least acceleration in the integral squared sense against weave maneuvers, issues guidance commands proportional to the zero effort miss and inversely proportional to the square of time to go until intercept. Chapter 20 showed that the compensated weave guidance law is given by

$$n_c = \frac{N'}{t_{go}^2}\left[y + \dot{y}t_{go} + \left(\frac{1 - \cos \omega t_{go}}{\omega^2}\right)\ddot{y}_T + \left(\frac{\omega t_{go} - \sin \omega t_{go}}{\omega^3}\right)\dddot{y}_T - n_L T^2(e^{-x} + x - 1)\right]$$

where in this case x (not to be confused with x in Kalman-filter equations) is given by

$$x = \frac{t_{go}}{T}$$

In the preceding equation t_{go} is the time to go until intercept, and T is defined as the approximate time constant of the flight control system. Again ω is the target weave frequency, which is assumed to be known. The effective navigation ratio in the compensated weave guidance law is time varying and is given by

$$N' = \frac{6x^2(e^{-x} - 1 + x)}{2x^3 + 3 + 6x - 6x^2 - 12xe^{-x} - 3e^{-2x}}$$

The compensated weave guidance law consists of five terms: the first two terms are related to the line-of-sight rate, the third term proportional to the target acceleration, a fourth term is proportional to target jerk, and a fifth term is proportional to the achieved missile acceleration. The first two terms in the compensated weave guidance law are proportional navigation. The third and fourth terms in the guidance law compensate for the sinusoidal motion of the target while the fifth term compensates for the dynamics in the missile flight control system. Details of the derivation of the compensated weave guidance law can also be found in Chapter 20.

It was also demonstrated in Chapter 25 that the four-state linear weave Kalman filter in conjunction with the compensated weave guidance law worked very well if the target weave frequency was known as shown in Fig. 27.3. Here we can see that we have virtually perfect estimates of the target acceleration after a very brief period of time.

However, if knowledge of the target weave frequency is in error the filter's estimate of the target acceleration will deteriorate. For example, consider the case where the actual target weave frequency is 2 r/s but the filter thinks the target weave frequency is 1 r/s. In this case we are underestimating the target weave frequency. Figure 27.4 shows that the filter estimate of the target acceleration does not track the actual target acceleration very well when the target weave frequency is underestimated. Figure 27.5 also shows that when the target weave frequency is overestimated at 4 r/s the Kalman filter's estimate of target acceleration also deteriorates.

Lack of knowledge of the target weave frequency will not only yield poorer state estimates but will also influence system performance as measured by the rms miss distance. A 50-run Monte Carlo miss distance experiment was repeated from Chapter 25 in which the flight times ranged from 0.5 to 10 s in steps of 0.5 s

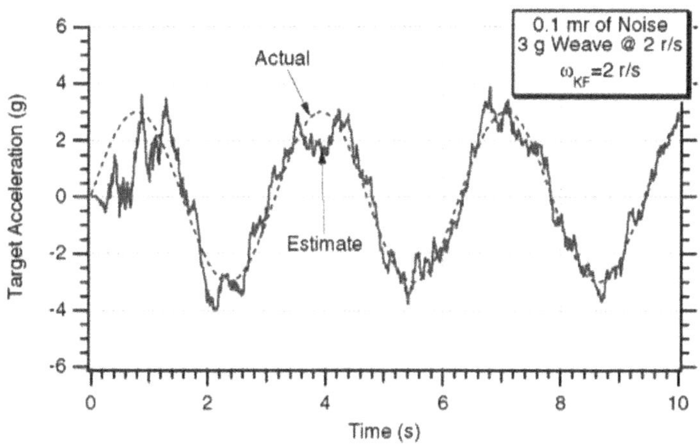

Fig. 27.3 Linear weave four-state Kalman filter works well if target weave frequency is known.

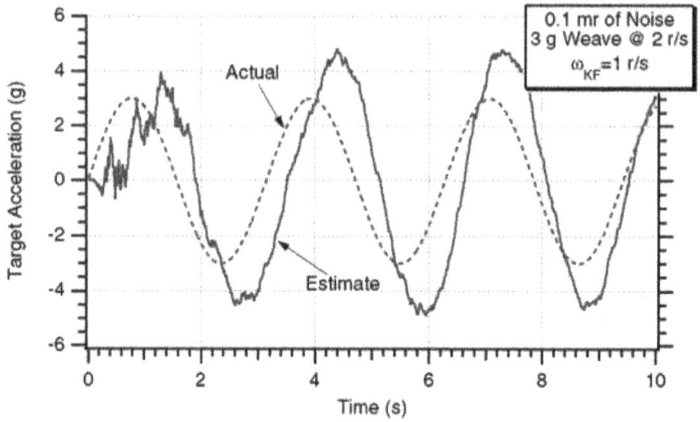

Fig. 27.4 Four-state linear weave Kalman-filter's estimate of target acceleration deteriorates when we underestimate target weave frequency.

using a modified version of Listing 25.2. Figure 27.6, which is identical to Fig. 25.17 (except scales have changed), shows that the price paid for the lack of knowledge of the target weave frequency is increased rms miss distance. When the target weave frequency is known, the maximum rms miss distance is 5 ft. However, when knowledge of the target weave frequency is in error on the low side by 1 r/s the maximum rms miss distance is 20 ft. Similarly, when knowledge of the target weave frequency is in error on the high side by 2 r/s the

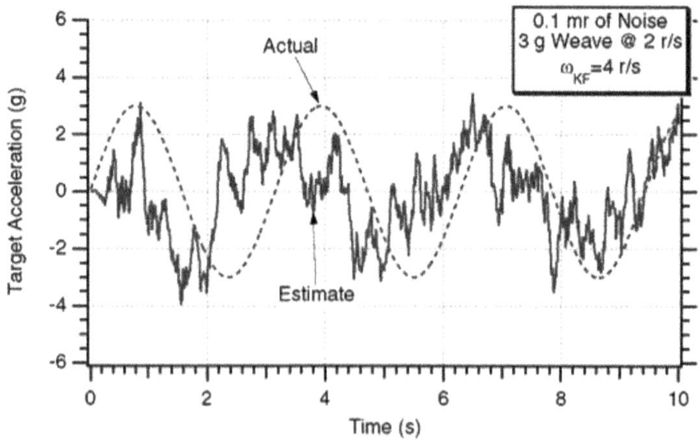

Fig. 27.5 Four-state linear weave Kalman-filter's estimate of target acceleration deteriorates when we overestimate target weave frequency.

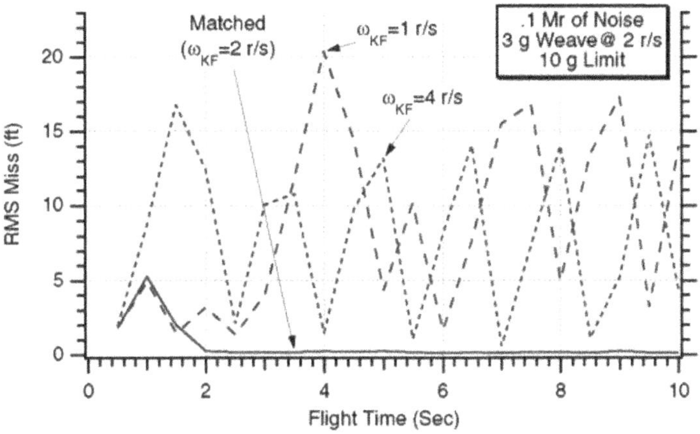

Fig. 27.6 Miss distance performance of four-state linear weave Kalman filter deteriorates when target weave frequency estimate is in error.

maximum rms miss distance is 15 ft. Clearly we need a better way of estimating the acceleration of a weaving target when the target weave frequency is unknown.

FILTER BANK METHODOLOGY

The fixed multiple model adaptive estimator or MMAE technique for determining an unknown parameter within a bank of Kalman filters was first developed by D. T. Magill in 1965 [2]. With this technique several Kalman filters are run in parallel with each one being tuned to a different parameter value. The likelihood function and Bayes' rule are used to determine the probability that a filter is the correct one. Estimates from each of the filters are weighted by the probability that it is the correct one and combined to form a resultant estimate [3–5].

The basic idea of this chapter is to use a bank of linear four-state weave Kalman filters, each one assuming a different target weave frequency, operating in parallel (that is, all filters receive the same measurements). Each filter is totally independent of every other filter in the filter bank. It is postulated that by using the fixed MMAE approach at every guidance update we can determine the probability that a given filter is the correct one (that is, tuned to the correct target weave frequency). The resultant estimates are then obtained by adding the state estimates of each filter multiplied by the probability that the filter is the one tuned to the correct frequency.

With the fixed MMAE approach each filter is totally independent of every other filter in the filter bank. The fixed MMAE approach makes use of the residual RES and covariance of the residual C of each filter in the filter bank. For the problem concerning the linear four-state weave Kalman filter, the residual and residual covariance are scalars. The ith filter residual and covariance at the kth instant are given by

$$\mathrm{RES}_k(i) = z_k - \mathbf{H}\boldsymbol{\Phi}_\mathbf{k}(i)\hat{\mathbf{x}}_{\mathbf{k-1}}(i) - \mathbf{H}\mathbf{G}_\mathbf{k}\mathbf{u}_{\mathbf{k-1}}$$

$$\sigma^2_{\mathrm{RES}_k}(i) = C_k(i) = \mathbf{H}\mathbf{M}_\mathbf{k}(i)\mathbf{H}^\mathrm{T} + \mathbf{R}_\mathbf{k}$$

It can also be shown that there is something called the likelihood function of the residual f, which is computed using the filter residual and covariance. The likelihood function is also a scalar for the four-state linear weave Kalman filter. The likelihood function for the ith filter at the kth instant is given by the scalar

$$f_k(i) = \frac{1}{\sqrt{2\pi C_k(i)}} e^{-0.5\mathrm{RES}_k^2(i)/C_k(i)}$$

The probability $p_k(i)$ that the ith filter is the correct one at the kth instant can be computed according to Bayes' rule as

$$p_k(i) = \frac{f_k(i)p_{k-1}(i)}{\sum_{i=1}^{r} f_k(i)p_{k-1}(i)}$$

$$= \frac{f_k(i)p_{k-1}(i)}{f_k(1)p_{k-1}(1) + f_k(2)p_{k-1}(2) + f_k(3)p_{k-1}(3) + \cdots + f_k(r)p_{k-1}(r)}$$

Note that the preceding equation is recursive and therefore needs initial estimates for the probability that each filter is the correct one. If we assume that it is equally likely that any one of r filters in the filter bank is the correct one, then we can say that [6]

$$p_0(i) = \frac{1}{r}$$

Thus at each instant of time we have a set of r probabilities (numbers ranging from zero to unity) telling us the likelihood that any filter is the correct one. The r probabilities add up to unity. The estimated target weave frequency to be used by the compensated weave guidance law can be found by weighting the frequency assigned to each filter by the probability that the filter is the correct one or

$$\hat{\omega}_\mathbf{k} = p_k(1)\omega(1) + p_k(2)\omega(2) + p_k(3)\omega(3) + \cdots + p_k(r)\omega(r)$$

where $\omega(1)\ldots\omega(r)$ are the tuned frequencies assumed by each of the r filters. The resultant state estimates to also be used by the compensated weave guidance law are obtained by using the state estimates of each individual filter weighted by the probability that the filter is correct or

$$\hat{\mathbf{x}}_\mathbf{k} = p_k(1)\hat{\mathbf{x}}_\mathbf{k}(1) + p_k(2)\hat{\mathbf{x}}_\mathbf{k}(2) + p_k(3)\hat{\mathbf{x}}_\mathbf{k}(3) + \cdots + p_k(r)\hat{\mathbf{x}}_\mathbf{k}(r)$$

where $\hat{\mathbf{x}}_\mathbf{k}(1)\ldots\hat{\mathbf{x}}_\mathbf{k}(r)$ are the state estimates of each of the r filters.

THREE FILTER BANK EXAMPLE

To better understand the mechanics of the fixed MMAE approach, it is best to first consider a numerical example in which there are only three filters in the filter bank. In this example it is assumed that one of the three filters is actually the correct one (meaning tuned to the correct target weave frequency). If three four-state linear weave Kalman filters are run in parallel, each one tuned to a different target weave frequency, the equations of the preceding section simplify. The likelihood function for each of the three filters at the kth instant is given by

$$f_k(1) = \frac{1}{\sqrt{2\pi C_k(1)}} e^{-0.5\mathrm{RES}_k^2(1)/C_k(1)}$$

$$f_k(2) = \frac{1}{\sqrt{2\pi C_k(2)}} e^{-0.5\mathrm{RES}_k^2(2)/C_k(2)}$$

$$f_k(3) = \frac{1}{\sqrt{2\pi C_k(3)}} e^{-0.5\mathrm{RES}_k^2(3)/C_k(3)}$$

where each of the covariances $C_k(i)$ can be found from each filters set of Ricatti equations according to

$$C_k(1) = \mathbf{HM_k}(1)\mathbf{H}^\mathrm{T} + \mathbf{R_k}$$

$$C_k(2) = \mathbf{HM_k}(2)\mathbf{H}^\mathrm{T} + \mathbf{R_k}$$

$$C_k(3) = \mathbf{HM_k}(3)\mathbf{H}^\mathrm{T} + \mathbf{R_k}$$

and the residuals can be found from the error signal within the filter as

$$\mathrm{RES}_k(1) = z_k - \mathbf{H\Phi_k}(1)\hat{\mathbf{x}}_{\mathbf{k-1}}(1) - \mathbf{HG_k u_{k-1}}$$
$$\mathrm{RES}_k(2) = z_k - \mathbf{H\Phi_k}(2)\hat{\mathbf{x}}_{\mathbf{k-1}}(2) - \mathbf{HG_k u_{k-1}}$$
$$\mathrm{RES}_k(3) = z_k - \mathbf{H\Phi_k}(3)\hat{\mathbf{x}}_{\mathbf{k-1}}(3) - \mathbf{HG_k u_{k-1}}$$

The probability that any given filter of the three filters is the correct one can be found from Bayes' rule as

$$p_k(1) = \frac{f_k(1)p_{k-1}(1)}{f_k(1)p_{k-1}(1) + f_k(2)p_{k-1}(2) + f_k(3)p_{k-1}(3)}$$

$$p_k(2) = \frac{f_k(2)p_{k-1}(2)}{f_k(1)p_{k-1}(1) + f_k(2)p_{k-1}(2) + f_k(3)p_{k-1}(3)}$$

$$p_k(3) = \frac{f_k(3)p_{k-1}(3)}{f_k(1)p_{k-1}(1) + f_k(2)p_{k-1}(2) + f_k(3)p_{k-1}(3)}$$

We can see from the preceding set of equations that at any instant of time

$$p_k(1) + p_k(2) + p_k(3) = 1$$

Again, $p_k(1)$ represents the probability that the first filter is correct at the kth instant. To start the preceding recursion for calculating the probabilities that a filter is correct, we assume initially each filter is equally likely to be the correct filter or

$$p_0(1) = \frac{1}{3}$$

$$p_0(2) = \frac{1}{3}$$

$$p_0(3) = \frac{1}{3}$$

At each instant of time, we calculate the probability that each filter is correct and then weigh the assumed frequency and estimates of each of the three filters according to their probabilities as

$$\hat{\omega}_k = p_k(1)\omega(1) + p_k(2)\omega(2) + p_k(3)\omega(3)$$
$$\hat{\mathbf{x}}_\mathbf{k} = p_k(1)\hat{\mathbf{x}}_\mathbf{k}(1) + p_k(2)\hat{\mathbf{x}}_\mathbf{k}(2) + p_k(3)\hat{\mathbf{x}}_\mathbf{k}(3)$$

Listing 27.1 implements the fixed MMAE approach for three four-state linear weave Kalman filters along with the inputs of Table 27.1. This listing is based upon the single four-state linear weave Kalman filter approach of Listing 25.2. However in Listing 27.1 we only consider the compensated weave guidance law. In addition, in Listing 27.1 the actual target weave frequency is 2 r/s. One filter in the filter bank is tuned to 1 r/s, the other is tuned to 2 r/s, and the third filter is tuned to 4 r/s. Therefore the correct filter in this example is the one tuned to 2 r/s. The additional code in going from the single filter approach of Listing 25.2 to the fixed MMAE approach of Listing 27.1 is highlighted in bold.

LISTING 27.1 THREE-FILTER FIXED MMAE APPROACH TO WEAVING TARGET PROBLEM

```
clear
count=0;
TAU=.5;
ORDER=4;
VC=9000.;
XLIM=322.;
W1=1.;
W2=2.;
W3=4.;
```

WREAL=2.;
XNT=96.6;
XNTREAL=96.6;
TS=.01;
YIC=0.;
VM=3000.;
HEDEG=0.;
HEDEGFIL=20.;
XNP=3.;
SIGRIN=.0001;
TF=10.;;
PHASE=0./57.3;
X1=W1*TS;
X2=W2*TS;
X3=W3*TS;
Y=YIC;
YD=-VM*HEDEG/57.3;
PHIS1=W1*W1*XNT*XNT/TF;
PHIS2=W2*W2*XNT*XNT/TF;
PHIS3=W3*W3*XNT*XNT/TF;
RTM=VC*TF;
SIGNOISE=SIGRIN;
SIGPOS=RTM*SIGNOISE;
SIGN2=SIGPOS^2;
PHI1=zeros([4,4]);
P1=zeros([4,4]);
Q1=zeros([4,4]);
IDNP=eye(4);
PHI2=zeros([4,4]);
P2=zeros([4,4]);
Q2=zeros([4,4]);
PHI3=zeros([4,4]);
P3=zeros([4,4]);
Q3=zeros([4,4]);

PHI1(1,1)=1;
PHI1(1,2)=TS;
PHI1(1,3)=(1-cos(X1))/(W1*W1);
PHI1(1,4)=(X1-sin(X1))/(W1*W1*W1);
PHI1(2,2)=1;
PHI1(2,3)=sin(X1)/W1;
PHI1(2,4)=(1-cos(X1))/(W1*W1);
PHI1(3,3)=cos(X1);
PHI1(3,4)=sin(X1)/W1;
PHI1(4,3)=-W1*sin(X1);
PHI1(4,4)=cos(X1);

```
PHI2(1,1)=1;
PHI2(1,2)=TS;
PHI2(1,3)=(1-cos(X2))/(W2*W2);
PHI2(1,4)=(X2-sin(X2))/(W2*W2*W2);
PHI2(2,2)=1;
PHI2(2,3)=sin(X2)/W2;
PHI2(2,4)=(1-cos(X2))/(W2*W2);
PHI2(3,3)=cos(X2);
PHI2(3,4)=sin(X2)/W2;
PHI2(4,3)=-W2*sin(X2);
PHI2(4,4)=cos(X2);

PHI3(1,1)=1;
PHI3(1,2)=TS;
PHI3(1,3)=(1-cos(X3))/(W3*W3);
PHI3(1,4)=(X3-sin(X3))/(W3*W3*W3);
PHI3(2,2)=1;
PHI3(2,3)=sin(X3)/W3;
PHI3(2,4)=(1-cos(X3))/(W3*W3);
PHI3(3,3)=cos(X3);
PHI3(3,4)=sin(X3)/W3;
PHI3(4,3)=-W3*sin(X3);
PHI3(4,4)=cos(X3);
Q1(1,1)=PHIS1*(.333*X1^3-2*sin(X1)+2*X1*cos(X1)+.5*X1-...
                  .25*sin(2*X1))/(W1^5);
Q1(1,2)=PHIS1*(.5*X1*X1-X1*sin(X1)+.5*sin(X1)*...
                  sin(X1))/(W1^4);
Q1(2,1)=Q1(1,2);
Q1(1,3)=PHIS1*(sin(X1)-X1*cos(X1)-.5*X1+...
                  .25*sin(2*X1))/(W1^3);
Q1(3,1)=Q1(1,3);
Q1(1,4)=PHIS1*(cos(X1)+X1*sin(X1)-.5*sin(X1)*...
                  sin(X1)-1)/(W1*W1);
Q1(4,1)=Q1(1,4);
Q1(2,2)=PHIS1*(1.5*X1-2*sin(X1)+.25*sin(2*X1))/(W1^3);
Q1(2,3)=PHIS1*(1-cos(X1)-.5*sin(X1)*sin(X1))/(W1*W1);
Q1(3,2)=Q1(2,3);
Q1(2,4)=PHIS1*(sin(X1)-.5*X1-.25*sin(2*X1))/W1;
Q1(4,2)=Q1(2,4);
Q1(3,3)=PHIS1*(.5*X1-.25*sin(2*X1))/W1;
Q1(3,4)=.5*PHIS1*sin(X1)*sin(X1);
Q1(4,3)=Q1(3,4);
Q1(4,4)=W1*PHIS1*(.5*X1+.25*sin(2*X1));

Q2(1,1)=PHIS2*(.333*X2^3-2*sin(X2)+2*X2*cos(X2)...
                   +.5*X2-.25*sin(2*X2))/(W2^5);
```

```
Q2(1,2)=PHIS2*(.5*X2*X2-X2*sin(X2)+.5*sin(X2)...
                        *sin(X2))/(W2^4);
Q2(2,1)=Q2(1,2);
Q2(1,3)=PHIS2*(sin(X2)-X2*cos(X2)-.5*X2+.25*...
                        sin(2*X2))/(W2^3);
Q2(3,1)=Q2(1,3);
Q2(1,4)=PHIS2*(cos(X2)+X2*sin(X2)-.5*sin(X2)*...
                        sin(X2)-1)/(W2*W2);
Q2(4,1)=Q2(1,4);
Q2(2,2)=PHIS2*(1.5*X2-2*sin(X2)+.25*sin(2*X2))/(W2^3);
Q2(2,3)=PHIS2*(1-cos(X2)-.5*sin(X2)*sin(X2))/(W2*W2);
Q2(3,2)=Q2(2,3);
Q2(2,4)=PHIS2*(sin(X2)-.5*X2-.25*sin(2*X2))/W2;
Q2(4,2)=Q2(2,4);
Q2(3,3)=PHIS2*(.5*X2-.25*sin(2*X2))/W2;
Q2(3,4)=.5*PHIS2*sin(X2)*sin(X2);
Q2(4,3)=Q2(3,4);
Q2(4,4)=W2*PHIS2*(.5*X2+.25*sin(2*X2));

Q3(1,1)=PHIS3*(.333*X3^3-2*sin(X3)+2*X3*cos(X3)+...
                        .5*X3-.25*sin(2*X3))/(W3^5);
Q3(1,2)=PHIS3*(.5*X3*X3-X3*sin(X3)+.5*sin(X3)*...
                        sin(X3))/(W3^4);
Q3(2,1)=Q3(1,2);
Q3(1,3)=PHIS3*(sin(X3)-X3*cos(X3)-.5*X3+.25*...
                        sin(2*X3))/(W3^3);
Q3(3,1)=Q3(1,3);
Q3(1,4)=PHIS3*(cos(X3)+X3*sin(X3)-.5*sin(X3)*...
                        sin(X3)-1)/(W3*W3);
Q3(4,1)=Q3(1,4);
Q3(2,2)=PHIS3*(1.5*X3-2*sin(X3)+.25*sin(2*X3))/(W3^3);
Q3(2,3)=PHIS3*(1-cos(X3)-.5*sin(X3)*sin(X3))/(W3*W3);
Q3(3,2)=Q3(2,3);
Q3(2,4)=PHIS3*(sin(X3)-.5*X3-.25*sin(2*X3))/W3;
Q3(4,2)=Q3(2,4);
Q3(3,3)=PHIS3*(.5*X3-.25*sin(2*X3))/W3;
Q3(3,4)=.5*PHIS3*sin(X3)*sin(X3);
Q3(4,3)=Q3(3,4);
Q3(4,4)=W3*PHIS3*(.5*X3+.25*sin(2*X3));

P1(1,1)=SIGN2;
P1(2,2)=(VM*HEDEGFIL/57.3)^2;
P1(3,3)=XNT*XNT;
P1(4,4)=W1*W1*XNT*XNT;

P2(1,1)=SIGN2;
```

```
P2(2,2)=(VM*HEDEGFIL/57.3)^2;
P2(3,3)=XNT*XNT;
P2(4,4)=W2*W2*XNT*XNT;
P3(1,1)=SIGN2;
P3(2,2)=(VM*HEDEGFIL/57.3)^2;
P3(3,3)=XNT*XNT;
P3(4,4)=W3*W3*XNT*XNT;
HMAT=[1 0 0 0];
HT=HMAT';
PHIT1=PHI1';
PHIT2=PHI2';
PHIT3=PHI3';
T=0.;
H=.001;
S=0.;
XNC=0.;
XNL=0.;
XLAM=Y/RTM;
YTDD=XNTREAL*sin(WREAL*T);
YTDDD=XNTREAL*WREAL*cos(WREAL*T);

YH1=0.;
YDH1=0.;
YTDDH1=0.;
YTDDDH1=0.;

YH2=0.;
YDH2=0.;
YTDDH2=0.;
YTDDDH2=0.;

YH3=0.;
YDH3=0.;
YTDDH3=0.;
YTDDDH3=0.;

PROB1=.333;
PROB2=.333;
PROB3=.333;
while T<=(TF-.0001)
	S=S+H;
	YOLD=Y;
	YDOLD=YD;
	XNLOLD=XNL;
	STEP=1;
	FLAG=0;
```

```
while STEP <=1
        if FLAG==1
                STEP=2;
                Y=Y+H*YD;
                YD=YD+H*YDD;
                XNL=XNL+H*XNLD;
                T=T+H;
        end
        TGO=TF-T+.000001;
        RTM=VC*TGO;
        XLAM=Y/(VC*TGO);
        YTDD=XNTREAL*sin(WREAL*T);
        XNLD=(XNC-XNL)/TAU;
        YDD=YTDD-XNL;
        FLAG=1;
end
FLAG=0;
Y=.5*(YOLD+Y+H*YD);
YD=.5*(YDOLD+YD+H*YDD);
XNL=.5*(XNLOLD+XNL+H*XNLD);
if S>=(TS-.00001)
        S=0.;
        TGO=TF-T+.000001;
        RTM=VC*TGO;
        SIGPOS=RTM*SIGNOISE;
        SIGN2=SIGPOS^2;
        PHIP1=PHI1*P1;
        PHIPPHIT1=PHIP1*PHIT1;
        M1=PHIPPHIT1+Q1;
        HM1=HMAT*M1;
        HMHT1=HM1*HT;
        HMHTR1=HMHT1(1,1)+SIGN2;
        HMHTRINV1(1,1)=1./HMHTR1;
        MHT1=M1*HT;
        GAIN1=MHT1*HMHTRINV1;
        KH1=GAIN1*HMAT;
        IKH1=IDNP-KH1;
        P1=IKH1*M1;

        PHIP2=PHI2*P2;
        PHIPPHIT2=PHIP2*PHIT2;
        M2=PHIPPHIT2+Q2;
        HM2=HMAT*M2;
        HMHT2=HM2*HT;
        HMHTR2=HMHT2(1,1)+SIGN2;
        HMHTRINV2(1,1)=1./HMHTR2;
```

```
MHT2=M2*HT;
GAIN2=MHT2*HMHTRINV2;
KH2=GAIN2*HMAT;
IKH2=IDNP-KH2;
P2=IKH2*M2;

PHIP3=PHI3*P3;
PHIPPHIT3=PHIP3*PHIT3;
M3=PHIPPHIT3+Q3;
HM3=HMAT*M3;
HMHT3=HM3*HT;
HMHTR3=HMHT3(1,1)+SIGN2;
HMHTRINV3(1,1)=1./HMHTR3;
MHT3=M3*HT;
GAIN3=MHT3*HMHTRINV3;
KH3=GAIN3*HMAT;
IKH3=IDNP-KH3;
P3=IKH3*M3;
CPZ1=HMHTR1;
CPZ2=HMHTR2;
CPZ3=HMHTR3;

YTDD=XNTREAL*sin(WREAL*T);
YTDDD=XNTREAL*WREAL*cos(WREAL*T);
XLAMNOISE=SIGNOISE*randn;
YSTAR=RTM*(XLAM+XLAMNOISE);

RES1=YSTAR-YH1-TS*YDH1-(1-cos(X1))*YTDDH1/...
        (W1*W1)-(X1-sin(X1))*YTDDDH1/(W1*W1*W1)...
        +.5*TS*TS*XNL;
YH1=YH1+TS*YDH1+(1-cos(X1))*YTDDH1/(W1*W1)+...
        (X1-sin(X1))*YTDDDH1/(W1*W1*W1)+GAIN1(1,1)*RES1-...
        .5*TS*TS*XNL;
YDH1=YDH1+sin(X1)*YTDDH1/W1+(1-cos(X1))*YTDDDH1/(W1*W1)...
        +GAIN1(2,1)*RES1-TS*XNL;
YTDDHNEW1=cos(X1)*YTDDH1+sin(X1)*YTDDDH1/W1+...
        GAIN1(3,1)*RES1;
YTDDDH1=-W1*sin(X1)*YTDDH1+cos(X1)*YTDDDH1+GAIN1(4,1)*RES1;
YTDDH1=YTDDHNEW1;

RES2=YSTAR-YH2-TS*YDH2-(1-cos(X2))*YTDDH2/(W2*W2)-...
        (X2-sin(X2))*YTDDDH2/(W2*W2*W2)+.5*TS*TS*XNL;
YH2=YH2+TS*YDH2+(1-cos(X2))*YTDDH2/(W2*W2)+...
        (X2-sin(X2))*YTDDDH2/(W2*W2*W2)+...
        GAIN2(1,1)*RES2-.5*TS*TS*XNL;
YDH2=YDH2+sin(X2)*YTDDH2/W2+(1-cos(X2))*...
        YTDDDH2/(W2*W2)+GAIN2(2,1)*RES2-TS*XNL;
```

```
YTDDHNEW2=cos(X2)*YTDDH2+sin(X2)*YTDDDH2/W2+...
        GAIN2(3,1)*RES2;
YTDDDH2=-W2*sin(X2)*YTDDH2+cos(X2)*YTDDDH2...
        +GAIN2(4,1)*RES2;
YTDDH2=YTDDHNEW2;

RES3=YSTAR-YH3-TS*YDH3-(1-cos(X3))*YTDDH3/...
        (W3*W3)-(X3-sin(X3))*YTDDDH3/(W3*W3*W3)...
        +.5*TS*TS*XNL;
YH3=YH3+TS*YDH3+(1-cos(X3))*YTDDH3/(W3*W3)+...
        (X3-sin(X3))*YTDDDH3/(W3*W3*W3)+...
        GAIN3(1,1)*RES3-.5*TS*TS*XNL;
YDH3=YDH3+sin(X3)*YTDDH3/W3+(1-cos(X3))*...
        YTDDDH3/(W3*W3)+GAIN3(2,1)*RES3-TS*XNL;
YTDDHNEW3=cos(X3)*YTDDH3+sin(X3)*YTDDDH3/W3+...
        GAIN3(3,1)*RES3;
YTDDDH3=-W3*sin(X3)*YTDDH3+cos(X3)*YTDDDH3...
        +GAIN3(4,1)*RES3;
YTDDH3=YTDDHNEW3;

F1=exp(-.5*RES1*RES1/CPZ1)/sqrt(6.28*CPZ1);
F2=exp(-.5*RES2*RES2/CPZ2)/sqrt(6.28*CPZ2);
F3=exp(-.5*RES3*RES3/CPZ3)/sqrt(6.28*CPZ3);

PROB1=PROB1*F1/(PROB1*F1+PROB2*F2+PROB3*F3);
PROB2=PROB2*F2/(PROB1*F1+PROB2*F2+PROB3*F3);
PROB3=PROB3*F3/(PROB1*F1+PROB2*F2+PROB3*F3);

WHPZ=W1*PROB1+W2*PROB2+W3*PROB3;
YHPZ=YH1*PROB1+YH2*PROB2+YH3*PROB3;
YDHPZ=YDH1*PROB1+YDH2*PROB2+YDH3*PROB3;
YTDDHPZ=YTDDH1*PROB1+YTDDH2*PROB2+YTDDH3*PROB3;
YTDDDHPZ=YTDDDH1*PROB1+YTDDDH2*PROB2+...
        YTDDDH3*PROB3;

XS=TGO/TAU;
TOP=6.*XS*XS*(exp(-XS)-1.+XS);
BOT1=2*XS*XS*XS+3.+6.*XS-6.*XS*XS;
BOT2=-12.*XS*exp(-XS)-3.*exp(-2.*XS);
XNPP=TOP/(.0001+BOT1+BOT2);
C1=XNPP/(TGO*TGO);
C2=XNPP/TGO;
C3=XNPP*(1.-cos(WHPZ*TGO))/(WHPZ*WHPZ*TGO*TGO);
C4=-XNPP*(exp(-XS)+XS-1.)/(XS*XS);
C5=XNPP*(WHPZ*TGO-sin(WHPZ*TGO))/(WHPZ*...
                WHPZ*WHPZ*TGO*TGO);
```

```
            XNC=C1*YHPZ+C2*YDHPZ+C3*YTDDHPZ+C4*XNL+...
                    C5*YTDDDHPZ;
            if XNC>XLIM
                XNC=XLIM;
            end
            if XNC<-XLIM
                XNC=-XLIM;
            end
            YTDDG=YTDD/32.2;
            YTDDHPZG=YTDDHPZ/32.2;
            count=count+1;
            ArrayT(count)=T;
            ArrayPROB1(count)=PROB1;
            ArrayPROB2(count)=PROB2;
            ArrayPROB3(count)=PROB3;
            ArrayWREAL(count)=WREAL;
            ArrayWHPZ(count)=WHPZ;
            ArrayYTDDG(count)=YTDDG;
            ArrayYTDDHPZG(count)=YTDDHPZG;
    end
end
output=[ArrayT',ArrayPROB1',ArrayPROB2',ArrayPROB3',...
            ArrayWREAL',ArrayWHPZ',ArrayYTDDG',...
            ArrayYTDDHPZG'];
save datfil.txt output /ascii
disp 'simulation finished'
clc
figure
plot(ArrayT,ArrayPROB1,ArrayT,ArrayPROB2,ArrayT,...
            ArrayPROB3),grid
xlabel('Time (s) ')
ylabel('Probability')
axis([0 10 0 1.2])
figure
plot(ArrayT,ArrayWREAL,ArrayT,ArrayWHPZ),grid
xlabel('Time (s) ')
ylabel('Frequency (r/s)')
axis([0 10 0 4])
figure
plot(ArrayT,ArrayYTDDG,ArrayT,ArrayYTDDHPZG),grid
xlabel('Time (s) ')
ylabel('Acceleration (g)')
axis([0 10 -6 6])
```

The nominal case of Listing 27.1 was run, and we can see from Fig. 27.7 that after 2 s the algorithm figures out that the correct filter is most likely the filter

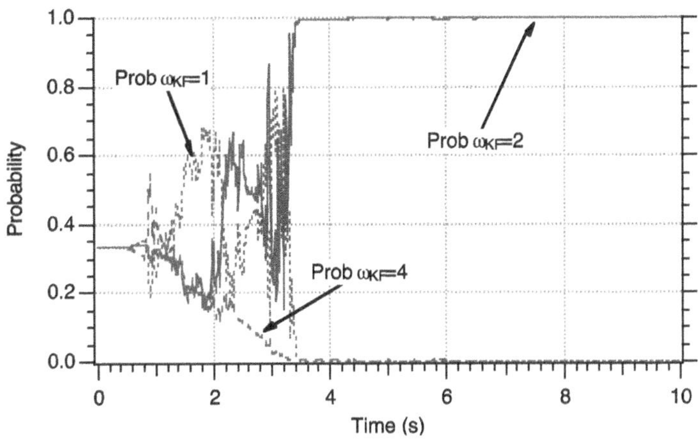

Fig. 27.7 Fixed MMAE approach identifies correct filter after 3 s.

tuned to 2 r/s. After 3.5 s the algorithm is almost certain that the 2 r/s filter is correct. Figure 27.8 shows that although it takes about 3.5 s to be absolutely certain which filter is correct, the interim estimates of the target weave frequency are quite close to the actual target weave frequency for the entire flight. Figure 27.9 shows that combining the state estimates of the three filters in a weighted manner yields excellent estimates of the target acceleration after only 2 s.

Listing 27.1 was modified to operate in the Monte Carlo mode. In addition, the flight time was varied from 0.5 to 10 s in steps of 0.5 s. For each flight time 50 runs were made to calculate the rms miss distance. We can see from Fig. 27.10 that the

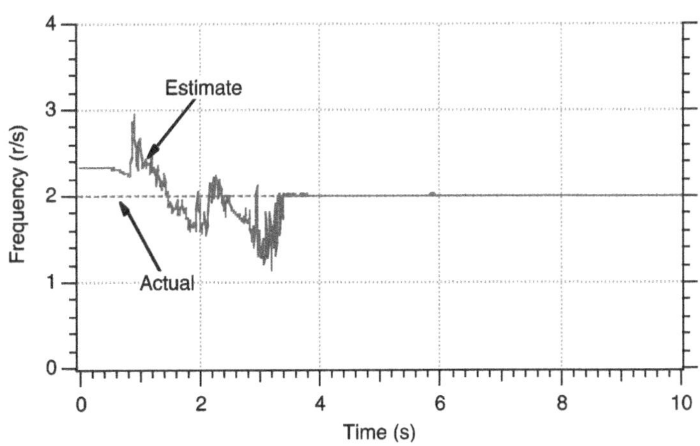

Fig. 27.8 Estimated target weave frequency is always quite close to actual target weave frequency.

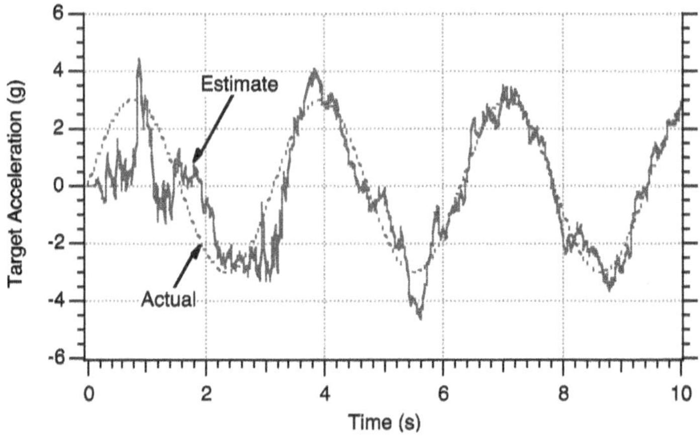

Fig. 27.9 Target acceleration estimate excellent after 3 s.

three-filter bank, which uses filter frequencies of 1, 2, and 4 r/s, yields rms miss distances that are usually less than 5 ft. Superimposed on Fig. 27.10 are the rms miss distance results for the mismatched four-state linear weave Kalman filters. The rms miss distances for the mismatched filters are usually several times larger than the filter bank results. Clearly the fixed multiple model or MMAE approach can yield significantly improved performance.

Miss-distance results can be further improved if there is less uncertainty in the target weave frequency (that is, more filters in the filter bank). Figure 27.11 shows

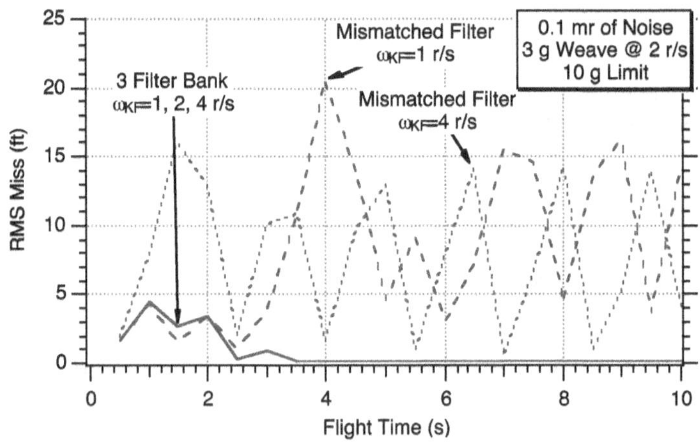

Fig. 27.10 Fixed MMAE approach yields improved performance when target weave frequency is unknown.

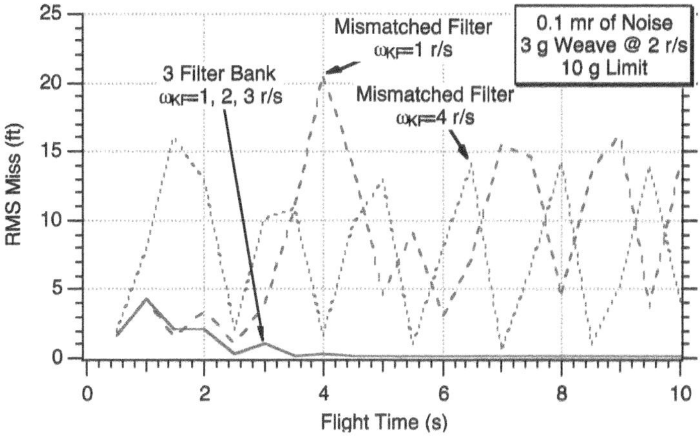

Fig. 27.11 Fixed MMAE performance is improved slightly when there is less uncertainty in actual target weave frequency.

that if the filter bank frequencies are 1, 2, and 3 r/s, respectively, the rms miss-distance performance improves slightly because at most we have a 1-r/s error in knowledge of the target weave frequency rather than 2 r/s as was the case in Fig. 27.10.

Figure 27.12 compares the optimal performance of the four-state linear weave Kalman filter (that is, when filter knows actual target weave frequency is 2 r/s), and the fixed multiple model results. Although there is room for improvement, we can see that the three-filter bank results are approaching the near-optimal

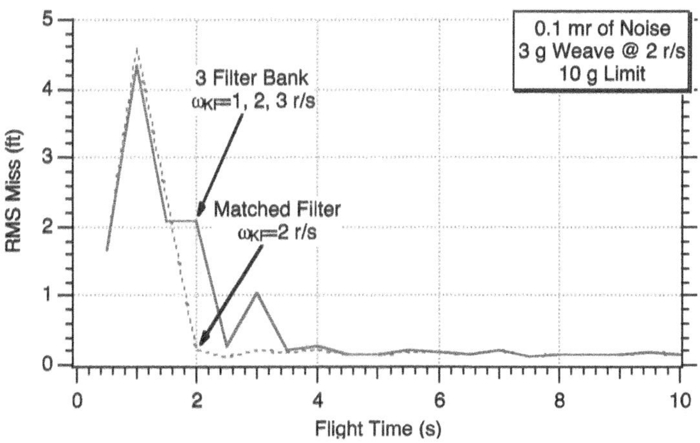

Fig. 27.12 Three-filter bank fixed MMAE approach results yield near-optimal performance.

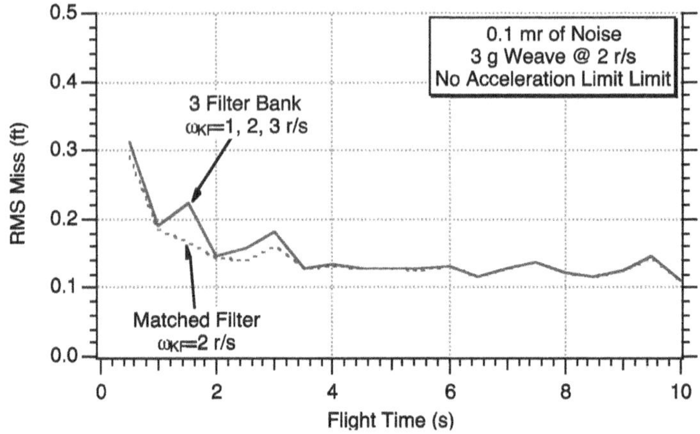

Fig. 27.13 Comparison of optimal weave Kalman filter and three filter-fixed MMAE approach when acceleration limit is removed.

results when at most there is a 1-r/s error in knowledge of the target weave frequency.

In the preceding cases it was assumed that the missile acceleration limit was 10 *g*. Another case was run in which the missile acceleration limit was removed. The performance of the single tuned Kalman filter should improve for flight times of less than 2 s. Figure 27.13 presents the comparison between the optimal performance of the single tuned Kalman filter and the filter bank approach. We can see that rms miss-distance results of the filter bank get even closer to the optimal single tuned filter results when the missile acceleration

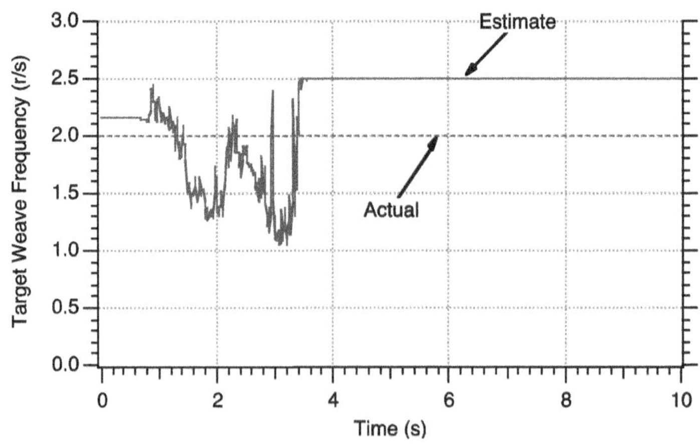

Fig. 27.14 Target weave frequency of filter closest to truth is selected.

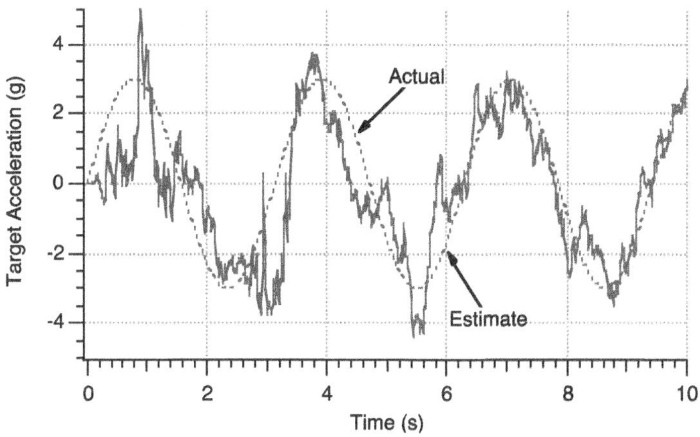

Fig. 27.15 Estimates of target acceleration are excellent even though none of the filters in filter bank are matched to truth.

limit is removed. This means that filter bank approach is yielding near optimal performance.

So far we have assumed that one of the filters in the filter bank is tuned to the actual target weave frequency. Actually this is a requirement for the fixed MMAE approach to work best. Let us see what happens when all three filters are tuned to the wrong frequency. A case was run where the actual target weave frequency was 2 r/s, but the three filters in the filter bank were tuned to 1, 2.5, and 3 r/s, respectively. We can see from Fig. 27.14 that the filter estimate closest to the truth is selected with the fixed MMAE approach. Figure 27.15 indicates that although we do not know the truth, the estimate of target acceleration is excellent after 2 s.

SUMMARY

It has been demonstrated in this chapter that a bank of linear four-state weave Kalman filters can be used for purposes of estimating the target weave frequency using the fixed MMAE technique. This filter bank approach makes use of each filter's likelihood function and Bayes' rule. It is shown that combining filter outputs in a probabilistic sense yields excellent estimates of the target weave frequency and yields small rms miss if the seeker measurement noise can be kept to the 0.1-mr level when the closing velocity is 9000 ft/s.

REFERENCES

[1] Zarchan, P., *Fundamentals of Kalman Filtering: A Practical Approach*, 2nd ed., Progress in Astronautics and Aeronautics, AIAA, Reston, VA, 2005, pp. 606–608, 647–675.

[2] Magill, D. T., "Optimal Adaptive Estimation of Sampled Stochastic Processes," *IEEE Transactions on Automatic Control*, Vol. 10, No. 4, 1965, pp. 434–439.

[3] Bar-Shalom, Y., Li, X. R., and Kirubarajan, T., *Estimation with Applications to Tracking and Navigation*, Wiley, New York, 2001, pp. 440–465.

[4] Shima, T., Oshman, Y., and Shinar, J., "Efficient Multiple Model Adaptive Estimation in Ballistic Missile Interception Scenarios," *Journal of Guidance, Control, and Dynamics*, Vol. 25, No. 4, 2002, pp. 667–675.

[5] Stengel, R., *Optimal Control and Estimation*, Dover, New York, 1994, pp. 402–407.

[6] Welch, G., and Bishop, G., "An Introduction to the Kalman Filter," SIGRAPH 2001, Aug. 2001, pp. 36–39.

[7] Marks, M., "Multiple Models Adaptive Estimation (MMAE) for Improving Guidance Performance Against Weaving Targets," AIAA Paper 2006-6697, Aug. 2006.

[8] Zarchan, P., and Alpert, J., "Using Filter Banks to Improve Performance Against Weaving Targets," AIAA Paper 2006-6700, Aug. 2006.

CHAPTER 28

Engagement Simulations in Three Dimensions

INTRODUCTION

So far all of the engagement simulations presented in this book have either been in one or two dimensions. This was done to make it easier for the reader to readily understand all of the concepts presented in the text. In this chapter we shall provide several examples on how to convert important elements of engagement simulation code to three dimensions in both the tactical and strategic worlds. Complete three-dimensional tactical and strategic engagement simulations will be presented to illustrate all important points discussed.

WEAVING TARGETS IN THREE DIMENSIONS

So far all of the guidance laws presented in the text have only been presented in one dimension. The extension of the guidance laws to three dimensions is quite straightforward and can most easily be done by making use of the zero effort miss concept. Recall that the zero effort miss is simply the miss distance that would result if the target continued to do what it is currently doing and the interceptor issued no further acceleration commands. We have previously demonstrated that all of the guidance laws presented in this text can be expressed in terms of the zero effort miss. The resultant guidance command is equal to the effective navigation ratio times the zero effort miss perpendicular to the line of sight divided by the square of time to go until intercept. The only difference between all of the guidance laws presented in this text is the way in which the zero effort miss is computed. In three dimensions we can define the relative position and velocity components between the missile and target as

$$R_{\mathrm{TM1}} = x_T - x_M$$
$$R_{\mathrm{TM2}} = y_T - y_M$$
$$R_{\mathrm{TM3}} = z_T - z_M$$

$$V_{TM1} = \dot{x}_T - \dot{x}_M$$
$$V_{TM2} = \dot{y}_T - \dot{y}_M$$
$$V_{TM3} = \dot{z}_T - \dot{z}_M$$

Let us start with the simplest guidance law, proportional navigation. For proportional navigation the components of the zero effort miss in three dimensions are simply

$$\mathrm{ZEM}_1 = R_{TM1} + V_{TM1} t_{go}$$
$$\mathrm{ZEM}_2 = R_{TM2} + V_{TM2} t_{go}$$
$$\mathrm{ZEM}_3 = R_{TM3} + V_{TM3} t_{go}$$

The weave guidance law on the other hand has a more complex expression for the zero effort miss that we shall deal with later in this section. As was mentioned earlier, for guidance purposes we are only interested in the component of the zero effort miss that is perpendicular to the line of sight. To find the component of the zero effort miss that is perpendicular to the line of sight, we must first find the component of the zero effort miss that is parallel to the line of sight. If the zero-effort-miss vector (**ZEM**) is defined as

$$\mathbf{ZEM} = \mathrm{ZEM}_1\mathbf{i} + \mathrm{ZEM}_2\mathbf{j} + \mathrm{ZEM}_3\mathbf{k}$$

then a unit vector along the line of sight can be expressed as

$$\mathbf{1}_{RTM} = \frac{R_{TM1}\mathbf{i} + R_{TM2}\mathbf{j} + R_{TM3}\mathbf{k}}{R_{TM}}$$

where the relative range between the missile and target is simply

$$R_{TM} = \sqrt{R_{TM1}^2 + R_{TM2}^2 + R_{TM3}^2}$$

The vector parallel to the line of sight has a magnitude equal to the dot product of the two vectors (zero effort miss and line of sight) and is along the same direction as the unit line-of-sight vector. Therefore the zero-effort-miss vector parallel to the line of sight is given by

$$\mathbf{ZEM_{PAR}} = \mathrm{ZEMDOTRTM}\left(\frac{R_{TM1}\mathbf{i} + R_{TM2}\mathbf{j} + R_{TM3}\mathbf{k}}{R_{TM}}\right)$$

where the quantity ZEMDOTRTM can be computed as

$$\mathrm{ZEMDOTRTM} = \frac{\mathrm{ZEM}_1 R_{TM1} + \mathrm{ZEM}_2 R_{TM2} + \mathrm{ZEM}_3 R_{TM3}}{R_{TM}}$$

The zero-effort-miss vector perpendicular to the line-of-sight $\mathbf{ZEM_{PER}}$ is simply the vector difference between the zero-effort-miss vector **ZEM** and the zero effort miss parallel to the line-of-sight $\mathbf{ZEM_{PAR}}$ or

$$\mathbf{ZEM_{PER}} = \mathbf{ZEM} - \mathbf{ZEM_{PAR}}$$

Therefore we can say that

$$\mathbf{ZEM_{PER}} = \mathrm{ZEM_{PER1}}\mathbf{i} + \mathrm{ZEM_{PER2}}\mathbf{j} + \mathrm{ZEM_{PER3}}\mathbf{k}$$

where the components of the zero effort miss perpendicular to the line of sight can be computed as

$$\mathrm{ZEM_{PER1}} = \mathrm{ZEM_1} - \frac{\mathrm{ZEMDOTRTM} * R_{\mathrm{TM1}}}{R_{\mathrm{TM}}}$$

$$\mathrm{ZEM_{PER2}} = \mathrm{ZEM_2} - \frac{\mathrm{ZEMDOTRTM} * R_{\mathrm{TM2}}}{R_{\mathrm{TM}}}$$

$$\mathrm{ZEM_{PER3}} = \mathrm{ZEM_3} - \frac{\mathrm{ZEMDOTRTM} * R_{\mathrm{TM3}}}{R_{\mathrm{TM}}}$$

As was mentioned earlier, the desired interceptor guidance commands are proportional to the zero effort miss perpendicular to the line of sight and inversely proportional to the square of time to go. Therefore the individual acceleration components of the guidance command are given by

$$a_{M1} = \frac{N'\mathrm{ZEMPER}_1}{t_{\mathrm{go}}^2}$$

$$a_{M2} = \frac{N'\mathrm{ZEMPER}_2}{t_{\mathrm{go}}^2}$$

$$a_{M3} = \frac{N'\mathrm{ZEMPER}_3}{t_{\mathrm{go}}^2}$$

where the effective navigation ratio N' is a constant set equal to three for proportional navigation.

The compensated weave guidance law, which is optimal in the sense that it requires the least acceleration against target weave maneuvers in the presence of a single-time-constant missile flight control system, also issues guidance commands proportional to the zero effort miss and inversely proportional to the square of time to go until intercept. In one dimension the compensated weave guidance law was shown in Chapter 20 to be

$$n_c = \frac{N'}{t_{go}^2}\left[y + \dot{y}t_{go} + \left(\frac{1 - \cos\omega t_{go}}{\omega^2}\right)\ddot{y}_T + \left(\frac{\omega t_{go} - \sin\omega t_{go}}{\omega^3}\right)\dddot{y}_T - n_L T^2(e^{-x} + x - 1)\right]$$

where x is given by

$$x = \frac{t_{go}}{T}$$

In the preceding expressions t_{go} is the time to go until intercept, and the bracketed quantity is the zero effort miss. The effective navigation ratio for the compensated weave guidance law is not a constant but is given by

$$N' = \frac{6x^2(e^{-x} - 1 + x)}{2x^3 + 3 + 6x - 6x^2 - 12xe^{-x} - 3e^{-2x}}$$

We can see that the compensated weave guidance law consists of five terms: the first two terms are related to the line-of-sight rate, the third term is proportional to the target acceleration, the fourth term is proportional to target jerk, and the fifth term is proportional to the achieved missile acceleration. The first two terms in the guidance law are proportional navigation, the third and fourth terms in the guidance law compensate for the sinusoidal motion of the target, and the fifth term compensates for the flight-control-system dynamics. Details of the derivation of the compensated weave guidance law can be found in Chapter 20.

Thus the zero effort miss for the compensated weave guidance law in three dimensions can be expressed as

$$\text{ZEM}_1 = R_{\text{TM1}} + V_{\text{TM1}}t_{go} + \left(\frac{1 - \cos\omega t_{go}}{\omega^2}\right)a_{T1} + \left(\frac{\omega t_{go} - \sin\omega t_{go}}{\omega^3}\right)\dot{a}_{T1} - T^2(e^{-x} + x - 1)a_{M1}$$

$$\text{ZEM}_2 = R_{\text{TM2}} + V_{\text{TM2}}t_{go} + \left(\frac{1 - \cos\omega t_{go}}{\omega^2}\right)a_{T2} + \left(\frac{\omega t_{go} - \sin\omega t_{go}}{\omega^3}\right)\dot{a}_{T2} - T^2(e^{-x} + x - 1)a_{M2}$$

$$\text{ZEM}_3 = R_{\text{TM3}} + V_{\text{TM3}} t_{\text{go}} + \left(\frac{1 - \cos \omega t_{\text{go}}}{\omega^2}\right) a_{T3}$$

$$+ \left(\frac{\omega t_{\text{go}} - \sin \omega t_{\text{go}}}{\omega^3}\right) \dot{a}_{T3} - T^2(e^{-x} + x - 1) a_{M3}$$

where a_{M1}, a_{M2}, and a_{M3} are the three components of the achieved missile accelerations. The target acceleration components are represented by a_{T1}, a_{T2}, and a_{T3}, whereas the components of target jerk are given by $\dot{a}_{T1}$, $\dot{a}_{T2}$, and $\dot{a}_{T3}$. The implementation of the compensated weave guidance law in three dimensions is the same as the implementation of proportional navigation. Therefore the only difference between the two guidance laws is in the expressions for the zero effort miss.

Now let us consider a numerical example to test our implementation of guidance laws in three dimensions. A weaving or spiraling target is chosen as the threat. The acceleration equations of a three-dimensional spiraling target are given by

$$a_{T1} = a_T \sin \omega t$$
$$a_{T2} = a_T \cos \omega t$$
$$a_{T3} = 0$$

where the subscript 1 indicates the downrange direction, the subscript 2 denotes the altitude direction, and the subscript 3 denotes the cross-range direction. Thus in this example the target is traveling in the cross-range direction and spiraling in the downrange and altitude direction. Because the compensated weave guidance law requires target jerk, we can calculate it exactly by differentiating the preceding equations. The resultant components of the target jerk are given by

$$\dot{a}_{T1} = a_T \omega \cos \omega t$$
$$\dot{a}_{T2} = -a_T \omega \sin \omega t$$
$$\dot{a}_{T3} = 0$$

A three-dimensional tactical engagement simulation, utilizing the preceding equations, appears in Listing 28.1. From Listing 28.1 we can see that the target is traveling at 1000 ft/s and is spiraling at 3 r/s with an acceleration level of 6 g. The missile is traveling at 3000 ft/s and has an infinite acceleration limit. There is a single-lag time constant of 1 s representing the dynamics of the flight control system in each of the three channels of the guidance system. From Listing 28.1 we can see that if QPN is set to 1 the missile uses proportional navigation for guidance, and if QPN is set to 0 the missile uses the compensated weave guidance law. The program is set up to run a number of cases—each of which has a different initial missile-target separation. Each separation corresponds to a different flight time. In this way we can generate adjoint-type curves using the brute-force approach as we did in Chapter 3 for the two-dimensional nonlinear engagement simulation. At the end of each flight, the components of the miss, along with the total miss, are printed out as a function of the engagement time. The only source of error in the simulation is the spiraling target maneuver.

LISTING 28.1 THREE-DIMENSIONAL TACTICAL ENGAGEMENT SIMULATION WITH SPIRALING TARGET

```
clear
count=0;
QPN=1;
TAU=1.;
W=3.;
AT=193.2;
VT=1000.;
VM=3000.;
XNP=3.;
XNCLIM=9999999999.;
for RT3IC=40000:-500:500
        RM1=0.;
        RM2=10000.;
        RM3=0.;
        RT1=0.;
        RT2=10000.;
        RT3=RT3IC;
        VT1=-AT/W;
        VT2=0.;
        VT3=-VT;
        T=0.;
        S=0.;
        RTM1=RT1-RM1;
        RTM2=RT2-RM2;
        RTM3=RT3-RM3;
        RTM=sqrt(RTM1^2+RTM2^2+RTM3^2);
        VM1=0.;
        VM2=0.;
        VM3=VM;
        VTM1=VT1-VM1;
        VTM2=VT2-VM2;
        VTM3=VT3-VM3;
        VC=-(RTM1*VTM1+RTM2*VTM2+RTM3*VTM3)/RTM;
        AM1=0.;
        AM2=0.;
        AM3=0.;
        while VC>=0
```

```
if RTM<1000
H=.0002;
else
H=.01;
end
RT1OLD=RT1;
RT2OLD=RT2;
RT3OLD=RT3;
RM1OLD=RM1;
RM2OLD=RM2;
RM3OLD=RM3;
VM1OLD=VM1;
VM2OLD=VM2;
VM3OLD=VM3;
VT1OLD=VT1;
VT2OLD=VT2;
VT3OLD=VT3;
AM1OLD=AM1;
AM2OLD=AM2;
AM3OLD=AM3;
STEP=1;
FLAG=0;
while STEP<=1
        if FLAG==1
                STEP=2;
                RT1=RT1+H*VT1;
                RT2=RT2+H*VT2;
                RT3=RT3+H*VT3;
                RM1=RM1+H*VM1;
                RM2=RM2+H*VM2;
                RM3=RM3+H*VM3;
                VM1=VM1+H*AM1;
                VM2=VM2+H*AM2;
                VM3=VM3+H*AM3;
                VT1=VT1+H*AT1;
                VT2=VT2+H*AT2;
                VT3=VT3+H*AT3;
                AM1=AM1+H*AM1D;
                AM2=AM2+H*AM2D;
                AM3=AM3+H*AM3D;
                T=T+H;
        end
        RTM1=RT1-RM1;
        RTM2=RT2-RM2;
        RTM3=RT3-RM3;
        RTM=sqrt(RTM1^2+RTM2^2+RTM3^2);
```

```
VTM1=VT1-VM1;
VTM2=VT2-VM2;
VTM3=VT3-VM3;
VC=-(RTM1*VTM1+RTM2*VTM2+RTM3*VTM3)/RTM;
TGO=RTM/VC;
AT1=AT*sin(W*T);
AT2=AT*cos(W*T);
AT3=0.;
if QPN==1
        ZEM1=RTM1+VTM1*TGO;
        ZEM2=RTM2+VTM2*TGO;
        ZEM3=RTM3+VTM3*TGO;
else
        AT1D=AT*W*cos(W*T);
        AT2D=-AT*W*sin(W*T);
        AT3D=0.;
        X=TGO/TAU;
        TOP=6.*X*X*(exp(-X)-1.+X);
        BOT1=2*X*X*X+3.+6.*X-6.*X*X;
        BOT2=-12.*X*exp(-X)-3.*exp(-2.*X);
        XNP=TOP/(.0001+BOT1+BOT2);
        ZEM1=RTM1+VTM1*TGO+AT1*(1.-cos(W*TGO))/W^2+...
                (AT1D*(W*TGO-sin(W*TGO))/W^3)-...
                AM1*TAU*TAU*(exp(-X)+X-1.);
        ZEM2=RTM2+VTM2*TGO+AT2*(1.-cos(W*TGO))/W^2+...
                (AT2D*(W*TGO-sin(W*TGO))/W^3)-...
                AM2*TAU*TAU*(exp(-X)+X-1.);
        ZEM3=RTM3+VTM3*TGO+AT3*(1.-cos(W*TGO))/W^2+...
                (AT3D*(W*TGO-sin(W*TGO))/W^3)-...
                AM3*TAU*TAU*(exp(-X)+X-1.);
end
ZEMDOTRTM=(ZEM1*RTM1+ZEM2*RTM2+ZEM3*RTM3)/RTM;
ZEMPER1=ZEM1-ZEMDOTRTM*RTM1/RTM;
ZEMPER2=ZEM2-ZEMDOTRTM*RTM2/RTM;
ZEMPER3=ZEM3-ZEMDOTRTM*RTM3/RTM;
AM1P=XNP*ZEMPER1/(TGO*TGO);
AM2P=XNP*ZEMPER2/(TGO*TGO);
AM3P=XNP*ZEMPER3/(TGO*TGO);
AM1D=(AM1P-AM1)/TAU;
AM2D=(AM2P-AM2)/TAU;
AM3D=(AM3P-AM3)/TAU;
if AM1>XNCLIM
        AM1=XNCLIM;
end
if AM1<-XNCLIM
        AM1=-XNCLIM;
```

```
                end
                if AM2>XNCLIM
                        AM2=XNCLIM;
                end
                if AM2<-XNCLIM
                        AM2=-XNCLIM;
                end
                if AM3>XNCLIM
                        AM3=XNCLIM;
                end
                if AM3<-XNCLIM
                        AM3=-XNCLIM;
                end
                XNCG=sqrt(AM1^2+AM2^2+AM3^2)/32.2;
                FLAG=1;
        end
        FLAG=0;
        RT1=.5*(RT1OLD+RT1+H*VT1);
        RT2=.5*(RT2OLD+RT2+H*VT2);
        RT3=.5*(RT3OLD+RT3+H*VT3);
        RM1=.5*(RM1OLD+RM1+H*VM1);
        RM2=.5*(RM2OLD+RM2+H*VM2);
        RM3=.5*(RM3OLD+RM3+H*VM3);
        VM1=.5*(VM1OLD+VM1+H*AM1);
        VM2=.5*(VM2OLD+VM2+H*AM2);
        VM3=.5*(VM3OLD+VM3+H*AM3);
        VT1=.5*(VT1OLD+VT1+H*AT1);
        VT2=.5*(VT2OLD+VT2+H*AT2);
        VT3=.5*(VT3OLD+VT3+H*AT3);
        AM1=.5*(AM1OLD+AM1+H*AM1D);
        AM2=.5*(AM2OLD+AM2+H*AM2D);
        AM3=.5*(AM3OLD+AM3+H*AM3D);
        S=S+H;
        if S>=.09999
                S=0.;
                RT1K=RT1/1000.;
                RT2K=RT2/1000.;
                RT3K=RT3/1000.;
                RM1K=RM1/1000.;
                RM2K=RM2/1000.;
                RM3K=RM3/1000.;
        end
end
count=count+1;
ArrayT(count)=T;
ArrayRTM1(count)=RTM1;
```

```
        ArrayRTM2(count)=RTM2;
        ArrayRTM3(count)=RTM3;
        ArrayRTM(count)=RTM;
end
output=[ArrayT',ArrayRTM1',ArrayRTM2',ArrayRTM3',ArrayRTM'];
save datfil.txt output /ascii
disp 'simulation finished'
clc
figure
plot(ArrayT,ArrayRTM1),grid
xlabel('Time (s) ')
ylabel('RTM1 (ft)')
axis([0 10 -30 30])
figure
plot(ArrayT,ArrayRTM2),grid
xlabel('Time (s) ')
ylabel('RTM2 (ft)')
axis([0 10 -30 30])
figure
plot(ArrayT,ArrayRTM3),grid
xlabel('Time (s) ')
ylabel('RTM3 (ft)')
axis([0 10 -30 30])
figure
plot(ArrayT,ArrayRTM),grid
xlabel('Time (s) ')
ylabel('RTM (ft)')
axis([0 10 0 30])
```

The nominal case of Listing 28.1 was run using the proportional navigation guidance law. The three components of the miss vs flight time are shown in Figs. 28.1, 28.2, and 28.3. We can see that the miss oscillates with flight time in Figs. 28.1 and 28.2, and the amplitude of the miss agrees with the single plane results of Fig. 20.7. This should not be surprising because identical inputs were used. We can see from Fig. 28.3 that the miss is virtually zero in the cross-range direction because there is no target maneuver in that direction. However we can see from Fig. 28.4 that the total miss is virtually a constant vs flight time after an initial transient period. This should not be surprising because the target weave was a sine wave in one direction and a cosine wave in the other direction, which yields circular motion. Apparently the total miss is also along a circle in the downrange and altitude direction. Therefore we have to be careful when looking at single plane results for a weaving target. One gets the impression from single plane results that the miss can be large or small depending on the flight time. In actuality the peak miss distance for the single plane results will define the radius of a circle or the three-dimensional miss distance for the weaving target.

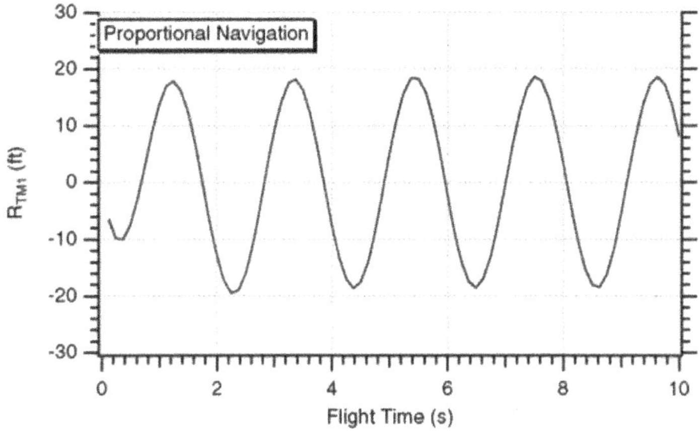

Fig. 28.1 Downrange component of miss oscillate at same frequency as 3-r/s weaving target maneuver.

Listing 28.1 was rerun using the compensated weave guidance law (QPN = 0). Figure 28.5 displays the total miss vs flight time for the same case of the 3-r/s spiraling 6-*g* target and the flight-control-system time constant of 1 s. We can see that with the compensated weave guidance law the miss is reduced from nearly 20 ft to virtually zero when we move to three dimensions, as predicted by theory and the planar results.

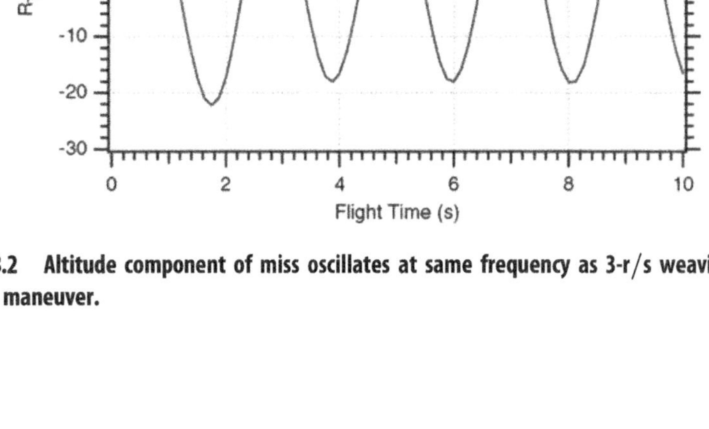

Fig. 28.2 Altitude component of miss oscillates at same frequency as 3-r/s weaving target maneuver.

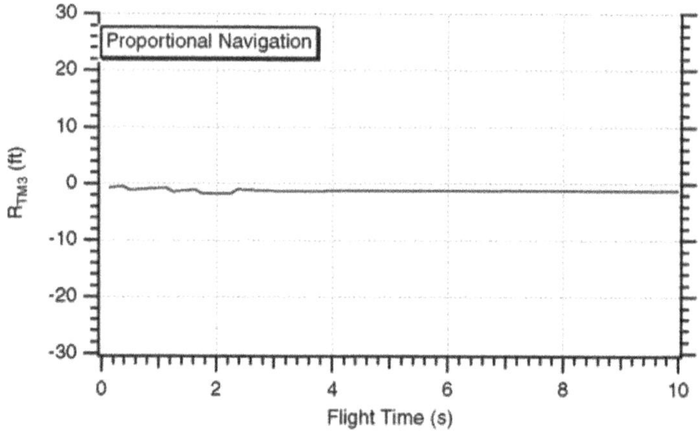

Fig. 28.3 Cross-range component of miss is virtually zero because target does not maneuver in that channel.

BALLISTIC TARGET TRAJECTORY GENERATOR IN THREE DIMENSIONS

Let us now consider a three-dimensional example of a ballistic target trajectory. For simplicity let us consider an example in which a ballistic target is impulsively launched from Monte Carlo to Las Vegas. Because we are neglecting the boost phase of the target, only gravity acts on the ballistic target. As was shown in

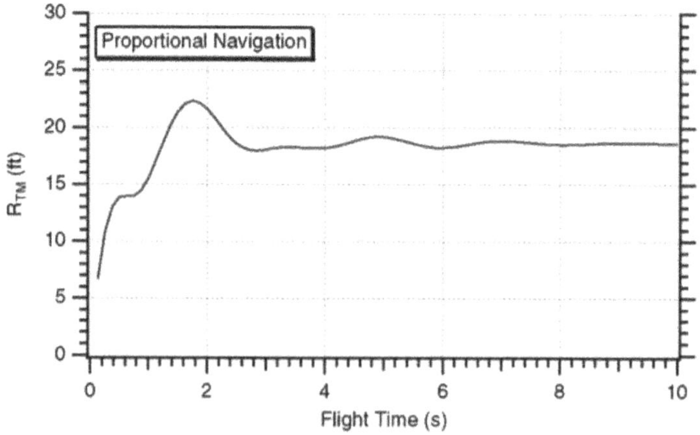

Fig. 28.4 Total three-dimensional miss caused by 3-r/s weaving target approaches a constant in steady state.

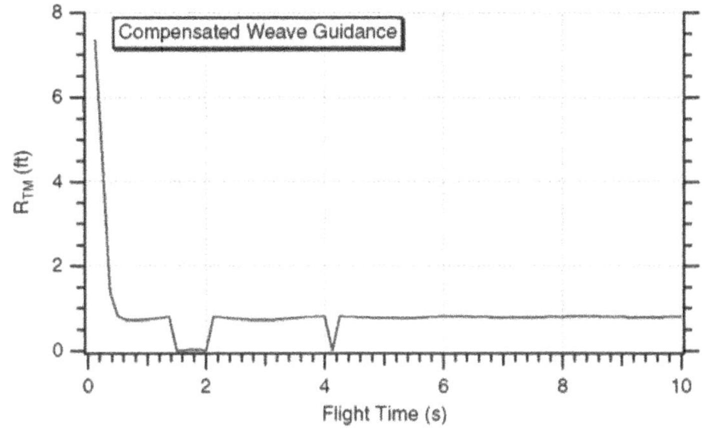

Fig. 28.5 Compensated weave guidance also dramatically reduces the total miss in three dimensions.

Chapter 11, a convenient coordinate system for the simulation of our strategic engagements is an Earth-centered Cartesian coordinate system as shown in Fig. 28.6. Because this coordinate system is fixed in inertial space (even though the Earth rotates), all missile acceleration differential equations can be integrated directly to yield velocity and position, without having to worry about Coriolis effects.

Because only gravity acts on the ballistic missile of our example, the three differential equations describing the acceleration of a target in a gravity field can be derived from Newton's law of universal gravitation in the Earth-centered Cartesian coordinate system as [2]

$$\ddot{x} = \frac{-gm\,x}{(x^2 + y^2 + z^2)^{1.5}}$$

$$\ddot{y} = \frac{-gm\,y}{(x^2 + y^2 + z^2)^{1.5}}$$

$$\ddot{z} = \frac{-gm\,z}{(x^2 + y^2 + z^2)^{1.5}}$$

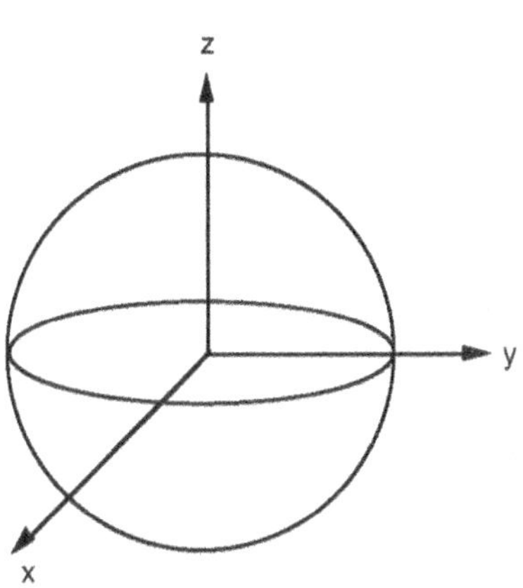

Fig. 28.6 Earth-centered coordinate system.

where x_T, y_T, and z_T are component distances to the ballistic missile measured from the center of the Earth and gm is the gravitational parameter with value

$$gm = 1.4077 * 1016\,\mathrm{ft}^3/\mathrm{s}^2$$

in the English system of units.

We would like to relate our inertial Earth-centered system (also known as the ECI coordinate system), where we will integrate the target equations of motion to a system in which we can draw maps (with longitude and latitude) using tools such as the Mapping Toolbox in MATLAB. It is convenient to use an Earth coordinate system for the drawing of maps (also known as the Earth-centered Earth-fixed or ECEF coordinate system). Our Earth coordinate system (x_e, y_e, z_e) is related to the inertial coordinate system (x, y, z) through the rotation of the Earth ω as shown in Fig. 28.7 [2]. In Fig. 28.7 the Earth rotates at ω rad/s, and t is time.

By inspection of Fig. 28.7, we can see that we can convert ECI coordinates to ECEF coordinates by using the trigonometric relationships

$$x_e = x \cos \omega t + y \sin \omega t$$
$$y_e = -x \sin \omega t + y \cos \omega t$$
$$z_e = z$$

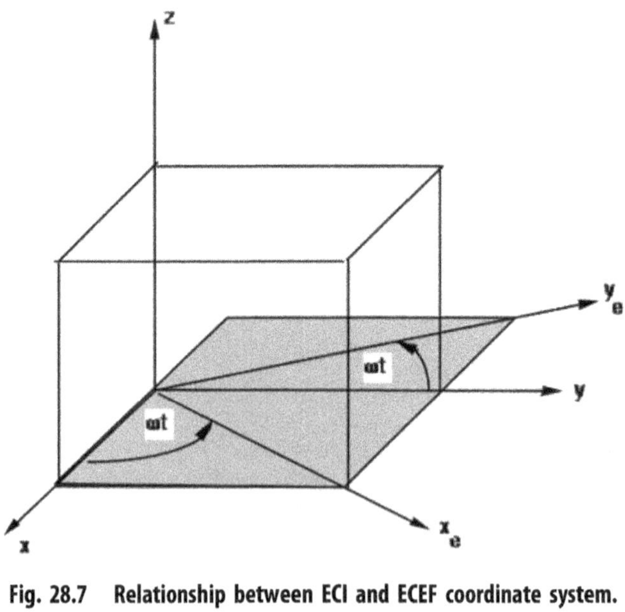

Fig. 28.7 Relationship between ECI and ECEF coordinate system.

Because the Earth rotates once per day, Earth rotation is given approximately by

$$\omega \approx \frac{360^\circ}{1\ \text{day}} = \frac{6.28\ \text{rad}}{86{,}400\ \text{s}} = 7.27 * 10^{-5}\ \text{rad/s}$$

Therefore an object can be expressed either in ECI or ECEF coordinates. An object that is in ECEF coordinates can also be converted to longitude, latitude, and altitude according to Fig. 28.8 [1].

If an object is in space, the distance from the center of the Earth to the object is simply the radius of the Earth plus the altitude of the object or

$$r = a + \text{alt} = \sqrt{x_e^2 + y_e^2 + z_e^2}$$

Longitude and latitude can be derived from Fig. 28.8 as

$$\text{long} = \tan^{-1}\left(\frac{y_e}{x_e}\right)$$

$$\text{lat} = \tan^{-1}\left(\frac{z_e}{\sqrt{x_e^2 + y_e^2}}\right)$$

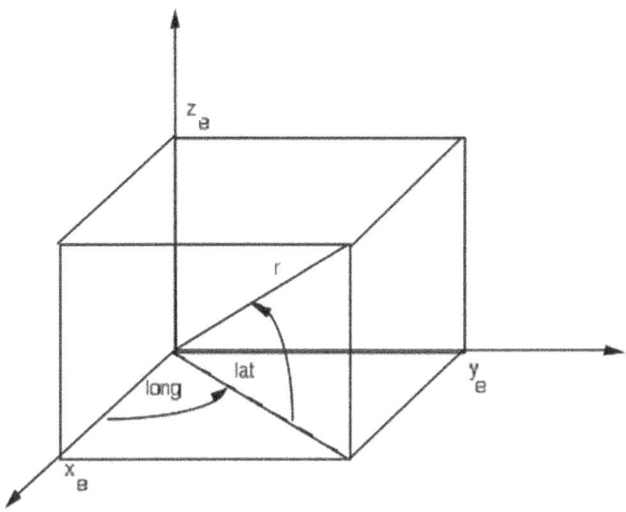

Fig. 28.8 Relationship between ECEF and mapping coordinates.

In addition, if an object is expressed in mapping coordinates, we can convert longitude and latitude to ECEF coordinates according to

$$x_e = r\cos(\text{lat})\cos(\text{long})$$
$$y_e = r\cos(\text{lat})\sin(\text{long})$$
$$z_e = r\sin(\text{lat})$$

We now have enough information to simulate an impulsively launched ballistic missile. All that we are missing is the three-dimensional Lambert routine to provide the initial velocity of the ballistic missile. The new three-dimensional Lambert routine is shown in Listing 28.2. We can see that it is a straightforward extension of the already derived two-dimensional efficient Lambert routine in Chapter 13. Reference 3 is another good source for understanding the code. In Listing 28.2 we enter the longitude and latitude of the target launch point (Monte Carlo) and its final destination (Las Vegas). It is specified that it will take the ballistic missile 2000 s to reach its destination. The missile starting and ending points are converted to ECI coordinates. We can see from the listing that the missile's destination must be modified in ECI coordinates because the Earth rotates significantly in 2000 s. As was already mentioned, all integrations are carried out in ECI coordinates. The inertial missile trajectory outputs are first converted to ECEF coordinates and then to longitude and latitude. In addition, the downrange (from missile launch point) and altitude of the missile are computed. The longitude and latitude outputs of the missile can be supplied to the MATLAB Mapping Toolbox in order to obtain geographical context.

LISTING 28.2 MATLAB BALLISTIC MISSILE TRAJECTORY GENERATOR

```
count=0;
SWITCH1=0;
XLONGTDEG=7.42;
XLATTDEG=43.75;
XLATFDEG=36.175;
XLONGFDEG=-115.136;
TF=2000.;
A=2.0926E7;
GM=1.4077E16;
W=-6.283185/86400.;
XLONGF=XLONGFDEG/57.3;
XLATF=XLATFDEG/57.3;
XLONGT=XLONGTDEG/57.3;
XLATT=XLATTDEG/57.3;
XLONGF=XLONGF-W*TF;
XF=A*cos(XLATF)*cos(XLONGF);
```

```
YF=A*cos(XLATF)*sin(XLONGF);
ZF=A*sin(XLATF);
XT=A*cos(XLATT)*cos(XLONGT);
YT=A*cos(XLATT)*sin(XLONGT);
ZT=A*sin(XLATT);
[VRX,VRY,VRZ]=LAMBERT3D(XT,YT,ZT,TF,XF,YF,ZF,SWITCH1);
XTD=VRX;
YTD=VRY;
ZTD=VRZ;
XTINIT=XT;
YTINIT=YT;
ZTINIT=ZT;
T=0.;
H=.001;
S=0.;
ALTNM=(sqrt(XT^2+YT^2+ZT^2)-A)/6076.;
while ALTNM>-1
          XTOLD=XT;
          YTOLD=YT;
          ZTOLD=ZT;
          XTDOLD=XTD;
          YTDOLD=YTD;
          ZTDOLD=ZTD;
          STEP=1;
          FLAG=0;
          while STEP<=1
                    if FLAG==1
          STEP=2;
                              XT=XT+H*XTD;
                              YT=YT+H*YTD;
                              ZT=ZT+H*ZTD;
                              XTD=XTD+H*XTDD;
                              YTD=YTD+H*YTDD;
                              ZTD=ZTD+H*ZTDD;
          T=T+H;
        end
        TEMPBOTT=(XT^2+YT^2+ZT^2)^1.5;
                    XTDD=-GM*XT/TEMPBOTT;
                    YTDD=-GM*YT/TEMPBOTT;
                    ZTDD=-GM*ZT/TEMPBOTT;
                    ALTNM=(sqrt(XT^2+YT^2+ZT^2)-A)/6076.;
                    FLAG=1;
          end
          FLAG=0;
          XT=.5*(XTOLD+XT+H*XTD);
          YT=.5*(YTOLD+YT+H*YTD);
```

```
        ZT=.5*(ZTOLD+ZT+H*ZTD);
        XTD=.5*(XTDOLD+XTD+H*XTDD);
        YTD=.5*(YTDOLD+YTD+H*YTDD);
        ZTD=.5*(ZTDOLD+ZTD+H*ZTDD);
        S=S+H;
        if S>=9.9999
                S=0.;
                XTE=XT*cos(W*T)-YT*sin(W*T);
                YTE=XT*sin(W*T)+YT*cos(W*T);
                ZTE=ZT;
                XLATT= atan2(ZTE, sqrt(XTE^2+YTE^2));
                XLATTDEG=57.3*XLATT;
                XLONGT= atan2(YTE, XTE);
                XLONGTDEG=57.3*XLONGT;
                DISTRTNM=distance3d(XTE,YTE,ZTE,XTINIT,YTINIT,ZTINIT);
                count=count+1;
                ArrayT(count)=T;
                ArrayDISTRTNM(count)=DISTRTNM;
                ArrayALTNM(count)=ALTNM;
                ArrayXLONGTDEG(count)=XLONGTDEG;
                ArrayXLATTDEG(count)=XLATTDEG;
        end
end
figure
plot(ArrayDISTRTNM,ArrayALTNM),grid
xlabel('Downrange (Nmi)')
ylabel('Altitude (Nmi) ')
clc
output=[ArrayT',ArrayDISTRTNM',ArrayALTNM'];
save datfil.txt output /ascii
output=[ArrayT',ArrayXLONGTDEG',ArrayXLATTDEG'];
csvwrite('trajfil.txt',output)

disp 'simulation finished'

function DISTKM=distance3d(XT,YT,ZT,XF,YF,ZF)
R=sqrt(XT^2+YT^2+ZT^2);
RF=sqrt(XF^2+YF^2+ZF^2);
A=2.0926E7;
CBETA=(XT*XF+YT*YF+ZT*ZF)/(R*RF);
if CBETA<=1.
        BETA=acos(CBETA);
        DISTKM=A*BETA/3280.;
else
        DISTKM=(XF-XT)/3280.;
end
```

```
function [VRX,VRY,VRZ]=LAMBERT3D(XT,YT,ZT,TF,XF,YF,ZF,SWITCH1)
PI=3.1415926535898;
DEG_PER_RAD=57.3;
EARTH_RADIUS=2.0926E7;
GM=1.4077E16;
HALFPI=PI/2.;
FT_PER_KM=3280.;
RAD_PER_SEC=1/57.3;
RF0X=XF-XT;
RF0Y=YF-YT;
RF0Z=ZF-ZT;
R0DOTRF=XT*XF+YT*YF+ZT*ZF;
R0DOTRF0=XT*RF0X+YT*RF0Y+ZT*RF0Z;
R0MAG=sqrt(XT^2+YT^2+ZT^2);
RFMAG=sqrt(XF^2+YF^2+ZF^2);
RF0MAG=sqrt(RF0X^2+RF0Y^2+RF0Z^2);
RATIO=R0MAG/RFMAG;
GMDIVR0=GM/R0MAG;
COS_T=R0DOTRF/(R0MAG*RFMAG);
VNUMER=GMDIVR0*(1.-COS_T);
T_MIN=0.;
if  SWITCH1==0
          G_MIN=HALFPI-acos(R0DOTRF0/(R0MAG*RF0MAG));
          G_MAX=HALFPI;
          THETA=acos(COS_T);
else
          G_MIN=-HALFPI;
          G_MAX=-HALFPI+acos(R0DOTRF0/(R0MAG*RF0MAG));
          THETA=2.*PI-acos(COS_T);
end
SIN_T=sin(THETA);
COT_HALFT=1./tan(THETA/2.);
GAMMA=(G_MAX+G_MIN)/2.;
GOLD=G_MIN;
TOLD=0.;
T=0.;
ITERS=1;

S=.5*(R0MAG+RFMAG+RF0MAG);
BL=sqrt(R0MAG*RFMAG)*cos(THETA/2.)/S;
BT=sqrt(8.*GM/(S*S*S))*TF;

while (abs(TF-T)>(.00000001*TF))
          SIN_G=sin(GAMMA);
          COS_G=cos(GAMMA);
          TAN_G=SIN_G/COS_G;
```

```
COS_TPLUSG=cos(THETA+GAMMA-2.*PI);
TERM1=(RATIO*COS_G-COS_TPLUSG)*COS_G;
RV0MAG=sqrt(VNUMER/TERM1);
LAMBDA=RV0MAG*RV0MAG/GMDIVR0;
if LAMBDA<1.9999999
  TERM0=sqrt(2./LAMBDA-1.);
  TERM1=(TAN_G*(1.-COS_T)+(1.-LAMBDA)*SIN_T)/...
              ((2.-LAMBDA)*RATIO);
  TERM2=(COS_G+COS_G)/(LAMBDA*TERM0*TERM0*TERM0);
  TERM3= atan2(TERM0, (COS_G*COT_HALFT-SIN_G));
  T=(R0MAG/(RV0MAG*COS_G))*(TERM1+TERM2*TERM3);
elseif  LAMBDA>2.0000001
  TERM0=sqrt(1.-2./LAMBDA);
  TERM1=(TAN_G*(1.-COS_T)+(1.-LAMBDA)*SIN_T)/...
              ((2.-LAMBDA)*RATIO);
  TERM2=COS_G/(LAMBDA*TERM0*TERM0*TERM0);
  TERM3=SIN_G-COS_G*COT_HALFT;
  TERM3=log((TERM3-TERM0)/(TERM3+TERM0));
  T=(R0MAG/(RV0MAG*COS_G))*(TERM1-TERM2*TERM3);
else
  TERM0=COS_G*COT_HALFT;
  TERM1=TERM0-SIN_G;
  TERM0=(3*TERM0*TERM1+1.)/(TERM1*TERM1*TERM1);
  T=TERM0*(2.*R0MAG)/(3.*RV0MAG);
end
if T>TF & GAMMA<G_MAX
  G_MAX=GAMMA;
end

if T<0. & GAMMA<G_MAX
  G_MAX=GAMMA;
end
if  T<TF & GAMMA>G_MIN
  G_MIN=GAMMA;
  T_MIN=T;
end
if (T<0.)
  NEXT=(G_MIN+G_MAX)/2.;
  GOLD=G_MIN;
  TOLD=T_MIN;
else
  NEXT=GAMMA+(TF-T)*(GAMMA-GOLD)/(T-TOLD);
  if NEXT>=G_MAX
    NEXT=(GAMMA+G_MAX)/2.;
  elseif NEXT<=G_MIN
```

```
                NEXT=(GAMMA+G_MIN)/2.;
              end
              GOLD=GAMMA;
              TOLD=T;
            end
            GAMMA=NEXT;
            ITERS=ITERS+1;
      if ITERS>100
            break
      end
end
if SWITCH1==0
                  GAMMA=GAMMA;
          ANGLE=HALFPI-GAMMA;
          SINA=sin(ANGLE);
          COSA=cos(ANGLE);
          V1X=XT;
          V1Y=YT;
          V1Z=ZT;
          V2X=XF;
          V2Y=YF;
          V2Z=ZF;
          MAG1=sqrt(V1X*V1X+V1Y*V1Y+V1Z*V1Z);
          DOTMAG=V1X*V2X+V1Y*V2Y+V1Z*V2Z;
          CROSSX=V1Y*V2Z-V1Z*V2Y;
          CROSSY=V1Z*V2X-V1X*V2Z;
          CROSSZ=V1X*V2Y-V1Y*V2X;
          CROSSMAG=sqrt(CROSSX*CROSSX+CROSSY*CROSSY+CROSSZ*CROSSZ);
          C2=MAG1*SINA/CROSSMAG;
          C1=COSA/MAG1-DOTMAG*C2/(MAG1*MAG1);
          RTEMPX=C1*V1X;
          RTEMPY=C1*V1Y;
          RTEMPZ=C1*V1Z;
          VUNITX=C2*V2X;
          VUNITY=C2*V2Y;
          VUNITZ=C2*V2Z;
          VUNITX=VUNITX+RTEMPX;
          VUNITY=VUNITY+RTEMPY;
          VUNITZ=VUNITZ+RTEMPZ;
else
          ANGLE=GAMMA-HALFPI;
          SINA=sin(ANGLE);
          COSA=cos(ANGLE);
          V1X=XT;
          V1Y=YT;
```

```
        V1Z=ZT;
        V2X=XF;
        V2Y=YF;
        V2Z=ZF;
        MAG1=sqrt(V1X*V1X+V1Y*V1Y+V1Z*V1Z);
        DOTMAG=V1X*V2X+V1Y*V2Y+V1Z*V2Z;
        CROSSX=V1Y*V2Z-V1Z*V2Y
        CROSSY=V1Z*V2X-V1X*V2Z;
        CROSSZ=V1X*V2Y-V1Y*V2X;
        CROSSMAG=sqrt(CROSSX*CROSSX+CROSSY*CROSSY+CROSSZ*CROSSZ);
        C2=MAG1*SINA/CROSSMAG;
        C1=COSA/MAG1-DOTMAG*C2/(MAG1*MAG1);
        RTEMPX=C1*V1X;
        RTEMPY=C1*V1Y;
        RTEMPZ=C1*V1Z;
        VUNITX=C2*V2X;
        VUNITY=C2*V2Y;
        VUNITZ=C2*V2Z;
        VUNITX=VUNITX+RTEMPX;
        VUNITY=VUNITY+RTEMPY;
        VUNITZ=VUNITZ+RTEMPZ;
end
VRX=RVOMAG*VUNITX;
VRY=RVOMAG*VUNITY;
VRZ=RVOMAG*VUNITZ;
```

The nominal case of Listing 28.2 was run and Fig. 28.9 presents a planar view of the actual target trajectory. We can see that the distance from Monte Carlo to Las Vegas is nearly 5000 n miles. The target apogee is approximately 900 n miles.

INTERCEPT POINT PREDICTION FOR BALLISTIC TARGETS

Soon, in this chapter we shall attempt to write a three-dimensional engagement simulation for a missile intercepting a ballistic target. To initially aim the missile, we will have to make a prediction of the intercept point. In other words we want to know the future location of the target at a desired intercept time. One way of finding the intercept point is by integrating the ballistic target equations of motion forward until the desired intercept time. This method certainly works but can be time consuming—especially if a small integration step size is used and the desired intercept time is a large number. A much better method, which is only applicable to ballistic targets (targets that are only influenced by gravity), is to use the numerical solution to Kepler's problem. With the Kepler method the initial target states (position and velocity) are known, and we desire to predict the target states at some given time in the future.

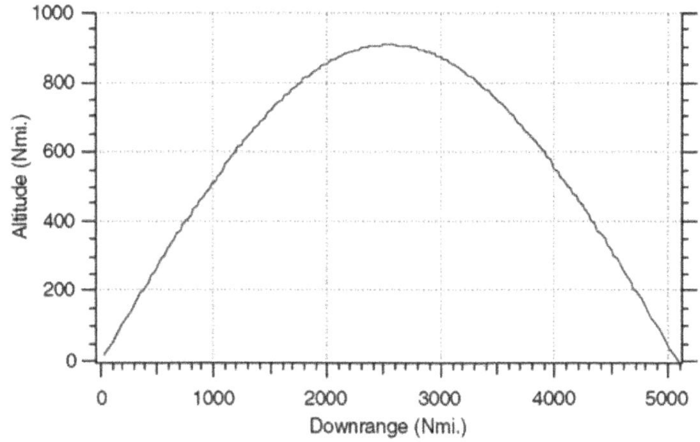

Fig. 28.9 Ballistic target trajectory.

Many algorithms exist for solving Kepler's problem, and the one used in this text was chosen simply because it appeared in the open literature [4] and not because it is the best. In this section we will not derive the Kepler routine but simply demonstrate that it works.

Let us consider a ballistic target prediction problem in which the initial conditions of the ballistic target are identical to that of the previous ballistic target simulation. We desire to predict the location and velocity of the ballistic target 1000 s after target launch. Listing 28.3 directly integrates the target equations of motion for 1000 s and prints the final states of the target. In addition, at the beginning of Listing 28.3 the target's initial conditions (converted to km and km/s) are fed into a Kepler routine to also predict the final states of the target. Quantities related to the Kepler subroutine are highlighted in bold.

The sample case of Listing 28.3 was run, and comparisons between numerical integration and the use of the Kepler subroutine are presented in Table 28.1. We can see that the difference in answers between the two methods is very small. Thus in the future we shall use the Kepler subroutine for the prediction of the intercept point of a ballistic target.

TABLE 28.1 COMPARISON IN PREDICTED INTERCEPT POINT BETWEEN DIRECT INTEGRATION AND KEPLER

Subroutine	x, km	y, km	z, km	$\dot{x}$, km/s	$\dot{y}$, km/s	$\dot{z}$, km/s
Integration	2547.4175	−3583.1082	6758.2423	−3.558504	−3.245617	−0.379447
Kepler	2547.4148	−3583.1074	6758.2384	−3.558509	−3.245615	−0.379455

LISTING 28.3 COMPARING DIRECT INTEGRATION WITH KEPLER PROPOGATION

```
%       Obtained from output of Listing C28L2
XT=14990432.9744621;
YT=1952093.10305573;
ZT=14469752.1663352;
XTD=996.773566434768;
YTD=-14954.8124604715;
ZTD=17528.1931768263;
TF=1000.;
A=2.0926E7;
GM=1.4077E16;
W=-6.283185/86400.;
T=0.;
H=.001;
S=0.;
T0=0.;
T1=TF;
X0(1)=XT/3280.;
X0(2)=YT/3280.;
X0(3)=ZT/3280.;
X0(4)=XTD/3280.;
X0(5)=YTD/3280.;
X0(6)=ZTD/3280.;
[X1]=KEPLER1(X0,T0,T1);
while T<=TF
        XTOLD=XT;
        YTOLD=YT;
        ZTOLD=ZT;
        XTDOLD=XTD;
        YTDOLD=YTD;
        ZTDOLD=ZTD;
        STEP=1;
        FLAG=0;
        while STEP<=1
                if FLAG==1
        STEP=2;
                        XT=XT+H*XTD;
                        YT=YT+H*YTD;
                        ZT=ZT+H*ZTD;
                        XTD=XTD+H*XTDD;
                        YTD=YTD+H*YTDD;
                        ZTD=ZTD+H*ZTDD;
        T=T+H;
      end
        TEMPBOTT=(XT^2+YT^2+ZT^2)^1.5;
```

```
                XTDD=-GM*XT/TEMPBOTT;
                YTDD=-GM*YT/TEMPBOTT;
                ZTDD=-GM*ZT/TEMPBOTT;
                FLAG=1;
        end
        FLAG=0;
        XT=.5*(XTOLD+XT+H*XTD);
        YT=.5*(YTOLD+YT+H*YTD);
        ZT=.5*(ZTOLD+ZT+H*ZTD);
        XTD=.5*(XTDOLD+XTD+H*XTDD);
        YTD=.5*(YTDOLD+YTD+H*YTDD);
        ZTD=.5*(ZTDOLD+ZTD+H*ZTDD);
        S=S+H;
        if S>=9.9999
                S=0.;
        end
end
XTKM=XT/3280.;
YTKM=YT/3280.;
ZTKM=ZT/3280.;
XTDKM=XTD/3280.;
YTDKM=YTD/3280.;
ZTDKM=ZTD/3280.;
ERRX=XTKM-X1(1);
ERRY=YTKM-X1(2);
ERRZ=ZTKM-X1(3);
ERRXD=XTDKM-X1(4);
ERRYD=YTDKM-X1(5);
ERRZD=ZTDKM-X1(6);
XTKM
X1(1)
YTKM
X1(2)
ZTKM
X1(3)
XTDKM
X1(4)
YTDKM
X1(5)
ZTDKM
X1(6)
ERRX
ERRY
ERRZ
ERRXD
ERRYD
ERRZD
```

```
function [X1]=KEPLER1(X0,T0,T1)
GMX=398923.;
REX=6380.;
TLIMIT=1.E-10;
KN=10;
MUQR = 1.;
DT = T1 - T0;
DX=10.;
if (abs(DT) <TLIMIT)
        for I=1:6
        X1(I) = X0(I);
   end
 end

TIME_FACTOR = sqrt(REX^3/GMX);
VEX = REX/TIME_FACTOR;
DT = DT/TIME_FACTOR;
for I=1:3
        R0(I) = X0(I)/REX;
        V0(I) = X0(I+3)/VEX;
end
R0MAG = sqrt(R0(1)^2 +R0(2)^2 +R0(3)^2);
V0MAG = sqrt(V0(1)^2 +V0(2)^2 +V0(3)^2);
D0 = R0(1)*V0(1) +R0(2)*V0(2) +R0(3)*V0(3);
SIGMA0 = D0/MUQR;
ALP0 = 2./R0MAG - V0MAG*V0MAG;
if ALP0 == 0.
       A0 = 1.E30;
else
       A0 = 1./ALP0;
end

X = ALP0*DT;
if ALP0 <= 0.
       X = .1*DT/R0MAG;
end
for K=1:KN
        if ALP0<0.
                Y = ALP0*X*X;
        YQR = sqrt(-Y);
        CY = (1.-cosh(YQR))/Y;
        SY = (sinh(YQR) -YQR)/(YQR^3);
       elseif ALP0==0.
        Y = 0.;
        CY = .5;
```

```
            SY = 1./6.;
          else
            Y = ALP0*X*X;
            YQR = sqrt(Y);;
            CY = (1.-cos(YQR))/Y;
            SY = (YQR -sin(YQR))/(YQR^3);
         end
            U1 = X*(1.-Y*SY);
            U2 = X*X*CY;
            U3 = X*X*X*SY;
            FX = R0MAG*U1 +SIGMA0*U2 +U3 -DT*MUQR;
            DFX = SIGMA0*U1 +(1. -ALP0*R0MAG)*U2 +R0MAG;
            DFX2 = SIGMA0*(1. -Y*CY) +(1. -ALP0*R0MAG)*U1;
            SDFX = DFX/(abs(DFX));
            DX0 = 16.*DFX*DFX;
            DX1 = 20.*FX*DFX2;
            DX2 = 16.*DFX*DFX - 20.*FX*DFX2;
            if DX2 > 0.
            DX = 5.*FX/(DFX +SDFX*sqrt(DX2));
            else
            DX = .5*X;
            end
            X =X -DX;
end

RMAG =DFX;
F = 1. -U2/R0MAG;
G = DT -U3/MUQR;
DF = -MUQR*U1/(RMAG*R0MAG);
DG = 1. -U2/RMAG;
for I=1:3
        X1(I) = (F*R0(I) +G*V0(I))*REX;
        X1(I+3) = (DF*R0(I) +DG*V0(I))*VEX;
end
```

STRATEGIC MISSILE-TARGET ENGAGEMENT SIMULATION

We now have enough information to write an engagement simulation. All we have to do is to add the missile equations to the existing target simulation of Listing 28.2. We will assume that the missile is also impulsively launched (TLAUNCH seconds after the target is launched) toward the predicted intercept point. The nearly exact predicted intercept point is obtained from the Kepler subroutine of the preceding section. In addition, we will add the ability to introduce deterministic intercept point prediction errors (PREDERR set to desired intercept point prediction error). Because the missile is initially stationary (before it is launched) but is on a rotating Earth, care must be taken on how to integrate the missile differential equations. A careful examination of the code reveals that the missile differential equations are different before and after launch. At some point near the end of the flight (200 s before intercept in this code), the missile guidance system is turned on to take out any remaining errors.

Once the missile guidance system is turned on, we have assumed that proportional navigation is used as the homing guidance law. As shown at the beginning of this chapter, we can express proportional navigation in terms of the zero effort miss perpendicular to the line of sight, and an example of its implementation in three dimensions has already been presented in Listing 28.1. Recall that the individual missile guidance acceleration components for proportional navigation are given by

$$a_{\text{XMGUID}} = \frac{N'\text{ZEMPER}_1}{t_{\text{go}}^2}$$

$$a_{\text{YMGUID}} = \frac{N'\text{ZEMPER}_2}{t_{\text{go}}^2}$$

$$a_{\text{ZMGUID}} = \frac{N'\text{ZEMPER}_3}{t_{\text{go}}^2}$$

If the missile is not yet in the homing phase of flight, the individual guidance acceleration components are zero. Therefore the acceleration differential equations for the impulsive missile after launch become

$$\ddot{x}_M = \frac{-gm\, x_M}{(x_M^2 + y_M^2 + z_M^2)^{1.5}} + a_{\text{XMGUID}}$$

$$\ddot{y}_M = \frac{-gm\, y_M}{(x_M^2 + y_M^2 + z_M^2)^{1.5}} + a_{\text{YMGUID}}$$

$$\ddot{z}_M = \frac{-gm\, z_M}{(x_M^2 + y_M^2 + z_M^2)^{1.5}} + a_{\text{ZMGUID}}$$

Before missile launch the acceleration differential equations are set to zero so that the missile remains on the ground. The missile-target engagement simulation appears in Listing 28.4. All equations that are either related to the missile or to the relative equations are highlighted in bold. Latitude and longitude information for the missile and target trajectories are written to the comma-delimited file TRAJFIL.TXT. The missile attempting to intercept the threat from Monte Carlo (and thereby protecting Las Vegas) is launched from Atlantic City.

LISTING 28.4 THREE-DIMENSIONAL STRATEGIC MISSILE–TARGET ENGAGEMENT SIMULATION (LAMBERT AND KEPLER SUBROUTINES NOT INCLUDED)

```
count=0;
SWITCH1=0;
SWITCHM=0;
TLAUNCH=200.;
XLONGTDEG=7.42;
XLATTDEG=43.75;
XLATFDEG=36.175;
XLONGFDEG=-115.136;
XLONGMDEG=-74.423;
XLATMDEG=39.364;
PREDERR=10.*6076.;
XNCLIM=161.;
XLATMDEGIC=XLATMDEG;
XLONGMDEGIC=XLONGMDEG;
TFTOT=2000.;
TF=1000.;
A=2.0926E7;
GM=1.4077E16;
W=-6.283185/86400.;
QBOOSTM=1;
XLONGF=XLONGFDEG/57.3;
XLATF=XLATFDEG/57.3;
XLONGT=XLONGTDEG/57.3;
XLATT=XLATTDEG/57.3;
XLONGM=XLONGMDEG/57.3;
XLATM=XLATMDEG/57.3;
XLONGF=XLONGF-W*TF;
XF=A*cos(XLATF)*cos(XLONGF);
YF=A*cos(XLATF)*sin(XLONGF);
ZF=A*sin(XLATF);
XT=A*cos(XLATT)*cos(XLONGT);
YT=A*cos(XLATT)*sin(XLONGT);
ZT=A*sin(XLATT);
[XTD,YTD,ZTD]=LAMBERT3D(XT,YT,ZT,TFTOT,XF,YF,ZF,SWITCH1);
XTINIT=XT;
YTINIT=YT;
ZTINIT=ZT;
XM=A*cos(XLATM)*cos(XLONGM);
YM=A*cos(XLATM)*sin(XLONGM);
ZM=A*sin(XLATM);
XMD=A*W*cos(XLATM)*sin(XLONGM);
YMD=-A*W*cos(XLATM)*cos(XLONGM);
ZMD=0.;
```

```
RTM1=XT-XM;
RTM2=YT-YM;
RTM3=ZT-ZM;
RTM=sqrt(RTM1^2+RTM2^2+RTM3^2);
VTM1=XTD-XMD;
VTM2=YTD-YMD;
VTM3=ZTD-ZMD;
VC=-(RTM1*VTM1+RTM2*VTM2+RTM3*VTM3)/RTM;
T=0.;
H=.001;
T0=0.;
T1=TF;
X0(1)=XT/3280.;
X0(2)=YT/3280.;
X0(3)=ZT/3280.;
X0(4)=XTD/3280.;
X0(5)=YTD/3280.;
X0(6)=ZTD/3280.;
[X1]=KEPLER1(X0,T0,T1);
XTF=X1(1)*3280.;
YTF=X1(2)*3280.;
ZTF=X1(3)*3280.;
XTF=XTF+PREDERR;
S=0.;
DELV=0.;
ALTNM=(sqrt(XT^2+YT^2+ZT^2)-A)/6076.;
while VC>0
        if RTM>5000.
                H=.01;
        else
                H=.00001;
        end
        XTOLD=XT;
        YTOLD=YT;
        ZTOLD=ZT;
        XTDOLD=XTD;
        YTDOLD=YTD;
        ZTDOLD=ZTD;
        XMOLD=XM;
        YMOLD=YM;
        ZMOLD=ZM;
        XMDOLD=XMD;
        YMDOLD=YMD;
        ZMDOLD=ZMD;
        DELVOLD=DELV;
        STEP=1;
```

```
  FLAG=0;
  while STEP<=1
        if FLAG==1
  STEP=2;
              XT=XT+H*XTD;
              YT=YT+H*YTD;
              ZT=ZT+H*ZTD;
              XTD=XTD+H*XTDD;
              YTD=YTD+H*YTDD;
              ZTD=ZTD+H*ZTDD;
              if T<TLAUNCH & QBOOSTM==1
                    XM=XM;
                    YM=YM;
                    ZM=ZM;
                    XMD=XMD;
                    YMD=YMD;
                    ZMD=ZMD;
              else
                    XM=XM+H*XMD;
                    YM=YM+H*YMD;
                    ZM=ZM+H*ZMD;
                    XMD=XMD+H*XMDD;
                    YMD=YMD+H*YMDD;
                    ZMD=ZMD+H*ZMDD;
              end
              DELV=DELV+H*DELVD;
  T=T+H;
 end
 TEMPBOTT=(XT^2+YT^2+ZT^2)^1.5;
        XTDD=-GM*XT/TEMPBOTT;
        YTDD=-GM*YT/TEMPBOTT;
        ZTDD=-GM*ZT/TEMPBOTT;
        ALTNM=(sqrt(XT^2+YT^2+ZT^2)-A)/6076.;
        RTM1=XT-XM;
        RTM2=YT-YM;
        RTM3=ZT-ZM;
        VTM1=XTD-XMD;
        VTM2=YTD-YMD;
        VTM3=ZTD-ZMD;
        RTM=sqrt(RTM1^2+RTM2^2+RTM3^2);
        VC=-(RTM1*VTM1+RTM2*VTM2+RTM3*VTM3)/RTM;
        TGO=RTM/VC;
        if TGO<200. & T>(TLAUNCH+50.)
              ZEM1=RTM1+VTM1*TGO;
              ZEM2=RTM2+VTM2*TGO;
              ZEM3=RTM3+VTM3*TGO;
```

```
                ZEMDOTRTM=(ZEM1*RTM1+ZEM2*RTM2+ZEM3*RTM3)/RTM;
                ZEMPER1=ZEM1-ZEMDOTRTM*RTM1/RTM;
                ZEMPER2=ZEM2-ZEMDOTRTM*RTM2/RTM;
                ZEMPER3=ZEM3-ZEMDOTRTM*RTM3/RTM;
                ZEMPERLOSKM=sqrt(ZEMPER1^2
                        +ZEMPER2^2+ZEMPER3^2)/3280.;
                AXMGUID=3.*ZEMPER1/(TGO^2);
                AYMGUID=3.*ZEMPER2/(TGO^2);
                AZMGUID=3.*ZEMPER3/(TGO^2);
        else
                AXMGUID=0.;
                AYMGUID=0.;
                AZMGUID=0.;
        end
        if AXMGUID>XNCLIM
                AXMGUID=XNCLIM;
        elseif AXMGUID<-XNCLIM
                AXMGUID=-XNCLIM;
        end
        if AYMGUID>XNCLIM
                AYMGUID=XNCLIM;
        elseif AYMGUID<-XNCLIM
                AYMGUID=-XNCLIM;
        end
        if AZMGUID>XNCLIM
                AZMGUID=XNCLIM;
        elseif AZMGUID<-XNCLIM
                AZMGUID=-XNCLIM;
        end
        if T>TLAUNCH
                TEMPBOTM=(XM^2+YM^2+ZM^2)^1.5;
                XMDD=-GM*XM/TEMPBOTM+AXMGUID;
                YMDD=-GM*YM/TEMPBOTM+AYMGUID;
                ZMDD=-GM*ZM/TEMPBOTM+AZMGUID;
        else
                XMDD=0.;
                YMDD=0.;
                ZMDD=0.;
        end
        DELVD=sqrt(AXMGUID^2+AYMGUID^2+AZMGUID^2);
        FLAG=1;
end
FLAG=0;
XT=.5*(XTOLD+XT+H*XTD);
YT=.5*(YTOLD+YT+H*YTD);
ZT=.5*(ZTOLD+ZT+H*ZTD);
```

```
XTD=.5*(XTDOLD+XTD+H*XTDD);
YTD=.5*(YTDOLD+YTD+H*YTDD);
ZTD=.5*(ZTDOLD+ZTD+H*ZTDD);
if T<TLAUNCH & QBOOSTM==1
        XM=A*cos(XLATMDEGIC/57.3)*cos(XLONGMDEGIC/57.3-W*T);
        YM=A*cos(XLATMDEGIC/57.3)*sin(XLONGMDEGIC/57.3-W*T);
        ZM=A*sin(XLATMDEGIC/57.3);
        XMD=A*W*cos(XLATMDEGIC/57.3)*sin(XLONGMDEGIC/57.3-W*T);
        YMD=-A*W*cos(XLATMDEGIC/57.3)*cos(XLONGMDEGIC/57.3-W*T);
        ZMD=0.;
else
        XM=.5*(XMOLD+XM+H*XMD);
        YM=.5*(YMOLD+YM+H*YMD);
        ZM=.5*(ZMOLD+ZM+H*ZMD);
        XMD=.5*(XMDOLD+XMD+H*XMDD);
        YMD=.5*(YMDOLD+YMD+H*YMDD);
        ZMD=.5*(ZMDOLD+ZMD+H*ZMDD);
end
DELV=.5*(DELVOLD+DELV+H*DELVD);
TGOM=TF-T;
if T>=TLAUNCH & QBOOSTM==1
        [XMD,YMD,ZMD]=LAMBERT3D(XM,YM,ZM,TGOM,XTF,YTF,ZTF,SWITCHM);
        QBOOSTM=0;
        XMDOLD=XMD;
        YMDOLD=YMD;
        ZMDOLD=ZMD;
end
S=S+H;
if S>=9.9999
        S=0.;
        XTE=XT*cos(W*T)-YT*sin(W*T);
        YTE=XT*sin(W*T)+YT*cos(W*T);
        ZTE=ZT;
        XLATT=atan2(ZTE, sqrt(XTE^2+YTE^2));
        XLATTDEG=57.3*XLATT;
        XLONGT=atan2(YTE, XTE);
        XLONGTDEG=57.3*XLONGT;
        DISTRTNM=distance3d(XTE,YTE,ZTE,XTINIT,YTINIT,ZTINIT);
        XME=XM*cos(W*T)-YM*sin(W*T);
        YME=XM*sin(W*T)+YM*cos(W*T);
        ZME=ZM;
        XLATM=atan2(ZME, sqrt(XME^2+YME^2));
        XLATMDEG=57.3*XLATM;
        XLONGM=atan2(YME, XME);
        XLONGMDEG=57.3*XLONGM;
        DISTRMNM=distance3d(XME,YME,ZME,XTINIT,YTINIT,ZTINIT);
```

```
                ALTMNM=(sqrt(XM^2+YM^2+ZM^2)-A)/6076.;
                AXMGUIDG=sqrt(AXMGUID^2+AYMGUID^2+AZMGUID^2)/32.2;
                count=count+1;
                ArrayT(count)=T;
                ArrayDISTRTNM(count)=DISTRTNM;
                ArrayALTNM(count)=ALTNM;
                ArrayDISTRMNM(count)=DISTRMNM;
                ArrayALTMNM(count)=ALTMNM;
                ArrayAXMGUIDG(count)=AXMGUIDG;
                ArrayXLONGTDEG(count)=XLONGTDEG;
                ArrayXLATTDEG(count)=XLATTDEG;
                ArrayXLONGMDEG(count)=XLONGMDEG;
                ArrayXLATMDEG(count)=XLATMDEG;
        end
end
XTE=XT*cos(W*T)-YT*sin(W*T);
YTE=XT*sin(W*T)+YT*cos(W*T);
ZTE=ZT;
XLATT=atan2(ZTE, sqrt(XTE^2+YTE^2));
XLATTDEG=57.3*XLATT;
XLONGT=atan2(YTE, XTE);
XLONGTDEG=57.3*XLONGT;
DISTRTNM=distance3d(XTE,YTE,ZTE,XTINIT,YTINIT,ZTINIT);
XME=XM*cos(W*T)-YM*sin(W*T);
YME=XM*sin(W*T)+YM*cos(W*T);
ZME=ZM;
XLATM=atan2(ZME, sqrt(XME^2+YME^2));
XLATMDEG=57.3*XLATM;
XLONGM=atan2(YME, XME);
XLONGMDEG=57.3*XLONGM;
DISTRMNM=distance3d(XME,YME,ZME,XTINIT,YTINIT,ZTINIT);
ALTMNM=(sqrt(XM^2+YM^2+ZM^2)-A)/6076.;
AXMGUIDG=sqrt(AXMGUID^2+AYMGUID^2+AZMGUID^2)/32.2;
count=count+1;
ArrayT(count)=T;
ArrayDISTRTNM(count)=DISTRTNM;
ArrayALTNM(count)=ALTNM;
ArrayDISTRMNM(count)=DISTRMNM;
ArrayALTMNM(count)=ALTMNM;
ArrayAXMGUIDG(count)=AXMGUIDG;
ArrayXLONGTDEG(count)=XLONGTDEG;
ArrayXLATTDEG(count)=XLATTDEG;
ArrayXLONGMDEG(count)=XLONGMDEG;
ArrayXLATMDEG(count)=XLATMDEG;
RTM
DELV
```

```
figure
plot(ArrayDISTRTNM,ArrayALTNM,ArrayDISTRMNM,ArrayALTMNM),grid
xlabel('Downrange (Nmi)')
ylabel('Altitude (Nmi) ')
figure
plot(ArrayT,ArrayAXMGUIDG),grid
xlabel('Time (s))')
ylabel('Acceleration (g) ')
axis([0 1000 0 .5])
clc
output=[ArrayT',ArrayDISTRTNM',ArrayALTNM',ArrayDISTRMNM',...
            ArrayALTMNM',ArrayAXMGUIDG'];
save datfil.txt output /ascii
output=[ArrayT',ArrayXLONGTDEG',ArrayXLATTDEG',ArrayXLONGMDEG',...
            ArrayXLATMDEG'];
csvwrite('trajfil.txt',output)
disp 'simulation finished'
% 3d Kepler, distance  and Lambert routines previously presented in this Chapter
```

The nominal case of Listing 28.4 was run in which there was 10 n miles of intercept point prediction error. Figure 28.10 displays the missile and target trajectories. We can see that intercept takes place about 2500 n miles from the target launch site at approximately 900 n miles altitude. Figure 28.11 plots the total commanded acceleration required for the missile to hit the target. Because homing guidance started at 800 s (200 s before intercept), the acceleration command was zero before that. A maximum acceleration of 0.1 g was required to take out the prediction error in 200 s.

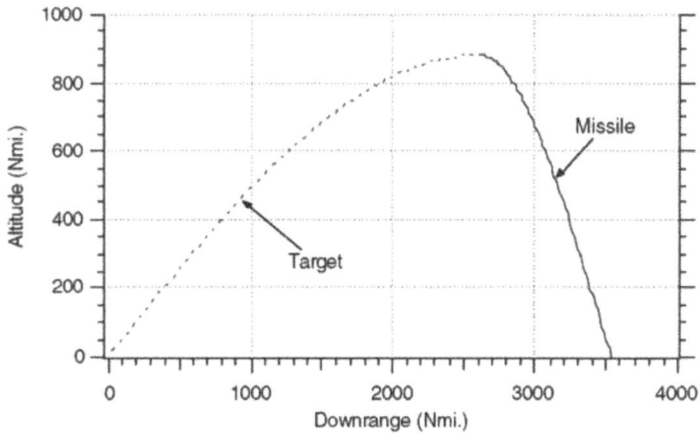

Fig. 28.10 Missile and target trajectories.

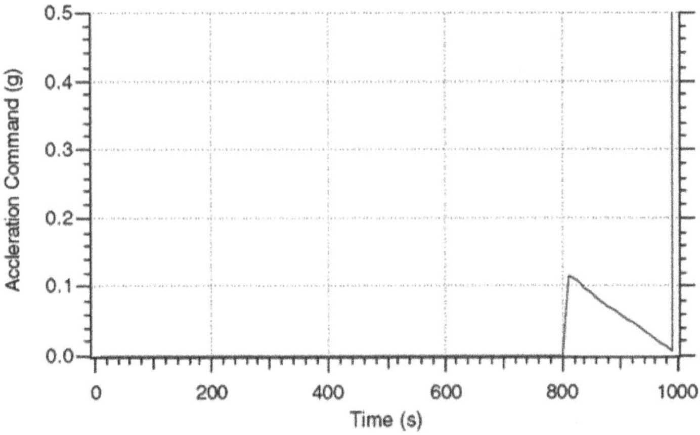

Fig. 28.11 Acceleration required to take out prediction error.

SUMMARY

This chapter has shown how three-dimensional simulations can be used to convey important information. In the tactical world we have seen how new insights against spiraling targets can be gained in three dimensions. We saw that the miss caused by a weaving target does not oscillate in three dimensions as it does in one or two dimensions but in fact approaches steady state as the flight time increases. Code has been provided to extend the Lambert routine to three dimensions in our strategic engagement simulations. New code for a Kepler subroutine has been provided so that we could do intercept point prediction for three dimensional ballistic targets without having to resort to numerical integration.

REFERENCES

[1] Miller, R., and Reddy, F., "Mapping the World in Pascal," *BYTE*, Vol. 12, Dec. 1987, pp. 329–334.

[2] Bate, R. R., Mueller, D. D., and White, J. E., *Fundamentals of Astrodynamics*, Dover, New York, 1971.

[3] Nelson, S. L., and Zarchan, P., "Alternative Approach to the Solution of Lambert's Problem," *Journal of Guidance, Control, and Dynamics*, Vol. 15, No. 4, 1992, pp. 1003–1009.

[4] Vinti, J. P., *Orbital and Celestial Mechanics*, G. J. Der, and N. L. Bonavito, eds., Progress in Astronautics and Aeronautics Series, AIAA, Reston, VA, 1998.

CHAPTER 29

Advanced Adjoint Applications

INTRODUCTION

Throughout this book we have discussed adjoints and their application to analyzing missile guidance systems. We have found that adjoints can be extremely valuable from both practical and computational points of view if the homing loop is linear (that is, without missile saturation). Single-run adjoint rms miss distance results have been shown to be equivalent to thousands of Monte Carlo runs generated from a forward engagement simulation. The value of adjoints has been not only demonstrated for continuous systems that could be described by a set of linear differential equations but also for mixed continuous discrete systems that could be described by a set of linear differential and difference equations.

In this chapter we shall demonstrate the utility of adjoints in two additional applications for the missile homing loop. In both of these applications we shall concentrate on systems that have a digital noise filter so that mixed continuous discrete adjoints must be used. In the first example we shall consider how to implement an adjoint if there are two samplers in the missile guidance system, each operating at different sampling rates. In this example a two-state, constant-gain digital fading memory filter is used as the noise filter. In the other example we shall consider a mixed continuous guidance system with a single sampler—but this time the digital noise filter is a three-state Kalman filter. Since the gains of the Kalman filter and associated guidance laws are both flight time dependent and time varying, a new interpretation of adjoint results is required.

MULTIPLE SAMPLING RATE ADJOINT

In some applications information is available more often than can be handled by the digital noise filter because of flight computer throughput considerations. In these applications a decision has to be made on what to do with the extra

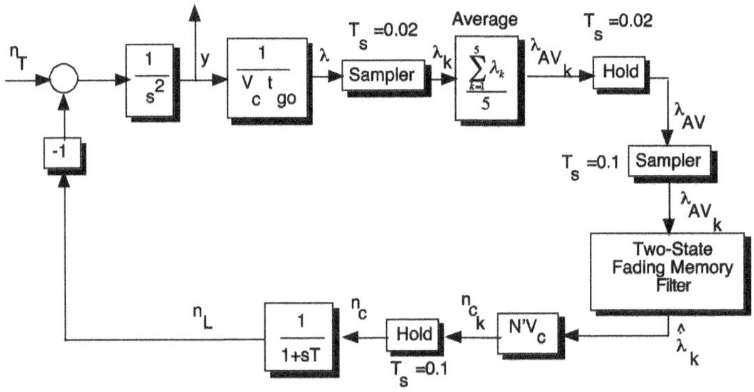

Fig. 29.1 General form of homing loop with two sampling rates and a fading memory filter.

measurements that cannot be utilized by the digital noise filter. One technique for handling the extra data is to preprocess the extra data [1] as shown in Fig. 29.1. In this example the normal sampling rate for the two-state digital fading memory noise filter is 10 Hz ($T_s = 0.1\,\text{s}$) but data is available at 50 Hz ($T_s = 0.02$ s). In other words we have five times as much data as can be used by the digital noise filter. In Fig. 29.1 the extra measurements are averaged by adding up five measurements and then dividing by five. The next time a measurement is available, the oldest measurement in the summation is dropped out while the new measurement is added so that the most recent five measurements are always being averaged. Note that the homing loop in Fig. 29.1 is similar to the mixed continuous discrete homing loop displayed in Fig. 7.1, except for the preprocessing of the measurement data and for the single time constant representation of the flight control system. The only source of error in the guidance system of Fig. 29.1 is a constant target maneuver n_T.

Figure 29.2 presents a more detailed block diagram of the homing loop with the two-state fading memory digital noise filter and two samplers—each operating at a different sampling rate. The input to the fading memory filter in the block diagram *pz* is the measured line-of-sight angle after it has been preprocessed and the output of the fading memory filter is estimated line-of-sight angle $\hat{\lambda}$ and estimated line-of-sight rate $\hat{\dot{\lambda}}$. The proportional navigation guidance law is implemented using the estimated line-of-sight rate according to

$$n_c = N'V_c\hat{\dot{\lambda}}$$

where N$'$ is the effective navigation ratio and V_c is the closing velocity. In addition to the fading memory filter, the only other lag that exists within the guidance system in this example is the single time constant representation of the flight control system with time constant T. The two gains of the fading memory filter

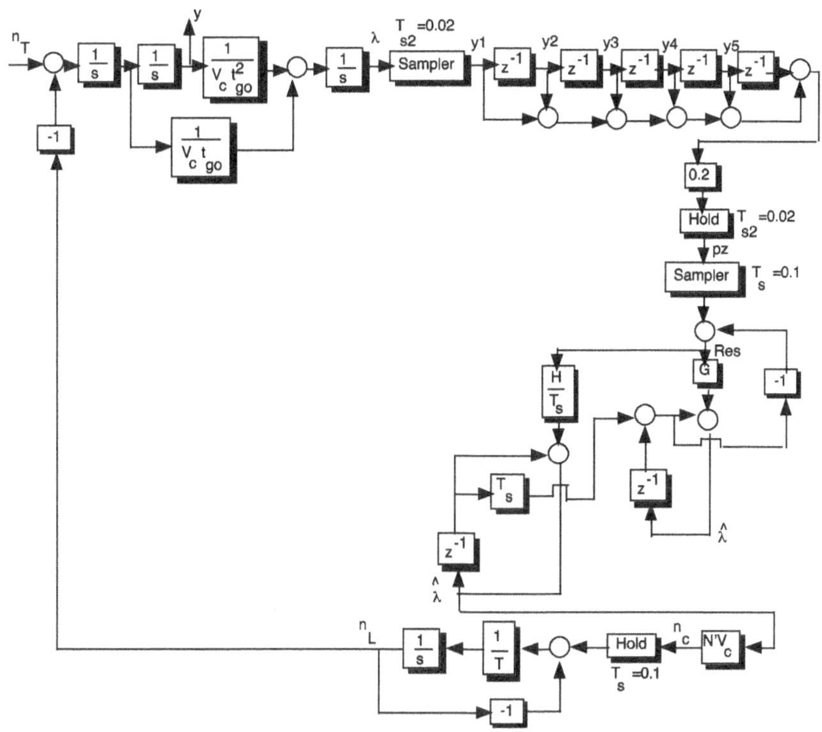

Fig. 29.2 Forward model of homing loop with two samplers operating at different rates.

G and H determine the bandwidth of the filter and guidance system. As was shown in Chapter 7 the two filter gains are computed from

$$G = 1 - \beta^2$$

$$H = (1 - \beta)^2$$

where β is a constant between 0 and 1 chosen by the designer. Higher values of β will result in a lower filter bandwidth or a more sluggish guidance system. Also note that in Fig. 29.2 that the line-of-sight angle has been computed in a laborious way. This is done to have an extra integrator in the system so that when we take the adjoint of the sampler (which will result in a differentiator) a simple cancellation of an integrator and differentiator can take place. The adjoints of sampler and hold blocks can be found in Chapter 7. To remind readers that the line-of-sight angle can be obtained in the laborious way indicated in Fig. 29.2 we start from the small angle definition of the line-of-sight angle which is given by

$$\lambda = \frac{y}{R_{TM}}$$

Differentiation of the preceding expression using the quotient rule from calculus yields

$$\dot{\lambda} = \frac{R_{TM}\dot{y} - y\dot{R}_{TM}}{R_{TM}^2}$$

Because the range from missile to target is given by

$$R_{TM} = V_c t_{go} = V_c(t_F - t)$$

the line-of-sight rate can be expressed as

$$\dot{\lambda} = \frac{V_c t_{go}\dot{y} + yV_c}{V_c^2 t_{go}^2} = \frac{y}{V_c t_{go}^2} + \frac{\dot{y}}{V_c t_{go}}$$

Integration of the preceding equation yields the line-of-sight angle as shown in Fig. 29.2. The line-of-sight angle is sampled every 0.02 s (50-Hz rate), passed through a preprocessing section and then passed through a zero-order hold. The output of the zero-order hold is then sampled at 0.1 s (10 Hz rate) and passed through the fading memory filter whose gains are calculated using $\beta = 0.8$. The term z^{-1} in Fig. 29.2 represents a pure delay. Again, it is important to emphasize that the forward model of the homing loop is presented in this way to facilitate the generation of an adjoint block diagram.

The system parameters for the guidance system of Fig. 29.2 appears in Table 29.1. The forward model of Fig. 29.2 is programmed and appears in Listing 29.1. However in the forward model simulation the line-of-sight angle

TABLE 29.1 SYSTEM INPUTS FOR MULTIPLE SAMPLING RATE SYSTEM

Symbol	Definition	Value
V_c	Closing velocity	4000 ft/s
T	Flight control system time constant	0.2 s
n_T	Target maneuver amplitude	96.6 ft/s^2 (3 g)
β	Fading memory filter tuning parameter	0.8
N'	Effective navigation ratio	3
T_s	Filter sampling time	0.1 s
T_{s2}	System sampling time	0.02 s

does not have to be computed in the laborious way of Fig. 29.2 but instead can be is computed more easily from the definition of the line-of-sight angle as

$$\lambda = \frac{y}{R_{TM}}$$

It is important to note that Listing 29.1 is nearly identical to Listing 7.2 except for the two samplers, the preprocessing section, and the single time constant representation of the flight control system. In Listing 7.2 there was only one sampling time set at 0.1 s and there were two sources of error: random target maneuver and measurement noise. In Listing 29.1 we only consider a deterministic step target maneuver that always starts at the beginning of the flight and whose amplitude is always 3 g (96.6 ft/s^2). Because there is no noise disturbance in Listing 29.1, it does not have to be run in the Monte Carlo mode as was done for Listing 7.2. However, 100 runs are made in the forward model of Listing 29.1—each with different flight times. The flight times vary between 0.1 s and 10 s in steps of 0.1 s. The miss distances for each of the flight times are written to a file. It is important to note that in this forward model simulation the high data rate sampler is encountered first. Thus the difference equations from the high data rate sampler are programmed first.

LISTING 29.1 SIMULATION OF FORWARD MODEL OF HOMING LOOP WITH TWO SAMPLERS OPERATING AT DIFFERENT RATES

```
count=0;
VC=4000.;
TAP=.2;
XNT=96.6;
BETA=.8;
XNP=3.;
TS=.1;
TS2=.02;
for TF=.1:.1:10.0,
	Y=0.;
	YD=0.;
	T=0.;
	H=.001;
	S=0.;
	S2=0.;
	GFILTER=1.-BETA^2;
	HFILTER=(1.-BETA)^2;
	XLAMHOLD=0.;
	XLAMDHOLD=0.;
	Y1OLD=0.;
```

```
Y2OLD=0.;
Y3OLD=0.;
Y4OLD=0.;
Y5OLD=0.;
XNC=0.;
XNL=0.;
Y1NEW=0.;
PZ=0.;
XLAM=0.;
while T<=(TF - 1e-5)
    YOLD=Y;
    YDOLD=YD;
    XNLOLD=XNL;
    STEP=1;
    FLAG=0;
    while STEP<=1
        if FLAG==1
                Y=Y+H*YD;
                YD=YD+H*YDD;
                XNL=XNL+H*XNLD;
                T=T+H;
                STEP=2;
        end;
        TGO=TF-T+.00001;
        RTM=VC*TGO;
        XLAM=Y/(VC*TGO);
        XLAMD=(RTM*YD+Y*VC)/(RTM^2);
        XNLD=(XNC-XNL)/TAP;
        YDD=XNT-XNL;
        FLAG=1;
    end;
    FLAG=0;
    Y=.5*(YOLD+Y+H*YD);
    YD=.5*(YDOLD+YD+H*YDD);
    XNL=.5*(XNLOLD+XNL+H*XNLD);
    S=S+H;
    S2=S2+H;
    if S2>=(TS2 - 1e-5)
        S2=0.;
        Y1NEW=XLAM;
        Y2NEW=Y1OLD;
        Y3NEW=Y2OLD;
        Y4NEW=Y3OLD;
        Y5NEW=Y4OLD;
        PZ=.2*(Y5OLD+Y5NEW+Y4NEW+Y3NEW+Y2NEW+XLAM);
        Y1OLD=Y1NEW;
```

```
            Y2OLD=Y2NEW;
            Y3OLD=Y3NEW;
            Y4OLD=Y4NEW;
            Y5OLD=Y5NEW;
        end;
        if S>=(TS - 1e-5)
            S=0.;
            RES=PZ-(XLAMHOLD+TS*XLAMDHOLD);
            XLAMHNEW=GFILTER*RES+XLAMHOLD+TS*XLAMDHOLD;
            XLAMDHNEW=HFILTER*RES/TS+XLAMDHOLD;
            XNC=XNP*VC*XLAMDHNEW;
            XLAMHOLD=XLAMHNEW;
            XLAMDHOLD=XLAMDHNEW;
        end;
    end;
    count=count+1;
    ArrayTF(count)=TF;
    ArrayY(count)=Y;
end;
figure
plot(ArrayTF',ArrayY'),grid
title('Standard miss for various flight times')
xlabel('Flight Time (S)')
ylabel('Miss (Ft) ')
axis([00,10,-40,100])
clc
output=[ArrayTF',ArrayY'];
save datfil.txt output /ascii
disp('Simulation Complete')
```

Figure 29.3 presents the adjoint model of the homing loop with two samplers operating at different rates derived from the forward model of Fig. 29.2. Figure 29.3 is constructed following the adjoint rules discussed in Chapters 3 and 7. As was previously mentioned, the adjoints of each sampler and hold are taken according to the rules of adjoints of Chapter 7. Recall that the adjoint of a sampler yields a difference equation and a differentiator. Numerical differentiation is avoided in Fig. 29.3 because there is cancellation of the differentiator with an integrator.

The adjoint of the homing loop with two samplers operating at different rates is programmed in Listing 29.2. Because the signal flow is reversed in the adjoint model, the low data rate sampler will be encountered first. This is the opposite of what happened in the forward model. Therefore in the adjoint listing the difference equations for the low data rate sampler are programmed first. Also note that there is no "for loop" in the adjoint listing and so all the miss distance results are generated in one computer run.

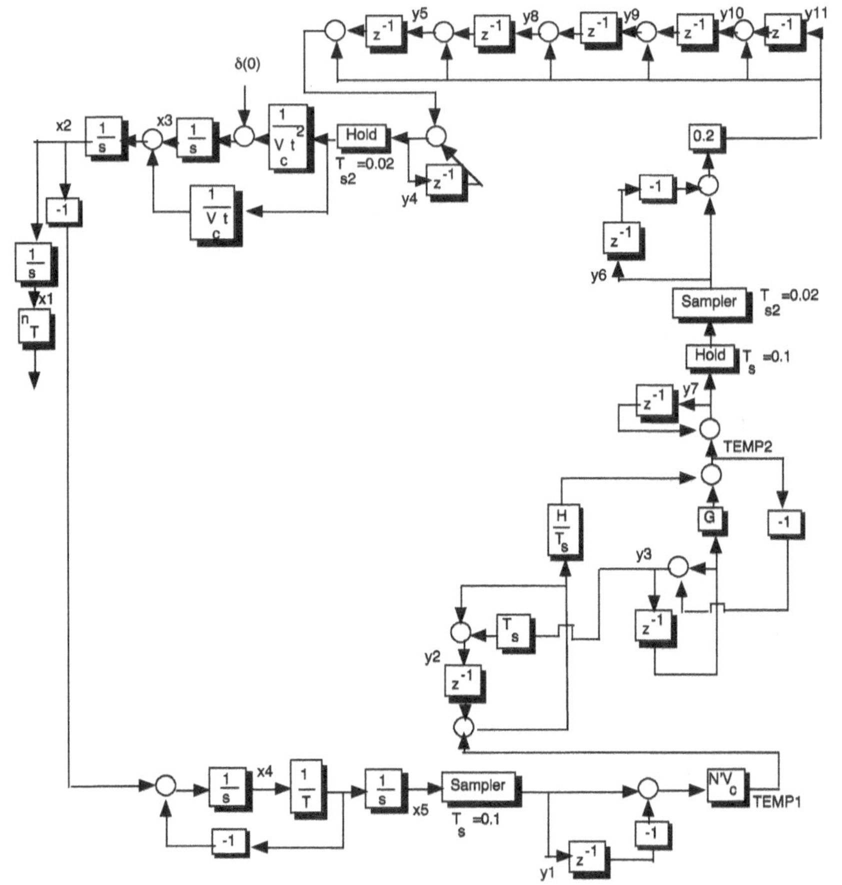

Fig. 29.3 Adjoint model of homing loop with two samplers operating at different rates.

LISTING 29.2 SIMULATION OF ADJOINT MODEL OF HOMING LOOP WITH TWO SAMPLERS OPERATING AT DIFFERENT RATES

```
count=1;
XNT=96.6;
XNP=3;
TF=10.;
TS=.1;
BETA=.8;
TAP=.2;
VC=4000.;
T=0.;
```

```
S=0.;
S2=0.;
TS2=.02;
TP=T+.00001;
X1=0;
X2=0;
X3=1;
X4=0.;
X5=0.;
Y1OLD=0.;
Y2OLD=0.;
Y3OLD=0.;
Y4OLD=0.;
Y5OLD=0.;
Y6OLD=0.;
Y7OLD=0.;
Y8OLD=0.;
Y9OLD=0.;
Y10OLD=0.;
Y11OLD=0.;
Y11NEW=0.;
Y10NEW=0.;
Y9NEW=0.;
Y7NEW=0.;
Y6NEW=0.;
Y8NEW=0.;
Y4NEW=0.;
Y1NEW=0.;
Y2NEW=0.;
Y3NEW=0.;
Y5NEW=0.;
H=.001;
GFILTER=1.-BETA^2;
HFILTER=(1.-BETA)^2;
XMNT=0.;
while TP<=(TF-1e-5)
        S=S+H;
        S2=S2+H;
        X1OLD=X1;
        X2OLD=X2;
        X3OLD=X3;
        X4OLD=X4;
        X5OLD=X5;
        STEP=1;
        FLAG=0;
        while STEP<=1
```

```
        if FLAG==1
            STEP=2;
            X1=X1+H*X1D;
            X2=X2+H*X2D;
            X3=X3+H*X3D;
            X4=X4+H*X4D;
            X5=X5+H*X5D;
            TP=TP+H;
        end
        TGO=TP;
        X1D=X2;
        X2D=X3+Y4NEW/(VC*TGO);
        X3D=(Y4NEW)/(VC*TGO*TGO);
        X4D=-X2-X4/TAP;
        X5D=X4/TAP;
        FLAG=1;
    end
    FLAG=0;
    X1=(X1OLD+X1)/2+.5*H*X1D;
    X2=(X2OLD+X2)/2+.5*H*X2D;
    X3=(X3OLD+X3)/2+.5*H*X3D;
    X4=(X4OLD+X4)/2+.5*H*X4D;
    X5=(X5OLD+X5)/2+.5*H*X5D;
    if S>=(TS-.0001)
        S=0.;
        Y1NEW=X5;
        TEMP1=(Y1NEW-Y1OLD)*XNP*VC;
        TEMP2=HFILTER*(Y2OLD+TEMP1)/TS+GFILTER*Y3OLD;
        Y2NEW=TEMP1+Y2OLD+TS*(Y3OLD-TEMP2);
        Y3NEW=Y3OLD-TEMP2;
        Y7NEW=TEMP2+Y7OLD;
        Y1OLD=Y1NEW;
        Y2OLD=Y2NEW;
        Y3OLD=Y3NEW;
        Y7OLD=Y7NEW;
    end
    if S2>=(TS2-.0001)
        S2=0.;
        Y6NEW=Y7NEW;
        Y11NEW=.2*(Y6NEW-Y6OLD);
        Y10NEW=Y11OLD+Y11NEW;
        Y9NEW=Y10OLD+Y11NEW;
        Y8NEW=Y9OLD+Y11NEW;
        Y5NEW=Y8OLD+Y11NEW;
        Y4NEW=Y4OLD+Y5OLD+Y11NEW;
        Y4OLD=Y4NEW;
```

```
            Y6OLD=Y6NEW;
            Y5OLD=Y5NEW;
            Y8OLD=Y8NEW;
            Y9OLD=Y9NEW;
            Y10OLD=Y10NEW;
            Y11OLD=Y11NEW;
            XMNT=XNT*X1;
            count=count+1;
            ArrayTP(count)=TP;
            ArrayXMNT(count)=XMNT;
      end
end
figure
plot(ArrayTP,ArrayXMNT),grid
xlabel('Flight Time (Sec)')
ylabel('Target Maneuver Miss (Ft)')
clc
output=[ArrayTP',ArrayXMNT'];
save datfil.txt output -ascii
disp 'simulation finished'
```

The nominal cases of Listing 29.1 (equivalent to 100 runs) and of Listing 29.2 (one run) were run and the results for the miss due to a constant 3-g maneuver for different flight times is displayed in Fig. 29.4. We can see that there is exact agreement between the adjoint and forward simulations indicating that we have

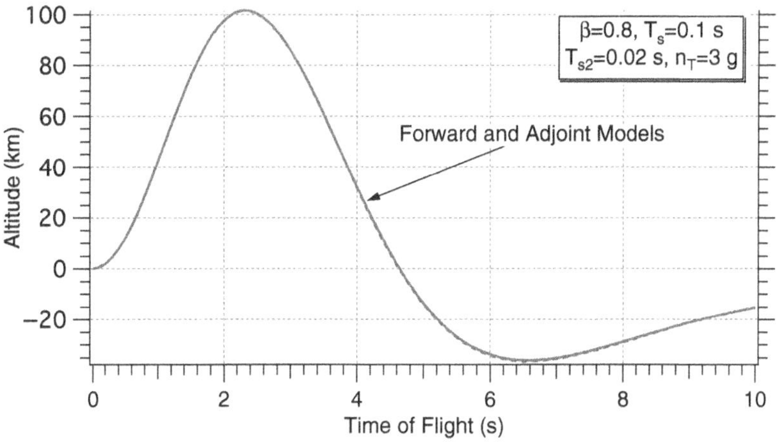

Fig. 29.4 Forward and adjoint models agree for homing loop with two samplers operating at different sampling rates.

correctly taken the adjoint of the homing loop with multiple samplers. We can see from Fig. 29.4 that the miss distances can be quite large due to the sluggishness of the fading memory filter ($\beta = 0.8$). Figure 29.5 shows that when the bandwidth of the fading memory filter is increased ($\beta = 0.3$) the guidance system performance improves and that there is still excellent agreement between the adjoint and forward models, again indicating that the adjoint has been taken correctly.

In the forward model the 3-g target maneuver always started at the beginning of flight and the flight time was varied from 0.1 s to 10 s in steps of 0.1 s. If we modify Listing 29.1 so that the time of flight is always 10 s but the time to go at which the 3-g maneuver occurs is varied from 0.1 s to 10 s in steps of 0.1 s, we can see from Fig. 29.6 that we get exactly the same results as that of Fig. 29.4. Thus we can see that adjoint time can either be interpreted as time of flight for a maneuver occurring at the beginning of flight or the time to go at which the maneuver occurs for a fixed time of flight.

ADJOINT OF DISCRETE LINEAR KALMAN FILTER

So far in this book whenever adjoints have been taken the noise filter was either a low pass filter which could be represented by a differential equation or a simple fading memory filter that could be represented by a set of difference equations. For more advanced guidance systems a Kalman filter is normally required as the noise filter. Although the Kalman filter has the same structure as the fading memory filter, its gains are not constant and must be computed from the Ricatti equations.

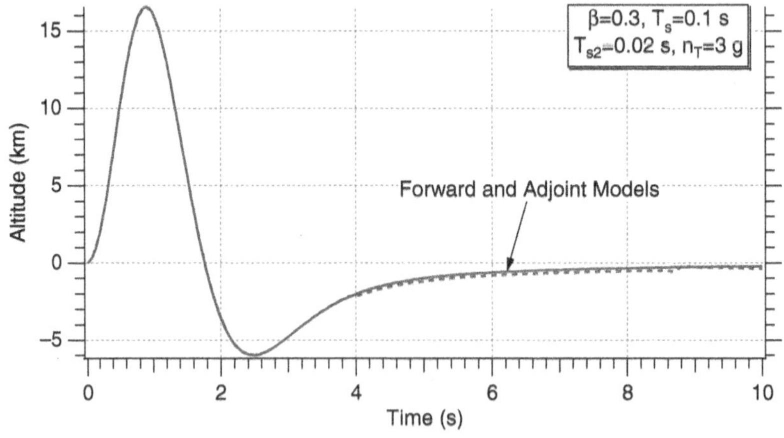

Fig. 29.5 Increasing sampling rate of first sampler causes miss distance to decrease.

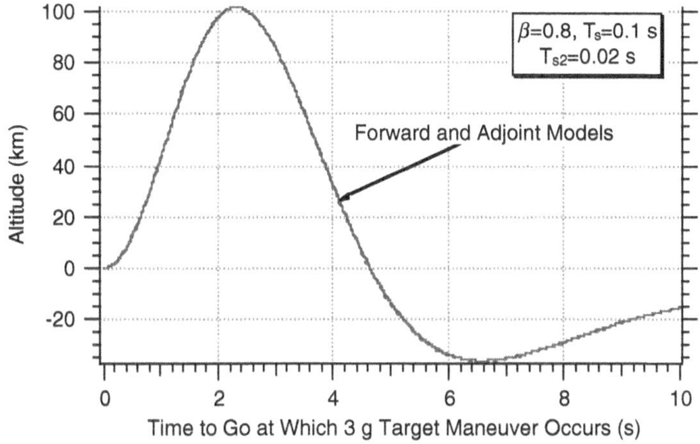

Fig. 29.6 Agreement between adjoint and forward models when target maneuver starting time is varied.

More advanced guidance laws can be used if a three-state Kalman filter is part of the guidance system. As the states of the three-state Kalman filter generally used in this text are relative position, relative velocity, and target acceleration possible guidance laws must be expressed in a slighly different way. As was shown in Chapters 8 and 9, the various possible guidance laws can be expressed in terms of the Kalman filter state estimates and control gains according to

$$n_c = C_1\hat{y}_k + C_2\hat{\dot{y}}_k + C_3\hat{n}_{T_k} + C_4 n_{L_k}$$

where $\hat{y}_k$ is the relative position estimate, $\hat{\dot{y}}_k$ is the relative velocity estimate, $\hat{n}_{T_k}$ is the target acceleration estimate and n_{L_k} is the missile achieved acceleration. The gains C_1, C_2, C_3, and C_4 are known as control gains. For proportional navigation the control gains are

$$C_1 = \frac{N'}{t_{\text{go}}^2}$$

$$C_2 = \frac{N'}{t_{\text{go}}}$$

$$C_3 = 0$$

$$C_4 = 0$$

For augmented proportional navigation the control gains become

$$C_1 = \frac{N'}{t_{\rm go}^2}$$

$$C_2 = \frac{N'}{t_{\rm go}}$$

$$C_3 = 0.5N'$$

$$C_4 = 0$$

With the two preceding guidance laws N′ was a constant. If we use optimal guidance the control gains become

$$C_1 = \frac{N'}{t_{go}^2}$$

$$C_2 = \frac{N'}{t_{go}}$$

$$C_3 = 0.5N'$$

$$C_4 = \frac{-N'T^2(e^{-x} + x - 1)}{t_{go}^2}$$

where T is the time constant of the flight control system and N' is no longer a constant but is given by

$$N' = \frac{6x^2(e^{-x} - 1 + x)}{2x^3 + 3 + 6x - 6x^2 - 12xe^{-x} - 3e^{-2x}}$$

and x is defined as

$$x = \frac{t_{go}}{T}$$

Figure 29.7 presents a block diagram of the homing loop with a three-state linear Kalman filter as the digital noise filter. In this diagram the line-of-sight angle measurement λ_k is multiplied by range to convert the actual measurement to a pseudo measurement of relative position y_k^*. The Kalman filter estimates relative position $\hat{y}_k$, relative velocity $\hat{\dot{y}}_k$, and target acceleration $\hat{n}_{T_k}$. In this homing loop there is also a single-lag representation of the flight control system with time constant T. Both the Kalman gains K_1, K_2 and K_3 and the control gains C_1, C_2, C_3, and C_4 are functions of time. As was previously mentioned the control gains determine which guidance law is used in conjunction with the Kalman filter.

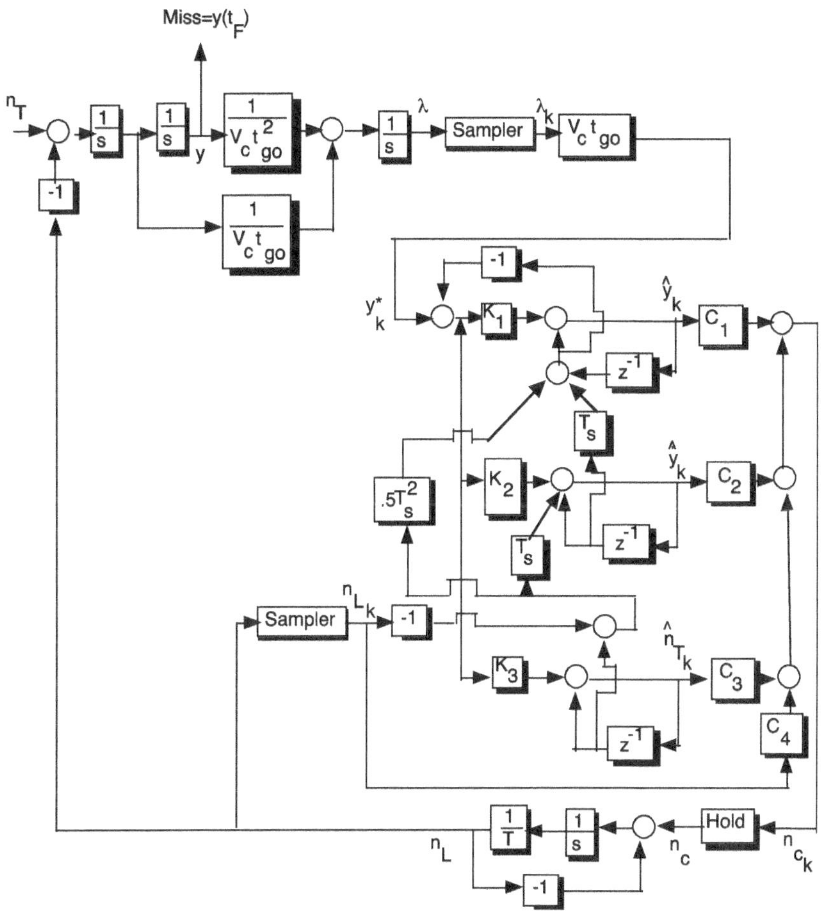

Fig. 29.7 Forward model of homing loop with discrete three-state Kalman filter.

The forward model of the Kalman filter in the homing loop of Listing 29.3, like Listing 29.1, has only one source of error—constant target maneuver. However, here the flight time is fixed at 10 s and the time to go at which the 3-*g* target maneuver begins is made a parameter in the "for loop." One hundred runs are made in the forward model with the maneuver starting time varying between 0.1 s and 10 s in steps of 0.1 s. This means that the time to go at which the maneuver occurs is varying between 9.9 and 0 s in steps of 0.1 s. The miss distances for each of the times to go at which the maneuver starts for the 10 s flight are written to a file. Note that in this forward model the sampling time is 0.1 s. For each run we solve for the Kalman gains via the Ricatti equations, although this is not necessary as the flight time is fixed and the gains will be the same from run to run.

LISTING 29.3 SIMULATION OF FORWARD MODEL OF HOMING LOOP WITH THREE-STATE KALMAN FILTER AND VARIOUS GUIDANCE LAW OPTIONS

```
count=0;
VM=3000.;
VC=4000.;
XNT=96.6;
YIC=0.;
HEDEGF=20.;
XNP=3.;
SIGNOISE=.001;
TS=.1;
TAU=.5;
APN=0;
TF=10.;
TS2=TS*TS;
TS3=TS2*TS;
TS4=TS3*TS;
TS5=TS4*TS;
PHIN=XNT*XNT/TF;
for TSTART=.1:.1:10.0,
	Y=YIC;
	YD=0.;
	T=0.;
	H=.01;
	S=0.;
	RTM=VC*TF;
	SIGPOS=RTM*SIGNOISE;
	SIGN2=SIGPOS^2;
	P11=SIGN2;
	P12=0.;
	P13=0.;
	P22=(VM*HEDEGF/57.3)^2;
	P23=0.;
	P33=XNT*XNT;
	YH=0.;
	YDH=0.;
	XNTH=0.;
	XNC=0.;
	XNL=0.;
	while T<=(TF - 1e-5)
		YOLD=Y;
		YDOLD=YD;
		XNLOLD=XNL;
		STEP=1;
		FLAG=0;
```

```
while STEP <=1
    if FLAG==1
            Y=Y+H*YD;
            YD=YD+H*YDD;
            XNL=XNL+H*XNLD;
            T=T+H;
            STEP=2;
    end;
    TGO=TF-T+.00001;
    RTM=VC*TGO;
    XLAM=Y/(VC*TGO);
    XLAMD=(RTM*YD+Y*VC)/(RTM^2);
    XNLD=(XNC-XNL)/TAU;
    if T>TSTART
            YDD=XNT-XNL;
    else
            YDD=0.;
    end;
    FLAG=1;
end;
FLAG=0;
Y=.5*(YOLD+Y+H*YD);
YD=.5*(YDOLD+YD+H*YDD);
XNL=.5*(XNLOLD+XNL+H*XNLD);
S=S+H;
if S>=(TS - 1e-5)
    S=0.;
    TGO=TF-T+.000001;
    RTM=VC*TGO;
    SIGPOS=RTM*SIGNOISE;
    SIGN2=SIGPOS^2;
    M11=P11+TS*P12+.5*TS2*P13+TS*(P12+TS*P22+.5*TS2*P23);
    M11=M11+.5*TS2*(P13+TS*P23+.5*TS2*P33)+TS5*PHIN/20.;
    M12=P12+TS*P22+.5*TS2*P23+TS*(P13+TS*P23+.5*TS2*P33)+TS4*PHIN/8.;
    M13=P13+TS*P23+.5*TS2*P33+PHIN*TS3/6.;
    M22=P22+TS*P23+TS*(P23+TS*P33)+PHIN*TS3/3.;
    M23=P23+TS*P33+.5*TS2*PHIN;
    M33=P33+PHIN*TS;
    K1=M11/(M11+SIGN2);
    K2=M12/(M11+SIGN2);
    K3=M13/(M11+SIGN2);
    P11=(1.-K1)*M11;
    P12=(1.-K1)*M12;
    P13=(1.-K1)*M13;
    P22=-K2*M12+M22;
    P23=-K2*M13+M23;
```

```
                P33=-K3*M13+M33;
                XLAMNOISE=0.;
                YSTAR=RTM*(XLAM+XLAMNOISE);
                RES=YSTAR-YH-TS*YDH-.5*TS*TS*(XNTH-XNL);
                YH=K1*RES+YH+TS*YDH+.5*TS*TS*(XNTH-XNL);
                YDH=K2*RES+YDH+TS*(XNTH-XNL);
                XNTH=K3*RES+XNTH;
                XLAMDH=(YH+YDH*TGO)/(VC*TGO*TGO);
                if APN==0
                        C1=XNP/TGO^2;
                        C2=XNP/TGO;
                        XNC=C1*YH+C2*YDH;
                elseif APN==1
                        C1=XNP/TGO^2;
                        C2=XNP/TGO;
                        C3=.5*XNP;
                        XNC=C1*YH+C2*YDH+C3*XNTH;
                else
                        X=TGO/TAU;
                        TOP=6.*X*X*(exp(-X)-1.+X);
                        BOT1=2*X*X*X+3.+6.*X-6.*X*X;
                        BOT2=-12.*X*exp(-X)-3.*exp(-2.*X);
                        XNPP=TOP/(.0001+BOT1+BOT2);
                        XNEW=XNPP*XNL*(exp(-X)+X-1.)/(X*X);
                        C1=XNPP/TGO^2;
                        C2=XNPP/TGO;
                        C3=.5*XNPP;
                        C4=-XNPP*(exp(-X)+X-1.)/(X*X);
                        XNC=C1*YH+C2*YDH+C3*XNTH+C4*XNL;
                end;
            end;
        end;
        TGOS=TF-TSTART;
        count=count+1;
        ArrayTGOS(count)=TGOS;
        ArrayY(count)=Y;
end;
figure
plot(ArrayTGOS',ArrayY'),grid
title('Miss for various tgo maneuver starting times')
xlabel('Time to go at which maneuver occurs (S)')
ylabel('Miss (Ft) ')
clc
output=[ArrayTGOS',ArrayY'];
save datfil.txt output /ascii
disp('Simulation Complete')
```

As was demonstrated in the previous section when adjoint results were generated for the fading memory filter, adjoint time can either be interpreted as missile flight time or time to go at which the target maneuver started. If we model a Kalman filter using the adjoint technique we must reverse all time varying gains (meaning Kalman gains and control gains) according to the rules of adjoints. Because the Kalman gains for a 10 s flight will be different than for a 5 s flight we must fix the flight time when modeling the adjoint of a Kalman filter. Therefore in this case adjoint time can only be interpreted as the time to go at which the disturbance occurs. Following the rules of adjoints shown in Chapter 7 for a mixed continuous discrete system we obtain Fig. 29.8.

The adjoint of the homing loop with the three-state Kalman filter and various guidance law options is programmed in Listing 29.4. Note that there is no "for loop" in the adjoint listing and so all the miss distance results will be generated

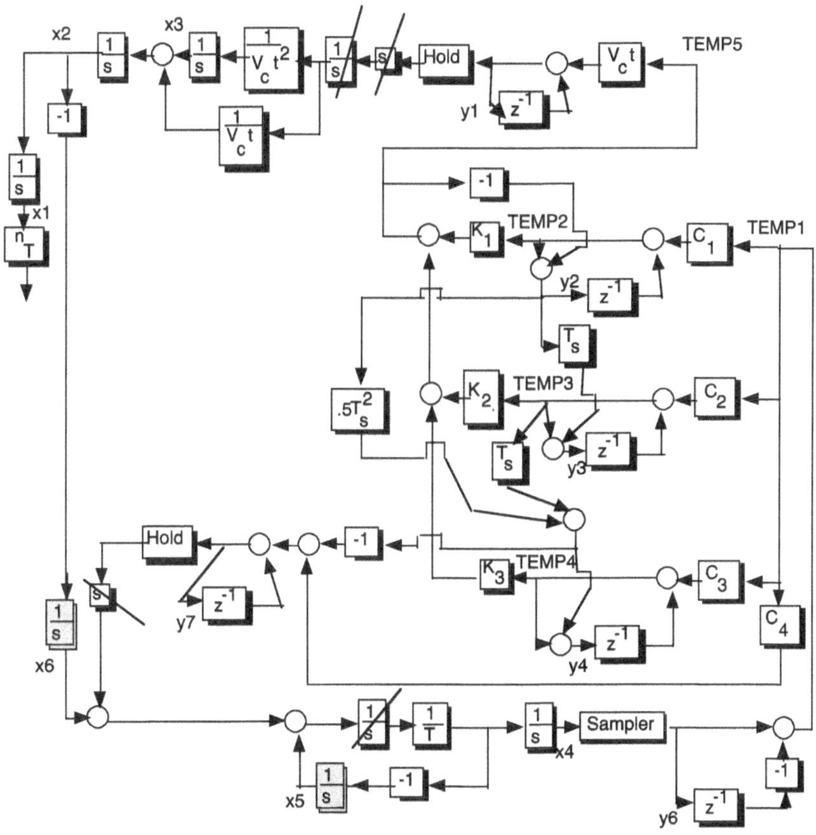

Fig. 29.8 Adjoint model of homing loop with discrete three-state Kalman filter.

in one computer run. However, as was mentioned previously adjoint time must be interpreted as time to go at which the constant maneuver starts for a 10 s flight. Also note that that the Kalman gains for a 10 s flight are first computed and stored as an array. After the Ricatti equations are finished the gains are reversed in time and also stored as an array. This array is used to provide the Kalman gains for each adjoint time.

LISTING 29.4 AJOINT OF HOMING LOOP WITH THREE-STATE KALMAN FILTER AND VARIOUS GUIDANCE LAW OPTIONS

```
count=0;
XNT=96.6;
XNP=3.;
TF=10;
TS=.1;
APN=0;
VM=3000.;
VC=4000.;
HEDEGF=20.;
SIGNOISE=.001;
TS2=TS*TS;
TS3=TS2*TS;
TS4=TS3*TS;
TS5=TS4*TS;
PHIN=XNT*XNT/TF;
RTM=VC*TF;
SIGPOS=RTM*SIGNOISE;
SIGN2=SIGPOS^2;
P11=SIGN2;
P12=0.;
P13=0.;
P22=(VM*HEDEGF/57.3)^2;
P23=0.;
P33=XNT*XNT;
C=0;
for T=TS:TS:TF
    TGO=TF-T+.000001;
    RTM=VC*TGO;
    SIGPOS=RTM*SIGNOISE;
    SIGN2=SIGPOS^2;
    M11=P11+TS*P12+.5*TS2*P13+TS*(P12+TS*P22+.5*TS2*P23);
    M11=M11+.5*TS2*(P13+TS*P23+.5*TS2*P33)+TS5*PHIN/20.;
    M12=P12+TS*P22+.5*TS2*P23+TS*(P13+TS*P23+.5*TS2*P33)+TS4*PHIN/8.;
    M13=P13+TS*P23+.5*TS2*P33+PHIN*TS3/6.;
    M22=P22+TS*P23+TS*(P23+TS*P33)+PHIN*TS3/3.;
```

```
        M23=P23+TS*P33+.5*TS2*PHIN;
        M33=P33+PHIN*TS;
        K1=M11/(M11+SIGN2);
        K2=M12/(M11+SIGN2);
        K3=M13/(M11+SIGN2);
        P11=(1.-K1)*M11;
        P12=(1.-K1)*M12;
        P13=(1.-K1)*M13;
        P22=-K2*M12+M22;
        P23=-K2*M13+M23;
        P33=-K3*M13+M33;
        C=C+1;
        U(C)=K1;
        V(C)=K2;
        W(C)=K3;
end;
% Modification
% --------------------
ICOUNT=round((TF/TS));
E(ICOUNT)=0;
F(ICOUNT)=0;
G(ICOUNT)=0;
ICOUNT=ICOUNT-1;
% --------------------
for I=1:1:ICOUNT
        REV=ICOUNT-I+1;
        E(REV)=U(I);
        F(REV)=V(I);
        G(REV)=W(I);
end;
TAP=.5;
VC=4000.;
T=0.;
S=0.;
TP=T+.00001;
X1=0;
X2=0;
X3=1;
X4=0.;
X5=0.;
X6=0.;
Y1OLD=0.;
Y2OLD=0.;
Y3OLD=0.;
Y4OLD=0.;
Y5OLD=0.;
```

```
Y6OLD=0.;
Y7OLD=0.;
Y7NEW=0.;
Y6NEW=0.;
Y8NEW=0.;
Y4NEW=0.;
Y1NEW=0.;
Y2NEW=0.;
Y3NEW=0.;
Y5NEW=0.;
H=.01;
XMNT=0.;
I=1;
while TP <= (TF - 1e-5)
	X1OLD=X1;
	X2OLD=X2;
	X3OLD=X3;
	X4OLD=X4;
	X5OLD=X5;
	X6OLD=X6;
	STEP=1;
	FLAG=0;
	while STEP<=1
		if FLAG==1
			X1=X1+H*X1D;
			X2=X2+H*X2D;
			X3=X3+H*X3D;
			X4=X4+H*X4D;
			X5=X5+H*X5D;
			X6=X6+H*X6D;
			TP=TP+H;
			STEP=2;
		end;
		TGO=TP;
		X1D=X2;
		X2D=X3+Y1NEW/(VC*TGO);
		X3D=(Y1NEW)/(VC*TGO*TGO);
		X4D=(X5+Y7NEW+X6)/TAP;
		X5D=-X4D;
		X6D=-X2;
		FLAG=1;
	end
	FLAG=0;
	X1=(X1OLD+X1)/2+.5*H*X1D;
	X2=(X2OLD+X2)/2+.5*H*X2D;
	X3=(X3OLD+X3)/2+.5*H*X3D;
```

```
    X4=(X4OLD+X4)/2+.5*H*X4D;
    X5=(X5OLD+X5)/2+.5*H*X5D;
    X6=(X6OLD+X6)/2+.5*H*X6D;
  S=S+H;
    if S>(TS-.0001)
        S=0.;
        K1=E(I);
        K2=F(I);
        K3=G(I);
    [TP I E(I) TS]
        I=I+1;
        if APN==0
            C1=XNP/TP^2;
            C2=XNP/TP;
            C3=0.;
            C4=0.;
        elseif APN==1
            C1=XNP/TP^2;
            C2=XNP/TP;
            C3=.5*XNP;
            C4=0.;
        else
            X=TP/TAP;
            TOP=6.*X*X*(exp(-X)-1.+X);
            BOT1=2*X*X*X+3.+6.*X-6.*X*X;
            BOT2=-12.*X*exp(-X)-3.*exp(-2.*X);
            XNPP=TOP/(.0001+BOT1+BOT2);
            C1=XNPP/TP^2;
            C2=XNPP/TP;
            C3=.5*XNPP;
            C4=-XNPP*(exp(-X)+X-1.)/(X*X);
    end
        TEMP1=X4-Y6OLD;
        TEMP2=C1*TEMP1+Y2OLD;
        TEMP3=C2*TEMP1+Y3OLD;
        TEMP4=C3*TEMP1+Y4OLD;
        TEMP5=K1*TEMP2+K2*TEMP3+K3*TEMP4;
        Y1NEW=Y1OLD+TEMP5*VC*TP;
        Y2NEW=TEMP2-TEMP5;
        Y3NEW=TEMP3+TS*Y2NEW;
        Y4NEW=TEMP4+TS*TEMP3+.5*TS*TS*Y2NEW;
        Y5=-(TS*TEMP3+.5*TS*TS*Y2NEW);
        Y7NEW=Y7OLD+C4*TEMP1+Y5;
        Y6NEW=X4;
        XMNT=XNT*X1;
        Y1OLD=Y1NEW;
```

```
            Y2OLD=Y2NEW;
            Y3OLD=Y3NEW;
            Y4OLD=Y4NEW;
            Y6OLD=Y6NEW;
            Y7OLD=Y7NEW;
        count=count+1;
        ArrayTP(count)=TP;
        ArrayXMNT(count)=XMNT;
        ArrayK1(count)=K1;
    end
end
figure
plot(ArrayTP',ArrayXMNT'),grid
title('Miss for various tgo maneuver starting times')
xlabel('Time to go at which maneuver occurs (S)')
ylabel('Miss (Ft) ')
clc
output=[ArrayTP',ArrayXMNT',ArrayK1'];
save datfil.txt output /ascii
disp('Simulation Complete')
```

The nominal cases of Listing 29.3 (equivalent to 100 runs) and of Listing 29.4 (one run) were run for each of the guidance law options (APN = 0, 1, and 2) and the results are displayed in Figs. 29.9 through 29.11. First we can see that there is exact agreement between the adjoint and forward simulations

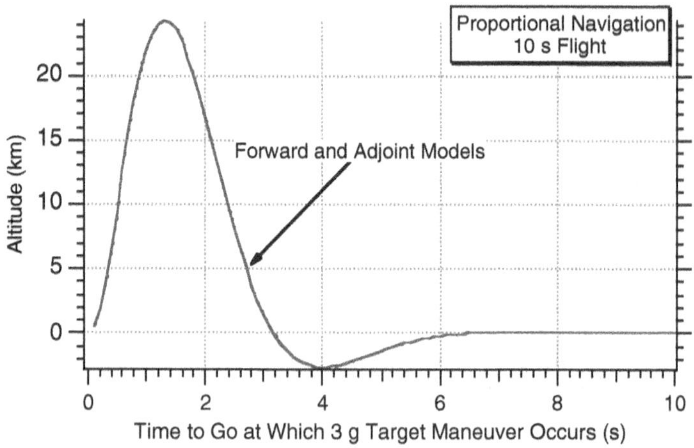

Fig. 29.9 Forward and adjoint models agree for homing loop with thee-state Kalman filter when proportional navigation is used.

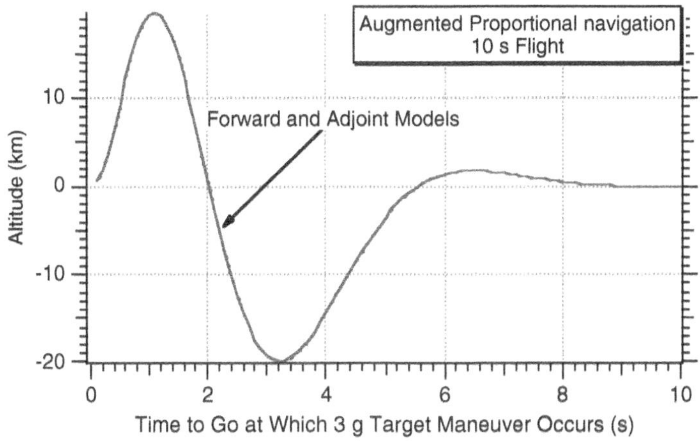

Figure 29.10 Forward and adjoint models agree for homing loop with thee-state Kalman filter when augmented proportional navigation is used

in all cases indicating that we have correctly taken the adjoint of the homing loop with the discrete three-state Kalman filter. We can also see from the three cases that the miss distance results get better as the guidance law becomes more advanced.

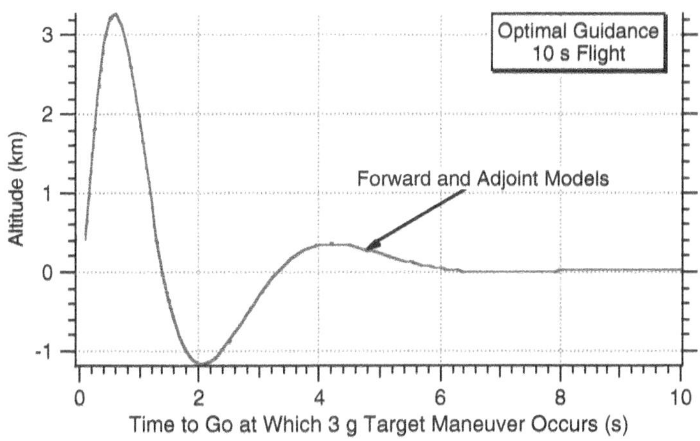

Fig. 29.11 Forward and adjoint models agree for homing loop with thee-state Kalman filter when optimal guidance is used.

SUMMARY

In this chapter we have shown two new applications of the method of adjoints for mixed continuous discrete systems. The first application involved multiple samplers used in a system in which measurement data was first preprocessed at one sampling rate and then used for input to a fading memory filter operating at another sampling rate. The second application involved taking the adjoint of a three-state discrete Kalman filter in the homing loop. In both examples experiments were conducted with a forward model to ensure that the adjoint was taken correctly.

REFERENCE

[1] Zarchan, P., *Fundamentals of Kalman Filtering A Practical Approach*, 3rd ed, Progress in Astronautics and Aeronautics, AIAA, Reston, VA, 2009, pp. 690–701.

CHAPTER 30

Miscellaneous Tactical Missile Guidance Topics

INTRODUCTION

In this chapter we shall discuss two important topics that have not yet been presented. The first topic covers the special case of shaping the interceptor trajectory against a stationary target without requiring time to go information. This new method will be compared to optimal guidance techniques for shaping the trajectory that require time to go information. The next topic will be in finding the minimum achievable miss distance with a radar homing missile. Formulas will be developed that depend on error sources beyond the control of the missile designer for establishing the minimum rms miss distance. Techniques for achieving the minimum possible rms miss are discussed and examples demonstrating their utility are presented.

BIASED PROPORTIONAL NAVIGATION FOR TRAJECTORY SHAPING AGAINST STATIONARY TARGETS [1]

In Chapter 24 we showed how optimal guidance techniques could be used against maneuvering targets to shape the trajectory of the interceptor so that it could hit the target with a desired final impact angle using minimum acceleration. However, as with other advanced guidance laws, the implementation of optimal guidance requires knowledge of the time to go until intercept. Time to go information might be available in radar homing missiles but may not be available for infrared homing missiles. In this section we shall show that for the special case of a stationary target, the missile trajectory can also be shaped by using biased proportional navigation. The form of biased proportional navigation discussed only requires line-of-sight rate information and knowledge of the missile's own velocity for the successful implementation of the guidance law. More complete details on using biased proportional navigation for hitting a stationary target can be found in Ref. 1.

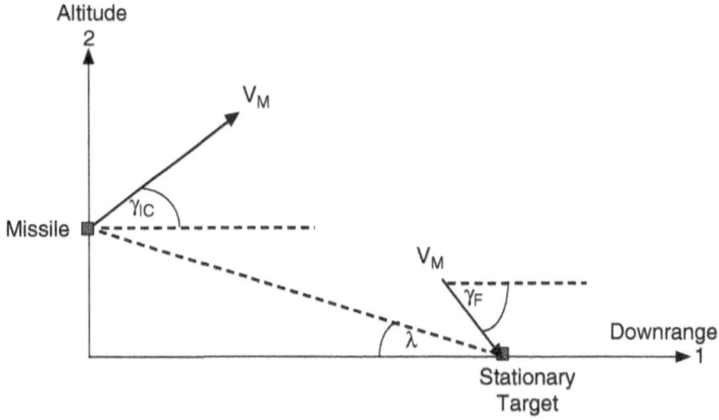

Fig. 30.1 Engagement geometry for stationary target.

The missile-target engagement geometry for a stationary target is depicted in Fig. 30.1. Here we can see that the missile is at an arbitrary initial flight path angle γ_{IC}. We desire to hit the stationary target at a final flight path angle γ_F using reasonable amounts of acceleration.

In Chapter 24 we showed that the optimal trajectory guidance law in which we hit the target, minimize the integral of the commanded acceleration squared, and attain a final line-of-sight angle λ_F is given by

$$n_c = 4V_c\dot{\lambda} + n_T + \frac{2V_c(\lambda - \lambda_F)}{t_{go}}$$

where n_c is the commanded missile acceleration, V_c is the closing velocity, n_T is the target maneuver, λ is the line-of-sight angle, $\dot{\lambda}$ is the line-of-sight rate and t_{go} is the time to go until intercept. If the target is stationary, there is no target maneuver and the final line-of-sight angle is the final flight path angle or

$$n_T = 0$$
$$\gamma_F = \lambda_F$$

Under these conditions the optimal guidance law against a stationary target simplifies to

$$n_c = 4V_c\dot{\lambda} + \frac{2V_c(\lambda - \gamma_F)}{t_{go}}$$

The optimal guidance law, in addition to hitting the target and achieving a final flight path angle, is suppose to minimize the integral of the acceleration squared or

$$\text{Performance Index} = \int_0^{t_F} n_c^2 dt$$

In addition, we shall use the preceding performance index to see how efficient biased proportional navigation is when compared to optimal guidance. Proportional navigation, which does not make use of closing velocity information, can be written in terms of the misile flight path rate $\dot{\gamma}$ as [2]

$$\dot{\gamma} = N\dot{\lambda}$$

where $\dot{\lambda}$ is the line-of-sight rate and N is the navigation ratio. It is important to note that N in general is not equal to the effective navigation ratio N' previously discussed in Chapter 2. If the missile acceleration command is perpendicular to the velocity vector, we can say that

$$\dot{\gamma} = \frac{n_c}{V_M}$$

Therefore we can say that this implementation of proportional navigation, where closing velocity information is not required, can also be written as

$$n_c = V_M\dot{\gamma} = NV_M\dot{\lambda}$$

If we equate this form of proportional navigation, where the acceleration command is perpendicular to the velocity vector, to the traditional form of proportional navigation where the acceleration command is perpendicular to the line-of-sight we obtain

$$n_c = NV_M\dot{\lambda} = N'V_c\dot{\lambda}$$

which means that the navigation ratio and the effective navigation ratio are related according to

$$N = N'\frac{V_c}{V_M}$$

For a stationary target the closing velocity is approximately the missile velocity, so in this particular case the two navigation ratios are approximately equal. Biased proportional navigation is proportional navigation plus an extra term and can be expressed in terms of the flight path rate as

$$\dot{\gamma} = N\dot{\lambda} + \text{bias}$$

where the bias can be zero for different portions of the flight.

Now let us see how we can choose a value for the bias that will help us shape the missile trajectory. If we integrate the preceding equation from 0 to t_F we obtain

$$\gamma_F - \gamma_{IC} = N(\lambda_F - \lambda_{IC}) + \text{bias}\Delta t$$

where Δt is the amount of time the bias is on (that is, has a non-zero value). For a stationary target

$$\lambda_F = \gamma_F$$

we can use substitution to yield

$$\gamma_F - \gamma_{IC} = N(\gamma_F - \lambda_{IC}) + \text{bias}\Delta t$$

Thus we can solve the preceding equation for the bias yielding

$$\text{bias} = \frac{-\gamma_F(N-1) + N\lambda_{IC} - \gamma_{IC}}{\Delta t}$$

In other words if the missile is on an initial flight path angle γ_{IC} and we wish to hit the target with a final flight path angle γ_F, we can solve for the bias by selecting a value for the amount of time Δt the bias is on.

Listing 24.2, which contains the optimal trajectory shaping guidance law, has been modified for the case of the stationary target and appears in Listing 30.1. Listing 30.1 provides the option of either using optimal trajectory shaping guidance law (IGUID $= 0$) or biased proportional navigation (IGUID $= 1$). We can see that the units in Listing 30.1 have been converted to the metric system and the nominal case is one in which the missile is attempting to hit a stationary target 10 km downrange with a final flight path angle of -90 deg. The missile is launched with a 30-deg heading error (or initial flight path angle of 30 deg) and the initial missile velocity is 250 m/s. So far in our work to date the missile acceleration command has always been perpendicular to the line of sight. In this simulation we have the additional capability of seeing what happens if the acceleration command is perpendicular to the missile velocity vector (ICHOICE $= 1$) or to the line of sight (ICHOICE $= 0$). If biased proportional navigation is used (IGUID $= 1$) the nominal case indicates that the bias is turned on at time zero (TBEG $= 0$) and turned off 30 s later (DELT $= 30$). The performance index (X $=$ integral of commanded missile acceleration squared at end of flight), flight path angle, and missile velocity are written to a file so that results can be compared. Differences between Listing 24.2 and Listing 30.1 are highlighted in bold.

LISTING 30.1 TWO-DIMENSIONAL ENGAGEMENT SIMULATION WITH OPTIMAL TRAJECTORY SHAPING AND BIASED PROPORTIONAL NAVIGATION GUIDANCE OPTIONS

```
n=0;
IGUID=0;
ICHOICE=0;
```

```
GAMDEG=30. ;
GAMIC=GAMDEG/57.3 ;
DELT=30. ;
TBEG=0. ;
TEND=TBEG+DELT ;
XNP=3. ;
RM1IC=0. ;
RM2IC=0. ;
RT1IC=10000.*3.28 ;
RT2IC=0. ;
VM=250.*3.28 ;
XNCLIMG=10. ;
GAMFDEG=-90. ;
H=.0001 ;
XNCLIM=32.2*XNCLIMG ;
RM1=RM1IC ;
RM2=RM2IC ;
RT1=RT1IC ;
RT2=RT2IC ;
VT1=0. ;
VT2=0. ;
T=0. ;
S=0. ;
RTM1=RT1-RM1 ;
RTM2=RT2-RM2 ;
RTM=sqrt(RTM1^2+RTM2^2) ;
XLAM=atan2(RTM2,RTM1) ;
VM1=VM*cos(GAMIC) ;
VM2=VM*sin(GAMIC) ;
VTM1=VT1-VM1 ;
VTM2=VT2-VM2 ;
VC=-(RTM1*VTM1+RTM2*VTM2) / RTM ;
GAMF=GAMFDEG/57.3 ;
BIASDEG=(-GAMFDEG* (XNP-1.) + XNP*XLAM*57.3-GAMDEG)/DELT ;
BIAS=BIASDEG/57.3 ;
X=0. ;
while VC >=0
   if RTM <1000
      H= .00001 ;
   else
      H= .0001 ;
   end
   RM1OLD=RM1 ;
   RM2OLD=RM2 ;
   VM1OLD=VM1 ;
   VM2OLD=VM2 ;
```

```
XOLD=X ;
STEP=1 ;
FLAG=0 ;
while STEP <=1
   if FLAG==1
         STEP=2 ;
         RM1=RM1+H*VM1 ;
         RM2=RM2+H*VM2 ;
         VM1=VM1+H*AM1 ;
         VM2=VM2+H*AM2 ;
         X=X+H*XD ;
         T=T+H ;
   end
   GAM=atan2 (VM2,VM1) ;
   VM=sqrt (VM1^2+VM2^2) ;
   RTM1=RT1-RM1 ;
   RTM2=RT2-RM2 ;
   RTM=sqrt(RTM1^2+RTM2^2) ;
   VTM1=VT1-VM1 ;
   VTM2=VT2-VM2 ;
   VC= -(RTM1*VTM1+RTM2*VTM2) / RTM ;
   XLAM=atan2(RTM2,RTM1) ;
   XLAMD=(RTM1*VTM2-RTM2*VTM1)/(RTM*RTM) ;
   if IGUID==0
      TGO=RTM/VC ;
      XNC=4. *VC*XLAMD+2. *VC*(XLAM-GAMF) / TGO ;
   else
      if T<TBEG
         GAMD=XNP*XLAMD ;
         XNC=VM*GAMD ;
      elseif T<TEND
         GAMD=XNP*XLAMD+BIAS ;
         XNC=VM*GAMD ;
      else
         GAMD=XNP*XLAMD ;
         XNC=VM*GAMD ;
      end
   end
   if XNC>XNCLIM
      XNC=XNCLIM ;
   end
   if XNC<-XNCLIM
      XNC=-XNCLIM ;
   end
   if ICHOICE==0
      AM1=-XNC*sin (XLAM) ;
```

```
        AM2=XNC*cos(XLAM) ;
      else
        AM1=-XNC*sin(GAM) ;
        AM2=XNC*cos(GAM) ;
      end
      XD=XNC*XNC ;
      FLAG=1 ;
    end
    FLAG=0 ;
    RM1= .5* (RM1OLD+RM1+H*VM1) ;
    RM2= .5* (RM2OLD+RM2+H*VM2) ;
    VM1= .5* (VM1OLD+VM1+H*AM1) ;
    VM2= .5* (VM2OLD+VM2+H*AM2) ;
    X= .5*(XOLD+X+H*XD) ;
    S=S+H ;
    if S>= .09999
      S=0. ;
      n=n+1 ;
      RT1K=RT1/3280. ;
      RT2K=RT2/3280. ;
      RM1K=RM1/3280. ;
      RM2K=RM2/3280. ;
      XNCG=XNC/32.2 ;
      GAMDEG=GAM*57.3 ;
      VMM=VM/3.28 ;
      XM=X/(3.28*3.28) ;
      ArrayT (n) =T ;
      ArrayRT1K (n) =RT1K ;
      ArrayRT2K (n) =RT2K ;
      ArrayRM1K (n) =RM1K ;
      ArrayRM2K (n) =RM2K ;
      ArrayXNCG (n) =XNCG ;
      ArrayGAMDEG (n) =GAMDEG ;
      ArrayXM (n) =XM ;
    end
end
figure
plot(ArrayRT1K, ArrayRT2K, ArrayRM1K, ArrayRM2K), grid
title( ' Engagement Geometry ' )
xlabel( ' Downrange (Kft) ' )
ylabel( ' Altitude (Kft) ' )
figure
plot(ArrayT,ArrayXNCG),grid
title( ' Commanded Acceleration ' )
xlabel( ' Time (Sec) ' )
ylabel( ' XNC (G) ' )
```

```
figure
plot(ArrayT, ArrayGAMDEG), grid
title( ' Flight Path Angle ' )
xlabel( ' Time (Sec) ' )
ylabel( ' GAM (Deg) ' )
clc
output=[ArrayT',ArrayRT1K',ArrayRT2K',ArrayRM1K',ArrayRM2K',...
                    ArrayXNCG',ArrayGAMDEG'] ;
save datfil.txt output /ascii
disp ' *** Simulation Complete '
RTM=sqrt(RTM1^2+RTM2^2)
VM=sqrt(VM1^2+VM2^2)
```

For a quick review the nominal case of Listing 30.1 was run in which the optimal trajectory shaping guidance law was used (IGUID = 0) and the missile acceleration command was perpendicular to the line of sight (ICHOICE = 0). Another case was run in which the same optimal guidance law was used but with the missile acceleration command perpendicular to the missile velocity vector (ICHOICE = 1). We can see from Fig. 30.2 that in both cases the missile trajectories are nearly identical and the missile hits the stationary target near vertically. However, Fig. 30.3 shows that the commanded acceleration profiles are vastly different. The maximum value of the acceleration command that is perpendicular to the line of sight is much smaller than when the acceleration command is perpendicular to the velocity vector and the flight time is much longer. Figure 30.4 shows that both guidance command implementations result in the flight path angle approaching −90 deg at intercept. Most importantly Fig. 30.5 shows that dramatic trajectory shaping causes a severe missile velocity

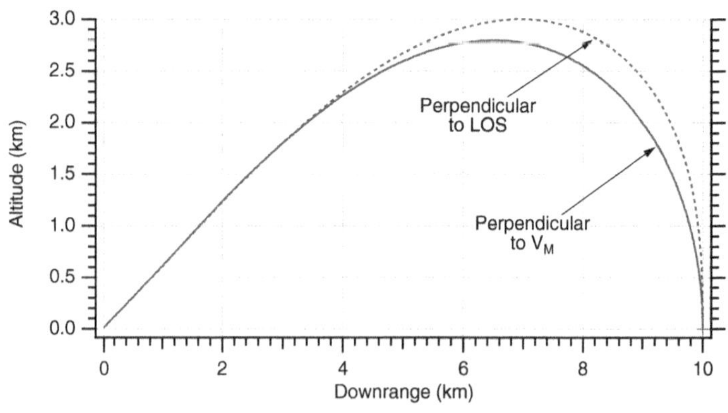

Fig. 30.2 Possible trajectories using optimal guidance for −90-deg impact against stationary target.

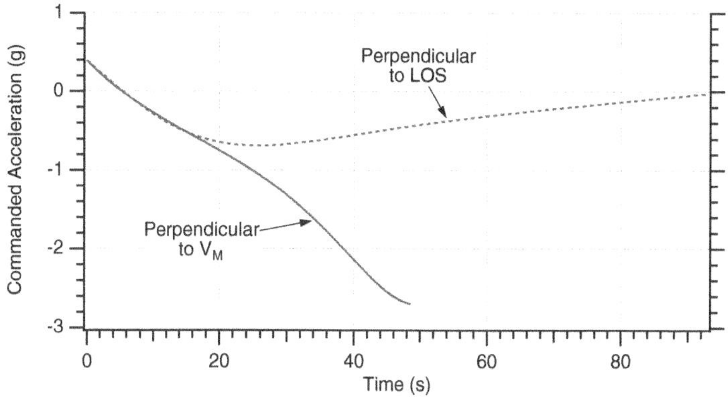

Fig. 30.3 Possible commanded acceleration profiles using optimal guidance for -90-deg impact against stationary target.

loss when the missile acceleration command is perpendicular to the line of sight and no velocity loss at all when the missile acceleration command is perpendicular to the velocity vector. From a practical point of view the velocity loss in this example cannot be tolerated because the actual acceleration capability of the missile diminishes as the velocity of the missile decreases. In future studies of trajectory shaping in this section we shall assume that the acceleration command is perpendicular to the velocity vector (ICHOICE = 1).

Next the optimal trajectory shaping guidance law was compared to biased proportional navigation for the case in which the acceleration commands were perpendicular to the velocity vector. Biased proportional navigation was considered

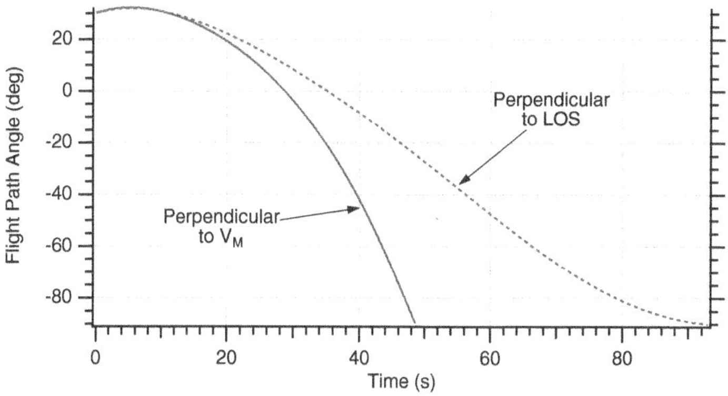

Fig. 30.4 Desired final flight path angle is achieved with optimal trajectory shaping guidance law.

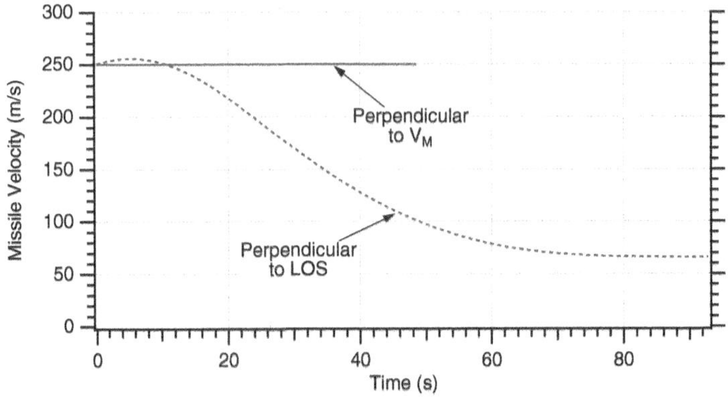

Fig. 30.5 Acceleration command perpendicular to line of sight causes missile velocity to decrease significantly when optimal trajectory shaping guidance law is used.

when the bias started at time zero and lasted for 30 s. In addition another case was considered when the bias started at 10 s and lasted for 30 s. We can see from Fig. 30.6 that both guidance laws result in missile trajectories that hit the stationary target near vertically. However the optimal guidance law trajectory is much tighter. We can also see from Fig. 30.6 that starting the bias later at 10 s also tightens the trajectory of the missile using biased proportional navigation. Figure 30.7 shows that the commanded acceleration profiles for the various guidance approaches are quite different but the maximum accelerators are similar. As expected, the flight time is much shorter for the tighter trajectories (optimal guidance case and biased proportional navigation when bias starts at 10 s). Figure 30.8 shows that both guidance laws achieve success in that the flight path angle approaches −90 deg at intercept. Figure 30.9 shows that although

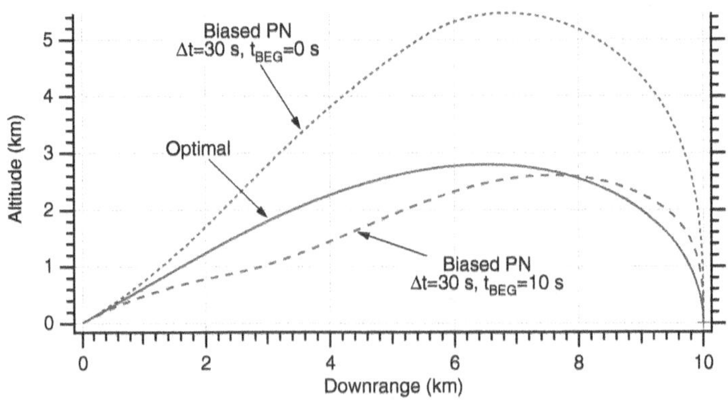

Fig. 30.6 Trajectories for both guidance laws are different.

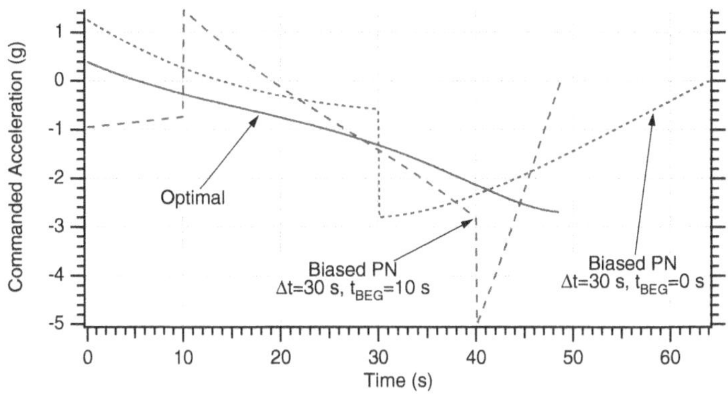

Fig. 30.7 Maximum acceleration command can be similar for both guidance laws.

the performance index is smaller for the case in which the optimal guidance law is used but the performance index for biased proportional navigation starting at time zero is only 33% larger. These results indicate that the acceleration requirements for biased proportional navigation against a stationary target should not be significantly higher than that of optimal guidance.

Figure 30.10 demonstrates that when the initial flight path angle is 30 deg, biased proportional navigation hits the stationary target for a variety of final impact angles. However Fig. 30.11 shows that more missile acceleration is required as the desired final impact angle increases.

In this section we have shown that for the problem of hitting a stationary target with a desired impact angle, biased propotional navigation is competitive with the optimal trajectory shaping guidance law. Optimal guidance does requires a smaller performance index (smaller integral of acceleration squared) than biased

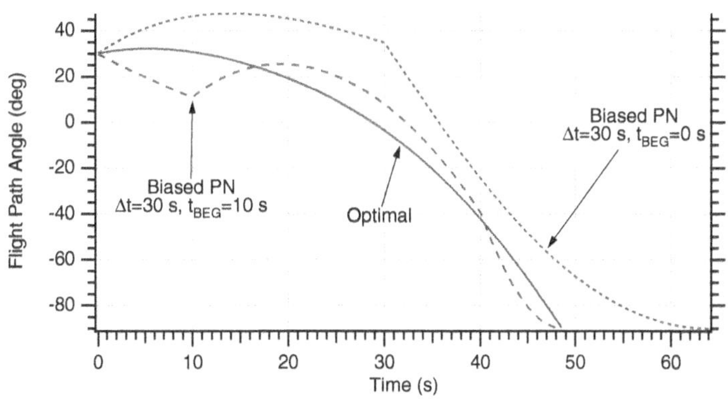

Fig. 30.8 Desired flight path angle at end is achieved with both guidance laws.

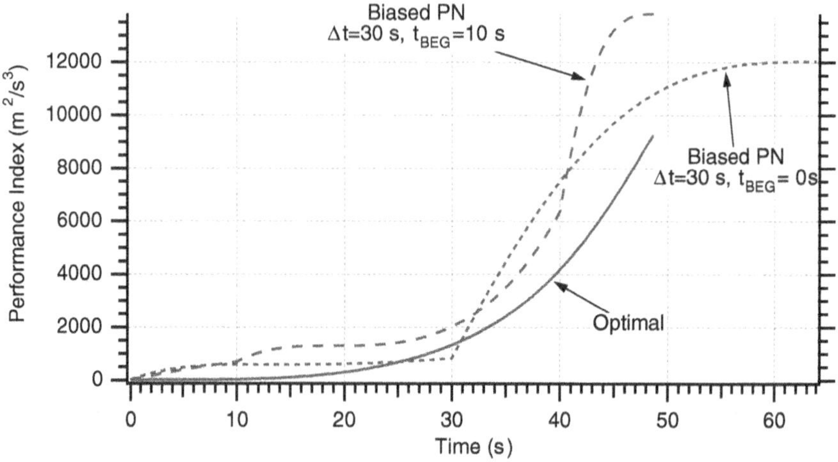

Fig. 30.9 Optimal guidance minimizes performance index.

proportional navigation. However, the main advatage of biased proportional navigation over optimal guidance in the case of a stationary target is that time-to-go information is not required.

SMALLEST POSSIBLE MISS DISTANCE FOR A RADAR HOMING MISSILE

In preliminary analysis it is sometimes of interest to find out the smallest possible miss distance a radar homing missile can achieve [3]. This calculation can be done by assuming the missile has infinite acceleration capability and that the two main

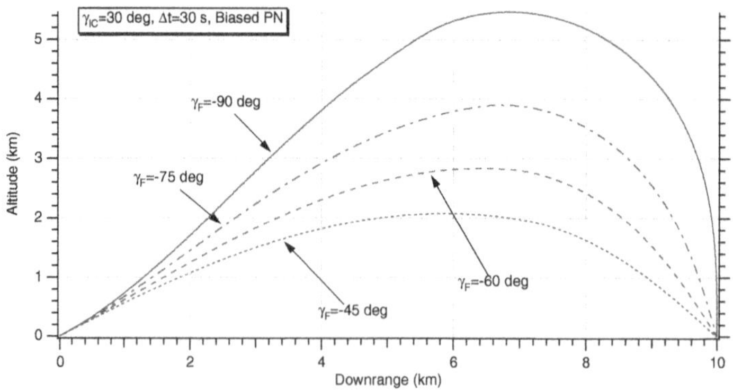

Fig. 30.10 Trajectory changes with changing final flight path angle when using biased PN.

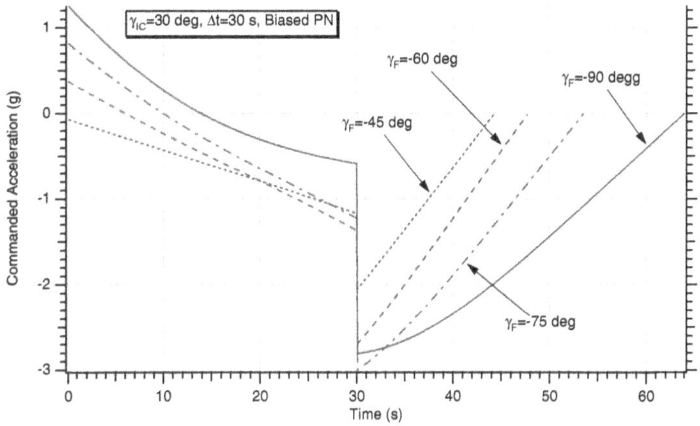

Fig. 30.11 Acceleration requirements increase as final flight path angle increases for biased PN.

sources of error are glint noise and target maneuver. The missile designer has no control over these error sources because they depend on the target. The standard deviation of the glint noise σ_{GL} is approximately one fifth of the target length that is perpendicular to the line of sight. This noise is highly correlated and can be modeled as white noise through a low-pass filter where the time constant of the low-pass filter T_{GL} can range from 0.1 s to 0.25 s [4].

However if a frequency agile radar is available the glint can simply be approximated as white noise as far as the guidance system is concerned [2]. If frequency agility is applied at the rate f_s then

$$f_s = \frac{1}{T_s}$$

where T_s is the sampling time of the frequency agile radar. The spectral density of the white glint noise in units squared per Hz is related to the standard deviation of the glint according to

$$\Phi_{\mathrm{GL}} = \sigma_{\mathrm{GL}}^2 T_s$$

The preceding relationship is identical to the one we used in Chapter 4 with the sampling time T_s being replaced by the integration interval.

Two factors will determine system performance if the missile has infinite acceleration capability. The first factor is the Kalman filter used to estimate the target states. If the Kalman filter process noise model is matched to the expected target maneuver and if the filter measurement noise model is matched to the actual measurement noise, then the filter is optimal. For the three-state Kalman filter discussed throughout this text, the first state is relative position and the square root of the first diagonal element of the covariance matrix represents the smallest possible rms value of the error in the estimate of position. As we cannot control the missile any better than we can estimate the states, this quantity also represents the smallest possible rms miss distance. The second factor is the guidance law. If we use an optimal guidance law (also matched to the shape of the expected maneuver) we should be able to achieve the minimum possible rms miss provided we have adequate missile acceleration capability and that the flight time is long enough so that system transients, such as heading error do not contribute to the miss.

If the sampling time of the guidance system is sufficiently small, the performance of a discrete Kalman filter will approach that of a continuous Kalman filter. In this section we shall show that the steady state, closed-form solutions for the diagonal elements of the covariance matrix of a continuous Kalman filter that depends on the glint noise and target maneuver levels can be found.

The original three-state linear Kalman filter of Chapter 9 was derived based on the homing loop model of Fig. 30.12. Recall that in this guidance system model we measured noisy relative position y* and were attempting to estimate relative position, relative velocity, and target acceleration. As was the case in Chapter 9, the achieved missile acceleration n_L was assumed to be known, and the target acceleration was considered to be modeled as a white noise through an integrator. It was shown in Chapter 4 that the shaping filter equivalent of a target maneuver with constant amplitude and random starting time (or uniformly distributed target maneuver) is mathematically equivalent in terms of second-order statistics to white noise through an integrator.

According to the results of Chapter 4, the spectral density of the white noise source u_s depicted in Fig. 30.12 is shown to be

$$\Phi_s = \frac{n_{T\text{MAX}}^2}{t_F}$$

where $n_{T\text{MAX}}$ is the assumed maximum target maneuver level magnitude and t_F the flight time over which the starting time of the maneuver is equally likely to occur. In this example u_n is the measurement noise which is assumed to be

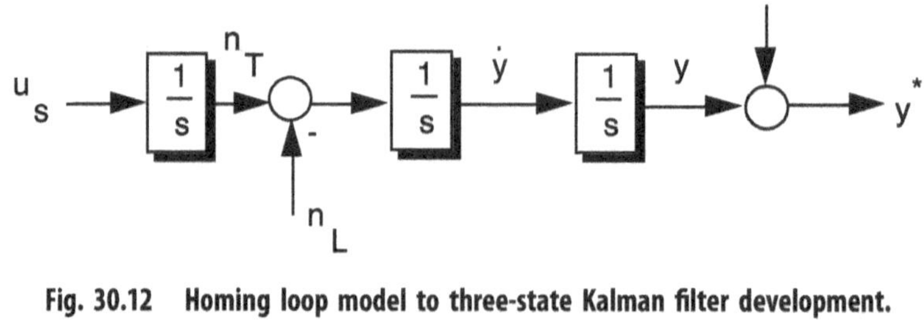

Fig. 30.12 Homing loop model to three-state Kalman filter development.

white glint noise. The model of Fig. 30.12 can be expressed in state space form as

$$\begin{bmatrix} \dot{y} \\ \ddot{y} \\ \dot{n}_T \end{bmatrix} = \begin{bmatrix} 0 & 1 & 0 \\ 0 & 0 & 1 \\ 0 & 0 & 0 \end{bmatrix} \begin{bmatrix} y \\ \dot{y} \\ n_T \end{bmatrix} + \begin{bmatrix} 0 \\ -1 \\ 0 \end{bmatrix} n_L + \begin{bmatrix} 0 \\ 0 \\ u_s \end{bmatrix}$$

Because the systems dynamics matrix of the preceding equation is given by

$$\mathbf{F} = \begin{bmatrix} 0 & 1 & 0 \\ 0 & 0 & 1 \\ 0 & 0 & 0 \end{bmatrix}$$

and the continuous process noise matrix can be found from

$$Q = E\left\{ \begin{bmatrix} 0 \\ 0 \\ u_s \end{bmatrix} \begin{bmatrix} 0 & 0 & u_s \end{bmatrix} \right\} = \begin{bmatrix} 0 & 0 & 0 \\ 0 & 0 & 0 \\ 0 & 0 & \Phi_s \end{bmatrix} = \frac{n_{\mathrm{TMAX}}^2}{t_F} \begin{bmatrix} 0 & 0 & 0 \\ 0 & 0 & 0 \\ 0 & 0 & 1 \end{bmatrix}$$

The measurement equation can be seen from Fig. 30.12 to be

$$y^* = y + u_n = \begin{bmatrix} 1 & 0 & 0 \end{bmatrix} \begin{bmatrix} y \\ \dot{y} \\ n_T \end{bmatrix} + u_n$$

Therefore the measurement matrix can be written by inspection of the previous equation as

$$H = \begin{bmatrix} 1 & 0 & 0 \end{bmatrix}$$

and the measurement noise matrix is a scalar given by

$$R = E[u_n u_n^T] = \Phi_n$$

The differential equation for the covariance matrix of a continuous Kalman filter is given by [5]

$$\dot{P} = -PH^T R^{-1} HP + PF^T + FP + Q$$

where P is the covariance matrix and has the property of being symmetric. We can solve the preceding equation in the steady state by setting the derivative of the covariance matrix to zero or

$$0 = -PH^T R^{-1} HP + PF^T + FP + Q$$

Substitution of the appropriate matrices into the preceding equation yields

$$\begin{bmatrix} 0 & 0 & 0 \\ 0 & 0 & 0 \\ 0 & 0 & 0 \end{bmatrix} = -\begin{bmatrix} P_{11} & P_{12} & P_{13} \\ P_{12} & P_{22} & P_{23} \\ P_{13} & P_{23} & P_{33} \end{bmatrix} \begin{bmatrix} 1 \\ 0 \\ 0 \end{bmatrix} \frac{1}{\Phi_n} [1 \quad 0 \quad 0] \begin{bmatrix} P_{11} & P_{12} & P_{13} \\ P_{12} & P_{22} & P_{23} \\ P_{13} & P_{23} & P_{33} \end{bmatrix}$$

$$+ \begin{bmatrix} P_{11} & P_{12} & P_{13} \\ P_{12} & P_{22} & P_{23} \\ P_{13} & P_{23} & P_{33} \end{bmatrix} \begin{bmatrix} 0 & 0 & 0 \\ 1 & 0 & 0 \\ 0 & 1 & 0 \end{bmatrix} + \begin{bmatrix} 0 & 1 & 0 \\ 0 & 0 & 1 \\ 0 & 0 & 0 \end{bmatrix}$$

$$\times \begin{bmatrix} P_{11} & P_{12} & P_{13} \\ P_{12} & P_{22} & P_{23} \\ P_{13} & P_{23} & P_{33} \end{bmatrix} + \begin{bmatrix} 0 & 0 & 0 \\ 0 & 0 & 0 \\ 0 & 0 & \Phi_s \end{bmatrix}$$

After taking symmetry into account and multiplying out the equations we get the following six scalar algebraic equations.

$$\begin{aligned} P_{11}^2 &= 2P_{12}\Phi_n \\ P_{12}^2 &= 2P_{23}\Phi_n \\ P_{13}^2 &= \Phi_s\Phi_n \\ P_{11}P_{12} &= \Phi_n(P_{22} + P_{13}) \\ P_{11}P_{13} &= P_{23}\Phi_n \\ P_{12}P_{13} &= P_{33}\Phi_n \end{aligned}$$

After much algebra we find that

$$P_{11} = 2\Phi_s^{1/6}\Phi_n^{5/6}$$

Because the square root of P_{11} represents the error in the estimate of relative position, taking the square root of both sides of the preceding equation yields the best the filter can estimate relative position or

$$\sqrt{P_{11}} = \sqrt{2\Phi_s^{1/6}\Phi_n^{5/6}}$$

Because we cannot control the missile any better than we can estimate, the preceding equation represents also the smallest possible achievable rms miss.

In order to illustrate the use of the preceding formula let us assume we have a guidance system in which there is a 0.01-s sampling time, a standard deviation of

10 ft of glint noise and a 5-g target maneuver whose starting time is equally likely to occur anywhere (that is, uniformly distributed) during a 10-s flight. Therefore the power spectral density (in units squared per Hertz) of the target maneuver and glint noise are given by

$$\Phi_s = \frac{n_{TMAX}^2}{t_F} = \frac{161^2}{10} = 2592$$

$$\Phi_N = \sigma_{GL}^2 T_s = 10^2 {*} 0.01 = 1$$

Therefore the standard deviation of the first diagonal element of the covariance matrix or rms value of the smallest possible miss is

$$\sqrt{P_{11}} = \sqrt{2\Phi_s^{1/6}\Phi_n^{5/6}} = \sqrt{2{*}2592^{1/6}{*}1^{5/6}} = 2.72 \text{ ft}$$

In the real word we want to build a discrete three-state Kalman filter. As was shown in Chapter 9, the gains for such a filter are obtained by numerically solving the recursive matrix Ricatti difference equations by iteration. As was shown in Chapter 9, the difference equations to be solved are

$$M_k = \Phi_k P_{k-1} \Phi_k^T + Q_k$$

$$K_k = M_k H^T [H M_k H^T + R_k]^{-1}$$

$$P_k = (I - K_k H) M_k$$

where P_k is the covariance matrix after an update, Φ_k is the fundamental matrix, Q_k is the discrete process noise matrix, M_k is the covariance matrix before an update, R_k is the discrete noise matrix and K_k is the Kalman gain matrix. In the preceding difference equations we need to find R_k, Φ_k and Q_k so that we can iterate and solve for the discrete covariance matrix P_k. It was shown in Chapter 9 that the discrete fundamental matrix can easily be derived from the systems dynamics matrix **F** as

$$\Phi_k = \begin{bmatrix} 0 & T_s & .5T_s^2 \\ 0 & 1 & T_s \\ 0 & 0 & 1 \end{bmatrix}$$

The discrete measurement noise R_k is simply the variance of the glint noise or

$$R_k = \sigma_{GL}^2$$

Finally Chapter 9 showed that the discrete process noise matrix can be found from the continuous process noise matrix as

$$Q_k = \int_0^{T_s} \Phi(\tau) Q \Phi^T(\tau) d\tau$$

$$= \int_0^{T_s} \begin{bmatrix} 1 & \tau & 0.5\tau^2 \\ 0 & 1 & \tau \\ 0 & 0 & 1 \end{bmatrix} \begin{bmatrix} 0 & 0 & 0 \\ 0 & 0 & 0 \\ 0 & 0 & \Phi_s \end{bmatrix} \begin{bmatrix} 1 & 0 & 0 \\ \tau & 1 & 0 \\ 0.5\tau^2 & \tau & 1 \end{bmatrix} d\tau$$

$$= \Phi_s \begin{bmatrix} \frac{T_s^5}{20} & \frac{T_s^4}{8} & \frac{T_s^3}{6} \\ \frac{T_s^4}{8} & \frac{T_s^3}{3} & \frac{T_s^2}{2} \\ \frac{T_s^3}{6} & \frac{T_s^2}{2} & T_s \end{bmatrix}$$

Listing 30.2 presents a simulation of the discrete Ricatti equations solved for different times of flight for the case of a 5-*g* uniformly distributed target maneuver and 10 ft of glint noise. The formula, obtained by solving the steady-state continuous Ricatti equations, for the square root of the first diagonal element is presented for comparison. In theory as the sampling time gets smaller, the continuous and discrete answers for the square root of the first diagonal element of the covariance matrix should agree.

LISTING 30.2 SOLVING DISCRETE RICATTI EQUATIONS

```
n=0;
VC=5.*3280.;
XNTIC=161.;
VM=3000.;
HEDEG=20.;
SIGNOISE=10.;
TS=.01;
for TF=.2:.2:10.0
        TS2=TS*TS;
        TS3=TS2*TS;
        TS4=TS3*TS;
        TS5=TS4*TS;
        PHIS=XNTIC*XNTIC/TF;
        PHIN=SIGNOISE*SIGNOISE*TS;
        SIGPOS=SIGNOISE;
        SIGN2=SIGPOS^2;
```

```
        P11=SIGN2;
        P12=0.;
        P13=0.;
        P22=(VM*HEDEG/57.3)^2;
        P23=0.;
        P33=XNTIC*XNTIC;
        T=0.;
        for T=TS:TS:TF
                TGO=TF-T+.000001;
                RTM=VC*TGO;
                SIGPOS=SIGNOISE;
                SIGN2=SIGPOS^2;
                M11=P11+TS*P12+.5*TS2*P13+TS*(P12+TS*P22+.5*TS2*P23);
                M11=M11+.5*TS2*(P13+TS*P23+.5*TS2*P33)+TS5*PHIS/20.;
                M12=P12+TS*P22+.5*TS2*P23+TS*(P13+TS*P23+.5*TS2*P33)
                M12=M12+TS4*PHIS/8.;
                M13=P13+TS*P23+.5*TS2*P33+PHIS*TS3/6.;
                M22=P22+TS*P23+TS*(P23+TS*P33)+PHIS*TS3/3.;
                M23=P23+TS*P33+.5*TS2*PHIS;
                M33=P33+PHIS*TS;
                K1=M11/(M11+SIGN2);
                K2=M12/(M11+SIGN2);
                K3=M13/(M11+SIGN2);
                P11=(1.-K1)*M11;
                P12=(1.-K1)*M12;
                P13=(1.-K1)*M13;
                P22=-K2*M12+M22;
                P23=-K2*M13+M23;
                P33=-K3*M13+M33;
                SP11=sqrt(P11);
        end;
        FORM=sqrt(2.*(PHIS^.16667)*(PHIN^.83333))
        n=n+1;
        ArrayT(n)=T;
        ArrayFORM(n)=FORM;
        ArraySP11(n)=SP11;
end;
figure
plot(ArrayT',ArrayFORM',ArraySP11),grid
title('RMS miss for various flight times')
xlabel('Flight Time (S)')
ylabel('RMS MISS (Ft) ')
clc
output=[ArrayT',ArrayFORM',ArraySP11'];
save datfil.txt output /ascii
disp('Simulation Complete')
```

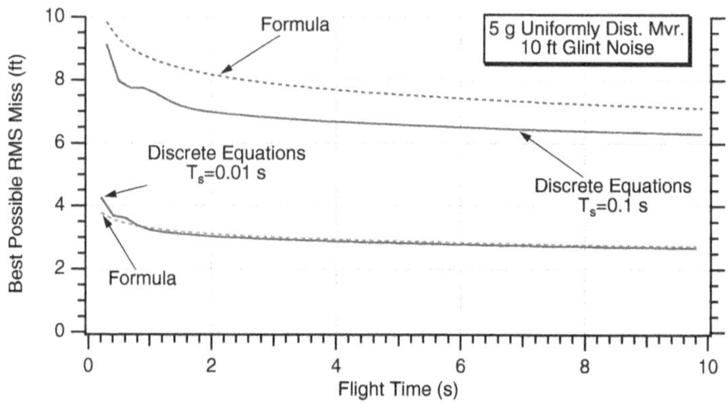

Fig. 30.13 Formula agrees with simulation results when sampling time is small.

Listing 30.2 was run for the cases in which the sampling time was 0.1 s and 0.01 s. We can see from Fig. 30.13 that the formula and discrete solutions are virtually identical when the sampling time is 0.01 s. However, when the sampling time is 0.1 s there is more of a difference between the formula and the solution obtained by solving the discrete Ricatti equations. As expected, we can also see that better performance can be obtained in a homing guidance system with smaller sampling times.

A homing loop, based on the linearized guidance system model used in Chapter 9 is shown in Fig. 30.14. In this homing loop the optimal guidance law, derived in Chapter 8, is used against the random uniformly distributed target maneuver. The homing loop includes the three-state Kalman filter that is

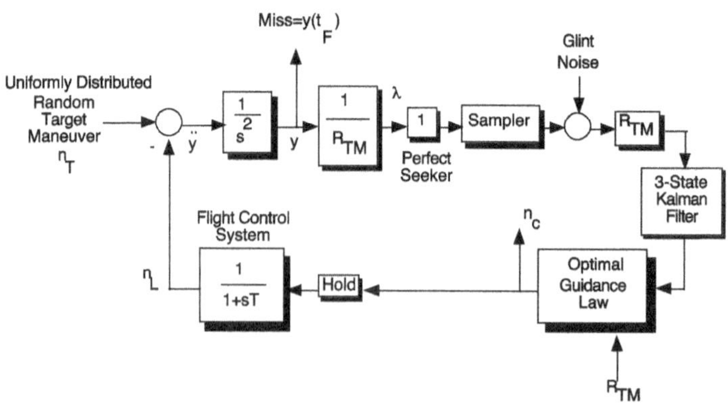

Fig. 30.14 Homing loop with random error sources, a three-state Kalman filter, and optimal guidance law.

required to estimate the states required by the optimal guidance law. In addition, there is a single time constant representation of the flight control system that indicates that there is a delay between the commanded and achieved missile acceleration. The transfer function of the flight control system is given by

$$\frac{n_L}{n_c} = \frac{1}{1 + sT}$$

where T is the flight control system time constant. In Fig. 30.14 we can also see that the acceleration command is not limited. As was previously mentioned, the two sources of error in this homing loop are glint noise and a uniformly distributed target maneuver.

In Chapter 8 it was shown the optimal guidance law for a single time constant guidance system where the target is executing a constant maneuver is given by

$$n_{c_{OG}} = \frac{N'}{t_{go}^2}[y + \dot{y}t_{go} + 0.5n_T t_{go}^2 - n_L T^2(e^{-x} + x - 1)]$$

where the effective navigation ratio can be expressed as

$$N' = \frac{6x^2(e^{-x} - 1 + x)}{2x^3 + 3 + 6x - 6x^2 - 12xe^{-x} - 3e^{-2x}}$$

and x is defined as

$$x = \frac{t_{go}}{T}$$

Listing 30.3 models the homing loop of Fig. 30.14 with the uniformly distributed target maneuver and the uncorrelated or white glint noise. Since there is measurement noise and a random target maneuver modeled in Listing 30.3, the simulation has to be run in the Monte Carlo mode (repeated simulation trials, each of which has a different maneuver starting time and different sign of the maneuver amplitude) in order to calculate the rms miss distance. The flight time is also varied in this simulation and 100 run Monte Carlo sets are made for flight times ranging from 0.2 s to 10 s in steps of 0.2 s. Therefore each case studied consists of 5000 runs (100 runs times 50 flight times). We can see from Listing 30.3 that the 3-state Kalman filter and optimal guidance law are also modeled.

LISTING 30.3 MONTE CARLO SIMULATION OF THREE-STATE KALMAN FILTER, OPTIMAL GUIDANCE LAW, UNCORRELATED GLINT NOISE AND UNIFORMLY DISTRIBUTED TARGET MANEUVER IN HOMING LOOP

```
%This simulation runs very slowly due to the small integration interval
%and large number of runs
```

```
%Preallocation
Z=zeros(size(1:1000));
I=zeros(size(1:50));
TF=zeros(size(1:50));
count=0;
VC=5.*3280.;
XNTIC=161.;
YIC=0.;
VM=3000.;
HEDEG=20.;
XNP=3.;
SIGNOISE=10.;
TS=.01;
TAU=.2;
RUN=100;
AMAXG=99999999.;
PZ1=.0001;
AMAX=AMAXG*32.2;
PHIN=SIGNOISE*SIGNOISE*TS;
for TF=.2:.2:10.0,
        Z1=0.;
        for JJ=1:RUN
                SUM=rand(1);
                TSTART=TF*SUM;
                PZ=rand(1);
                PZ=PZ-.5;
                if PZ > 0
                        COEF=1;
                else
                        COEF=-1;
                end;
                Y=0.;
                YD=0.;
                TS2=TS*TS;
                TS3=TS2*TS;
                TS4=TS3*TS;
                TS5=TS4*TS;
                PHIS=XNTIC*XNTIC/TF;
                RTM=VC*TF;
                SIGPOS=SIGNOISE;
                SIGN2=SIGPOS^2;
                P11=SIGN2;
                P12=0.;
                P13=0.;
                P22=(VM*HEDEG/57.3)^2;
                P23=0.;
```

```
P33=XNTIC*XNTIC;
T=0.;
H=.001;
S=0.;
YH=0.;
YDH=0.;
XNTH=0.;
XNC=0.;
XNL=0.;
while T <= (TF - 1e-5)
        YOLD=Y;
        YDOLD=YD;
        XNLOLD=XNL;
        STEP=1;
        FLAG=0;
        while STEP <=1
                if FLAG==1
                        Y=Y+H*YD;
                        YD=YD+H*YDD;
                        XNL=XNL+H*XNLD;
                        T=T+H;
                        STEP=2;
                end;
                if T<TSTART
                        XNTC=0.;
                else
                        XNTC=COEF*XNTIC;
                end;
                TGO=TF-T+.00001;
                RTM=VC*TGO;
                XLAM=Y/(VC*TGO);
                XNLD=(XNC-XNL)/TAU;
                XNT=XNTC;
                YDD=XNT-XNL;
                FLAG=1;
        end;
        FLAG=0;
        Y=.5*(YOLD+Y+H*YD);
        YD=.5*(YDOLD+YD+H*YDD);
        XNL=.5*(XNLOLD+XNL+H*XNLD);
        S=S+H;
        if S>=(TS - 1e-5)
        S=0.;
                TGO=TF-T+.000001;
                RTM=VC*TGO;
                SIGPOS=SIGNOISE;
```

```
                    SIGN2=SIGPOS^2;
                    M11=P11+TS*P12+.5*TS2*P13+TS*(P12+TS*P22
                            +.5*TS2*P23);
                    M11=M11+.5*TS2*(P13+TS*P23+.5*TS2*P33)
                            +TS5*PHIS/20.;
                    M12=P12+TS*P22+.5*TS2*P23+TS*(P13+TS*P23
                            +.5*TS2*P33);
                    M12=M12+TS4*PHIS/8.;
                    M13=P13+TS*P23+.5*TS2*P33+PHIS*TS3/6.;
                    M22=P22+TS*P23+TS*(P23+TS*P33)+PHIS*TS3/3.;
                    M23=P23+TS*P33+.5*TS2*PHIS;
                    M33=P33+PHIS*TS;
                    K1=M11/(M11+SIGN2);
                    K2=M12/(M11+SIGN2);
                    K3=M13/(M11+SIGN2);
                    P11=(1.-K1)*M11;
                    P12=(1.-K1)*M12;
                    P13=(1.-K1)*M13;
                    P22=-K2*M12+M22;
                    P23=-K2*M13+M23;
                    P33=-K3*M13+M33;
                    YNOISE=SIGNOISE*randn;
                    YSTAR=Y+YNOISE;
                    RES=YSTAR-YH-TS*YDH-.5*TS*TS*(XNTH-XNL);
                    YH=K1*RES+YH+TS*YDH+.5*TS*TS*(XNTH-XNL);
                    YDH=K2*RES+YDH+TS*(XNTH-XNL);
                    XNTH=K3*RES+XNTH;
                    X=TGO/TAU;
                    ZEM2H=YH+YDH*TGO-XNL*TAU*TAU*(exp(-X)+X-1.)...
                           +.5*XNTH*TGO*TGO;
                    TOP=6.*X*X*(exp(-X)-1.+X);
                    BOT1=2*X*X*X+3.+6.*X-6.*X*X;
                    BOT2=-12.*X*exp(-X)-3.*exp(-2.*X);
                    XNPP=TOP/(PZ1+BOT1+BOT2);
                    XNEW=XNPP*XNL*(exp(-X)+X-1.)/(X*X);
                    XNC=XNPP*ZEM2H/TGO^2;
                    if XNC>AMAX
                           XNC=AMAX;
                    end;
                    if XNC<-AMAX
                           XNC=-AMAX;
                    end;
          end;
end;
SP11=sqrt(P11);
Z(JJ)=Y;
```

```
            Z1=Z(JJ)+Z1;
            XMEAN=Z1/JJ;
    end;
    SIGMA=0.;
    Z1=0.;
    Z2=0.;
    for JJ=1:RUN,
            Z1=(Z(JJ)-XMEAN)^2+Z1;
            Z2=Z(JJ)^2+Z2;
            if JJ==1,
                    SIGMA=0.;
                    RMS=0.;
            else
                    SIGMA=sqrt(Z1/(JJ-1));
                    RMS=sqrt(Z2/(JJ-1));
            end;
    end;
    FORM=sqrt(2.*(PHIS^.16667)*(PHIN^.8333));
    count=count+1;
    ArrayTF(count)=TF;
    ArrayRMS(count)=RMS;
    ArraySP11(count)=SP11;
    ArrayFORM(count)=FORM;
end;
figure
plot(ArrayTF',ArrayRMS',ArrayTF',ArrayFORM',ArrayTF',ArraySP11'),grid
title('RMS miss for various flight times')
xlabel('Flight Time (S)')
ylabel('RMS MISS (Ft) ')
clc
output=[ArrayTF',ArrayRMS',ArraySP11',ArrayFORM'];
save datfil.txt output /ascii
disp('Simulation Complete')
```

The nominal case of Listing 30.3 was run and rms miss versus flight time results from the Monte Carlo simulation are presented in Fig. 30.15. Superimposed on the graph are the predictions from the discrete matrix Ricatti equations (square root of the first diagonal element of the covariance matrix). Also superimposed on Fig. 30.15 are the smallest possible rms miss results obtained from the formula

$$\sqrt{P_{11}} = \sqrt{2\Phi_s^{1/6}\Phi_n^{5/6}}$$

We can see that after the transients have died out, the Monte Carlo rms miss distance results are in very close agreement with both the predictions from the

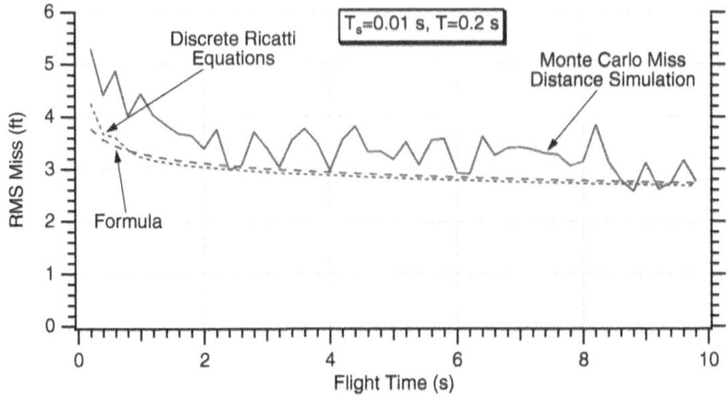

Fig. 30.15 Optimal guidance is nearly able to achieve smallest possible rms miss distance.

discrete and continuous Ricatti equations. This close rms miss distance agreement means that we are controlling the missile via the optimal guidance law as well as we can estimate the relative position state via the three-state Kalman filter. *This means that, in the absence of missile acceleration saturation effects, we can do no better with any other guidance law against the uniformly distributed target maneuver.*

Listing 30.3 was rerun again with the flight control system time constant increased from 0.2 s to 1 s. Figure 30.16 shows that the Monte Carlo rms miss distance results are now larger than the theoretical projections of the Ricatti equations. As the optimal guidance law is suppose to cancel out the single-lag flight control system dynamics perfectly, something appears to be wrong because the rms miss should not increase in the absence of missile acceleration saturation effects.

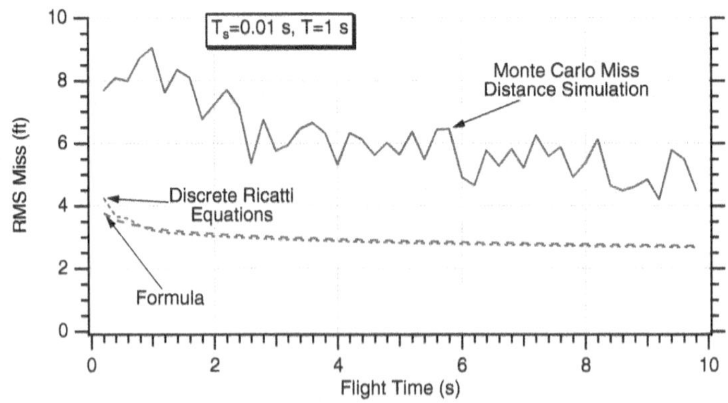

Fig. 30.16 Increasing flight control system time constant increases rms miss.

In order to understand what happened when the flight control system time constant was increased, let us review the expression for the navigation ratio in the optimal guidance law. As was previously mentioned, the expression for the optimal navigation ratio is given by

$$N' = \frac{6x^2(e^{-x} - 1 + x)}{2x^3 + 3 + 6x - 6x^2 - 12xe^{-x} - 3e^{-2x}}$$

where

$$x = \frac{t_{go}}{T}$$

If we examine the denominator of the formula for N' we can see that as t_{go} or x goes to zero, the denominator goes to zero. A careful examination of Listing 30.3 indicates that a division by zero was avoided by adding the term γ to the denominator in order to prevent the expression from blowing up as t_{go} approached zero or

$$N' = \frac{6x^2(e^{-x} - 1 + x)}{\gamma + 2x^3 + 3 + 6x - 6x^2 - 12xe^{-x} - 3e^{-2x}}$$

The term γ was set to 0.0001 and the optimal guidance law appeared to work rather well in the experiments conducted so far.

The miss distance experiment of Fig. 30.16 was rerun, except this time γ was reduced from 0.0001 to 0.00000001. We can see from Fig. 30.17 that making γ closer to zero enabled the optimal guidance law to yield rms miss distance results that were much closer to the theoretical predictions of the Ricatti equations.

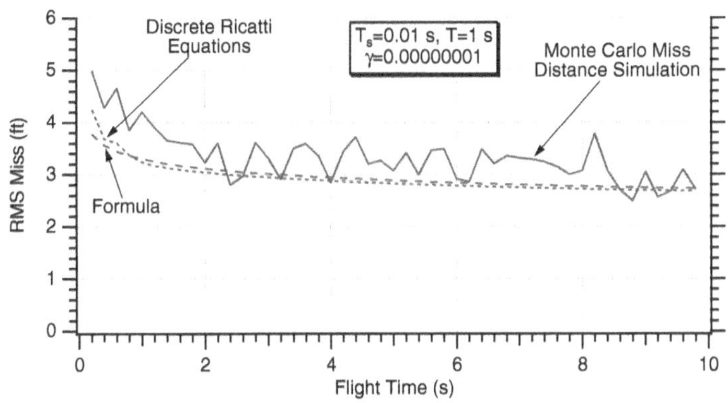

Fig. 30.17 Reducing γ enables optimal guidance to achieve smallest possible rms miss when flight control system time constant is large.

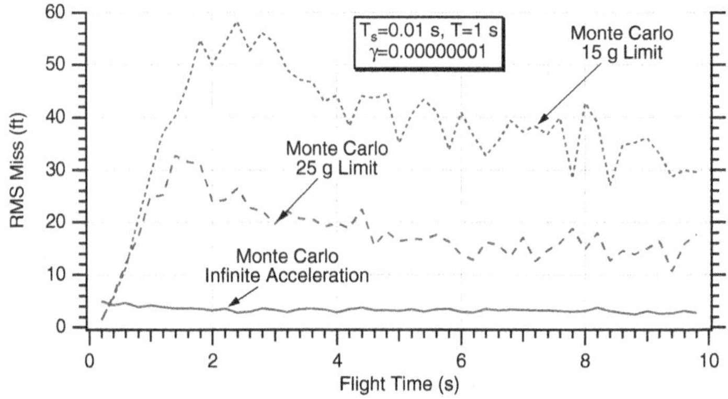

Fig. 30.18 Acceleration saturation effects influence our ability to compensate for large flight control system time constants.

It appears from Fig. 30.17 that miraculous performance improvement can be obtained by reducing γ in the formula for the effective navigation ratio when the flight control system time constant is large. However reducing γ has the effect of increasing the effective navigation ratio and missile acceleration near the end of the flight. An increase in missile acceleration near the end of the flight can lead to acceleration saturation and increased miss distance. Figure 30.18 shows that for the case of a large flight control system time constant ($T = 1$ s) we only get very small miss distances when the missile acceleration limit is infinite. For a 3 to 1 (15 g missile acceleration limit) and a 5 to 1 (25 g missile acceleration limit) missile acceleration advantages over the target the miss distances can be quite large.

Let us go back to the original nominal case in which the flight control system time constant was 0.2 s and the value of γ used in the optimal guidance law was 0.0001. Figure 30.19 first shows that the rms miss distances are much smaller for a 5 to 1 and a 3 to 1 missile to target acceleration advantage than they were in Fig. 30.18. Next, Fig. 30.19 shows that with a 5 to 1 missile to target acceleration advantage we are not too far from the smallest possible rms miss.

In this section we have developed formulas for the minimimum possible rms miss distance for the case in where there is white glint noise, a uniformly distributed target maneuver, and infinite missile acceleration capability. As these error sources are up to the target rather than the missile designer, they place a lower limit on the best achievable performance of a radar homing missile that has infinite acceleration capability. Finite missile acceleration capability plus other sources of error will only serve to increase the miss distance. In the next chapter we shall investigate a guidance law that can achieve small miss distances when the missile to target acceleration advantage is less than three.

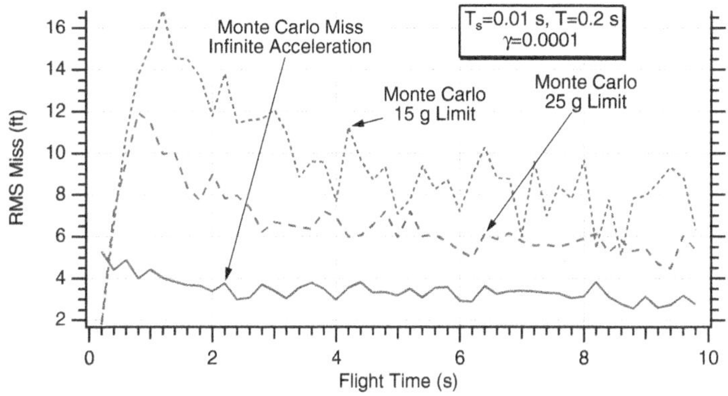

Fig. 30.19 Reducing the flight control system time constant allows to get closer to lowest possible theoretical rms miss distance.

REFERENCES

[1] Erer, K. S., and Merttopcuoglu, O., "Indirect Control of Impact Angle Against Stationary Targets Using Biased PPN," AIAA Guidance, Navigation and Control Conference, August 2010, Toronto, Canada, accepted for publication in *Journal of Guidance, Control and Dynamics.*

[2] Jerger, J., *Systems Preliminary Design*, D. Van Nostrand Company, Inc. Princeton, N. J., 1960, pp. 182-187.

[3] Rusnak, I., "Bounds on the RMS Miss of Radar-Guided Missiles," *Journal of Guidance, Control and Dynamics*, Vol. 3, No. 6, Nov.-Dec. 2010, pp. 1718-1723.

[4] Garnell, P., and East, D. J., *Guided Weapon Control Systems*, Pergamon Press, Oxford, England, 1977, pp. 15-17.

[5] Zarchan, P., and Musoff, H., *Fundamentals of Kalman Filtering: A Practical Approach*, 3rd ed, AIAA Progress Series, Reston, VA, 2009, pp. 219-255.

CHAPTER 31

Comparison of Differential Game Guidance With Optimal Guidance

INTRODUCTION

In Chapter 30 we showed that the minimum possible rms miss distance could be achieved if the Kalman filter was matched to the real world, the missile had infinite acceleration capability, the sampling time was small and the missile employed an optimal guidance law that was matched to the shape of the target maneuver assuming that it was known. In the example presented in Chapter 30, the three-state discrete Kalman filter knew (via the process noise) and the guidance law knew that the target was performing a random uniformly distributed constant maneuver. It was shown that when there was infinite missile acceleration available the rms miss distances that could be achieved approached the estimation capability of the optimal filter. However, when there was only a 3 to 1 missile to target acceleration advantage, rms miss distance performance started to degrade. In this chapter we shall see if other guidance approaches can yield better performance than optimal guidance when the Kalman filter is not matched to the target maneuver and there is a low missile-to-target acceleration advantage.

Pursuit-evasion differential game theory can be used to derive a guidance law that does not depend on on the knowledge of the future target maneuver. An important family of pursuit-evasion games assume linear kinematics and bounded controls [1–8]. However, usually differential game theory guidance laws with bounded controls have not been considered for practical application in endoatmospheric missiles because of their bang-bang nature (meaning the missile guidance command is either at plus or minus the maximum missile acceleration capability) which may lead to excessive induced drag or severe actuator requirements. In this chapter we shall initially disregard the practical objections to differential game guidance with bounded controls in order to see if there might be performance benefits to this unusual missile guidance approach. If there are substantial performance benefits to differential game guidance with bounded controls we will see if something simple can be done to make this type of guidance law more practical.

In this chapter we shall first review the various types of stressing target maneuvers we shall consider for the missile guidance law comparison. Then, we shall the use the three-state Kalman filter of Chapter 9 (its derivation based on a constant amplitude target maneuver and starting time uniformly distributed over the flight time). The three-state Kalman filter will be used for estimating states of a variety of other challenging target maneuvers. Under these circumstances the Kalman filter will no longer be optimal but should still work. The Kalman filter will estimate the states required for the different guidance laws considered and a performance comparison will be made. The performance comparison will be made under very stressing conditions when there is only a 2 to 1 missile-to-target acceleration advantage.

TRADITIONAL GUIDANCE LAW REVIEW

Proportional navigation is probably the world's most popular guidance law because of its robustness and ease of implementation. The guidance law can be expressed as

$$n_{c_{\rm PN}} = N' V_c \dot{\lambda}$$

where $n_{c_{\rm PN}}$ is the missile acceleration command perpendicular to the line of sight, N' is a designer-chosen gain usually in the range of 3 to 5, V_c is the closing velocity, and $\dot{\lambda}$ is the line-of-sight rate. Here it is assumed that the line-of-sight rate and closing velocity can be measured by the seeker. Proportional navigation can also be written in terms of the zero effort miss ZEM (meaning the miss that would result if the target continued to do what it was doing and the missile did not issue acceleration commands) as

$$\mathrm{ZEM_{PN}} = y + \dot{y} t_{\rm go}$$

$$n_{c_{\rm PN}} = N' V_c \dot{\lambda} = \frac{N' \mathrm{ZEM_{PN}}}{t_{\rm go}^2}$$

It can be shown that proportional navigation is an optimal guidance law if the missile has ideal dynamics (that is, zero-lag guidance system), the target does not maneuver, and $N' = 3$. Here optimal means that under ideal circumstances proportional navigation can yield zero miss distance for the least amount of missile acceleration (that is, the integral of the acceleration squared over the flight time is minimized). If the target is maneuvering, proportional navigation still works but another guidance law might do better.

The augmented proportional navigation guidance law assumes that the target is executing a constant maneuver of magnitude n_T. The guidance law is proportional navigation plus an extra term to account for the target maneuver and is given by

$$n_{c_{\rm APN}} = N' V_c \dot{\lambda} + 0.5 N' n_T$$

where it is assumed that the target acceleration can be estimated. It can be shown that augmented proportional navigation is an optimal guidance law (that is, zero miss and the integral of the acceleration squared is minimized) if the target executes a constant maneuver and N′ = 3. Augmented proportional navigation can also be written in terms of the zero effort miss as

$$n_{c_{\text{APN}}} = \frac{N'\text{ZEM}_{\text{APN}}}{t_{\text{go}}^2}$$

where

$$\text{ZEM}_{\text{APN}} = y + \dot{y}t_{\text{go}} + 0.5n_T t_{\text{go}}^2$$

Again, if the target executes another type of maneuver, augmented proportional navigation still works but another guidance law might do better.

It was shown in Chapter 8 that missile flight control system dynamics can cause miss distance if the target maneuvers a short time before intercept. The flight control system dynamics cause an unwanted delay between the commanded acceleration n_c and the achieved missile acceleration n_L. Let us assume that the dynamics of the missile flight control system can be represented by the single-lag network

$$\frac{n_L}{n_c} = \frac{1}{1 + sT}$$

where T is the approximate time constant of the missile flight control system. Since large miss distances can result if the time constant is big, it is desirable to keep the flight control system time constant as small as possible. If we also assume that the target is executing a constant maneuver and the time constant of the missile flight control system is known, then it can be shown that the optimal guidance law which compensates for the flight control system dynamics is given by

$$n_{c_{\text{OG}}} = \frac{N'}{t_{\text{go}}^2}\left[y + \dot{y}t_{\text{go}} + 0.5n_T t_{\text{go}}^2 - n_L\, T^2(e^{-x} + x - 1)\right]$$

where the effective navigation ratio is no longer constant and can be expressed as

$$N' = \frac{6x^2(e^{-x} - 1 + x)}{2x^3 + 3 + 6x - 6x^2 - 12xe^{-x} - 3e^{-2x}}$$

and x is defined as

$$x = \frac{t_{\text{go}}}{T}$$

The effective navigation ratio for the optimal guidance law is displayed in Fig. 31.1. We can see that at the beginning of the flight (long time to go) the

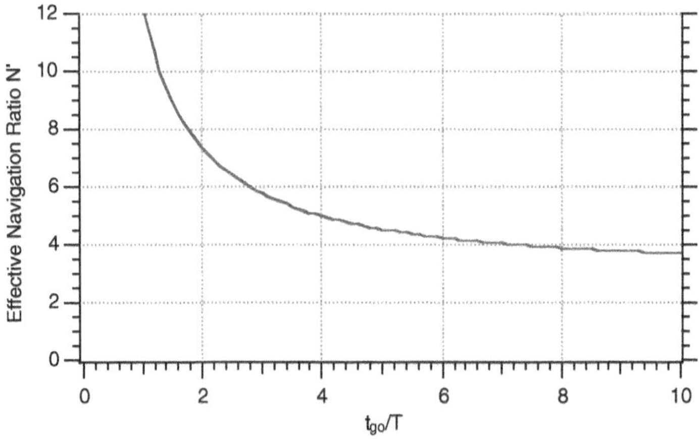

Fig. 31.1 Normalized effective navigation ratio for optimal guidance law.

effective navigation ratio is approximately constant and is approaching 3. As we get closer to intercept (small time to go), the effective navigation ratio grows considerably.

DIFFERENTIAL GAME GUIDANCE LAW

Much work has been done in the past 30-plus years on guidance laws based on differential game theory with bounded controls [1–8]. The concept of differential pursuit-evasion games is based on assuming that the target maneuvering strategy is the best for maximizing the miss distance. Against such target maneuver the optimal stategy of the interceptor missile is the guidance law based on the solution of a differential game. This guidance law is robust with respect to the actual target maneuver by guaranteeing the smallest miss distance against any bounded target maneuver. This guidance law, like the guidance laws of the previous section, is also based on the zero effort miss concept but is bang-bang in nature (missile executes either maximum positive or negative acceleration) because the control effort is not penalized. The bang-bang nature of this guidance law may not make it suitable for endoatmospheric application (possibly too much induced drag or excessive actuator requirements), but for the moment we will put this practical concern aside. The differential game guidance law with bounded controls considered in this chapter is based on the zero effort miss from the optimal guidance law, assuming ideal target dynamics, is given by

$$n_{c_{DG}} = n_{MAX} sign(ZEM_{DG})$$

where the zero effort miss *does not depend on the future target maneuver* and is given by

$$\text{ZEM}_{\text{DG}} = y + \dot{y}t_{\text{go}} - n_L T^2(e^{-x} + x - 1)$$

where

$$x = \frac{t_{\text{go}}}{T}$$

We can see that the differential game guidance zero effort miss calculation *does not depend on any assumptions concerning the target maneuver type or even the magnitude of the current target acceleration.* The fact that this guidance law does not require knowledge of the target maneuver means that this guidance law should perform equally as well (or as poorly) against all types of target maneuvers.

In the preceding guidance law it is assumed that the target dynamics are ideal in the sense that the target can execute its maneuver instantaneously without a lag. If there are target dynamics that can be represented by a first-order lag with time constant T_T and these dynamics are known to the pursuing interceptor then the zero effort miss becomes

$$\text{ZEM}_{\text{DG}} = y + \dot{y}t_{\text{go}} - n_L T^2(e^{-x} + x - 1) + n_T T_T^2(e^{-x_T} + x_T - 1)$$

where

$$x_T = \frac{t_{\text{go}}}{T_T}$$

If target maneuver dynamics are considered then the target maneuver level must be estimated with a Kalman filter. If target maneuver dynamics are neglected ($T_T = 0$), then we can see from the zero effort miss calculation that the target maneuver level does not have to be estimated.

TARGET MANEUVERS

For the guidance law comparison it is desireable to pick a set of random target maneuvers of varying degrees of difficulty for the pursuing interceptor. The first target maneuver considered, and one that has been used extensively in this text, is the uniformly distributed target maneuver. This constant maneuver's starting time is random and equally likely to occur anywhere during the flight (uniformly distributed over the flight time). The magnitude of the maneuver is fixed but for any given flight the maneuver is equally likely to be either positive or negative. An example of a 5-*g* uniformly distributed target maneuver is shown in Fig. 31.2.

Another maneuver that is often used in missile guidance system analysis is the Poisson (or random telegraph signal) target maneuver. This target maneuver,

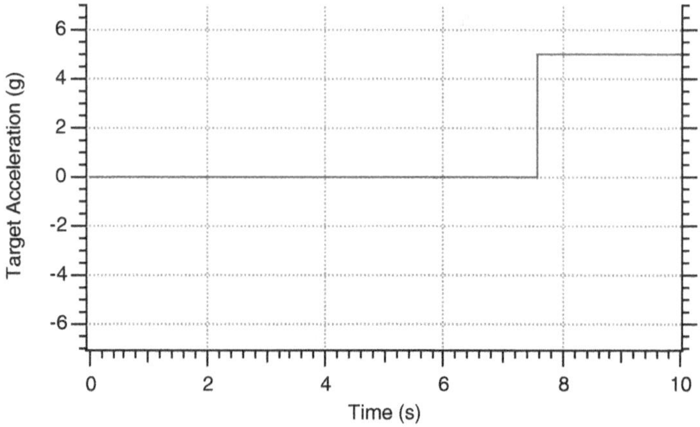

Fig. 31.2 Sample uniformly distributed constant 5-*g* target maneuver.

while not the most realistic, is probably the most challenging to intercept of all possible target maneuvers because it stresses the missile guidance system. The Poisson square wave or random telegraph signal is defined as a maneuver of amplitude $+\beta$ or $-\beta$. The length of time for which the maneuver $n_T(t)$ remains in either the positive or negative direction is random. In particular, the number of times the maneuver changes sign (zero crossings) during one second is given by the Poisson distribution. If $P(k)$ represents the probability of k sign changes in t_F seconds, then we can say that

$$P(k) = \frac{(\upsilon t_F)^k e^{-\upsilon t_F}}{k!}$$

where k is the number of sign changes and υ is the average number of zero crossings per second. Details on how to model the Poisson maneuver both in the forward and adjoint time domains can be found in the appendix. An example of a 5-g Poisson target maneuver with an average of 0.5 zero crossings per second is displayed in Fig. 31.3. Generally speaking, if good performance can be obtained against the Poisson maneuver, even better performance will be obtained against other types of target maneuvers.

A random weave maneuver, originally discussed in Chapter 20 is a sinusoidal maneuver of constant frequency with a starting time that is uniformly distributed over the flight time and whose phase is uniformly distributed between 0 deg and 360 deg. A sample 5-g, 2-r/s random weave maneuver is displayed in Fig. 31.4.

The random vertical-S maneuver is one in which the target is always at maximum positive or negative acceleration but the sign of the acceleration is periodically reversed. The starting time of the maneuver is uniformly distributed over the flight time. A sample random 5-g, 2-r/s vertical-S maneuver is displayed in Fig. 31.5.

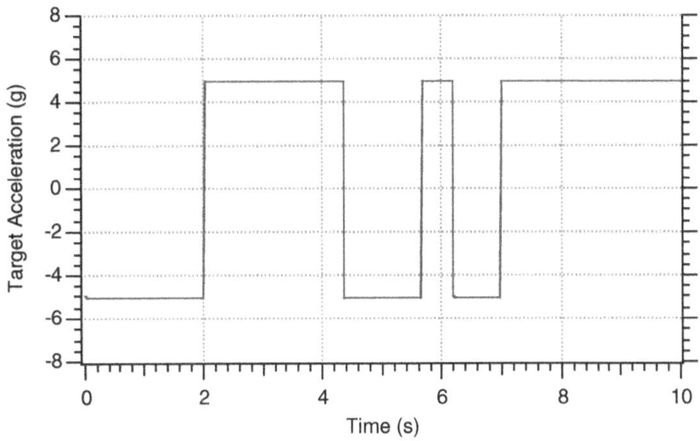

Fig. 31.3 Sample 5-*g* Poisson target maneuver with an average of 0.5 zero crossings per second.

GUIDANCE LAW COMPARISON

Figure 31.6 shows the homing loop when measurement angle noise and random target maneuver are considered as the two sources of error that can cause miss distance. Here the three-state Kalman filter, which was originally derived in Chapter 9, is used to estimate the states required for guidance.

Using the three-state Kalman filter, sample estimates of each of the target maneuver types are presented in Figs. 31.7–31.10. In the following examples

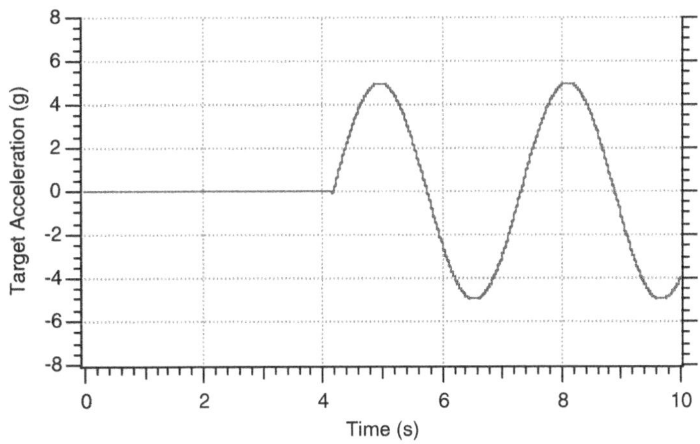

Fig. 31.4 Sample random 5-*g*, 2-r/s target weave maneuver.

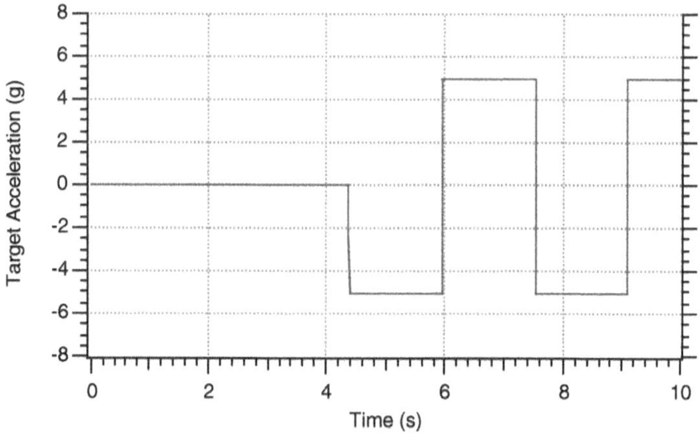

Fig. 31.5 Sample 5-*g*, 2-r/s random vertical-S target maneuver.

the angle measurement noise standard deviation is 0.1 mr and the sampling time is 0.01 s. The assumed process noise adjusts the bandwidth of the Kalman filter. The value of the continuous process noise matrix (used in deriving the discrete process noise matrix) in the English system of units was assumed to be

$$Q_c = \begin{bmatrix} 0 & 0 & 0 \\ 0 & 0 & 0 \\ 0 & 0 & \frac{n_{TMAX}^2}{t_F} \end{bmatrix} = \begin{bmatrix} 0 & 0 & 0 \\ 0 & 0 & 0 \\ 0 & 0 & \frac{161^2}{10} \end{bmatrix}$$

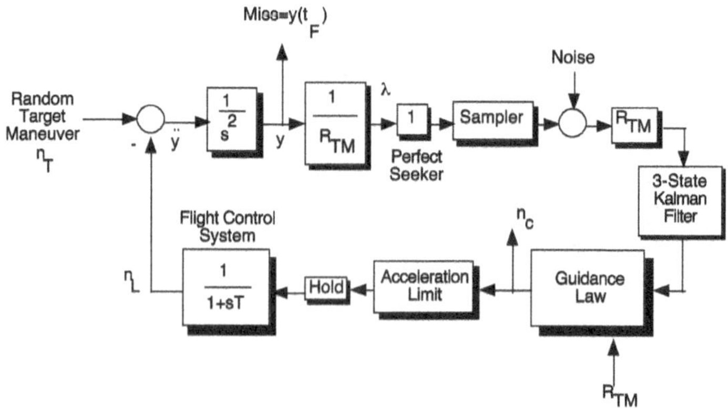

Fig. 31.6 Homing loop with measurement noise and three-state Kalman filter.

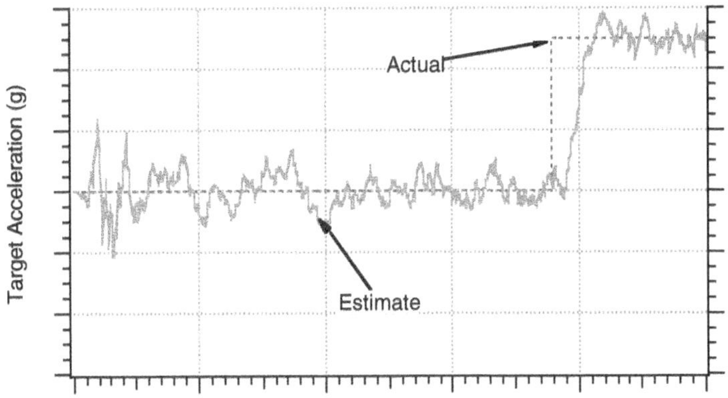

Fig. 31.7 Estimating random constant target maneuver with three-state Kalman filter.

We can see from Fig. 31.7 that the estimate of the random constant target maneuver is excellent. This should not be surprising because the three-state Kalman filter is optimal for this type of target maneuver. The worst estimate of the target maneuver is for the Poisson maneuver case of Fig. 31.8. In this case the filter cannot keep up with the zero crossings of the target. More process noise or a wider bandwidth filter would have allowed the filter to follow the target motion better at the expense of much noisier estimates. Figures 31.9 and 31.10 show that the filter can provide reasonable estimates of both the 2-r/s weaving and vertical-S target maneuvers. However, the filter estimates in these two examples lag the actual target maneuver.

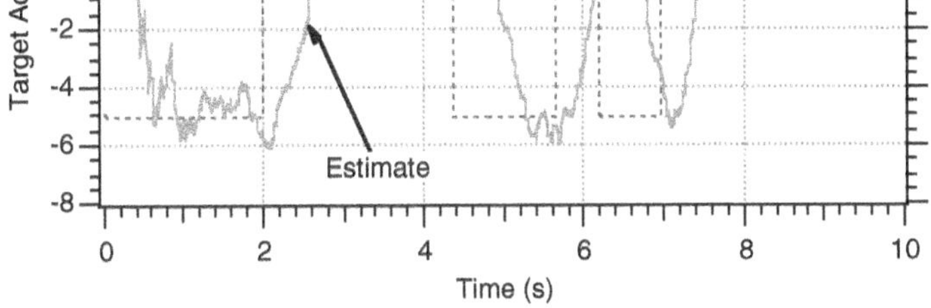

Fig. 31.8 Estimating Poisson target maneuver with three-state Kalman filter.

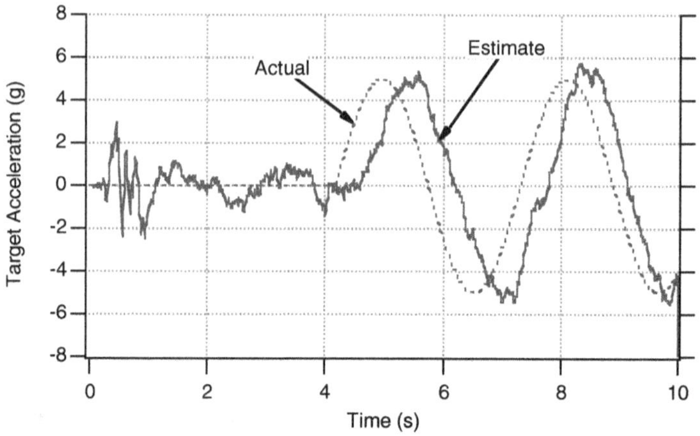

Fig. 31.9 Estimating random weave target maneuver with three-state Kalman filter.

The system parameters of our homing loop shown in Fig. 31.6 are defined below in Table 31.1. We can immediately see that system performance will be a challenge since there is only a 2 to 1 missile-to-target acceleration advantage.

A Monte Carlo simulation of Fig. 31.6 is used to generate rms miss distance performance and appears in Listing 31.1. Nominally 50-run Monte Carlo sets are run for each of 100 flight times (ranging from 0.1 s to 10 s in steps of 0.1 s) in order to generate guidance law comparisons.

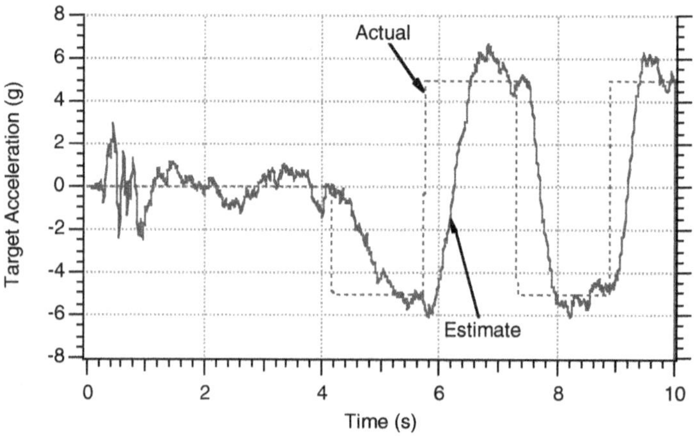

Fig. 31.10 Estimating random vertical-S target maneuver with three-state Kalman filter.

TABLE 31.1 GUIDANCE SYSTEM PARAMETERS

Parameter	Definition	Value
n_T	Target maneuver magnitude	5 g
σ_{Noise}	Measurement noise standard deviation	0.0001 r
T	Flight control system time constant	0.2 s
T_s	Kalman filter sampling time	0.01 s
a_{MAX}	Missile acceleration limit	10 g
V_c	Closing velocity	5 km/s

LISTING 31.1 LINEARIZED MONTE CARLO SIMULATION FOR COMPARING GUIDANCE LAW EFFECTIVENESS WHEN NOISE AND FILTERING ARE CONSIDERED

```
% Runs very slowly because of small integration interval
%Preallocation
Z=zeros(size(1:1000));
I=zeros(size(1:50));
TF=zeros(size(1:50));
count=0;
TSW=1;
VC=5.*3280.;
XNTIC=161;
YIC=0.;
VM=3000.;
HEDEG=20.;
XNP=3.;
SIGNOISE=.0001;
TS=.01;
TAU=.2;
NOISE=1;
RUN=50;
% TYPE OF GUIDANCE (0=PN,1=OG,2=DG,3=HYBRID)
APN=1;
XLIM=322;
XNU=.5;
W=2;
% TYPE OF MANEUVER (1=POISSON,2=UNIF CONST,3=RANDOM SINE,4=RANDOM VS
ICONSTANT=2;
for TF=.1:.1:10.0,
        Z1=0;
        for I=1:RUN
                SUM=rand(1);
```

```
        TSTART=TF*SUM;
        PZ=rand(1);
        PZ=PZ-.5;
        if PZ > 0
            COEF=1;
        else
            COEF=-1;
        end;
SUM=rand(1);
PHASE=6.28*SUM;
        Y=YIC;
        YD=0;
        TS2=TS*TS;
        TS3=TS2*TS;
        TS4=TS3*TS;
        TS5=TS4*TS;
        PHIN=XNTIC*XNTIC/10;
        RTM=VC*TF;
        SIGPOS=RTM*SIGNOISE;
        SIGN2=SIGPOS^2;
        P11=SIGN2;
        P12=0.;
        P13=0.;
        P22=(VM*HEDEG/57.3)^2;
        P23=0.;
        P33=XNTIC*XNTIC;
        T=0.;
        H=.001;
        S=0.;
        YH=0.;
        YDH=0.;
        XNTH=0.;
        XNC=0.;
        XNL=0.;
        BETA=XNTIC;
        QFIRST=1;
        SIG=1./sqrt(2.*XNU);
        XNOISE=randn;
        if XNOISE > 0
            XNTP=BETA;
        else
            XNTP=-BETA;
        end;
        DELT=9999.;
        TNOW=0.;
        while T <= (TF - 1e-5)
```

```
if QFIRST==1
        XNOISE1=SIG*randn;
        XNOISE2=SIG*randn;
        DELT=XNOISE1^2+XNOISE2^2;
        QFIRST=0;
        TNOW=T;
end;
if T>= (DELT+TNOW)
        XNTP=-XNTP;
        QFIRST=1;
end;
YOLD=Y;
YDOLD=YD;
XNLOLD=XNL;
STEP=1;
FLAG=0;
while STEP <=1
        if FLAG==1
Y=Y+H*YD;
YD=YD+H*YDD;
XNL=XNL+H*XNLD;
T=T+H;
STEP=2;
        end;
        if ICONSTANT==1
                XNTC=XNTP;
        elseif ICONSTANT==2
                if T < TSTART
                        XNTC=0.;
                else
                        XNTC=COEF*XNTIC;
                end;
        elseif ICONSTANT==3
                if T < TSTART
                        XNTC=0.;
                else
                        XNTC=XNTIC*sin(W*T+PHASE);
                end;
        else
                if T < TSTART
                        XNTC=0.;
                else
                       XNTC=COEF*XNTIC*sign(sin(W*(T-TSTART)));
                end;
        end;
        TGO=TF-T+.00001;
```

```
        RTM=VC*TGO;
        XLAM=Y/(VC*TGO);
        XLAMD=(RTM*YD+Y*VC)/(RTM^2);
        XNLD=(XNC-XNL)/TAU;
        YDD=XNTC-XNL;
        FLAG=1;
end;
FLAG=0;
Y=.5*(YOLD+Y+H*YD);
YD=.5*(YDOLD+YD+H*YDD);
XNL=.5*(XNLOLD+XNL+H*XNLD);
S=S+H;
if S > =(TS - 1e-5)
        S=0.;
        TGO=TF-T+.000001;
        RTM=VC*TGO;
        SIGPOS=RTM*SIGNOISE;
        SIGN2=SIGPOS^2;
        M11=P11+TS*P12+.5*TS2*P13+TS*(P12+TS*P22
              +.5*TS2*P23);
        M11=M11+.5*TS2*(P13+TS*P23+.5*TS2*P33)
              +TS5*PHIN/20.;
        M12=P12+TS*P22+.5*TS2*P23+TS*
              (P13+TS*P23+.5*TS2*P33)...
              +TS4*PHIN/8.;
        M13=P13+TS*P23+.5*TS2*P33+PHIN*TS3/6.;
        M22=P22+TS*P23+TS*(P23+TS*P33)+PHIN*TS3/3.;
        M23=P23+TS*P33+.5*TS2*PHIN;
        M33=P33+PHIN*TS;
        K1=M11/(M11+SIGN2);
        K2=M12/(M11+SIGN2);
        K3=M13/(M11+SIGN2);
        P11=(1.-K1)*M11;
        P12=(1.-K1)*M12;
        P13=(1.-K1)*M13;
        P22=-K2*M12+M22;
        P23=-K2*M13+M23;
        P33=-K3*M13+M33;
                XLAMNOISE=SIGNOISE*randn;
        YSTAR=RTM*(XLAM+XLAMNOISE);
        RES=YSTAR-YH-TS*YDH-.5*TS*TS*(XNTH-XNC);
        YH=K1*RES+YH+TS*YDH+.5*TS*TS*(XNTH-XNC);
        YDH=K2*RES+YDH+TS*(XNTH-XNC);
        XNTH=K3*RES+XNTH;
        XLAMDH=(YH+YDH*TGO)/(VC*TGO*TGO);
                X=TGO/TAU;
```

```
                    ZEM1H=YH+YDH*TGO-XNL*TAU*TAU
                            *(exp(-X)+X-1.);
                    ZEM2H=YH+YDH*TGO-XNL*TAU*TAU
                            *(exp(-X)+X-1.)+...
                            .5*XNTH*TGO*TGO;
                if APN==0
                XNC=XNP*(YH+YDH*TGO)/TGO^2;
                elseif APN==1
                X=TGO/TAU;
                TOP=6.*X*X*(exp(-X)-1.+X);
                BOT1=2*X*X*X+3.+6.*X-6.*X*X;
                BOT2=-12.*X*exp(-X)-3.*exp(-2.*X);
                XNPP=TOP/(.0001+BOT1+BOT2);
                        XNC=XNPP*ZEM2H/TGO^2;
                    elseif APN==2
                        XNC=XLIM*sign(ZEM1H);
                    else
                        if TGO>TSW
                            TOP=6.*X*X*(exp(-X)-1.+X);
                            BOT1=2*X*X*X+3.+6.*X-6.*X*X;
                            BOT2=-12.*X*exp(-X)-3.*exp(-2.*X);
                            XNPP=TOP/(.0001+BOT1+BOT2);
                            XNEW=XNPP*XNL*(exp(-X)+X-1.)/
                                    (X*X);
                            XNC=XNPP*ZEM2H/TGO^2;
                            if XNC>XLIM
                                XNC=XLIM;
                            end
                            if XNC<-XLIM
                                XNC=-XLIM;
                            end
                        else
                            XNC=XLIM*sign(ZEM1H);
                        end;
                end;
                if XNC>XLIM
                    XNC=XLIM;
                elseif XNC<-XLIM
                    XNC=-XLIM;
                end;
            end;
        end;
        Z(I)=Y;
        Z1=Z(I)+Z1;
        XMEAN=Z1/I;
end;
```

```
        SIGMA=0;
        Z1=0;
        Z2=0.;
        for I=1:RUN,
        Z1=(Z(I)-XMEAN)^2+Z1;
                        Z2=Z(I)^2+Z2;
        if I==1,
                SIGMA=0;
                                RMS=0.;
        else
                SIGMA=sqrt(Z1/(I-1));
                                RMS=sqrt(Z2/(I-1));
        end;
        end;
        count=count+1;
        ArrayTF(count)=TF;
        ArraySIGMA(count)=SIGMA;
        ArrayXMEAN(count)=XMEAN;
        ArrayRMS(count)=RMS;
end;
figure
plot(ArrayTF',ArrayRMS'),grid
title('RMS miss for various flight times')
xlabel('Flight Time (S)')
ylabel('RMS MISS (Ft) ')
clc
output=[ArrayTF',ArrayRMS'];
save datfil.txt output /ascii
disp('Simulation Complete')
```

Listing 31.1 was first run in the Monte Carlo mode for the proportional navigation guidance law for the case of the 5-*g* uniformly distributed constant target maneuver when the missile acceleration limit was 15 g and 10 g. We can see from Fig. 31.11 that when the missile only has a 2 to 1 acceleration advantage over the target, the miss is unacceptable. As we are only considering missile to target acceleration advantages of 2 to 1 in this chapter, proportional navigation guidance will be eliminated from further consideration.

Next Monte Carlo runs were made for both the optimal guidance (APN = 1) and differential game (APN = 2) guidance laws with bounded controls in the case of a 5-*g* random constant ideal target maneuver (ICONSTANT = 2) and a 10-*g* missile acceleration limit. We can see from Fig. 31.12 that against the random constant target maneuver, both guidance laws yield similar performance for flight times greater than 4 s. However for smaller flight times differential game guidance with bounded controls yielded slightly smaller rms miss distances.

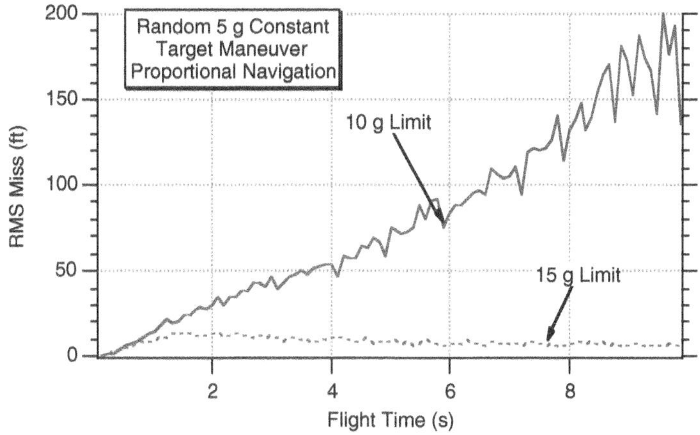

Fig. 31.11 Proportional navigation rms miss is too large in presence of 2 to 1 missile to target acceleration advantage.

Other Monte Carlo sets of runs were made for both optimal guidance and differential game guidance laws with bounded controls for the case of a 5-*g* Poisson target maneuver (ICONSTANT = 1) with an average of 0.5 zero crossings per second and a 10-*g* missile acceleration limit. From Fig. 31.13 we can see

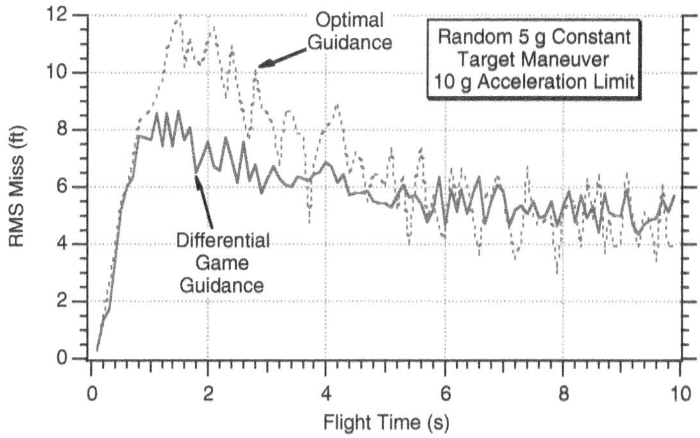

Fig. 31.12 Both optimal guidance and differential game guidance with bounded controls have similar performance against a random constant maneuver if there is only a 2 to 1 missile-to-target acceleration advantage.

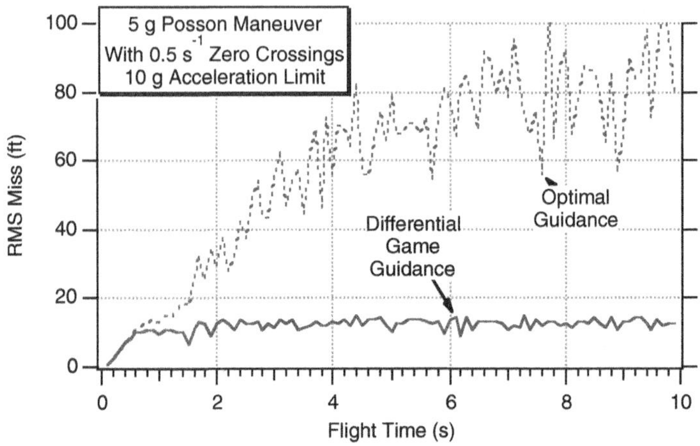

Fig. 31.13 Differential game guidance with bounded controls is more effective than optimal guidance against a Poisson target maneuver if there is only a 2 to 1 missile-to-target acceleration advantage.

that against this difficult target maneuver, differential game guidance with bounded controls performed much better than optimal guidance.

Another Monte Carlo sets of runs were made for both guidance laws for the case of a 5-g, 2-r/s random weaving target maneuver (ICONSTANT = 3) with a 10-g missile acceleration limit. From Fig. 31.14 we can see that against the weaving

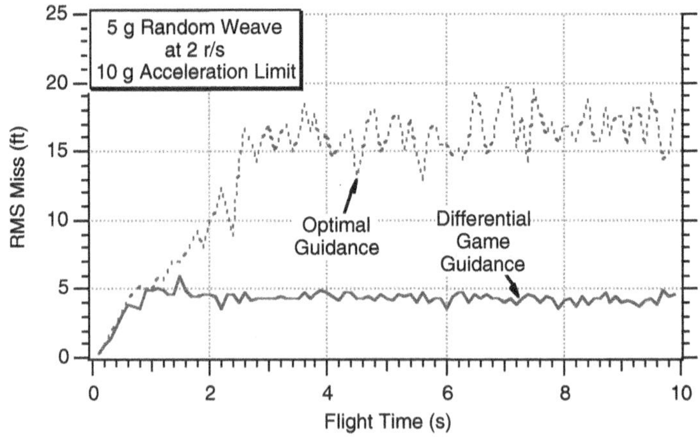

Fig. 31.14 Differential game guidance with bounded controls is more effective than optimal guidance against a random weaving target maneuver if there is only a 2 to 1 missile-to-target acceleration advantage.

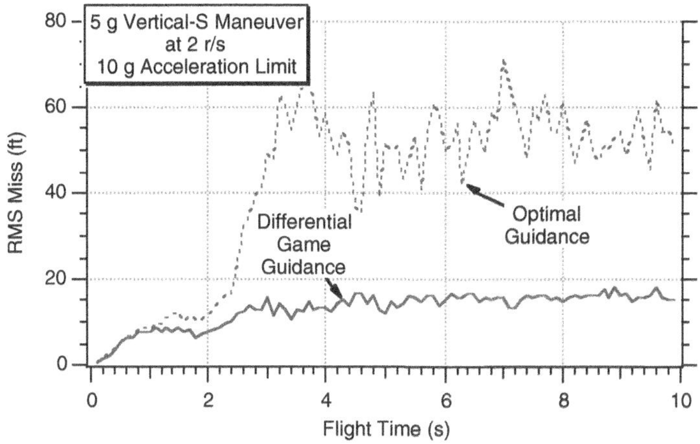

Fig. 31.15 Differential game guidance with bounded controls is more effective than optimal guidance against a random vertical-S target maneuver if there is only a 2 to 1 missile-to-target acceleration advantage.

target maneuver, the differential game guidance again performed much better than optimal guidance.

Other Monte Carlo sets of runs were made for different guidance laws for the case of a 5-*g*, 2-r/s random vertical-S target maneuver (ICONSTANT = 4) with a 10-*g* missile acceleration limit. From Fig. 31.15 we can see that against the vertical-S target maneuver, differential game guidance with bounded controls is again better than optimal guidance.

From the previous set of results we can see that if there was 0.1 mr of measurement noise and the sampling time was 0.01 s and there was a 2 to 1 missile-to target-acceleration advantage, the differential game guidance law with bounded controls generally yields better performance than the optimal guidance law against any of the four random target maneuvers considered. To see if the performance benefit is due to the low measurement noise or the small sampling time, let us now see how optimal guidance and differential game guidance with bounded controls compare when the measurement noise is increased by an order of magnitude against random weave maneuver. The missile-to-target acceleration limit is increased to 3 to 1. Figure 31.16 shows that when the measurement noise is increased by an order of magnitude the rms miss distances increases significantly but differential game guidance with bounded controls still yields better performance than optimal guidance. If the measurement noise is kept at 0.1 mr but the sampling time is increased to 0.1 s while the missile to target acceleration advantage is 3 to 1, Fig. 31.17 shows that when the sampling time is increased by an order of magnitude the rms miss distances increases significantly but differential game guidance with bounded controls still yields better performance than optimal guidance.

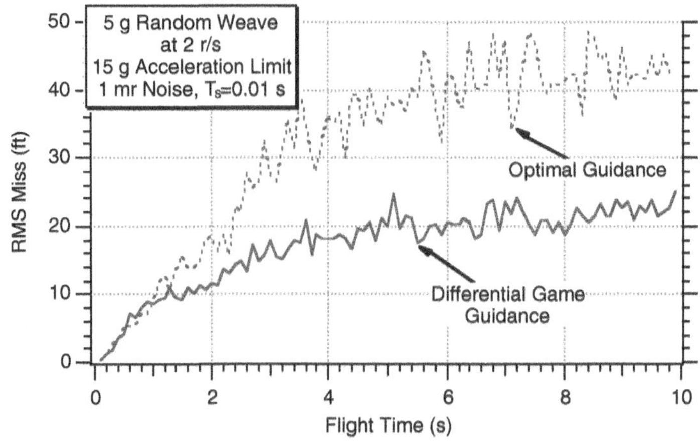

Fig. 31.16 Differential game guidance with bounded controls is more effective than optimal guidance against a random weave target maneuver when the measurement noise is increased.

MAKING DIFFERENTIAL GAME GUIDANCE MORE PRACTICAL

So far we have demonstrated that differential game guidance with bounded controls appears to be more effective than optimal guidance for a variety of very difficult target maneuvers when the missile-to-target acceleration advantage is low.

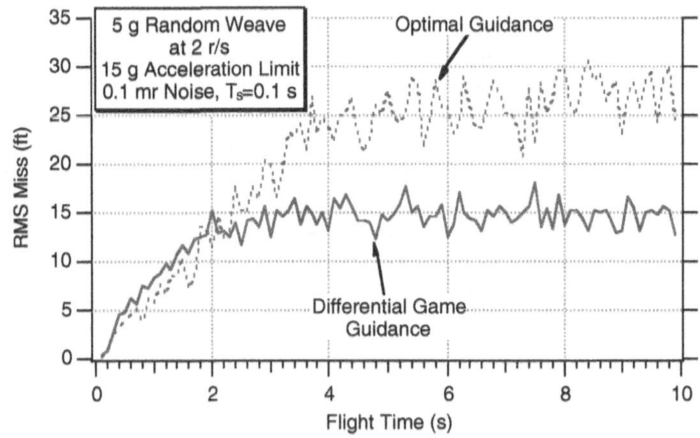

Fig. 31.17 Differential game guidance with bounded controls is more effective than optimal guidance against a random weave target maneuver when the sampling time is increased.

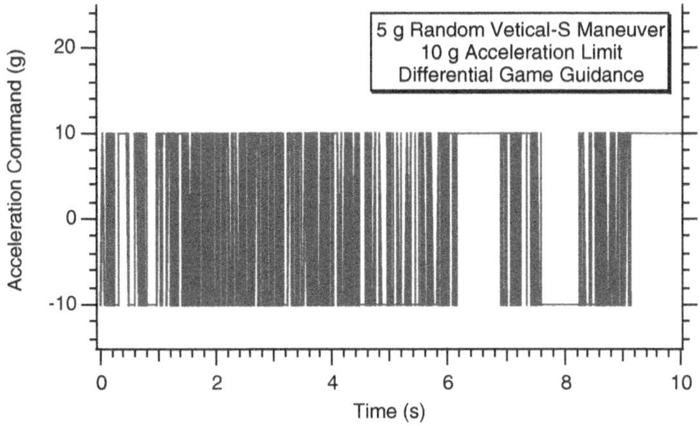

Fig. 31.18 Sample single-run missile acceleration command profile for differential game guidance with bounded controls indicates a great deal of chattering.

The advantages of differential game guidance with bounded controls has been well known for more than three decades by the academic community but has not been met with enthusiasm by practicing engineers. As was mentioned previously, one of the main practical objections to the implementation of differential game guidance with bounded controls has been the chattering back and fourth of the acceleration command as is illustrated in the single flight results of Fig. 31.18. Here it can be seen that the missile acceleration command due to the vertical-S target maneuver constantly switches back and fourth between maximum positive and negative acceleration. It has been thought that this chattering could lead to excessive induced drag or excessive actuator requirements, which would negate any of the idealized performance benefits illustrated in this chapter.

It is hypothesized that most of the miss distance improvements offered by the differential game guidance law with bounded controls occur in actions take near the very end of the flight. Therefore a possible solution to this potential induced drag problem is to limit the application of differential game guidance with bounded controls to the last few seconds of flight. In other words optimal guidance could be used for most of the flight and then TSW (parameter in Listing 31.1) seconds before the end of the flight differential game guidance with bounded controls could be used. This type of simple switching logic could be termed *hybrid guidance* and can be found in Listing 31.1 (APN = 3).

Figure 31.19 illustrates the use of hybrid guidance for a single flight against the vertical-S target maneuver with TSW = 2. Here we can see a gentler missile acceleration command profile when compared to Fig. 31.18. Chattering has been removed from 80% of the flight.

In order to see if comparable results to differential game guidance with bounded controls could be obtained with hybrid guidance a set of Monte Carlo

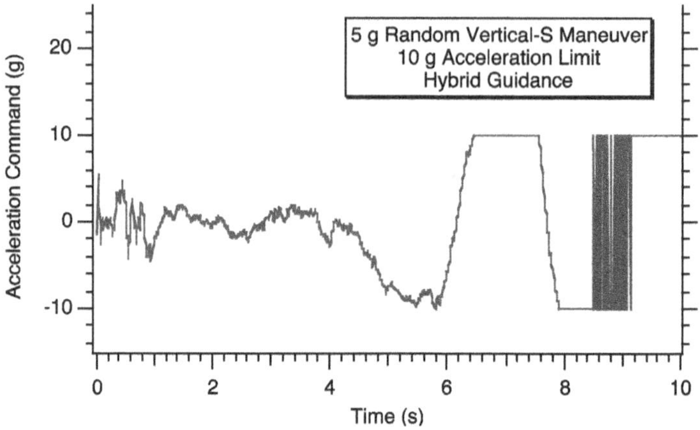

Fig. 31.19 Sample single-run missile acceleration command profile for hybrid guidance indicates most of the chattering has been eliminated.

results were generated for each of the four different random target maneuvers previously considered. The goal of the experiment was to see how small TSW could be made so that performance degradation due to chattering would either not result or would be considerably alleviated. Figures 31.20–31.23 show that hybrid guidance can yield performance very similar to that of differential game guidance with bounded controls with values of TSW ranging from 1 to 3 seconds. Thus, depending on the application, hybrid guidance may yield

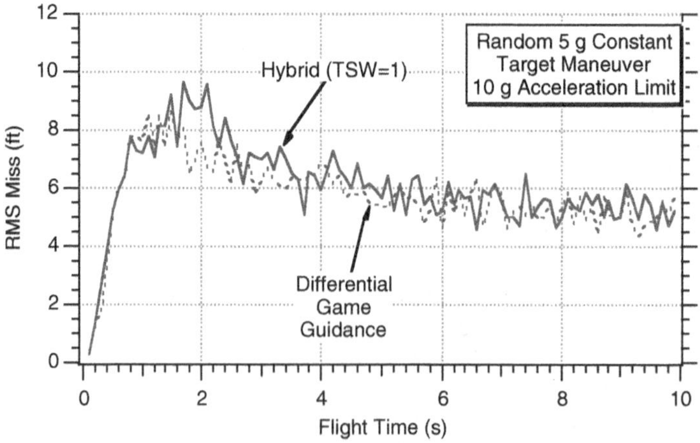

Fig. 31.20 Hybrid (TSW = 1 s) and differential game guidance with bounded controls yield similar performance against random constant target maneuver.

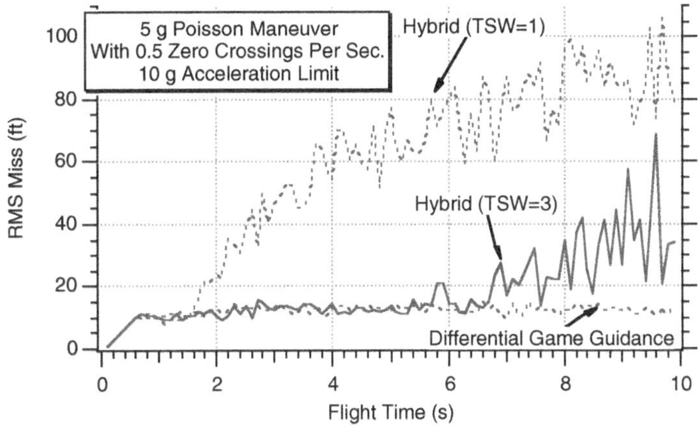

Fig. 31.21 Hybrid (TSW=3 s) and differential game guidance with bounded controls yield similar performance against Poisson target maneuver.

significant and achievable performance advantages over optimal guidance against very challenging target maneuvers.

TARGET DYNAMICS

It was previously mentioned in this chapter that the target could also be modeled as having a single time constant representation between the commanded and

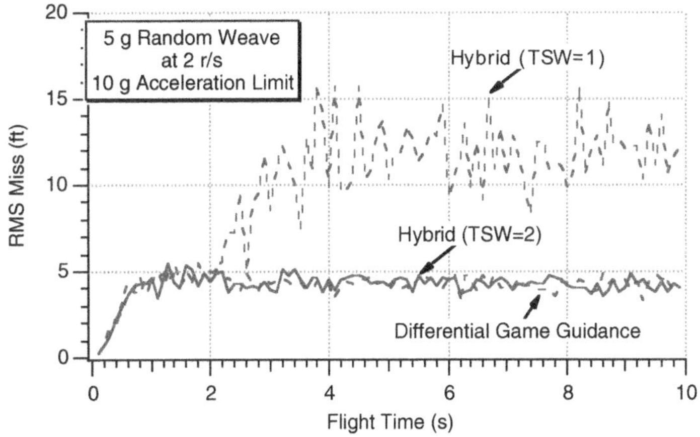

Fig. 31.22 Hybrid (TSW=2 s) and differential game guidance with bounded controls yield similar performance against random weave target maneuver.

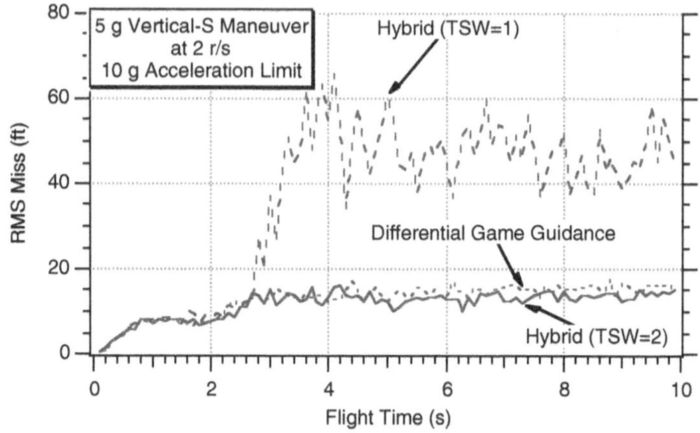

Fig. 31.23 Hybrid (TSW = 2 s) and differential game guidance with bounded controls yield similar performance against random vertical-S target maneuver.

achieved target acceleration. If the target dynamics were known, the zero effort miss would have an extra term to account for those dynamics as shown in the section which discussed the differential game guidance law with bounded controls. Figure 3.24 shows what happens to performance in the case of the random weave target maneuver when differential game guidance with bounded controls is used assuming target time constants of 0 and 0.2 s. We can see that

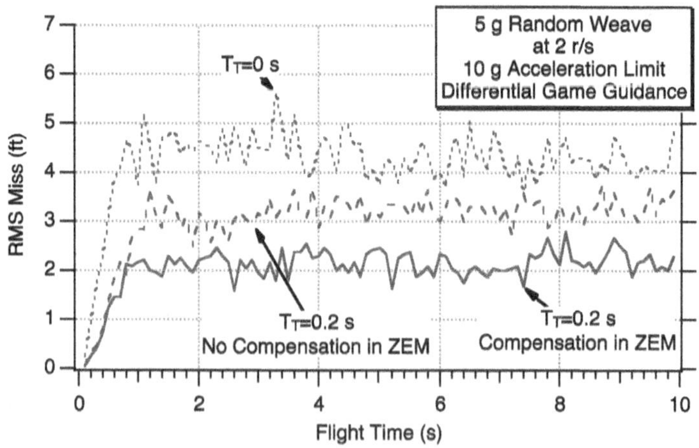

Fig. 31.24 Compensating for target dynamics can yield better system performance when the target time constant is 0.2 s.

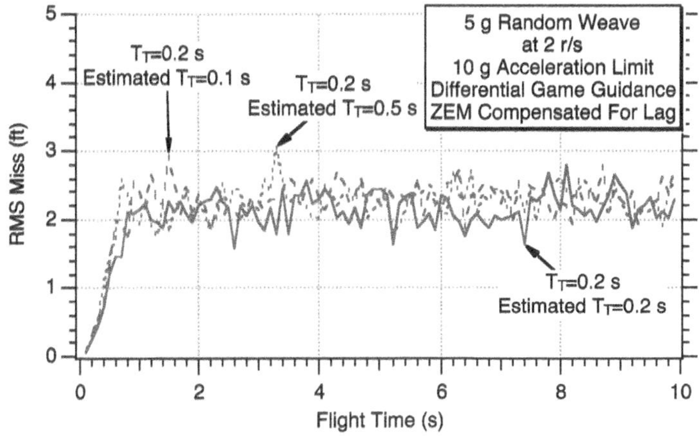

Fig. 31.25 Differential game guidance law with bounded controls does not have to know target time constant exactly.

miss distance performance improves slightly if there is a target time constant of 0.2 s—even though the guidance law is not compensating for the dynamics. However there is a more substantial improvement in performance when the guidance law compensates for the target dynamics.

In Fig. 31.24 it was assumed that when the target dynamics were known perfectly, the differential game guidance law with bounded controls could yield much smaller miss distances. Figure 31.25 shows that the differential game guidance law

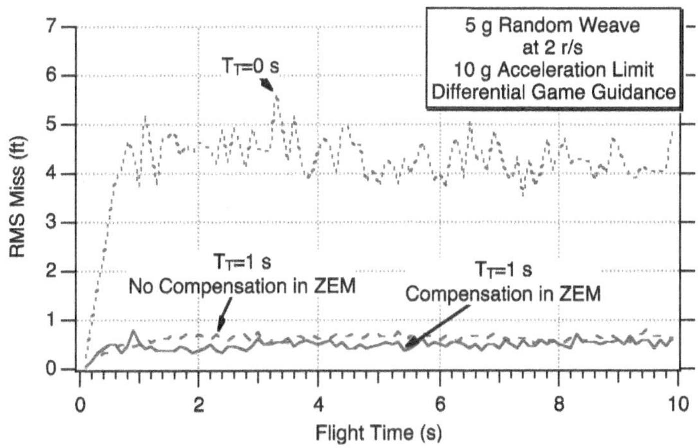

Fig. 31.26 Compensating for target dynamics is not as important when the target time constant is increased from 0.2 s to 1 s.

with bounded controls is not very sensitive to knowing the exact target time constant since nearly identical performance is achieved if we overestimate or underestimate the target time constant.

Finally Fig. 31.26 shows that if the target time constant is increased from 0.2 s to 1 s, the miss distances are much smaller and it is not as important to compensate for the target time constant.

SUMMARY

In this chapter the differential game guidance law with bounded controls has been introduced. In theory (that is, in noise free scenarios) this guidance law guarantees the smallest miss distance against *any* bounded target maneuver, including the entire family of unknown and/or random target maneuvers. It has been demonstrated that in realistic noisy scenarios with an estimator in the loop, the differential game guidance with bounded controls offers considerable performance improvements over optimal guidance when the missile-to-target acceleration advantage is low. The chapter also demonstrates how the chattering caused by the bang-bang nature of differential game guidance with bounded controls could be could be reduced dramatically without significant performance degradation by simply applying this guidance law in the last few seconds of flight. Other guidance laws that can potentially eliminate the typical chattering of the guidance laws based on differential game theory with bounded controls can be found in Ref. 9. Finally it has been shown that if the there is a target time constant, performance improvements could be achieved by compensating for the target dynamics in the computation of the zero effort miss.

REFERENCES

[1] Gutman, S., and Leitmann, G., "Optimal Strategies in the Neighborhood of a Collision Course," *AIAA Journal*, Vol. 14, No. 9, 1976, pp. 1210–1212.

[2] Gutman, S., "On Optimal Guidance for Homing Missiles," *Journal of Guidance and Control*, Vol. 3, No. 4, 1979, pp. 296–300.

[3] Shinar, J., and Gutman, S., "Three Dimensional Optimal Pursuit and Evasion with Bounded Control," *IEEE Transacactions on Automatic Control*, Vol. AC-25, No. 3, 1980, pp. 492–496.

[4] Shinar, J., "Solution Techniques for Realistic Pursuit-Evasion Games," *Advances in Control and Dynamic Systems, C.T. Leondes, ed.*, Vol. 17, Academic Press, N.Y. 1981, pp. 63–124.

[5] Shinar, J., "Interception of Maneuvering Targets in Theater Missile Defense,"*Theater Ballistic Missile Defense, B. Naveh. ed., Progress in Aeronautics and Astronautics,* Vol. 192, American Institute of Aeronautics and Astronautics, Washington D.C., 2000.

[6] Shima, T., and Shinar, J., "Time Varying Linear Pursuit-Evasion Game Models with Bounded Controls," *Journal of Guidance, Control and Dynamics*, Vol. 25, No. 3, 2002, pp. 425–432.

[7] Shinar, J., and Turetsky, V., "Missile Guidance Laws Based on Pursuit-Evasion Game Formulations," *Automatica*, Vol. 39, No. 4, 2003, pp. 607–618.

[8] Shinar, J., Turetsky, V., and Oshman, Y., "Integrated Estimation/Guidance Design Approach for Improved Homing Against Randomly Maneuvering Targets," *Journal of Guidance, Control and Dynamics*, Vol. 30, No. 1, Jan.–Feb. 2007, pp. 154–161.

[9] Shinar, J., and Turetsky, V., "Interceptor Guidance Laws—Robustness and Implementation," 50th Israel Annual Conference on Aviation and Astronautics, Tel Aviv–Haifa, Feb., 2010.

CHAPTER 32

Kinematics of Intercepting a Ballistic Target

In Chapter 13 we showed how a long-range ballistic target, using the principals of Lambert guidance, could fly to its intended destination. In this chapter we shall investigate the amount of velocity that is required by a pursuing interceptor to hit the target during the target's ballistic or midcourse portion of flight. In addition, standard graphical methods of conveying interceptor performance shall be introduced in this chapter. To simplify matters and to keep the discussion generic we shall assume that both the ballistic target and pursuing interceptor are impulsively launched, but not at the same time, so that they both get up to speed immediately. In addition, for simplicity we shall initially consider two-dimensional engagements on a round, non-rotating Earth where gravity is determined by Newton's Law of Universal Gravitation and atmospheric effects are neglected. It shall be assumed that the interceptor knows where the ballistic target will be at the desired intercept time [that is, predicted intercept point (PIP) is known perfectly]. From Chapter 11 we know that the impulsive ballistic target differential equations of motion in two dimensions are given by

$$\ddot{x}_T = -gm\frac{x_T}{(x_T^2 + y_T^2)^{1.5}}$$

$$\ddot{y}_T = -gm\frac{y_T}{(x_T^2 + y_T^2)^{1.5}}$$

where x_T and y_T are component distances from the center of the Earth to the target and gm is the gravitational parameter with value in the English system of units of

$$gm = 1.4077{*}10^{16}\,\frac{\text{ft}^3}{\text{s}^2}$$

In Chapter 28 a general three-dimensional Lambert routine was presented in Listing 28.2 showing how to calculate the initial velocity required by the ballistic target to reach its final destination at the desired time. The call to the Lambert routine was given by

```
[VRX,VRY,VRZ] = LAMBERT3D (XT,YT,ZT,TGOLAM,XF,YF,ZF,SWITCH);
```

where XT,YT,ZT was a three-dimensional vector describing the initial target location, TGOLAM was the allotted time to go for the target to reach its final destination and XF,YF,ZF was a three-dimensional vector describing the target's intended destination. The Lambert routine's output is VRX,VRY,VRZ, which is a three-dimensional vector describing the components of the required initial target velocity.

Listing 32.1, which is based on Listing 28.3, simulates an impulsive ICBM (Intercontinental Ballistic Missile) in two dimensions. The z coordinate in this simulation has been set to zero and all quantities that depend on latitude have been set to zero as well so that the two-dimensional problem will work with the three-dimensional Lambert routine. As was previously mentioned, the Lambert routine, which is *not* included in Listing 32.1, can be found in Listing 28.3. A careful examination of Listing 32.1 reveals that the time-of-flight formula for the minimum energy trajectory of the target has been programmed. The derivation of the formula for the time of flight for a minimum energy trajectory $t_{F_{\mathrm{ME}}}$ is derived in the appendix and is given by

$$t_{F_{\mathrm{ME}}} = 252 + 0.223DR_{\mathrm{km}} - 5.44{*}10^{-6}\mathrm{DR}_{\mathrm{km}}^2$$

where DR_{km} is the desired distance of travel in km. The trajectory can be lofted by adding TLOFT in units of seconds to $t_{F_{\mathrm{ME}}}$ or depressed by subtracting TLOFT from $t_{F_{\mathrm{ME}}}$. From a three-dimensional point of view the ICBM in Listing 32.1 is flying across the equator (latitude is zero) starting at 0-deg longitude and traveling 10,000 km or approximately a quarter of the way around the world.

LISTING 32.1 TWO-DIMENSIONAL TRAJECTORY SIMULATION OF IMPULSIVE TWO-DIMENSIONAL ICBM

```
count=0;
TS=1.;
RDESKM=10000.;
TLOFT=0.;
TFTOT=252.+.223*RDESKM-(5.44E-6)*RDESKM*RDESKM;
TFTOT=TFTOT+TLOFT;
A=2.0926E7;
GM=1.4077E16;
XLONGFDEG=57.3*RDESKM*3280./A;
XLONGTDEG=0.;
XLATTDEG=0.;
XLATFDEG=0.;
SWITCH=0;
T=0.;
S=0.;
XLONGF=XLONGFDEG/57.3;
XLATF=XLATFDEG/57.3;
```

```
XF=A*cos(XLATF)*cos(XLONGF);
YF=A*cos(XLATF)*sin(XLONGF);
ZF=0.;
XLONGT=XLONGTDEG/57.3;
XLATT=XLATTDEG/57.3;
XT=A*cos(XLATT)*cos(XLONGT);
YT=A*cos(XLATT)*sin(XLONGT);
ZT=0.;
XTINIT=XT;
YTINIT=YT;
ZTINIT=0.;
RTINIT=sqrt(XTINIT^2+YTINIT^2+ZTINIT^2);
H=.01;
ALTTKM=(sqrt(XT^2+YT^2)-A)/3280.;
TGOLAM=TFTOT-T;
[VRX,VRY,VRZ]=LAMBERT3D(XT,YT,ZT,TGOLAM,XF,YF,ZF,SWITCH);
XTD=VRX;
YTD=VRY;
ZTD=VRZ;
VBOT=sqrt(XTD^2+YTD^2)/3280.;
while ALTTKM > -1
     XTOLD=XT;
     YTOLD=YT;
     XTDOLD=XTD;
     YTDOLD=YTD;
     STEP=1;
     FLAG=0;
     while STEP <=1
          if FLAG==1
     STEP=2;
               XT=XT+H*XTD;
               YT=YT+H*YTD;
               XTD=XTD+H*XTDD;
               YTD=YTD+H*YTDD;
               T=T+H;
          end;
          TEMPBOTT=(XT^2+YT^2)^1.5;
          XTDD=-GM*XT/TEMPBOTT;
          YTDD=-GM*YT/TEMPBOTT;
          ALTTKM=(sqrt(XT^2+YT^2)-A)/3280.;
          FLAG=1;
     end
     FLAG=0;
     XT=.5*(XTOLD+XT+H*XTD);
     YT=.5*(YTOLD+YT+H*YTD);
     XTD=.5*(XTDOLD+XTD+H*XTDD);
```

```
        YTD=.5*(YTDOLD+YTD+H*YTDD);
        S=S+H;
        if S >=(TS-.0001)
                S=0.;
                ALTTKM=(sqrt(XT^2+YT^2)-A)/3280.;
                DISTRTKNM=distance3dkm(XT,YT,ZT,XTINIT,YTINIT,ZTINIT);
                count=count+1;
                ArrayT(count)=T;
                ArrayDISTRTKNM(count)=DISTRTKNM;
                ArrayALTTKM(count)=ALTTKM;
        end
end
figure
plot(ArrayDISTRTKNM,ArrayALTTKM),grid
xlabel('Downrange (km)')
ylabel('Altitude (km) ')
clc
output=[ArrayT',ArrayDISTRTKNM',ArrayALTTKM'];
save datfil.txt output /ascii
disp 'simulation finished'
% LAMBERT3D can be found in Listing 28.3
% distance3dkm can be found in Listing 28.3
```

Minimum energy (TLOFT = 0), lofted (TLOFT = 500), and depressed (TLOFT = −200) trajectories for the ballistic target were run using Listing 32.1 and the resultant trajectories are compared in Fig. 32.1. The initial velocity required for the minimum energy trajectory was 7.19 km/s, while 7.33 km/s was required for the lofted trajectory and 7.23 km/s was required for the depressed trajectory.

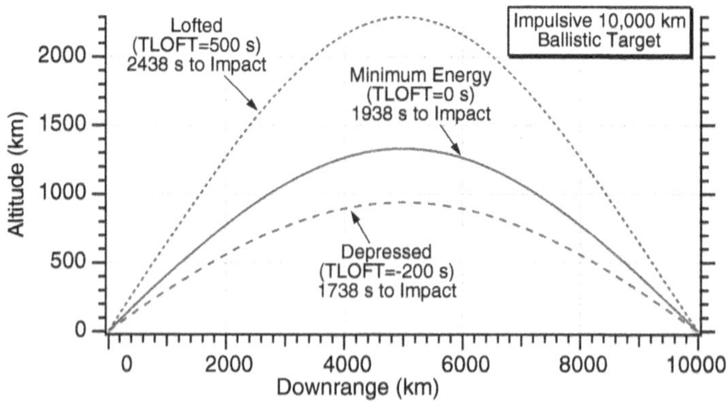

Fig. 32.1 Possible trajectories for 10,000-km impulsive ICBM.

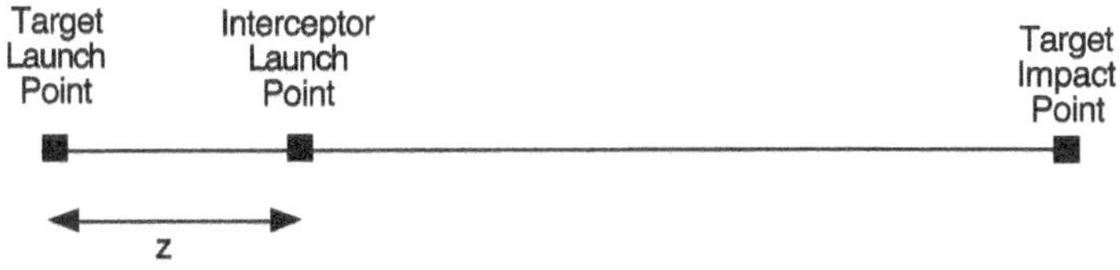

Fig. 32.2 Distance between interceptor and target launch points.

We can see from Fig. 32.1 that all the trajectories have different apogees and arrive at the 10,000-km destination at different times.

Now let us examine how much velocity would be required by a pursuing impulsive interceptor to hit the ballistic target at a desired intercept time t_F. It is assumed that the interceptor is launched some time after the target is launched and that the interceptor launch point is located z km downrange from the target launch point as shown in Fig. 32.2.

A multiple-run simulation in which the desired intercept time TF is varied for each run appears in Listing 32.2. From Listing 32.2 we can see that the interceptor launch time is nominally 300 s after the target is launched (TLAUNCH = 300) and the distance from the target launch point to the missile launch point is nominally 1000 km (XLONGMDEGICKM = 1000). The target is flying a 10,000-km (RDESKM = 10000) minimum energy trajectory (TLOFT = 0). A perfect prediction routine predict32.m tells the interceptor where the impulsively launched target will be at the desired intercept time TF. This prediction routine knows that the target is impulsive and takes the initial position and velocity of the target and integrates the target differential equations forward to the desired intercept time, thus yielding the predicted intercept point (XTFACT and YTFACT). At the desired interceptor launch time (TLAUNCH), a three-dimensional Lambert routine (with z component set to zero) calculates the required interceptor velocity vector so that it will arrive at the predicted intercept point at the desired intercept time. The equations of motion for the interceptor and target are numerically integrated forward and the engagement simulation is stopped at the point of closest approach, which is the miss distance. After each run the interceptor flight time (TF-TLAUNCH) and the required interceptor velocity (VMBO) are tabulated for interceptor burnout velocities less than 8 km/s. In this simulation, the precise miss distance is not calculated since the integration interval is not small enough in the main program nor the prediction routine to get a very accurate calculation of the miss. Decreasing the integration interval by an order of magnitude would have yielded very long running times. Under the condition of a large integration interval, a miss of several hundred feet will be considered a direct hit. If the integration interval was reduced by an order of magnitude, the miss distances would be near zero. Flights are considered a success if the required missile velocity is less than 8 km/s, the miss is less than 1000 ft, and the altitude of intercept is greater than 50 km. We can see from Listing 32.2 that one run involves cases in which the desired intercept time is varied from 60 s after the interceptor is launched to 20 s before the target would impact the ground in steps of 20 s.

LISTING 32.2 TWO-DIMENSIONAL KINEMATIC MULTIPLE-RUN ENGAGEMENT SIMULATION

```
count=0;
TLAUNCH=300.;
TF=1400.;
TS=1.;
XLONGMDEGICKM=1000.;
XLATMDEGICKM=0.;
RDESKM=10000.;
TFTOT=2000.;
ALTMKMIC=0.;
TLOFT=0.;
QEARTH=0;
TFTOT=252.+.223*RDESKM-(5.44E-6)*RDESKM*RDESKM;
TFTOT=TFTOT+TLOFT;
for TF=(TLAUNCH+60):20:(TFTOT-20),
        SWITCH=0;
        SWITCHM=0;
        ALTM=ALTMKMIC*3280.;
        TFTOTP=TFTOT;
        RDESKMP=RDESKM;
        A=2.0926E7;;
        GM=1.4077E16;
        XLONGFDEG=57.3*RDESKM*3280./A;
        XLONGMDEGIC=XLONGMDEGICKM/111.;
        XLATMDEGIC=0.;
        XLONGMDEG=XLONGMDEGIC;
        XLATMDEG=XLATMDEGIC;
        QGUID=0;
        QLAUNCH=0;
        QFIRST=1;
        XLONGTDEG=0.;
        XLATTDEG=0.;
        XLATFDEG=0.;
        TBO=0.;
        QBOOST=1;
        QBOOSTM=1;
        SWITCH=0;
        SWITCHM=0;
        T=0.;
        S=0.;
```

```
AXT=0.;
AYT=0.;
ATP=0.;
XLONGF=XLONGFDEG/57.3;
XLATF=XLATFDEG/57.3;
XF=A*cos(XLATF)*cos(XLONGF);
YF=A*cos(XLATF)*sin(XLONGF);
ZF=0.;
XLONGT=XLONGTDEG/57.3;
XLONGM=XLONGMDEG/57.3;;
XLATM=XLATMDEG/57.3;
XLATT=XLATTDEG/57.3;
XT=A*cos(XLATT)*cos(XLONGT);
YT=A*cos(XLATT)*sin(XLONGT);
ZT=0.;
XTINIT=XT;
YTINIT=YT;
ZTINIT=0.;
RTINIT=sqrt(XTINIT^2+YTINIT^2+ZTINIT^2);
XM=(A+ALTM)*cos(XLATM)*cos(XLONGM);
YM=(A+ALTM)*cos(XLATM)*sin(XLONGM);
ZM=0.;
XMINIT=XM;
YMINIT=YM;
ZMINIT=ZM;
RMINIT=sqrt(XMINIT^2+YMINIT^2+ZMINIT^2);
ATP=1.;
AXM=0.;
AYM=0.;
AMP=0.;
H=.01;
ALTTKM=(sqrt(XT^2+YT^2)-A)/3280.;
XMD=0.;
YMD=0.;
ACC=0.;
AXMGUID=0.;
AYMGUID=0.;
PREDERRKM=0.;
ZEM1=0.;
ZEM2=0.;
ALTMKM=(sqrt(XM^2+YM^2)-A)/3280.;
ALTTKM=(sqrt(XT^2+YT^2)-A)/3280.;
TGOLAM=TFTOT-T;
[VRX,VRY]=LAMBERT3D(XT,YT,ZT,TGOLAM,XF,YF,ZF,SWITCH);
XTD=VRX;
YTD=VRY;
```

```
	ZTD=0.;
[XTFACT,YTFACT]=predict32(T,XT,YT,XTD,YTD,TF,TFTOT,XF,YF);
	VBOT=sqrt(XTD^2+YTD^2)/3280.;
RTM1=XT-XM;
	RTM2=YT-YM;
	RTM=sqrt(RTM1^2+RTM2^2);
	VTM1=XTD-XMD;
	VTM2=YTD-YMD;
	VC=-(RTM1*VTM1+RTM2*VTM2)/RTM;
	while ~((T>(TLAUNCH+50.)) & VC<0. & RTM<10000.)
		if RTM<1000
			H=.0001;
		else
			H=.01;
		end
		XTOLD=XT;
		YTOLD=YT;
		XTDOLD=XTD;
		YTDOLD=YTD;
		XMOLD=XM;
		YMOLD=YM;
		XMDOLD=XMD;
		YMDOLD=YMD;
		STEP=1;
		FLAG=0;
		while STEP <=1
			if FLAG==1
				STEP=2;
				XT=XT+H*XTD;
				YT=YT+H*YTD;
				XTD=XTD+H*XTDD;
				YTD=YTD+H*YTDD;
				XM=XM+H*XMD;
				YM=YM+H*YMD;
				XMD=XMD+H*XMDD;
				YMD=YMD+H*YMDD;
				T=T+H;
			end
			TEMPBOTT=(XT^2+YT^2)^1.5;
			XTDD=-GM*XT/TEMPBOTT;
			YTDD=-GM*YT/TEMPBOTT;
			RTM1=XT-XM;
			RTM2=YT-YM;
			VTM1=XTD-XMD;
			VTM2=YTD-YMD;
			RTM=sqrt(RTM1^2+RTM2^2);
```

```
                VC=-(RTM1*VTM1+RTM2*VTM2)/RTM;
                TGO=RTM/VC;
                if T>TLAUNCH
                        TEMPBOTM=(XM^2+YM^2)^1.5;
                        XMDD=-GM*XM/TEMPBOTM;
                        YMDD=-GM*YM/TEMPBOTM;
                else
                        XMDD=0.;
                        YMDD=0.;
                end
                ALTMKM=(sqrt(XM^2+YM^2)-A)/3280.;
                ALTTKM=(sqrt(XT^2+YT^2)-A)/3280.;
                ALTT=sqrt(XT^2+YT^2)-A;
                FLAG=1;
        end
        FLAG=0;
        XT=.5*(XTOLD+XT+H*XTD);
        YT=.5*(YTOLD+YT+H*YTD);
        XTD=.5*(XTDOLD+XTD+H*XTDD);
        YTD=.5*(YTDOLD+YTD+H*YTDD);
        XM=.5*(XMOLD+XM+H*XMD);
        YM=.5*(YMOLD+YM+H*YMD);
        XMD=.5*(XMDOLD+XMD+H*XMDD);
        YMD=.5*(YMDOLD+YMD+H*YMDD);
        S=S+H;
        if T>=TLAUNCH
                TGOLAMM=TF-T;
                XTF=XTFACT;
                YTF=YTFACT;
                ZTF=0.;
                QLAUNCH=1;
        end
        TGOLAMM=TF-T;
        if ((T>=TLAUNCH) & QBOOSTM==1)
                QBOOSTM=0;
                [VRXM,VRYM,VRZM]=LAMBERT3D(XM,YM,ZM,TGOLAMM,
                          XTF,YTF,ZTF,SWITCHM);
                XMD=VRXM;
                YMD=VRYM;
                XMDOLD=VRXM;
                YMDOLD=VRYM;
                VBOM=sqrt(XMD^2+YMD^2)/3280.;
                QLAUNCH=1;
        end
        if S>=(TS-.0001)
                S=0.;
```

```
                    ALTTKM=(sqrt(XT^2+YT^2)-A)/3280.;
                    DISTRTKM=distance3dkm(XT,YT,ZT,XTINIT,YTINIT,ZTINIT);
                    ALTMKM=(sqrt(XM^2+YM^2)-A)/3280.;
                    DISTRMKM=distance3dkm(XM,YM,ZM,XTINIT,YTINIT,ZTINIT);
                    VTK=sqrt(XTD^2+YTD^2)/3280.;
                    ATG=sqrt(XTDD^2+YTDD^2)/32.2;
                    VMKM=sqrt(XMD^2+YMD^2)/3280.;
                    VTKM=sqrt(XTD^2+YTD^2)/3280.;
            end
        end
        if (VBOM<8. & RTM<1000. & ALTTKM>50.)
            count=count+1;
            ArrayTFTL(count)=TF-TLAUNCH;
            ArrayVBOM(count)=VBOM;
        end
end
figure
plot(ArrayTFTL,ArrayVBOM),grid
xlabel('Interceptor Flight Time (s)')
ylabel('Interceptor Velocity (km/s) ')
clc
output=[ArrayTFTL',ArrayVBOM'];
save datfil.txt output /ascii
disp 'simulation finished'

% This is m file for PREDICT32.M
function [xtf,ytf]=predict(tp,xtp,ytp,xtdp,ytdp,tf,tftot,xf,yf,xtinit,ytinit)
t=tp;
switch1=0;
xt=xtp;
yt=ytp;
zt=0.;
xtd=xtdp;
ytd=ytdp;
ztd=0.;
zf=0.;
a=2.0926E7;
gm=1.4077E16;
qboost=1;
h=.01;
s=0.;
axt=0.;
ayt=0.;
tgolam=tftot-t;
[vrx,vry,vrz]=LAMBERT3D(xt,yt,zt,tgolam,xf,yf,zf,switch1);
xtd=vrx;
```

```
ytd=vry;
ztd=0;
while t<=(tf-.00001)
        xtold=xt;
        ytold=yt;
        xtdold=xtd;
        ytdold=ytd;
        step=1;
        flag=0;
        while step <=1
                if flag==1
                        xt=xt+h*xtd;
                        yt=yt+h*ytd;
                        xtd=xtd+h*xtdd;
                        ytd=ytd+h*ytdd;
                        t=t+h;
                        step=2;
                end
                tembot=(xt^2+yt^2)^1.5;
                xtdd=-gm*xt/tembot;
                ytdd=-gm*yt/tembot;
                flag=1;
        end;
        flag=0;
        xt=(xtold+xt)/2+.5*h*xtd;
        yt=(ytold+yt)/2+.5*h*ytd;
        xtd=(xtdold+xtd)/2+.5*h*xtdd;
        ytd=(ytdold+ytd)/2+.5*h*ytdd;
end
xtf=xt;
ytf=yt;
% LAMBERT3D can be found in Listing 28.3
% distance3dkm can be found in Listing 28.3
```

The nominal case of Listing 32.2 was run in which the target is on a 10,000-km minimum energy trajectory and the interceptor is launched 300 s after the target is launched. Addition cases are also run in which the distance from the interceptor launch point to the target launch point was considered to be 1000 km, 5000 km, and 9000 km respectively. We can see from Fig. 32.3 that for very short interceptor flight times the required interceptor velocity can be quite large. However, depending on the launch point separation of the interceptor and target there appears to be a minimum required interceptor velocity. We can see from Fig. 32.3 that a 5.3 km/s interceptor is required if the interceptor is launched 9000 km from the target launch point (or 1000 km from the target impact point) and the flight time is approximately 1400 s.

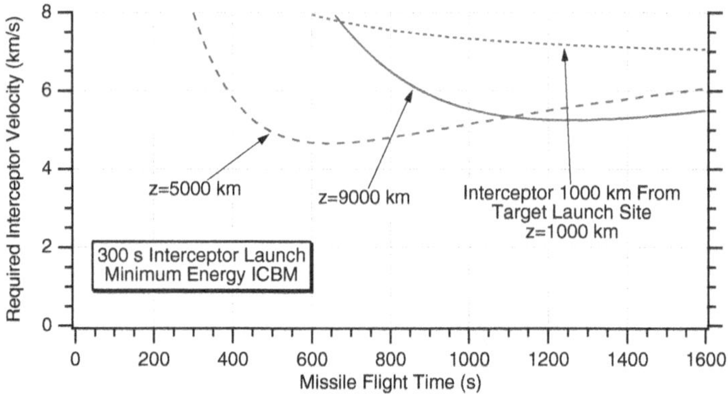

Fig. 32.3 Required initial interceptor velocity against a minimum energy ICBM depends on interceptor launch point location and flight time.

We can examine specific cases of the preceding example in more detail for the scenario in which the interceptor launch point is 9000 km downrange from the target launch point. Figure 32.4 presents the interceptor–target engagement geometries for an interceptor located 9000 km downrange of the target launch point—one in which the interceptor flight time is 800 s (6.5 km/s initial interceptor velocity required according to Fig. 32.3) and the other in which the interceptor flight time is 1400 s (5.3 km/s initial interceptor velocity is required according to Fig. 32.3). We can see that when the flight time is shorter the interceptor flies directly to the intercept point. However when the flight time is much longer, the interceptor has to fly a lofted trajectory in order to reach the intercept point

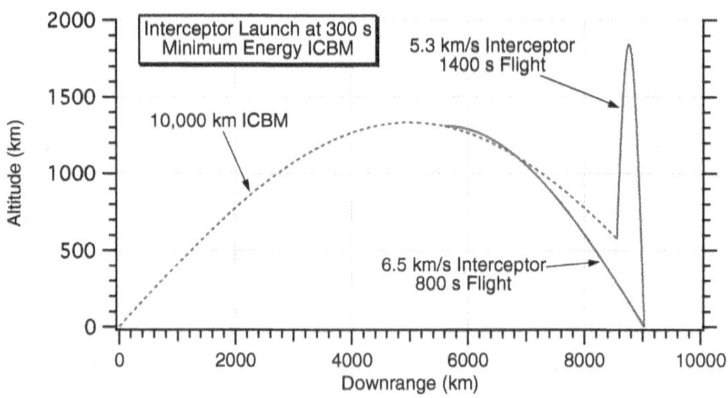

Fig. 32.4 Two sample engagement geometries when the interceptor launch point is 9000 km from the target launch point.

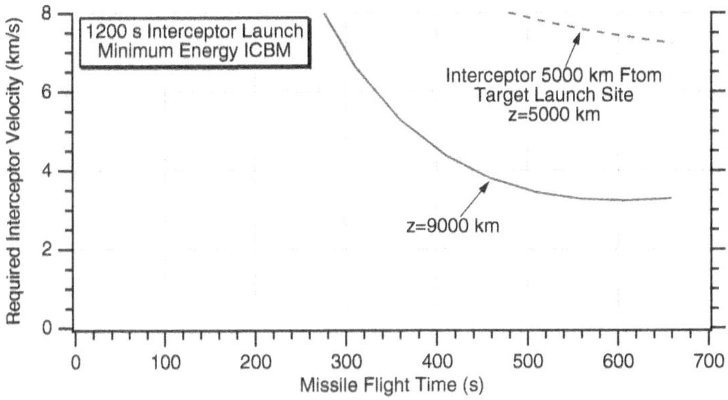

Fig.32.5 Increasing interceptor launch time requires interceptor launch point to be closer to target impact point.

under the influence of gravity only. In practice, homing sensor constraints would have to be checked to see if such a lofted trajectory is practical.

Figure 32.5 shows that if the interceptor launch time is increased to 1200 s from 300 s it is not possible for the interceptor launch site to be only 1000 km from the target launch site. In addition, we can see that lower interceptor velocities are only possible if the interceptor is near the target impact point ($z = 9000$ km).

Figure 32.6 examines two cases taken from Fig. 32.5 in more detail. The first case shows an interceptor launched at 1200 s and located 5000 km from the target launch site. In this case the interceptor has an initial velocity of 7.4 km/s and travels for 610 s (as indicated from Fig. 32.5). In this case Fig. 32.6 indicates

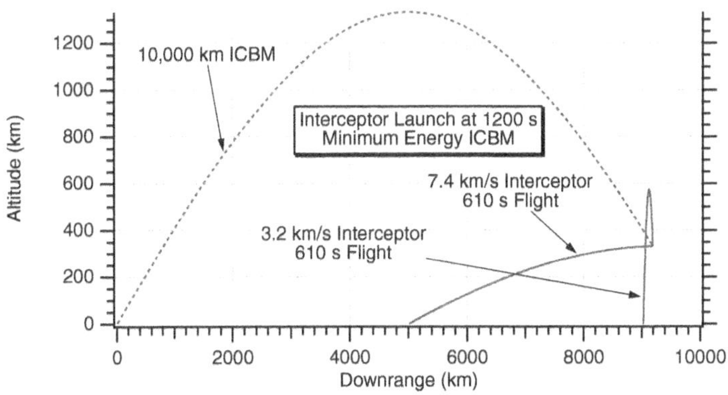

Fig. 32.6 Two sample engagement geometries for two different interceptor launch points with the same flight time.

that the interceptor flies directly to the target. The second case shows an interceptor launched at 1200 s and located 9000 km from the target launch site. In this case the interceptor has an initial velocity of 3.2 km/s but also travels for 610 s (as indicated from Fig. 32.5). In this case Fig. 32.6 indicates a lofted trajectory for the interceptor as the interceptor is very close to the impact point of the target and has a great deal of flight time.

OPERATIONAL AREA

Although considerable insight has been gained from the two-dimensional results of the previous section, even more insight can be gained by examining the three-dimensional case. In addition, we can speed up the simulation of Listing 32.2 significantly by eliminating the numerical integration of the differential equations for the interceptor and target. This can be done as both the interceptor and target are impulsive. The three-dimensional Kepler subroutine that first appeared in Listing 28.3 can be used to accurately predict where the target will be at the desired intercept time and the three-dimensional Lambert routine that first appeared in Listing 28.2 can be used to calculate the required interceptor velocity and direction so that the interceptor will arrive at the intercept point at the desired intercept time. The avoidance of numerical integration can speed up the simulation of Listing 32.2 by several orders of magnitude. The resultant computational savings allows us to vary more parameters and run many more cases in order to get a more complete picture. We can set up a program in which the target trajectory is fixed so we can figure out where the interceptor should be placed for a successful intercept. Displaying information in this fashion is known as an operational area.

Listing 32.3 develops an operational area for an interceptor with an initial velocity that cannot exceed VBOLIM against targets that can fly different types of trajectories (for example, TLOFT=0 yields minimum energy trajectory). In addition, checks are performed to ensure that intercepts are made above 50-km altitude. In this three-dimensional simulation, the impulsive target is flying a minimum energy 10,000-km trajectory along the equator. The interceptor is placed at different downrange locations specified by longitude and latitude. The initial interceptor downrange locations are longitude XLONGMDEG expressed in degrees and varied from 0 deg to 150 deg in steps of 5 deg (0 deg corresponds to the interceptor being at the target launch point while 150 deg corresponds to the interceptor being more than 16,000 km from the target launch point). The initial interceptor crossrange locations are latitude XLATMDEG expressed in degrees and vary from −40 deg to 40 deg in steps of 2.5 deg. The interceptor launch time is varied from 300 s after target launch to 1600 s after target launch in steps of 100 s. Latitude and longitude have been approximately converted to distances by assuming that 1 deg corresponds to 111 km (which is valid along the equator but overestimates distances as the latitude increases). The desired intercept time is varied from 60 s after interceptor launch to to the time the target would impact the ground (TFTOT) in steps of 50 s. If the required interceptor velocity is less than VBOLIM, the intercept is considered a success and the initial interceptor location is noted. In addition, cases are not considered in which the target velocity would be greater than 7.5 km/s.

LISTING 32.3 GENERATING THREE-DIMENSIONAL OPERATIONAL AREAS

```
count=0;
TLAUNCH=300.;
RDESKM=10000.;
ALTMKMIC=0.;
TLOFT=0.;
VBOLIM=5.;
I=1;
for XLONGMDEG=0:5:150,
        for XLATMDEG=-40:2.5:40,
                for TLAUNCH=300:100:1600,
                        ALTM=ALTMKMIC*3280.;
                        A=2.0926E7;
                        GM=1.4077E16;
                        XLONGFDEG=57.3*RDESKM*3280./A;
                        XLATFDEG=0.;
                        PI=3.14159;
                        XLONGTDEG=0.;
                        XLATTDEG=0.;
                        SWITCH=0;
                        SWITCHM=0;
                        T=0.;
                        XLONGF=XLONGFDEG/57.3;
                        XLATF=XLATFDEG/57.3;
                        XF=A*cos(XLATF)*cos(XLONGF);
                        YF=A*cos(XLATF)*sin(XLONGF);
                        ZF=A*sin(XLATF);
                        XLONGT=XLONGTDEG/57.3;
                        XLONGM=XLONGMDEG/57.3;
                        XLATM=XLATMDEG/57.3;
                        XLATT=XLATTDEG/57.3;
                        XT=A*cos(XLATT)*cos(XLONGT);
                        YT=A*cos(XLATT)*sin(XLONGT);
                        ZT=A*sin(XLATT);
                        XM=(A+ALTM)*cos(XLATM)*cos(XLONGM);
                        YM=(A+ALTM)*cos(XLATM)*sin(XLONGM);
                        ZM=(A+ALTM)*sin(XLATM);
                        DISTFKM=distance3dkm(XF,YF,ZF,XT,YT,ZT);
                        TFTOT=252.+.223*DISTFKM-(5.44E-6)*DISTFKM*DISTFKM;
```

```
                TFTOT=TFTOT+TLOFT;
                for TF=(TLAUNCH+60.):50:TFTOT,
                        TGOLAM=TFTOT-T;
% CALCULATE TARGET VELOCITY REQUIRED TO REACH ITS DESTINATION
                        [VRX,VRY,VRZ]=LAMBERT3D(XT,YT,ZT,TGOLAM,XF,
                                YF,ZF,SWITCH);
                        XTD=VRX;
                        YTD=VRY;
                        ZTD=VRZ;
% CALCULATE TARGET STATES AT DESIRED INTERCEPT TIME
                        T0=0.;
                        T1=TF;
                        X0(1)=XT/3280.;
                        X0(2)=YT/3280.;
                        X0(3)=ZT/3280.;
                        X0(4)=XTD/3280.;
                        X0(5)=YTD/3280.;
                        X0(6)=ZTD/3280.;
                        [X1]=KEPLER1(X0,T0,T1);
                        XTF=X1(1)*3280.;
                        YTF=X1(2)*3280.;
                        ZTF=X1(3)*3280.;
                        ALTFKM=(sqrt(XT^2+YTF^2+ZTF^2)-A)/3280.;
                        if ALTFKM<50.
                                break
                        end
% CALCULATE MISSILE VELOCITY REQUIRED TO INTERCEPT TARGET AT DESIRED
INTERCEPT TIME
                        TGOLAMM=TF-TLAUNCH;
                        [VRXM,VRYM,VRZM]=LAMBERT3D(XM,YM,ZM,TGOLAMM,
                                XTF,YTF,ZTF,SWITCHM);
                        XMD=VRXM;
                        YMD=VRYM;
                        ZMD=VRZM;
% CALCULATE MISSILE STATES AT DESIRED INTERCEPT TIME
                        T1=TF;
                        T0=TLAUNCH;
                        X0(1)=XM/3280.;
                        X0(2)=YM/3280.;
                        X0(3)=ZM/3280.;
                        X0(4)=XMD/3280.;
                        X0(5)=YMD/3280.;
                        X0(6)=ZMD/3280.;
                        [X1]=KEPLER1(X0,T0,T1);
                        XMF=X1(1)*3280.;
                        YMF=X1(2)*3280.;
```

```
                    ZMF=X1(3)*3280.;
                    XMDF=X1(4)*3280.;
                    YMDF=X1(5)*3280.;
                    ZMDF=X1(6)*3280.;
                    VBOM=sqrt(XMD^2+YMD^2+ZMD^2)/3280.;
                    VBOT=sqrt(XTD^2+YTD^2+ZTD^2)/3280.;
                    if (VBOM<VBOLIM & VBOT<7.5)
                         count=count+1;
                         ArrayTF(count)=TF;
                         ArrayTLAUNCH(count)=TLAUNCH;
                         ArrayXLONGM(count)=XLONGMDEG*111.;
                         ArrayXLATM(count)=XLATMDEG*111.;
                         ArrayVBOM(count)=VBOM;
                    end
               end
          end
     end
end
figure
plot(ArrayXLONGM,ArrayXLATM,'r+'),grid
xlabel('Downrange (km)')
ylabel('Crossrange (km) ')
clc
output=[ArrayTF',ArrayTLAUNCH',ArrayXLONGM',ArrayXLATM',ArrayVBOM'];
save datfil.txt output /ascii
disp 'simulation finished'
% LAMBERT3D can be found in Listing 28.2
% distance3dkm can be found in Listing 28.2
% KEPLER1 can be found in Listing 28.3
```

The nominal case of Listing 32.3 was run for a 6-km/s interceptor against the 10,000-km target flying a minimum energy trajectory. Figure 32.7 displays the resultant operational area, indicated by plus signs, and the target trajectory represented by the solid straight line. Thus, the operational area represents places from which an interceptor can be launched and hit the target at some point in its trajectory. Missing from the plot are the interceptor launch and intercept times that resulted in a success. Some plus signs correspond to many possible combinations of launch and intercept times while other plus signs may be the result of a unique combination. We can see that the operational area is very large because the interceptor speed can be as much as 6 km/s. We can also see that the operational area is symmetrical about the target trajectory. At first glance these results might appear to disagree with the two-dimensional results of Fig. 32.3 because the operational area of Fig. 32.7 at zero latitude extends from 3000 km to more than 14,000 km. However, recall in Fig. 32.3 we only considered three downrange cases ($z = 1000$ km, $z = 5000$ km and $z = 9000$ km). In

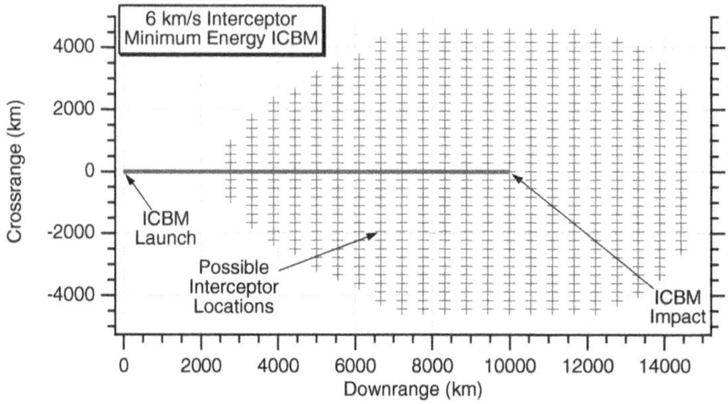

Fig. 32.7 Operational area for 6-km/s interceptor against 10,000-km minimum energy target.

Fig. 32.3 it was indicated that there were solutions for a 6-km/s interceptor at $z = 5000$ km and $z = 9000$ km, which is certainly consistent with Fig. 32.7.

Figure 32.8 shows that when the maximum interceptor velocity is reduced from 6 km/s to 5 km/s, the operational area shrinks considerably. When the interceptor had a 6 km/s capability the interceptor could be placed from 3000 km to 14,500 km from the target launch site. However when the interceptor velocity is reduced to 5 km/s we can see from Fig. 32.8 that the interceptor must be placed from 3500 km to 12,500 km from the target launch site.

Figure 32.9 shows that the operational area reduces even more dramatically when the interceptor velocity is further reduced to 4 km/s. Now the interceptor

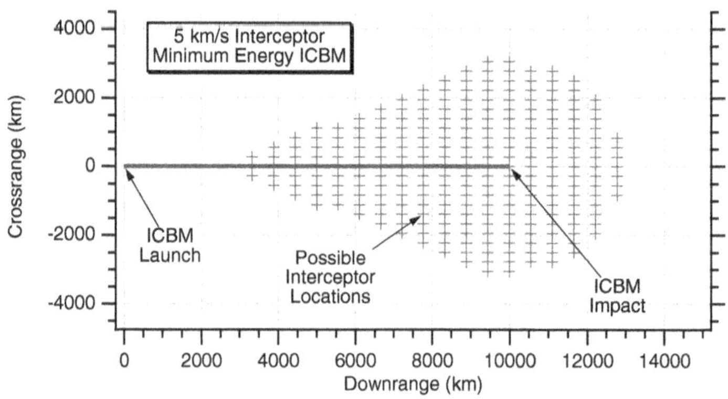

Fig. 32.8 Operational area for 5 km/s interceptor against 10,000-km minimum energy target.

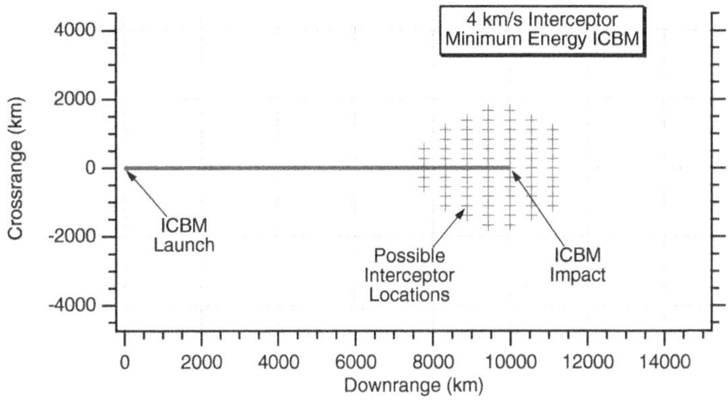

Fig. 32.9 Operational area for 4 km/s interceptor against 10,000-km minimum energy target.

must be placed 7500 km to 10,500 km from the target launch site for an intercept to be successful.

LAUNCH AREA DENIED

Successful target intercept information can be presented in a variety of ways. The operational area method of presentation, presented in the last section, assumed a fixed target trajectory that was previously represented by a thick straight line and possible interceptor launch locations that resulted in a successful engagement were represented by plus signs. Cases can also be run in which the target aimpoint and initial interceptor locations are fixed and the initial target launch site is varied. Graphics generated in this way are known as *launch area denied plots*. The operational area code of Listing 32.3 can be modified so that launch area denied results can be generated (shown in Listing 32.4). Listing 32.4 has the appropriate statements that generate launch area denied results and are highlighted in bold. In this simulation we can see that the interceptor is nominally placed at the equator at 60-deg longitude and the target launch point is varied from -100 to 200 deg in longitude (in steps of 5 deg) and from -60 deg to 60 deg in latitude (in steps of 2.5 deg). Cases are ruled out if the required impulsive target velocity exceeds 7.5 km/s.

LISTING 32.4 GENERATING THREE-DIMENSIONAL LAUNCH AREA DENIED RESULTS

```
count=0;
TLAUNCH=300.;
RDESKM=10000.;
```

```
ALTMKMIC=0.;
TLOFT=0.;
VBOLIM=5.;
XLONGMDEG=60.;
XLATMDEG=0.;
I=1;
for XLONGTDEG=-100:5:200,
        for XLATTDEG=-60:2.5:60,
                for TLAUNCH=300:100:1600,
                        ALTM=ALTMKMIC*3280.;
                        A=2.0926E7;
                        GM=1.4077E16;
                        XLONGFDEG=57.3*RDESKM*3280./A;
                        XLATFDEG=0.;
                        PI=3.14159;
                        SWITCH=0;
                        SWITCHM=0;
                        T=0.;
                        S=0.;
                        XLONGF=XLONGFDEG/57.3;
                        XLATF=XLATFDEG/57.3;
                        XF=A*cos(XLATF)*cos(XLONGF);
                        YF=A*cos(XLATF)*sin(XLONGF);
                        ZF=A*sin(XLATF);
                        XLONGT=XLONGTDEG/57.3;
                        XLONGM=XLONGMDEG/57.3;
                        XLATM=XLATMDEG/57.3;
                        XLATT=XLATTDEG/57.3;
                        XT=A*cos(XLATT)*cos(XLONGT);
                        YT=A*cos(XLATT)*sin(XLONGT);
                        ZT=A*sin(XLATT);
                        XM=(A+ALTM)*cos(XLATM)*cos(XLONGM);
                        YM=(A+ALTM)*cos(XLATM)*sin(XLONGM);
                        ZM=(A+ALTM)*sin(XLATM);
                        DISTFKM=distance3dkm(XF,YF,ZF,XT,YT,ZT);
                        TFTOT=252.+.223*DISTFKM-(5.44E-6)*DISTFKM*DISTFKM;
                        TFTOT=TFTOT+TLOFT;
                        for TF=(TLAUNCH+60.):50:TFTOT,
                                TGOLAM=TFTOT-T;
% CALCULATE TARGET VELOCITY REQUIRED TO REACH ITS DESTINATION
                                [VRX,VRY,VRZ]=LAMBERT3D(XT,YT,ZT,TGOLAM,XF,
                                        YF,ZF,SWITCH);
                                XTD=VRX;
                                YTD=VRY;
                                ZTD=VRZ;
% CALCULATE TARGET STATES AT DESIRED INTERCEPT TIME
                                T0=0.;
```

```
                              T1=TF;
                              X0(1)=XT/3280.;
                              X0(2)=YT/3280.;
                              X0(3)=ZT/3280.;
                              X0(4)=XTD/3280.;
                              X0(5)=YTD/3280.;
                              X0(6)=ZTD/3280.;
                              [X1]=KEPLER1(X0,T0,T1);
                              XTF=X1(1)*3280.;
                              YTF=X1(2)*3280.;
                              ZTF=X1(3)*3280.;
                              ALTFKM=(sqrt(XT^2+YTF^2+ZTF^2)-A)/3280.;
                              if ~(ALTFKM < 50.)
 %CALCULATE MISSILE VELOCITY REQUIRED TO INTERCEPT TARGET AT DESIRED
INTERCEPT TIME
                                    TGOLAMM=TF-TLAUNCH;
    [VRXM,VRYM,VRZM]=LAMBERT3D(XM,YM,ZM,TGOLAMM,XTF,YTF,ZTF,SWITCHM);
                                    XMD=VRXM;
                                    YMD=VRYM;
                                    ZMD=VRZM;
                                    VBOM=sqrt(XMD^2+YMD^2+ZMD^2)/3280.;
                                    VBOT=sqrt(XTD^2+YTD^2+ZTD^2)/3280.;
                                    if (VBOM < VBOLIM & VBOT<7.5)
                                          count=count+1;
                                          ArrayTF(count)=TF;
                                          ArrayTLAUNCH(count)=TLAUNCH;
                                          ArrayXLONGT(count)=XLONGTDEG*111.;
                                          ArrayXLATT(count)=XLATTDEG*111.;
                                          ArrayVBOM(count)=VBOM;
                                          ArrayVBOT(count)=VBOT;
                                    end
                              end
                        end
                  end
            end
end
figure
plot(ArrayXLONGT,ArrayXLATT,'r+'),grid
xlabel('Downrange (km)')
ylabel('Crossrange (km) ')
clc
output=[ArrayTF',ArrayTLAUNCH',ArrayXLONGT',ArrayXLATT',ArrayVBOM',ArrayVBOT'];
save datfil.txt output /ascii
disp 'simulation finished'
% LAMBERT3D can be found in Listing 28.2
% distance3dkm can be found in Listing 28.2
% KEPLER1 can be found in Listing 28.3
```

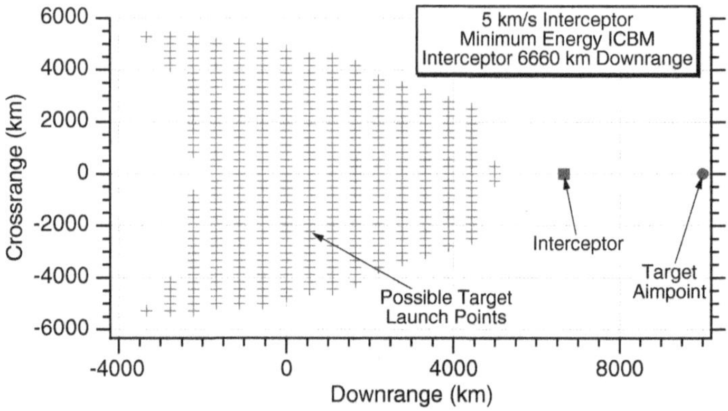

Fig. 32.10 Launch area denied for 5-km/s interceptor against minimum energy target.

The nominal case of Listing 32.4 was run for a 5-km/s interceptor against the impulsive ICBM target flying a minimum energy trajectory. Figure 32.10 displays possible target launch points that are denied, indicated by plus signs, because they can be successfully intercepted by a 5-km/s interceptor whose initial launch location is 6660 km downrange (60-deg longitude). We can see that the launch area denied is very large because the interceptor is located in a favorable location. Again, it is important to note that each plus sign in Fig. 32.10 corresponds to a different interceptor launch and intercept time that results in a successful intercept. Figure 32.11 shows the same case except this time the target trajectory is lofted (TLOFT = 500). We can see that the size of the launch area denied region shrinks considerably because the target is harder to reach since in may be higher in altitude. Figure 32.12 shows that when the target trajectory is depressed

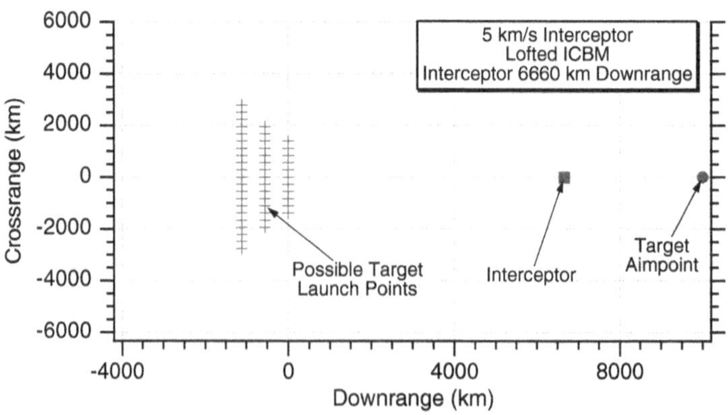

Fig. 32.11 Launch area denied for 5-km/s interceptor against a lofted target.

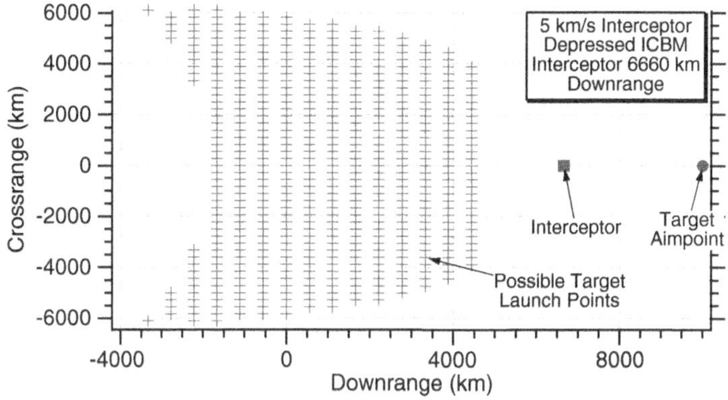

Fig. 32.12 Launch area denied for 5-km/s interceptor against a depressed target.

(TLOFT = −200) the launch area denied region is roughly the same size as it was when the target was flying a minimum energy trajectory.

DEFENDED AREA

As was mentioned in the previous sections, successful target intercept information can be presented in a variety of ways. The operational area method of presentation assumed a fixed target trajectory that was previously represented by thick straight line and possible interceptor launch locations, represented by plus signs, were varied. The previous section showed that launch area denied plots can be generated by having the target aimpoint and initial interceptor locations fixed and varying the initial target launch site. Finally, cases can also be run in which the target launch point and initial interceptor locations are fixed and the final target impact point is varied. Graphics generated in this way are known as *defended area plots.* The launch area denied code of Listing 32.4 can be modified so that defended area results can be generated and is shown in Listing 32.5. This listing has the appropriate statements that generate defended area results and are highlighted in bold. In this simulation we can see that the interceptor is nominally placed at the equator at 60-deg longitude and the target impact point is varied from 20 to 200 deg in longitude and from −60 deg to 60 deg in latitude. Cases are ruled out if the required target velocity exceeds 7.5 km/s.

LISTING 32.5 GENERATING THREE-DIMENSIONAL DEFENDED AREA RESULTS

```
count=0;
TLAUNCH=300.;
RDESKM=10000.;
```

```
ALTMKMIC=0.;
TLOFT=0.;
VBOLIM=5.;
XLONGMDEG=60.;
XLATMDEG=0.;
XLONGTDEGIC=0.;
XLATTDEGIC=0.;
I=1;
for XLONGFDEG=20:5:200,
        for XLATFDEG=-60:2.5:60,
                for TLAUNCH=300:100:1600,
                        ALTM=ALTMKMIC*3280.;
                        A=2.0926E7;
                        GM=1.4077E16;
                        XLONGTDEG=XLONGTDEGIC;
                        XLATTDEG=XLATTDEGIC;
                        PI=3.14159;
                        SWITCH=0;
                        SWITCHM=0;
                        T=0.;
                        S=0.;
                        XLONGF=XLONGFDEG/57.3;
                        XLATF=XLATFDEG/57.3;
                        XF=A*cos(XLATF)*cos(XLONGF);
                        YF=A*cos(XLATF)*sin(XLONGF);
                        ZF=A*sin(XLATF);
                        XLONGT=XLONGTDEG/57.3;
                        XLONGM=XLONGMDEG/57.3;
                        XLATM=XLATMDEG/57.3;
                        XLATT=XLATTDEG/57.3;
                        XT=A*cos(XLATT)*cos(XLONGT);
                        YT=A*cos(XLATT)*sin(XLONGT);
                        ZT=A*sin(XLATT);
                        XM=(A+ALTM)*cos(XLATM)*cos(XLONGM);
                        YM=(A+ALTM)*cos(XLATM)*sin(XLONGM);
                        ZM=(A+ALTM)*sin(XLATM);
                        DISTFKM=distance3dkm(XF,YF,ZF,XT,YT,ZT);
                        TFTOT=252.+.223*DISTFKM-(5.44E-6)*DISTFKM*DISTFKM;
                        TFTOT=TFTOT+TLOFT;
                        for TF=(TLAUNCH+60.):50:(TFTOT-50.),
                                TGOLAM=TFTOT-T;
% CALCULATE TARGET VELOCITY REQUIRED TO REACH ITS DESTINATION
                                [VRX,VRY,VRZ]=LAMBERT3D(XT,YT,ZT,TGOLAM,XF,
                                        YF,ZF,SWITCH);
                                XTD=VRX;
                                YTD=VRY;
```

```
                    ZTD=VRZ;
% CALCULATE TARGET STATES AT DESIRED INTERCEPT TIME
                    T0=0.;
                    T1=TF;
                    X0(1)=XT/3280.;
                    X0(2)=YT/3280.;
                    X0(3)=ZT/3280.;
                    X0(4)=XTD/3280.;
                    X0(5)=YTD/3280.;
                    X0(6)=ZTD/3280.;
                    [X1]=KEPLER1(X0,T0,T1);
                    XTF=X1(1)*3280.;
                    YTF=X1(2)*3280.;
                    ZTF=X1(3)*3280.;
                    ALTFKM=(sqrt(XTF^2+YTF^2+ZTF^2)-A)/3280.;
                    if ALTFKM<50.
                         break
                    end
% CALCULATE MISSILE VELOCITY REQUIRED TO INTERCEPT TARGET AT DESIRED
INTERCEPT TIME
                    TGOLAMM=TF-TLAUNCH;

[VRXM,VRYM,VRZM]=LAMBERT3D(XM,YM,ZM,TGOLAMM,XTF,YTF,ZTF,SWITCHM);
                    XMD=VRXM;
                    YMD=VRYM;
                    ZMD=VRZM;
                    VBOM=sqrt(XMD^2+YMD^2+ZMD^2)/3280.;
                    VBOT=sqrt(XTD^2+YTD^2+ZTD^2)/3280.;
                    if (VBOM<VBOLIM & VBOT<7.5)
                         count=count+1;
                         ArrayTF(count)=TF;
                         ArrayTLAUNCH(count)=TLAUNCH;
                         ArrayXLONGF(count)=XLONGFDEG*111.;
                         ArrayXLATF(count)=XLATFDEG*111.;
                         ArrayVBOM(count)=VBOM;
                    end
               end
          end
     end
end
figure
plot(ArrayXLONGF,ArrayXLATF,'r+'),grid
xlabel('Downrange (km)')
ylabel('Crossrange (km) ')
clc
output=[ArrayTF',ArrayTLAUNCH',ArrayXLONGF',ArrayXLATF',ArrayVBOM'];
```

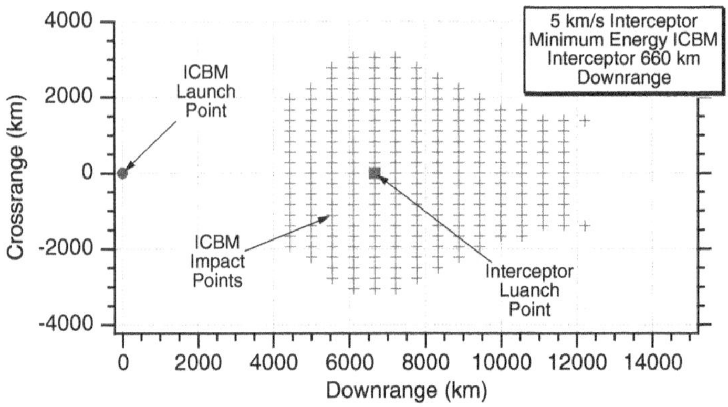

Fig. 32.13 Defended area for 5-km/s interceptor against minimum energy trajectory target.

```
save datfil.txt output /ascii
disp 'simulation finished'
% LAMBERT3D can be found in Listing 28.2
% distance3dkm can be found in Listing 28.2
% KEPLER1 can be found in Listing 28.3
```

The nominal case of Listing 32.5 was run for a 5-km/s interceptor against the impulsive ICBM target flying a minimum energy trajectory. Figure 32.13 displays possible target impact points that can be defended, indicated by plus signs, because they can be successfully intercepted by a 5-km/s interceptor located 6660 km downrange from the target launch point. Again, it is important to

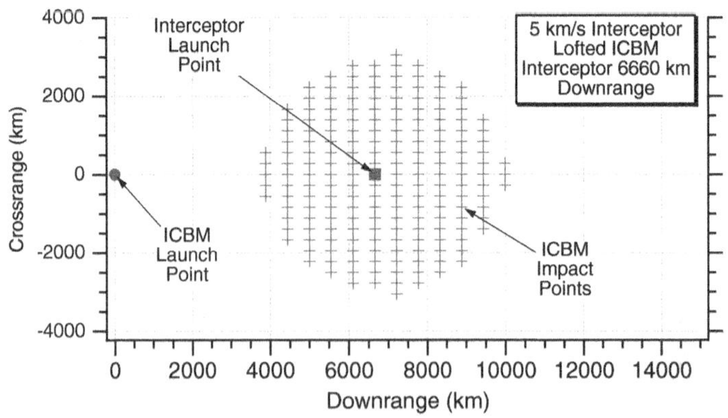

Fig. 32.14 Defended area for 5-km/s interceptor against a lofted trajectory target.

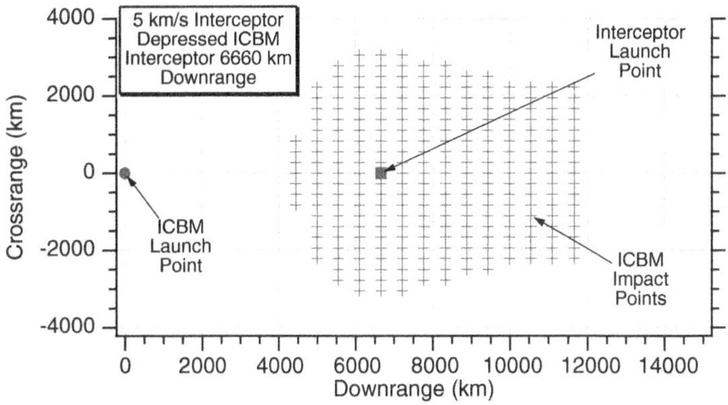

Fig. 32.15 Area for 5-km/s interceptor against a depressed trajectory target.

note that each plus sign in Fig. 32.13 corresponds to a different interceptor launch and intercept time that results in a successful intercept. Figure 32.14 shows the same case except this time the target trajectory is lofted (TLOFT = 500). We can see that the size of the defended area region shrinks because the target is harder to reach since its apogee is higher than that of a minimum energy trajectory and thus harder to reach with a 5-km/s interceptor. However the difference in results between the minimum energy and loft target trajectories is not as great as it was in the previous section because the interceptor is much closer to the target impact point. Figure 32.15 shows that when the target trajectory is depressed (TLOFT = −200) the defended area region is roughly the same size as it was when the target was flying a minimum energy trajectory.

SUMMARY

In this chapter we have seen how to present information on successful intercepts of an impulsive ballistic target being pursued by an impulsive interceptor. The operational area method of presentation assumes that the target trajectory is fixed and we figure out where the interceptor can be placed for a successful intercept. The launch area denied method of presentation assumes that the target aimpoint and initial interceptor locations are fixed and the initial target launch site is varied. Finally, the defended area method of presentation assumes that the target launch point and initial interceptor locations are fixed and the final target impact point is varied. Using the three methods of presentation cases were run in which the size of the area was influenced either by the missile velocity or the target trajectory type.

CHAPTER 33

Boost-Phase Filtering Options

INTRODUCTION

Intercepting ballistic missiles during their boost phase is attractive because ballistic missiles are easy to detect and track while they are thrusting and, if the intercept is successful, the raid size for subsequent missile defense layers is thinned. In addition, decoys and other countermeasures, which can be very important for midcourse phase intercepts (that is, when the target is only under the influence of gravity) are usually considered to be more difficult to devise against boosting targets. An important part of the boost-phase intercept problem is the problem of tracking the boosting target and predicting where it will be in the future. Robust filtering options that require a minimum of computation are highly desired for the boost-phase intercept application. This chapter [1] takes the approach of simplifying the boost-phase filtering problem down to its most basic levels so that two different fundamental approaches can be explored.

It is usually thought that the more *a priori* information one gives a Kalman filter, the better the Kalman filter will perform. This is certainly true if the *a priori* information is correct. If the *a priori* information has slight errors however, considerable filtering errors may result. In fact if the *a priori* information is incorrect, filter divergence may result. Process noise reflects how much confidence we have in the mathematical representation of the model of the real world embedded in the Kalman filter. Using large amounts of process noise in the filter design is an engineer's way of telling the filter that we have very little confidence in our model of the real world and that the filter should always pay attention to the measurements. Using small amounts or zero process noise in the filter design means that we are so confident in our model of the real world that the filter can eventually stop paying attention to the measurements (Kalman gains eventually go to zero with zero process noise). In order to make a Kalman filter robust to errors in *a priori* information, sufficient process noise must be added to the filter—sometimes massive amounts of process noise! Although increased process noise can often eliminate filter divergence, it also increases the errors in the state estimates. Sometimes it is possible that when *a priori* information is in error one might have been better off with a simpler Kalman filter, requiring less process noise, that does not require *a priori* information.

Although there is an extensive body of academic papers on high-order filters for tracking applications, many systems that are built use low-order decoupled filtering techniques [2, 3]. For both simplicity and ease of understanding we will only consider the low-order tracking filters in this chapter. Two filtering options for tracking a boosting intercontinental ballistic missile (ICBM) will be investigated in this chapter. First, the performance of a decoupled two-state Kalman filter, sometimes called a position-velocity filter, will be investigated assuming that the current ICBM acceleration magnitude and direction is known perfectly. Next, the degradation in performance of the two-state Kalman filter will be studied assuming that the current acceleration magnitude is known exactly but the current direction is not known and must be guessed. In this case the filter will assume that the ICBM is performing a gravity turn even though it is not. Next a simple linear three-state decoupled polynomial Kalman filter, sometimes called a position-velocity-acceleration filter, for tracking the boosting ICBM will also be considered. This type of filter, which does not require *a priori* information, is often used in the tactical missile world for tracking unpredictable maneuvering aircraft. Performance comparisons will be made between the two filtering options.

ICBM MODEL

To get at the heart of the matter and to avoid unnecessary complexity, the physics of the real world will be kept as simple as possible and an ICBM model from the open literature will be used. One such generic ICBM model appears in the American Physical Society (APS) report of Ref. 4 and is representative of a two-stage, liquid-fueled ICBM that is capable of traveling 12,000 km. The magnitude of the longitudinal acceleration profile of the APS two-stage liquid ICBM during its boost phase appears in Fig. 33.1.

Figure 33.1 indicates that the ICBM boost phase lasts 240 s and the discontinuities in the acceleration profile are due to staging events. At the end of the first-stage burn at 120 s the peak ICBM longitudinal acceleration is approximately 6 *g* and at the end of the second-stage burn at 240 s the peak acceleration is approximately 13 *g*. The job of the Kalman filter will be to estimate the position, velocity, and possibly the acceleration of the boosting ICBM.

ICBM GUIDANCE

Usually an ICBM flies straight up for a while and then performs a gravity turn while it is in the atmosphere to minimize loading and drag effects. Once the ICBM is out of the atmosphere or when the dynamic pressure falls below a

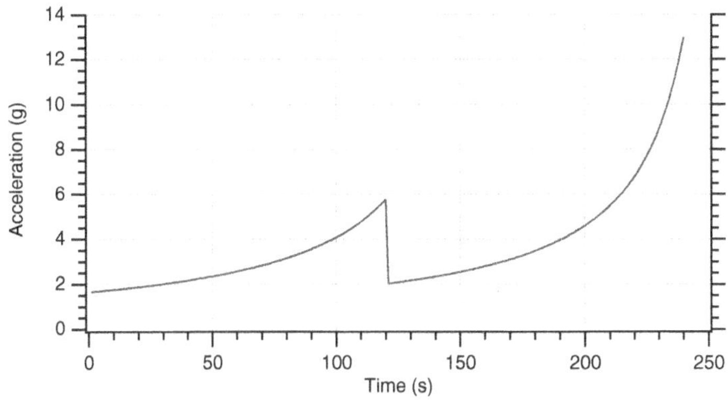

Fig. 33.1 Two-stage APS liquid ICBM acceleration profile.

certain level the ICBM can perform closed-loop guidance so that it can reach its intended target [5]. Lambert guidance is one possible method of steering a boosting ICBM to its intended target at the desired impact time. Lambert guidance involves the numerical solution to Lambert's problem and was discussed fully in Chapter 13 [6–8]. Essentially, at each instant of time if the ICBM knows where it is and where it wants to go and how long it should take to get to its destination, the solution to Lambert's problem tells the ICBM at each instant of time the magnitude and direction of the required velocity vector.

As was shown in Chapter 12, with a gravity turn the ICBM's thrust vector is aligned with the velocity vector in an attempt to drive the angle of attack to zero in order to minimize drag. The differential equations for the acceleration components of the ICBM while performing the gravity turn in an ECI coordinate system was shown to be given by

$$a_x = \frac{-gm\,x}{(x^2+y^2)^{1.5}} + a_T\frac{\dot{x}}{(x^2+y^2)^{.5}}$$

$$a_y = \frac{-gm\,y}{(x^2+y^2)^{1.5}} + a_T\frac{\dot{y}}{(\dot{x}^2+\dot{y}^2)^{.5}}$$

The first term in the preceding differential equation is due to gravity and the second term is the gravity turn portion of the boosting target's acceleration.

Listing 33.1, which is based on Listing 13.3, is a two-dimensional trajectory generator for the APS ICBM traveling on a 7000-km trajectory. Here the ICBM goes straight up for 20 s (TUPT = 20), then performs an open-loop gravity turn for the next 80 s (TGRAVEND = 100) and finally uses closed-loop Lambert guidance to get to its final destination. Note that the trajectory generator also

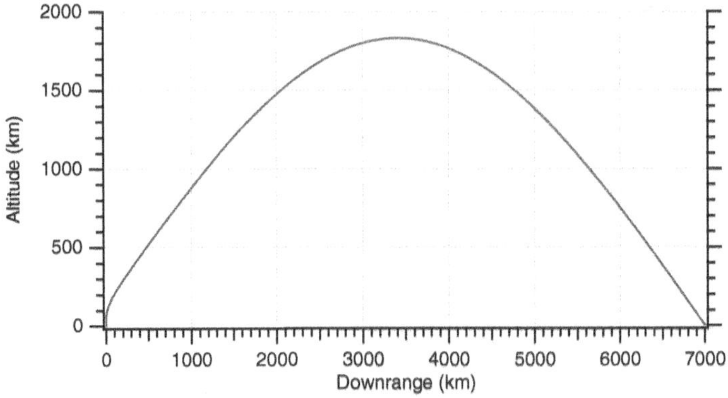

Fig. 33.2 Nominal 7000-km ICBM trajectory.

calculates the dynamic pressure Q of the ICBM. The dynamic pressure acting on the ICBM is given by the expression

$$Q = 0.5\rho V^2$$

where V is the booster's velocity in ft/s and ρ is air density in slug/ft^3. In the English system of units air density ρ was shown in Chapter 10 to be (above 30-kft altitude) approximated by

$$\rho = 0.0034e^{-\text{alt}/22000}$$

where alt is the booster altitude in feet. The nominal case of Listing 33.1 was run and the resultant lofted (TLOFT = 500) 7000-km ICBM trajectory appears in Fig. 33.2. We can see that the apogee of the trajectory is approximately 1800 km.

LISTING 33.1 TRAJECTORY GENERATOR FOR NOMINAL ICBM

```
count=0;
RDESKM=7000.;
TF=2000.;
TFINISH=999999.;
TLOFT=500.;
TGRAVEND=100.;
GAMDEGIC=89.8;
TUPT=20.;
RDESRKM=560.;
LEFT=1;
QBOOST=1;
QOOMPH=1;
```

```
QFINISH=1;
QZERO=0;
CW=0;
SWITCH=0;
QFIRST=1;
GAMDEG=GAMDEGIC;
H=.01;
T=0.;
S=0.;
A=2.0926E7;
GM=1.4077E16;
ALTNM=0.;
ALT=ALTNM*6076.;
ANGDEG=0.;
ANG=ANGDEG/57.3;
XLONGM=ANG;
X=(A+ALT)*cos(ANG);
Y=(A+ALT)*sin(ANG);
ALT=sqrt(X^2+Y^2)-A;
XFIRST=X;
YFIRST=Y;
X1=cos(1.5708-GAMDEG/57.3+ANG);
Y1=sin(1.5708-GAMDEG/57.3+ANG);
AXT=0.;
AYT=0.;
XLONGTDEG=57.3*RDESKM*3280./A;
XLONGRDEG=57.3*RDESRKM*3280./A;
TF=252.+.223*RDESKM-(5.44E-6)*RDESKM*RDESKM;
TF=TF+TLOFT;
XLONGT=XLONGTDEG/57.3;
XLONGR=XLONGRDEG/57.3;
XF=A*cos(XLONGT);
YF=A*sin(XLONGT);
XR=A*cos(XLONGR);
YR=A*sin(XLONGR);
AXT=0.;
AYT=0.;
Z=0;
ZF=0;
ZFIRST=0;
while ~(ALTNM<-1 | T>TFINISH)
        XOLD=X;
        YOLD=Y;
        X1OLD=X1;
        Y1OLD=Y1;
        STEP=1;
```

```
FLAG=0;
while STEP <=1
	if FLAG==1
		STEP=2;
		X=X+H*XD;
		Y=Y+H*YD;
		X1=X1+H*X1D;
		Y1=Y1+H*Y1D;
		T=T+H;
	end;
	if T<120.
		WGT=-2622*T+440660.;
		TRST=725850.;
	elseif T<240.
		WGT=-642.*T+168120.;
		TRST=182250.;
	else
		WGT=5500.;
		TRST=0.;
	end
	AT=32.2*TRST/WGT;
	XD=X1;
	YD=Y1;
	VEL=sqrt(XD^2+YD^2);
	TEMBOT=(X^2+Y^2)^1.5;
	X1D=-GM*X/TEMBOT+AXT;
	Y1D=-GM*Y/TEMBOT+AYT;
	ALT=sqrt(X^2+Y^2)-A;
	ACCG=sqrt(AXT^2+AYT^2)/32.2;
	FLAG=1;
end
FLAG=0;
X=(XOLD+X)/2+.5*H*XD;
Y=(YOLD+Y)/2+.5*H*YD;
X1=(X1OLD+X1)/2+.5*H*X1D;
Y1=(Y1OLD+Y1)/2+.5*H*Y1D;
S=S+H;
Z=0.;
ZF=0.;
TGOLAM=TF-T;
if (QBOOST==1 & T>TGRAVEND)
	TGOLAM=TF-T;
	[VRX,VRY,VRZ]=LAMBERT3D(X,Y,Z,TGOLAM,XF,YF,ZF,SWITCH);
	DELX=VRX-X1;
	DELY=VRY-Y1;
	DEL=sqrt(DELX^2+DELY^2);
```

```
		if (T<240 & DEL>500.)
			AXT=AT*DELX/DEL;
			AYT=AT*DELY/DEL;
		elseif DEL<500.
			TRST=0.;
			QBOOST=0;
			AXT=0.;
			AYT=0.;
			X1=VRX;
			Y1=VRY;
			X1OLD=X1;
			Y1OLD=Y1;
		else
			QBOOST=0;
			AXT=0.;
			AYT=0.;
		end
	elseif (T>=TUPT & T<= TGRAVEND & QFIRST==1)
		QFIRST=0;
		VEL=sqrt(XD^2+YD^2);
		X1=VEL*cos(1.5708-GAMDEGIC/57.3+ANG);
		Y1=VEL*sin(1.5708-GAMDEGIC/57.3+ANG);
		X1OLD=X1;
		Y1OLD=Y1;
		AXT=AT*X1/VEL;
		AYT=AT*Y1/VEL;
	elseif (T>=TUPT & T<=TGRAVEND)
		VEL=sqrt(XD^2+YD^2);
		AXT=AT*X1/VEL;
		AYT=AT*Y1/VEL;
	elseif T<=TUPT
		RTMAG=sqrt(X^2+Y^2);
		AXT=AT*X/RTMAG;
		AYT=AT*Y/RTMAG;
	end
	if S>=.9999
		S=0.;
		DISTNM=distance3dkm(X,Y,Z,XFIRST,YFIRST,ZFIRST);
		ALTNM=(sqrt(X^2+Y^2)-A)/3280.;
		RMAG=sqrt(X^2+Y^2);
		VMAG=sqrt(XD^2+YD^2);
		GAMDEG=90-57.3*acos((X*XD+Y*YD)/(RMAG*VMAG));
		RHO=.0034*exp(-ALT/22000.);
		Q=.5*RHO*VEL*VEL;
		RRMAG=sqrt(XR^2+YR^2);
		RRTMAG=sqrt((X-XR)^2+(Y-YR)^2);
```

```
			ELDEG=90.-57.3*acos((XR*(X-XR)+YR*(Y-YR))/(RRMAG*RRTMAG));
			if (ELDEG>2. & ELDEG<85)
				ISEE=1;
			else
				ISEE=0;
			end
			count=count+1;
			ArrayT(count)=T;
			ArrayDISTNM(count)=DISTNM;
			ArrayALTNM(count)=ALTNM;
			ArrayQ(count)=Q;
			ArrayISEE(count)=ISEE;
		end
end
figure
plot(ArrayDISTNM,ArrayALTNM),grid
xlabel('Downrange (km)')
ylabel('Altitude (km)')
clc
output=[ArrayT',ArrayDISTNM',ArrayALTNM',ArrayQ',ArrayISEE'];
save datfil.txt output /ascii
disp 'simulation finished'
% LAMBERT3D can be found in Listing 28.3
% distance3dkm can be found in Listing 28.3
```

For the boost-phase portion of the ICBM trajectory, or first 240 s of flight, the dynamic pressure, in the English system of units, is depicted in Fig. 33.3. As was previously mentioned, usually a gravity turn is performed while the ICBM is in the atmosphere to reduce drag and loading effects. However, when the dynamic pressure is low enough, closed-loop guidance can be begin. If we assume that

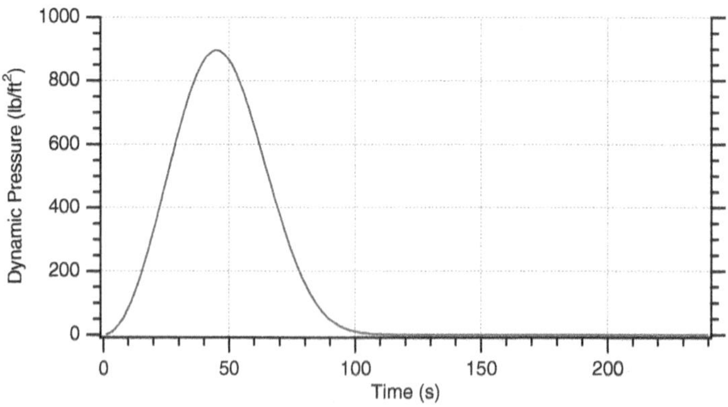

Fig. 33.3 Dynamic pressure for ICBM boost-phase portion of flight.

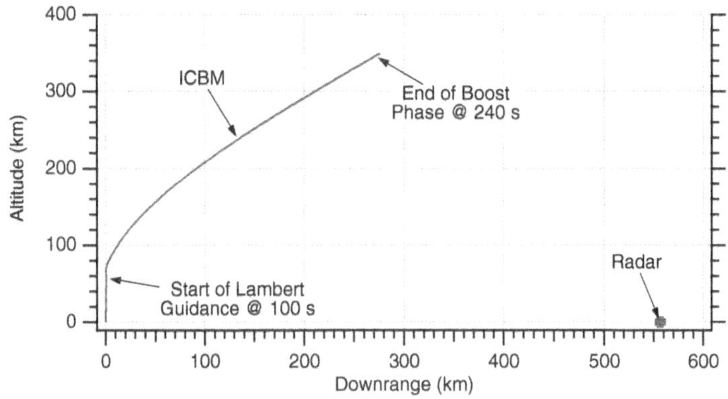

Fig. 33.4 Boost-phase portion of 7000-km ICBM trajectory.

ICBM guidance cannot begin until the dynamic pressure drops below 600 lb/ft^2 (that is, the actual number depends on booster design), then we can say that for this example closed-loop Lambert guidance can not begin until 65 s after the ICBM takes off. However, in order to be conservative, it shall be assumed that closed-loop guidance can begin at 100 s where the dynamic pressure is below 50 lb/ft^2.

The boost-phase portion of the ICBM 7000-km trajectory is shown in Fig. 33.4. In this example the boosting ICBM first goes straight up for 20 s, performs a gravity turn for the next 80 s and then switches to Lambert guidance at 100 s after launch. Also shown in Fig. 33.4 is a tracking radar located approximately 560 km from the ICBM launch site. The radar will not be able to see the target immediately due to the curvature of the Earth and radar elevation angle constraints. If a minimum radar elevation angle of 2 deg is required for acquisition then running Listing 33.1 also indicates that the radar will not be able to see the ICBM for the first 90 s of flight (that is, when ISEE = 1).

Figure 33.5 displays the downrange and altitude acceleration profiles of the boosting ICBM. The figure indicates that for the first 20 s the downrange acceleration is zero because the APS ICBM is going straight up. As was mentioned before, the ICBM goes straight up for 20 s, a gravity turn is employed from 20 s to 100 s, and Lambert guidance is used for the remainder of the flight. There is a staging event at 120 s.

FILTERING OPTIONS

For radar tracking applications range and angle measurements are available. A fully coupled extended Kalman filter (EKF) can be designed to estimate the position, velocity, and acceleration of the target. With this approach the filter will require accurate state and covariance initialization and have a certain computational burden. An alternative approach that has been used in older tracking

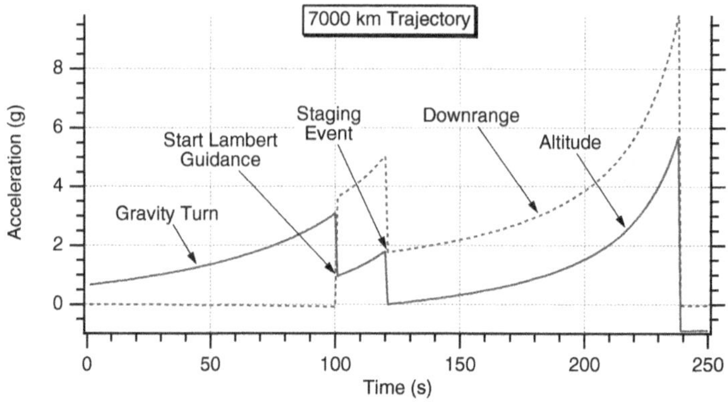

Fig. 33.5 ICBM acceleration components during boost phase for 7000-km trajectory.

systems [2, 3] is to transform the actual radar measurements to pseudo position measurements and then to build decoupled linear polynomial Kalman filters in each part of the coordinate system under consideration. Such an approach offers a considerable reduction in computation over the EKF approach, which means that many more targets can be tracked for a given flight computer throughput capacity. In addition, linear polynomial Kalman filters are very robust and are insensitive to filter initialization issues, which can be very important in military applications. Therefore, in each of the filtering approaches considered in this chapter, the filters will be considered to be decoupled and only the equations in one of the channels will be presented.

If *a priori* information concerning the current acceleration profile of the ICBM is available, a two-state Kalman filter can be designed to estimate the ICBM's position and velocity. Because the acceleration of the boosting ICBM is assumed to be known, it does not have to be estimated. If acceleration information, also known as a *template*, is not available a three-state polynomial Kalman filter can be designed to estimate the position, velocity, and acceleration of the boosting ICBM.

Figure 33.4 presented the geometry for a surface-based radar tracking an ICBM during its boost phase. In this generic model it is assumed that the radar measures range and angle to the ICBM and that the radar tracks the target with an angular accuracy of 1 mr and a range accuracy of 10 m. Measurements are taken every second. Due to previously mentioned radar horizon and elevation angle constraints, the radar sees the ICBM at 90 s. after target launch.

The range and angle from the radar to the target or ICBM is given by

$$R_T = \sqrt{(x_T - x_R)^2 + (y_T - y_R)^2}$$

$$\theta_T = \tan^{-1}\left(\frac{y_T - y_R}{x_R - x_T}\right)$$

where the coordinates of the radar x_R, y_R are assumed to be known and the target location x_T, y_T is unknown. As was previously mentioned, although the radar actually measures range and angle, one can pretend the radar measures downrange and altitude in order to avoid building an EKF. Creating such position pseudo measurements will allow us to build a linear decoupled polynomial Kalman filter.

The target location can be expressed in terms of the radar location, angle to the target and range to the target as

$$x_T = -R_T \cos\theta_T + x_R$$
$$y_T = R_T \sin\theta_T + y_R$$

Using the chain rule from calculus yields

$$\Delta x_T = \frac{\partial x_T}{\partial R_T}\Delta R_T + \frac{\partial x_T}{\partial \theta_T}\Delta\theta_T$$
$$\Delta y_T = \frac{\partial y_T}{\partial R_T}\Delta R_T + \frac{\partial y_T}{\partial \theta_T}\Delta\theta_T$$

After taking the partial derivatives we obtain

$$\Delta x_T = -\cos\theta_T \Delta R_T + R_T \sin\theta_T \Delta\theta_T$$
$$\Delta y_T = \sin\theta_T \Delta R_T + R_T \cos\theta_T \Delta\theta_T$$

Squaring both sides of the preceding two equations, ignoring cross-coupling efects, and taking expectations yields

$$\sigma_x^2 = \cos^2\theta_T \sigma_R^2 + R_T^2 \sin^2\theta_T \sigma_\theta^2$$
$$\sigma_y^2 = \sin^2\theta_T \sigma_R^2 + R_T^2 \cos^2\theta_T \sigma_\theta^2$$

where it has been assumed that

$$E(\Delta x_T^2) = \sigma_x^2$$
$$E(\Delta y_T^2) = \sigma_y^2$$
$$E(\Delta R_T^2) = \sigma_R^2$$
$$E(\Delta \theta_T^2) = \sigma_\theta^2$$

TWO-STATE TEMPLATED BASED FILTER

If cross coupling effects are neglected, two decoupled two-state linear polynomial Kalman filters can be built. One filter is in downrange and the other in altitude so that the position and velocity of the booster can be estimated. A template-based Kalman filter assumes that a perfect acceleration template of the boosting target is available in which the current components of the booster acceleration a_{Tx}, a_{Ty} are known exactly. Under these conditions the equations of the downrange template-based two-state linear polynomial Kalman filter become

$$\mathrm{Res}_k = x^*_{T_k} - \hat{x}_{T_{k-1}} - \hat{\dot{x}}_{T_{k-1}} T_s - 0.5 a_{T_{k-1}} T_s^2$$

$$\hat{x}_{T_k} = \hat{x}_{T_{k-1}} + \hat{\dot{x}}_{T_{k-1}} T_s + 0.5 a_{T_{k-1}} T_s^2 + K_{1_k} \mathrm{Res}_k$$

$$\hat{\dot{x}}_{T_k} = \hat{\dot{x}}_{T_{k-1}} + a_{T_{k-1}} T_s + K_{2_k} \mathrm{Res}_k$$

where Res stands for the filter residual. There is an identical set of equations for the Kalman filter in the altitude direction. The Kalman gains for the downrange and altitude filters will be different because the variance of the pseudo measurement noise in each channel is different. The process noise matrix for both decoupled filters is given by

$$\mathbf{Q}_k = \Phi_s \begin{bmatrix} \dfrac{T_s^3}{3} & \dfrac{T_s^2}{2} \\ \dfrac{T_s^2}{2} & T_s \end{bmatrix}$$

where Φ_s is the process noise power spectral density in units squared per Hz. In practice Φ_s is chosen by extensive computer experiments in which practical errors are introduced to ensure that the filter is robust.

THREE-STATE FILTER

Kalman filters that are not template-based do not have to be invented because they have been used in the tactical missile world for many decades for tracking unpredictable maneuvering aircraft targets. These filters simply take derivatives and are known as *linear polynomial Kalman filters.* For tracking an ICBM during the boost phase in a two-dimensional world, two decoupled three-state linear polynomial Kalman filters can also be built, one in downrange and the other in altitude, to estimate the position, velocity, and acceleration of the boosting ICBM. The equations of the downrange three-state linear polynomial Kalman filter are given by [9, 10]

$$\mathrm{Res}_k = x^*_{T_k} - \hat{x}_{T_{k-1}} - \hat{\dot{x}}_{T_{k-1}} T_s - 0.5 \hat{\ddot{x}}_{T_{k-1}} T_s^2$$

$$\hat{x}_{T_k} = \hat{x}_{T_{k-1}} + \hat{\dot{x}}_{T_{k-1}} T_s + 0.5 \hat{\ddot{x}}_{T_{k-1}} T_s^2 + K_{1_k} \mathrm{Res}_k$$

$$\hat{\dot{x}}_{T_k} = \hat{\dot{x}}_{T s_{k-1}} + \hat{\ddot{x}}_{T_{k-1}} T_s + K_{2_k} \mathrm{Res}_k$$

$$\hat{\ddot{x}}_{T_k} = \hat{\ddot{x}}_{T s_{k-1}} + K_{3_k} \mathrm{Res}_k$$

where Res again represents the filter residual. There is an identical set of equations in the altitude direction. The Kalman gains for the downrange and altitude filters will be different because the variance of the pseudo measurement noise is different in each channel. The process noise matrix for both decoupled filters is given by

$$Q = \Phi_s \begin{bmatrix} \frac{T_s^5}{20} & \frac{T_s^4}{8} & \frac{T_s^3}{6} \\ \frac{T_s^4}{8} & \frac{T_s^3}{3} & \frac{T_s^2}{2} \\ \frac{T_s^3}{6} & \frac{T_s^2}{2} & T_s \end{bmatrix}$$

where Φ_s is the process noise power spectral density in units squared per Hz. For all studies in this chapter with the three-state linear polynomial Kalman filter, the value of process noise is determined by experiment and is set to 260.

Listing 33.2 is a simulation of the ICBM trajectory generator with the just discussed two- and three-state linear decoupled polynomial Kalman filters. If IFILTER = 1, then the two-state template filter is used where the target acceleration is assumed to be known. In addition, we can see that the filter does not depend on knowledge of the target's initial position and velocity but initializes itself by using a two-state least squares filter for the first few measurements. If IFILTER = 2, then the three-state linear polynomial Kalman filter is used. In this case we can see that the filter also does not depend on knowledge of the target's initial position, velocity, and acceleration estimates but initializes itself by using a three-state least squares filter for the first few measurements [10]. The radar elevation angle is calculated and it is assumed that the radar cannot see the target unless the elevation angle is between 2 deg and 85 deg. Filtering does not start until the radar can see the target (ISEE = 1).

LISTING 33.2 SIMULATION OF TWO POSSIBLE LINEAR DECOUPLED POLYNOMIAL KALMAN FILTERS FOR TRACKING AN ICBM DURING BOOST PHASE

```
count=0;
IFILTER=1;
RDESKM=7000.;
TF=2000.;
TFINISH=240.;
TLOFT=500.;
TGRAVEND=100.;
GAMDEGIC=89.8;
TUPT=20.;
RDESRKM=560.;
SWITCH=0;
PHIS=0.;
PHIS1=260.;
ERR=0.;
```

```
TS=1.;
SIGTHET=.001;
SIGR=10.*3.28;
QGRAV=0;
ORDER =2;
LEFT=1;
QBOOST=1;
QFINISH=1;
QFIRST=1;
GAMDEG=GAMDEGIC;
HINT=.01;
XH=0.;
XDH=0.;
XDDH=0.;
YH=0.;
YDH=0.;
YDDH=0.;
PHI=zeros([2,2]);
P=zeros([2,2]);
Q=zeros([2,2]);
IDNPZ=eye(2);
P(1,1)=99999999999.;
P(2,2)=99999999999.;
PP(1,1)=99999999999.;
PP(2,2)=99999999999.;
PHI(1,1)=1;
PHI(1,2)=TS;
PHI(2,2)=1;
HMAT(1,1)=1.;
HMAT(1,2)=0.;
PHIT=PHI';
HT=HMAT';
Q(1,1)=PHIS*TS^3/3.;
Q(1,2)=PHIS*TS^2/2.;
Q(2,1)=Q(1,2);
Q(2,2)=PHIS*TS;
P11=99999999999.;
P12=0.;
P13=0.;
P22=99999999999.;
P23=0.;
P33=99999999999.;
P11P=99999999999.;
P12P=0.;
P13P=0.;
P22P=99999999999.;
```

```
P23P=0.;
P33P=99999999999.;
XN=0.;
T=0.;
S=0.;
A=2.0926E7;
GM=1.4077E16;
ALTNM=0.;
ALT=ALTNM*6076.;
ANGDEG=0.;
ANG=ANGDEG/57.3;
XLONGM=ANG;
X=(A+ALT)*cos(ANG);
Y=(A+ALT)*sin(ANG);
ALT=sqrt(X^2+Y^2)-A;
XFIRST=X;
YFIRST=Y;
X1=cos(1.5708-GAMDEG/57.3+ANG);
Y1=sin(1.5708-GAMDEG/57.3+ANG);
AXT=0.;
AYT=0.;
XLONGTDEG=57.3*RDESKM*3280./A;
XLONGRDEG=57.3*RDESRKM*3280./A;
TF=252.+.223*RDESKM-(5.44E-6)*RDESKM*RDESKM;
TF=TF+TLOFT;
XLONGT=XLONGTDEG/57.3;
XLONGR=XLONGRDEG/57.3;
XF=A*cos(XLONGT);
YF=A*sin(XLONGT);
XR=A*cos(XLONGR);
YR=A*sin(XLONGR);
AXT=0.;
AYT=0.;
Z=0;
ZF=0;
ZFIRST=0;
while ~(ALT< -1 | T>TFINISH)
        XOLD=X;
        YOLD=Y;
        X1OLD=X1;
        Y1OLD=Y1;
        STEP=1;
        FLAG=0;
        while STEP <=1
                if FLAG==1
                        STEP=2;
```

```
                X=X+HINT*XD;
                Y=Y+HINT*YD;
                X1=X1+HINT*X1D;
                Y1=Y1+HINT*Y1D;
                T=T+HINT;
        end
        if T<120.
                WGT=-2622*T+440660.;
                TRST=725850.;
        elseif T<240.
                WGT=-642.*T+168120.;
                TRST=182250.;
        else
                WGT=5500.;
                TRST=0.;
        end
        AT=32.2*TRST/WGT;
        XD=X1;
        YD=Y1;
        VEL=sqrt(XD^2+YD^2);
        TEMBOT=(X^2+Y^2)^1.5;
        X1D=-GM*X/TEMBOT+AXT;
        Y1D=-GM*Y/TEMBOT+AYT;
        ALT=sqrt(X^2+Y^2)-A;
        ACCG=sqrt(AXT^2+AYT^2)/32.2;
        FLAG=1;
end
FLAG=0;
X=(XOLD+X)/2+.5*HINT*XD;
Y=(YOLD+Y)/2+.5*HINT*YD;
X1=(X1OLD+X1)/2+.5*HINT*X1D;
Y1=(Y1OLD+Y1)/2+.5*HINT*Y1D;
S=S+HINT;
Z=0.;
ZF=0.;
ZR=0;
TGOLAM=TF-T;
if (QBOOST==1 && T>TGRAVEND)
        TGOLAM=TF-T;
        [VRX,VRY,VRZ]=LAMBERT3D(X,Y,Z,TGOLAM,XF,YF,ZF,SWITCH);
        DELX=VRX-X1;
        DELY=VRY-Y1;
        DEL=sqrt(DELX^2+DELY^2);
        if (T<240 & DEL>500.)
                AXT=AT*DELX/DEL;
                AYT=AT*DELY/DEL;
```

```
            elseif DEL<500.
                    TRST=0.;
                    QBOOST=0;
                    AXT=0.;
                    AYT=0.;
                    X1=VRX;
                    Y1=VRY;
                    X1OLD=X1;
                    Y1OLD=Y1;
            else
                    QBOOST=0;
                    AXT=0.;
                    AYT=0.;
    end
      elseif (T>=TUPT & T<=TGRAVEND & QFIRST==1)
            QFIRST=0;
            VEL=sqrt(XD^2+YD^2);
            X1=VEL*cos(1.5708-GAMDEGIC/57.3+ANG);
            Y1=VEL*sin(1.5708-GAMDEGIC/57.3+ANG);
            X1OLD=X1;
            Y1OLD=Y1;
            AXT=AT*X1/VEL;
            AYT=AT*Y1/VEL;
      elseif (T>=TUPT & T<=TGRAVEND)
            VEL=sqrt(XD^2+YD^2);
            AXT=AT*X1/VEL;
            AYT=AT*Y1/VEL;
      elseif T<=TUPT
            RTMAG=sqrt(X^2+Y^2);
            AXT=AT*X/RTMAG;
            AYT=AT*Y/RTMAG;
      end
      if S>=(TS-.00001)
            S=0.;
            DISTNM=distance3dkm(X,Y,Z,XFIRST,YFIRST,ZFIRST);
            ALTNM=(sqrt(X^2+Y^2)-A)/3280.;
            RMAG=sqrt(X^2+Y^2);
            VMAG=sqrt(XD^2+YD^2);
            GAMDEG=90-57.3*acos((X*XD+Y*YD)/(RMAG*VMAG));
            RHO=.0034*exp(-ALT/22000.);
            QPRES=.5*RHO*VEL*VEL;
            RRKM=sqrt((X-XR)^2+(Y-YR)^2)/3280.;
            DISTRKM=distance3dkm(XR,YR,ZR,XFIRST,YFIRST,ZFIRST);
            ALTRKM=(sqrt(XR^2+YR^2)-A)/3280.;
            RRMAG=sqrt(XR^2+YR^2);
            RRTMAG=sqrt((X-XR)^2+(Y-YR)^2);
```

```
ELDEG=90.-57.3*acos((XR*(X-XR)+YR*(Y-YR))/(RRMAG*RRTMAG));
if (ELDEG>2. & ELDEG<85)
        ISEE=1;
else
        ISEE=0;
end
if ISEE==1
        XN=XN+1;
        if IFILTER==1
                XK1=2.*(2.*XN-1.)/(XN*(XN+1));
                XK2=6./(XN*(XN+1)*TS);
        else
                XK1=3*(3*XN*XN-3*XN+2)/(XN*(XN+1)*(XN+2));
                XK2=18*(2*XN-1)/(XN*(XN+1)*(XN+2)*TS);
                XK3=60/(XN*(XN+1)*(XN+2)*TS*TS);
        end
        TS2=TS*TS;
        TS3=TS2*TS;
        TS4=TS3*TS;
        TS5=TS4*TS;
        THET=atan2(Y-YR,XR-X);
        R=sqrt((XR-X)^2+(Y-YR)^2);
        THETNOISE=SIGTHET*randn;
        RNOISE=SIGR*randn;
        RMEAS=R+RNOISE;
        THETMEAS=THET+THETNOISE;
        XTS=XR-RMEAS*cos(THETMEAS);
        YTS=YR+RMEAS*sin(THETMEAS);
        XTNOISE=X-XTS;
        SIGX=sqrt((cos(THET)*SIGR)^2+(R*sin(THET)*SIGTHET)^2);
        YTNOISE=Y-YTS;
        SIGY=sqrt((sin(THET)*SIGR)^2+(R*cos(THET)*SIGTHET)^2);
        if IFILTER==1
                RMAT(1,1)=SIGX^2;
                PHIP=PHI*P;
                PHIPPHIT=PHIP*PHIT;
                M=PHIPPHIT+Q;
                HM=HMAT*M;
                HMHT=HM*HT;
                HMHTR=HMHT+RMAT;
                HMHTRINV(1,1)=1./HMHTR;
                MHT=M*HT;
                K=MHT*HMHTRINV;
                KH=K*HMAT;
                IKH=IDNPZ-KH;
                P=IKH*M;
```

```
                if XK1>K(1,1)
                        XK1PZ=XK1;
                        XK2PZ=XK2;
                else
                        XK1PZ=K(1,1);
                        XK2PZ=K(2,1);
                end
                if QGRAV==0
                        XDDH=X1D*(1.+ERR);
                else
                        VELH=sqrt(XDH^2+YDH^2);
                        XDDH=AT*XDH/(VELH+.0001);
                end
                RES=XTS-XH-TS*XDH-.5*TS*TS*XDDH;
                XH=XH+XDH*TS+.5*TS*TS*XDDH+XK1PZ*RES;
                XDH=XDH+XDDH*TS+XK2PZ*RES;
                RMATP(1,1)=SIGY^2;
                PHIPP=PHI*PP;
                PHIPPHITP=PHIPP*PHIT;
                MP=PHIPPHITP+Q;
                HMP=HMAT*MP;
                HMHTP=HMP*HT;
                HMHTRP=HMHTP+RMATP;
                HMHTRINVP(1,1)=1./HMHTRP;
                MHTP=MP*HT;
                KP=MHTP*HMHTRINVP;
                KHP=KP*HMAT;
                IKHP=IDNPZ-KHP;
                PP=IKHP*MP;
                if XK1>K(1,1)
                        XK1PZP=XK1;
                        XK2PZP=XK2;
                else
                        XK1PZP=KP(1,1);
                        XK2PZP=KP(2,1);
                end
                if QGRAV==0
                        YDDH=Y1D*(1.+ERR);
                else
                        YDDH=AT*YDH/(VELH+.0001);
                end
                RESP=YTS-YH-TS*YDH-.5*TS*TS*YDDH;
                YH=YH+YDH*TS+.5*TS*TS*YDDH+XK1PZP*RESP;
                YDH=YDH+YDDH*TS+XK2PZP*RESP;
        else
           M11=P11+TS*P12+.5*TS2*P13+TS*(P12+TS*P22+.5*TS2*P23);
```

```
M11=M11+.5*TS2*(P13+TS*P23+.5*TS2*P33)+TS5*PHIS1/20.;
M12=P12+TS*P22+.5*TS2*P23+TS*(P13+TS*P23+.5*TS2*P33);
M12=M12+TS4*PHIS1/8.;
M13=P13+TS*P23+.5*TS2*P33+PHIS1*TS3/6.;
M22=P22+TS*P23+TS*(P23+TS*P33)+PHIS1*TS3/3.;
M23=P23+TS*P33+.5*TS2*PHIS1;
M33=P33+PHIS1*TS;
BOT=M11+SIGX*SIGX;
K1=M11/BOT;
K2=M12/BOT;
K3=M13/BOT;
FACT=1.-K1;
P11=FACT*M11;
P12=FACT*M12;
P13=FACT*M13;
P22=-K2*M12+M22;
P23=-K2*M13+M23;
P33=-K3*M13+M33;
if XK1>K1
        XK1PZ=XK1;
        XK2PZ=XK2;
        XK3PZ=XK3;
else
        XK1PZ=K1;
        XK2PZ=K2;
        XK3PZ=K3;
end
RES=XTS-XH-TS*XDH-.5*TS*TS*XDDH;
XH=XH+XDH*TS+.5*TS*TS*XDDH+XK1PZ*RES;
XDH=XDH+XDDH*TS+XK2PZ*RES;
XDDH=XDDH+XK3PZ*RES;
M11P=P11P+TS*P12P+.5*TS2*P13P+TS*(P12P+TS*P22P+.5*TS2*P23P);
M11P=M11P+.5*TS2*(P13P+TS*P23P+.5*TS2*P33P)+TS5*PHIS1/20.;
M12P=P12P+TS*P22P+.5*TS2*P23P+TS*(P13P+TS*P23P+.5*TS2*P33P);
M12P=M12P+TS4*PHIS1/8.;
M13P=P13P+TS*P23P+.5*TS2*P33P+PHIS1*TS3/6.;
M22P=P22P+TS*P23P+TS*(P23P+TS*P33P)+PHIS1*TS3/3.;
M23P=P23P+TS*P33P+.5*TS2*PHIS1;
M33P=P33P+PHIS1*TS;
BOTP=M11P+SIGY*SIGY;
K1P=M11P/BOTP;
K2P=M12P/BOTP;
K3P=M13P/BOTP;
FACTP=1.-K1P;
P11P=FACTP*M11P;
P12P=FACTP*M12P;
```

```
                    P13P=FACTP*M13P;
                    P22P=-K2P*M12P+M22P;
                    P23P=-K2P*M13P+M23P;
                    P33P=-K3P*M13P+M33P;
                    if XK1>K1
                        XK1PZP=XK1;
                        XK2PZP=XK2;
                        XK3PZP=XK3;
                    else
                        XK1PZP=K1P;
                        XK2PZP=K2P;
                        XK3PZP=K3P;
                    end
                    RESP=YTS-YH-TS*YDH-.5*TS*TS*YDDH;
                    YH=YH+YDH*TS+.5*TS*TS*YDDH+XK1PZP*RESP;
                    YDH=YDH+YDDH*TS+XK2PZP*RESP;
                    YDDH=YDDH+XK3PZP*RESP;
                end
            end
            if IFILTER==1
                ERRXTD=(X1-XDH);
                SP22=sqrt(P(2,2));
                ERRYTD=(Y1-YDH);
                SP22P=-SP22;
                count=count+1;
                ArrayT(count)=T;
                ArrayDISTNM(count)=DISTNM;
                ArrayALTNM(count)=ALTNM;
                ArrayERRXTD(count)=ERRXTD;
                ArraySP22(count)=SP22;
                ArraySP22P(count)=-SP22;
            else
                ERRXTD=(X1-XDH);
                SP22=sqrt(P22);
                ERRYTD=(Y1-YDH);
                SP22P=-SP22;
                count=count+1;
                ArrayT(count)=T;
                ArrayDISTNM(count)=DISTNM;
                ArrayALTNM(count)=ALTNM;
                ArrayERRXTD(count)=ERRXTD;
                ArraySP22(count)=SP22;
                ArraySP22P(count)=-SP22;
            end
        end
    end
```

```
figure
plot(ArrayDISTNM,ArrayALTNM),grid
xlabel('Downrange (km)')
ylabel('Altitude (km)')
figure
plot(ArrayT,ArrayERRXTD,ArrayT,ArraySP22,ArrayT,ArraySP22P),grid
xlabel('Time (s)')
ylabel('Velocity Error (f/s)')
axis([90 240 -500 500])
clc
output=[ArrayT',ArrayDISTNM',ArrayALTNM',ArrayERRXTD',ArraySP22',ArraySP22P'];
save datfil.txt output /ascii
disp 'simulation finished'
% LAMBERT3D can be found in Listing 28.3
% distance3dkm can be found in Listing 28.3
```

Listing 33.2 was first run assuming that the two-state template-based filter did not require any process noise (IFILTER = 1, PHIS = 0). Figure 33.6 shows single flight results that indicate when there is no process noise the velocity error of the template based filter starts to diverge from the covariance matrix predictions. Initially the results of Fig. 33.6 might seem strange because we have perfect knowledge of the target's current acceleration states. However, we do not have information on where the target is going and when it will arrive at its destination. From a filtering point of view this means that our predictions to the next measurement, via the fundamental matrix, are not exact. Making the process noise zero means that the filter will eventually stop paying attention to the measurements and filter divergence results as is indicated in Fig. 33.6. It is important to note that in this chapter we shall be making judgments based on single flight results. Normally a Monte Carlo analysis is required for precise estimates in filtering

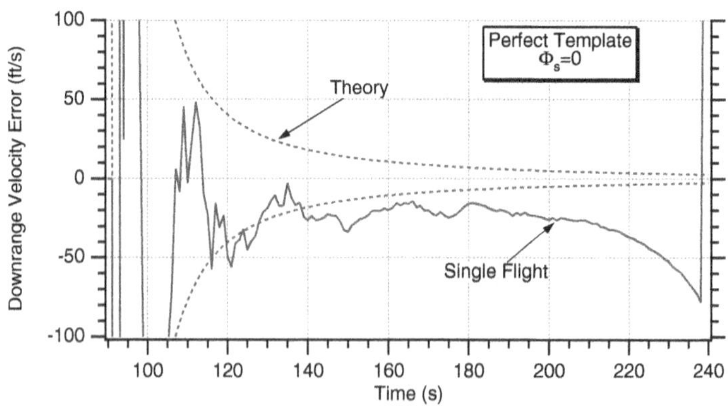

Fig. 33.6 Filter error in estimates diverge when there is zero process noise.

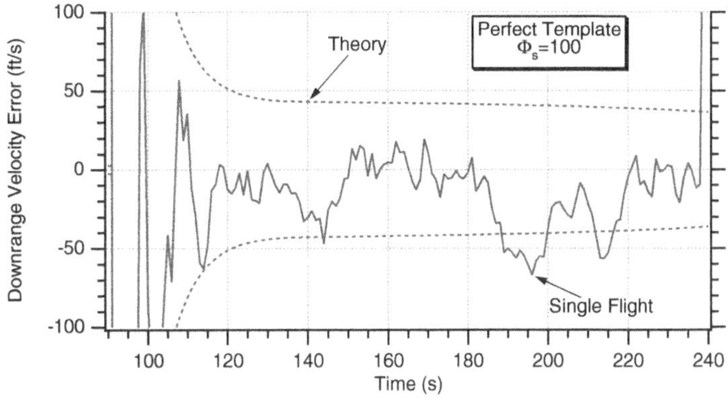

Fig. 33.7 With perfect acceleration template two-state Kalman filter yields excellent downrange velocity estimates.

work. However, usually disaster, such as filter divergence shown in Fig. 33.6, can be detected in a single run.

The simple engineering fix to filter divergence is to increase the filter process noise. Figure 33.7 indicates that divergence is eliminated and filter consistency is achieved when Φ_s is increased from 0 to 100. The figure indicates that, on a single flight basis, the error in the estimate of downrange velocity is consistent with the theoretical predictions of the covariance matrix obtained from the Riccati equations. Perfect knowledge of the current target's acceleration magnitude and direction in the template-based filter helps keep downrange velocity errors to less than 40 ft/s by the end of the ICBM boost phase.

Next, a 10% acceleration template error was introduced into our knowledge of the ICBM's acceleration magnitude ($\text{ERR} = 0.1$). It was still assumed that the current direction of the ICBM longitudinal acceleration was known perfectly. Figure 33.8 indicates that an acceleration template error is hardly visually noticeable.

However Fig. 33.9 shows that even a small 10% template error significantly influences two-state template-based Kalman filter consistency. The figure demonstrates that the downrange error in the velocity estimate diverges from the theoretical bounds for a 10% template error. Because our model of the real world has additional errors, more process noise is required by the filter to obtain filter consistency.

Figure 33.10 shows that when the filter process noise is increased by an order of magnitude (Φ_s increased from 100 to 1000) the two-state template-based Kalman filter becomes consistent and divergence is no longer an issue when there is a 10% filter template error. The price paid for increasing the process noise is that the theoretical errors in the velocity estimate at 240 s increases from 40 ft/s ($\Phi_s = 100$) to 90 ft/s ($\Phi_s = 1000$).

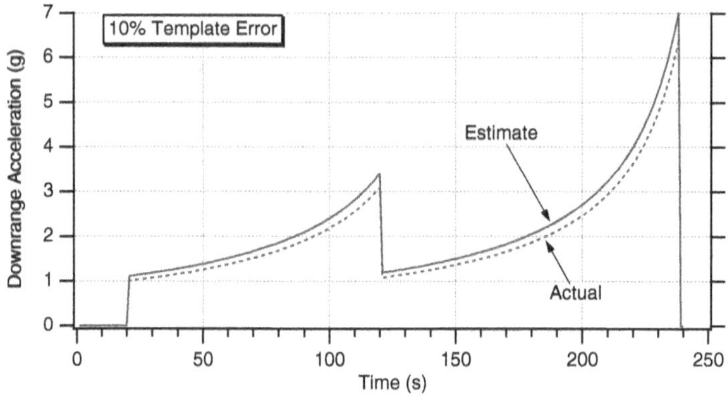

Fig. 33.8 10% errors slightly degrade downrange acceleration template.

So far all of the Kalman filter results have depended on a template in which the current booster acceleration direction was known exactly. Because in reality it is impossible to know where the ICBM is going and how it will get there, an assumption must be made on the direction of the future booster acceleration vector. One such popular assumption is to assume the ICBM acceleration direction is that of a gravity turn [9]. As was shown in the previous section, the gravity turn assumption implies that the booster thrust vector is always aligned with its velocity vector.

Many consider the gravity turn assumption to be reasonable because a gravity turn will minimize loading and drag while the booster is in the atmosphere. However a gravity turn is not a closed-loop guidance law that can enable an ICBM to reach its intended target in the desired time. An ICBM performs gravity turn type maneuvers in the atmosphere to minimize drag and loading

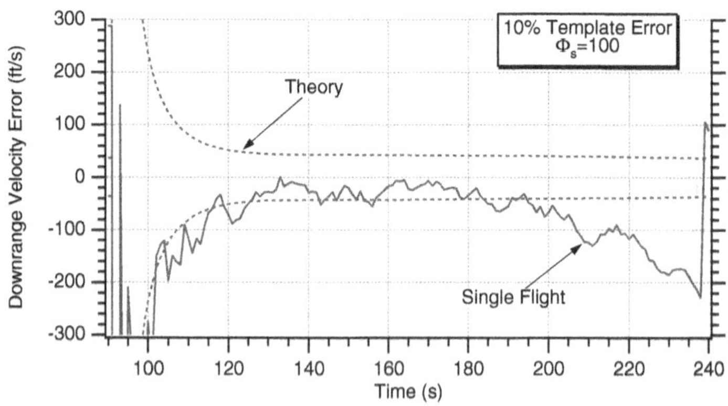

Fig. 33.9 10% template errors cause two-state Kalman filter to diverge.

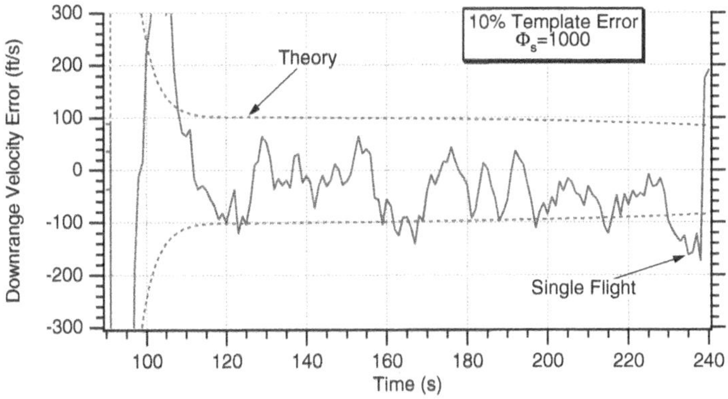

Fig. 33.10 More process noise is required by two-state Kalman filter to eliminate divergence when there is template error.

effects, or to provide gentle maneuvering if the dynamic pressure is too high. When the dynamic pressure is low enough, an ICBM must use a closed-loop guidance law (in our case the ICBM is guiding using Lambert guidance) to reach its intended target. Therefore there will be a model mismatch with the gravity turn assumption of the Kalman filter and what is happening in the real world when the dynamic pressure is low. In our real world model the target is performing a gravity turn until 100 s and then the target switches to Lambert guidance when the dynamic pressure for this example is close to zero.

Figure 33.11 shows that when the gravity turn assumption is incorporated in the two-state template-based Kalman filter (QGRAV = 1) there is divergence because there is not sufficient process noise ($\Phi_s = 1000$).

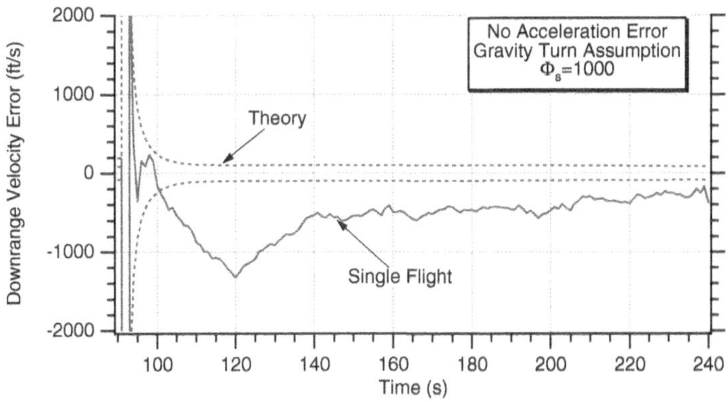

Fig. 33.11 Kalman filter diverges with gravity turn assumption because there is not sufficient process noise.

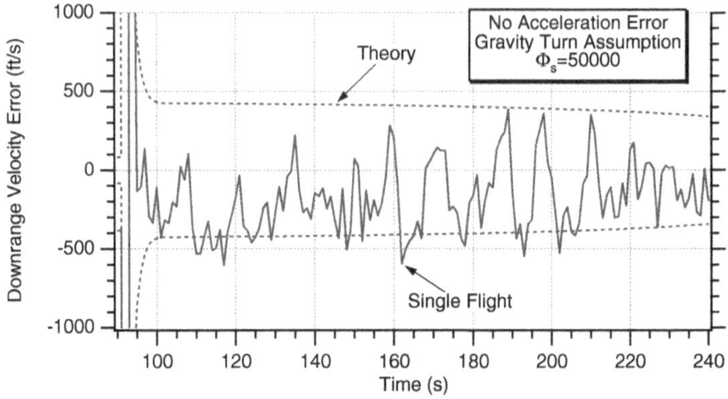

Fig. 33.12 Kalman filter is consistent with gravity turn assumption when process noise is increased by more than an order of magnitude.

Figure 33.12 shows that when the filter process noise is increased by more than an order of magnitude (Φ_s increased from 1000 to 50,000) the two-state template-based Kalman filter becomes consistent and divergence is no longer an issue with the gravity turn assumption. The price paid for increasing the process noise is that the theoretical errors in the velocity estimate at 240 s increases from 40 ft/s ($\Phi_s = 100$) to 350 ft/s ($\Phi_s = 50{,}000$).

Therefore it has been demonstrated that our template-based gravity turn two-state Kalman filter is very sensitive to modeling errors—even when knowledge of the current acceleration magnitude and direction is perfect. Large amounts of process noise are required to prevent filter divergence when the acceleration direction is not known. Errors in our downrange velocity estimates can increase by an order of magnitude when minor errors in our knowledge of the real world are introduced. Does a Kalman filter exist that is not template-based and is more robust than the Kalman filter just considered?

Listing 33.2 was run for the case of the three-state filter (IFILTER = 2) with a value of process noise that was determined by experiment (PHIS1 = 260). Figure 33.13 shows single flight results for the acceleration estimate. The figure indicates that there is no problem in estimating the target acceleration without a template with a three-state linear polynomial Kalman filter.

Figure 33.14 shows that for the nominal trajectory the filter is consistent and that the error in the estimate of downrange velocity for the three-state polynomial Kalman filter is approximately 200 ft/s. This error in the estimate is not as good as the two-state filter with a perfect template and perfect knowledge of the future intention of the target. However it is better than the two-state gravity turn template-based Kalman filter (see Fig. 33.12).

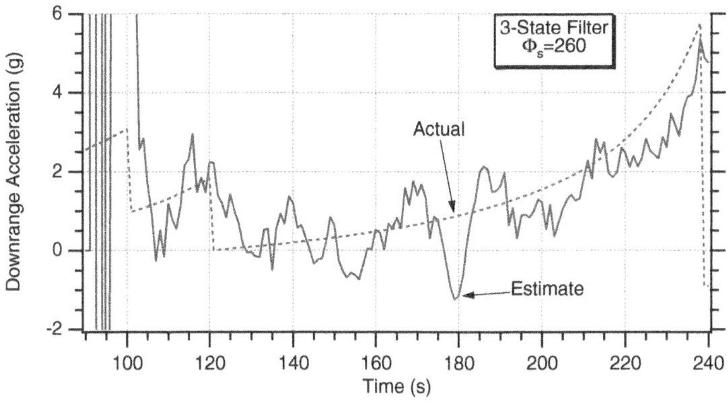

Fig. 33.13 Three-state linear polynomial Kalman filter does not require a template to estimate ICBM acceleration.

So far sufficient process noise has been added to the two Kalman filters considered in order to obtain filter consistency. With a consistent filter the covariance matrix of the Riccati equations can be used to obtain accurate performance projections concerning the errors in the estimate. Figure 33.15 presents the theoretical error in the downrange velocity estimate for both the three-state linear polynomial Kalman filter and the two-state, perfect template, gravity turn assumption Kalman filter when they both have sufficient process noise. The figure indicates that the resultant errors in the downrange velocity estimate for the three-state filter are nearly half of those for the two-state, perfect template, gravity turn assumption Kalman filter.

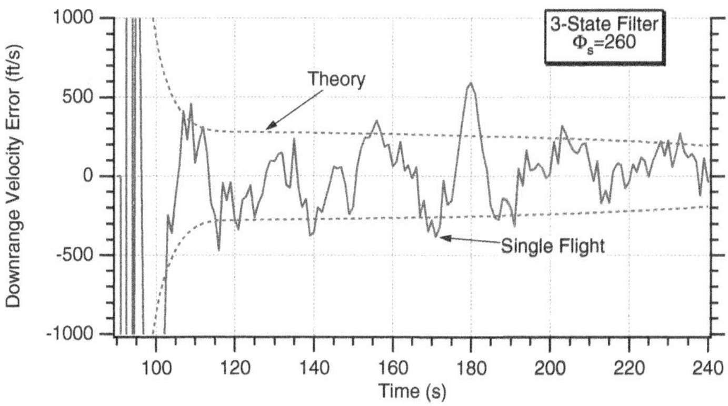

Figure 33.14 Three-state Kalman filter is consistant.

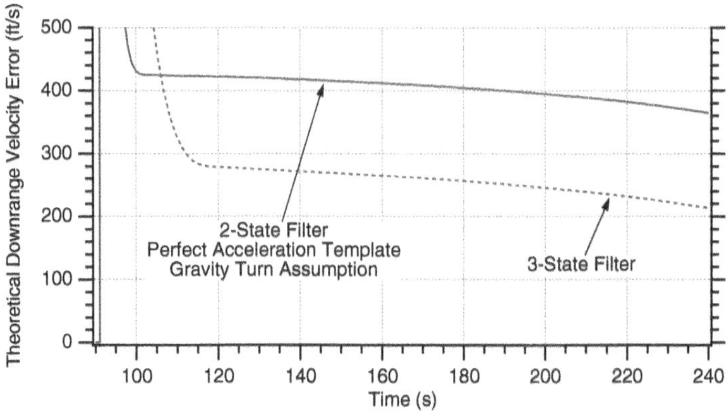

Fig. 33.15 Three-state Kalman filter is superior on a theoretical basis to two-state, perfect acceleration template, gravity turn Kalman filter.

SUMMARY

For the boost-phase intercept problem it has been demonstrated that when a template-based Kalman filter has perfect *a priori* information it offers significantly better performance than a linear three-state polynomial Kalman filter that does not require such information. However, when the two-state template-based filter has slight errors, this chapter shows that equivalent or sometimes better performance can be obtained with the simpler three-state polynomial Kalman filter.

REFERENCES

[1] Zarchan, P., "Boost Phase Filtering Options: Is Simpler Better?," *Journal of Guidance, Control, and Dynamics*, Vol. 33, Nov–Dec. 2010, pp. 1724–1731.

[2] Blackman, S. S., *Multiple Target Tracking With Radar Applications*, Artech House, Norwood, MA, 1986, pp. 19–44.

[3] Brookner, E., *Tracking and Kalman Filtering Made Easy*, John Wiley & Sons, Inc., New York, 1998, pp. 3–104.

[4] Kleppner, D., and Lamb, F. K. (eds.), *Boost Phase Intercept Systems for National Missile Defense*, American Physical Society, July 2003, pp. 263–278.

[5] White, J. E., "Guidance and Targeting for the Strategic Target System," *Journal of Guidance, Control, and Dynamics*, Vol. 15, Nov.–Dec. 1992, pp. 1313–1319.

[6] Battin, R. H., *An Introduction to the Mathematics and Methods of Astrodynamics*, AIAA Education Series, New York, 1987.

[7] Brand, T. J., "A New Approach to Lambert Guidance," Charles Stark Draper Laboratory, Rept. R-694, Cambridge, MA, June 1971.

[8] Nelson, S. L., and Zarchan, P., "Alternative Approach to the Solution of Lambert's Problem," *Journal of Guidance, Control, and Dynamics*, Vol. 15, July–Aug. 1992, pp. 1003–1009.

[9] Blackman, S., and Popoli, R., *Design and Analysis of Modern Tracking Systems*, Artech House, Boston, MA, 1999, pp. 241–250.

[10] Zarchan, P., *Fundamentals of Kalman Filtering A Practical Approach*, 3rd ed, Progress in Astronautics and Aeronautics, AIAA, Reston, VA 2009, pp. 159–165.

CHAPTER 34

Kill Vehicle Guidance and Control Sizing For Boost-Phase Intercept

INTRODUCTION

This chapter addresses some of the guidance and control issues involved in enabling an air-launched interceptor carrying a highly maneuverable kinetic kill vehicle (KKV) to perform an exoatmospheric intercept of a boosting threat target capable of traveling many thousands of kilometers [1]. The chapter takes the reader through part of the first iteration of the multi-iteration design process in order to get a preliminary estimate of how much divert and acceleration may be required by the kinetic kill vehicle to hit the target. Simplified examples are presented to indicate how conventional guidance and filtering techniques can be used as a starting point in the iterative design process for this important problem in missile defense. More advanced guidance and filtering techniques can be used in subsequent iterations to more accurately size the kinetic kill vehicle and improve system performance and robustness.

BACKGROUND

As was mentioned in Chapter 33, intercept of intercontinental ballistic missiles (ICBMs) during their boost phase has long been considered attractive. A few years ago the American Physical Society (APS) released a detailed report [2] that studied the use of surface-based interceptors for intercepting ICBMs during their boost phase. But the APS report found that the surface-based interceptors required for boost-phase intercept would have to be very heavy, due to both the high required burnout velocities and lateral divert requirements of the interceptor. In addition, in the case of an Iranian ICBM launch against the United States, for example, the APS report showed that the interceptors would have to be based in countries that might present a political challenge for the United States. Thus, in general, the APS report was pessimistic about the success of a terrestrial-based boost-phase intercept system.

AIR-LAUNCHED INTERCEPTOR APPROACH

An alternative approach to boost-phase intercept involves the use of airborne interceptors and was first considered in the open literature by Wilkening [3] and then expanded upon, with considerable practical detail, by Corbett [4]. At first glance this alternative approach to boost-phase intercept might be considered to be inferior to surface-based interceptors, since airborne interceptors would have even lower burnout velocities than surface-based interceptors due to aircraft payload weight constraints. However, in this alternative approach, stealthy aircraft, which would be used as both launch and sensor platforms, initially would be manned but in the future would be unmanned using a platform such as the Naval Unmanned Combat Air System Carrier (N-UCAS) [5]. Stealthiness would enable the aircraft to penetrate enemy territory to get much closer to ICBM launch sites than would be possible with surface-based interceptors located near the borders of an enemy nation. Having the defensive interceptor launch platform closer to an enemy launch site means that the required burnout velocity of an air-launched interceptor can be much less than that of a surface-launched interceptor.

The key elements in Corbett's boost-phase intercept system construct are stealthy fighter aircraft with infrared search-and-track (IRST) systems to detect and track the target and airborne interceptors with highly maneuverable kinetic kill vehicles for exoatmospheric intercepts of the enemy missiles. In this system construct pairs of stealthy aircraft travel in oval racetracks in opposite directions over enemy territory. Their combined IRST systems have 360-deg coverage and can search for, detect, and track enemy ICBMs and intermediate range ballistic missiles (IRBMs) autonomously during their boost phase. When the IRST system of one aircraft detects a threat, it can cue another off-board IRST system to establish an additional angles-only track on the target so that the position, velocity, and acceleration of the target can be estimated. When sufficient track accuracy of the target states are obtained, a prediction is made of the target's position at the desired intercept time. This prediction will be imperfect as it is impossible to know a boosting target's future intentions. The launch aircraft turns so that it can fire its interceptor directly at the predicted intercept point (PIP). The interceptor's thrust is not only used to increase the speed of the interceptor but as the PIP is constantly changing, the interceptor thrust vector must also be steered in order for the interceptor to hit the latest and most refined estimate of the PIP. When the interceptor burns out, the PIP will still be in considerable error. Therefore, additional fuel and guidance is required so that a KKV, which separates from the interceptor after the interceptor burns out, will hit the target using its lateral divert engines for responding to guidance commands outside of the Earth's atmosphere.

GUIDANCE AND CONTROL ISSUES

The purpose of this chapter is to illustrate how some key guidance and control issues influence the amount of fuel and acceleration the KKV must have so that it can successfully engage both IRBMs and ICBMs during their boost phase. Sample trade-offs will be conducted using conventional guidance and filtering methods to illustrate the first step of an iterative design process that must take place for all practical designs. Subsequent steps in the design process may consider more advanced guidance and filtering techniques which in turn might reduce the KKV divert requirements derived in this chapter.

Lambert guidance was shown in Chapter 13 to be an effective method of guidance when the control authority of the interceptor is in the axial direction while augmented proportional navigation (APN) was shown in Chapter 8 to be appropriate when the control authority of the KKV was in the lateral direction. Therefore, Lambert guidance can be used while the interceptor is thrusting and APN can be used afterwards by the KKV's lateral divert engines. The interceptor must be sized for both adequate burnout velocity and sufficient fuel and acceleration for the KKV's divert engines. When the KKV gets close enough to the target, its seeker can acquire the target plume. The seeker software must be capable of distinguishing the target hard body from the plume and enable the KKV to hit that target's warhead.

This chapter starts out by first considering the effects of apparent target maneuver and guidance law (assuming zero PIP error) in a noise-free, one-dimensional engagement environment. Formulas will be developed showing how KKV divert requirements are related to target maneuver and KKV action or homing time. Next engagement experiments in two dimensions are conducted in a noise-free environment to see how simulation results compare with closed-form solutions. Finally it is demonstrated that sensor noise and filtering effects also play an important role in establishing KKV lateral divert requirements.

ONE-DIMENSIONAL MODEL FOR UNDERSTANDING GUIDANCE

Figure 34.1 presents the classical interceptor homing loop for understanding guidance. Here n_T represents the apparent target acceleration, as seen by the pursuing interceptor, of a boosting threat. That portion of the target's axial acceleration that is perpendicular to the KKV-target line of sight will appear as a target maneuver to the KKV. In this example we want to ensure that the KKV has adequate acceleration capability (does not saturate near the end of the flight) so that it can hit the target. For exoatmospheric intercepts the kill vehicle time constants are so small that they can be neglected in a preliminary analysis. In addition, we will mainly be concerned about the amount of KKV fuel or lateral divert required for a successful intercept.

Under worst-case geometrical conditions, all of the threats axial acceleration will be seen by the KKV as an apparent target maneuver. A typical acceleration

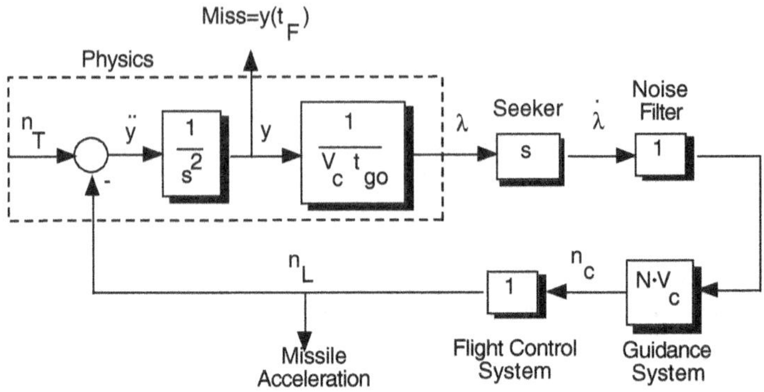

Fig. 34.1 One-dimensional guidance system model for initial analysis.

profile for a generic one-stage, 180-s burn IRBM is displayed as the solid curve in Fig. 34.2. Here it can be seen that the threat acceleration increases with increasing time because as the IRBM propellant burns, the weight of the IRBM decreases. In this example the maximum acceleration of the IRBM is approximately 9 g at 180 s. For academic and analytical purposes the generic IRBM acceleration can be approximated by a parabola with zero acceleration initially and $n_{T\text{MAX}}$ acceleration finally. The parabolic approximation is given by

$$n_{T_{\text{Parabola}}} = n_{T_{\text{MAX}}} \left(\frac{t}{t_F} \right)^2$$

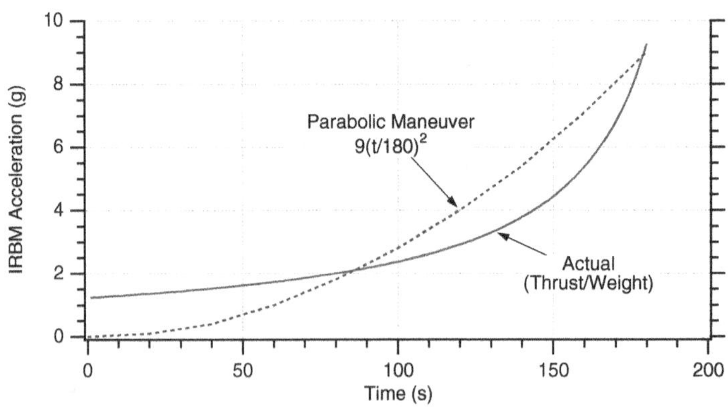

Fig. 34.2 For simplicity the axial acceleration of an IRBM will be approximated by a parabola.

where t_F is the IRBM burnout time. It will be shown later that the real purpose of the parabolic approximation is to enable us to rapidly estimate the KKV divert and acceleration requirements on hypothetical threats when only a minimal amount of information is available.

Let us simulate the one-dimensional guidance system of Fig. 34.1, assuming the proportional navigation (PN) guidance law with an effective navigation ratio of 3, for the academic case in which the IRBM target of Fig. 34.2 and the interceptor are both launched at time zero and intercept occurs at target burnout (180 s). The linearized one-dimensional engagement simulation of Listing 2.2 was modified so that the effects on the guidance system due to a parabolic target maneuver could be compared to the actual apparent target maneuver (that is, thrust divided weight). The modified code appears in Listing 34.1. The changes to the original code are highlighted in bold. Here the guidance law can be changed from proportional navigation (APN = 0) to augmented proportional navigation (APN = 1). In addition the effects of the parabolic target maneuver (OPTION = 0) on the guidance system can be compared to the actual apparent target maneuver (OPTION = 1). The code for the IRBM's thrust and weight profiles (ITGT = 1) appears with the differential equations before the FLAG=1 statement. The code also considers an ICBM threat (ITGT = 2) that will be discussed later in this chapter.

LISTING 34.1 ONE-DIMENSIONAL ENGAGEMENT SIMULATION BASED ON LINEARIZED GEOMETRY

```
count=0;
IOPTION=0;
ITGT=1;
XNTAV=117.6;
if ITGT==1
        TF=180.;
else
        TF=240.;
end
PRED=0.*3280.;
VM=9000.;
VC=18000.;
XNTMAX=9.*32.2;
XNCMAX=966.;
APN=0.;
HEDEG=-57.3*PRED/(VM*TF);
YD=-VM*HEDEG/57.3;
Y=0.;
XNP=3.;
T=0.;
H=.001;
```

```
S=0.;
DELV=0.;
SUM=0.;
XN=0.;
while T<(TF-.0001)
        YOLD=Y;
        YDOLD=YD;
        DELVOLD=DELV;
        STEP=1;
        FLAG=0;
        while STEP<=1
                if FLAG==1
                        STEP=2;
                        Y=Y+H*YD;
                        YD=YD+H*YDD;
                        DELV=DELV+H*DELVD;
                        T=T+H;
                end;
                TGO=TF-T+.00001;
                if ITGT==1
                        if IOPTION==0
                                XNT=XNTMAX*(T/TF)^2;
                        else
                                if T<180.
                                        WGT=-212.*T+44000.;
                                        TRST=54100.;
                                else
                                        WGT=3300.;
                                        TRST=0.;
                                end
                                        XNT=32.2*TRST/WGT;
                        end
                else
                        if IOPTION==0
                                XNT=XNTAV;
                        else
                                if T<120.
                                        WGT=-2622*T+440660.;
                                        TRST=725850.;
                                elseif T<240.
                                        WGT=-642.*T+168120.;
                                        TRST=182250.;
                                else
                                        WGT=5500.;
                                        TRST=0.;
                                end
```

```
                    XNT=32.2*TRST/WGT;
                end
            end
            XLAMD=(Y+YD*TGO)/(VC*TGO*TGO);
            XNC=XNP*VC*XLAMD+.5*APN*XNP*XNT;
            if XNC>XNCMAX
                XNC=XNCMAX;
            end
            if XNC<-XNCMAX
                XNC=-XNCMAX;
            end
            DELVD=abs(XNC);
            YDD=XNT-XNC;
            FLAG=1;
        end
        FLAG=0;
        Y=.5*(YOLD+Y+H*YD);
        YD=.5*(YDOLD+YD+H*YDD);
        DELV=.5*(DELVOLD+DELV+H*DELVD);
        S=S+H;
        if S>=.09999
            S=0.;
            SUM=SUM+XNT;
            XN=XN+1.;
            count=count+1;
            ArrayT(count)=T;
            ArrayXNT(count)=XNT/32.2;
            ArrayXNC(count)=XNC/32.2;
            ArrayDELV(count)=DELV/3.28;
        end
end
figure
plot(ArrayT,ArrayXNC),grid
xlabel('Missile Flight Time (s)')
ylabel('KKV Acceleration (g)')
clc
output=[ArrayT',ArrayXNC',ArrayDELV'];
save datfil.txt output /ascii
disp 'simulation finished'
Y
DELV/3.28
```

The nominal case of Listing 34.1 was run in which the parabolic target maneuver (OPTION = 0) and actual target maneuver (OPTION = 1) were used when the proportional navigation guidance law (APN = 0) was employed. Figure 34.3 shows that in this case the actual KKV acceleration required by the PN guidance law at

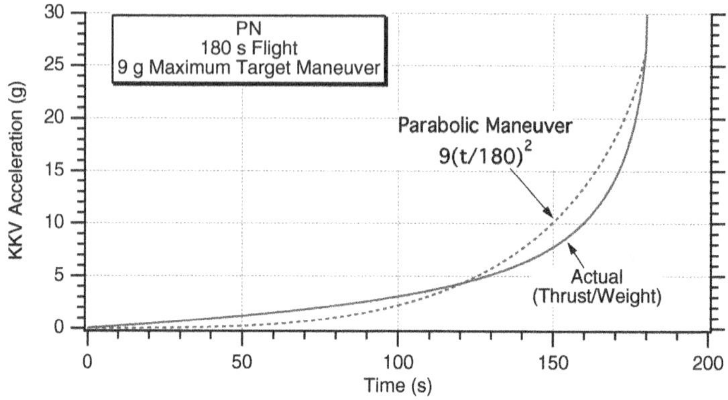

Fig. 34.3 For PN guidance system, parabolic maneuver approximation yields accurate performance projections.

the end of the flight is 30 g or three times the maximum acceleration capability of the target. This KKV acceleration requirement is the same as would be the case if there was a constant target maneuver as has been shown in Chapter 2. It can also be seen from Fig. 34.3 that the parabolic target maneuver approximation yields nearly identical KKV acceleration requirements at the end of flight indicating that the parabolic target maneuver approximation might be a good approximation for a boosting target being pursued by a KKV employing PN guidance.

The one-dimensional simulation experiment was also repeated for the case in which the interceptor guidance law was changed to augmented proportional navigation (APN = 1) with an effective navigation ratio of 3. The solid curve of Fig. 34.4 represents the actual acceleration required by the KKV using the augmented proportional navigation guidance law against the boosting IRBM, whereas the dashed curve represents the required KKV acceleration due to a parabolic target maneuver approximation. From the solid curve of Fig. 34.4 it can be seen that the maximum acceleration required by the KKV against the actual boosting target is now only 3 g or about $1/3$ of the maximum axial acceleration capability of the target, whereas the parabolic approximation indicates a 4.5-g maximum KKV acceleration or one half the maximum acceleration capability of the target. In addition, for both the actual and parabolic target acceleration models, the maximum acceleration no longer occurs at the end of flight. This means that should the KKV acceleration saturate, there is a chance that it will come out of saturation and not cause miss distance. *Thus, the maximum KKV acceleration required by the augmented proportional navigation guidance law is nearly an order of magnitude smaller than that required by the proportional navigation guidance law!* This means that the performance improvement with augmented proportional navigation is so great that we shall not even consider

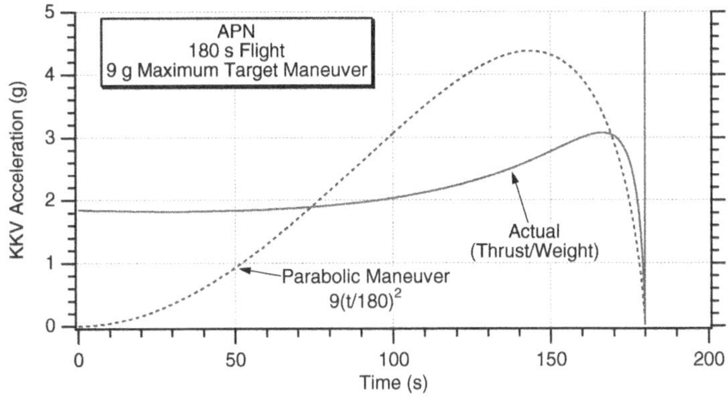

Fig. 34.4 For augmented proportional navigation guidance system, parabolic maneuver approximation to IRBM boosting target overestimates KKV acceleration requirements.

using proportional navigation guidance against a boosting target. Figure 34.4 also indicates that the parabolic approximation to the target maneuver is not as good in predicting interceptor performance as it was in the previous example where proportional navigation guidance was used by the KKV. However, the parabolic approximation still indicates that there is a dramatic performance improvement with augmented proportional navigation guidance. It can be seen that using the parabolic approximation for the boosting IRBM tends to overestimate the KKV acceleration requirements. Therefore, for a conservative starting point the parabolic approximation to the boosting target might be appropriate for analysis because of its simplicity.

DEVELOPING FORMULAS FOR DIVERT DUE TO BOOSTING TARGET AND PIP ERRORS

Generally speaking, an exoatmospheric interceptor has to be used against a long-range boosting target because most of the target's flight during its boost phase is outside the atmosphere. An endoatmospheric interceptor may be required against short-range ballistic missiles (SRBMs) because their apogees are very low. In this chapter only long-range IRBMs and ICBMs are considered. The KKV part of the interceptor has lateral divert engines for implementing guidance commands outside of the atmosphere. Since the divert engines burn propellant to implement the guidance law, guidance terminates when the propellant is expended because the KKV can no longer maneuver. The amount of propellant required by the KKV is related to the lateral divert ΔV through the rocket equation. Thus the amount of lateral divert required for an intercept is an important measure of interceptor performance. As was shown in Chapter 14, the lateral divert is simply the integral of the absolute value of the interceptor acceleration or

$$\Delta V = \int_0^{t_F} |n_c| \, dt$$

Figure 34.1 was evaluated, using Listing 34.1, for a 10-g parabolic target maneuver and the APN guidance law for the missile or kinetic kill vehicle (KKV) flight times ranging from 10 s to 50 s in steps of 10 s. The simulation results of Fig. 34.5 indicate that the amount of lateral divert required increases linearly with KKV flight time. We can also see from Fig. 34.5 that the simulation results can be curve fitted with a straight line and an empirical formula can be developed for the KKV lateral divert due to a parabolic target maneuver as

$$\Delta V_{\text{MVR}} = 0.25 n_{\text{T}_{\text{MAX}}} t_F$$

where $n_{T_{\text{MAX}}}$ is the maximum value of the parabolic maneuver in units of m/s^2 and ΔV_{MVR} is in units of m/s. The quantity t_F represents the KKV action time or the amount of time the KKV is maneuvering.

Up to this point it has been assumed that the only disturbance entering the KKV guidance system was an apparent target maneuver. Another important error source is the PIP error. This error source is due to the fact that the location of the boosting target at the desired intercept time is unknown. A prediction of the intercept point must be made and this prediction will have errors. The errors will be the same whether a PN or APN guidance law is used.

It is important to note that some people believe that the PIP errors for a boosting target will be small because *a priori* information concerning the threat will be

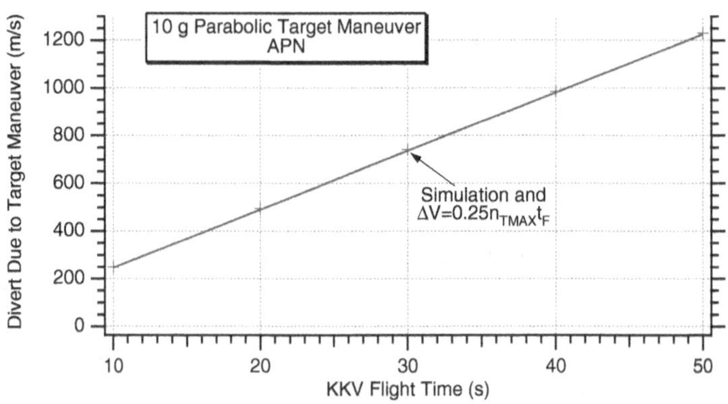

Fig. 34.5 Formula for KKV divert due to parabolic target maneuver in APN guidance system can be developed.

available. The point of view taken in this chapter is that even if the thrust-weight profile of the threat was known perfectly the future direction of the acceleration vector is unknown. In other words, we do not know where the target is going or when it will get there based on past information. Therefore, the conservative point of view taken in this chapter is that the system must work when information concerning the threat is not available or denied and that the lateral divert of the KKV must be sized accordingly. It was shown in Chapter 14 that for the model of Fig. 34.1, the KKV lateral divert due to the PIP error for either the PN or APN guidance laws with an effective navigation ratio of 3 is given by

$$\Delta V_{\mathrm{PIP}} = 1.5\frac{\mathrm{PIP}}{t_F}$$

Here it can be observed that for a given PIP error, more divert will be required for shorter KKV action times t_F. Thus the total divert required, under worse case conditions, is simply given by

$$\Delta V_{\mathrm{TOT_{APN}}} = \frac{1.5*\mathrm{PIP}}{t_F} + 0.25*n_{TMAX}t_F$$

INTERCEPTOR-IRBM ENGAGEMENTS

The previously mentioned single-stage generic IRBM model was put in a two-dimensional nonlinear simulation, assuming round Earth and Newton's Law of Universal Gravitation, based on Listing 13.3. The simulation was modified for a sample 2000-km IRBM lofted trajectory and the modified simulation appears in Listing 34.2. Although Listing 34.2 is a two-dimensional simulation, three-dimensional Lambert and distance routines (obtained from Chapter 28) are used by setting the z-components of various quantities to zero. As was mentioned previously, an ICBM target (ITGT = 2) option also appears in Listing 34.2 and will be discussed later in this chapter.

LISTING 34.2 TARGET TRAJECTORY GENERATOR

```
count=0;
RDESKM=2000.;
% 1=IRBM,2=ICBM
ITGT=1;
TLOFT=200.;
TUPT=15.;
if ITGT==1
   TPZ=180.;
else
   TPZ=240.;
```

```
end
QMIN=1;
TF=2000.;
TFINISH=3000.;
LEFT=1;
QBOOST=1;
QOOMPH=1;
CW=0;
SWITCH=0;
GAMDEG=89.99;
H=.01;
T=0.;
S=0.;
A=2.0926E7;
GM=1.4077E16;
ALT=0;
ANGDEG=0.;
ANG=ANGDEG/57.3;
XLONGM=ANG;
X=(A+ALT)*cos(ANG);
Y=(A+ALT)*sin(ANG);
Z=0;
ALT=sqrt(X^2+Y^2)-A;
XFIRST=X;
YFIRST=Y;
ZFIRST=Z;
X1=cos(1.5708-GAMDEG/57.3+ANG);
Y1=sin(1.5708-GAMDEG/57.3+ANG);
AXT=0.;
AYT=0.;
XLONGTDEG=57.3*RDESKM*3280./A;
TF=252.+.223*RDESKM-(5.44E-6)*RDESKM*RDESKM;
TF=TF+TLOFT;
XLONGT=XLONGTDEG/57.3;
XF=A*cos(XLONGT);
YF=A*sin(XLONGT);
ZF=0;
while ALT>-1
   XOLD=X;
   YOLD=Y;
   X1OLD=X1;
   Y1OLD=Y1;
   STEP=1;
   FLAG=0;
   while STEP <=1
      if FLAG==1
```

```
            STEP=2;
            X=X+H*X1;
            Y=Y+H*Y1;
            X1=X1+H*X1D;
            Y1=Y1+H*Y1D;
            T=T+H;
        end
        if ITGT==1
            if T<180.
                WGT=-212.*T+44000.;
                TRST=54100.;
            else
                WGT=3300.;
                TRST=0.;
            end
        else
            if T<120
                WGT=-2622*T+440660.;
                TRST=725850.;
            elseif T<240.
                WGT=-642.*T+168120.;
                TRST=182250.;
            else
                WGT=5500.;
                TRST=0.;
            end
        end
        AT=32.2*TRST/WGT;
        TEMBOT=(X^2+Y^2)^1.5;
        X1D=-GM*X/TEMBOT+AXT;
        Y1D=-GM*Y/TEMBOT+AYT;
        ALT=sqrt(X^2+Y^2)-A;
        FLAG=1;
    end
    FLAG=0;
    X=(XOLD+X)/2+.5*H*X1;
    Y=(YOLD+Y)/2+.5*H*Y1;
    X1=(X1OLD+X1)/2+.5*H*X1D;
    Y1=(Y1OLD+Y1)/2+.5*H*Y1D;
    S=S+H;
    if QBOOST==1
        TGOLAM=TF-T;
        [VRX,VRY,VRZ]=LAMBERT3D(X,Y,Z,TGOLAM,XF,YF,ZF,SWITCH);
        DELX=VRX-X1;
        DELY=VRY-Y1;
        DEL=sqrt(DELX^2+DELY^2);
```

```
        if T<TPZ&DEL>500
           AXT=AT*DELX/DEL;
           AYT=AT*DELY/DEL;
        elseif DEL<500
           TRST=0.;
           QBOOST=0;
           AXT=0.;
           AYT=0.;
           X1=VRX;
           Y1=VRY;
           X1OLD=X1;
           Y1OLD=Y1;
        else
           QBOOST=0;
           AXT=0.;
           AYT=0.;
        end
        if T<TUPT
           RTMAG=sqrt(X^2+Y^2);
           AXT=AT*X/RTMAG;
           AYT=AT*Y/RTMAG;
        end
    end
    if S>=.99999
        S=0.;
        DISTKM=distance3dkm(X,Y,Z,XFIRST,YFIRST,ZFIRST);
        ALTKM=(sqrt(X^2+Y^2)-A)/3280.;
        VELK=sqrt(X1^2+Y1^2)/3280.;
        count=count+1;
        ArrayT(count)=T;
        ArrayDISTKM(count)=DISTKM;
        ArrayALTKM(count)=ALTKM;
        ArrayVELK(count)=VELK;
    end
end
figure
plot(ArrayDISTKM',ArrayALTKM'),grid
xlabel('Downrange (km)')
ylabel('Altitude (km) ')
clc
output=[ArrayT',ArrayDISTKM',ArrayALTKM'];
save datfil.txt output /ascii
disp 'simulation finished'
% LAMBERT3D can be found in Listing 28.3
% distance3dkm can be found in Listing 28.3
```

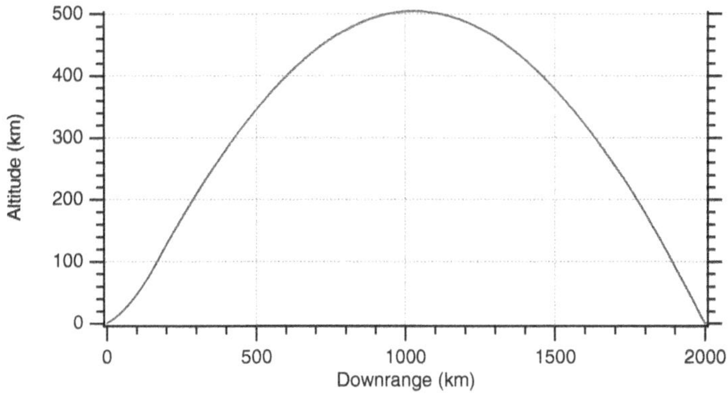

Fig. 34.6 Sample 2000-km IRBM trajectory.

The nominal case of Listing 34.2 was run (TLOFT = 200, TUPT = 200) and the resultant IRBM trajectory is presented in Fig. 34.6. We can see that the IRBM travels 2000 km downrange with an apogee of approximately 500 km.

The boost-phase portion or the first 180 s of the IRBM trajectory is presented with 10-s time tics in Fig. 34.7. If it is assumed that there is cloud cover until 7-km altitude then the IRBM can be seen by airborne IRST sensors at 80 s. Of course, on a clear day the target can be seen much sooner.

Listing 34.3 presents an engagement simulation in which an impulsively launched interceptor pursues the just discussed IRBM. Here we have two options for calculating the predicted intercept point (PIP) at the desired intercept time TF. The first option assumes that PIP is known perfectly (QPERFECT = 1). With this option the routine predict34.m integrates the target equations of

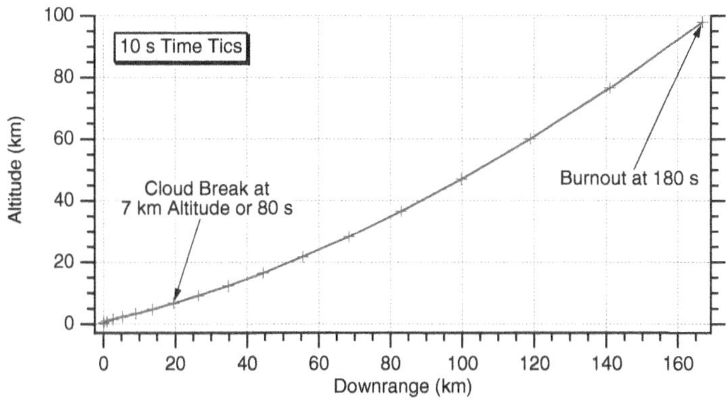

Fig. 34.7 Boost-phase portion of IRBM trajectory.

motion forward to calculate the exact location of the target at the desired intercept time. Another option for calculating the PIP, assuming no *a priori* information is available, is to use a three-term Taylor series (QPERFECT = 0). At the time of interceptor launch the required velocity vector of the interceptor is calculated given the location of the interceptor, its launch time, the PIP, and desired intercept time. It is important to note that a "for loop" is included in Listing 34.3 to find the earliest possible intercept time for an impulsive interceptor whose speed is less than 4 km/s. Augmented proportional navigation guidance begins at a user-specified time TGUID. Again, we can see from Listing 34.3 that the three-dimensional Lambert and distance routines are used in the two-dimensional engagement simulation with the required z-component inputs set to zero.

LISTING 34.3 TWO-DIMENSIONAL NONLINEAR ENGAGEMENT SIMULATION

```
count=0;
TLAUNCH=90.;
TF=170.;
TS=1.;
XLONGMDEGICKM=400.;
RDESKM=2000.;
GAMDEG=89.99;
TFTOT=2000.;
TUPT=15.;
TGUID=110.;
XNCLIM=322.;
XNP=3.;
QPERFECT=1;
TLOFT=200.;
ALTMKMIC=15.;
QTAYLOR=1;
PIPERRKM=0.;
QGUID=1;
ITGT=1;
SWITCH1=0;
SWITCHM=0;
DELTF=0.;
QFIX=1;
if ITGT==1
        TPZ=180.;
else
        TPZ=240.;
end
ALTM=ALTMKMIC*3280.;
```

```
TFTOT=252.+.223*RDESKM-(5.44E-6)*RDESKM*RDESKM;
TFTOT=TFTOT+TLOFT;
A=2.0926E7;
GM=1.4077E16;
W=0.;
XLONGFDEG=57.3*RDESKM*3280./A;
XLONGMDEGIC=XLONGMDEGICKM/111.;
XLONGMDEG=XLONGMDEGIC;
QLAUNCH=0;
QFIRST=1;
XLONGTDEG=0.;
QBOOST=1;
QOOMPH=1;
T=0.;
S=0.;
AXT=0.;
AYT=0.;
ATP=0.;
XLONGF=XLONGFDEG/57.3;
XLONGF=XLONGF-W*TFTOT;
PIPERR=0.;
XF=A*cos(XLONGF);
YF=A*sin(XLONGF);
ZF=0;
XLONGT=XLONGTDEG/57.3;
XLONGM=XLONGMDEG/57.3;
XT=A*cos(XLONGT);
YT=A*sin(XLONGT);
XTINIT=XT;
YTINIT=YT;
RTINIT=sqrt(XTINIT^2+YTINIT^2);
XM=(A+ALTM)*cos(XLONGM);
YM=(A+ALTM)*sin(XLONGM);
ZM=0;
XFIRST=XT;
YFIRST=YT;
ZT=0;
ZFIRST=0;
DISTRTKMIC=distance3dkm(XT,YT,ZT,XFIRST,YFIRST,ZFIRST);
XMINIT=XM;;
YMINIT=YM;
RMINIT=sqrt(XMINIT^2+YMINIT^2);
XTD=cos(1.5708-GAMDEG/57.3);
YTD=sin(1.5708-GAMDEG/57.3);
ATP=1.;
AXM=0.;
```

```
AYM=0.;
AMP=0.;
H=.01;
ALTTKM=(sqrt(XT^2+YT^2)-A)/3280.;
XMD=0.;
YMD=0.;
RTM1=XT-XM;
RTM2=YT-YM;
RTM=sqrt(RTM1^2+RTM2^2);
VTM1=XTD-XMD;
VTM2=YTD-YMD;
VC=-(RTM1*VTM1+RTM2*VTM2)/RTM;
DELV=0.;
ACC=0.;
AXMGUID=0.;
AYMGUID=0.;
PREDERRKM=0.;
ZEM1=0.;
ZEM2=0.;
ALTMKM=(sqrt(XM^2+YM^2)-A)/3280.;
TBOT=0.;
DELVELM=0.;
PIPKMBO=0.;
ZEMPERPTOT=0.;
if QFIX==1
% FIND EXACT LOCATION OF TARGET AT DESIRED INTERCEPT TIME EXACT PIP)
        [XTFACT,YTFACT]=predict34(T,XT,YT,XTD,YTD,TF,TFTOT,TUPT,XF,YF,ITGT);
        ZTFACT=0;
        TGOLAM=TF-TLAUNCH;
        [VRX,VRY,VRZ]=LAMBERT3D(XM,YM,ZM,TGOLAM,XTFACT,YTFACT,
                ZTFACT,SWITCHM);
        VMXRQD=VRX;
        VMYRQD=VRY;
        VMRQDKM=sqrt(VMXRQD^2+VMYRQD^2)/3280.;
else
        for TF=(TLAUNCH+30.):10:(TPZ-10.),
% FIND EXACT LOCATION OF TARGET AT DESIRED INTERCEPT TIME EXACT PIP)
        [XTFACT,YTFACT]=predict34(T,XT,YT,XTD,YTD,TF,TFTOT,TUPT,XF,YF,ITGT);
        TGOLAM=TF-TLAUNCH;
        [VRX,VRY,VRZ]=LAMBERT3D(XM,YM,ZM,TGOLAM,XTFACT,YTFACT,
                ZTFACT,SWITCHM);
        VMXRQD=VRX;
        VMYRQD=VRY;
        VMRQDKM=sqrt(VMXRQD^2+VMYRQD^2)/3280.;
        if VMRQDKM<4.
                break
```

```
		end
	end
end
TF=TF+DELTF;
while ~((T>(TF-10.))&VC<0.)
	if RTM<1000
		H=.00001;
	else
		H=.01;
	end
	XTOLD=XT;
	YTOLD=YT;
	XTDOLD=XTD;
	YTDOLD=YTD;
	XMOLD=XM;
	YMOLD=YM;
	XMDOLD=XMD;
	YMDOLD=YMD;
	DELVOLD=DELV;
	STEP=1;
	FLAG=0;
	while STEP <=1
		if FLAG==1
			STEP=2;
			XT=XT+H*XTD;
			YT=YT+H*YTD;
			XTD=XTD+H*XTDD;
			YTD=YTD+H*YTDD;
			XM=XM+H*XMD;
			YM=YM+H*YMD;
			XMD=XMD+H*XMDD;
			YMD=YMD+H*YMDD;
			DELV=DELV+H*DELVD;
			T=T+H;
		end
		if ITGT==1
			if T<180.
				WGT=-212.*T+44000.;
				TRST=54100.;
			else
				WGT=3300.;
				TRST=0.;
			end
		else
			if T<120.
				WGT=-2622*T+440660.;
```

```
				TRST=725850.;
			elseif T<240.
				WGT=-642.*T+168120.;
				TRST=182250.;
			else
				WGT=5500.;
				TRST=0.;
			end
		end
		ATP=32.2*TRST/WGT;
		TEMPBOTT=(XT^2+YT^2)^1.5;
		XTDD=-GM*XT/TEMPBOTT+AXT;
		YTDD=-GM*YT/TEMPBOTT+AYT;
		RTM1=XT-XM;
		RTM2=YT-YM;
		VTM1=XTD-XMD;
		VTM2=YTD-YMD;
		RTM=sqrt(RTM1^2+RTM2^2);
		VC=-(RTM1*VTM1+RTM2*VTM2)/RTM;
		TGO=RTM/VC;
		ACCDOTRTM=(XTDD*RTM1+YTDD*RTM2)/RTM;
		ACCPER1=XTDD-ACCDOTRTM*RTM1/RTM;
		ACCPER2=YTDD-ACCDOTRTM*RTM2/RTM;
		ACCPERPTOT=sqrt(ACCPER1^2+ACCPER2^2)/32.2;
		if T>TGUID
			TEMPBOTM=(XM^2+YM^2)^1.5;
			XMDDGRAV=-GM*XM/TEMPBOTM;
			YMDDGRAV=-GM*YM/TEMPBOTM;
			ZEM1=RTM1+VTM1*TGO+.5*(XTDD-XMDDGRAV)*TGO^2;
			ZEM2=RTM2+VTM2*TGO+.5*(YTDD-YMDDGRAV)*TGO^2;
			ZEMDOTRTM=(ZEM1*RTM1+ZEM2*RTM2)/RTM;
			ZEMPER1=ZEM1-ZEMDOTRTM*RTM1/RTM;
			ZEMPER2=ZEM2-ZEMDOTRTM*RTM2/RTM;
			ZEMPERPTOT=sqrt(ZEMPER1^2+ZEMPER2^2)/3280.;
			AXMGUID=XNP*ZEMPER1/(TGO^2);
			AYMGUID=XNP*ZEMPER2/(TGO^2);
			TGO=RTM/VC;
			if QGUID==0
				XNCLIM=0.;
			end
			if AXMGUID>XNCLIM
				AXMGUID=XNCLIM;
			elseif AXMGUID<-XNCLIM
				AXMGUID=-XNCLIM;
			end
			if AYMGUID>XNCLIM
```

```
                        AYMGUID=XNCLIM;
                elseif AYMGUID<-XNCLIM
                        AYMGUID=-XNCLIM;
                end
        else
                AXMGUID=0.;
                AYMGUID=0.;
        end
        if T>TLAUNCH
                TEMPBOTM=(XM^2+YM^2)^1.5;
                XMDD=-GM*XM/TEMPBOTM+AXMGUID;
                YMDD=-GM*YM/TEMPBOTM+AYMGUID;
        else
                XMDD=0.;
                YMDD=0.;
        end
        ACCNEW=sqrt(AXMGUID^2+AYMGUID^2);
        DELVD=ACCNEW;
        ALTMKM=(sqrt(XM^2+YM^2)-A)/3280.;
        FLAG=1;
end
FLAG=0;
XT=.5*(XTOLD+XT+H*XTD);
YT=.5*(YTOLD+YT+H*YTD);
XTD=.5*(XTDOLD+XTD+H*XTDD);
YTD=.5*(YTDOLD+YTD+H*YTDD);
XM=.5*(XMOLD+XM+H*XMD);
YM=.5*(YMOLD+YM+H*YMD);
XMD=.5*(XMDOLD+XMD+H*XMDD);
YMD=.5*(YMDOLD+YMD+H*YMDD);
DELV=.5*(DELVOLD+DELV+H*DELVD);
S=S+H;
if QBOOST==1
        TGOLAM=TFTOT-T;
        [VRX,VRY,VRZ]=LAMBERT3D(XT,YT,ZT,TGOLAM,XF,YF,ZF,SWITCH1);
        VTX=VRX;
        VTY=VRY;
        DELVXT=VTX-XTD;
        DELVYT=VTY-YTD;
        VELT=sqrt(XTD^2+YTD^2);
        DELVELT=sqrt(DELVXT^2+DELVYT^2);
        if (T<TPZ & DELVELT>500.)
                AXT=ATP*DELVXT/DELVELT;
                AYT=ATP*DELVYT/DELVELT;
        elseif DELVELT<500.
                TRST=0.;
```

```
			QBOOST=0;
			AXT=0.;
			AYT=0.;
			XTD=VTX;
			XTDOLD=XTD;
			YTD=VTY;
			YTDOLD=YTD;
			TBOT=T;
		else
			QBOOST=0;
			QOOMPH=0;
			AXT=0.;
			AYT=0.;
			TBOT=T;
		end
	end
	if T<TUPT
		RTMAG=sqrt(XT^2+YT^2);
		AXT=ATP*XT/RTMAG;
		AYT=ATP*YT/RTMAG;
	end
	if T>=TLAUNCH
		TGOLAMM=TF-T;
		if QPERFECT==1
			XTF=XTFACT;
			YTF=YTFACT;
		elseif (QPERFECT==0 & QTAYLOR==1)
			TGOM=TF-T;
			XTF=XT+XTD*TGOM+.5*XTDD*TGOM*TGOM;
			YTF=YT+YTD*TGOM+.5*YTDD*TGOM*TGOM;
		end
		ZTF=0;
		QLAUNCH=1;
		PIPERR=sqrt((XTF-XTFACT)^2+(YTF-YTFACT)^2)/3280.;
	end
	TGOLAMM=TF-T;
	if (T>=TLAUNCH & QFIRST==1)
		QFIRST=0;
		TGOPZ=TF-TLAUNCH;
		[VRX,VRY,VRZ]=LAMBERT3D(XM,YM,ZM,TGOPZ,XTF,YTF,ZTF,SWITCHM);
		VMXRQD=VRX;
		VMYRQD=VRY;
		VMRQDKM=sqrt(VMXRQD^2+VMYRQD^2)/3280.;
		XMD=VMXRQD;
		XMDOLD=XMD;
		YMD=VMYRQD;
```

```
            YMDOLD=YMD;
        end
        if S>=(TS-.0001)
            S=0.;
            ALTTKM=(sqrt(XT^2+YT^2)-A)/3280.;
            DISTRTKM=distance3dkm(XT,YT,ZT,XFIRST,YFIRST,ZFIRST);
            ALTMKM=(sqrt(XM^2+YM^2)-A)/3280.;
            DISTRMKM=distance3dkm(XM,YM,ZM,XFIRST,YFIRST,ZFIRST);
            VTK=sqrt(XTD^2+YTD^2)/3280.;
            ATG=sqrt(XTDD^2+YTDD^2)/32.2;
            VMKM=sqrt(XMD^2+YMD^2)/3280.;
            VTKM=sqrt(XTD^2+YTD^2)/3280.;
            XLAM=atan2(RTM2,RTM1);
            ATPLOS=-XTDD*sin(XLAM)+YTDD*cos(XLAM);
            ATPLOSG=ATPLOS/32.2;
            ZEMPLOS=-ZEM1*sin(XLAM)+ZEM2*cos(XLAM);
            ZEMPLOSG=ZEMPLOS/3280.;
            XNCPLOSG=(-AXMGUID*sin(XLAM)+AYMGUID*cos(XLAM))/32.2;
            DELVKM=DELV/3280.;
            ACCNEWG=ACCNEW/32.2;
            AXMGUIDG=AXMGUID/32.2;
            AYMGUIDG=AYMGUID/32.2;
            if T>TLAUNCH
                PIPKM=sqrt((XTFACT-XTF)^2+(YTFACT-YTF)^2)/3280.;
                PIPPLOS=-(XTFACT-XTF)*sin(XLAM)+(YTFACT-YTF)*cos(XLAM);
                PIPPLOSKM=PIPPLOS/3280.;
            else
                PIPKM=0.;
                PIPPLOSKM=0.;
            end
            count=count+1;
            ArrayT(count)=T;
            ArrayDISTRTKM(count)=DISTRTKM;
            ArrayALTTKM(count)=ALTTKM;
            ArrayDISTRMKM(count)=DISTRMKM;
            ArrayALTMKM(count)=ALTMKM;
            ArrayATPLOSG(count)=ATPLOSG;
            ArrayXNCPLOSG(count)=XNCPLOSG;
            ArrayPIPPLOSKM(count)=PIPPLOSKM;
        end
end
figure
plot(ArrayDISTRTKM,ArrayALTTKM,ArrayDISTRMKM,ArrayALTMKM),grid
xlabel('Downrange (km)')
ylabel('Altitude (km) ')
figure
```

```
plot(ArrayT,ArrayATPLOSG,ArrayT,ArrayXNCPLOSG),grid
xlabel('Time (s)')
ylabel('Acceleration (g) ')
clc
output=[ArrayT',ArrayDISTRTKM',ArrayALTTKM',ArrayDISTRMKM',ArrayALTMKM',...
        ArrayATPLOSG',ArrayXNCPLOSG'];
save datfil.txt output /ascii
disp 'simulation finished'
DELV/3.28

% predict34.m subroutine file
function [xtf,ytf]=predict(tp,xtp,ytp,xtdp,ytdp,tf,tftot,tupt,xf,yf,itgt)
if itgt==1
    tpz=180;
else
    tpz=240;
end
t=tp;
switch1=0;
xt=xtp;
yt=ytp;
zt=0.;
xtd=xtdp;
ytd=ytdp;
ztd=0.;
zf=0.;
a=2.0926E7;
gm=1.4077E16;
qboost=1;
h=.01;
s=0.;
axt=0.;
ayt=0.;
ztd=0;
while t<=(tf-.00001)
    xtold=xt;
    ytold=yt;
    xtdold=xtd;
    ytdold=ytd;
    step=1;
    flag=0;
    while step <=1
        if flag==1
            xt=xt+h*xtd;
            yt=yt+h*ytd;
            xtd=xtd+h*xtdd;
```

```
                ytd=ytd+h*ytdd;
                t=t+h;
                step=2;
        end
        tembot=(xt^2+yt^2)^1.5;
        xtdd=-gm*xt/tembot+axt;
        ytdd=-gm*yt/tembot+ayt;
        if itgt==1
                if t<180.
                        wgt=-212.*t+44000.;
                        trst=54100.;
                else
                        wgt=3300.;
                        trst=0.;
                end
        else
                if t<120
                        wgt=-2622*t+440660.;
                        trst=725850.;
                elseif t<240.
                        wgt=-642.*t+168120.;
                        trst=182250.;
                else
                        wgt=5500.;
                        trst=0.;
                end
        end
        atp=32.2*trst/wgt;
        flag=1;
end;
flag=0;
xt=(xtold+xt)/2+.5*h*xtd;
yt=(ytold+yt)/2+.5*h*ytd;
xtd=(xtdold+xtd)/2+.5*h*xtdd;
ytd=(ytdold+ytd)/2+.5*h*ytdd;
if qboost==1
        tgolam=tftot-t;
        [vrx,vry,vrz]=LAMBERT3D(xt,yt,zt,tgolam,xf,yf,zf,switch1);
        vtx=vrx;
        vty=vry;
        delvxt=vtx-xtd;
        delvyt=vty-ytd;
        delvelt=sqrt(delvxt^2+delvyt^2);
        if (t<tpz&delvelt>500.)
                axt=atp*delvxt/delvelt;
                ayt=atp*delvyt/delvelt;
```

```
                elseif delvelt<500.
                        trst=0.;
                        qboost=1;
                        axt=0.;
                        ayt=0.;
                        xtd=vtx;
                        xtdold=xtd;
                        ytd=vty;
                        ytdold=ytd;
                else
                        qboost=0;
                        qoomph=0;
                        axt=0.;
                        ayt=0.;
                end
                if t<tupt
                        rtmag=sqrt(xt^2+yt^2);
                        axt=atp*xt/rtmag;
                        ayt=atp*yt/rtmag;
                end
        end
end
xtf=xt;
ytf=yt;
% LAMBERT3D can be found in Listing 28.3
% distance3dkm can be found in Listing 28.3
```

Assuming that 10 s are required by the aircraft IRST sensors to establish a firm track on the IRBM, then the earliest an interceptor can be launched would be at 90 s (Fig. 34.7 shows cloud break occurring at 80 s). Figure 34.8 depicts an impulsively air-launched 3.4 km/s interceptor (that is, it takes zero seconds to get up to speed being launched at 90 s and 15-km altitude) at the PIP. In this scenario an intercept is to occur at 170 s or 10 s before the IRBM burns out. The case in which the PIP is known perfectly is done first in order to establish KKV divert requirements due to an apparent target maneuver. Next a post target burnout intercept, where there is no apparent target maneuver, is investigated in order to establish KKV divert requirements due to PIP error.

Let us first examine the case in which there is no PIP error. In this exercise the interceptor would not even require a KKV as it can fly directly toward the perfect PIP. However, as part of this academic exercise, the KKV guidance system, using the APN guidance law, is turned on at 110 s (20 s after interceptor launch to account for the fact that in the real world it might take 20 s for the interceptor to build up to speed). Since the portion of the IRBM acceleration that is perpendicular to the LOS appears as a target maneuver to the pursuing interceptor, the KKV will maneuver in order to hit the apparently maneuvering target. The KKV

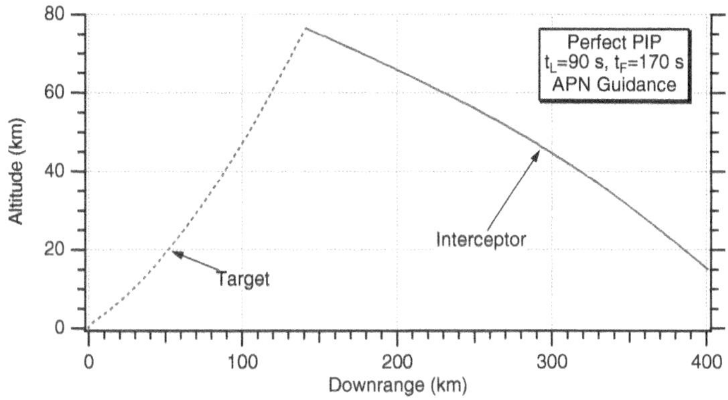

Fig. 34.8 Engagement with interceptor launched 10 s after cloud break and intercept occurring 10 s before the end of the IRBM boost phase.

maneuvering will occur even though the PIP is known perfectly! The interceptor-target engagement geometry for this example is displayed in Fig. 34.8.

The required KKV acceleration to hit the target along with the target acceleration perpendicular to the LOS is displayed in Fig. 34.9. Here it can be observed that for this engagement geometry the maximum target acceleration perpendicular to the LOS is approximately 3.6 g and the maximum KKV acceleration required by the APN guidance law is approximately 0.7 g or approximately five times less acceleration than the target. The amount of lateral divert required by the KKV, as indicated by the engagement simulation, is 208 m/s in this example.

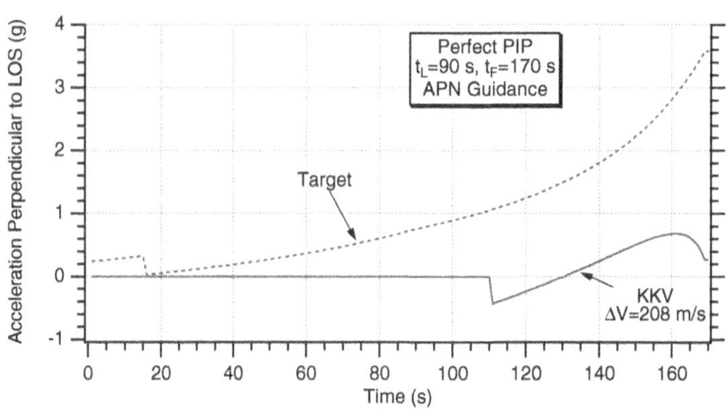

Fig. 34.9 For maximum range engagement KKV requires much less acceleration than target.

If the divert formula based on the parabolic target maneuver is utilized it can be seen that the theoretical KKV divert predicted is 540 m/s or

$$\Delta V_{\mathrm{MVR}} = 0.25 n_{\mathrm{TMAX}} t_F = 0.25 * 36 * (170 - 110) = 540\ \frac{\mathrm{m}}{\mathrm{s}}$$

As expected, the theoretical KKV divert prediction is observed to be very conservative since it is more than twice as high as the actual KKV divert required as indicated by the engagement simulation. Generally speaking, it was found that the KKV divert due to the apparently maneuvering one-stage IRBM target, as indicated by the engagement simulation, for different initial separations between both the interceptor and target launch points, was usually less than a few hundred m/s and is thus considered to be negligible.

A potentially more important error source against this one-stage IRBM target can be the PIP error. One simple way of predicting where the target will be at intercept is to use a three-term Taylor series based on the current position, velocity, and acceleration of the target or in two dimensions

$$x_F = x + \dot{x} t_{\mathrm{go}} + 0.5 \ddot{x} t_{\mathrm{go}}^2$$

$$y_F = y + \dot{y} t_{\mathrm{go}} + 0.5 \ddot{y} t_{\mathrm{go}}^2$$

where t_{go} is given by

$$t_{\mathrm{go}} = t_F - t$$

In other words the future location of the object at some time t_F is the current target position plus the current target velocity times the time to go plus one-half the current target acceleration times the time to go squared. The Taylor series method of prediction has many faults but its main virtue is that it does not require *a priori* information.

Figure 34.10 shows a case in which the PIP is not known as and is calculated from the preceding three-term Taylor series. In this example the PIP error perpendicular to the line of sight is 89 km when the interceptor starts to guide. In this academic exercise, the KKV guidance system is immediately turned on after the target burns out (so that none of the KKV divert is due to the apparent target maneuver), and an intercept geometry is set up so that the desired intercept time is also after target burnout at 230 s, but the achieved intercept time in the nonlinear engagement simulation turns out to be 223 s. To accommodate the postboost intercept time and to ensure that the intercept geometry was kinematically feasible the interceptor launch point was moved further downrange from the target launch point.

Figure 34.11 displays the KKV acceleration profile for the engagement with the 89-km PIP error. It can be seen that when the KKV guidance system turns on at 180 s there is an immediate step in KKV acceleration. After the initial step the KKV acceleration saturates for a brief period of time (simulation has

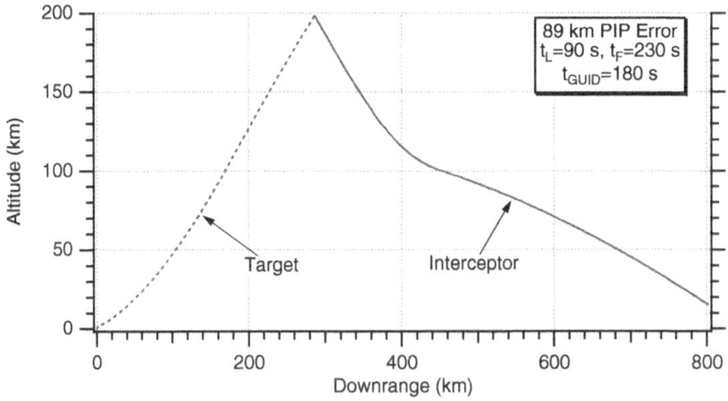

Fig. 34.10 Postboost phase engagement to test formula for PIP error.

10-g limit on KKV) and then linearly decreases to zero which is in accordance with theory. The lateral divert required by the KKV, as indicated by the engagement simulation, is 2.8 km/s and intercept occurs at 223 s. Had the KKV not saturated, the divert would have been slightly higher. The enormous amount of divert required for this hypothetical secenario was due to the fact that the interceptor was launched while the target was boosting. If the interceptor was launched after the target boost phase and the engagement was kinematically feasible, much less divert would be required for the intercept.

A formula was previously provided for the KKV divert due to PIP error. For the case in which there is 89 km of PIP error and the time to take it out is 43 s $(223 - 180 = 43)$, the theoretical divert formula indicates that 3.1 km/s of

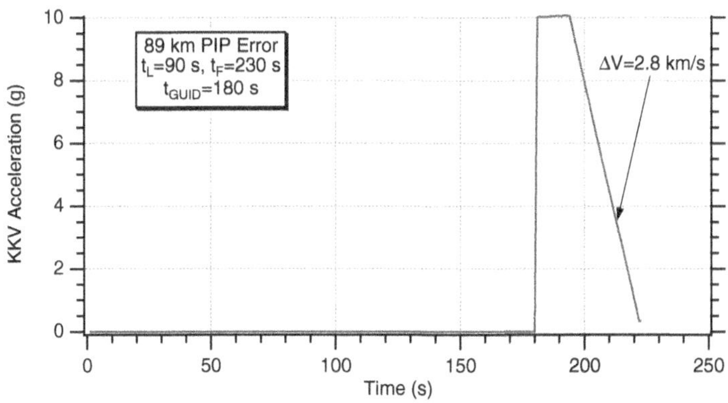

Fig. 34.11 Large initial KKV acceleration is required to take out PIP error in short time.

TABLE 34.1 DETAILS OF PIP ERROR EXPERIMENT

Distance (km)	t_F (s)	ΔV_{Sim} (m/s)	PIP (m)	$\Delta V_{Formula}$ (m/s)
400	170	570	20,400	510
350	160	410	12,500	375
300	150	290	7200	270
250	140	210	3900	195
200	130	140	1800	135
150	120	110	700	105
100	120	110	680	102

KKV lateral divert is required or

$$\Delta V_{\text{PIP}} = 1.5\frac{\text{PIP}}{t_F} = 1.5\frac{89{,}000}{(223 - 180)} = 3.1\ \frac{\text{km}}{s}$$

which is close to the engagement simulation results of Fig. 34.11.

The scenario of Fig. 34.8 was rerun for several cases where the interceptor launch point was gradually moved closer to the target launch point and intercept occurred during the target boost phase. Recall that the interceptor launch time is 90 s and KKV guidance starts at 110 s. In each case, the earliest boost-phase intercept time was selected, assuming the interceptor speed could be no greater than 4 km/s. The simulated and calculated divert results appear along with the PIP error perpendicular to the line of sight in Table 34.1. Figure 34.12 compares both the simulated and computed divert requirements for the KKV due to PIP

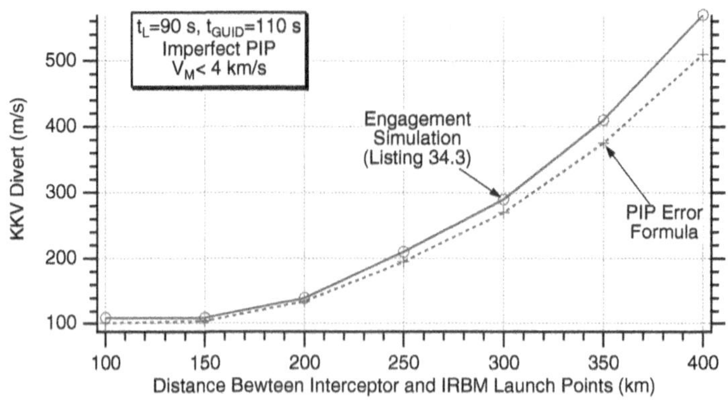

Fig. 34.12 Most of the KKV divert is due to PIP error for one-stage IRBM.

error. It is important to note that the simulated results include the effect of the apparent target maneuver, whereas the divert formula does not. We can see that the theoretical divert formula due to PIP error slightly underestimates the total required KKV lateral divert as indicated by the simulation. This means that for the one-stage IRBM example most of the KKV divert is due to the PIP error rather than the apparent target maneuver.

INTERCEPTOR-ICBM ENGAGEMENTS

Next a two-stage liquid ICBM threat with a 240-s burn time, as described in [2], was considered as the target for analysis. A sample lofted 10,000-km trajectory for the ICBM was generated using Listing 34.2 (ITGT = 2, TLOFT = 500, TUPT = 20) and is depicted in Fig. 34.13 where it can be seen that the apogee of this trajectory is approximately 2000 km. The boost-phase portion or first 240 s of the ICBM trajectory is displayed in Fig. 34.14 with 20-s time tics. It can be observed that the ICBM breaks the clouds, assuming a 7-km altitude cloud cover, at approximately 60 s.

The total axial acceleration of the ICBM is depicted in Fig. 34.15. Here it can be seen that the first staging event occurs at 120 s with a maximum acceleration of 6 *g* and the second staging event occurs at 240 s with a maximum acceleration of nearly 13 *g*. If the interceptor is launched early and the KKV guidance starts before 120 s, then the KKV guidance system will experience a large step in target acceleration at 120 s. It is apparent that this complex apparent acceleration cannot be represented by a single parabola.

An engagement was set up using Listing 34.3 in which the initial interceptor launch point is about 900 km from the target launch site. The interceptor is launched at 80 s and the KKV guidance system using the APN guidance law is

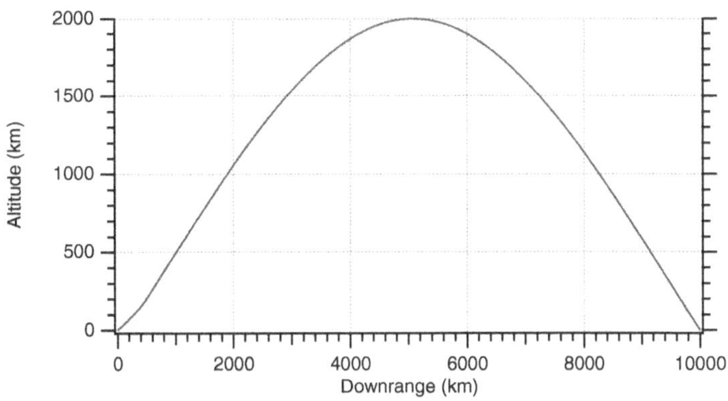

Fig. 34.13 Sample 10,000-km ICBM trajectory.

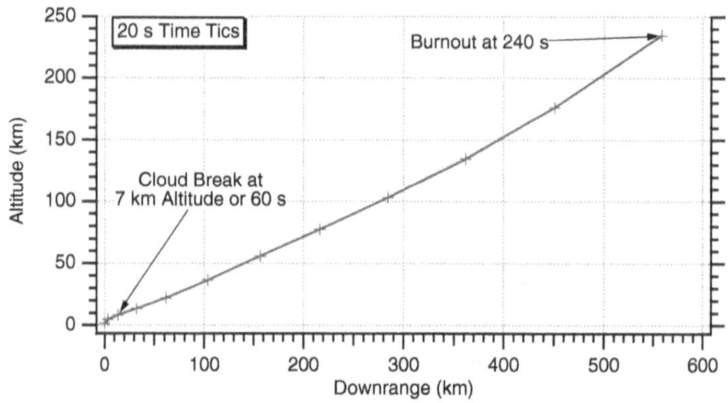

Fig. 34.14 Boost-phase portion of ICBM trajectory.

initiated at 100 s or about 20 s before the first target staging event. The engagement geometry is depicted in Fig. 34.16.

Figure 34.17 displays the target axial acceleration that is perpendicular to the LOS (dashed curve) for the engagement of Fig. 34.16. As was mentioned previously, this acceleration projection appears as a target maneuver to the KKV. The resultant KKV acceleration response to the apparent target maneuver (solid curve) is also displayed in Fig. 34.17. It can be seen that at first the KKV acceleration closely follows the apparent target maneuver and then changes abruptly when the target goes through the staging event at 120 s. The magnitude of the KKV acceleration after the staging event becomes a fraction of the actual target acceleration perpendicular to the LOS. The resultant lateral divert required by the KKV is shown in Fig. 34.17 to be 998 m/s.

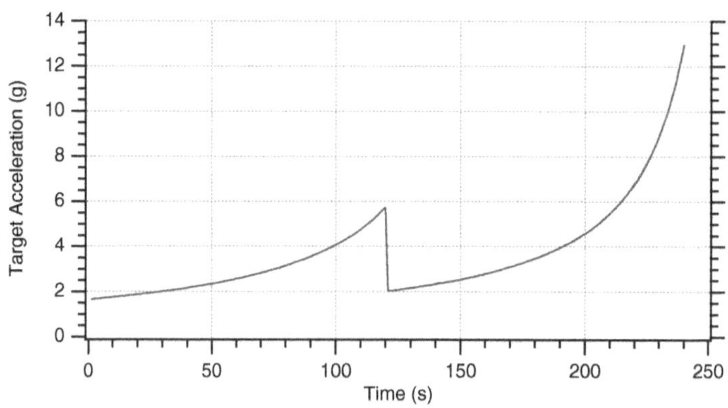

Fig. 34.15 ICBM has two staging events.

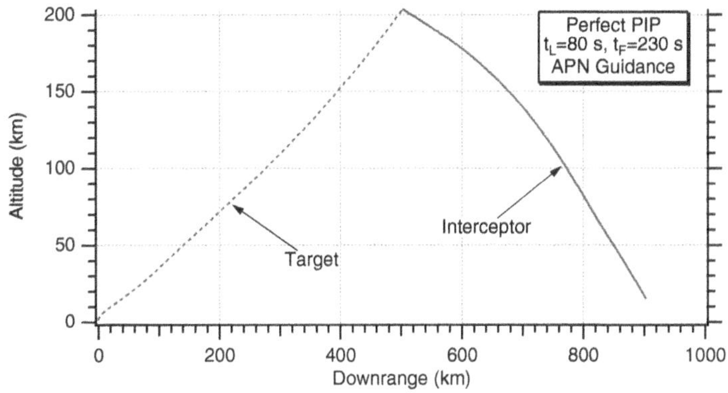

Fig. 34.16 ICBM engagement with interceptor launched 20 s after cloud break and intercept occurring 10 s before end of boost.

The average acceleration of the target maneuver while the KKV guidance system is activated can be computed from the target acceleration perpendicular to the LOS and is shown in Fig. 34.17 to be 1.3 *g*. If we pretend that the average target acceleration represents the complex apparent target maneuver shown in Fig. 34.17 then we can calculate the theoretical KKV divert from Chapter 14 (assuming APN guidance with an effective navigation ratio of 3) to be

$$\Delta V_{\mathrm{APN}} = 0.75 n_{\mathrm{TAV}}\, t_F = 0.75 * 13 * (230 - 100) = 1268\ \frac{\mathrm{m}}{\mathrm{s}}$$

which is approximately 25% greater than the value of 998 m/s shown in Fig. 34.17.

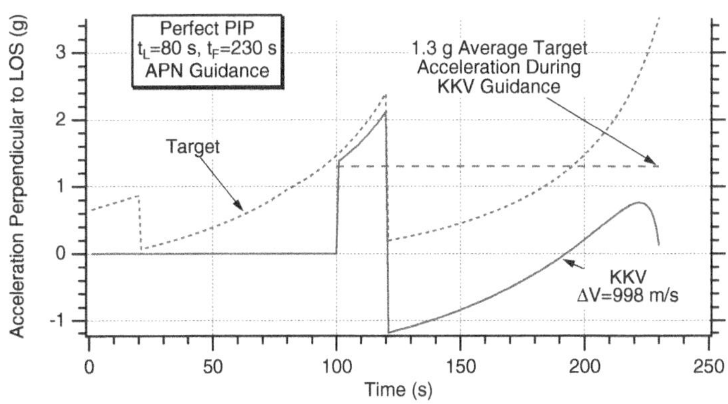

Fig. 34.17 For maximum range ICBM engagement apparent target acceleration is not a parabola.

An important purpose of the preceding divert formula is to qualitatively explain the simulation results and to suggest that the required KKV divert increases with increasing homing or action time. The formula also suggests that launching the interceptor earlier may increase the KKV divert requirements for a given intercept time. Thus an important purpose of the divert formula is to suggest future simulation experiments that must be conducted as part of the iterative design process.

The scenario of Fig. 34.16 was rerun for several cases where the interceptor launch point was gradually moved closer to the target launch point. Recall that the interceptor launch time is 80 s and KKV guidance starts at 100 s. *In each case, the earliest boost-phase intercept time was selected, assuming the interceptor speed could be no greater than 4 km/s.* Therefore for the 900-km downrange case it was found the intercept time could be reduced from 230 s (see Fig. 34.17) to 220 s. The simulated and calculated divert results appear both in Table 34.2 and Fig. 34.18. It is important to note that the calculated results are based on an average target acceleration of 1.3 *g*, which may not be accurate for all of the cases examined. However, it can be seen that the KKV divert trend is accurately captured with the simple divert formula appropriate for a constant average target maneuver.

Figure 34.19 shows another set of cases where the distance from the interceptor launch point to the target launch point is varied in the same way as was done in Fig. 34.18. However, this time the PIP is calculated from a three-term Taylor series rather than being perfect. It can be seen from the engagement simulation results of Fig. 34.19 that for the two-stage ICBM case the KKV divert requirements do not change significantly when PIP errors are considered. In fact sometimes the PIP errors reduce the miss distance because of the direction of the PIP error. Thus it can be concluded that for the ICBM case the apparent target maneuver is the major contributor to the KKV divert requirements.

TABLE 34.2 DETAILS OF BOOSTING ICBM EXPERIMENT

Downrange (km)	t_F (s)	ΔV_{SIM} (m/s)	$\Delta V_{Formula}$ (m/s)
900	220	896	1170
800	210	823	1072
700	190	727	878
600	180	689	780
500	160	557	585
400	150	463	487
300	130	177	292
200	120	141	195
100	110	2	98

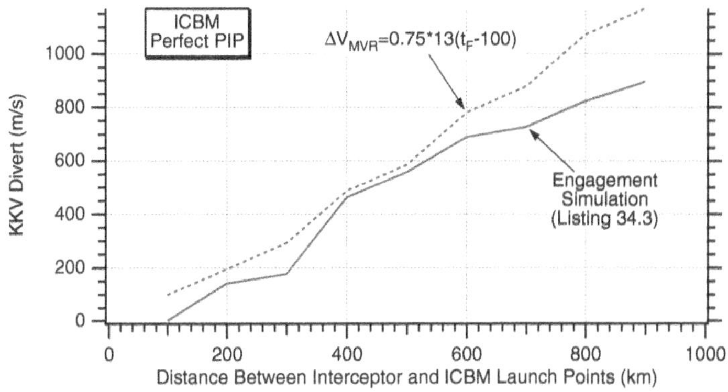

Fig. 34.18 Divert due to apparent constant target maneuver formula captures trends of actual KKV divert due to boosting two-stage ICBM.

NOISE AND FILTERING

Ideally we would like to have an aircraft sensor that could measure both range and angle to the target so that we can estimate the target states required for calculating the PIP and for implementing Lambert guidance for the boosting interceptor and APN guidance for the KKV. Unfortunately, an airborne radar that can see the target at the long distances required might be too heavy for airborne applications. Similarly, an airborne LADAR may also not work at required distances to see the target. On the other hand, an IRST sensor can see the boosting target at great distances but can only measure angle. For angle-only tracking of an unpredictable target, triangulation or stereo tracking is generally required to get target state estimates. In order to triangulate on the target two aircraft are required, each

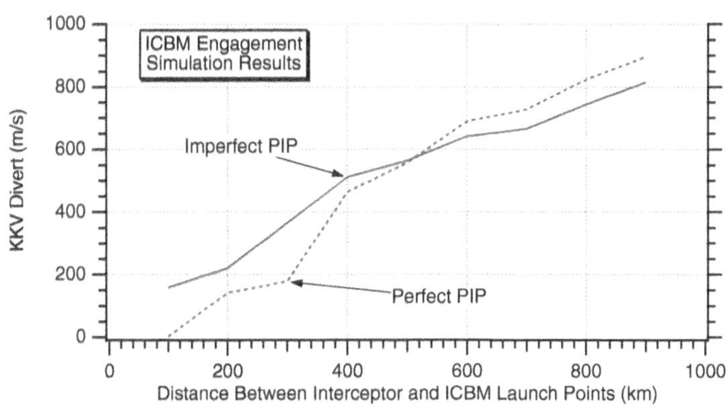

Fig.34.19 Influence of PIP errors on KKV lateral divert for ICBM intercept is small.

having an IRST sensor, separated by a large distance known as a baseline. From the angle-only measurements of the two sensors, filters can be designed to estimate the position, velocity, and acceleration of the target. The triangulation of the angle measurements is also known as stereo tracking. One logical choice for filtering would be to design an extended Kalman filter (EKF) for this stereo tracking application.

Another choice might be to design an even simpler filter for the first step of the iterative design process so that initial estimates of the increase in KKV divert requirements due to sensor noise can be rapidly obtained. Such a choice might be the use of linear decoupled polynomial three-state Kalman filters using pseudo measurements. Although decoupled linear polynomial three-state Kalman filters are not optimal in this stereo tracking application, they can easily and rapidly be designed by pretending the sensors are measuring distances to the target rather than angles from the sensor to the target. Later on, during subsequent stages of the iterative design and sizing process, more complex filters such as the EKF can be considered to see if they can reduce the resultant KKV divert requirements.

The basis for the pseudo measurements for the decoupled linear polynomial three-state Kalman filters in two dimensions can be derived from Fig. 34.20. Here we see two sensors measuring angles θ_1 and θ_2. It is assumed that the location of the sensors (x_{s1}, y_{s1} and x_{s2}, y_{s2}) are known, but the location of the target (x_T, y_T) is unknown.

From Fig. 34.20 one can express the two sensor measurements of the angles θ_1 and θ_2 as

$$\theta_1 = \tan^{-1}\left(\frac{y_{s1} - y_T}{x_{s1} - x_T}\right)$$

$$\theta_2 = \tan^{-1}\left(\frac{y_{s2} - y_T}{x_{s2} - x_T}\right)$$

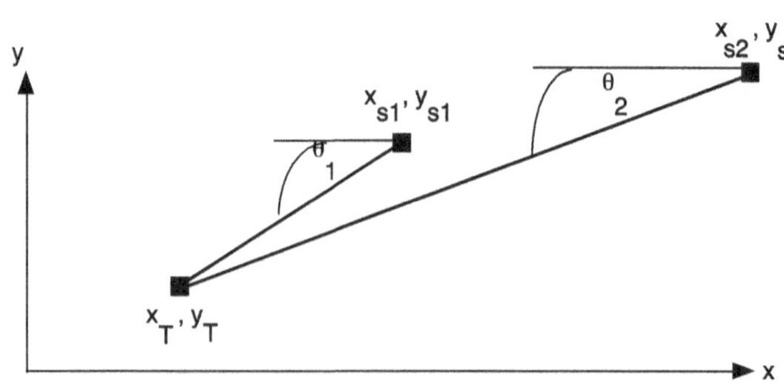

Fig. 34.20 Two angle-only sensors tracking a target.

Here we have two nonlinear equations with two unknowns. After some algebraic manipulation one can solve for the coordinates of the target in terms of the angle measurements and the sensor locations as

$$x_T^* = \frac{x_{s2}\tan(\theta_2) - x_{s1}\tan(\theta_1) + y_{s1} - y_{s2}}{\tan(\theta_2) - \tan(\theta_1)}$$

$$y_T* = y_{s1} - x_{s1}\tan(\theta_1) + \frac{x_{s2}\tan(\theta_2)\tan(\theta_1) - x_{s1}\tan^2(\theta_1) + (y_{s1} - y_{s2})\tan(\theta_1)}{\tan(\theta_2) - \tan(\theta_1)}$$

Thus the two pseudo measurements x_T^* and y_T^* will serve as inputs to the two decoupled three-state linear polynomial Kalman filters. One also has to develop formulas for the variance of the pseudo measurement noise to be used by the Riccati equations in the Kalman filter. The variance of the pseudo measurement noise on x_T can be found by using the chain rule from calculus. According to the chain rule

$$\Delta x_T = \frac{\partial x_T}{\partial \theta_1}\Delta\theta_1 + \frac{\partial x_T}{\partial \theta_2}\Delta\theta_2$$

By squaring and taking expectations of both sides of the preceding equation one can express the variance of the pseudo measurement noise in terms of the variances of each of the IRST sensors as

$$\sigma_{x_T}^2 = \left(\frac{\partial x_T}{\partial \theta_1}\right)^2 \sigma_{\theta_1}^2 + \left(\frac{\partial x_T}{\partial \theta_2}\right)^2 \sigma_{\theta_2}^2$$

where the partial derivatives are evaluated as

$$\frac{\partial x_T}{\partial \theta_1} = \frac{(x_{s2} - x_{s1})\tan(\theta_2) + y_{s1} - y_{s2}}{[(\cos(\theta_1)\tan(\theta_2) - \tan(\theta_1))]^2}$$

$$\frac{\partial x_T}{\partial \theta_2} = \frac{(x_{s1} - x_{s2})\tan(\theta_1) + y_{s2} - y_{s1}}{[(\cos(\theta_2)\tan(\theta_2) - \tan(\theta_1))]^2}$$

The variance of the pseudo measurement noise on y_T can be found in a similar way.

A Monte Carlo simulation was set up where it was assumed that angle measurements were taken 10 times per second. The interceptor and KKV were command guided (first with Lambert guidance and then APN) until 10 s before intercept based on the Kalman filter estimates. The two linear polynomial three-state Kalman filter initial state estimates were set to zero (in other words, "cold starting" the filter) and within 10 s accurate state estimates were obtained using least squares filter techniques [6]. During the last 10 s it was assumed that homing guidance took place and that near-perfect estimates of the target states were available at a 100-Hz rate so that APN guidance could be implemented with range estimates being uplinked from the aircraft to the KKV. The code appears below in Listing 34.4. Setting RUN=1 allows the simulation to run in the single flight mode while setting RUN = 50 allows the simulation to run in the Monte Carlo mode.

LISTING 34.4 TWO-DIMENSIONAL MONTE CARLO ENGAGEMENT SIMULATION USING STEREO TRACKING

```
%Runs very slowly in Monte Carlo mode
count=0;
PHIS=576;
TLAUNCH=80.;
TF=230.;
TS=.1;
XLONGMDEGICKM=400.;
XLONGS2DEGKM=500.;
RDESKM=10000.;
GAMDEG=89.99;
TFTOT=2000.;
TUPT=20.;
TGUID=100.;
XNCLIM=322.;
XNP=3.;
QPERFECT=0;
TLOFT=500.;
ALTMKMIC=15.;
QTAYLOR=1;
PIPERRKM=0.;
QGUID=1;
ITGT=2;
DELTF=0.;
QFIX=0;
SIGTHET1=.00005;
THOM=10.;
BIAS1=0.;
RUN=50;
VMRQDKMIC=4.;
if ITGT==1
        TPZ=180.;
else
        TPZ=240.;
end
SIGTHET2=SIGTHET1;
XLONGS1DEGKM=XLONGMDEGICKM;
ORDER=3;
ALTM=ALTMKMIC*3280.;
```

```
TFTOT=252.+.223*RDESKM-(5.44E-6)*RDESKM*RDESKM;
TFTOT=TFTOT+TLOFT;
for JJ=1:RUN,
        SWITCHM=0;
        SWITCH1=0;
        A=2.0926E7;
        GM=1.4077E16;
        W=0.;
        XLONGFDEG=57.3*RDESKM*3280./A;
        XLONGMDEGIC=XLONGMDEGICKM/111.;
        XLONGMDEG=XLONGMDEGIC;
        XLONGS1DEG=XLONGS1DEGKM/111.;
        XLONGS2DEG=XLONGS2DEGKM/111.;
        QLAUNCH=0;
        QFIRST=1;
        XLONGTDEG=0.;
        QBOOST=1;
        QOOMPH=1;
        T=0.;
        S=0.;
        AXT=0.;
        AYT=0.;
        ATP=0.;
        XLONGF=XLONGFDEG/57.3;
        XLONGF=XLONGF-W*TFTOT;
        PIPERR=0.;
        XF=A*cos(XLONGF);
        YF=A*sin(XLONGF);
        ZF=0;
        XLONGT=XLONGTDEG/57.3;
        XLONGM=XLONGMDEG/57.3;
        XLONGS1=XLONGS1DEG/57.3;
        XLONGS2=XLONGS2DEG/57.3;
        XT=A*cos(XLONGT);
        YT=A*sin(XLONGT);
        ZT=0;
        XTINIT=XT;
        YTINIT=YT;
        ZTINIT=0;
        RTINIT=sqrt(XTINIT^2+YTINIT^2);
        XM=(A+ALTM)*cos(XLONGM);
        YM=(A+ALTM)*sin(XLONGM);
        ZM=0;
        XS1=(A+ALTM)*cos(XLONGS1);
        YS1=(A+ALTM)*sin(XLONGS1);
        XS2=(A+ALTM)*cos(XLONGS2);
```

```
YS2=(A+ALTM)*sin(XLONGS2);
XFIRST=XT;
YFIRST=YT;
ZFIRST=0;
DISTRTKMIC=distance3dkm(XT,YT,ZT,XFIRST,YFIRST,ZFIRST);
XMINIT=XM;
YMINIT=YM;
RMINIT=sqrt(XMINIT^2+YMINIT^2);
XTD=cos(1.5708-GAMDEG/57.3);
YTD=sin(1.5708-GAMDEG/57.3);
ATP=1.;
AXM=0.;
AYM=0.;
AMP=0.;
H=.01;
ALTTKM=(sqrt(XT^2+YT^2)-A)/3280.;
XMD=0.;
YMD=0.;
RTM1=XT-XM;
RTM2=YT-YM;
RTM=sqrt(RTM1^2+RTM2^2);
VTM1=XTD-XMD;
VTM2=YTD-YMD;
VC=-(RTM1*VTM1+RTM2*VTM2)/RTM;
DELV=0.;
ACC=0.;
AXMGUID=0.;
AYMGUID=0.;
PREDERRKM=0.;
ZEM1=0.;
ZEM2=0.;
ALTMKM=(sqrt(XM^2+YM^2)-A)/3280.;
TBOT=0.;
DELVELM=0.;
PIPKMBO=0.;
ZEMPERPTOT=0.;
if QFIX==1
        [XTFACT,YTFACT]=predict34(T,XT,YT,XTD,YTD,TF,TFTOT,TUPT,XF,YF,ITGT);
        ZTFACT=0;
        TGOLAM=TF-TLAUNCH;
        XLONGM=atan2(YM,XM);
        XLONGT=atan2(YTFACT,XTFACT);
        [VRX,VRY,VRZ]=LAMBERT3D(XM,YM,ZM,TGOLAM,XTFACT,
                YTFACT,ZTFACT,SWITCHM);
        VMXRQD=VRX;
        VMYRQD=VRY;
```

```
		VMRQDKM=sqrt(VMXRQD^2+VMYRQD^2)/3280.
	else
		for TF=(TLAUNCH+30.):10:(TPZ-10.),
			[XTFACT,YTFACT]=predict34(T,XT,YT,XTD,YTD,TF,TFTOT,TUPT,
				XF,YF,ITGT);
			ZTFACT=0;
  TGOLAM=TF-TLAUNCH;
			XLONGM=atan2(YM,XM);
			XLONGT=atan2(YTFACT,XTFACT);
			[VRX,VRY,VRZ]=LAMBERT3D(XM,YM,ZM,TGOLAM,XTFACT,
				YTFACT,ZTFACT,SWITCHM);
			VMXRQD=VRX;
			VMYRQD=VRY;
			VMRQDKM=sqrt(VMXRQD^2+VMYRQD^2)/3280.;
			if VMRQDKM<VMRQDKMIC
				break
			end
		end
	end
	TF=TF+DELTF;
	XH=0.;
	XDH=0.;
	XDDH=0.;
	YH=0.;
	YDH=0.;
	YDDH=0.;
	PHI=zeros([3,3]);
	P=zeros([3,3]);
	Q=zeros([3,3]);
	IDNPZ=eye(3);
	P(1,1)=99999999999.;
	P(2,2)=99999999999.;
	P(3,3)=99999999999.;
	PP(1,1)=99999999999.;
	PP(2,2)=99999999999.;
	PP(3,3)=99999999999.;
	PHI(1,1)=1;
	PHI(1,2)=TS;
	PHI(1,3)=.5*TS*TS;
	PHI(2,2)=1;
	PHI(2,3)=TS;
	PHI(3,3)=1;
	HMAT(1,1)=1.;
	HMAT(1,2)=0.;
	HMAT(1,3)=0.;
	PHIT=PHI';
```

```
HT=HMAT';
Q(1,1)=PHIS*TS^5/20;
Q(1,2)=PHIS*TS^4/8;
Q(1,3)=PHIS*TS^3/6;
Q(2,1)=Q(1,2);
Q(2,2)=PHIS*TS^3/3;
Q(2,3)=PHIS*TS*TS/2;
Q(3,1)=Q(1,3);
Q(3,2)=Q(2,3);
Q(3,3)=PHIS*TS;
XN=0.;
while ~((T>(TF-10.)) & VC<0.)
        if RTM<1000
                H=.00001;
        else
                H=.01;
        end
        XTOLD=XT;
        YTOLD=YT;
        XTDOLD=XTD;
        YTDOLD=YTD;
        XMOLD=XM;
        YMOLD=YM;
        XMDOLD=XMD;
        YMDOLD=YMD;
        DELVOLD=DELV;
        STEP=1;
        FLAG=0;
        while STEP <=1
                if FLAG==1
                        STEP=2;
                        XT=XT+H*XTD;
                        YT=YT+H*YTD;
                        XTD=XTD+H*XTDD;
                        YTD=YTD+H*YTDD;
                        XM=XM+H*XMD;
                        YM=YM+H*YMD;
                        XMD=XMD+H*XMDD;
                        YMD=YMD+H*YMDD;
                        DELV=DELV+H*DELVD;
                        T=T+H;
                end
                if ITGT==1
                        if T<180
                                WGT=-212.*T+44000.;
                                TRST=54100.;
```

```
		else
			WGT=3300.;
			TRST=0.;
		end
	else
		if T<120.
			WGT=-2622*T+440660.;
			TRST=725850.;
		elseif T<240.
			WGT=-642.*T+168120.;
			TRST=182250.;
		else
			WGT=5500.;
			TRST=0.;
		end
	end
	ATP=32.2*TRST/WGT;
	TEMPBOTT=(XT^2+YT^2)^1.5;
	XTDD=-GM*XT/TEMPBOTT+AXT;
	YTDD=-GM*YT/TEMPBOTT+AYT;
	RTM1=XT-XM;
	RTM2=YT-YM;
	VTM1=XTD-XMD;
	VTM2=YTD-YMD;
	RTM=sqrt(RTM1^2+RTM2^2);
	VC=-(RTM1*VTM1+RTM2*VTM2)/RTM;
	TGO=RTM/VC;
	ACCDOTRTM=(XTDD*RTM1+YTDD*RTM2)/RTM;
	ACCPER1=XTDD-ACCDOTRTM*RTM1/RTM;
	ACCPER2=YTDD-ACCDOTRTM*RTM2/RTM;
	ACCPERPTOT=sqrt(ACCPER1^2+ACCPER2^2)/32.2;
	if (T>TGUID & TGO<THOM)
		TEMPBOTM=(XM^2+YM^2)^1.5;
		XMDDGRAV=-GM*XM/TEMPBOTM;
		YMDDGRAV=-GM*YM/TEMPBOTM;
		ZEM1=RTM1+VTM1*TGO+.5*(XTDD-XMDDGRAV)*TGO^2;
		ZEM2=RTM2+VTM2*TGO+.5*(YTDD-YMDDGRAV)*TGO^2;
		ZEMDOTRTM=(ZEM1*RTM1+ZEM2*RTM2)/RTM;
		ZEMPER1=ZEM1-ZEMDOTRTM*RTM1/RTM;
		ZEMPER2=ZEM2-ZEMDOTRTM*RTM2/RTM;
		ZEMPERPTOT=sqrt(ZEMPER1^2+ZEMPER2^2)/3280.;
		AXMGUID=XNP*ZEMPER1/(TGO^2);
		AYMGUID=XNP*ZEMPER2/(TGO^2);
		TGO=RTM/VC;
		if QGUID==0
			XNCLIM=0.;
```

```
                end
                if AXMGUID>XNCLIM
                        AXMGUID=XNCLIM;
                elseif AXMGUID<-XNCLIM
                        AXMGUID=-XNCLIM;
                end
                if AYMGUID>XNCLIM
                        AYMGUID=XNCLIM;
                elseif AYMGUID<-XNCLIM
                        AYMGUID=-XNCLIM;
                end
        end
        if T<=TGUID
                AXMGUID=0.;
                AYMGUID=0.;
        end
        if T>TLAUNCH
                TEMPBOTM=(XM^2+YM^2)^1.5;
                XMDD=-GM*XM/TEMPBOTM+AXMGUID;
                YMDD=-GM*YM/TEMPBOTM+AYMGUID;
        else
                XMDD=0.;
                YMDD=0.;
        end
        ACCNEW=sqrt(AXMGUID^2+AYMGUID^2);
        DELVD=ACCNEW;
        ALTMKM=(sqrt(XM^2+YM^2)-A)/3280.;
        FLAG=1;
end
FLAG=0;
XT=.5*(XTOLD+XT+H*XTD);
YT=.5*(YTOLD+YT+H*YTD);
XTD=.5*(XTDOLD+XTD+H*XTDD);
YTD=.5*(YTDOLD+YTD+H*YTDD);
XM=.5*(XMOLD+XM+H*XMD);
YM=.5*(YMOLD+YM+H*YMD);
XMD=.5*(XMDOLD+XMD+H*XMDD);
YMD=.5*(YMDOLD+YMD+H*YMDD);
DELV=.5*(DELVOLD+DELV+H*DELVD);
S=S+H;
if QBOOST==1
        TGOLAM=TFTOT-T;
        XLONGM=atan2(YT,XT);
        XLONGT=atan2(YF,XF);
        [VRX,VRY,VRZ]=LAMBERT3D(XT,YT,ZT,TGOLAM,XF,YF,ZF,SWITCH1);
        VTX=VRX;
```

```
        VTY=VRY;
        DELVXT=VTX-XTD;
        DELVYT=VTY-YTD;
        VELT=sqrt(XTD^2+YTD^2);
        DELVELT=sqrt(DELVXT^2+DELVYT^2);
        if (T<TPZ & DELVELT>500.)
                AXT=ATP*DELVXT/DELVELT;
                AYT=ATP*DELVYT/DELVELT;
        elseif DELVELT<500.
                TRST=0.;
                QBOOST=0;
                AXT=0.;
                AYT=0.;
                XTD=VTX;
                XTDOLD=XTD;
                YTD=VTY;
                YTDOLD=YTD;
                TBOT=T;
        else
                QBOOST=0;
                QOOMPH=0;
                AXT=0.;
                AYT=0.;
                TBOT=T;
        end
end
if T<TUPT
        RTMAG=sqrt(XT^2+YT^2);
        AXT=ATP*XT/RTMAG;
        AYT=ATP*YT/RTMAG;
end
if T>=TLAUNCH
        TGOLAMM=TF-T;
        if QPERFECT==1
                XTF=XTFACT;
                YTF=YTFACT;
        elseif (QPERFECT==0 & QTAYLOR==1)
                TGOM=TF-T;
                XTF=XT+XTD*TGOM+.5*XTDD*TGOM*TGOM;
                YTF=YT+YTD*TGOM+.5*YTDD*TGOM*TGOM;
        end
        ZTF=0;
        QLAUNCH=1;
        PIPERR=sqrt((XTF-XTFACT)^2+(YTF-YTFACT)^2)/3280.;
end
TGOLAMM=TF-T;
```

```
if (T>=TLAUNCH & QFIRST==1)
        QFIRST=0;
        TGOPZ=TF-TLAUNCH;
        [VRX,VRY,VRZ]=LAMBERT3D(XM,YM,ZM,TGOPZ,XTF,YTF,
                ZTF,SWITCHM);
        VMXRQD=VRX;
        VMYRQD=VRY;
        VMRQDKM=sqrt(VMXRQD^2+VMYRQD^2)/3280.;
        XMD=VMXRQD;
        XMDOLD=XMD;
        YMD=VMYRQD;
        YMDOLD=YMD;
end
if S>=(TS-.0001)
        S=0.;
        THET1=atan2(YS1-YT,XS1-XT);
        THET2=atan2(YS2-YT,XS2-XT);
        THET1NOISE=SIGTHET1*randn;;
        THET2NOISE=SIGTHET2*randn;;
        THET1S=THET1+THET1NOISE+BIAS1;
        THET2S=THET2+THET2NOISE;
        TOP1=XS2*tan(THET2S)-XS1*tan(THET1S)+YS1-YS2;
        XTS=TOP1/(tan(THET2S)-tan(THET1S));
        TOP2=XS2*tan(THET2S)*tan(THET1S)-XS1*tan(THET1S)*
                tan(THET1S)...
        +tan(THET1S)*(YS1-YS2);
        YTS=YS1-XS1*tan(THET1S)+TOP2/(tan(THET2S)-tan(THET1S));
        XTNOISE=XT-XTS;
        YTNOISE=YT-YTS;
        DXDT1=(tan(THET2)*(XS2-XS1)+YS1-YS2)/((cos(THET1)*
                (tan(THET2)...
                        -tan(THET1)))^2);
        DXDT2=(tan(THET1)*(XS1-XS2)+YS2-YS1)/((cos(THET2)*
                (tan(THET2)-...
        tan(THET1)))^2);
        SIGX=sqrt((DXDT1*SIGTHET1)^2+(DXDT2*SIGTHET2)^2);
        DYDT1=-XS1/(cos(THET1)*cos(THET1));
        DYDT1=DYDT1+(XS2*tan(THET2)*tan(THET2)-2.*XS1*
                tan(THET1)*...
                        tan(THET2)+(YS1-YS2)*tan(THET2)+XS1*
                            tan(THET1)...
                        *tan(THET1))/((cos(THET1)*(tan(THET2)...
                            -tan(THET1)))^2);
        DYDT2=(tan(THET1)*tan(THET1)*(XS1-XS2)-(YS1-YS2)*...
        tan(THET1))/((cos(THET2)*(tan(THET2)-tan(THET1)))^2);
        SIGY=sqrt((DYDT1*SIGTHET1)^2+(DYDT2*SIGTHET2)^2);
```

```
XN=XN+1.;
XK1=3*(3*XN*XN-3*XN+2)/(XN*(XN+1)*(XN+2));
XK2=18*(2*XN-1)/(XN*(XN+1)*(XN+2)*TS);
XK3=60/(XN*(XN+1)*(XN+2)*TS*TS);
RMAT(1,1)=SIGX^2;
PHIP=PHI*P;
PHIPPHIT=PHIP*PHIT;
M=PHIPPHIT+Q;
HM=HMAT*M;
HMHT=HM*HT;
HMHTR=HMHT+RMAT;
HMHTRINV(1,1)=1./HMHTR(1,1);
MHT=M*HT;
K=MHT*HMHTRINV;
KH=K*HMAT;
IKH=IDNPZ-KH;
P=IKH*M;
if XN<10.
        XK1PZ=XK1;
        XK2PZ=XK2;
        XK3PZ=XK3;
else
        XK1PZ=K(1,1);
        XK2PZ=K(2,1);
        XK3PZ=K(3,1);
end
RES=XTS-XH-TS*XDH-.5*TS*TS*XDDH;
XH=XH+XDH*TS+.5*TS*TS*XDDH+XK1PZ*RES;
XDH=XDH+XDDH*TS+XK2PZ*RES;
XDDH=XDDH+XK3PZ*RES;
RMATP(1,1)=SIGY^2;
PHIPP=PHI*PP;
PHIPPHITP=PHIPP*PHIT;
MP=PHIPPHITP+Q;
HMP=HMAT*MP;
HMHTP=HMP*HT;
HMHTRP=HMHTP+RMATP;
HMHTRINVP(1,1)=1./HMHTRP(1,1);
MHTP=MP*HT;
KP=MHTP*HMHTRINVP;
KHP=KP*HMAT;
IKHP=IDNPZ-KHP;
PP=IKHP*MP;
if XN<10.
        XK1PZP=XK1;
        XK2PZP=XK2;
```

```
        XK3PZP=XK3;
else
        XK1PZP=KP(1,1);
        XK2PZP=KP(2,1);
        XK3PZP=KP(3,1);
end
RESP=YTS-YH-TS*YDH-.5*TS*TS*YDDH;
YH=YH+YDH*TS+.5*TS*TS*YDDH+XK1PZP*RESP;
YDH=YDH+YDDH*TS+XK2PZP*RESP;
YDDH=YDDH+XK3PZP*RESP;
if (T>TGUID & TGO>THOM)
        RTM1H=XH-XM;
        RTM2H=YH-YM;
        RTMH=sqrt(RTM1H^2+RTM2H^2);
        VTM1H=XDH-XMD;
        VTM2H=YDH-YMD;
        VCH=-(RTM1H*VTM1H+RTM2H*VTM2H)/RTMH;
        TGOH=RTMH/VCH;
        TEMPBOTM=(XM^2+YM^2)^1.5;
        XMDDGRAV=-GM*XM/TEMPBOTM;
        YMDDGRAV=-GM*YM/TEMPBOTM;
        ZEM1H=RTM1H+VTM1H*TGOH+.5*(XDDH-XMDDGRAV)*
                TGOH^2;
        ZEM2H=RTM2H+VTM2H*TGOH+.5*(YDDH-YMDDGRAV)*
                TGOH^2;
        ZEMDOTRTMH=(ZEM1H*RTM1H+ZEM2H*RTM2H)/RTMH;
        ZEMPER1H=ZEM1H-ZEMDOTRTMH*RTM1H/RTMH;
        ZEMPER2H=ZEM2H-ZEMDOTRTMH*RTM2H/RTMH;
        AXMGUID=XNP*ZEMPER1H/(TGOH^2);
        AYMGUID=XNP*ZEMPER2H/(TGOH^2);
        if QGUID==0
                XNCLIM=0.;
        end
        if AXMGUID>XNCLIM
                AXMGUID=XNCLIM;
        elseif AXMGUID<-XNCLIM
                AXMGUID=-XNCLIM;
        end
        if AYMGUID>XNCLIM
                AYMGUID=XNCLIM;
        elseif AYMGUID<-XNCLIM
                AYMGUID=-XNCLIM;
        end
end
if RUN==1
```

```
                    end
               end
               if RUN==1
                    ERRXTDD=(XTDD-XDDH)/32.2;
                    SP33X=sqrt(P(3,3))/32.2;
                    SP33XP=-SP33X;
                    ERRYTDD=(YTDD-YDDH)/32.2;
                    SP33Y=sqrt(PP(3,3))/32.2;
                    SP33YP=-SP33Y;
                    count=count+1;
                    ArrayT(count)=T;
                    ArrayERRXTDD(count)=ERRXTDD;
                    ArraySP33X(count)=SP33X;
                    ArraySP33XP(count)=SP33XP;
               end
          end
          count=count+1
          ArrayJJ(count)=count;
          ArrayRTM(count)=RTM;
          ArrayDELV(count)=DELV/3.28;
end
if RUN==1
   figure
   plot(ArrayT,ArrayERRXTDD,ArrayT,ArraySP33X,ArrayT,ArraySP33XP),grid
   xlabel('Time (s)')
   ylabel('Acceleration Error (f/s) ')
   axis([0 240 -10 10])
   clc
   output=[ArrayT',ArrayERRXTDD',ArraySP33X',ArraySP33XP'];
   save datfil.txt output /ascii
else
   figure
   plot(ArrayJJ,ArrayRTM),grid
   xlabel('Run')
   ylabel('Miss (ft) ')
   figure
   plot(ArrayJJ,ArrayDELV),grid
   xlabel('Run')
   ylabel('Divert (m/s) ')
   clc
   output=[ArrayJJ',ArrayRTM',ArrayDELV'];
   %save ('datfil.txt', 'output', '-ascii')
        save datfil.txt output /ascii
   RTMSORT=sort(ArrayRTM)
   DELVSORT=sort(ArrayDELV)
   RTM90=RTMSORT(45)
```

```
    DELV90=DELVSORT(45)
end
disp 'simulation finished'

% LAMBERT3D can be found in Listing 28.3
% distance3dkm can be found in Listing 28.3
% predict34 can be found in Listing 34.3
```

A case was run against the ICBM target in which the shooter aircraft (with the first sensor) was 800-km downrange of the target launch point and the second sensor was 900-km downrange of the target launch point. Both sensors were 15 km in altitude and the intercept time was set at 230 s for the case where there were no PIP errors. In addition it was assumed that the angular accuracy of both sensors was 50 μr and that the filter sampling time was 0.1 s. Figures 34.21 and 34.22 show how the linear polynomial Kalman filter is able to estimate the target acceleration for different values of filter process noise (Φ_s). Figure 34.22 shows that the filter with less process noise yields a much smoother estimate of target acceleration. However, by comparing the acceleration estimates to that of Fig. 34.21, one can see that the price paid for the smoother estimate is that the estimated target acceleration lags the actual target acceleration.

Frequently, in investigating Kalman filter performance, it is popular to look at the error in the estimates as was shown in Chapter 9. Figure 34.23 shows how the single run error in the acceleration estimate for the filter with larger process noise ($\Phi_s = 576$) compares to the theoretical predictions provided by the Riccati equations. It can be seen that the single run results fall within the theoretical bounds, which indicates that the filter is consistent. On the other hand, Fig. 34.24 shows that when the process noise is reduced ($\Phi_s = 9$) the filter is no

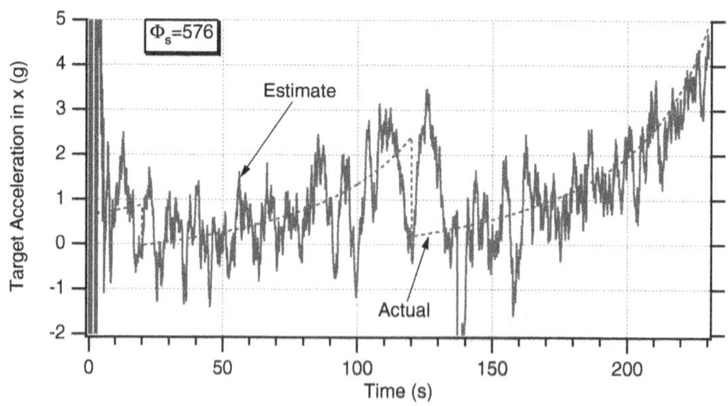

Fig. 34.21 Large Kalman filter process noise yields noisy estimates of target acceleration.

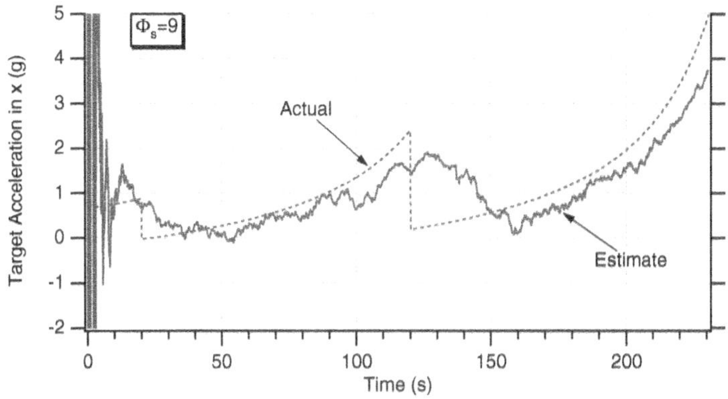

Fig. 34.22 Reducing Kalman filter process provides smoother estimates of target acceleration.

longer consistent. However, one can also see that the errors in the acceleration estimate of the filter with less process noise is smaller than when the process noise is increased. We shall soon see from a performance point of view which filter is better for this particular boost-phase intercept problem.

INTERCEPTOR ENGAGEMENTS WITH NOISE AND FILTERING

Fifty run Monte Carlo sets were run for both the ICBM and IRBM targets to see how much the inclusion of noise and filtering increased the KKV lateral divert requirements. The 90% point (that is, 90% of the flights had required divert

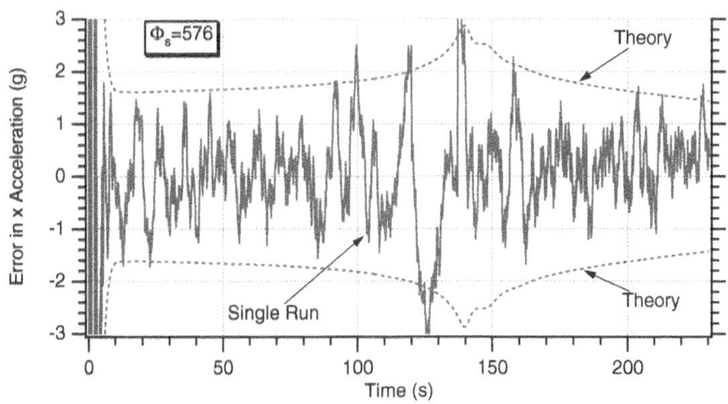

Fig. 34.23 Filter is consistent when process noise is high but estimation errors are large.

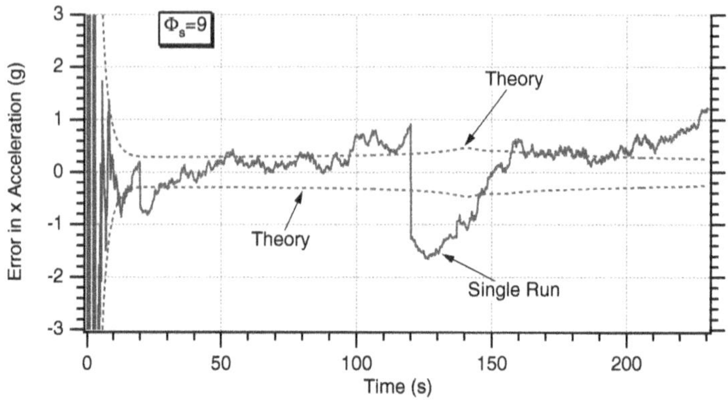

Fig. 34.24 Filter is not consistent when process noise is low but estimation errors are much lower.

that was less than the amount shown in the following figures) in a Monte Carlo set was used as the figure of merit. Figure 34.25 shows that the KKV divert requirements still increase with the initial distance from the interceptor launch point to the target launch site increasing. One can also see that sensor noise may increase the KKV lateral divert requirements substantially over the case where the state estimates are perfect. Figure 34.25 also shows that the selection of the amount of process noise that is used in the filter is important. *Lower amounts of process noise may reduce the interceptor divert requirements—even though the filter is not consistent!* However, even with a small value of process noise the divert requirements can still increase more than 500 m/s over the case without noise

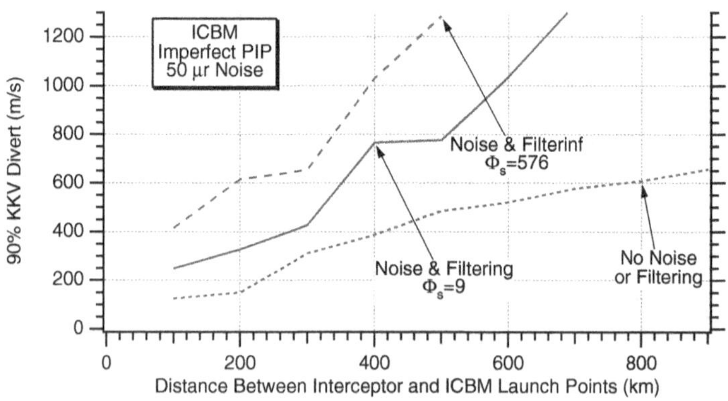

Fig. 34.25 Choice of process noise in Kalman filter can be very important in setting KKV divert requirements.

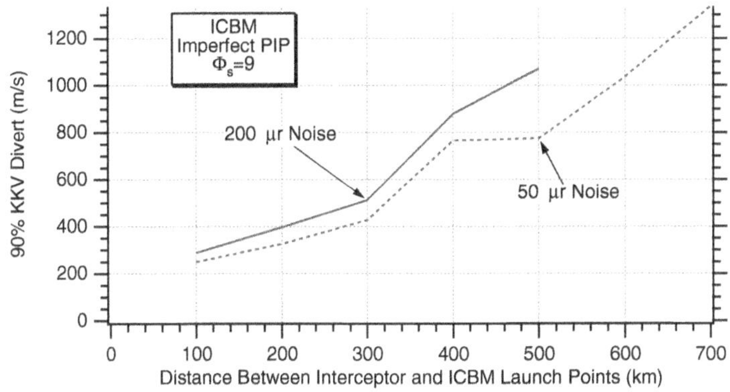

Fig. 34.26 KKV divert requirements increase with increasing measurement noise for the ICBM engagement.

or filtering. In addition, the inclusion of noise and filtering limits the maximum distance that the interceptor launch point can be from the target launch site. Thus, the inclusion of sensor noise and filtering is a very important part of the design process in setting KKV divert requirements. Subsequent iterations in the design and sizing process must include alternative filtering approaches such as the EKF to see if KKV divert requirements can be reduced.

The amount of sensor measurement noise is also very important. Figure 34.26 shows that if the sensor noise increases from its nominal value of 50 μr to 200 μr the lateral divert requirements can sometimes increase by more than 200 m/s at the longer ranges. More importantly, the maximum distance the missile launch point can be from the target launch point is reduced when the measurement noise is increased. Thus, one can see that the guidance and control engineer must interact with the sensor designer to establish reasonable sensor requirements to achieve a balanced system design.

The IRBM engagement results were also repeated for the case in which noise and filtering were considered. One can see from Fig. 34.27 that including these realistic effects also increase the KKV lateral divert requirements. In practice, a wide variety of threats and possible trajectories would have to be considered in deriving KKV lateral divert requirements.

It has been mentioned that there are many steps in the iterative design process. In later steps other sources of error must be considered to highlight potential problems and to suggest design work that must be performed. For example, in the analysis conducted so far we have not considered measurement angle bias errors. Let us repeat the results of Fig. 34.25, where there is 50 μr of measurement noise and no bias errors, to now include a 100 μr angle bias error on the first IRST sensor. Figure 34.28 shows that the bias error increases the KKV lateral divert requirements. Methods for improving the KKV divert performance have to be

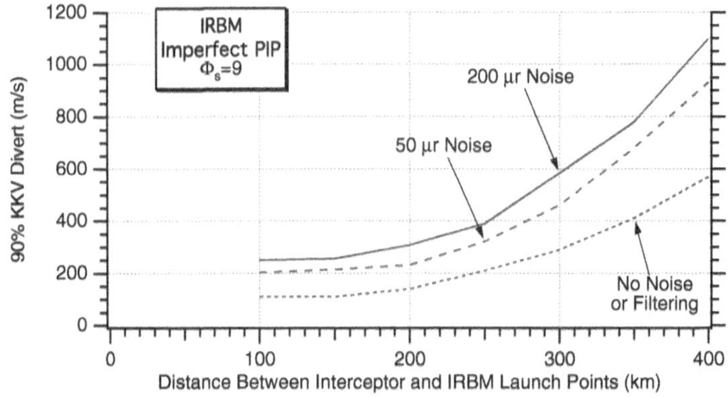

Fig. 34.27 KKV divert requirements also increase with increasing measurement noise for the IRBM engagement.

explored. Thus alternative filter structures must also be examined in subsequent iterations of the design process to see if the influence of biases can be alleviated. Included for study in later iterations of the design process should be such traditional approaches such as redesigning the Kalman filter, increasing the homing time to avoid biases, placing more stringent design requirements on the IRST sensor, or including a star tracker on the aircraft. If no acceptable solutions to the bias error can be found, we can see that the effective range of the interceptor will be diminished. It is important to point out that problems are a normal part of the design process and that the search for solutions to these problems often leads to engineering innovation if the right people are involved.

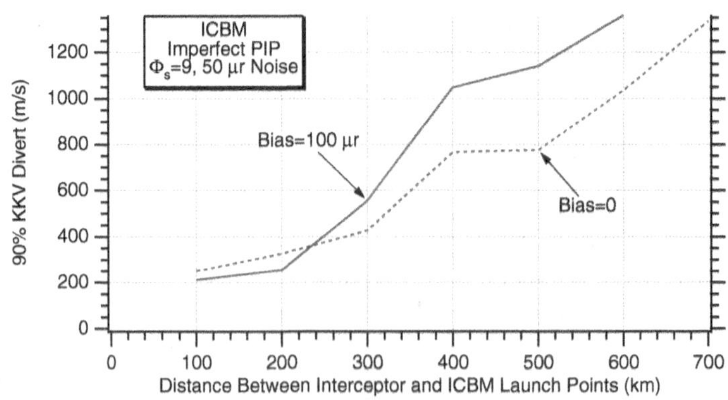

Fig. 34.28 Angle bias errors can increase KKV divert requirements or reduce maximum effective range of interceptor.

SUMMARY

This chapter illustrates, through simplified examples, how airborne interceptor guidance and filtering technology are important in determining KKV lateral divert requirements against both boosting IRBM and ICBM threats. It was shown that with 50 μr of aircraft IRST sensor noise and 2 km/s of KKV, divert boost-phase intercepts of both IRBM and ICBM targets could be achieved if the interceptor launching aircraft could be within 300 km to 800 km of the target launch site. Traditional methods of guidance and filtering were employed to achieve these results. These methods were used to illustrate how sensor noise, prediction error, and apparent target maneuver work together in setting the KKV's lateral divert requirements. Simple formulas have been developed that can be used to understand and explain engagement simulation results. Alternative guidance and filtering techniques have to be explored in subsequent iterations of the design process to see if the KKV divert and acceleration requirements can be reduced.

REFERENCES

[1] Zarchan, P., "Kill Vehicle Guidance and Control Sizing for Boost-Phase Intercept," *Journal of Guidance, Control and Dynamics*, March-April 2010.

[2] Kleppner, D., and Lamb, F. K. (eds.), *Boost-Phase Intercept Systems For National Defense*, American Physical Society, July 2003.

[3] Wilkening, D. A., "Airborne Boost-Phase Ballistic Missile Defense," *Science and Global Security*, June 2004, pp. 1-67.

[4] Corbett, M., and Zarchan, P., "The Role of Air Power in Active Missile Defense," *U.S. Air and Space Power Journal*, Summer 2010, pp. 57-71.

[5] Ehrhard, T. P., and Work, R. O., "Range, Persistence, Stealth, and Networking: The Case For a Carrier-Based Unmanned Combat Air System," Center for Strategic and Budgetary Assessments, 2008.

[6] Zarchan, P., *Fundamentals of Kalman Filtering A Practical Approach*, 3rd ed, Progress in Astronautics and Aeronautics, AIAA, Reston, VA 2009, pp. 91-128.

Additional Examples

INTRODUCTION

In this appendix additional examples will be presented showing the interested reader how to use the source code listings found in the supporting materials for this book. The listings are often slightly modified so that the reader will understand examples of specific changes that can be made in order to explore issues beyond the scope of the text.

SOFTWARE DETAILS

To facilitate learning, source code that is formatted for both IBM and Macintosh computers containing all of the text's MATLAB listings are included on the AIAA website. The equivalent FORTRAN source code listings can also be found on this website.

The MATLAB and FORTRAN source code should run, as is, with either MATLAB R2010b software or Version 11.05 of the Absoft Macintosh Pro Fortran compiler. The software has been tested on a MacPro Intel Mac Computer. Use of different computers or compilers in either the Macintosh- or IBM-compatible world may require some slight modification of the source code.

The naming conventions of the source code files for both languages are slightly different. The MATLAB naming convention is CxLy.M where x corresponds to Chapter number and y corresponds to listing number. In other words C4L2.M corresponds to MATLAB Listing 4.2 of the text (that is, Chapter 4, Listing 2). The FORTRAN naming convention is C4L2.F.

SENSITIVITY OF OPTIMAL GUIDANCE TO TIME TO GO ERRORS

In evaluating the performance of the optimal guidance law, we have seen tremendous performance benefits over proportional navigation in the presence of guidance system dynamics. It has been assumed that the time to go information, required by optimal guidance, was perfect. Adjoint Listing 8.2 was modified to

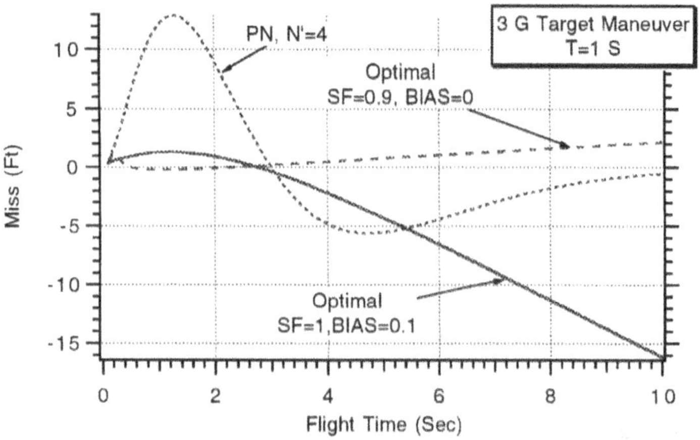

Fig. A.1 Time to go must be known accurately for optimal guidance to yield performance benefits.

include scale factor SF and bias BIAS errors on the estimated time to go TGOH when the optimal guidance option was used (APN = 2). An error-free time to go case would require SF = 1 and BIAS = 0. Time to go errors are not introduced into the proportional navigation option (APN = 0) because in practice this guidance law does not require time to go but works directly on the line-of-sight rate. Statements that have been modified from the original Listing 8.2 are highlighted in boldface in Listing A.1.

Cases were rerun for proportional navigation (APN = 0), optimal guidance with a time to go scale factor error of 0.9 (APN = 2, SF = 0.9, BIAS = 0), and optimal guidance with a time to go bias error of 0.1 s (APN = 2, SF = 1, BIAS = 0.1). The miss distance sensitivity to a 3-*g* target maneuver when the guidance system time constant is 1 s is displayed in Fig. A.1. We can see that both scale factor and bias errors degrade the optimal guidance performance. Figure A.1 shows that the performance of optimal guidance with a 0.9 scale factor error is worse than that of proportional navigation for flight times greater than 8 s and with a 0.1 s bias error is worse than that of proportional navigation for flight times greater than 5.5 s. [1]. Therefore time to go must be known accurately in order for optimal guidance to perform better than proportional navigation.

LISTING A.1 ADJOINT SIMULATION OF OPTIMAL GUIDANCE SYSTEM (MODIFIED LISTING 8.2)

```
XNT=96.6;
XNP=4.;
TAU=1.;
TF=10.;
```

```
VM=3000.;
HEDEG=-20.;
APN=2;
BIAS=.1;
SF=1;
T=0.;
S=0.;
TP=T+.00001;
X1=0.;
X2=0.;
X3=1.;
X4=0.;
XNPP=0.;
H=.01;
HE=HEDEG/57.3;
n=0.;
while TP<=(TF-1e-5)
    X1OLD=X1;
    X2OLD=X2;
    X3OLD=X3;
    X4OLD=X4;
    STEP=1;
    FLAG=0;
    while STEP<=1
        if FLAG==1
        STEP=2;
        X1=X1+H*X1D;
        X2=X2+H*X2D;
        X3=X3+H*X3D;
        X4=X4+H*X4D;
        TP=TP+H;
        end
        TGO=TP+.00001;
        if APN==0
        C1=XNP/(TGO*TGO);
        C2=XNP/TGO;
        C3=0.;
        C4=0.;
        elseif APN==1
        C1=XNP/(TGO*TGO);
        C2=XNP/TGO;
        C3=.5*XNP;
        C4=0.;
        else
                TGOH=SF*TGO+BIAS;
                if TGOH<0
```

```
                        TGOH=.0001;
                    end
            X=TGOH/TAU;
            TOP=6.*X*X*(exp(-X)-1.+X);
            BOT1=2*X*X*X+3.+6.*X-6.*X*X;
            BOT2=-12.*X*exp(-X)-3.*exp(-2.*X);
            XNPP=TOP/(.0001+BOT1+BOT2);
            C1=XNPP/(TGOH*TGOH);
            C2=XNPP/TGOH;
            C3=.5*XNPP;
            C4=-XNPP*(exp(-X)+X-1.)/(X*X);
            end
            X1D=X2+C3*X4/TAU;
            X2D=X3+C2*X4/TAU;
            X3D=C1*X4/TAU;
            X4D=-X4/TAU-X2+C4*X4/TAU;
            FLAG=1;
        end
        FLAG=0;
        X1=(X1OLD+X1)/2+.5*H*X1D;
        X2=(X2OLD+X2)/2+.5*H*X2D;
        X3=(X3OLD+X3)/2+.5*H*X3D;
        X4=(X4OLD+X4)/2+.5*H*X4D;
        S=S+H;
        if S>=.0999
            S=0.;
            n=n+1;
            ArrayTP(n)=TP;
            ArrayXMNT(n)=XNT*X1;
            ArrayXMHE(n)=-VM*HE*X2;
        end
end
figure
plot(ArrayTP,ArrayXMNT),grid
xlabel('Flight Time (Sec)')
ylabel('Target Maneuver Miss (Ft)')
clc
output=[ArrayTP',ArrayXMNT',ArrayXMHE'];
save datfil.txt output /ascii
disp 'simulation finished'
```

SIMULATING AN IMPULSE

So far, when the adjoint simulation technique has been used the impulse required for the implementation of the adjoint method has been simulated by finding the

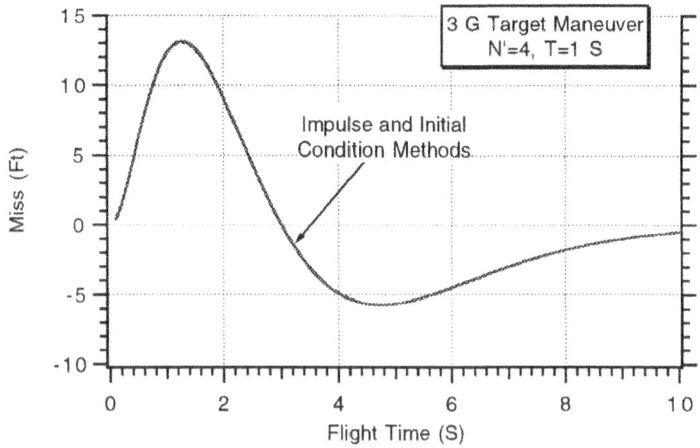

Fig. A.2 Both impulse and initial condition methods are equivalent for target maneuver disturbance.

appropriate initial conditions on the various integrators in the system. In some applications, in which the adjoint method is being automated, it may be advantageous to actually simulate the impulse rather than trying to develop the logic in finding the appropriate initial conditions [2]. Listing 3.1 has been modified to show how the impulse can be simulated. Changes to the original source code have been highlighted in boldface. We can see that before the STEP = 1 statement the height of the impulse is chosen so that the simulated impulse has unit area. The width of the impulse is half an integration interval since the differential equations are called twice with the second-order Runge–Kutta integration technique. Before the STEP = 1 statement we can see that the impulse is added only once to the derivative of $x3$. When we come back to this section of code at other times the value of the simulated impulse will be zero.

We can see from Figs. A.2 and A.3 that the miss due to target maneuver and heading error is identical whether we are using Listing 3.1 where initial conditions are used for the impulse or Listing A.2 where the impulse is actually simulated. However it is important to note that when the impulse is simulated the answers are more sensitive to the integration step size than when the initial condition method is used. This means that the answers in Listing A.2 will start to diverge from the true answers sooner than the answers from Listing 3.1 if the integration interval is made larger.

LISTING A.2 SIMULATING AN IMPULSE RATHER THAN USING INITIAL CONDITIONS FOR USE IN THE ADJOINT METHOD (EQUIVALENT TO LISTING 3.1)

```
n=0.;
XNP=4;
```

```
XNT=96.6;
TAU=1;
TF=10;
VM=3000;
HEDEG=-20;
T=0.;
S=0.;
TP=T+.00001;
X1=0;
X2=0;
X3=0;
X4=0;
H=.01;
HE=HEDEG/57.3;
while TP<=(TF-1e-5)
        S=S+H;
        X1OLD=X1;
        X2OLD=X2;
        X3OLD=X3;
        X4OLD=X4;
        if TP<H/2
                IMPULSE=2./H;
        else
                IMPULSE=0;
        end
        STEP=1;
        FLAG=0;
        while STEP<=1
                if FLAG==1
                        STEP=2;
                        X1=X1+H*X1D;
                        X2=X2+H*X2D;
                        X3=X3+H*X3D;
                        X4=X4+H*X4D;
                        TP=TP+H;
                end
                X1D=X2;
                X2D=X3;
                Y1=(X4-X2)/TAU;
                TGO=TP+.00001;
                if STEP==2
                        IMPULSE=0;
                end
                X3D=XNP*Y1/TGO+IMPULSE;
                X4D=-Y1;
                FLAG=1;
```

```
        end
        FLAG=0;
        X1=(X1OLD+X1)/2+.5*H*X1D;
        X2=(X2OLD+X2)/2+.5*H*X2D;
        X3=(X3OLD+X3)/2+.5*H*X3D;
        X4=(X4OLD+X4)/2+.5*H*X4D;
        if S>=.0999
                S=0.;
n=n+1;
ArrayTP(n)=TP;
                ArrayXMNT(n)=XNT*X1;
                ArrayXMHE(n)=-VM*HE*X2;
        end
end
figure
plot(ArrayTP,ArrayXMNT),grid
xlabel('Flight Time (Sec)')
ylabel('Target Maneuver Miss (Ft)')
figure
plot(ArrayTP,ArrayXMHE),grid
xlabel('Flight Time (Sec)')
ylabel('Heading Error Miss (Ft)')
clc
output=[ArrayTP',ArrayXMNT',ArrayXMHE'];
save datfil.txt output /ascii
disp 'simulation finished'
```

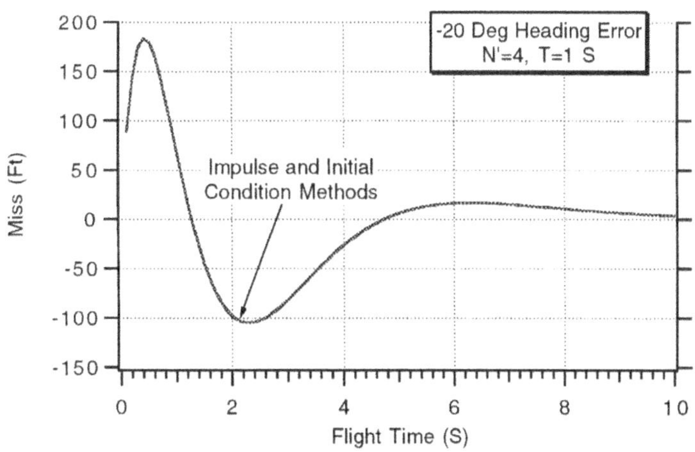

Fig. A.3 Both impulse and initial condition methods are equivalent for heading error disturbance.

DIFFERENT GUIDANCE SYSTEM DISTRIBUTIONS

So far in the text we have chosen the binomial distribution as our canonic guidance system form (that is, all equal time constants). The canonic guidance system form was chosen for simplicity since only the total guidance system time constant had to be specified. It was never demonstrated how the binomial guidance system compared to other canonic guidance system forms in terms of the resultant performance projections.

In this section we will consider two additional guidance system forms. The only feature that the various guidance systems will have in common with the binomial guidance system is that the total guidance system time constant will be the same. The fifth-order binomial guidance system adjoint block diagram of Fig. 6.5 was modified so that other guidance system configurations could be studied by input changes and the resultant adjoint block diagram appears in Fig. A.4. We can see that if $QD = 0$ and all the time constants are the same (i.e., $T_1 = T_2 = T_3 = T_4 = T_5$) we have the fifth-order binomial guidance system that we have already studied. However, now we have the flexibility to make the individual time constants different from one another. If $QD = 1$ in Fig. A.4, we have three real poles and a quadratic distribution for the guidance system. Again the real poles can all have different values. Note that the only disturbance in the guidance system of Fig. A.4 is a step target maneuver.

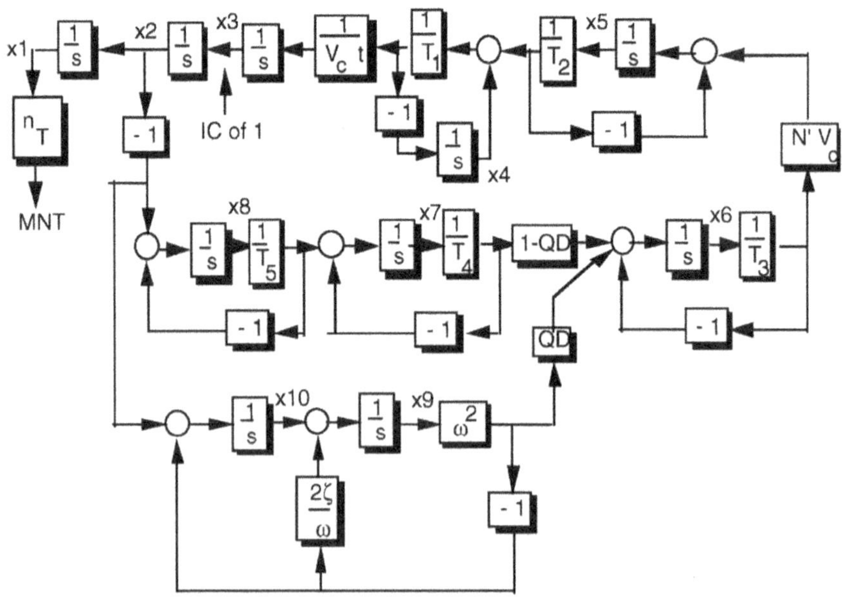

Fig. A.4 Canonic guidance system adjoint block diagram.

The adjoint simulation of Listing 6.1 was modified to correspond to Fig. A.4, and the resultant simulation appears in Listing A.3. The statements that have changed from Listing 6.1 have been highlighted in boldface. We can see from the listing that the only disturbance to the guidance system is a 1-*g* target maneuver.

LISTING A.3 CANONIC GUIDANCE SYSTEM ADJOINT SIMULATION

```
n=0;
QD=0.;
XNT=32.2;
XNP=4.;
T1=.0667;
T2=.133;
T3=.2;
T4=.267;
T5=.333;
W=10;
Z=.7;
TF=10;
VC=4000;
T=0.;
S=0;
TP=T+.00001;
X1=0;
X2=0;
X3=1;
X4=0;
X5=0;
X6=0;
X7=0;
X8=0;
X9=0;
X10=0;
H=.01;
while TP<=(TF-1e-5)
        S=S+H;
        X1OLD=X1;
        X2OLD=X2;
        X3OLD=X3;
        X4OLD=X4;
        X5OLD=X5;
        X6OLD=X6;
        X7OLD=X7;
        X8OLD=X8;
```

```
X9OLD=X9;
X10OLD=X10;
STEP=1;
FLAG=0;
while STEP<=1
        if FLAG==1
        STEP=2;
                        X1=X1+H*X1D;
                        X2=X2+H*X2D;
                        X3=X3+H*X3D;
                        X4=X4+H*X4D;
                        X5=X5+H*X5D;
                        X6=X6+H*X6D;
                        X7=X7+H*X7D;
                        X8=X8+H*X8D;
                        X9=X9+H*X9D;
                        X10=X10+H*X10D;
                        TP=TP+H;
                end
                X1D=X2;
                X2D=X3;
                TGO=TP+.00001;
                X3D=(X4+X5/T2)/(VC*TGO*T1);
                X4D=-(X4+X5/T2)/T1;
                X5D=-X5/T2+XNP*VC*X6/T3;
                X6D=-X6/T3+QD*W*W*X9+(1.-QD)*X7/T4;
                X7D=-X7/T4+X8/T5;
                X8D=-X8/T5-X2;
                X9D=X10-2.*Z*W*X9;
                X10D=-W*W*X9-X2;
                FLAG=1;
end
FLAG=0;
X1=(X1OLD+X1)/2+.5*H*X1D;
X2=(X2OLD+X2)/2+.5*H*X2D;
X3=(X3OLD+X3)/2+.5*H*X3D;
X4=(X4OLD+X4)/2+.5*H*X4D;
X5=(X5OLD+X5)/2+.5*H*X5D;
X6=(X6OLD+X6)/2+.5*H*X6D;
X7=(X7OLD+X7)/2+.5*H*X7D;
X8=(X8OLD+X8)/2+.5*H*X8D;
X9=(X9OLD+X9)/2+.5*H*X9D;
X10=(X10OLD+X10)/2+.5*H*X10D;
if S>=.0999
        S=0;
        XMNT=XNT*X1;
```

```
            n=n+1;
        ArrayTP(n)=TP;
        ArrayXMNT(n)=XNT*X1;
        end
end
figure
plot(ArrayTP,ArrayXMNT),grid
xlabel('Flight Time (Sec)')
ylabel('Target Maneuver Miss (Ft)')
clc
output=[ArrayTP',ArrayXMNT'];
save datfil.txt output /ascii

disp 'simulation finished'
```

For reference the first guidance system studied is the fifth-order binomial guidance system in which all the individual time constants are 0.2 s or

$$\frac{n_L}{\dot{\lambda}} = \frac{N'V_c}{(1+0.2\,\mathrm{s})(1+0.2\,\mathrm{s})(1+0.2\,\mathrm{s})(1+0.2\,\mathrm{s})(1+0.2\,\mathrm{s})}$$

We can see that the total time constant of the preceding expression is 1 s since

$$T = 0.2 + 0.2 + 0.2 + 0.2 + 0.2 = 1\,\mathrm{s}$$

The next guidance system under consideration has five unequal time constants where the second time constant is twice as big as the first, the third time constant is three times as big as the first, the fourth time constant is four times bigger than the first, and the fifth time constant is five times bigger than the first or

$$\frac{n_L}{\dot{\lambda}} = \frac{N'V_c}{(1+0.0667\,\mathrm{s})(1+0.133\,\mathrm{s})(1+0.2\,\mathrm{s})(1+0.267\,\mathrm{s})(1+0.333\,\mathrm{s})}$$

Again we can see that the total time constant of the preceding expression is also 1 s since

$$T = 0.0667 + 0.133 + 0.2 + 0.267 + 0.333 = 1\,\mathrm{s}$$

The last fifth-order guidance system configuration studied is the one with three real poles and a quadratic distribution or

$$\frac{n_L}{\dot{\lambda}} = \frac{N'V_c}{(1+0.1\,\mathrm{s})(1+0.2\,\mathrm{s})(1+0.56\,\mathrm{s})[1+(2*0.7/10)\mathrm{s}+(\mathrm{s}^2/10^2)]}$$

Again we can see that the total time constant of the preceding expression is 1 s since

$$T = 0.1 + 0.2 + 0.56 + 2*0.7/10 = 1\,\mathrm{s}$$

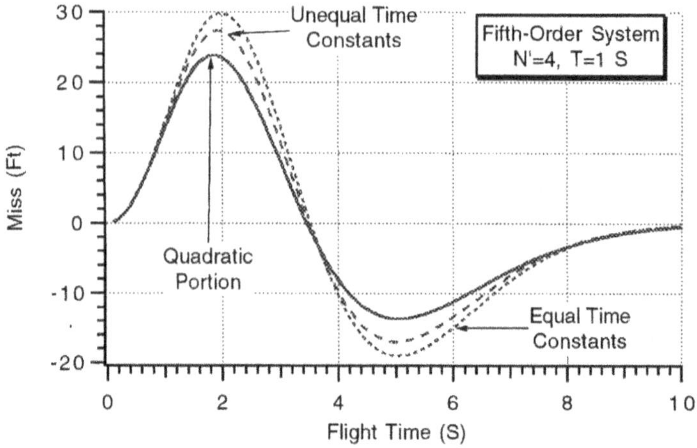

Fig. A.5 All fifth-order guidance system configurations yield approximately same answers if total guidance system time constant is same.

The adjoint simulation results appear in Fig A.5 where the miss distance due to a 1-g step target maneuver disturbance is presented as a function of flight time We can see that all of the miss distance curves are close to one another, indicating that the performance of all of the guidance system configurations considered are approximately the same since the total guidance system time constant of each guidance systems is 1 s. This means that the value of the total guidance system time constant is far more important than the guidance system pole distribution.

SAMPLING EXPERIMENTS

In Chapters 7 and 9 we conducted simplified data rate studies with both the digital fading memory and Kalman noise filters. We concluded that the miss distance due to noise and target maneuver tended to decrease as the data rate increased (that is, sampling time decreased). For simplicity, in the data rate studies, the measurement noise standard deviation was held constant as the data rate changed. In many systems, when one gets into the details of the signal processing, it becomes readily apparent that the data rate and measurement noise standard deviation are *not* independent. In these systems the measurement noise spectral density Φ remains constant, which means that the standard deviation of the simulated digital measurement noise is proportional to the square root of the data rate (in other words, inversely proportional to the square root of the sampling time T_s) or

$$\sigma = \sqrt{\frac{\Phi}{T_s}}$$

Therefore if we double the data rate (that is, sampling time halved), we must also increase the standard deviation of the simulated digital measurement noise by 41.4% ($2^{.5} = 1.414$). In this section we shall repeat the experiments of Chapters 7 and 9 to see if the miss due to digital measurement noise still decreases with increasing data rate when the measurement noise spectral density is held constant.

The miss distance adjoint program of Listing 7.3 represents the adjoint of a digital two-state fading memory filter in the homing loop. The program was modified so that a change in the data rate would cause a change in the standard deviation of the measurement noise according to the preceding relationship under the constant spectral density assumption. It was assumed that a sampling time of 0.1 s (10 Hz data rate) corresponded to 1 mr of measurement noise. Adjoint runs were made in which the sampling time was considered a parameter. Figure A.6 shows that the standard deviation of the miss distance due to digital measurement noise still decreases with increasing data rate, although not as dramatically as was the case in Fig. 7.22.

The Monte Carlo simulation used to generate Fig. 9.12 was also modified so that the equivalent spectral density of the measurement noise would remain constant and the standard deviation of the digital measurement noise would vary with data rate. Fifty run Monte Carlo sets were run for 20 different values of flight time at data rates of 2 Hz ($T_s = 0.5$ s), 10 Hz ($T_s = 0.1$ s), and 20 Hz ($T_s = 0.05$ s). We can see from Fig. A.7 that, although the noise miss distance dependence on data rate is not as dramatic as in Fig. 9.13, the miss distance still decreases with increasing data rate (decreasing sampling time).

In summary, we can say that even in cases in which it is appropriate to hold the measurement noise spectral density constant when data rate studies are conducted, the noise induced miss still decreases with increasing data rate when either

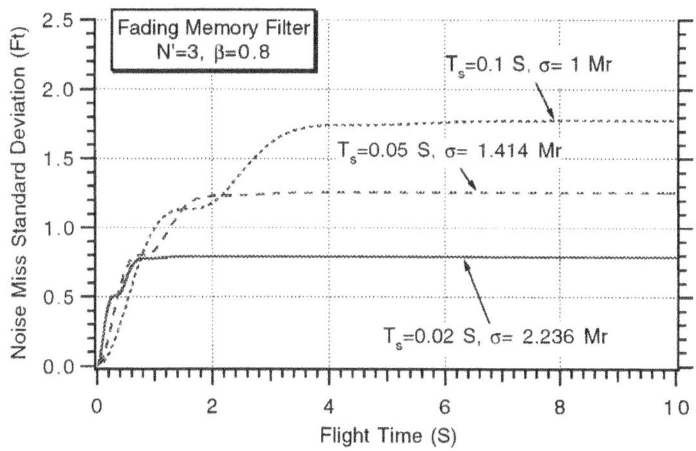

Fig. A.6 Increasing data rate still reduces miss due to digital measurement noise.

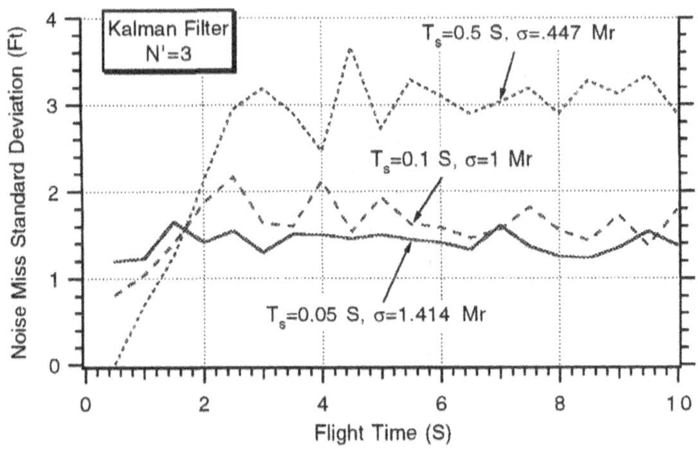

Fig. A.7 Measurement noise miss still increases with decreasing sampling rate.

a fading memory or Kalman filter is used. Both of these digital noise filters take into account information concerning the data rate so that they can both achieve good performance when the data rate changes.

BRUTE FORCE FREQUENCY RESPONSE [3]

We have seen in Chapter 22 how we can find the frequency response of a linear system by first analytically deriving the open-loop transfer function of the system under consideration and then finding its magnitude and phase as a function of frequency. Because a great deal of algebraic manipulation of the system under consideration is involved in this procedure, it is easy to make an error. This section will show that an independent check of the frequency response can be made by using a brute force simulation approach on the system under consideration.

Recall that when we are finding the frequency response of a system we are essentially finding the amplitude and phase of the steady-state sinusoidal output of a linear system driven by a sinusoidal input, as can be seen in Fig. A.8. It is important to note that for a frequency response the sinusoidal output magnitude and phase is found for different sinusoidal input frequencies.

Fig. A.8 Model for brute force frequency response method.

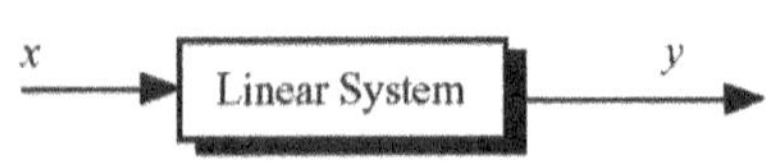

Here, the sinusoidal input to the linear system of Fig. A.8 is given by

$$x = \sin \omega t$$

where ω is the sinusoidal input frequency. Because the system under consideration is linear, the output in the steady state (that is, after transients have died out) must also be a sine wave and can be expressed as

$$y = A \sin(\omega t + \phi)$$

where A is the amplitude of the sinusoidal output and ϕ is the phase angle. Because the system is linear, the frequency of the system output is the same as the frequency of the input. If we multiply the steady-state system output by a sine wave of the same frequency as the input and integrate the result over a period, we obtain P or

$$P = \int_0^{2\pi/\omega} y \sin \omega t \, \mathrm{d}t = \int_0^{2\pi/\omega} A \sin(\omega t + \phi) \sin \omega t \, \mathrm{d}t = \frac{A\pi}{\omega} \cos \phi$$

Similarly, if we multiply the steady-state system output by a cosine wave of the same frequency as the input and integrate the result over a period, we obtain Q or

$$Q = \int_0^{2\pi/\omega} y \cos \omega t \, \mathrm{d}t = \int_0^{2\pi/\omega} A \sin(\omega t + \phi) \cos \omega t \, \mathrm{d}t = \frac{A\pi}{\omega} \sin \phi$$

From the two preceding equations we can see that P and Q are related to the magnitude and phase of the system output according to

$$\sqrt{P^2 + Q^2} = \frac{A\pi}{\omega}$$

$$\phi = \tan^{-1} \frac{Q}{P}$$

In other words, information concerning the magnitude and phase of the system's steady-state sinusoidal output because of a sinusoidal input can be obtained from P and Q. To make the two preceding relationships useful we must first figure out a way to determine when we are in steady state (that is, transients have died out). If we integrate over a period and evaluate P_0, and then integrate again over another period and evaluate P_1, the difference is

$$\Delta P_1 = P_1 - P_0$$

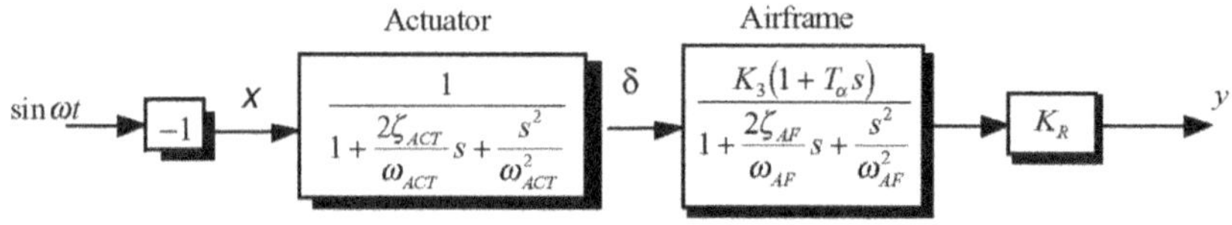

Fig. A.9 Computing frequency response by brute force.

If the difference is very small or zero we know we are in steady state. If not, we evaluate other differences until we are in steady state or

$$\Delta P_2 = P_2 - P_1$$

$$\vdots$$

$$\Delta P_n = P_n - P_{n-1}$$

To see if the brute-force frequency response technique works, let us again consider the rate gyro flight control system that was originally presented in Fig. 22.11 of Chapter 22. If we break the loop at the actuator (as was done in Fig. 22.15) the open-loop transfer function can be obtained by brute force from Fig. A.9. In this diagram, we will first choose a sinusoidal input frequency and then evaluate P and Q when steady state is reached. Another input frequency will be chosen and the process will be repeated. Enough input frequencies will be chosen to compute a proper frequency response.

To simulate Fig. A.9, we must first convert the transfer functions of the block diagram to differential equations. If

$$x = -\sin \omega t$$

we have already shown that by using the chain rule from calculus that the transfer function for the actuator can be converted to the differential equation

$$\ddot{\delta} = \omega_{ACT}^2 \left(x - \delta - \frac{2\zeta_{ACT}}{\omega_{ACT}} \dot{\delta} \right)$$

while the airframe transfer function becomes

$$\ddot{e} = \omega_{AF}^2 \left(\delta - e - \frac{2\zeta_{AF}}{\omega_{AF}} \dot{e} \right)$$

and the system output becomes

$$y = K_R K_3 (e + T_\alpha \dot{e})$$

The computerized method for finding the brute-force frequency response appears in Listing A.4. Notice that the integration interval is small as in other simulations of the rate gyro flight control system. Error criteria are set to ensure that P is in steady-state. For example, we assume that it takes at least 20 s for transients to die out. At the lower frequencies it will take longer for the transients to die out, and therefore we have an additional error criteria (that is, test when differences in P over a period are sufficiently small). We can see from Listing A.4 that the "For loop" increments the sinusoidal input frequency in an intelligent manner. The resultant magnitude and phase of the system output for each input frequency is printed out and also written to a file.

LISTING A.4 BRUTE-FORCE FREQUENCY RESPONSE PROGRAM

```
% Program runs very slowly
n=0;
ZACT=.7;
WACT=150;
K3=-1.89;
TA=.457;
ZAF=.058;
WAF=25.3;
KR=.1;
PI=3.1416;
H=.0001;
for I=2:160
        W=10^(.025*I-1);
        PERIOD=2.*PI/W;
        T=0.;
        S=0.;
        E=0.;
        ED=0.;
        DEL=0.;
        DELD=0.;
        P=0.;
        Q=0;
        PPREV=0;
        QPREV=0;
        DELP=0;
        DELQ=0;
        DELPOLD=0;
        DELQOLD=0;
        DELDELP=100;
        DELDELQ=100;
        while ~(T>20. & abs(DELDELP)<.0001)
                EOLD=E;
                EDOLD=ED;
                DELOLD=DEL;
                DELDOLD=DELD;
```

```
        POLD=P;
        QOLD=Q;
        STEP=1;
        FLAG=0;
        while STEP<=1
        if FLAG==1
        STEP=2;
                        E=E+H*ED;
                        ED=ED+H*EDD;
                        DEL=DEL+H*DELD;
                        DELD=DELD+H*DELDD;
                        P=P+H*PD;
                        Q=Q+H*QD;
                        T=T+H;
                end
                X=-sin(W*T);
                DELDD=WACT*WACT*(X-DEL-2.*ZACT*DELD/WACT);
                EDD=WAF*WAF*(DEL-E-2.*ZAF*ED/WAF);
                Y=KR*K3*(E+TA*ED);
                PD=Y*sin(W*T);
                QD=Y*cos(W*T);
                FLAG=1;
        end
        FLAG=0;
        E=.5*(EOLD+E+H*ED);
        ED=.5*(EDOLD+ED+H*EDD);
        DEL=.5*(DELOLD+DEL+H*DELD);
        DELD=.5*(DELDOLD+DELD+H*DELDD);
        P=.5*(POLD+P+H*PD);
        Q=.5*(QOLD+Q+H*QD);
        S=S+H;
        if (S>=(PERIOD-.0001))
        S=0.;
                DELP=P-PPREV;
                DELQ=Q-QPREV;
                PPREV=P;
                QPREV=Q;
                DELDELP=DELPOLD-DELP;
                DELDELQ=DELQOLD-DELQ;
                DELPOLD=DELP;
                DELQOLD=DELQ;
        end
end
PHASE=57.3*atan2(DELQ,DELP);
if PHASE>90.
        PHASE=PHASE-360;
```

```
        end
        GAIN=10.*log10((DELP^2+DELQ^2)*W*W/(PI*PI));
        n=n+1;
        ArrayW(n)=W;
ArrayPHASE(n)=PHASE;
ArrayGAIN(n)=GAIN;
end
figure
plot(ArrayW,ArrayGAIN),grid
xlabel('Frequency (r/s)')
ylabel('Gain (db)')
figure
plot(ArrayW,ArrayPHASE),grid
xlabel('Frequency (r/s)')
ylabel('Phase (deg)')
clc
output=[ArrayW',ArrayGAIN',ArrayPHASE'];
save datfil.txt output /ascii
disp 'simulation finished'
```

Listing A.4 was run for the nominal case, and Fig. A.10 displays the open-loop frequency response obtained by the method of brute force. By comparing this open-loop response to that of Fig. 22.16, we can see that both responses are identical in magnitude and phase. However, the analytic frequency response method of Chapter 22 (namely, Listing 22.2) yields the answers faster because numerical integration techniques are not involved. It is important to note that when using the analytical frequency response method, it is easier to make mistakes in setting up the program. On the other hand, the brute-force method has a longer computer running time (the running time in MATLAB is more than

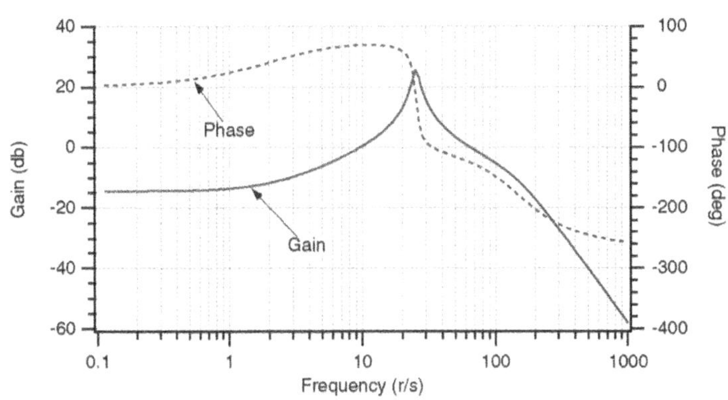

Fig. A.10 Brute-force method results are identical to those of Fig. 22.16.

TABLE A.1 BOTH METHODS ARE NUMERICALLY IDENTICAL

	Frequency domain		Time domain	
ω, r/s	Gain, db	Phase, deg	Gain, db	Phase, deg
0.112	-14.46	2.85	-14.46	2.85
0.501	-14.25	12.5	-14.25	12.5
3.981	-7.91	58.0	-7.91	58.0
33.4966	11.53	-100.43	11.53	-100.45
149.624	-11.30	-179.5	-11.30	-179.4
1000	-58.1	-257.85	-58.1	-258.53

15 minutes while in FORTAN it is less than a minute) because numerical integration is involved for each input frequency, but there is less chance of making a mistake in setting up the program because it is similar to simulation.

Table A.1 shows, for selected frequencies, how the frequency and time domain approaches compare. It can be assumed that the frequency domain answers are exact and the time domain answers are approximate. We can see that the gain is identical for both methods, but at the higher frequencies the phase angle obtained by the time domain method is in slight error.

In summary, we can say that the brute-force method for finding the open-loop frequency response of a system can be used as a useful check on analytically derived answers.

MINIMUM ENERGY TRAJECTORIES

We derived the hit equation in Chapter 11. The formula for the required velocity for an impulsive missile to travel a certain distance was given by

$$V = \sqrt{\frac{gm(1 - \cos\phi)}{a\cos\gamma[\cos\gamma - \cos(\phi + \gamma)]}}$$

where ϕ was related to the distance to be traveled, gm was a gravitational constant, a was the radius of the Earth, and γ was the flight path angle.

A minimum energy trajectory is one in which the velocity to travel a certain distance is minimized. To minimize the velocity in the preceding expression, we set the derivative of the velocity with respect to the flight path angle to zero or

$$\frac{dV}{d\gamma} = 0$$

After much algebra one can show that the flight path angle that yields minimum energy trajectories γ_{ME} is given by [4]

$$\gamma_{ME} = \frac{\pi}{4} - \frac{\phi}{4}$$

Recall that when we use Lambert guidance we specify where we are, where we want to go, and the amount of time it takes to get there. Therefore, in future uses of Lambert guidance for minimum energy trajectories we would like to specify the flight time that yields the appropriate flight path angle γ_{ME}. Recall in Chapter 11 we showed that the formula for the flight time was given by

$$t_F = \frac{a}{V\cos\gamma}\left\{\frac{\tan\gamma(1-\cos\phi)+(1-\lambda)\sin\phi}{(2-\lambda)\left[\dfrac{1-\cos\phi}{\lambda\cos^2\gamma}+\dfrac{\cos(\gamma+\phi)}{\cos\gamma}\right]} + \frac{2\cos\gamma}{\lambda\left(\dfrac{2}{\lambda}-1\right)^{1.5}}\tan^{-1}\left[\frac{\sqrt{\frac{2}{\lambda}-1}}{\dfrac{\cos\gamma}{\tan\frac{\phi}{2}}-\sin\gamma}\right]\right\}$$

Listing A.5 programs the preceding formulas for the minimum energy flight path angle, velocity, and flight time. We can see that a loop appears in which the downrange distance to be traveled is varied from 100 to 20,000 km in steps of 100 km. Listing A.12 then calculates the flight time that corresponds to each minimum energy trajectory.

The simulation of Listing A.5 was run, and the minimum energy flight time corresponding to each downrange to be traveled is displayed in Fig. A.11. We

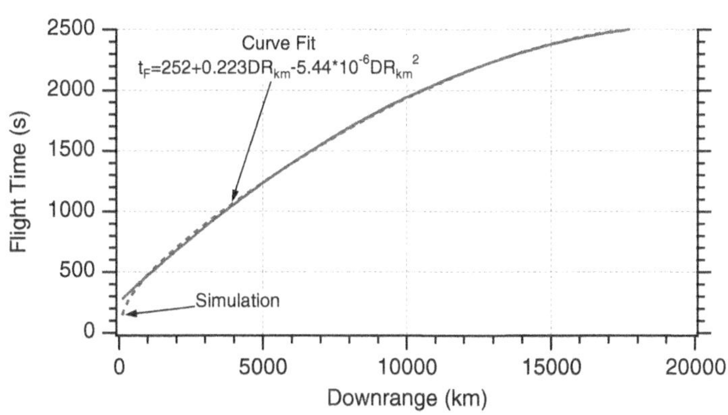

Fig. A.11 Deriving flight time formula for minimum energy trajectories.

can see that it appears that the flight time for a minimum energy trajectory is a linear function of downrange (that is, distance to be traveled in kilometers). Superimposed on the graph is the best linear least-squares curve fit to the simulation results. We can see that if the downrange to be traveled is expressed in kilometers then the minimum energy flight time in units of seconds can be expressed as

$$t_{F_{ME}} = 252 + 0.223DR_{\text{km}} - 5.44 * 10^{-6}DR_{\text{km}}^2$$

LISTING A.5 CALCULATING FLIGHT TIME FOR A MINIMUM ENERGY TRAJECTORY

```
n=0;
GM=1.4077E16;
A=2.0926E7;
CONST=sqrt(GM/A);
for DISTKM=100:100:20000,
        PHI=DISTKM*3280./A ;
        GAM=3.14159/4.-PHI/4.;
        GAMDEG=57.3*GAM;
        TOP=GM*(1.-cos(PHI));
        TEMP=A*cos(GAM)/A-cos(PHI+GAM);
        BOT=A*cos(GAM)*TEMP;
        V=sqrt(TOP/BOT);
        XLAM=A*V*V/GM;
        TOP1=tan(GAM)*(1-cos(PHI))+(1-XLAM)*sin(PHI);
        BOT1P=(1-cos(PHI))/(XLAM*cos(GAM)*cos(GAM));
        BOT1=(2-XLAM)*(BOT1P+cos(GAM+PHI)/cos(GAM));
        TOP2=2*cos(GAM);
        BOT2=XLAM*((2/XLAM-1)^1.5);
        TOP3=sqrt(2/XLAM-1);
        BOT3=cos(GAM)/tan(PHI/2)-sin(GAM);
        TEMP=(TOP2/BOT2)*atan2(TOP3,BOT3);
        TF=A*(TOP1/BOT1+TEMP)/(V*cos(GAM));
        n=n+1;
        ArrayDISTKM(n)=DISTKM;
        ArrayTF(n)=TF;
end
figure
plot(ArrayDISTKM,ArrayTF),grid
xlabel('Distance (km)')
ylabel('Flight Time (s)')
clc
output=[ArrayDISTKM',ArrayTF'];
save datfil.txt output /ascii
disp 'simulation finished'
```

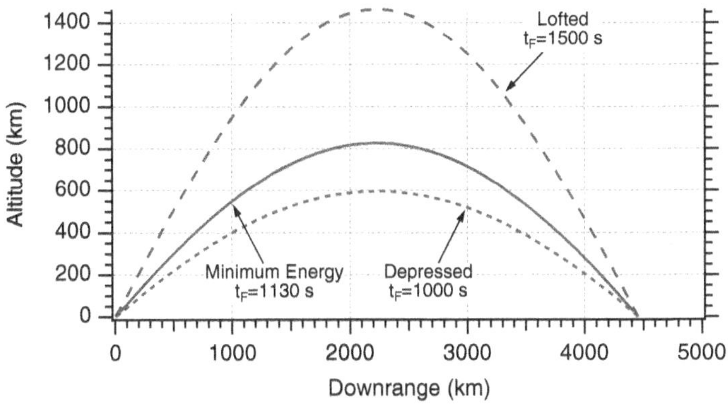

Fig. A.12 Minimum energy trajectory compared with lofted and depressed trajectories.

To test the minimum energy formula for flight time, Listing 13.3 was modified to include the new formula. We can see that the distance to be traveled is 40 deg (namely, 70 deg − 30 deg) or approximately 4500 km. In this simulation, a two-stage booster flies to its intended target according to Lambert guidance.

The first case run was for a minimum energy trajectory. In this case, the flight time turned out to be 1130 s. Trajectories corresponding to flight times of 1000 s and 1500 s were also run. We can see from Fig. A.12 that decreasing the flight time from the minimum energy value depresses the trajectory, whereas increasing the flight time from its minimum energy value lofts the trajectory.

To ensure that a flight time of 1130 s corresponds to a minimum energy trajectory, a comparison was made of the velocity profiles for each of the three trajectories. Figure A.13 shows that the final velocity for the 1000 s trajectory is

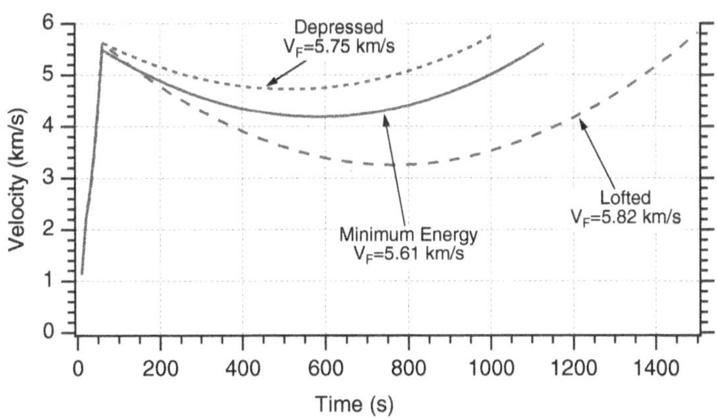

Fig. A.13 Minimum energy trajectory has least velocity at the end of the flight.

5.75 km/s, the final velocity for the 1500 s trajectory is 5.82 km/s, and the final velocity for the minimum energy trajectory is 5.61 km/s. As expected, the minimum energy trajectory yields the smallest final velocity. However, it is important to note that the velocity differential between all three trajectories is not very large.

TRAJECTORY SHAPING GUIDANCE IN THREE DIMENSIONS

In Chapter 24 we showed that the trajectory shaping guidance law in one dimension was given by

$$n_c(t) = \frac{6y + 4\dot{y}t_{\text{go}} + n_T t_{\text{go}}^2 + 2\dot{y}(t_F)t_{\text{go}}}{t_{\text{go}}^2}$$

where y was the relative position between the target and missile, $\dot{y}$ was the relative velocity between target and missile, n_T was the target acceleration, and $\dot{y}(t_F)$ was the desired relative velocity at intercept between the target and missile. In Chapter 28 we showed how several of the guidance laws presented in the text could be programmed in three dimensions. In this section we shall also show how we can convert the trajectory shaping guidance law to three dimensions. Before we can convert the preceding guidance law to three dimensions, it is first necessary to write the trajectory shaping guidance law in a slightly different form than was presented in Chapter 24. Expanding the final relative velocity term of the trajectory shaping guidance law yields

$$n_c = \frac{6y + 4\dot{y}t_{\text{go}} + n_T t_{\text{go}}^2 + 2[\dot{y}_T(t_F) - \dot{y}_M(t_F)]t_{\text{go}}}{t_{\text{go}}^2}$$

If we assume that the target velocity does not change very much during the flight, then we can say that

$$\dot{y}_T(t_F) \approx \dot{y}_T$$

and the trajectory shaping guidance law can then be rewritten as

$$n_c = \frac{6y + 4\dot{y}t_{\text{go}} + n_T t_{\text{go}}^2 + 2[\dot{y}_T - \dot{y}_M(t_F)]t_{\text{go}}}{t_{\text{go}}^2}$$

Because we know that

$$\dot{y} = \dot{y}_T - \dot{y}_M$$

we can also say that

$$n_c = \frac{6y + 4\dot{y}t_{go} + n_T t_{go}^2 + 2[\dot{y} + \dot{y}_M - \dot{y}_M(t_F)]t_{go}}{t_{go}^2}$$

or

$$n_c = \frac{6y + 6\dot{y}t_{go} + n_T t_{go}^2 + 2[\dot{y}_M - \dot{y}_M(t_F)]t_{go}}{t_{go}^2}$$

LISTING A.6 TRAJECTORY SHAPING GUIDANCE LAW IN THREE-DIMENSIONAL SIMULATION

```
n=0;
XNTG=4.;
VT=1000.;
VM=3000.;
RM1=0.;
RM2=10000.;
RM3=-1000.;
RT1=30000.;
RT2=10000.;
RT3=0.;
GAMFPDEG= -30.;
GAMFYDEG= 20.;
XNT=32.2*XNTG;
BETA=0.;
VT1=-VT*cos(BETA);
VT2=VT*sin(BETA);
VT3=0;
GAMFP=GAMFPDEG/57.3;
GAMFY=GAMFYDEG/57.3;
[VM1,VM2,VM3,TF]=LAUNCHLOGIC(RM1,RM2,RM3,RT1,RT2,RT3,VT1,VT2,...
                                                  VT3,VM);
H=.0001;
T=0.;
S=0.;
RTM1=RT1-RM1;
RTM2=RT2-RM2;
RTM3=RT3-RM3;
RTM=sqrt(RTM1^2+RTM2^2+RTM3^2);
VTM1=VT1-VM1;
VTM2=VT2-VM2;
VTM3=VT3-VM3;
```

```
VC=-(RTM1*VTM1+RTM2*VTM2+RTM3*VTM3)/RTM;
VM1F=VM*cos(GAMFP)*cos(GAMFY);
VM2F=VM*sin(GAMFP);
VM3F=VM*cos(GAMFP)*sin(GAMFY);
while ~(VC<0)
        if RTM<1000.
                H=.00001;
        else
                H=.0001;
        end
        BETAOLD=BETA;
        RT1OLD=RT1;
        RT2OLD=RT2;
        RT3OLD=RT3;
        RM1OLD=RM1;
        RM2OLD=RM2;
        RM3OLD=RM3;
        VM1OLD=VM1;
        VM2OLD=VM2;
        VM3OLD=VM3;
        STEP=1;
        FLAG=0;
        while STEP<=1
                if FLAG==1
                        STEP=2;
                        BETA=BETA+H*BETAD;
                        RT1=RT1+H*VT1;
                        RT2=RT2+H*VT2;
                        RT3=RT3+H*VT3;
                        RM1=RM1+H*VM1;
                        RM2=RM2+H*VM2;
                        RM3=RM3+H*VM3;
                        VM1=VM1+H*AM1;
                        VM2=VM2+H*AM2;
                        VM3=VM3+H*AM3;
                        T=T+H;
                end
                RTM1=RT1-RM1;
                RTM2=RT2-RM2;
                RTM3=RT3-RM3;
                RTM=sqrt(RTM1^2+RTM2^2+RTM3^2);
                VTM1=VT1-VM1;
                VTM2=VT2-VM2;
                VTM3=VT3-VM3;
                VC=-(RTM1*VTM1+RTM2*VTM2+RTM3*VTM3)/RTM;
                TGO=RTM/VC;
```

```
        BETAD=XNT/VT;
        XNT1=XNT*sin(BETA);
        XNT2=XNT*cos(BETA);
        XNT3=0.;
        VT1=-VT*cos(BETA);
        VT2= VT*sin(BETA);
        VT3=0.;
        VM=sqrt(VM1*VM1+VM2*VM2+VM3*VM3);
        XNC1=((6.*RTM1+6.*VTM1*TGO)/TGO^2) +2.0*(VM1-VM1F)/(TGO)+XNT1;
        XNC2=((6.*RTM2+6.*VTM2*TGO)/TGO^2) +2.0*(VM2-VM2F)/(TGO)+XNT2;
        XNC3=((6.*RTM3+6.*VTM3*TGO)/TGO^2) +2.0*(VM3-VM3F)/(TGO)+XNT3;
        XNCG=sqrt(XNC1^2+XNC2^2+XNC3^2)/32.2;
        XNCDOTVM=(XNC1*VM1+XNC2*VM2+XNC3*VM3)/VM;
        AM1=XNC1-XNCDOTVM*VM1/VM;
        AM2=XNC2-XNCDOTVM*VM2/VM;
        AM3=XNC3-XNCDOTVM*VM3/VM;
        GAMYDEG=57.3*atan2(VM3,VM1);
        GAMPDEG=57.3*atan2(VM2,sqrt(VM1^2+VM3^2));
        FLAG=1;
    end
    FLAG=0;
    BETA=.5*(BETAOLD+BETA+H*BETAD);
    RT1=.5*(RT1OLD+RT1+H*VT1);
    RT2=.5*(RT2OLD+RT2+H*VT2);
    RT3=.5*(RT3OLD+RT3+H*VT3);
    RM1=.5*(RM1OLD+RM1+H*VM1);
    RM2=.5*(RM2OLD+RM2+H*VM2);
    RM3=.5*(RM3OLD+RM3+H*VM3);
    VM1=.5*(VM1OLD+VM1+H*AM1);
    VM2=.5*(VM2OLD+VM2+H*AM2);
    VM3=.5*(VM3OLD+VM3+H*AM3);
    S=S+H;
    if S>=.0999
        S=0.;
        n=n+1;
        RT1K=RT1/1000.;
        RT2K=RT2/1000.;
        RT3K=RT3/1000.;
        RM1K=RM1/1000.;
        RM2K=RM2/1000.;
        RM3K=RM3/1000.;
        ArrayT(n)=T;
        ArrayRM1K(n)=RM1K;
        ArrayRM2K(n)=RM2K;
        ArrayRM3K(n)=RM3K;
        ArrayRT1K(n)=RT1K;
```

```
                ArrayRT2K(n)=RT2K;
                ArrayRT3K(n)=RT3K;
                ArrayGAMPDEG(n)=GAMPDEG;
                ArrayGAMFPDEG(n)=GAMFPDEG;
                ArrayGAMYDEG(n)=GAMYDEG;
                ArrayGAMFYDEG(n)=GAMFYDEG;
                ArrayXNCG(n)=XNCG;
   end
end
n=n+1;
                RT1K=RT1/1000.;
                RT2K=RT2/1000.;
                RT3K=RT3/1000.;
                RM1K=RM1/1000.;
                RM2K=RM2/1000.;
                RM3K=RM3/1000.;
                ArrayT(n)=T;
                ArrayRM1K(n)=RM1K;
                ArrayRM2K(n)=RM2K;
                ArrayRM3K(n)=RM3K;
                ArrayRT1K(n)=RT1K;
                ArrayRT2K(n)=RT2K;
                ArrayRT3K(n)=RT3K;
                ArrayGAMPDEG(n)=GAMPDEG;
                ArrayGAMFPDEG(n)=GAMFPDEG;
                ArrayGAMYDEG(n)=GAMYDEG;
                ArrayGAMFYDEG(n)=GAMFYDEG;
                ArrayXNCG(n)=XNCG;
figure
plot(ArrayRM1K,ArrayRM2K,ArrayRT1K,ArrayRT2K),grid
xlabel('Downrange (kft)')
ylabel('Altitude (kft)')
figure
plot(ArrayRM1K,ArrayRM3K,ArrayRT1K,ArrayRT3K),grid
xlabel('Downrange (kft)')
ylabel('Crossrange (kft)')
figure
plot(ArrayT,ArrayGAMPDEG,ArrayT,ArrayGAMFPDEG),grid
xlabel('Time (Sec)')
ylabel('Pitch Reentry Angle (deg)')
figure
plot(ArrayT,ArrayGAMYDEG,ArrayT,ArrayGAMFYDEG),grid
xlabel('Time (Sec)')
ylabel('Yaw Reentry Angle (deg)')
figure
plot(ArrayT,ArrayXNCG),grid
```

```
xlabel('Time (Sec)')
ylabel('Commanded Acceleration (g)')
axis([00,10,00,60])
clc
output=[ArrayT',ArrayRM1K',ArrayRM2K',ArrayRM3K',ArrayRT1K',...
        ArrayRT2K',ArrayRT3K',ArrayGAMPDEG',ArrayGAMFPDEG',...
        ArrayGAMYDEG',ArrayGAMFYDEG'];
save datfil.txt output -ascii
disp 'simulation finished'
RTM

function [VM1,VM2,VM3,TF]=LAUNCHLOGIC(RM1,RM2,RM3,RT1,RT2,RT3,...
                                              VT1,VT2,VT3,VM)
for TF=.1:.1:10,
        RTM1=RT1-RM1;
        RTM2=RT2-RM2;
        RTM3=RT3-RM3;
        RT1F=RT1+VT1*TF;
        RT2F=RT2+VT2*TF;
        RT3F=RT3+VT3*TF;
        THET=asin((RT2F-RM2)/(VM*TF));
        PSI=atan2(RT3F-RM3,RT1F-RM1);
        VM1=VM*cos(THET)*cos(PSI);
        VM2=VM*sin(THET);
        VM3=VM*cos(THET)*sin(PSI);
        RM1F=RM1+VM1*TF;
        RM2F=RM2+VM2*TF;
        RM3F=RM3+VM3*TF;
        RTM1F=RT1F-RM1F;
        RTM2F=RT2F-RM2F;
        RTM3F=RT3F-RM3F;
        RTMF=sqrt(RTM1F^2+RTM2F^2+RTM3F^2);
        VTM1=VT1-VM1;
        VTM2=VT2-VM2;
        VTM3=VT3-VM3;
        VC=-(RTM1F*VTM1+RTM2F*VTM2+RTM3F*VTM3)/RTMF;
        if VC<0
            break
        end
end
```

An examination of the term in brackets of the preceding expression indicates that we are attempting to make the missile velocity $\dot{y}_M$ reach a specified value at the end of the flight. Soon we will show that this is equivalent to controlling the missile flight-path angle. By duplicating the expression for the trajectory shaping guidance law in each of the Earth's inertial coordinates, we can express the guidance law in three dimensions by inspection as

$$n_{c_1} = \frac{6R_{TM1} + 6V_{TM1}t_{\text{go}} + n_{T1}t_{\text{go}}^2 + 2[V_{M1} - V_{M1}(t_F)]t_{\text{go}}}{t_{\text{go}}^2}$$

$$n_{c_2} = \frac{6R_{TM2} + 6V_{TM2}t_{\text{go}} + n_{T2}t_{\text{go}}^2 + 2[V_{M2} - V_{M2}(t_F)]t_{\text{go}}}{t_{\text{go}}^2}$$

$$n_{c_3} = \frac{6R_{TM3} + 6V_{TM3}t_{\text{go}} + n_{T3}t_{\text{go}}^2 + 2[V_{M3} - V_{M3}(t_F)]t_{\text{go}}}{t_{\text{go}}^2}$$

where R_{TM} and V_{TM} are relative position and velocity, respectively. In the preceding expressions, 1, 2, and 3 represent downrange, altitude, and cross range, respectively, in the Earth or inertial coordinate system. Thus the guidance commands are in each of those directions. For the trajectory shaping guidance law we want to make the total acceleration perpendicular to the missile velocity vector. This will ensure that the missile velocity will remain constant throughout the flight. A similar approach was taken in [5].

If we want to make the missile hit the target at desired flight-path angles γ_{PF} and γ_{YF}, we can say that the desired missile velocity components at the end of the flight are given by

$$V_{M1}(t_F) = V_M \cos\gamma_{\text{PF}} \cos\gamma_{\text{YF}}$$

$$V_{M2}(t_F) = V_M \sin\gamma_{\text{PF}}$$

$$V_{M3}(t_F) = V_M \cos\gamma_{\text{PF}} \sin\gamma_{\text{YF}}$$

where V_M is the total missile velocity. The trajectory shaping guidance law was implemented in the three-dimensional simulation of Listing A.6. In the nominal case of the simulation, the target is executing a constant 4-g maneuver perpendicular to the target velocity vector in the altitude-downrange plane. In addition, it is desired that the missile hit the target with a pitch flight-path angle of −30 deg and a yaw flight-path angle of 20 deg. The simulation calculates the instantaneous pitch and yaw flight-path angles, as well as the miss distance, to see if the trajectory shaping guidance law meets its objectives. A missile launch logic subroutine, valid for flight times of less than 10 s, is included to place the missile on a collision triangle (assuming no target maneuver) with the target.

The nominal case of Listing A.6 was run. The resultant altitude-downrange and crossrange-downrange trajectories of Figs A.14 and A.15 indicate that the missile is hitting the target. In addition we can see that there is much curvature to the missile, trajectory indicating that a great deal of trajectory shaping has taken place.

Figures A.16 and A.17 present the missile pitch and yaw flight-path angle profiles, respectively. We can see from both figures that the missile is meeting the

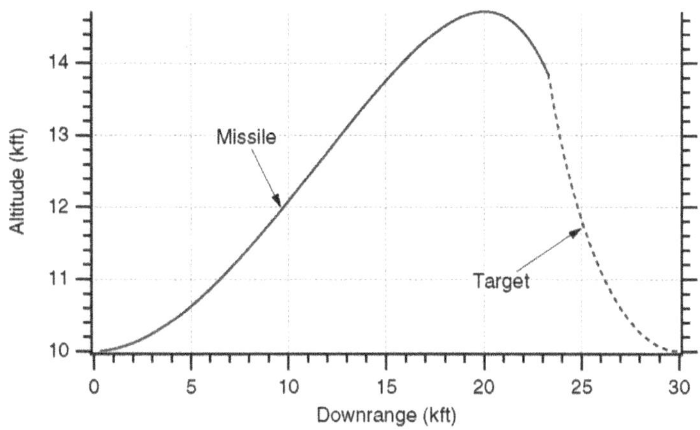

Fig. A.14 Engagement viewed in altitude-downrange plane.

desired pitch and yaw flight-path-angle objectives at the end of the flight. However we also can see from Fig. A.18 that the price paid for the trajectory shaping is a very large commanded missile acceleration. Chapter 24 goes into detail in showing how the acceleration requirements of trajectory shaping guidance can be reduced.

MODELING POISSON TARGET MANEUVER

So far in this text we have mainly considered the uniformly distributed step or weave target maneuvers for evaluating guidance system performance. We were

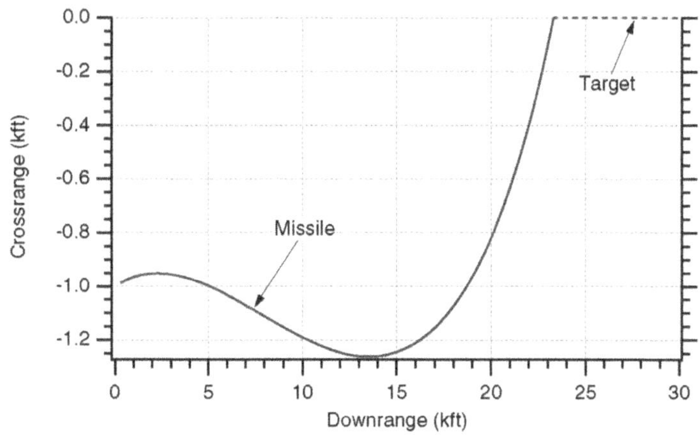

Fig. A.15 Engagement viewed in crossrange-downrange plane.

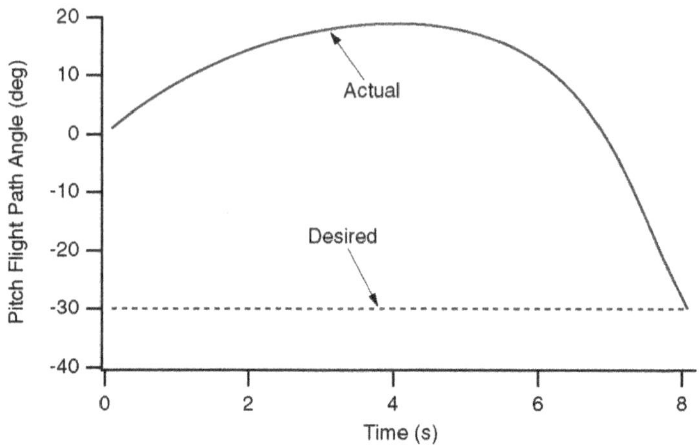

Fig. A.16 Trajectory shaping guidance law enables missile to achieve pitch flight-path-angle objective.

able to find shaping filter equivalents for these maneuvers so that the method of adjoints could be utilized to efficiently evaluate guidance system performance. Another maneuver that is often used in missile guidance system analysis is the Poisson (or random telegraph signal) target maneuver. The Poisson square wave is defined as a maneuver of amplitude $+\beta$ or $-\beta$. The length of time for which the maneuver $n_T(t)$ remains in either position is random [6]. In particular,

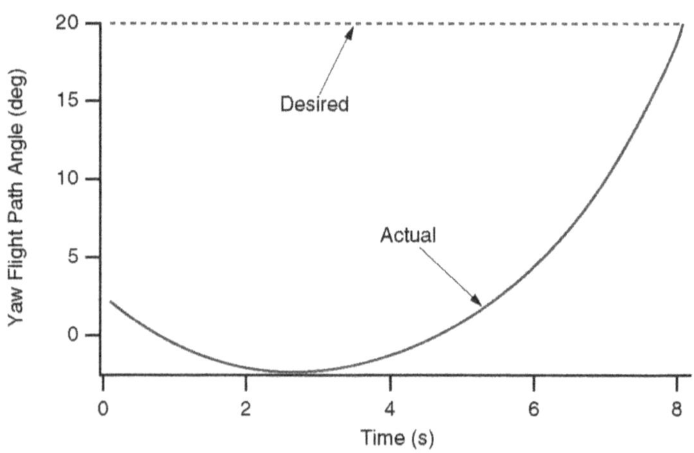

Fig. A.17 Trajectory shaping guidance law enables missile to achieve yaw flight-path-angle objective.

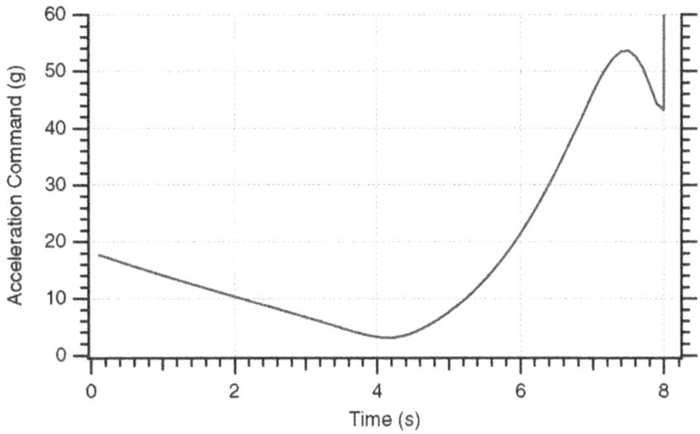

Fig. A.18 Great deal of acceleration is required to make trajectory shaping guidance law work.

the number of times the maneuver changes sign (zero crossings) is given by the Poisson distribution. If $P(k)$ represents the probability of k sign changes in t_F seconds, then we can say that

$$P(k) = \frac{(\nu t_F)^k e^{-\nu t_F}}{k!}$$

where k is the number of sign changes and ν is the average number of zero crossings per second. This type of maneuver can be very stressing to a missile guidance system and is similar in shape to the vertical-S maneuver of Chapter 6. A typical realization of a 3-g Poisson target maneuver ($\upsilon = 0.5\ \mathrm{s}^{-1}$, $\beta = 3\ g$, and $t_F = 20\ \mathrm{s}^{-1}$) is shown in Fig. A.19.

The two steps in simulating a Poisson square wave are in a forward simulation are as follows:

1. Determine the initial sign of the square wave from a Gaussian random number generator with zero mean and unity standard deviation.
2. The length of time between sign changes $\Delta\tau$ is determined by squaring and adding two Gaussian distributions with zero mean and standard deviation given by

$$\sigma = \frac{1}{\sqrt{2\upsilon}}$$

The second step for making the Poisson target maneuver is displayed in Fig. A.20.

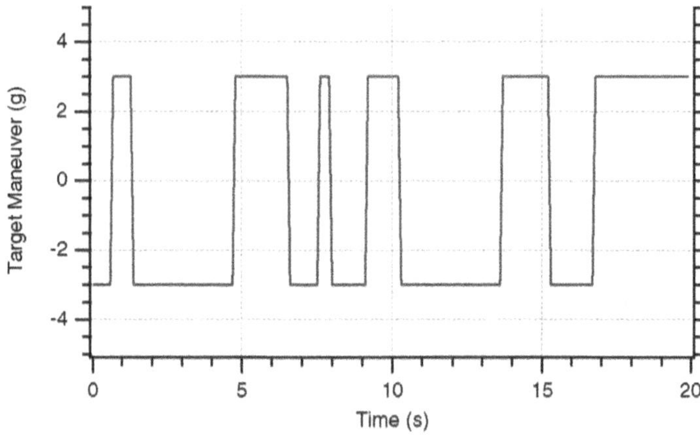

Fig. A.19 Poisson square wave target maneuver.

In Chapter 4, Fig. 4.15 modeled a single-time-constant proportional navigation guidance system whose only error source was a uniformly distributed target maneuver. A Monte Carlo simulation of that guidance system appeared in Listing 4.5. Listing A.7 modifies Listing 4.5 by replacing the uniformly distributed step target maneuver with the Poisson target maneuver. Statements that pertain to the modeling of the Poisson target maneuver have been highlighted in boldface. In addition we can see from Listing A.7 that 1000 runs are made (rather than 50 runs in Listing 4.5) for each flight time and that the flight time

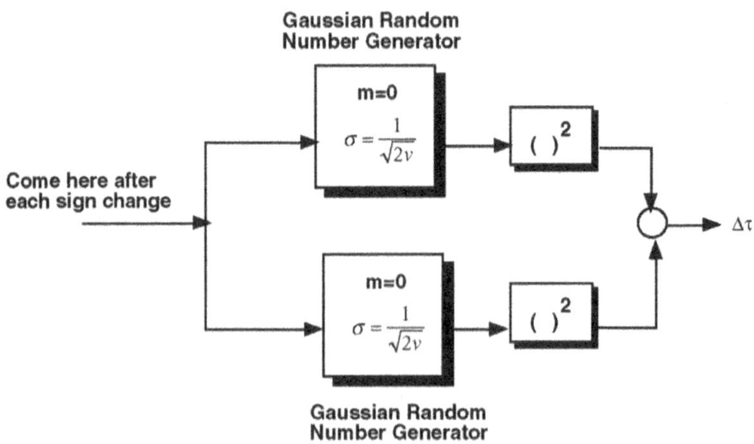

Fig. A.20 Simulating Poisson target maneuver.

is incremented every 0.2 s (rather than every second in Listing 4.5) in order to get better accuracy.

LISTING A.7 MONTE CARLO SIMULATION OF SINGLE-TIME-CONSTANT GUIDANCE SYSTEM DRIVEN BY POISSON TARGET MANEUVER

```
count=0;
VC=4000;
XNT=96.6;
VM=3000;
XNP=3;
TAU=1;
RUN=1000;
BETA=96.6;
XNU=.5;
for TF=.2:.2:10,
        Z1=0;
        for I=1:RUN
                QFIRST=1;
                SIG=1./sqrt(2.*XNU);
                PZ=uniform;
                PZ=PZ-.5;
                if PZ > 0
                        COEF=1;
                else
                        COEF=-1;
                end;
                XNT=COEF*BETA;
                DELT=9999.;
                TNOW=0;
        Y=0;
        YD=0;
        T=0;
        H=.01;
        S=0;
        XNC=0;
        XNL=0;
        while T<=(TF - 1e-5)
        if QFIRST==1
                                XNOISE1=SIG*randn;
                                XNOISE2=SIG*randn;
                                DELT=XNOISE1^2+XNOISE2^2;
                                QFIRST=0;
                                TNOW=T;
                        end;
```

```
                    if T>=(DELT+TNOW)
                      XNT=-XNT;
                      QFIRST=1;
                    end
        YOLD=Y;
        YDOLD=YD;
        XNLOLD=XNL;
        STEP=1;
        FLAG=0;
        while STEP <=1
        if FLAG==1
        Y=Y+H*YD;
        YD=YD+H*YDD;
        XNL=XNL+H*XNLD;
        T=T+H;
        STEP=2;
  end
  TGO=TF-T+.00001;
  RTM=VC*TGO;
  XLAMD=(RTM*YD+Y*VC)/(RTM^2);
  XNC=XNP*VC*XLAMD;
  XNLD=(XNC-XNL)/TAU;
  YDD=XNT-XNL;
  FLAG=1;
 end;
 FLAG=0;
 Y=.5*(YOLD+Y+H*YD);
 YD=.5*(YDOLD+YD+H*YDD);
 XNL=.5*(XNLOLD+XNL+H*XNLD);
 S=S+H;
end;
Z(I)=Y;
Z1=Z(I)+Z1;
XMEAN=Z1/I;
        end;
        SIGMA=0;
        Z1=0;
Z2=0.;
        for I=1:RUN
        Z1=(Z(I)-XMEAN)^2+Z1;
 Z2=Z(I)^2+Z2;
        if I==1
        SIGMA=0;
        RMS=0.;
        else
        SIGMA=sqrt(Z1/(I-1));
```

```
        RMS=sqrt(Z2/(I-1));
        end
        end;
        count=count+1;
        ArrayTF(count)=TF;
        ArrayRMS(count)=RMS;
end;
figure
plot(ArrayTF,ArrayRMS)
title('Shaping filter Monte Carlo results')
xlabel('Time')
ylabel('RMS Miss (ft)')
clc
output=[ArrayTF',ArrayRMS'];
save datfil.txt output -ascii
disp 'simulation finished'
```

The shaping filter equivalent of a Poisson target maneuver can be represented by white noise u_s through a low-pass filter with time constant $1/2\upsilon$ as shown in Fig. A.21 [6]. In this figure the white-noise input has spectral density Φ_s given by

$$\Phi_s = \frac{\beta^2}{\nu}$$

and the initial condition on the integrator has value β (to ensure that the standard deviation of the shaping filter output is β at all times).

Because the network of Fig. A.21 is driven by both an impulse and white noise, we can take its adjoint. The resultant adjoint block diagram of the Poisson target maneuver appears in Fig. A.22. In this figure we are using the same notation as the adjoint of the single-time-constant homing loop of Fig. 4.17. The outputs

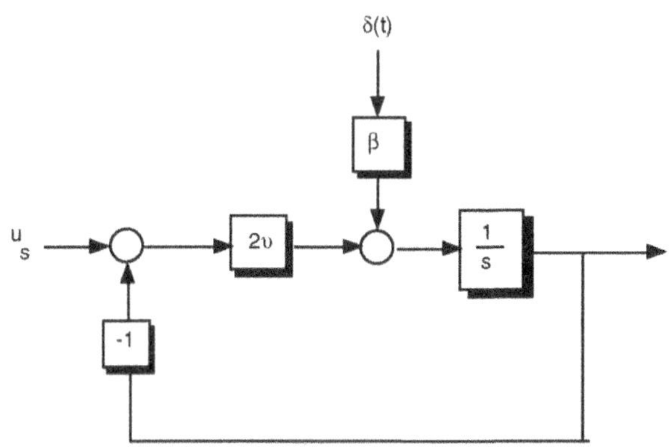

Fig. A.21 Shaping filter equivalent of Poisson target maneuver.

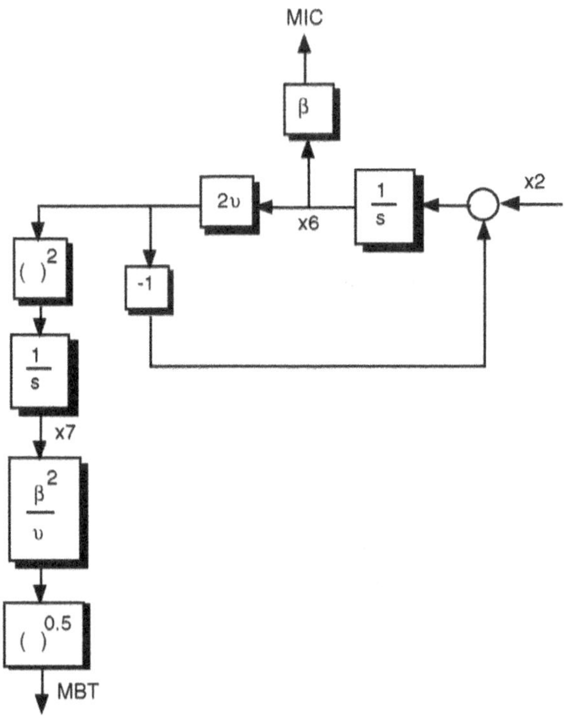

Fig. A.22 Adjoint of Poisson target maneuver.

MBT and MIC of Fig. A.22 represent the adjoint outputs of the miss caused by noise and miss caused by random initial condition on the shaping filter, respectively. Therefore the total rms miss caused by the Poisson target maneuver is given by

$$\mathrm{RMS} = \sqrt{\mathrm{MBT}^2 + \mathrm{MIC}^2}$$

The adjoint of the single-time-constant proportional navigation guidance system was taken, and the resultant adjoint simulation appears in Listing A.8. Only a few modifications to the original adjoint simulation of Listing 4.6 were required, and these statements are highlighted in boldface.

The nominal cases of Listing A.7 (Monte Carlo code) and Listing A.8 (adjoint code) were run, and the rms miss distance vs flight results as a result of the Poisson target maneuver are displayed in Fig. A.23. We can see that the Monte Carlo results, which required 50,000 runs (50 flight times multiplied by 1000 runs per flight time), are identical to the single-run adjoint results.

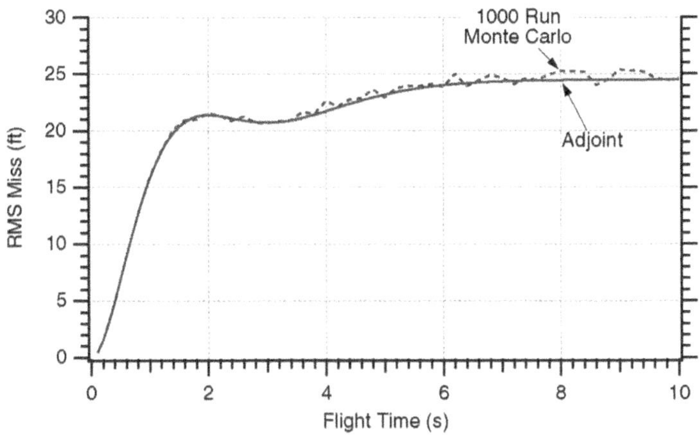

Fig. A.23 Adjoint and Monte Carlo models agree for Poisson target maneuver disturbance.

LISTING A.8 ADJOINT OF SINGLE-TIME-CONSTANT GUIDANCE SYSTEM DRIVEN BY POISSON TARGET MANEUVER

```
count=0;
XNT=96.6;
XNP=3;
TAU=1;
TF=10;
T=0;
S=0;
TP=T+.00001;
X1=0;
X2=0;
X3=1;
X4=0;
X5=0.;
X6=0;
X7=0;
H=.01;
XNU=.5;
BETA=96.6
while TP <= (TF - 1e-5)
        STEP=1;
        FLAG=0;
        S=S+H;
        X1OLD=X1;
        X2OLD=X2;
```

```
X3OLD=X3;
X4OLD=X4;
X5OLD=X5;
X6OLD=X6;
X7OLD=X7;
while STEP <=1
        if FLAG==1
                STEP=2;
                X1=X1+H*X1D;
                X2=X2+H*X2D;
                X3=X3+H*X3D;
                X4=X4+H*X4D;
                X5=X5+H*X5D;
                X6=X6+H*X6D;
                X7=X7+H*X7D;
                TP=TP+H;
        end;
        X1D=X2;
        X2D=X3;
        Y1=(X4-X2)/TAU;
        TGO=TP+.00001;
        X3D=XNP*Y1/TGO;
        X4D=-Y1;
        X5D=X1*X1;
        X6D=X2-2.*XNU*X6;
        X7D=(2.*XNU*X6)^2;
        FLAG=1;
end;
FLAG=0;
X1=(X1OLD+X1)/2+.5*H*X1D;
X2=(X2OLD+X2)/2+.5*H*X2D;
X3=(X3OLD+X3)/2+.5*H*X3D;
X4=(X4OLD+X4)/2+.5*H*X4D;
X5=(X5OLD+X5)/2+.5*H*X5D;
X6=(X6OLD+X6)/2+.5*H*X6D;
X7=(X7OLD+X7)/2+.5*H*X7D;
S=S+H;
if S>=.000999
        S=0.;
                XMBT=BETA*sqrt(X7/XNU);
                XIC=BETA*X6;
                RMS=sqrt(XMBT^2+XIC^2);
        count=count+1;
        ArrayTP(count)=TP;
        ArrayRMS(count)=RMS;
end;
```

```
end
figure
plot(ArrayTP, ArrayRMS),grid
title('Adjoint model using shaping filter approach')
xlabel('Flight Time (S)')
ylabel('RMS Miss (Ft)')
%axis([00,10,00,30])
clc
output=[ArrayTP',ArrayRMS'];
save datfil.txt output /ascii
disp('Simulation Complete')
```

REFERENCES

[1] Nesline, F. W., and Zarchan, P., "A New Look at Classical Versus Modern Homing Guidance," *Journal of Guidance and Control*, Vol. 4, No. 1, 1981, pp. 78–85.

[2] Bucco, D., "Adjoints Revisited: A Software Tool to Facilitate Their Application," *Proceedings of 1997 AIAA Guidance and Control Conference*, Washington, DC.

[3] *Advanced Continuous Simulation Language (ACSL) Reference Manual*, Mitchell and Gauthier Associates (MGA), Inc., Concord, MA, 1991, pp. A62–A69.

[4] Regan, F. J., and Anandakrishnan, J. M., *Dynamics of Atmospheric Re-Entry*, AIAA, Washington, DC, 1993, pp. 154–156.

[5] Ohlmeyer, E. J., and Phillips, C. A., "Generalized Vector Explicit Guidance," *Journal of Guidance, Control, and Dynamics*, Vol. 29, No. 2, 2006, pp. 261–268.

[6] Schwartz, M., *Information Transmission, Modulation, and Noise*, McGraw–Hill, New York, 1959, pp. 441–448.

INDEX

Note: Page numbers with "f" represent figures. Page numbers with "t" represent tables. Page numbers with "L" represent listings.

SUPPORTING MATERIALS

To access the MATLAB and FORTRAN source code that accompany this work, please go to www.aiaa.org/books and click on the "Supporting Materials" link. Select this work from the list provided and enter the password

routine

when prompted.

As was mentioned in the preface, some material has been deleted from the fifth edition in the sixth edition. The fifth edition PDFs for Chapters 10, 16, 26, 27 and all of the appendices can be found on AIAA's website.

Many of the topics introduced in this book are discussed in more detail in other AIAA publications. For a complete listing of titles in the Progress in Astronautics and Aeronautics series, as well as other AIAA publications, please visit www.aiaa.org.

AIAA is committed to devoting resources to the education of both practicing and future aerospace professionals. In 1996, the AIAA Foundation was founded. Its programs enhance scientific literacy and advance the arts and sciences of aerospace. For more information, please visit www.aiaafoundation.org.